Rhizoctonia Species:
Taxonomy, Molecular Biology,
Ecology, Pathology and Disease Control

Rhizoctonia Species:
Taxonomy, Molecular Biology, Ecology, Pathology and Disease Control

Edited by

Baruch Sneh
Tel Aviv University,
Ramat Aviv, Tel Aviv, Israel

Suha Jabaji-Hare
McDonald Campus, McGill University,
Quebec, Canada

Stephen Neate
CSIRO, Division of Soils,
Adelaide, South Australia

and

Gerda Dijst
Research Institute for Plant Protection (IPO-DLO),
Wageningen, The Netherlands

KLUWER ACADEMIC PUBLISHERS
DORDRECHT / BOSTON / LONDON

Library of Congress Cataloging-in-Publication Data

Rhizoctonia species : taxonomy, molecular biology, ecology, pathology, and disease control / edited by Baruch Sneh ... [et al.].
p. cm.
Papers from the Second International Symposium on Rhizoctonia, held in Noordwijkerhout, Netherlands, June 27-30, 1995.
Includes index.

1. Rhizoctonia--Congresses. I. Sneh, Buruch. II. International Symposium on Rhizoctonia (2nd : 1995 : Noordwijkerhout, Netherlands)
QK625.A48R48 1996
589.2'4--dc20 96-30327

ISBN 978-90-481-4597-3

Published by Kluwer Academic Publishers,
P.O. Box 17, 3300 AA Dordrecht, The Netherlands.

Kluwer Academic Publishers incorporates
the publishing programmes of
D. Reidel, Martinus Nijhoff, Dr W. Junk and MTP Press.

Sold and distributed in the U.S.A. and Canada
by Kluwer Academic Publishers,
101 Philip Drive, Norwell, MA 02061, U.S.A.

In all other countries, sold and distributed
by Kluwer Academic Publishers Group,
P.O. Box 322, 3300 AH Dordrecht, The Netherlands.

Printed on acid-free paper

Printed in the Netherlands

TABLE OF CONTENTS

LIST OF CONTRIBUTORS

Gerard C Adams Department of Botany and Plant Pathology, Michigan State University, East Lansing, Michigan 48824 USA.
Fax: + +1 (517) 353 1926, E-Mail: gadams@pilot.msu.edu

Torben F Andersen Botanical Laboratory, University of Copenhagen, Gothersgade 140, DK-1123 Copenhagen K, Denmark.
Fax: + +45 33 13 91 04, E-Mail: torbena@bot.ku.dk

Sabine Banniza International Mycological Institute, Bakeham Lane, Egham, Surrey TW20 9TY, United Kingdom .
Fax: + +44 1784 470909, E-Mail: imi@cabi.org

Gilbert J Banville Ministere de l'Agriculture des Pecheries et de l'Alimentation du Quebec, Station de recherches Les Buissons C.P. 455, Les Buissons, Quebec G0H 1H0, Canada.
Fax: + +1 (418) 567-8791. E-Mail: otrysko@quebectel.com

Michael D Benson Department of Plant Pathology, North Carolina State University Raleigh, NC 27695-7629 USA,
Fax: + +1 (919) 515 5657, E-mail: mike_benson@ncsu.edu

Mark Bolda Department of Plant Pathology, University of California, Davis CA 95616, USA,
Fax: + +1 (916) 752 5674

Timothy B Brenneman Department of Plant Pathology, University of Georgia, Coastal Plain Experiment Station, Tifton, GA 31793-0748, USA,
Fax: + +1 (912) 386 7285.

Paul D Bridge International Mycological Institute, Bakeham Lane, Egham, Surrey TW20 9TY, United Kingdom,
Fax: + +44 1784 470909, E-Mail: pbridge@cabi.org

Leon L Burpee Department of Plant Pathology, University of Georgia, Georgia Station, Griffin, GA 30223 USA
Fax: + +1 (404) 412 4010, E-Mail: lburpee@gaes.griffin.peachnet.edu

Donald E Carling University of Alaska Fairbanks, 533 E. Fireweed, Palmer, Alaska 99645 USA.
Fax: + +1 (907) 746-2677, E-mail: pfdec@orion.alaska.edu

D Kelly Cartwright Department of Plant Pathology, North Carolina State University Raleigh, NC 27695-7629 USA.
Fax: + +1 (919)-515-5657, E-mail: kelly_cartwright@ncsu.edu

Kan-Fa Chang Crop Diversification Centre-South, SS #4, Brooks, Alberta, Canada T1R 1E6
Fax: + +1 (403) 362-2554

Yih-Chang Chang Taiwan Agricultural Research Institute 189 Chung-Cheng Road, Wu-Feng 413 Taichung, Taiwan Republic of China
Fax: + + 88 64 333 8126

R James Cook USDA-ARS, Root Disease & Biological Control Unit, 365 Johnson Hall, WSU, Pullman, WA 99164 USA
Fax: + +1 (509) 335 7674, E-Mail: cookrj@wsuvm1.csc.wsu.edu

Ben JC Cornelissen Section of Plant Pathology Institute for Molecular Cell Biology BioCentrum Amsterdam Kruislaan 318 1098 SM Amsterdam the Netherlands
Fax: + +31 20 525 7934, E-Mail: cornelissen@bio.uva.nl

Marc A Cubeta Department of Plant Pathology, North Carolina State University, Vernon G. James Research and Extension Center, 207 Research Station Road, Plymouth, NC 27962 USA
Fax + +1 (919) 684 5412, E-Mail: cub@plymouth.ces.ncsu.edu

Earl A Curl Department of Plant Pathology, Auburn University, AL 36849 USA.
Fax: + +1 (205) 844 1947, E-Mail: biones@ag.auburn.edu

James W Deacon Institute of Cell & Molecular Biology, University of Edinburgh, Mayfield Road, Edinburgh EH9 3JH, United Kingdom
Fax: + +31 131 650 5392, E-Mail: Jdeacon@srvo.bio.edinburgh.ac.uk

Gerda Dijst DLO Institute for Plant Protection Research (IPO-DLO), P.O.Box 9060, 6700 GW Wageningen, the Netherlands
Fax: + +31 8370 10113, E-Mail: dijst@ipo.dlo.nl

Mirjam P Does Section of Plant Pathology Institute for Molecular Cell Biology BioCentrum Amsterdam Kruislaan 318 1098 SM Amsterdam the Netherlands,
Fax: + +31 20 525 7934,

Yigal Elad Deptement of Plant Pathology, Agricultural Research Organization, The Volcani Center, P.O. Box 6, Bet Dagan 50250, Israel
Fax: + +972 3 968 3543, E-Mail: dani@agri.huji.ac.il

Ulrich Gisi Agrobiological Research Station, Sandoz Agro Limited, CH-4108 Witterswil, Switzerland

Dolores Gonzalez Instituto de Ecologia, A.C., 91000 Xalapa, Vera Cruz, Mexico.

Niklaus J Grünwald Department of Plant Pathology, University of California, Davis CA 95616, USA.
Fax: + +1 (916) 752 5674,

Teruyoshi Hashiba Faculty of Agriculture, Tohoku University , 1-1 Tsutsumidori-Amamiyamachi Sendai 981, Japan
Fax: + +81-22 22 721 870

Leonard J Herr Department of Plant Pathology, The Ohio State University, Ohio Agricultural Research and Development Center, Wooster, Ohio 44691, USA
Fax: + +1 (216) 263 3841, E-Mail: beke.1@osu.edu

Ari M Hietala Finnish Forest Research Institute, P.O. Box 18, FIN-01301 Vantaa, Finland

Mark Holderness International Mycological Institute, Bakeham Lane, Egham, Surrey TW20 9TY, United Kingdom
Fax: + +44 1784 470 909, E-Mail: imi@cabi.org

Yoshihisa Homma Hokkaido National Agricultural Experiment Station, Hitsujigaoka-1, Toyohira, Sapporo, 062 Japan
Fax: + +81 11-851-4953, E-mail: hommay@cryo.affrc.go.jp

Ronald J Howard Crop Diversification Centre-South, SS #4, Brooks, Alberta, Canada T1R 1E6.
Fax: + +1 (403) 362 2554.

Ting-Fang Hsieh Taiwan Agricultural Research Institute 189 Chung-Cheng Road, Wu-Feng 413 Taichung, Taiwan Republic of China
Fax: + + 88 64 333 8126

Don M Huber Department of Botany & Plant Pathology, Purdue University, West Lafayette, IN 47907-1155 USA
Fax: + +1 (317) 494 0363, E-Mail: huber@btny.purdue.edu

Sheau-Fung Hwang Alberta Environmental Centre, Bag 4000, Vegreville, Alberta, Canada T9C 1T4.
Fax: + +(403) 632-8379, E-Mail: hwang@aec.env.gov.ab.ca

Mitsuro Hyakumachi Faculty of Agriculture, Gifu University, Yanagido 1-1 Gifu 501-11, Japan
Fax: + +81 582 30 1712, E-Mail: hyakumac@cc.gifu-u.ac.jp

Suha Jabaji-Hare Department of Plant Science, Macdonald Campus of McGill University, Ste.-Anne-de-Bellevue, Quebec, Canada, H9X 3V9
Fax: + +1 514 398 7897,. E-Mail: share@nrs.mcgill.ca

Jaacov Katan Department of Plant Pathology and Microbiology, The Hebrew University of Jerusalem, Faculty of Agriculture, Rehovot 76100, Israel.
Fax: + +972-8-9466794, E-mail: gamliel@agri.huji.ac.il

Hans R Kataria International Rice Research Institute, IRRI-India Liaison Office, C-18 Friends Colony (East), New Delhi 110 065, INDIA. Present address: Institute for Phytomedicine University of Hohnheim, D-70593 Stutgart, Germany
Fax: + +49 711 459 2408.

Jaap Keijer DLO-Research Institute for Plant Protection (IPO-DLO), Wageningen, The Netherlands and DLO-State Institute for Quality Control of Agricultural Products (RIKILT-DLO). P.O. Box 230, 6700 AE Wageningen, the Netherlands
Fax + +31 317 417 717, E-Mail: j.keijer@rikilt.dlo.nl

Takashi Kobayashi Faculty of Agriculture, Tohoku University , 1-1 Tsutsumidori-Amamiyamachi Sendai 981, Japan
Fax: + +81-2222721870

Martin M. Kulik Soybean and Alfalfa Research Laboratory Agricultural Research Service U. S. Department of Agriculture Beltsville, Maryland, 20705 USA.
Fax: + +1 (301) 504 5358, E-Mail: mkulik@asrr.arsusda.gov

Shiro Kuninaga Health Science University of Hokkaido, Ishikari-Tohbetu, Hokkaido, 061-02, Japan
Fax: + +81 1332 3 1669

Dilip K Lakshman University of Maine, Plant Biology and Pathology Department, Orono, Maine 04469, USA.
Fax: + +1 (207) 581 2986, E-mail: stellos@maine.maine edu

Robert T Lartey Department of Biochemistry and Cell Biology, State University of New York, Stony Brook, NY 11794 USA.
Fax: + +1 (516) 632 8575, E-Mail: lartey@mcbsgi.bio.sunysb.edu

Jack A Lewis Biocontrol of Plant Diseases Laboratory, Agricultural Research Service USDA Beltsville, Maryland 20705, USA.
Fax: ++1 (301) 504 5968, E-Mail: rlumsden@asrr.arsusda.gov

Gordon C Macnish School of Biological & Environmental Sciences, Murdoch University, Murdoch, WA, 6150, Australia
Fax: ++61 9 310 7084. E-Mail: gmacnish@central.murdoch.edu.au

Bruce Martin Clemson University, Pee Dee Research and Education Center, Route 1, Box 531, Florence, SC 29501-9603 USA

Mark Mazzola USDA-ARS, Tree Fruit Research Lab, 1104 N. Western Ave., Wenatchee WA 98801 USA.
Fax: ++1 (509) 664 2287, E-Mail: mazzola@tfrl.ars.usda.gov

Leo S Melchers Mogen International nv Einsteinweg 97 2333 CB Leiden, the Netherlands.
Fax: ++31 20 525 7934,

TW Mew Division of Entomology & Plant Pathology International Rice Research Institute, P.O. Box 933, 1099 Manila, Philippines
Fax ++63 2 817 8470, E-Mail: c.salonga@cgnet.com

Royall T Moore School of Applied Biological and Chemical Sciences University of Ulsterat Coleraine Northern Ireland BT52 1SA:UK
Fax: ++44 1265 324906, E-Mail: rt.moore@ulst.ac.uk

J Elizabeth M Mordue International Mycological Institute, Bakeham Lane, Egham, Surrey TW20 9TY, United Kingdom
Fax: ++44 1784 470 909, E-Mail: e.mordue@cabi.org

Shigeo Naito Tohoku National Agricultural Experiment Station, Akahira 4, Shimo-kuriyagawa, Morioka, 020-01 Japan
Fax: ++81-196-41-7794, E-mail: nai@tnaes.affrc.go.jp

Stephen M Neate CSIRO Division of Soils/CRC for Soil and Land Management, P.B. No 2 Glen Osmond, 5064, Australia
Fax: ++61 8 303 8550, E-Mail: neates@adl.soils.csiro.au

Philip A O'brien School of Biological and Environmental Sciences, Western Australian State Agricultural Biotechnology Center, Murdoch University, Murdoch WA 6150, Australia.
Fax: ++61 9 310 7084, E-mail: obrien@murdoch.edu.au

Akira Ogoshi Faculty of Agriculture, Hokkaido University, Kita-ku, Sapporo 060, Japan,
Fax. ++81 11 706 4938, E-Mail: sakino@res.agr.hokudai.ac.jp

Barbara E Otrysko Ministere de l'Agriculture des Pecheries et de l'Alimentation du Quebec, Station de recherches Les Buissons C.P. 455, Les Buissons, Quebec G0H 1H0, Canada.
Fax: ++1 (418) 567-8791, E-mail: otrysko@quebectel.com

Lee Panella USDA-ARS Sugarbeet Research Crops Research Laboratory 1701 Centre Ave. Fort Collins, Colorado 80526 USA.
Fax: ++1 (970) 482-2909, E-mail: lpanella@lamar.colostate.edu

Hanne N Rasmussen Research Centre Aarslev, Danish Institute of Plant and Soil Science, Kirstinebjergvej 10, DK-5792 Aarslev, Denmark.
E-mail: h.rasmussen@aarslev.ou.dk

Craig S Rothrock Department of Plant Pathology, University of Arkansas Fayetteville Arkansas 72701, USA
Fax: ++ (501) 575 7601, E-mail: c.rothrock@comp.uark.edu

Albert D Rovira CSIRO Division of Soils, Private Bag No. 2, Glen Osmond, SOUTH Australia 5064.
Fax: ++61 8303 8500

Victor Rubio Centro Nacional de Biotecnologia (CSIC-UAM). Campus Cantoblanco, Universidad Autonoma de Madrid. 28049 Madrid, SPAIN.
Fax: ++34 1 585 4506, E-mail: ffvr@cnbvx3.cnb.uam.es

Earl G Ruppel USDA-ARS Sugarbeet Research Crops Research Laboratory 1701 Centre Ave. Fort Collins, Colorado 80526 USA
Fax: ++1 (970) 482 2909, E-mail: eruppel@lamar.colostate.edu

Michael A Rutherford International Mycological Institute, Bakeham Lane, Egham, Surrey TW20 9TY, United Kingdom
Fax: ++44 1784 470 909, E-Mail: m.rutherford@cabi.org.

S Savary Division of Entomology & Plant Pathology International Rice Research Institute, P.O. Box 933, 1099 Manila, Philippines
Fax: ++63 2 817 8470, E-mail: ssavary@irri.cgnet.com

Johannes HM Schneider IPO-DLO, P.O. Box 9060, 6700 GW Wageningen, the Netherlands

Robin Sen Department of Biosciences, Division of General Microbiology, P.O. Box 56, FIN-00014 University of Helsinki, Finland
Fax: ++35 8 0 7085 9262 E-mail: rsen@cc.helsinki.fi

James B Sinclair University of Illinois, 1101 W. Peabody, Urbana, IL 61801-4723 USA.
Fax: ++1 (217) 244 1707, E-Mail: sinclair1@ux1.cso.uiuc.edu

Richard W Smiley Oregon State University, Columbia Basin Agricultural Research Center, P.O. Box 370, Pendleton, OR 97801 USA
Fax ++1 (503) 278 4188, E-Mail: smileyr@bcc.orst.edu

Baruch Sneh Department of Botany and Institute for Nature conservation Research, The George S. Wise Faculty of Life Sciences, Tel Aviv University Ramat Aviv, Israel 69978,
Fax: ++972 3 640 9380, E-mail: snehba@post.tau.ac.il

Joost A Stalpers CBS , Oosterstraat 1, 3742 SK Baarn, the Netherlands
Fax ++31 355 416 142, E-Mail: stalpers@pc.cbsc.nl

Donald R Sumner Department of Plant Pathology, University of Georgia, Coastal Plain Experiment Station, Tifton, GA 31793-0748, USA
Fax: ++1 (912) 386 7285, E-Mail: path2%tifton.bitnet@uga.cc.uga.edu

Mark W Sweetingham Agriculture Western Australia, Baron-Hay Court, South Perth, 6151, Australia.
Fax: ++61 9 367 2625, E-mail: msweeting@infotech.agric.wa.gov.au

Stellos M Tanvantzis University of Maine, Plant Biology and Pathology Department, Orono, Maine 04469, USA,
Fax: + +1 (207) 581 2986, E-mail: stellos@maine.maine edu

Chin-Chyu Tu Taiwan Agricultural Research Institute 189 Chung-cheng Road, Wu-Feng 413 Taichung, Taiwan Republic of China
Fax: + + 88 64 333 8126

Ariena HC Van Bruggen Department of Plant Pathology, University of California, Davis CA 95616, USA
Fax: + +1 (916) 752 5674, E-Mail: ahvanbruggen@ucdavis.edu

Paul HJF Van Den Boogert DLO Research Institute for Plant Protection, P.O. Box 9060, 6700GW Wageningen, the Netherlands
Fax: + +31 317.410113, E-mail: boogert@ipo.dlo.nl

Prithwi Raj Verma Agriculture and Agri-Food Canada Saskatoon Research Cenetr 107 Science Place Saskatoon, Saskatchewan Canada S7N 0X2,
Fax: + +1 (306) 956 7247,

Rytas Vilgalys Department of Botany, Duke University, Durham, NC 27708 USA,
Fax: + +1 (919) 684 5412, E-Mail: fungi@acpub.duke.edu

Albert R Weinhold University of California, 151 Hilgard Hall, Berkeley, CA 94720, USA,
Fax: + +1 (510) 643 5098, E-mail: aweinhol@nature.Berkeley.EDU

PREFACE

Species of *Rhizoctonia* are ubiqutous and variable soil inhabitant fungi. Many different *Rhizoctonia* isolates cause worldwide economically imporant diseases on most of the world's important crop plants, such as: cereals, cotton, sugarbeet potato, vegetables, field crops, turfgrasses, ornamentals, fruit trees and forest trees. On the other hand, many isolates of *Rhizoctonia* are saprophytic. Some are mycorrhizal with orchids and other plants, while some are biocontrol agents.

The first book on *Rhizoctonia,* "Biology and Pathology of *Rhizoctonia solani*" was edited by Parmeter (1970). It was the result of the first symposium on *Rhizoctonia,* held in Florida in 1965. Since then, there have been major advances in our understanding of the taxonomy, ecology, pathology, epidemiology and disease control. The second book: "Identification of *Rhizoctonia* Speceis" written by Sneh, Burpee and Ogoshi (1991), which was focused on classification and selected methodologies in *Rhizoctonia* research, served as an introduction and a trigger for us to continue the task and embark upon the challenge of writing the third book on *Rhizoctonia*. The aim of this book was to summarize the vast amount of information accumulated in the literature on the various aspects of *Rhizoctonia* research since the book of Parmeter was published in 1970.

The process was facilitated by the establishment of an International *Rhizoctonia* Committee (IRC) in the International Society of Plant Pathology (ISPP) during the International Congress of Plant Pathology held in Montreal (August 1993). During this meeting the IRC decided to organize the Second International Symposium on *Rhizoctonia* and to edit a comprehensive book on the subject. The Second International Sympoisum on *Rhizoctonia* (ISR-95) was held in Noordwijkerhout, The Netherlands during 27-30 June 1995. Thirty of the current book chapters, are extended reveiws of papers presented by the invited speakers during this symposium

The book chapters were written by the world's most reputable experts in their respective fields of *Rhizoctonia* research, summarizes years of research in the various aspects of the ubiquitous complex of the soil-borne fungi belonging to the anamorph genus *Rhizoctonia*. The subject reviews covered in this book include classical as well as modern molecular approaches to *Rhizoctonia* research in: Taxonomy and evolution, genetics and pathogenicity, plant-*Rhizoctonia* interactions, Ecology, population and disease dynamics, disease occurrence and management in various crops, cultural control, biological control, germplasm for resistance, molecular use in breeding strategies for resistance, chemical and integrated control. This publication is aimed to be an important source book on *Rhizoctonia* for researchers, plant pathologists, students, extension specialists, crop producers and companies interested in plant disease control.

We would like to express our sincere thanks to all who contributed in the various stages of the work on this book. Without their invested efforts this project could not have been accomplished. To all of the authors and co-authors (see list of contributors) who wrote their respective chapters and improved the quality of the other chapters by reviewing each other's manuscripts. The Dutch National Organization Committee of the International *Rhizoctonia* Symposium for establishing the background for the book. To Yair Hadas for bringing the manuscripts into the camera-ready format and indexing.

April 1996

Baruch Sneh,
Suha Jabaji-Hare,
Stephen Neate
Gerda Dijst.

INTRODUCTION - THE GENUS *RHIZOCTONIA*

AKIRA OGOSHI

Faculty of Agriculture, Hokkaido University,
Kita-ku, Sapporo 060, JAPAN

I. *RHIZOCTONIA* AND *R. SOLANI*

The genus *Rhizoctonia* was described by DeCandolle in 1815. He designated *R. crocorum* (Pers.) DC. as the type species, while *R. solani*, the most important species of *Rhizoctonia*, was described by Kühn in 1858. The history of *R. solani* is almost as long as the history of plant pathology. It was published around the time when the potato late blight broke out in Europe, which caused the noteworthy potato famine in Ireland. Since these initial reports, the pathology, taxonomy, ecology, and disease control of *Rhizoctonia* have been extensively studied.

It was interesting for me to scan the literature reports published on *Rhizoctonia*. To estimate the number of reports, I tried to count them in the Review of Applied Mycology (Volumes 1-48) and the Review of Plant Pathology (Volumes 49-73) from 1922 to 1994. In the first volume, 24 reports of *Rhizoctonia* were listed, whereas in the last volume 241 citations on *Rhizoctonia* were listed. *Rhizoctonia* reports comprised 3% of the total reports on plant pathology. Thus, from this review we may extrapolate that worldwide, the number of reports on *Rhizoctonia* might have been 10,000, taking into consideration that not all of the reports have been reviewed in these periodicals.

II. EVENTS IN THE STUDIES ON *RHIZOCTONIA* SPP.

Major events in the history of studies on *Rhizoctonia* are listed below in a chronological order:

A. 1815 - 1900

Following the establishment of the form genus *Rhizoctonia*, 32 taxa were reported in this period. However, most did not belong to *Rhizoctonia* and some were synonyms of earlier described species. Only two species from the nineteenth century, *R. crocorum* and *R. solani,* are presently recognized. It is reasonable to assume that both technology and the flow of information were the major obstacles for research in those days, but it emphasizes the need for correct and accurate information as the basis for a proper identification of *Rhizoctonia*.

B. 1901 - 1930

Early in the 20th century, physiological (Matsumoto *et al.*, 1923) as well as comparative studies were carried out with regard to several taxa related to *Rhizoctonia* (Duggar, 1915). The first report of Rhizoctonia mycorrhizal associations with orchids was published by Bernard (1909)

C. 1931 - 1940

The number of studies related to *R. solani* were increased and anastomosis groups were reported. Anastomosis groups were based on the studies of Matsumoto *et al.* (1932) and Schultz (1936).

B. Sneh et al. (eds.),
Rhizoctonia Species: Taxonomy, Molecular Biology, Ecology, Pathology and Disease Constrol, 1–9.

D. 1941 - 1950

Studies on *Rhizoctonia* were scarce, possibly, due to the 2nd World War.

E. 1951 - 1960

During the 1950s, the number and quality of studies related to *Rhizoctonia* increased considerably. Anastomosis groups (Richter and Schneider, 1953), forms (Exner, 1953) and several types (Ito *et al.*, 1955) *of R. solani*, perfect states (Donk, 1956; Olive, 1957), and ecology in soil (Sanford, 1952; Herzog and Wartenberg, 1958; Rao, 1959) were studied. Development of qualitative isolation methods for *R. solani* propagules from soil and quantitative methods for estimating saprophytic activity and population density of *R. solani* in soil were important particularly for studies regarding the epidemiology, ecology and disease control. Comparative studies of isolation techniques were also reported (Papavizas and Davey, 1959; Boosalis and Scharen, 1959). Consequently, isolates of *Rhizoctonia* spp. could be easily collected from many different regions. The importance of identification and classification of *Rhizoctonia* spp. isolates was again emphasized.

Seasonal changes in population densities of *R. solani* in cultivated field soils were determined. Populations of *Rhizoctonia* were found to be increased during the spring time, to decrease somewhat during the summer and to increase again during the autumn. Presently, it is not known whether this behavioral pattern is common among other isolates of *Rhizoctonia* spp., as their cultural features, pigmentation, growth rate, abundance of aerial mycelia, pathogenicity, competitive saprophytic ability etc. are quite diverse. It is very difficult to categorize these characteristics, as there are always many isolates with intermediate characteristics. Handy and appropriate methods were needed to identify isolates more efficiently thus enabling us to establish more accurate concepts for estimating population dynamics of *Rhizoctonia* spp. in soil.

F. 1961 - 1970

During this decade, the number of studies regarding the physiology, ecology and biological control of *Rhizoctonia* spp. was in focus. Beginning in the 1960s, many potential biological control agents against *R. solani* were identified. Some of these were not directly antagonistic to the pathogen, but non pathogenic or hypovirulent strains of *Rhizoctonia* spp. were shown to function sometimes also through cross protection (see chapter VI.B4).

Among the bacterial biocontrol agents *Bacillus* spp. were studied extensively (Olsen and Baker, 1968), later followed by studies of *Pseudomonas* spp. (Savithiry and Gnanamanickam, 1987). The fungal antagonists belonging to *Trichoderma* spp. already recognized by Weindling, (1932) were predominantly studied in this period. In more recent years other genera of fungal antagonists like *Gliocladium* spp. (Jager and Velvis, 1983) have been extensively studied.

The publication of the book: "*Rhizoctonia solani*, Biology and Pathology" edited by Parmeter (1970) was a highly significant event. Every *Rhizoctonia* researcher is familiar with this book as an introduction to *Rhizoctonia solani* and other *Rhizoctonia* species. Anastomosis grouping and publication of this book are thought to be the beginning of the current research of *Rhizoctonia* spp. After this book the only one book on *Rhizoctonia* spp.: "Identification of *Rhizoctonia* Species" (Sneh *et al.*, 1991) has been published.

One of the epoch-making areas of study was related to the role of anastomosis grouping in the subdivision of *R. solani*. Matsumoto *et al.* (1932) introduced hyphal fusion to differentiate the rice sheath blight pathogen from the other groups of *R. solani*. They divided the hyphal fusion phenomenon into three categories. Namely, perfect fusion, imperfect fusion and contact fusion. They found that the hyphae of the rice sheath blight isolates could fuse with each other by either perfect or imperfect fusion, but could not fuse with isolates of other groups of *Rhizoctonia* spp. Schultz (1936) reported the existence of 5 AGs in *R. solani*, although one of these groups. *R. solani* var. *fuchsiae* is not considered to be *R. solani*, but belongs to a binucleate *Rhizoctonia* species. (Ogoshi *et al.*, 1983). The importance of anastomosis groups (AGs) was emphasized once again by Richter and Schneider in (1953). They reported the existence of 6 AGs. Again, Gruppe E, is not *R. solani* It also belongs to a binucleate *Rhizoctonia* species.

Even after these studies anastomosis grouping was still not used routinely to identify *Rhizoctonia* isolates. Parmeter *et al.* (1969) reported the existence of 4 groups within *R. solani*, and they indicated the existence of "bridging" isolates, as well as predicting the existence of additional AGs. This is a very important paper, as it may be considered as a rediscovery of the importance of anastomosis grouping in *R. solani*. Indeed, Kuninaga *et al.*, (1978) confirmed the existence of "bridging" isolates, and additional AGs having later been described. (Ogoshi, 1976; Kuninaga *et al.*, 1978; Homma *et al.*, 1983; Rovira *et al.*, 1986; Carling *et al.*, 1987; 1994; Ogoshi *et al.*, 1990). Identification and classification of *R. solani* isolates have been markedly improved due to the use of anastomosis grouping.

Anastomosis grouping may be used directly for ecological studies of *Rhizoctonia spp.*, e.g. for the investigation of number of clones existing in a certain field and how these clones are distributed in this field. Using this method, a clone can be identified by the perfect fusion of the isolates with the representative clone. Many clones may thus be found even in a small rice field (Ogoshi and Ui, 1983). Although some clones are widely distributed, others may be more limited in their distribution patterns. Periodical recoveries of different clones can provide valuable information on their differential patterns of distribution in the same field and at different seasons, as shown in potato and sugar beet fields.

G. 1971 - 1980

During this decade, the number of AGs was increased from 4 to 7 (Ogoshi, 1976; Kuninaga *et al.*, 1978), and subgroups of some AGs were reported. New techniques were introduced into the research of *R. solani,* such as ultrastructure and host-parasite interactions of *R. solani,* which were studied with the electron microscope (Naiki and Ui, 1974; Lisker *et al.*, 1976).

H. 1981 - 1990

The number of *R. solani* AGs increased to 11. Currently, there are 12 groups, as AG 11 was reported recently in Australia and in the USA (Carling *et al.* 1994). Analysis of *R. solani* DNA and DNA base sequence homologies were used to compare relatedness of isolates within and among AGs (Kuninaga and Yokosawa, 1982). The genetic basis of anastomosis groups was thus established. Further molecular techniques were employed to compare AGs, subgroups, isolates, as well as for other purposes (Vilgalys, 1988; Neate *et al.*, 1988; Vilgalys and Gonzalez, 1990; Liu *et al.*, 1990; Jabaji-Hare *et al.*, 1990).

I. 1991-1995

The number of *R. solani* subgroups continue to increase. *Rhizoctonia* researchers can now be expected to identify the AGs and subgroups of their isolates at the beginning of their studies. Biotechnological studies are increasing. The development and use of transgenic resistant plants (Broglie *et al.*, 1991), detection by ELISA (Benson, 1992), use of ribosomal DNA polymorphism (Liu *et al.*, 1993), RAPD assay (Duncan, 1993) and other molecular techniques were reported.

The term intraspecific group (ISG) was proposed by Ogoshi (1987) as a name for subgroups of AG of *R. solani*. It may mean that AGs could be considered as species while their subgroup could be interpreted as subgroups. The term ISG was subsequently used in the literature resulting in some confusion. No doubt that some of this confusions been created because AGs have not been formally recognized as taxonomic species. Recognizing this, I would like to indicate that use of the term ISG should be discontinued. The term subgroup should be a more suitable one to use instead.

III. Taxa related to *R. solani* and other *Rhizoctonia* spp.

Rhizoctonia spp. consists of a very diverse collection of teleomorphs referred to different families orders, and even classes, which are only considered together in this book due to certain vegetative features distinguishing the form genus. There are presently about 120 species, varieties or formae speciales described within *Rhizoctonia* spp. Some of these were clearly miss-identified and are not *Rhizoctonia*, while others are considered to e synonyms of earlier described species. Recently, Andersen and Stalpers (1994) have made a thorough review of *Rhizoctonia* taxonomy, which will facilitate later taxonomical work in this form genus. Due to diversity of teleomorphs, the recognized species have been referred to five different form genera (Moore 1987) and two more are erected later in this book to bring the anamorph taxonomy in line with the teleomorphs (see chapters I.A1, I.A3). The genera with anamorphs referred to *Rhizoctonia* are *Helicobasidium, Thanatephorus, Ceratobasidium, Waitea, Tulasnella and Sebacina,* which are treated in more detail in chapter I.A3. *Helicobasidium,* only contains the type species of the form genus *R. crocorum*, which by now is fairly well known, and thus no longer extensively studied.

Of the teleomprphs, the closely related *Thanatephorus* (with *R. solani*) and *Ceratobasidium* are by far the most studied, as several of these are severely pathogenic to various crops throughout the world. For the determination of distribution, host-plant relations and pathogenicity, the sub-division into AGs and subgroups has been the focus of the last decades. Also the binucleate species of *Ceratobasidium* have been divided into AGs (Burpee *et al.*, 1980; Ogoshi *et al.*, 1983; Sneh *et al.*, 1991). Particularly biochemical and molecular studies such as DNA base sequence relatedness (Kuninaga and Yokosawa, 1985), rDNA polymorphism (Vilgalys and Gonzales, 1990) and electrophoretic isozyme analyses (Neate *et al.*, 1988; Liu *et al.*, 1990) have been powerful tools. *R. solani* has thus been divided into at least 28 groups and subgroups and most likely several more will be reported in the future. Recently, uninucleate *Rhizoctonia* have been reported, probably referable to *Ceratobasidium bicorne* (Hall, 1986; Hietala *et al.*, 1994) (see also chapter V.11).

Several strains can be referred to *Waitea circinata* (Parmeter and Whitney, 1970); Ogoshi *et al.*, 1985), which has a multinucleate *Rhizoctonia* anamorph, *R. zeae*. The strains are very different in morphology from *R. solani* and *Waitea* may not belong to the family ceratobasidiaceae, but rather to the corticaceae.

Members of the *Tulasnella* and *Sebacina,* as well as *Thanatephorus* and *Ceratobasidium* spp. are known to form mycorrhizal associations with orchids (see also chapter V.14). The two former genera are mainly nonpathogenic and live saprophytically.

IV. HOST RANGE AND DISTRIBUTION OF *RHIZOCTONIA* SPP. AGS

The host plants infected by *Thanatephorus cucumeris* reported in Japan alone belong to 35 orders, 52 families, 125 genera and comprise more than 142 species from the Cycadopsida to the Monocotyledoneae.

R. solani and binucleate *Rhizoctonia* spp. AGs have been identified in relation to many diseases. Isolates of certain AGs were found to infect many host plants, while those of other AGs have a rather limited host ranges. AG 1 and its subgroups infect many different crops. The most important disease caused by isolates of this AG is the rice sheath blight. Isolates of AG 2 and its subgroups infect rice, sugar beet and corn. Isolates of AG 3 cause the black scurf disease on potato tubers. Isolates of AG 4 have a wide host range, including cotton and onion. Isolates of AG 8 cause the severe "bare patch" disease on wheat and barley. Isolates of other anastomosis groups may be weakly pathogenic on several plants or not pathogenic at all such as AG 6, which is considered to be mainly non pathogenic. Certain isolates are capable to form mycorrhizal associations with orchids. (Masuhara *et al*, 1993; see chapter V.14).

Anastomosis groups of *R. solani*, binucleate *Rhizoctonia* spp. or *Waitea* species have been reported in 47 countries or districts; but more likely at least some AGs are present in every country. Distribution of anastomosis groups depends for pathogenic groups on the host plant distribution and on the climate and vegetation types for saprophytic groups. However, they are basically cosmopolitans, since they have a relatively wide host range and a flexible capability for adaptation.

V. SUGGESTED FURTHER RESEARCH DIRECTIONS FOR *RHIZOCTONIA SPP.*

Rhizoctonia spp. have been extensively studied in recent years, due to the introduction of new techniques and new ideas. The following aspects deserve special attention and continued basic research:

1. As opposed to other groups, isolates of AGs 5, 6, 7, 9, 10 and BI of *R. solani* have been reported to be weakly pathogenic or not pathogenic. Are all of their isolates really saprophytes? Isolates of AG 7 have recently been reported as a severe pathogen on carnation (Lo *et al.*, 1990) in Taiwan and on soybean in Indonesia (Naito *et al.*, 1991). Therefore, the possibility that isolates of AGs 6, 9, 10 and BI are severe pathogens on some plants under certain environmental conditions should not be outruled.

2. Genetic studies regarding the pathogenicity of *Rhizoctonia* species and host-parasite relationships are relatively scarce compared with other pathogens. *R. solani* is basically multinucleate, and its hyphae are primarily heterokaryotic. Multinucleate heterokaryotic cells are less suitable for genetic studies. Therefore, the use of uninucleate basidiospores as inoculum and consecutive infection on host plants, should enable researchers to obtain more useful information of the genetic basis of pathogenicity, similar to genetic studies performed with airborne

pathogens. Transformations of isolated genes from one isolate to the other could be expected to contribute to the understanding of the genetics of pathogenicity and physiology of *Rhizoctonia* spp.

3. It would be interesting to find out the destiny of fused protoplasts obtained from isolates of different AGs or different subgroups. Protoplasts obtained from isolates of different *R. solani* AGs are capable of fusing with each other. Can these fused protoplasts survive and establish new strains? What characteristics are displayed by isolates originating from these fused protoplasts? Such investigations may help elucidate characteristics and relationships of AGs and subgroups.

4. How does an isolate recognize the hyphae of members of its own AG? The first step of the recognition process must include hyphal attraction. A hypha of one isolate is attracted only by a hypha of another one which belongs to the same AG. Different chemotactic substances emanating from hyphae of the different AGs probably play an important role in the first steps of recognition prior to hyphal fusion. Therefore the number of attracting substances may be the same as that of AGs.

5. The main mycorrhizal fungi of orchids are binucleate *Rhizoctonia (R. repens, R. anaticula* and others) *R. solani* has also been recorded to form mycorrhizal associations with orchids. However, relationships between AGs of *R. solani* and orchids have not been studied in detail. AG 5 and AG 6 have been isolated from orchids (Uetake *et al.*, unpublished) and it has been shown that all AGs of *R. solani* can potentially form mycorrhizal associations (Masuhara *et al.*, 1993). Furthermore, several new isolates of multinucleate *Rhizoctonia* may be found to be involved in orchid mycorrhizal associations (Uetake *et al.*, unpublished). There are 3 anastomosis groups of binucleate *Rhizoctonia* which can not be assigned to the known groups (Uetake *et al.*, unpublished data). Additional work is needed to establish whether they belong to *Thanatephorus, Ceratobasidium, Tulasnella, Sebacina* or others.

6. In Australia, AG 8 has been divided into 5 groups according to the zymogram patterns of the isolates (MacNish *et al* 1993). It is quite possible that additional groups will be found in AG 8 in future. Cultural appearances are diverse among isolates from USA, Australia, and Scotland and may deserve special attention in this respect.

7. AG 8 is known as a bridging group. Isolates of this group were found to be capable to fuse with isolates of AGs 2-1, 2-2 and BI (Rovira *et al.*, 1986). The bridging behavior has not been sufficiently studied.

8. In South Africa, the "Crater rot" disease of wheat has been known for a long time (Scott *et al.* 1979). It should be determined to which AG the isolates causing this disease, belong. Researchers in South Africa claim that this pathogen is *R. solani,* but the isolates do not fuse with any representative tester isolates of the known AGs (Lubbe and Meyer, 1995).

9. It should be determined how to consider *Thanatephorus praticola* (AG 4 of *T. cucumeris*). Should it be included in *T. cucumeris*, or retained as *T. praticola,* an independent species. This fungus has at least 3 features which may differentiate it from other *R. solani* AGs. The number of nuclei in young hyphae is ca. four in average (Ogoshi, 1976), the number of sterigmata on a basidium is three in

average (Kotila, 1929), and hyphal width is narrower than that of isolates belonging to the other AGs (Ogoshi, 1976). Numbers of *Rhizoctonia* spp. show 4 different hyphal nuclear numbers, *C. bicorne* is uninucleate (Hietala *et al.*, 1994), *C. cornigerum* and other *Ceratobasidium* spp. are binucleate (Burpee *et al.*, 1980), *T. praticola* is tetranucleate (Ogoshi, 1976) and *T. cucumeris* is multinucleate (Parmeter and Whitney, 1970). Using this criterion, *T. praticola* could be considered as an independent species.

10. Classification and naming of anastomosis groups and subgroups of *R. solani* (or *T. cucumeris*) requires thorough consideration and collaborative efforts. Based on DNA base sequence homology and other studies, each AG may be considered as an independent species. However, in that case, *T. cucumeris* alone would comprise 11 or 12 species. It is not convenient to have so many species which have essentially the same perfect state may be with the exception of AG 4 (*T. praticola*). It is more convenient to introduce variety and form, as suggested by Schultz (1936), such as *R. solani* var. *brassicae* for AG 2-1 and *R. solani* var. *typica* for AG 3 (Sneh et al., 1991). Such suggestions should be considered in studies and discussions among the taxonomical specialists in future.
11. International distribution of *Rhizoctonia* isolates is basically not acceptable for quarantine reasons. Over the years, I have sent our isolates to 34 countries, 77 locations and 85 researchers. In a sense this is dangerous with regard to dissemination and infestation of the pathogens. I would like to suggest to establish a system, where at least one *Rhizoctonia* researcher in every country, district or continent will assume the responsibility to maintain and distribute *Rhizoctonia* spp. isolates. The key researchers should keep isolates of *R. solani*, binucleate *Rhizoctonia* spp. and *Waitea* AGs from their own countries or districts and determine their AGs with the known authentic isolates carefully. The local isolates could then be made available to researchers in the country, district or continent as tester strains. in the country, district or continent.

It is forecasted that within 50 years the world's human population may reach 10 billion people, and that food production will be the most serious problem facing the human race in the next century. *Rhizoctonia* is one of the important pathogens which are obstacles to food production. Our role in this context should be to improve the efficiency of plant protection against plant pathogens and thus contribute to the reduction of crop losses, which means to increase the world's capability to produce food required for the forecasted population size.

VI. ACKNOWLEDGMENT

I would like to express my sincere thanks to Dr. TF Andersen, for his critical review of the manuscript and for his valuable comments.

VII. References

Andersen TF & Stalpers JA (1994) A check-list of Rhizoctonia epiphytes. Mycotaxon 51:437-457.

Benson DM (1992) Detection by enzyme-linked immunosorbent assay of *Rhizoctonia* species on poinsettia stem cutting. Plant Dis. 76:578-581.

Bernard N (1909) L'evolution dans la symbiose des orchidéés et leur champignons commensaux. Ann. Sci. Nat. Paris 9. sér. 9:1-196

Boosalis MG & Scharen AL (1959) Methods for microscopic detection of *Aphanomyces euteiches* and *Rhizoctonia solani* associated with plant debris. Phytopathology 49:192-198.

Broglie K, Chet I, Holliday M, Cressman R, Biddle P, Knowlton S, Mauvais CJ & Broglie R (1991) Transgenic plants with enhanced resistance to the fungal pathogen *Rhizoctonia solani*. Science (Washington) 254:1194-1197.

Burpee LL, Sanders PL, Cole H Jr & Sherwood RT (1980) Anastomosis groups among isolates of *Ceratobasidium cornigerum* and related fungi. Mycologia 72:689-701.

Carling DE, Leiner RH & Kebler KM (1987) Characterization of a new anastomosis group (AG 9) of *Rhizoctonia solani*. Phytopathology 77:1609-1612.

Carling DE, Rothrock CS, MacNish GC, Sweetingham MW & Brainard KA (1994) Characterization of anastomosis group 11 (AG 11) of *Rhizoctonia solani*. Phytopathology 84:1387-1393.

De Candolle AP (1815) Flore Française ed. 3, Vol. V. Desray, France.

Donk MA (1956) Notes on resupinate Hymenomycetes - 2. The tulasnelloid fungi. Reinwardtia 3:363-379.

Duncan S, Barton JE & O'Brien PA (1993) Analysis of variation in isolates of *Rhizoctonia solani* by random amplified polymorphic DNA assay. Mycol. Res. 97:1075-1082.

Duggar BM (1915) *Rhizoctonia crocorum* (Pers.) D.C and *R. solani* Kühn (*Corticium vagum* B & C), with notes on other species. Ann. Missouri Botanical Garden 2:403-458.

Exner B (1953) Comparative studies on four *Rhizoctonias* occurring in Louisiana. Mycologia 45:698-719.

Hall G (1986) A species of *Rhizoctonia* with uninucleate hyphae isolated from roots of winter wheat. Trans. Brit. Mycol. Soc. 87:466-471.

Herzog W & Wartenberg H (1958) Untersuchungen uber die Lebensdauer der Sklerotien von *Rhizoctonia solani* Kühn im Boden. Phytopathol. Z. 33:291-315.

Hietala AM, Sen R & Lilja A (1994) Anamorphic and teleomorphic characteristics of a uninucleate *Rhizoctonia* sp. isolated from the roots of nursery grown conifer seedlings. Mycol. Res. 98:1044-1050.

Homma Y, Yamashita Y & Ishii M (1983) A new anastomosis group (AG 7) of *Rhizoctonia* solani Kühn from Japanese radish fields. Ann. Phytopath. Soc. Japan 49:184-190.

Ito K, Kontani S & Kondo H (1955) Web-blight fungus of Japanese larch seedlings. Bull. Forestry Experiment Station (Japan) 79:43-70.

Jabaji-Hare SH, Meller Y, Gill S & Charest PM (1990) Investigation of genetic relatedness among anastomosis groups of *Rhizoctonia solani* using cloned DNA probes. Can. J. Plant Pathol. 12:393-404.

Jager G & Velvis H (1983) Suppression of *Rhizoctonia solani* in potato fields. 1. Occurrence. Neth. J. Pl. Path. 89: 21-29.

Kotila J (1929) A study of the biology of a new spore forming *Rhizoctonia*, *Corticium praticola*. Phytopathology 19:1059-1099.

Kuninaga S & Yokosawa R (1982) DNA base sequence homology in *Rhizoctonia* solani Kühn. I. Genetic relatedness within anastomosis group 1. Ann. Phytopath. Soc. Japan 48:559-667.

Kuninaga S & Yokosawa R (1985) DNA base sequence homology in *Rhizoctonia solani* Kühn. VI. Genetic relatedness among seven anastomosis groups. Ann. Phytopath. Soc. Japan 51:127-132.

Kuninaga S, Yokosawa R & Ogoshi A (1978) Anastomosis grouping of *Rhizoctonia solani* Kühn isolated from non-cultivated soils. Ann. Phytopath. Soc. Japan 44:591-598.

Kühn JG (1858) Die Krankheiten der Kulturegewachse, ihre Ursachen und ihre Verhutung. Gustav Bosselmann, Berlin. 312 pp.

Liu ZL, Domier LL & Sinclair JB (1993) ISG-specific ribosomal DNA polymorphism of the *Rhizoctonia solani* species complex. Mycologia 85:795-800.

Liu ZL, Nicksrent DL & Sinclair JB (1990) Genetic relationships among isolates of *Rhizoctonia solani* anastomosis group-2 based on isozyme analysis. Can. J. Plant Pathol. 12:376-382.

Lisker N, Katan J & Henis Y (1976) Scanning electron microscopy of the infection of beans by *Rhizoctonia solani* propagules. Ann. Bot. 40:625-629.

Lo CT, Tu CC & Tsai WH (1990) Anastomosis groups and pathogenicity of *Rhizoctonia* spp. from carnation. Plant Prot. Bull. 32:158-161 (Taiwan).

Lubbe EM & Meyer L (1995) Crater disease of wheat in South Africa: an overview. Abstract of International Symposium of *Rhizoctonia*. the Netherlands. pp 59.

Masuhara G, Katsuya K & Yamaguchi K (1993) Potential for symbiosis of *Rhizoctonia solani* and binucleate *Rhizoctonia* with seeds of *Spiranthes sinensis* var. *amoena* in vitro. Mycol. Res. 97:746-752.

Matsumoto T, Yamamoto W & Hirane S (1932) Physiology and parasitology of the fungi generally referred to as *Hypochnus Sasakii* Shirai. I. Differentiation of the strains by means of hyphal fusion and culture in differential media. J. Soc. Trop. Agric. 4:370-388.

MacNish GC & Sweetingham MW (1993) Evidence of stability of pectic zymogram groups within *Rhizoctonia solani* AG 8. Mycol. Res. 97:1056-1058.

Naiki T & Ui T (1974) Ultrastructures of the hyphal call walls and the sclerotia of *Rhizoctonia solani* Kühn. Trans. Mycol Soc. Japan 15:113-120.

Naito S, Djaeni M, Nasution A, Purwanti H, Sutoyo, Amir M, Kardin K, Sudjadi M, Hifni HR, Achmad NH & Lidya B (1991) Soil-borne diseases affecting yield loss of soybean. Proceeding of Final Seminar of the Strengthening of Pioneering Research for PALAWIJA Crop Production (ATA:378), Indonesia. pp 24-37.

Neate SM, Cruikshank RH & Rovira AD (1988) Pectic enzyme patterns of *Rhizoctonia* solani isolates from agricultural soils in South Australia. Trans. Brit. Mycol. Soc. 90:37-42.

Ogoshi A (1976) Studies on the grouping of *Rhizoctonia solani* Kühn with hyphal anastomosis and on the perfect stages of groups. Bull. Nat. Inst. Agr. Sci. Ser. C. No 30:1-63.

Ogoshi A & Ui T (1983) Diversity of clones within an anastomosis group of *Rhizoctonia* solani in a field. Ann. Phytopath. Soc. Japan 49:239-245.

Ogoshi A, Oniki M, Araki T & Ui (1983) Studies on the anastomosis groups of binucleate *Rhizoctonia* and their perfect states. J. Fac. Agr. Hokkaido Univ. 61:244-260.

Ogoshi A, Cook RJ & Bassett EN (1990) *Rhizoctonia* species and anastomosis groups causing root rot of wheat and barley in the pacific northwest. Phytopathology 80:784-788.

Olive LS (1957) Two new genera of the Ceratobasidiaceae and their phylogenetic significance. Amer. J. Bot. 44:429-435.

Olsen CM & Baker KF (1968) Selective heat treatment of soil, and its effect on the inhibition of *Rhizoctonia solani* by *Bacillus subtilis*. Phytopathology 58:79-87.

Oniki M, Ogoshi A, Araki T, Sakai R & Tanaka S (1985) The perfect state of *Rhizoctonia oryzae* and *R. zeae*, and the anastomosis groups of *Waitea circinata*. Trans. Mycol. Soc. Japan 26:189-198.

Papavizas GC & Davey CB (1959) Isolation of *Rhizoctonia solani* Kühn from naturally infested and artificially inoculated soils. Plant Dis. Reptr. 43:404-410.

Parmeter JR Jr (ed.) (1970) Biology and Pathology of *Rhizoctonia solani*. University of California Press, Berkeley. 255 pp.

Parmeter JR Jr & Whitney HS (1970) Taxonomy and nomenclature of the imperfect state. In: Parmeter JR Jr (ed.) Biology and Pathology of *Rhizoctonia solani*. (pp 7-19) University of California Press, Berkeley.

Parmeter JR Jr, Sherwood, RT & Platt WD (1969) Anastomosis grouping among isolates of *Thanatephorus cucumeris*. Phytopathology 59:1270-1278.

Rao AS (1959) A comparative study of competitive saprophytic ability in twelve root-infecting fungi by an agar plate method. Trans. Brit. Mycol. Soc. 42:97-111.

Richter H & Schneider R (1953) Untersuchungen zur morphologischen und biologischen Differenzierung von *Rhizoctonia solani* K. Phytopathol. Z. 20:264-271.

Rovira AD, Ogoshi A & McDonald HJ (1986) Characterization of isolates of *Rhizoctonia solani* from cereal roots in South Australia and New South Wales. Phytopahtology 28:233-246.

Sanford GB (1952) Persistence of *Rhizoctonia solani* Kühn in soil. Can. J. Botany 30:652-664.

Savithiry S & Gnanamanickam SS (1987) Bacterization of peanut with *Pseudomonas fluorescens* for biological control of *Rhizoctonia solani* and for enhanced yield. Plant and Soil 102:11-15.

Schultz H (1936) Vergleichende Untersuchungen zur Oekologie, Morphologie und Systematik des 'Vermehrungspilzes'. Arb. K. Biol. Reichsanst. Land und Forstw. 22:1-41.

Scott DB, Visser CPN & Rufenacht EMC (1979) Crater disease of summer wheat in African drylands. Plant Dis. Reptr. 63:836-840.

Sneh B, Burpee B & Ogoshi A (1991) Identification of *Rhizoctonia* Species. APS Press, St. Paul, USA. 133 pp.

Vilgalys R (1988) Genetic relatedness among anastomosis groups in *Rhizoctonia* as measured by DNA/DNA hybridization. Phytopathology 78:698-702.

Vilgalys R & Gonzalez D (1990) Ribosomal DNA restriction fragment length polymorphism in *Rhizoctonia solani*. Phytopathology 80:151-158.

Weindling R (1932) *Trichoderma lignorum* as a parasite of other soil fungi. Phytopathology 22:837-845.

Section I.

TAXONOMY AND EVOLUTION OF *RHIZOCTONIA* SPP.

I.A Classical Methods

I.B Biochemical and Molecular Methods

I.A1 THE DOLIPORE/PARENTHESOME SEPTUM IN MODERN TAXONOMY

ROYALL T. MOORE

School of Applied Biological and Chemical Sciences University of Ulster at Coleraine Northern Ireland BT52 1SA:UK

I. INTRODUCTION

Fungal septa are a consequence of opposing layers of wall material being deposited between opposite faces of a centripetally invaginating membrane. Septa are entire in the aquatic fungi and most of the zygomycetes, but are generally perforate in the higher fungi (Moore, 1994). It is not known how cytokinesis terminates in a central pore. Ascomycete septa are homogeneous, electron translucent; the pores have rounded edges, are of a size to allow nuclear migration, and are accompanied by one or more Woronin bodies on either side that are large enough to block the pore (Buller, 1933; Markham and Collinge, 1987). Basidiomycete septa, however, are tripartite and appear light-dark-light layered when viewed electron microscopically (e.g., Bracker and Butler, 1963). The essential compositional difference between ascomycete and basidiomycete septal morphology was first noted by Kreger-van Rij and Veenhuis (1971) and is now recognized to be a fundamental indicator of the two subkingdoms (Table 1). This dichotomy also occurs in the walls and is most useful in determining the taxonomic affinities of yeast phases where the distinction is diagnostic during budding (see Moore, 1989c; 1996b).

II. BASIDIOMYCETE CLASSIFICATION

The dichotomy of the basidiomycetes into the classes of Homobasidiomycetes and Heterobasidiomycetes dates from Patouillard's 1900 *Essai taxonomique des hyménomycètes* and, though obsolete, is still commonly used in one form or another (e.g., Jülich, 1981; Wells, 1994). Contemporary difficulties with this dichotomy are taxonomic and nomenclatural. The nomenclatural problems are a consequence of deciding what the major groups of fungi are, the ranks they are to be given, and the International Code of Botanical Nomenclature which specifies suffixes for taxa and subtaxa but makes no provision for supertaxa (Moore, 1974). Treating the basidiomycetes as a unitary assemblage, as class Basidiomycetes (Lowy, 1971; Alexopoulos and Mims, 1979) or as subdivision or subphylum Basidiomycotina (Talbot, 1971; Kendrick, 1985; Carlile and Watkinson, 1994), disallows formal expression of considered taxonomic judgments above the ordinal level. Many authors choose to ignore these niceties and use in their schemes such class-mimicking terms (sometimes formed as "subclasses") as Gastero-, Hemibasidio-, Holobasidio-, Hymeno-, Phragmobasidio-, and Teliomycetes. The kingdom Fungi (Table 1) is now known to be a natural assemblage comprising the chytrids and all the terrestrial taxa (Bruns *et al.*, 1991). Superordinal nomenclatural space is opened out, therefore, by excluding the mycetozoa from the mycological hierarchy, by omitting higher and lower fungi as designated taxa, and by formally recognizing subkingdoms (Table 1; Moore, 1994).

B. Sneh et al. (eds.),
Rhizoctonia Species: Taxonomy, Molecular Biology, Ecology, Pathology and Disease Constrol, 13–35.

The subkingdom Basidiomycetia comprises two phyla: the Basidiomycota and the Ustomycota (Table 1). The phylum Basidiomycota, the basidiomycetes proper, is characterized by 5S rRNA in Walker's Group 5 (Walker and Doolittle, 1983) and, except for the rusts and *Eocronartium*, the dolipore/parenthesome septal pore complex (Figs. 1-8). The phylum Ustomycota (Moore, 1988a; 1996a, b), ustomycetes, however, has other 5S rRNA (see Moore, 1988a) and nanoscopic pores with no paraseptal cytoplasmic differentiation (Oberwinkler and Bauer, 1989; Moore, 1996a; Fig. 9). This deeper, contemporary dichotomy makes previously used class names (Hemibasidio-, Heterobasidio-, Holobasidio-, Homobasidio-, Phragmobasidio-, and Teliomycetes) inapplicable and they are, therefore, rejected. (Parentheticaly, but very important for clarity, terms ending in '-mycetes' are classes and singular if the prefix starts with a capital letter but are vernacular and plural if the prefix starts with a small letter; the singular '-mycete' is always vernacular and must start with a lowercase letter).

TABLE 1. Major taxa of the kingdom fungi and subkingdom basidiomycetia

SUBKINGDOM MASTIGOMYCETIA		
PHYLUM OOMYCOTA [guest taxon]		
PHYLUM HYPHOCHYTRIDIOMYCOTA [guest taxon]		
PHYLUM CHYTRIDIOMYCOTA		
SUBKINGDOM ZYGOMYCETIA		
PHYLUM ZYGOMYCOTA		
PHYLUM TRICHOMYCOTA		
SUBKINGDOM ASCOMYCETIA		
PHYLUM ASCOMYCOTA		
PHYLUM HEMIASCOMYCOTA		
SUBKINGDOM BASIDIOMYCETIA		
PHYLUM BASIDIOMYCOTA		
Class Hymenomycetes		
Order Agaricales	Order Coltriciales	Order Polyporales
Order Amanitales	Order Hericiales	Order Russulales
Order Boletales	Order Meruliales	Order Sistotrematales
Order Cantharellales	Order Phaeolales	Order Stereales
Order Ceratobasidiales	Order Pluteales	Order Thelephorales
Class Clavariomycetes		
Order Botryobasidiales	Order Lentinales	Order Dacrymycetales
Order Clavariales	Order Tulasnellales	Order Hyphodontiales
Order Hymenochaetales	Order Clavuliciales	Order Hirschioporales
Class Gasteromycetes		
Class Exidiomycetes		
Order Exidiales	Order Tremellales	Order Filobasidiales
Class Auriculariomycetes		
Order Auriculariales	Order Eocronartiales	Order Pucciniales
PHYLUM USTOMYCOTA		
Order Agaricostilbales	Order Tilletiales	Order Heterogastridiales
Order Atractiellales	Order Platygloeoales	Order Ustilaginales
Order Exobasidiales	Order Septobasidiales	
Order Graphiolales	Order Sporidiales	
SUBKINGDOM DEUTEROMYCETIA		
PHYLUM DEUTEROMYCOTA		
PHYLUM BLASTOMYCOTA		

III. THE DOLIPORE/PARENTHESOME SEPTUM

The dolipore/parenthesome (d/p) septum (Fig. 1) is composed of three elements (Moore, 1985): the dolipore (d), its occlusions (o), and the paraseptal parenthesomes (p). All dolipores have a similar underlying morphology that can be imagined as arising when the cleaving invaginating membrane (Orlovich and Ashford, 1994) encounters some sort of a dense central material and partially flows around it before growth ceases (Patton and Marchant, 1978a; Moore, 1985). The two types of wall material are variously distributed in the resulting 'barrel-shaped' mold (Moore, 1985). The pressure of the closing diaphragm on the central core appears to divide it, resulting in a pair of slightly separated, base-to-base champagne-cork-shaped moieties occupying the dolipore channel. The outer portion of an occlusion occupies the dolipore orifice and can have one of two cross-sectional morphologies (Fig. 2): either that of a dense granule (O_1) or that of a striated band (O_2). There are three principal *parenthesome* morphologies (Fig. 2): regularly perforate (P_1), imperforate (P_2), though a single, minuscule central perforation may occur (Fig. 5), and vesiculate (P_{3-5}); they may also be absent (P_6). It is not known how parenthesomes form (Orlovich and Ashford, 1994). Parenthesomes of the P_1 and P_2 type, associated with O_1 occlusions, are dome-shaped organelles differentiated from the endoplasmic reticulum; at the juncture of the differentiation the intermembrane space widens slightly, an unexplained central line/layer/membrane is interposed, and the lumen is filled with a material that is more electron dense than the cytoplasm. There have been proposals to use perceived differences in lumen morphology, pore size, and pore density (Patton and Marchant, 1978b) as taxonomic criteria but, to date, there is no systematic study that demonstrates such subtilities in this organelle, a structure whose ultracytobiology we do not understand: the problems of size and morphology of the organelle, fixation, state of material under different growth conditions/ages, biochemistry and molecular biology, etc. (see Orlovich and Ashford, 1994).

Each O_1 type dolipore has a pair of parenthesomes, one on each side, delimiting organelle-free cytoplasm (Figs. 3-7). In septa with O_2 type occlusions, Tremellales *sensu stricto* (see below) and Filobasidiales, each parenthesome, if present, comprises an aggregate of cuneiform to ampulliform vesicles 'tethered' by filaments to the striate band of the occlusion (Fig. 8). The morphology of a vesicle can be pictured as a long-stemmed goblet with transparent film covering the top, the stem and the glass represent the tether and the vesicle, which has the typical parenthesome cross-sectional morphology, while the film across the abseptal end represents the continuation of the outer vesicle membrane across the opening.

IV. THE MAIN ASSEMBLAGES OF THE BASIDIOMYCOTA

The exploitation of septal ultrastructure as a major taxonomic character has been inhibited by the absence of a comprehensive inventory (Tables 3, 4, 5) because much of the information occurs widely and casually in the literature. This has led to a general impression that hymenomycetes have perforate parenthesomes, that those of the jelly fungi are imperforate, and that the imperforate occurrences in certain aphyllophores are exceptional and anomalous (Moore, 1994).

The several septal morphologies resulting from combinations of the septal components are, however, taxononomically significant (Fig. 2). They, in combination with basidial structure, allow a clearer definition of the major taxa of the

Basidiomycota (Table 2). The following recommendations are proposed towards a more felicitous taxonomy of these fungi.

1. All species within a genus should have the same basic parenthesome type, e.g., *Hirschioporus/Trichaptum* (these genera may be synonymous but the parenthesome morphology of the types species of *Trichaptum* has yet to be demonstrated; Moore, 1988b; 1994).
2. All genera comprising a family and families comprising an order should be of the same basic parenthesome type (Tables 3, 4, 5).
3. Classes should be defined by a combination of basidial and septal morphologies (Table 2). Five such classes are proposed in the phylum Basidiomycota: Auriculariomycetes, Clavariomycetes, Exidiomycetes, Gasteromycetes, and Hymenomycetes (if, in the future, genuine and consistant differences are found within these taxa, they could be the basis for recognizing subclasses).

TABLE 2. Types of basidia and septa in major taxa

Taxon	**Basidia**[a]			**Septa**[b]
	O	$\ominus$	$\otimes$	
HYMENOMYCETES	√			O_1P_1
GASTEROMYCETES	√			O_1P_1
CLAVARIOMYCETES	√			O_1P_2
AURICULARIOMYCETES				
Auriculariales		√		O_1P_2
Eocronartiales		√		rust-type
Uredinales		√		rust-type
EXIDIOMYCETES				
Exidiales			√	O_1P_2
Tremellales			√	$O_2P_{3,4}$
Filobasidiales	√			$O_2P_{3\text{-}6}$

[a]Key to basidial morphology: O = non-septate; $\ominus$ = transverse septate; $\otimes$ = cruciate septate.
[b]Key to septal morphology: O = dolipore Occlusion; P =Parenthesome.

A. Hymenomycetes and Clavariomycetes

1. Discussion The fundamental dichotomy between these two classes is whether, respectively, the parenthesomes are perforate or imperforate (Tables 1, 2). Several aggregate taxa contain species with contradictory parenthesomes; the traditional order Aphyllophorales, in particular, is a large, heterogeneous omnium gatherum. The order, however, is no longer tenable, besides being heterogeneous, it is also illegitimate and a synonym of the Polyporales (Jülich, 1981). The much reduced Polyporales can be retained for species with P_1 parenthesomes that remain unassigned to other orders in the Hymenomycetes.

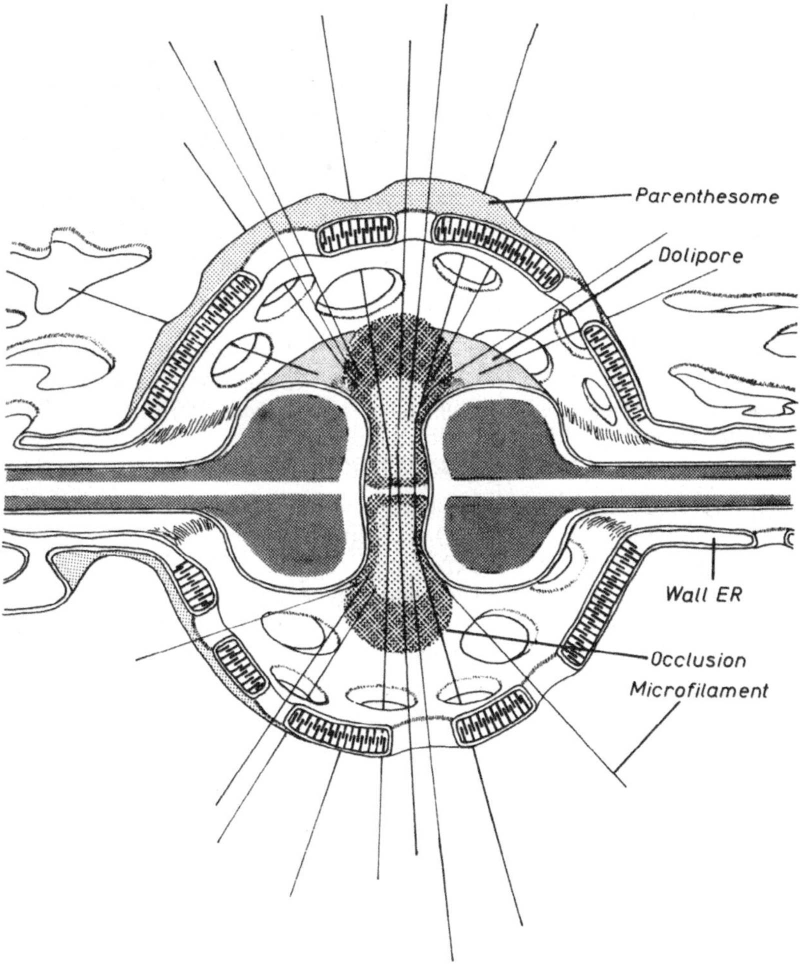

FIGURE 1. Perspective interpretation of a dolipore septum with perforate parenthesomes showing the champagne-cork-shape type of occlusions characteristic of Hymenomycetes and Gasteromycetes. Note that the parenthesomes are differentiated from the endoplasmic reticulum (ER) and that each has an interposed 'line'. (From Moore & Marchant, 1972.)

The corticoid fungi and other resupinate aphyllophores comprise a taxonomic archipeligo through which a number of authors have striven to draw a line between forms they perceive as being either "Homobasidiomycetes" or "Heterobasidiomycetes"; they have, as well, attempted to define the limits of the taxa they perceive comprising each "class". The simple basidiomata of these fungi - often little more than thin layers of mycelium bearing basidia that may be more or less tulasnelloid - have been variously interpreted, either as primitive or as advanced and reduced. Efforts to systematically structure this melange range from broadly complimentary to widely divergent (see Jülich, 1981; Parmasto, 1986; Ginns and

Lefebvre, 1993). Yet these and other authors, past and present, are all using the same visual and light microscopic data (sometimes enhanced by scanning electron micrographs). The consequent result is a plethora of higher taxa from which to choose. The several schemes proposed for ordering this complex of fungi have employed a number of morphological characters but none has exploited septal morphology.

The disposition of the orders and families of the Hymenomycetes (Table 3) generally follows the scheme of Ginns and Lefebvre (1993). The colligation of the Clavariomycetes (Table 4) selects, as much as possible, from available taxa; in the absence of these, new taxa are proposed that are based on species whose ultrastructure is known.

2. **Nomenclature** **HYMENOMYCETES** Fries 1821-1832 emend. RT Moore (Figs. 1, 2)
Phylum Basidiomycota. Septa O_1/P_1 type; basidiomata hemiangiogarpic to gymnocarpic; metabasidia becoming autobasidia (Corner, 1948).

Coltriciales R.T. Moore, new order. (Figs. 5, 6)
Class Hymenomycetes. Type: Coltriciaceae Jülich 1981.

CLAVARIOMYCETES R.T. Moore, new class
Phylum Basidiomycota. Septa O_1/P_2 type; basidiomata hemiangiocarpic to gymnocarpic; metabasidia remaining entire, ranging in form from hymenomycete-like to tulasnelloid (forming inflated sterigmata that are constricted at the base and which may become deciduous). Type: Clavariales R.T. Moore.

Clavariales R.T. Moore, new order.
Class Clavariomycetes. Type: Clavariaceae Chev. 1826.

Clavuliciales R.T. Moore, new order.
Class Clavariomycetes. Type: **Clavuliciaceae** R.T. Moore, fam. nov. (Inter se aequales *Clavulicium* Boidin 1957, Type).

Hirschioporales R.T. Moore, new order (Fig. 5)
Class Clavariomycetes. Type: **Hirschioporaceae** (Bond. & Sing.) R.T. Moore, comb. nov. (type,Hirschioporae Bondarzew and Singer 1941).

Hyphodontiales R.T. Moore, new order.
Class Clavariomycetes. Type: **Hyphodontiaceae** (Parm.) R.T. Moore, comb. nov. (type,Hyphodontieae Parmasto 1986).

Lentinales R.T. Moore, new order.
Class Clavariomycetes. Type: Lentinaceae Jülich 1981.

Paullicorticiales R.T. Moore, new order.
Class Clavariomycetes. Type: **Paullicorticiaceae** R.T. Moore, fam. nov. (Inter se aequales *Paullicorticium* Eriksson 1958, Type).

B. **Gasteromycetes Fries 1821-1832.**
Phylum Basidiomycota. Basidiomata angiocarpic; metabasidia becoming apobasidia forming passively shed spores.

Only a few species of these fungi have been examined by transmission electron microscopy but those that have been studied have perforate parenthesomes (O_1/P_1 type septa): *Calvatia* (Beneke, 1963); *Clathrus* (Eymé and Parriaud, 1970); *Lycoperdon* (Marchant, 1969; Flegler *et al.*, 1976); *Nidularia* (Patton and Marchant, 1978b); *Nia*

(Brooks, 1975); *Phallus* (Patton and Marchant, 1978b); *Pisolithus* (Orlovich and Ashford, 1994); *Sphaerobolus* (Dykstra, 1982).

C. Auriculariomycetes

1. Discussion The three orders comprising the Auriculariomycetes are a natural group united by transversely septate epibasidia. The order Auriculariales is characterized by saprobic habit, prominent basidiomata, and O_1/P_2 type septa (Moore, 1978; Tu and Kimbrough, 1978; Oberwinkler, 1985; Lü and McLaughlin, 1991).

The Eocronartiales and Uredinales are parasitic and have a reduced septal morphology composed of simple pores with sharp edges and, frequently, an included pulley-shaped, electron-dense granule; the paraseptal cytoplasm is denser, organelle free, and delimited by domes of simple vesicles (Coffey *et al.*, 1972; Littlefield and Heath, 1979; Khan *et al.*, 1982; Boehm and McLaughlin, 1989).

Jülich (1982) established the monotypic family Eocronartiacaeae for the single species *Eocronartium muscicola* (Fr.) Fitz. and stated that its position was in the Uredinales rather than the Auriculariales. Boehm and McLaughlin (1988; 1989), however, stated that it could not be lodged among the rusts because of its type of biotrophy and the micromorphology of nuclear division and branching. For these reasons it is established in its own order.

2. Nomenclature AURICULARIOMYCETES (Jülich) R.T. Moore, comb. nov.
Phylum Basidiomycota. Type: Auriculariinae Jülich 1981.

Auriculariales Schroeter emend. R.T. Moore

Class Auriculariomycetes, with the additional character of O_1/P_2 type septa.

Eocronartiales R.T. Moore, new order

Class Auriculariomycetes. Type: Eocronartiaceae Jülich 1982 emend. R.T. Moore, with the additional character of the septal complex composed of simple pores with sharp edges (frequently with an included pulley-shaped, electron-dense granule) and the paraseptal cytoplasm delimited by domes of simple vesicles.

Pucciniales Clements and Shear emend. R.T. Moore

Class Auriculariomycetes, with the additional character of the septal complex composed of simple pores with sharp edges (frequently with an included pulley-shaped, electron-dense granule) and the paraseptal cytoplasm delimited by domes of simple vesicles.

D. Exidiomycetes

1. Discussion The three orders comprising the Exidiomycetes (Table 2) are a natural group united by interlocking character sets (see Moore, 1978; 1985; 1988c; 1996b; Wells, 1994). The order Exidiales is characterized by saprobic habit, large basidiomata, cruciately septate basidia, hyphal monokaryons, and O_1P_2 type septa (Table 5).

The order Tremellales sens. str. (Bandoni, 1988) is characterized by parasitic habit (Zugmaier *et al.*, 1994; Roberts, 1995), generally large basidiomata, cruciately septate basidia, a yeast haplophase, and $O_2P_{3\text{-}4}$ type septa.

The order Filobasidiales (Julich, 1981) is characterized by basidiomata that are absent or stillboid, non-septate basidia, a yeast haplo-phase (except for two homothallic species) that may be parasitic, and septa that are $O_2P_{4\text{-}6}$ type. The family Filobasidiaceae (Kwon-Chung, 1988) includes two genera *Filobasidium* and

Filobasidiella with O_2P_4 or 6 type septa; *Cystofilobasidium*, a ustomycete, order Sporidiales, is excluded (Moore, 1996a,b). The family Wallemiaceae is monotypic. *Wallemia sebi* has O_2P_5 type septa and has been interpreted as a teleomorph (Moore, 1986).

2. Nomenclature EXIDIOMYCETES R.T. Moore, new class (Figs. 7, 8)

Phylum Basidiomycota. Type: **Exidiales** R.T. Moore, new order (Type, Exidiaceae R.T. Moore 1978).

Tremellales Rea emend. R.T. Moore (Fig. 9)

Class Exidiomycetes. Type: Tremellineae R.T. Moore 1978.

Wallemiaceae R.T. Moore, fam. nov.

Ordo Filobasidiales. Idem *Wallemia* Johan-Olsen atque septum doliporus cum parenthesoma vesiculata (O_2P_5 forma).

TABLE 3. Class Hymenomycetes: Taxa with P_1 parenthesomes
[ ] = Septal structure cited but not illustrated; { } = Taxonomic reference

AGARICACEAE	
Agaricus	Manocha, 1965; Craig *et al.*, 1979
Lepiota	Gow & Gad, 1995
AMANITACEAE	
Amanita	Flegler *et al.*, 1976
ATHELIACEAE	
Athelia	Tu & Kimbrough, 1978; Tu *et al.*, 1977
BOLBITIACEAE	
Agrocybe	Gull, 1976
BOLETACEAE	
Boletus	Patrignani & Pellegrini, 1986; Becket *et al.*, 1974
CANTHARELLACEAE	
Cantharellus	Danell *et al.*, 1993
CERATOBASIDIACEAE	
Ceratobasidium	Tu *et al.*, 1977; Patton & Marchant, 1978b; Tu & Kimbrough, 1978; Currah & Sherburne, 1992;Wells, 1994; {Warcup & Talbot,1966; Warcup, 1975; Ginns, 1983;Moore, 1987; 1989b}
Oliveonia	{Warcup, 1975}
Thanetephorus	Bracker & Butler, 1963; Setliff *et al.*, 1972; Tu & Kimbrough, 1978; Tu *et al.*, 1977; Curah & Sherburne, 1992; Langer (G.), 1994; {Warcup, 1975; Parmesto, 1986}
Tofispora	Langer(G.), 1994
Uthatobasidium	Tu *et al.*, 1977; Tu & Kimbrough, 1978; Oberwinkler, 1985; Langer (G.), 1994; {Parmesto, 1986}
Waitea	Tu & Kimbrough, 1978; Tu *et al.*, 1977; {Moore, 1987}
COLTRICIACEAE	{Jülich, 1981}
Coltricia	Moore, 1980

TABLE 3. cont'd

CONIOPHORACEAE	
Coniophora	Langvad, 1971; {Ginns, 1982a}
COPRINACEAE	
Coprinus	Valk & Marchant, 1978; Henderson & Ross, 1983
CORTINARIACEAE	
Galerina	Besson & Froment, 1968
DIGITATISPORACEAE	
Digitatispora	Brooks, 1975
FISTULINACEAE	
Fistulina	Patrignani & Pellegrini, 1986
GANODERMATACEAE	
Ganoderma	Mims & Seabury, 1989
HERICIACEAE	{Jülich, 1981}
Hericium	Flegler, *et al.*, 1976
HYPHODERMATACEAE	{Jülich, 1981}
Hyphoderma	Langer(E.) & Oberwinkler, 1993
Hypochnicium	Langer(E.) & Oberwinkler, 1993
LACHNOCLADIACEAE	
Scytinostroma	Besson & Froment, 1968; {Rattan, 1974}
MERULIACEAE	
Chondrostereum	Moore, 1988b; Spiers & Hopcroft, 1988
Gloeoporus	Moore, 1988b
Merulius	Moore, 1988b
PHAEOLACEAE	{Jülich, 1981}
Phaeolus	Moore, 1980
PHANEROCHAETACEAE	
Cora	Roskin, 1970
Dictyonema	Slocum, 1980
PLEUROTACEAE	
Pleurotus	Moore, 1977; 1989a; Sl'zec, 1986
PLUTEACEAE	
Volvariella	Flegler, *et al.*, 1976
POLYPORACEAE	
Favolus	Flegler, *et al.*, 1976
Fomes	Moore, 1980
Lenzites	Hyde & Walkinshaw, 1966
Polyporus	Wilsenach & Kessel, 1965; Moore & Marchant, 1972; Patton & Marchant, 1978b; Moore, 1975; 1980; 1985;
Poria	Setliff *et al.*, 1972; Figs. 3,4
Trametes	Aylmore *et al.*, 1984
SCHIZOPHYLLACEAE	
Schizophyllum	Marchant & Wessels, 1973
SISTOTREMATACEAE	{Jülich, 1981}
Galzinia	Wells, 1994
Sistotrema	Currah & Sherburne, 1992; Langer (G.), 1994
SPARASSIDACEAE	
Sparassis	Patrignani & Pellegrini, 1986

TABLE 3. cont'd

STEREACEAE	
Stereum	Patrignani & Pellegrini, 1986
STROPHARIACEAE	{Jülich, 1981}
Naematoloma	Khan & Kimbrough, 1979
Pholiota	Wells, 1978; Arita, 1979
Psilocybe	Flegler, *et al.*, 1976; Tu & Kimbrough, 1978
Stropharia	Blanz, 1978
THELEPHORACEAE	{Stalpers, 1993}
Thelephora	Patrignani & Pellegrini, 1986; Langer (G.), 1994
Tomentella	Calonge, 1969
TRICHOLOMATACEAE	
Armillaria	Berliner & Duff, 1965; Moore, 1965; Schmid & Liese, 1970
Clitocybe	Patrignani & Pellegrini, 1986
Collybia	Foerster *et al.*, 1965
Crinipellis	[Moore, 1985]
Flammulina	Williams *et al.*, 1985
Panellus	Lingle, 1989
TYPHULACEAE	
Typhula	Metzler, 1984
VUILLEMINIACEAE	
Corticium	Duncan, 1974; Patton & Marchant, 1978b; {Warcup, 1975}
Laetisaria	Hoch & Howard, 1981
TAXA INCERTAE SEDIS	
Dendrosporomyces	Nawawi *et al.*, 1977
Digitatispora	Brooks, 1975
Moniliophthora	Evans *et al.*, 1978
Moniliopsis	Currah & Sherburne, 1992
Riessia	Goos & Tubaki, 1973
Sclerotium	Tu *et al.*, 1977; Punter *et al.*,1984; Hanlin & Tortolero, 1989; Mordue, 1983
Spiniger	Hanlin, 1978
(ant fungi)	Angeli-Papa & Eymé ,1979
(*Pseudotsuga* mycorrhzia)	Strullu, 1979
(mycangial fungus)	Happ *et al.*, 1976
(raspberry root nodules)	Harris & Mackenzie, 1982

TABLE 4. Class Clavariomycetes Hymenomycetous taxa with P_2 parenthesomes [ ] = Septal structure cited but not illustrated; { } = Taxonomic reference

BOTRYOBASIDIACEAE	
Botryobasidium	Langer (G.), 1994
CLAVARIACEAE	
Clavaria	Peterson *et al.*, 1980; Patrignani & ellegrini, 1986;{Seviour *et al.*, 1973}
CLAVULICIACEAE	
Clavulicium	[Oberwinkler, 1985]; {Parmesto, 1986}

TABLE 4. cont'd

DACRYMYCETACEAE	
Calocera	Blanz, 1978; Patton & Marchant, 1978b; Tu & Kimbrough, 1978
Cerinomyces	Maekawa, 1987; Wells, 1994; {Ginns, 1982b; Parmesto, 1986}
Dacrymyces	Moore, 1965; 1978; Tu & Kimbrough, 1978; Flegler *et al.*, 1976
Guepiniopsis	Wells, 1994
HIRSCHIOPORACEAE	
Hirschioporus	Moore, 1988; Traquair & McKeen, 1978; {Macrae & Aoshima, 1966}; Fig. 5
Trichaptum	Alexander *et al.*. 1989; Rajchenberg & Bianchinotti, 1991; Rajchenberg & Bianchinotti, 1992, {Ryvarden, 1978}
HYMENOCHAETACEAE	
Hymenochaete	Oberwinkler, 1985
HYPHODONTIACEAE	
Hyphodontia	Langer (E.), *et al.*, 1992; Langer (E.) & Oberwinkler, 1993
Basidioradulum	Langer (E.) & Oberwinkler, 1993; {Parmesto, 1986}
Schizopora	Langer (E.) & Oberwinkler, 1993
LENTINACEAE	
Lentinus	Ando, 1972; Nakai & Ushiyama, 1974; {Hibbert *et al.*, 1994}
PAULLICORTICIACEAE	
Paullicorticium	[Oberwinkler, 1985]
PHELLINACEAE	{Fiasson, 1983; Fiasson & David, 1983; Jülich, 1981}
Inonotus	Moore, 1980
Onnia	Moore, 1980; Setliff *et al.*, 1972
Phellinus	Moore, 1980
TULASNELLACEAE	
Monosporonella	Oberwinkler & Ryvarden, 1991;
Tulasnella	Bandoni & Oberwinkler,1982; Khan & Talbot,1976; Moore, 1978; Langer (G.), 1994; {War cup & Talbot, 1967; Warcup, 1975}; Fig.6
Stilbotulasnella	Bandoni & Oberwinkler, 1982
TAXA INCERTAE SEDIS	
Dactylaria	Saikawa *et al.*, 1994
Pneumatospora	Fisher & Wakeley, 1986
(*Calluna* endosymbiont)	Bonfante-Fasolo, 1980

TABLE 5. Class Exidiomycetes, Exidiales: taxa with P_2 parenthesomes
[ ] = Septal structure cited but not illustrated; { } = Taxonomic reference

EXIDIACEAE	
Basidiodendron	Wells, 1994
Ductifera	[Oberwinkler, 1985]
Eichleriella	[Oberwinkler, 1985]
Exidia	Moore, 1978; Patton & Marchant, 1978b; Wells, 1994
Exidiopsis	Wells, 1994
HYALORIACEAE	
Elmeria	Wells, 1994
Heterochaetella	Wells, 1994
Myxarium	[Oberwinkler, 1985]
Phlogiotis	Patrignani & Pellegrini, 1986;
Pseudohydnum	[Oberwinkler, 1985]; Fig. 7
Protodontia	Wells, 1994
Stypella	[Oberwinkler, 1985]
SEBACINACEAE	
Sebacina	Williams & Thilo, 1989; Currah & Sherburne, 1992; {Warcup, 1975}
Tremelloscypha	[Oberwinkler, 1985]
TREMELLODENDROPSIDACEAE	
Tremellodendropsis	Wells, 1994

V. CONCERNING TAXA ASCRIBED TO *RHIZOCTONIA*

The partitioning of the basidiomycetes into the phyla Basidiomycota and Ustomycota and the reformatting of the basidiomycetes proper into new classes provides a contemporary context within which to classify forms ascribed to *Rhizoctonia*. These forms are anamorphic agonomycetes lacking clear cut morphological characters. They are associated with roots, as parasites or mycorrhiza but some types are mycoparasites and others are saprobes, and occur in both phyla of the basidiomycetes. The generally ambiguous nature of the hyphae has led to characterizing isolates by anastomosis groups (Table 6) and the description of a large number of species. Andersen and Stalpers (1994) analysed 119 epithets described in *Rhizoctonia* and reduced them to seven taxa considered to be validly published (see also Sneh *et al.*, 1991 and chapter V.14). The current respective taxa can be summarized as follows:

- ***Ceratorhiza*** R.T. Moore.

Teleomorph: *Ceratobasidium* Rogers; Hymenomycetes, Ceratobasidiales, Ceratobasidiaceae.

- ***Chrysorhiza*** Andersen and Stalpers.

Teleomorph: *Waitea* Warcup and Talbot; Hymenomycetes, Ceratobasidiales, Ceratobasidiaceae.

- ***Epulorhiza*** R.T. Moore emend T.F. Andersen and R.T. Moore.

Rhizoctonia similis, cellularis moniliformis formans, sed a eadum ultrastructura pori septalis differens. Ultrastructura pori enim septum doliporum, minus, cum parenthesomate imperforato vel pauciperforato. Lumen parenthesomatis tristratum, stratum centrale translucens, strata distalia densa. Parenthesomata ad marginem saepissime recurvata. Teleomorphosa *Tulasnella* Schröter (Fig. 6). Species typica *Rhizoctonia repens* Bernard.

Epulorhiza is similar to *Rhizoctonia* in the production of monilioid cells, but differs with respect to the septal pore ultrastructure. The pore ultrastructure is likewise a dolipore structure, but it is smaller and has imperforate or pauciperforate parenthesomes. The images of pauciperforate parenthesomes show a small central area narrowing to a minuscule pore that may appear as a discontinuity. The parenthesome lumen is three-layered, with a central electron transparent layer, surrounded by two electron dense layers. The margin of the parenthesomes is almost always recurved. The teleomorphs belong to the genus *Tulasnella* Schröeter (Clavariomycetes). Type species of *Epulorhiza* is *Rhizoctonia repens* Bernard.

- ***Epulorhiza repens*** (Bernard) R.T. Moore emend R.T. Moore.
Idem *Epulorhiza repens* atque teleomorphosa *Tulasnella deliquescens* (Juel) Juel. **Neotypus**: Exsicatte of CBS 573.83 (J.H. Warcup 062). In herb. **C** no. 33549, isotypes in **L, K**.

- ***Opadorhiza*** T.F. Andersen and R.T. Moore gen. nov.
Epulorhiza similis, cellularis moniliformis formans, sed a eadum ultrastructura pori septalis differens. Ultrastructura pori septalis differens. Ultrastructura pori enim septum doliporum, minute. Parenthesomata imperforatum vel pauciperforatum, lumen tristratum, stratum centrale densum, strata distalia translucentia. Parenthesomata ad marginem nunquam recurvata. Teleomorphosa *Sebacina* Tul. Species typica *Rhizoctonia globularis* Saksena and Vaartaja emend. T.F. Andersen and R.T. Moore. Etym. *opados* Gr., attending, accompanying, *rhiza* Gr., root.

Opidorhiza is similar to *Epulorhiza* with respect to the production of monilioid cells, but differs with respect to the septal pore ultrastructure. The septal pore ultrastructure is likewise a dolipore, which is imperforate or pauciperforate. It is, however, much smaller and the three-layered lumen consists of a central electron dense layer surrounded by two electron transparent layers. Furthermore, the margin of the parenthesomes are never recurved. The teleomorphs belong to the genus *Sebacina* (Exidiomycetes, Sebacinaceae). The type species is *Rhizoctonia globularis* Saksena and Vaartaja emend T.F. Andersen and R.T. Moore.

- ***Opadorhiza globularis*** (Saksena and Vaartaja) T.F. Andersen and R.T. Moore comb. nov.
Basionym: *Rhizoctonia globularis* Saksena and Vaartaja (1960) Can. J. Bot 38:939. Excl. H.K. Saksena no. 2616. **Neotypus**: Exsiccate of CBS 262.60 (H.K. Saksena no. 2287). In herb. **C** no. 33550, isotypes in **L, K**.

- ***Rhizoctonia*** DC (proposed for conservation over *Moniliopsis* Ruhland; see chapter I.A3).
Teleomorph: *Thanatephorus* Donk.; Hymenomycetes, Ceratobasidiales, Ceratobasidiaceae.

-***Thanatophytum*** Nees.
Teleomorph: *Helicobasidium* Pat. (Fig. 9); Ustomycetes, Platygloeales, Platygloeaceae (Moore, 1990; 1994, resp.).

TABLE 6. Summary of taxa by teleomorph (anamorph) ascribed to *Rhizoctonia* with anastomosis groups ATCC accession numbers provided within (#). Herb. C. (#[a]).

BINUCLEATE		MLTINUCLEATE	
CERATOBASIDIUM **(*Ceraorhiza*): benign "spp."**		***THANATEPHORUS*** **(*Rhizoctonia* nom. cons.): *R. solani***	
AG-A [=CAG 2]	(76133)	AG 1-IA	(76121)
AG-Ba	(76134)	AG 1-IB	(76122)
AG-Bb	(76135)	AG 1-IC	(76123)
AG-Bo	(76136)	AG 2-1	(76168)
AG-C	(76137)	AG 2-t	( )
AG-D [=CAG 1]	(76138)	AG 2-2IIIB	(76124)
AG-E [=CAGs 3,6]	(76139)	AG 2-2IV	(76125)
AG-F [=CAG 4]	(76140)	AG 2-3	(200225)
AG-G	(76141)	AG 3	(76167)
AG-H	(76132)	AG 4 HG-I	(76126)
AG-I	(76143)	AG 4 HG-II	(76127)
AG-J	(76432)	AG 5	(76128)
AG-K	(76145)	AG 6 HG-I	(76129)
AG-L	(76146)	AG 6 GV	(76130)
AG-M	(lost)	AG 7	(76131)
AG-N	(76148)	AG 8	(76106)
AG-O	(76149)	AG 9 TP	(90332)
AG-P	(76150)	AG 9 TX	(90331)
AG-Q	(76151)	AG 10	(76106)
AG-R [=CAG 5]	(76144)	AG 11	(90857)
AG-S [=CAG 7]	(76147)	AG BI	(76132)
TULASNELLA **(*Epulorhiza*): *E. anaticula, E. repens***		***WAITEA*** **(*Chrysorhiza*): *C. oryzae, C. zeae***	
E. repens	(33549[a])		
Rr1	(76152)	*C. oryzae* WAG-O	(76154)
Rr2	(76153)	*C. zeae* WAG-Z	(76155)
SEBACINA **(*Opadorhiza*): *O. globularis*** (33550[a])			

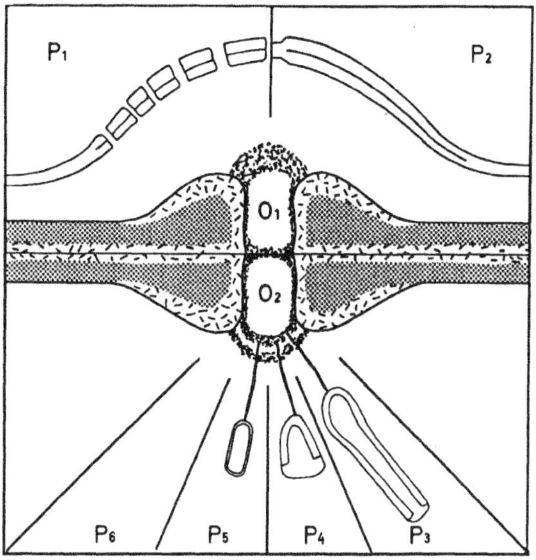

FIGURE 2. Summary diagram of basic dolipore/parenthesome (d/p) septal configurations (from Moore, 1985). The diagram is divided horizontally between the two types of occlusions found in the dolipore orifice: O_1 septa appear to have a granule while O_2 septa appear to have a a striated band. Associated with these septal types are several forms of parenthesome: P_1 regularly perforate; P_2 imperforate, $P_{3\text{-}5}$ vesiculate, P_6 absent. See Table 2 for major taxa characterized by combinations of these characters.

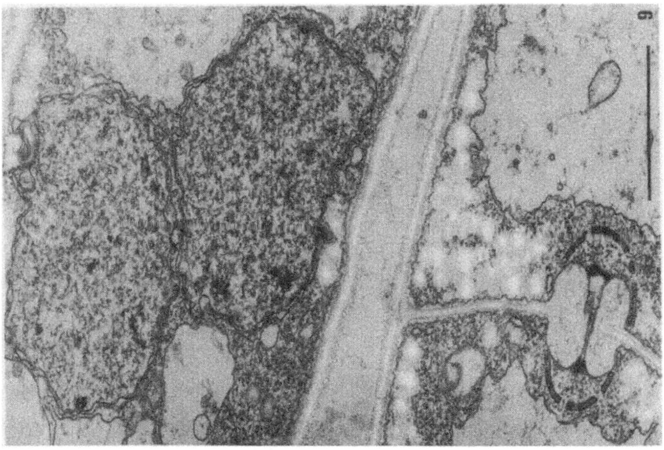

FIGURE 3. Portions of two dikaryotic hyphae of *Polyporus biennis* (Gull.:Fr.) Fries (Hymenomycetes). The left hypha shows the intimate association of the paired nuclei while the right hypha shows a representative O_1P_1 d/p septum. Scale bar = 1.0μm. Note that the septal complex is considerably smaller than the nuclei (about 1 x 2μm) and is not discernible by light microscopy. (From Moore, 1989a.)

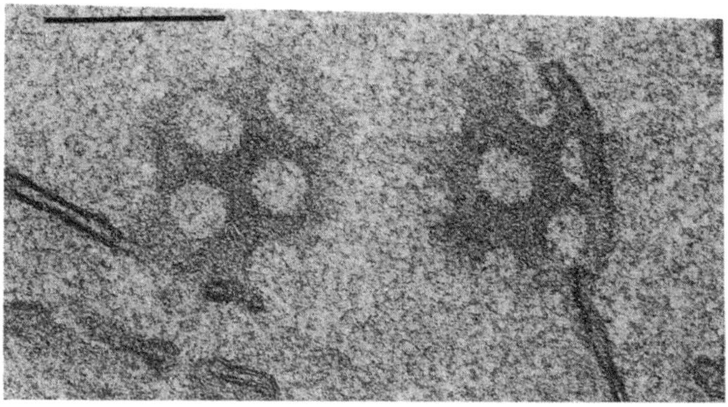

FIGURE 4. Tangential section of a P_1 parenthesome of *P. biennis* showing typical pores and their distribution and the abrupt transition with undifferentiated ER. Scale line = 0.25 1m. (From Moore, 1975.)

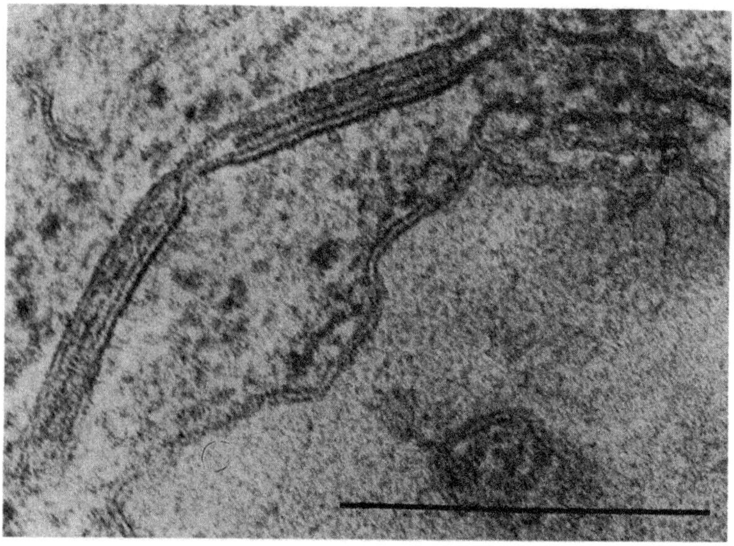

FIGURE 5. Imperforate parenthesome (O_1P_2) of *Hirschioporus abietinus* (Dickson ex Fr.) Donk (Clavariomycetes) showing the central discontinuity in the middle layer that gives this type of parenthesome a characteristic tented profile; a perfect medium section might have shown a minuscule pore. Note the ER-parenthesome continuity. Scale line = 0.5 1m. (From Moore, 1985.)

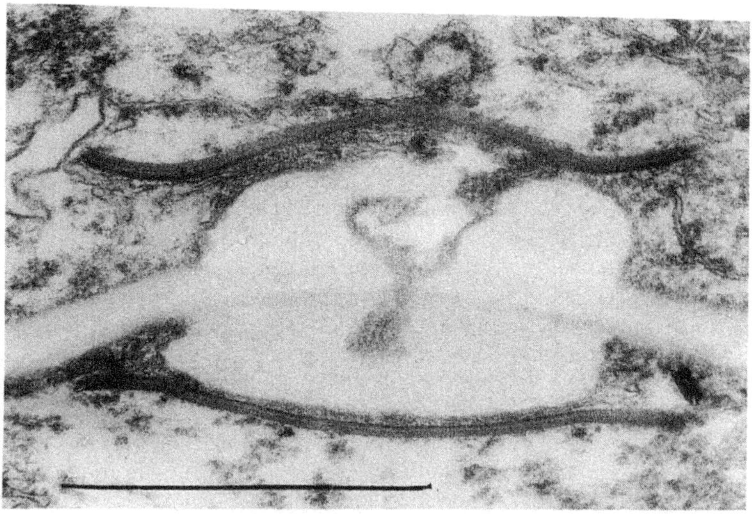

FIGURE 6. Imperforate parenthesomes (O_1P_2) of *Tulasnella fuscoviolacea* Bres. (Clavariomycetes) showing low arch profiles with recurved margins. Scale line = 0.75 1m. (See Moore, 1978.)

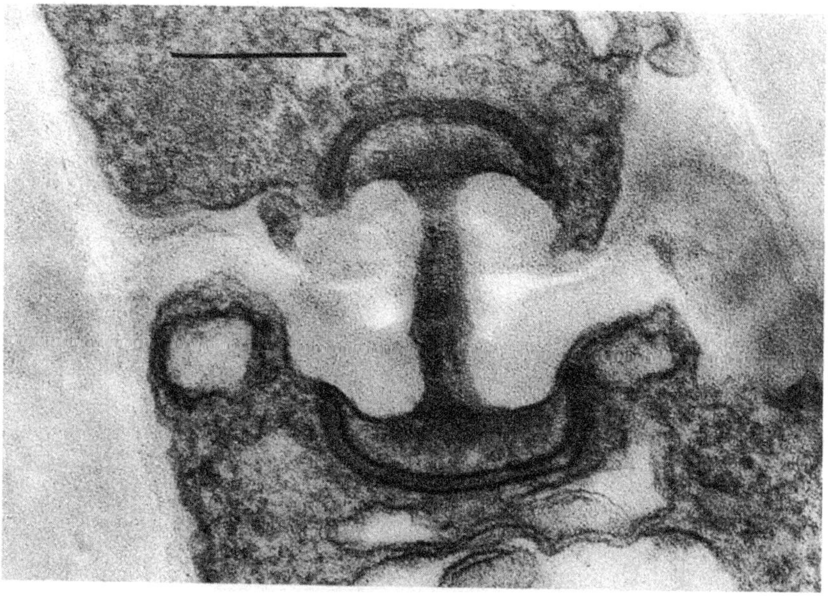

FIGURE 7. Imperforate parenthesomes (O_1P_2) of *Pseudohydnum gelatinosum* (Scop.) Karst. (Exidiomycetes, Exidiales) showing high arch profiles. Scale line = 0.5 1m. (Original micrograph from field collected material.)

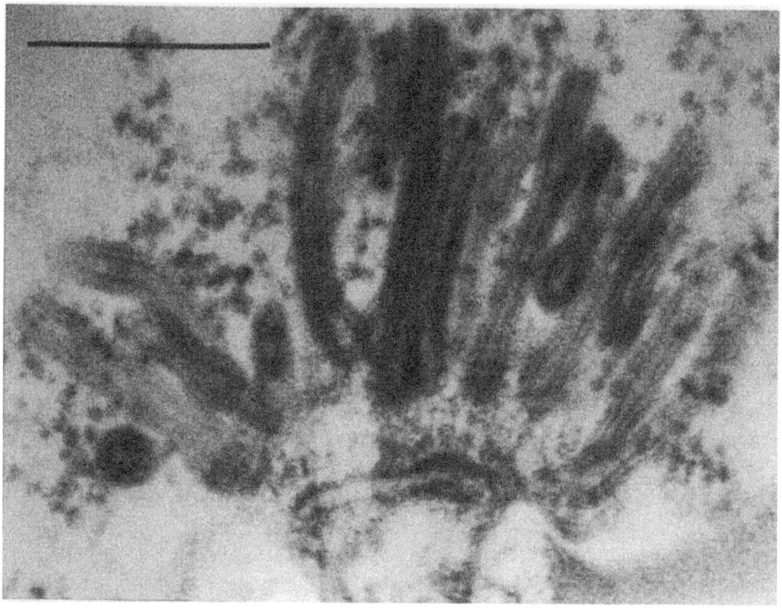

FIGURE 8. O_2P_3 septum of *Sirobasidium magnum* Boedjin (Exidiomycetes, Tremellales) with ampuliform parenthesome vesicules tethered by microfibrils to the striated band occlusion. Scale line = 0.25 1m. (From Moore, 1979.)

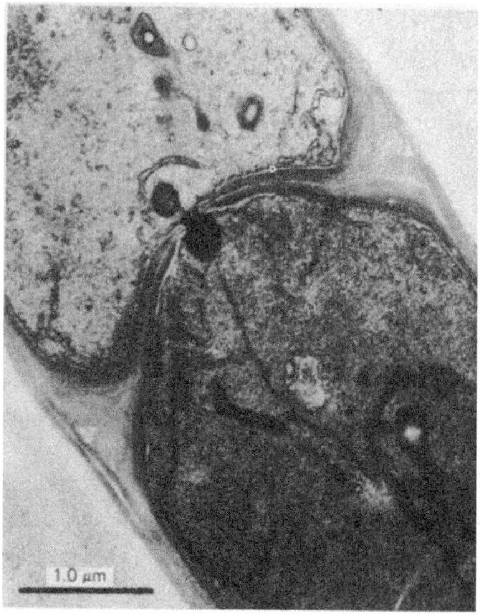

FIGURE 9. Ustomycete type septum of *Helicobasidium mompa* Tanaka (Ustomycota, Platygloeales). Scale line = 1.0 1m. (From Moore, 1985.)

VI. ACKNOWLEDGEMENTS

It is a particular pleasure to recognize the extensive and patient help of Dr. Torben Andersen for the Latin diagnoses and other assistance in the section on the nomenclature of the *Rhizoctonia* complex, to thank Dr. Baruch Sneh for the data in Table 6, and to acknowledge both Dr. Donald Carling for his review of the manuscript and Dr. Suha Jubadi-Hare for her exemplary editing of the final manuscript.

VII. REFERENCES

Alexander J, Welden AL & Ovrebo CL (1989) *Trichaptum sector* (Ehrenb.: Fr.) Kreisel (Polyporaceae s.l.). Mycologia Helvetica 3:291-302.

Alexopoulos CJ & Mims CW (1979) Introductory Mycology, ed. 3. John Wiley, New York.

Andersen TF & Stalpers JA (1994) A check-list of *Rhizoctonia* epithets. Mycotaxon 51:437-457.

Ando M (1972) Fines structure of the hyphae of *Lentinus edodes* Sing. Trans. Mycol. Soc. Japan 13:191-195.

Angeli-Papa J & Eymé J (1979) Le champignon cultivé par la "Fourmi-manioc" *Acromyrmex octinospinosus* Reich en Guadeloupe; résultats préliminaires sur le mycélium en culture pure et sur l'infrastructure des hyphes. CR Acad.Sc. Paris 289:21-24.

Arita I (1979) Cytological studies on *Pholiota*. Repts. Tottori Mycol. Inst. 17:1-118

Aylmore RC, Wakley GE & Todd NK (1984) Septal sealing in the basidiomycete *Coriolus versicolor*. J. Gen. Microbiol. 130:2975-2982.

Bandoni RJ (1988) Taxonomic overview of the Tremellales. In: de Hoog GS, Smith MTh & Weijman ACM (eds.) The Expanding Realm of Yeast-like Fungi (pp 87-110) Elsevier, Amsterdam.

Bandoni RJ & Oberwinkler F (1982) *Stilbotulasnella*:a new genus in the Tulasnellaceae. Can. J. Bot. 60:1875-1879.

Beckett, A Heath JM & McLaughlin DJ (1974) An Atlas of Fungal Ultrastructure. Longman, London.

Beneke ES (1963) *Calvatia*, calvacin, & cancer. Mycologia 55:257-270.

Berliner MD & Duff RH (1965) Ultrastucture of *Armillaria mellea* hyphae. Can. J. Bot. 43:171-172

Besson M & Froment A (1968) Observation d'un capuchon septal de type polypore hors des polyporacées. Bull. Soc. Mycol. France 84:485-488.

Blanz PA (1978) Über die systematische Stellung der Exobasidiales. Z. Mykol. 44:91-107.

Boehm EWA & McLaughlin DJ (1988) *Eocronartium muscicola*: a basidiomycetous moss parasite exploiting gametophytic transfer cells. Can. J. Bot. 66:762-770

Boehm EWA & McLaughlin DJ (1989) Phylogeny and ultrastructure in *Eocronartium muscicola*: meiosis and basidial development. Mycologia 81:98-114.

Bondarzew A & Singer R (1941) Zur Systematik der Polyporaceae. Ann. Mycol. 39:43-65.

Bonfante-Fasolo P (1980) Occurrence of a basidiomycete in living cells of mycorrhizal hair roots of *Calluna vulgaris*. Trans. Brit. Mycol. Soc. 75:320-325.

Bracker CE & Butler EE (1963) The ultrastructure and development of septa in hyphae of *Rhizoctonia solani*. Mycologia 55:35-58.

Brooks RD (1975) The presence of dolipore septa in *Nia vibrissa* and *Digitatispora marina*. Mycologia 67:172-174.

Bruns TJ, White TJ & Taylor JW (1991) Fungal molecular systematics. Annu. Rev. Ecol. Syst. 22:525-564.

Buller AHR (1933) Woronin bodies and their movements. Researches on Fungi 5:127-130.

Calonge FD (1969) Electron microscope studies on *Tomentella*. I. Ultrastructure of the vegetative hyphae. Arch. Mikrobiol. 65:136-145.

Carlile MJ & Watkinson SC (1994) The Fungi. Academic Press, London.

Coffey MD, Palevitz BA & Allen PJ (1972) The fine structure of two rust fungi *Puccinia helianthi* and *Melampsora lini*. Can. J. Bot. 50:231-240.

Corner EJH (1948) Studies in the basidium. I. The ampoule effect with a note on nomenclature. New Phytol. 47:22-51.

Craig GD, Newsam RJ, Gull K & Wood DA (1994) An ultrastructural and autoradiographic study of stipe elongation in *Agaricus bisporus*. Protoplasma 98:15-29.

Currah RS & Sherburne R (1992) Septal ultrastructure of some fungal endophytes from boreal orchid mycorrhizas. Mycol. Res. 96:583-587.

Danell E, Alström S & Terström A (1993) *Pseudomonas fluorescens* in association with fruit bodies of the ectomycorrhizal mushroom *Cantharellus cibarius*. Mycol. Res. 97:1148-1152.

Duncan EJ (1974) Studies on the vegetative hyphae of *Corticium areolatum*. Trans. Brit. Mycol. Soc. 63:115-120.

Dykstra MJ (1982) A cytological examination of *Sphaerobolus stellatus* fruiting bodies. Mycologia 74:44-53.

Evans HC, Stalpers HA, Samson RA & Benny GL (1978) On the taxonomy of *Monilia rorei*, an important pathogen of *Theobroma cacao* in South America. Can. J. Bot. 56:2528-2532.

Eymé J & Parriaud H (1970) Au sujet de l'infrastructure des hyphes de *Clathrus cancellatus* Tournefort Champignon gast romycète. CR Acad. Sci. Paris Série D 270:1890-1892.

Fiasson J-L (1983) Taxonomie phylétique des hyménochétacées porées d'Europe spécialement du genre *Phellinus*. Lyon: Docteur Sciences Naturelles thése Université Claude Bernard.

Fiasson J.-L & David A (1983) Approche taximétrique de hyménochétacées porées d'Europe (champignons Aphyllophorales). Can. J. Bot. 61:442-451.

Fisher PJ & Wakeley G (1986) Acetylenes and ultrastructure as aids in classification of *Pneumatospora obcoronata*. Trans. Brit. Mycol. Soc. 86:507-508.

Flegler SL, Hooper GR & Fields WG (1976) Ultrastructural and cytochemical changes in the basidiomycete dolipore septum associated with fruiting. Can. J. Bot. 54:2243-2253.

Foerster GE, Behrens PG & Airth RL (1965) Bioluminescence and other characteristics of *Collybia velutipes*. Am. J. Bot. 52:487-495.

Ginns J (1982a) A monograph of the genus *Coniophora* (Aphyllophorales basidiomycetes). Opera Botanica 61:1-61.

Ginns J (1982b) *Cerinomyces ceraceus* sp. nov. and the similar *C. grandinioides* & *Clagerheimii*. Can. J. Bot. 60:519-524.

Ginns J (1983) "*Ceratobasidium fibulatum*" an invalid name. Mycotaxon 18:439-442.

Ginns J & Lefebvre MNL (1993) Lignicolous Corticioid Fungi (Basidiomycota) of North America: Systematics Distribution & Ecology. American Phytopathological Society: St. Paul.

Goos RD & Tubaki K (1973) Conidium ontogeny in *Riessia semiophora*. Can. J. Bot. 51:1439-1442.

Gow NAB & Gadd GM (1995) The Growing Fungus. Chapman and Hall: London.

Gull K (1976) Differentiation of septal ultrastructure according to cell type in the basidiomycete *Agrocybe praecox*. J. Ultrastruc. Res. 54:89-94.

Hanlin RT (1978) Septum structure in *Spiniger meineckellus*. Amer. J. Bot. 65:471-476.

Hanlin RT & Tortolero O (1989) Morphology of *Sclerotium coffeicola*, a tropical foliar pathogen. Can. J. Bot. 67:1852-1860.

Happ GM, Happ CM & Barras SJ (1976) Bark beetle - fungal symbiosis. II. Fine structure of a basidiomycetous ectosymbiont of the southern pine beetle. Can. J. Bot. 54:1049-1062.

Harris DD & Mackenzie KAD (1982) Nodular bodies of an unidentified basidiomycete on raspberry roots. Trans. Brit. Mycol. Soc. 79:530-534.

Henderson LE & Ross IK (1983) Ultrastructural studies of vegetative and fruiting mycelia of *Coprinus congregatus*. Mycologia 75:634-647.

Hoch HH & Howard RJ (1981) Conventional chemical fixations induce artifactual swelling of dolipore septa. Exp. Mycol. 5:167-172.

Holm L (1987) The terminology of rusts: suggestions for a compromise. Notes Roy. Bot. Gard. Edinborough 44:433-435.

Jülich W(1981) Higher Taxa of Basidiomycetes. Cramer: Vaduz.

Jülich W(1982) Notes on some basidiomycetes (Aphyllophorales and Heterobasidiomycetes). Persoonia 11:421-428.

Hibbett DS, Tsuneda A & Murakami S (1994) The secotioid form of *Lentinus tigrinus*: genetics and developments of a fungal morphological innovation. Amer. J. Bot. 81:466-778.

Hyde JM & Walkinshaw CH (1966) Ultrastructure of basidiospores and mycelium of *Lenzites saepiaria*. J. Bacteriol. 92:1218-1227.

Kendrick B (1985) The Fifth Kingdom. Mycologue, Waterloo, Canada.

Khan SR & Kimbrough JW (1979) Ultrastructure of septal pore apparatus in the lamellae of *Naematoloma puigsgarii*. Can. J. Bot. 57:1064-2070.

Khan SR & Talbot PHB (1976) Ultrastructure of septa in hyphae and basidia of *Tulasnella*. Mycologia 68:1027-1036.

Khan SR, Kimbrough JW & Webb PG (1982) The fine structure of septa and haustoria of *Cronartium quercuum* formae speciales *fusinforme* on *Quercus rubra*. Mycologia 74:808-819.

Kreger-van Rij NJW & Veenhuis M (1971) A comparative study of the cell wall structure of basidiomycetous and related yeasts. J. Gen. Microbiol. 68:87-95

Kwon-Chung KJ (1988) Filobasidiaceae,a taxonomic survey. In: de Hoog GS, Smith MTh & Weijman ACM (eds.) The Expanding Realm of Yeast-like Fungi (pp 75-85) Elsevier: Amsterdam.

Langer E & Oberwinkler F (1993) Corticioid basidiomycetes. I. Morphology and ultrastructure. Windahlia 20:1-28.

Langer E, Langer G, Oberwinkler F & Tschen J (1992) A new *Hyphodontia* species from Taiwan. Trans. Mycol. Soc. Japan 33:401-408.

Langer G (1994) Die Gattung *Botryobasidium* Donk (Corticiaceae Basidiomycetes). Bibliotheca Mycologica 158:1-459.

Langvad F (1971) New structures in the basidiomycete *Coniophora cerebella*. J. Bacteriology 106:679-682.

Lingle, WL (1989) Enhanced staing of the basidiomycete *Panellus stypticus* prepared for transmission microscopy by freeze-substitution. Cryptogamic Bot. 1:236-242.

Littlefield LJ & Heath MC (1979) Ultrastructure of Rust Fungi. Academic Press, New York.

Lowy B (1968) Taxonomic problems in the Heterobasidiomycetes. Taxon 17:118-127.

Lowy B (1971) Tremellales. Flora Neotropica 6:1-153.

Lü H & McLaughlin DJ (1991) Ultrastructure of the septal pore apparatus and early septum initiation in *Aucicularia auricula-judae*. Mycologia 83:322-334.

Macrae R & Aoshima K (1966) *Hirschioporus* [*Lenzites*] *larcinus* and its synonyms: *L. abietis*, *L. ambigua*, *L. pinicola*. Mycologia 58:912-926.

Maekawa N (1987) A new species of the genus *Cerinomyces*. Can. J. Bot. 65:583-588.

Manocha MS (1965) Fine structure of the *Agaricus* carpophore. Can. J. Bot. 43:1329-1333.

Marchant R (1969) The fine structure and development of the fructification of *Lycoperdon perlatum*. Trans. Brit. Mycol. Soc. 53:63-68.

Marchant R & Wessels JGH (1973) Septal structure in normal and modified strains of *Schizophyllum commune* carrying mutations affecting septal dissolution. Arch. Mikrobiol. 9:35-45.

Markham P & Collinge AJ (1987) Woronin bodies of filamentous fungi. FEMS Microbiol. Rev. 46:1-11.

Metzler B (1984) Der Entwicklungszyklus von *Typhula incarnata* Lasch ex Fr-- Feinstrukturelle und Funktionelle aspekte. Doktors thesis Eberhard-Karls-Universität: Tübingen

Mims CW & Seabury F (1989) Ultrastructure of tube formation and basidiospore development in *Ganodema lucidum*. Mycologia 81:754-764.

Moore RT (1965) The ultrastructure of fungal cells. In: Ainsworth GC & Sussman AS (eds.) The Fungi. An Advanced Treatise. Vol. 1 (pp 95-118) Academic Press, New York.

Moore RT (1974) Proposal for the recognition of super ranks. Taxon 23:650-652.

Moore, RT (1975) Early ontogeneic stages in dolipore/parenthesome formation in *Polyporus biennis*. J. Gen. Microbiol. 87:251-259.

Moore RT (1977) Dolipore disjunction in *Antromycopsis broussonetiae* Pat. Exp. Mycol. 1:92-101.

Moore RT (1978) Taxonomic significance of septal ultrastructure with particular reference to the jelly fungi. Mycologia 70:1007-1024.

Moore, RT (1979) Septal ultrastructure in *Sirobasidium magnum* and its taxonomic implications. Antonie van Leeuwenhoek 45:113-118.

Moore RT (1980) Taxonomic significance of septal ultrastructure in the genus *Onnia* Karsten (Polyporineae/ Hymenochaetaceae). Bot. Not. 133:169-175.

Moore RT (1985) The challenge of the dolipore/ parenthesome septum. In: Moore D, Casselton LA, Wood DA & Frankland J (eds.). Developmental Biology of Higher Fungi (pp 175-212). Cambridge University Press, Cambridge.

Moore RT (1986) A note on *Wallemia sebi*. Antonie van Leeuwenhoek 52:183-187.

Moore RT (1987) The genera of *Rhizoctonia*-like fungi: *Ascorhizoctonia Ceratorhiza* gen. nov. *Epulorhiza* gen. nov. *Moniliopsis* and *Rhizoctonia*. Mycotaxon 29:91-99.

Moore, RT (1988a) A reconnaissance of the division Ustomycota (Basidiomycotera). In: Moriarty C (ed.) Taxonomy,Putting Plants and Animals in their Place. (pp 63-91) Royal Irish Academy: Dublin.

Moore RT (1988b) Taxonomic implications of septal ultrastructure in the Aphyllophorales. In: Wolkinger F (ed) Internationales Symposium über Aphyllorphorales (Fungi Basidiomycetes) Eisenstadt (1982). (pp 55-63) Austrian Academy of Sciences: Graz.

Moore RT (1988c) Micromorphology of yeasts and yeast-like fungi and its taxonomic implications. In: de Hoog GS, Smith MTh & Weijman ACM (eds.) The Expanding Realm of Yeast-like Fungi (pp 203-226) Elsevier, Amsterdam.

Moore RT (1989a) Alicean taxonomy,Small characters made large. Bot. J. Linn. Soc. 99:59-79.

Moore RT (1989b) *Ceratorhiza oryzae-sativae* a new combination for the anamorph of *Ceratobasidium setariae*. Antonie van Leeuwenhoek 55:393-395.

Moore RT (1989c) Observations on budding in basidiomycetous yeasts. Yeast 5:S305-S308.

Moore RT (1990) Order Platygloeales ord. nov. Mycotaxon 39:245-248.

Moore RT (1994) Third order morphology: TEM in the service of taxonomy. In: Hawksworth DL (ed.) Identification and Characterization of Pest Organisms. (pp 259-249). Wallingford: CAB International.

Moore RT (1996a) An Inventory of the Ustomycota. Mycotaxon 59:1-31.

Moore RT (1996b) Cytology and ultrastructure. In: Kurtzman CP and Fell J (eds.) The Yeasts a Taxonomic Study, Ed. 4 [in press]. Elsevier: Amsterdam.

Moore RT & Marchant R (1972) Ultrastructural characterization of the basidiomycete septum of *Polyporus biennis*. Can. J. Bot. 50:2463-2469.

Mordue JEM (1983) Dolipore septa in *Sclerotium hydrophilum*. Trans. Brit. Mycol. Soc. 81:654-655.

Nakai Y & Ushiyama R (1974) Fine structure of shiitake *Lentinus edodes* (Berk.) Sing. II. Development of basidia and basidiospores. Report of the Tottori Mycological Institute (Japan) 11:7-15.

Nawawi A, Webster J & Davey RA (1977) *Dendrosporomyces prolifer* gen. et sp. nov., a basidiomycete with branched conidia. Trans. Brit. Mycol. Soc. 68:59-63.

Oberwinkler F (1985) Ammerkungen zur Evolution und Systematik der Basidiomyceten. Bot. Jahrb. Syst. 107:541-580.

Oberwinkler F & Bauer R (1989) The systematics of gastroid auricularioid heterobasidiomycetes. Sydowia 41:224-256.

Oberwinkler F & Ryvarden L (1991) *Monosporella* a new genus in the Tulasnellaceae, basidiomycetes. Mycol. Res. 95:377-379.

Orlovich DA & Ashford AE (1994) Structure and development of the dolipore septum in *Pisolithus tinctorius*. Protoplasma 178:66-80.

Parmesto E (1986) On the origin of the hymenomycetes. (What are the corticioid fungi?). Windahlia 16:3-19.

Patrignani G & Pellegrini S (1986) Fine structures of the fungal septa on varieties of basidiomycetes. Caryologia 39:239-250.

Patton AM & Marchant R (1978a) An ultrastructural study of septal development in hyphae of *Polyporus biennis*. Arch. Microbiol. 118:271-277.

Patton AM & Marchant R (1978b) A mathematical analysis of dolipore/parenthesome structure in basidiomycetes. J. Gen. Microbiol. 109:335-349.

Peterson TA, Mueller WC & Englander L (1980) Anatomy and ultastructure of a *Rhododendron* root-fungus association. Can. J. Bot. 58:2421-2433.

Punter D, Reid J & Hopkin AA (1984) Notes on sclerotium-forming fungi from *Zizania aquatica* (wildrice) and other hosts. Mycologia 76:722-732.

Rajchenberg M & Bianchinotti V (1991) *Trametes fumoso-avellanea* (Aphyllophorales): a taxonomic study. Nord. J. Bot. 11:225-230.

Rajchenberg M & Bianchinotti V (1992) On the presence of imperforate parenthesomes in *Daedalea sprucei* and its taxonomical disposition. Mycol. Res. 96:956-958.

Rattan SS (1974) *Scytinostroma* in India with notes on extralimital species. Trans. Brit. Mycol. Soc. 63:1-12.

Roberts P (1995) British *Tremella* species I: *Tremella aurantia* & *T. mesenterica* Mycologist 9:110-114.

Roskin PA (1970) Ultrastructure of the host-parasite interaction in the basidiolichen *Cora pavonia* (Web.) EFries. Arch. Mikrobiol. 70:176-182.

Ryvarden L (1978) The Polyporaceae of North Europe 2:438. Fungiflora, Oslo.

Saikawa M, Baba N & Aoki Y (1994) Electron microscopy of amoeba-capturing *Dactylella tylopaga* showing the morphology of a basidiomycete. Mycologia 86:474-477.

Schmid R & Liese W (1970) Feinstruktur der Rhizomorphen von *Armillaria mellea*. Phytopath. Zeit. 68:221-231.

Setliff EC, McDonald WL & Patton RF (1972) Fine structure of the septal pore apparatus in *Polyporus tomentosus*, *Poria latemarginata* and *Rhizoctoia solani*. Can. J. Bot. 50:2559-2563.

Seviour RJ, Willing RR & Chilvers GA (1973) Basidiocarps associated with ericoid mycorrhizas. New Phytol. 72:381-385.

Slézec A.-M (1986) Les étapes de la méiose che le pleurotes des ombellifères. Cryptogam. Mycol. 7:235-265.

Slocum RD (1980) Light and electron microscopic investigations in the Dictyonemataceae (basidiolichens). II. *Dictyonema irpicinum*. Can. J. Bot. 58) 1005-1015.

Sneh B, Burpee LL & Ogoshi A (1991) Identification of *Rhizoctonia* Species. APS Press, St. Paul Minnesota. 133pp.

Spiers AG & Hopcroft DH (1988) Ultrastructural studies of basidial and basidiospore development and basidiospore release in *Chondrostereum purpureum*. European J. Forest Pathology 18:367-381.

Stalpers JA (1993) The aphyllophoraceous fungi I. Keys to the species of the Thelephorales. Studies in Mycology 35:1-168.

Strullu DG (1979) Ultrastructure et représsentation spatiale du manteau fongique des ectomycorhizes. Can. J. Bot. 57:23(19-2324.

Talbot PHB (1971) Principles of Fungal Taxonomy. Macmillan, London.

Traquair JA & McKeen WE (1978) Ultrastructure of the dolipore/parenthesome septum in *Hirschioporus pargamenus* (Polyporaceae). Can. J. Microbiol. 24:767-771.

Tu CC & Kimbrough JW (1978) Systematics and phylogeny of the *Rhizoctonia* complex. Bot. Gaz. 139:454-466.

Tu CC, Kimbrough JW & Aldrich HC (1977) Cytology and ultrastructure of *Thanatephorus cucumeris* and related taxa of the *Rhizoctonia* complex. Can. J. Bot. 55:2419-2436.

Valk P van der & Marchant R (1978) Hyphal ultrastructure in fruit-body primordia of the basidiomycetes *Schizophyllum commune* and *Coprinus cinereus*. Protoplasma 95:57-72.

Walker, WF & Doolittle, WF (1983) 5S rRNA sequences from eight basidiomycetes and fungi imperfecti. Nucleic Acids Res. 11:7625-7630.

Warcup JH (1975) Factors affecting symbiotic germination of orchid seed. In: Sanders, FE, Mosse, B & Tinker, PB (eds.). Endomycorrhizas (pp 87-104) Academic Press.

Warcup JH & Talbot PHB (1966) Perfect states of some rhizoctonias. Trans. Brit. Mycol. Soc. 49:427-435.

Warcup JH & Talbot PHB (1967) Perfect states of rhizoctonias associated with orchids. New Phytol. 66:631-641.

Wells K (1978) The fine structure of septal pore apparatus in the lamellae of *Pholiota terrestris*. Can. J. Bot. 56:2915-2924

Wells K (1994) Jelly fungi then and now! Mycologia 86:18-48.

Williams MAJ, Beckett A & Read ND (1985) Ultrastructural aspects of fruit body differentiation in *Flammulina velutipes*. In: Moore D, Casselton LA, Wood DA & Frankland J (eds.). Developmental Biology of Higher Fungi (pp 429-450) Cambridge University Press.

Williams PG & Thilo E (1989) Ultrastructural evidence for the identity of some multinucleate rhizoctonias. New Phytol. 112:513-518.

Wilsenach R & Kessel M (1965) On the function and structure of the septal pore of *Polyporus rugulosus*. J. Gen. Microbiol. 40:397-400.

Zugmaier W, Bauer R & Oberwinkler F (1994) Mycoparasitism of some *Tremella* species. Mycologia 86:49-56.

I.A2 GROUPING IN *RHIZOCTONIA SOLANI* BY HYPHAL ANASTOMOSIS REACTION.

DONALD E. CARLING

University of Alaska Fairbanks, 533
E. Fireweed, Palmer Alaska 99645, USA

I. INTRODUCTION

Grouping of strains of *Rhizoctonia solani* Kühn (teleomorph: *Thanatephorus cucumeris* [(Frank) Donk] based on hyphal anastomosis reactions has been described as the principle most helpful to plant pathologists in studies of *R. solani* during a 15 year period beginning in 1965 (Anderson, 1982). Similarly, Vilgalys and Cubeta (1994) state that "Since its inception, the anastomosis group concept has represented the single most important advance toward understanding genetic diversity in *Rhizoctonia*,". Indeed, the anastomosis group concept has been and is a powerful tool to use as we seek to unravel the complex nature of this very important plant pathogenic fungus. Grouping based on hyphal anastomosis of *Rhizoctonias* other than *R. solani*, including the binucleates (Ogoshi *et al.*, 1983) and those with *Waitea* teleomorphs (Oniki *et al.*, 1985), is also successfully practiced although hyphal anastomosis reactions in *Aquathanatephorus pendulus* (Tu and Kimbrough, 1978) have not been studied. This chapter will focus only on hyphal anastomosis reactions in *R. solani*.

Reports on the occurrence of hyphal anastomosis reactions among compatible strains of *R. solani* date back more than 60 years to the early papers of Matsumoto (1921) and Matsumoto *et al.* (1932). In the latter report, Matsumoto *et al.* (1932) first spoke of using what we now call hyphal anastomosis reactions to differentiate strains of *Hypochnus sasakii* Shirai (=*R. solani*) from one another. Schultz (1936) first reported grouping isolates (the anastomosis group concept) based on anastomosis reactions. Other reports added to and refined the findings of these initial studies (Flentje and Stretton, 1964; Richter and Schneider, 1953; Parmeter *et al.*, 1969) (Table 1) and by 1970 the hyphal anastomosis reaction was beginning to gain some general recognition as a useful method for dividing strains of *R. solani* into more homogeneous, and therefore more comprehensible, groups.

However, doubts lingered as to the meaning and usefulness of grouping strains of *R. solani* based on hyphal anastomosis reactions. Little was known about the process of anastomosis and less was known about genetic factor(s) controlling its success or failure. Some isolates that were very similar in behavior and appearance would not anastomose with one another and many isolates would not fit into the groups that had been characterized at that time (Richter and Schneider, 1953; Baker, 1970). Furthermore, grouping by anastomosis reaction did not always correspond to grouping by pathogenicity, colony morphology, or other physical features. Parmeter and Whitney (1970) rightly pointed out that failure to anastomose does not provide adequate information for the exclusion of isolates from *R. solani*. Although grouping of strains based on hyphal anastomosis reactions was being put to some practical use,

B. Sneh et al. (eds.),
Rhizoctonia Species: Taxonomy, Molecular Biology, Ecology, Pathology and Disease Constrol, 37–47.

the consensus in 1970 seemed to be that additional study was required before assigning any type of formal taxonomic status to anastomosis groups. The debate about how or if the AG is to be formalized in a taxonomic context continues today.

TABLE 1. Historical review of subdividing *Rhizoctonia solani* based on hyphal anastomosis reactions.

Year	Author	Group proposed	Present day equivalent
1936	Schultz	1. Gruppe I (*hortensis*)	(AG 1)
		2. Gruppe II (*brassicae*)	(AG 2)
		3. Gruppe III (*typica*)	(AG 3)
		4. Gruppe IV (*chicorii-endiviae*)	(AG 4)
		5. Gruppe V (*fuchsiae*)	("binucleate")
1953	Richter & Schneider	1. Fusion gruppe A	(AG 1)
		2. Fusion gruppe B	(AG 5)
		3. Fusion gruppe C	(AG 4)
		4. Fusion gruppe D "*cruciferen*"	(AG 2)
		5. Fusion gruppe E	("binucleate")
		6. Fusion gruppe F "*kartoffel*"	(AG 3)
1969	Parmeter *et al.*	1. AG 1	(AG 1)
		2. AG 2	(AG 2)
		3. AG 3	(AG 3)
		4. AG 4	(AG 4)

Anastomosis in *R. solani* is defined as a manifestation of somatic, or vegetative, incompatibility between hyphae of different but related strains (Anderson, 1982). The anastomosis group (AG) is a collection of closely related isolates grouped together based on their capability to anastomose with one another. The anastomosis reaction frequently results in death (the so called "killing reaction") of anastomosing and immediately adjacent hyphal cells. Less frequently, anastomosis reactions are not accompanied by a killing of hyphal cells, such as self anastomosis reactions and hyphal reactions between clonal isolates. Although dead hyphal cells often are present at points where anastomosis has occurred, and are easy to locate in microscope fields and therefore are helpful in identifying potential sites of anastomosis, cell death is neither the sole nor the primary criterion used to define anastomosis points (Carling *et al.*, 1988; MacNish *et al.*, 1993).

II. Confrontation of isolates

Placement of an isolates in one or another of the anastomosis groups requires that pairs of isolates be confronted followed by evaluation and interpretation of the cytological reactions that are observed in the "zone of confrontation". The zone of confrontation includes the area where hyphae of the two isolates have the opportunity to come in contact with one another. Confrontations can be created by pairing isolates on water agar in petri plates (Ogoshi, 1975; Rovira *et al.*, 1986), on cellophane overlaying agar (Parmeter *et al.*, 1969; Castro, 1982; Castro *et al.*, 1988; Carling and Sumner, 1992), or on glass slides (Kronland and Stanghellini, 1988). Each of these approaches is known to work well but one thing to keep in mind is that the quality and observibility of anastomosis reactions is enhanced when nutrients, especially nitrogen, are below optimal supplies (Yokoyama and Ogoshi, 1988).

III. Categorization of Hyphal Anastomosis Reactions

Hyphal anastomosis reactions observed within the zone of confrontation are then categorized. Different terminologies have been created for characterizing these reactions (Table 2) but, unfortunately, each is lacking the specific cytological detail required to properly understand what criteria are being used to define each category. Matsumoto *et al.*, (1932) labelled the reactions they observed, from most closely to least closely related, as perfect fusion, imperfect fusion, contact fusion and no reaction. Perfect fusion was not defined but rather was described as that reaction observed when hyphae of the same parent mycelium are confronted (i.e. self anastomosis). Unfortunately, not all self anastomosis reactions are similar, the obvious exception being the non-self anastomosing isolates. Imperfect fusion is the reaction most specifically defined by Matsumoto *et al.* (1932); a reaction that includes the incomplete dissolution of the membrane, no cytoplasmic mixing, and no cell death. Contact fusion is defined as involving contact between participating hyphae but no fusion. Matsumoto *et al.*, (1932) do not make clear if contact fusion is limited to hyphae merely touching one another or if the participating hyphae actually connect and stick together. However, they do compare contact fusion to the relationship between an appressorium and its host surface, where growth of a participating hypha seems to be interrupted as it approaches its host surface. It is interesting to note that Matsumoto *et al.*, (1932) made no mention of cell death as a part of any of the hyphal reactions they observed, and the reason for this may be the short period of time they allowed between confrontation and assessment of reaction. This omission does illustrate, however, that categorization of hyphal anastomosis reactions can be done without using cell death as a criterion.

Flentje and Stretton (1964) categorized the same continuum of hyphal reactions using different terminology (Table 2). They do not cite the paper of Matsumoto *et al.* (1932) and the terms perfect, imperfect and contact are not mentioned. Instead the self (S), killing (K), wall fusion (WF), and no reaction (NR) hyphal reactions are defined and cell death is a key criterion in the system. The S reaction involved no cell death and although wall and membrane fusion were clearly implied they do not provide detailed descriptions of the cytological changes observed. They do, however, make the point that S reactions were observed between confronted pairs of different parent mycelia, thereby indicating the existence of very closely related, though not necessarily identical, individuals in the field. In the K reaction, cytoplasmic anastomosis occurred followed by death of anastomosed cells plus other adjacent cells. In some more recent reports, the K reaction sometimes is equated to the imperfect reaction of Matsumoto *et al.* (1932), but by definition this is not possible because cell death was not reported in the imperfect reaction. In the WF reaction, Flentje and Stretton (1964) reported wall anastomosis but no cytoplasmic anastomosis. Flentje proved this condition by very meticulously puncturing or attempting to tease apart connected cells. They found that if they punctured one cell of a WF reaction, the other would not leak, proving the absence of a cytoplasmic connection. They also found that WF cells were so firmly stuck together that they could not be teased apart without disrupting the cells. It is tempting to equate WF with contact fusion, but again the definitions do not permit it. In the NR category, hyphae are in close proximity and the opportunity for reaction exists, but no reaction occurs.

TABLE 2. Summarization of terminologies used to describe hyphal anastomosis reaction in *Rhizoctonia solani*.[1]

Matsumoto *et al.*[2] (1932)	**Flentje & Stretton[3] (1964)**	**Parmeter *et al.*[4] (1969)**	**Carling *et al.* (1988)**
Perfect	**S**	**2 (Perfect)**	**'C3'**
Hyphae from same parent mycelia	No cell death (-wall and membrane fusion implied)	Wall and cytoplasmic fusion Cell death	
Imperfect	**K**	**2 (Imperfect)**	**'C2'**
Membrane incompletely dissolved, no cytoplasmic mixing	Cytoplasmic anastomosis Death of anastomosed cells plus others	Wall but no cytoplasmic fusion Cell death	
Contact	**WF**	**1**	**'C1'**
Contact but no fusion	Wall anastomosis but no cytoplasmic anastomosis	No fusion No cell death	
No reaction	**NR**	**0**	**'C0'**
No reaction	No reaction	No reaction	

[1] Terms at the top of the table indicate close relationships, whereas terms at the bottom indicate distant relationships. Terms in rows are not necessarily equivalent.
[2] Did not mention the cell death reaction.
[3] Did not cite Matsumoto, *et al.*
[4] Parmeter, *et al.* did not mention the terminology of Flentje & Stretton and they redefined the terminology of Matsumoto, *et al.*

Parmeter *et al.* (1969), in their descriptions of anastomosis reactions, use the terminology of Matsumoto *et al.* (1932) and, though citing the work of Flentje and Stretton (1964), do not make reference to their terminology. The use by Parmeter *et al.* (1969) of the terminology of Matsumoto *et al.* (1932) creates confusion because they indicate cell death occurs in both perfect and imperfect reactions whereas, by the original definitions, cell death was not mentioned or recorded. Structural studies of perfect and imperfect anastomosis reactions, as defined by Parmeter *et al.*, (1969), have been done (Yokoyama *et al.*, 1983, 1985; Yokoyama and Ogoshi, 1986), and many other researchers also appear to use the terms perfect, imperfect and contact in ways that are inconsistent with the original definitions of Matsumoto *et al.* (1932). In addition to using Matsumoto's terms, Parmeter *et al.* (1969) also presented their own brief cytological descriptions of three classes of hyphal anastomosis reactions observed between paired isolates of AG 2. These three classes of reaction were 0, 1, and 2, where 0 = no anastomosis, 1 = hyphal contact but neither fusion nor cell death, and 2 = cell wall and, often, cytoplasmic fusion followed by cell death.

In attempting to overcome differences and deficiencies in definitions among the three systems described above, a system of four categories (C3-C0) for describing hyphal anastomosis reactions was constructed (Carling *et al.*, 1988; MacNish *et al.*, 1993), taking information from each of the three previous systems and adding supplemental detail to definitions when needed (Table 3). C3 represents a very close relationship and indicates the same AG, the same vegetatively compatible population (VCP), and possibly the same individual. C2 represents related isolates that are in the same AG but in different VCPs. C1 represents a more distant relationship such as may be found within highly heterogeneous AGs including AGs 2, 8 or BI; or between AGs where bridging relationship exists. C0 indicates different AGs and the lack of an intimate vegetative relationship.

The term "anastomosis" was not used by Matsumoto *et al.*, (1932) in his first description of the different hyphal anastomosis reactions and indeed it may not be the best term for the job. Some dictionary definitions, i.e. "the joining of one hollow or tubular organ to another", illustrate that "anastomosis" is a perfectly suitable term. However, the C0 reaction is no reaction at all and C1 and C2 actually are failed attempts at anastomosis. Only the C3 reaction is what truly can be considered a successful anastomosis reaction. The placement of isolates into AGs most often is based on the observation of C2 and C1 reactions, both of which are actually failed anastomosis reactions. Therefore it is more accurate and may be more appropriate to call the resulting groups failed anastomosis groups (FAG). Equally correct but perhaps more acceptable for consideration as alternatives to AGs include hyphal reaction group (HRG) and vegetative incompatibility group (VIG). Recognizing that anastomosis and AGs are probably here to stay, these alternatives are presented only as food for thought.

IV. GROUPING OF STRAINS BASED ON HYPHAL ANASTOMOSIS REACTIONS

The use of the term anastomosis to describe these hyphal reactions and the concept of grouping strains based on these reactions has evolved over time. In their initial work, Matsumoto *et al.* (1932) did not use the term anastomosis and did not formally propose grouping isolates based on hyphal reactions. They did report the types of reactions they observed between confronted pairs of isolates and spoke of these reactions as indicating levels of relationship between isolates. They did not include cell death as a descriptive criterion for any of the hyphal reactions they observed, but the reactions they described illustrated their recognition of the complexity of hyphal anastomosis and laid the groundwork for the anastomosis group concept.

Schultz (1936) proposed "gruppe" 1-5 of *Hypochnus solani* (=*R. solani*) based on hyphal reactions and later Richter and Schneider (1953) proposed "fusion gruppe" A-F based on hyphal reaction criteria similar to those of Schultz. In 1969, (Parmeter *et al.*, 1969) proposed anastomosis groups 1-4 and the formalized use of the term "anastomosis group" began. These early systems of grouping isolates based on hyphal reactions along with current AGs equivalents are summarized in Table 1. Since 1969, eight additional AGs have been characterized including AGs 5 (Ogoshi, 1975), 6 and BI (Kuninaga *et al.*, 1979), 7 (Homma *et al.*, 1983), 8 (Neate and Warcup, 1985; Rovira *et al.*, 1986), 9 (Carling *et al.*, 1987), 10 (MacNish *et al.*, 1995), and 11 (Carling *et al.*, 1994). Ogoshi (1975) presents the two major subgroups of AG 2 (AGs 2-1 and 2-2) as independent anastomosis groups but Adams (1988) argues that AGs 2-

1, 2-2, and BI belong together in AG 2, illustrating that some differences of opinion exist as to boundaries of the various AGs.

TABLE 3. Categories of hyphal anastomosis reactions in *Rhizoctonia* solani[1]

Category	Relatedness	Description of interaction
'C3'	CLOSELY RELATED (same anastomosis groups) [same vegetatively compatible population (VCP)] (same isolate)	Walls fuse; membranes fuse; anastomosis point frequently not obvious; diameter of anastomosis point equal or nearly equal hyphal diameter; anastomosing cells and adjacent cells may die but generally do not.
'C2'	RELATED (same anastomosis groups but different VCP)	Wall connection obvious membrane contact uncertain; location of reaction site obvious; diameter of anastomosis point less than hyphal diameter; anastomosing and adjacent cells always die.
'C1'	DISTANTLY RELATED (different anastomosis groups or same anastomosis groups)	Contact between hyphae; apparent connection of walls but no evidence of wall penetration or membrane - membrane contact; occasionally one or both anastomosing cells and adjacent cells die.
'C0'	NOT RELATED (different anastomosis groups)	No interaction

[1] Categorization based upon reactions and cytological changes occurring at the point of anastomosis between two hyphae. These reactions and cytological changes are observable at the light microscopic level.

V. ANASTOMOSIS GROUPS (AG), AND AG SUBGROUPS [INTRASPECIFIC GROUPS (ISG)]

The term intraspecific group (ISG) was proposed by Ogoshi (1987) as a name for subgroups of AGs of *R. solani*. The term ISG was accepted and immediately began to appear in the literature but not always in the context created. No doubt, some of the confusions arose because AGs have not been formally recognized as taxonomic species. Recognizing this, Ogoshi (see introduction chapter) now has indicated that use of the term ISG should be discontinued and many, including this author, agree with that point of view.

Still it is possible, and always useful, to identify subgroups of AGs when they are known. This can be done without reference to ISG or an ISG designation, i.e.: AG 1 with subgroups IA, IB, and IC (subgrouping based on colony morphology and pathogenicity), AG 4 with subgroups HG-I and HG-II (subgroupings based on DNA-DNA complementarity), and AG 8 with subgroups ZGs 1, 2, 3, 4 and 5 (subgroupings based on zymogram patterns). Presently, all AGs except AG 5, 7, 10 and BI have identifiable subgroups and further research probably will identify subgroups in all AGs. It is important to remember, however, that subgroups of AGs, with the possible exception of AG 2-1 and AG 2-2, are not distinguishable on the basis of hyphal anastomosis reactions from other members of their AGs and therefore AGs subgroups are not AG per se. Subgroups of AGs should not be thought of or described as AGs.

VI. BRIDGING GROUPS AND BRIDGING RELATIONSHIPS

Using hyphal anastomosis reactions, one can place nearly all isolates of *R. solani* into one AG or another, however, the existence of bridging isolates sometimes makes placement of an isolate into the proper AG a difficult task. Bridging isolates are isolates that can produce a hyphal anastomosis reaction (generally the C1 reaction) with isolates from more than one AG. By definition, AG BI can contain only isolates that will bridge with members of one or more other AGs. At this time, no isolates of AGs 1, 4, 5, 7, 9, or 10 are known to bridge with isolates of any other AG. At least some isolates of each of the remaining six AGs (Fig. 1) are known to bridge with other AGs. Some isolates of AG 8 and AG BI are known to bridge with some isolates of five other AGs (Fig. 1) whereas some isolates of AGs 2, 3, 6, and 11 are known to bridge with some isolates from two to four other AGs. Although AGs are often described as being genetically isolated from one another, the existence of bridging isolates illustrate that these separations are not always definitive. For this reason, it may be argued that AGs perhaps do not qualify as species.

VII. NON-SELF ANASTOMOSING ISOLATES (NSAI)

It is generally observed that isolates of *R. solani* will anastomose with themselves although this is not always the case. Parmeter *et al.*, (1969) reported that some isolates that would not anastomose with testers of any known group also would not anastomose with themselves and Kuninaga (1980) reported that some single basidiospore isolates in his studies behaved in a similar way. Hyakumachi and Ui (1987) found some non-self anastomosing isolates (NSAI) of AG 2-2 amongst collections from sugar beet fields that had similar growth rates on culture media to self anastomosing isolates (SAI) of AG 2-2 from the same fields. Contrary to the SAI, however, NSAI produced abundant aerial mycelium, would not grow in non-sterile field soil, did not contain melanin, and would not produce sclerotia. NSAI representing AGs 1, 3, 4, and 5 also have been found (Hyakumachi and Ui, 1987), and more recently Carling and Brainard (unpublished) have found NSAI in a collection of strains of AG 3 from tobacco. Further searching likely will reveal that other, and perhaps all, AGs of *R. solani* have NSAI amongst their populations.

NSAI will anastomose neither with themselves nor with other NSAI although they will anastomose with self anastomosing isolates from their respective AGs (Hyakumachi and Ui, 1987), permitting placement of NSAI within an AG. Frequency of anastomosis between NSAI and SAI, however, is considerably less than frequency of anastomosis between SAI. Although the behavior of NSAI is not well understood, it appears that they are missing a factor required for anastomosis to occur. Also, based on the occurrence of anastomosis reactions between NSAI and SAI, it seems clear that only one of a pair of confronted isolates needs to possess the factor(s) required for anastomosis for a reaction to occur. In time, NSAI may prove useful in studies aimed at understanding the genetic systems controlling anastomosis in *R. solani*.

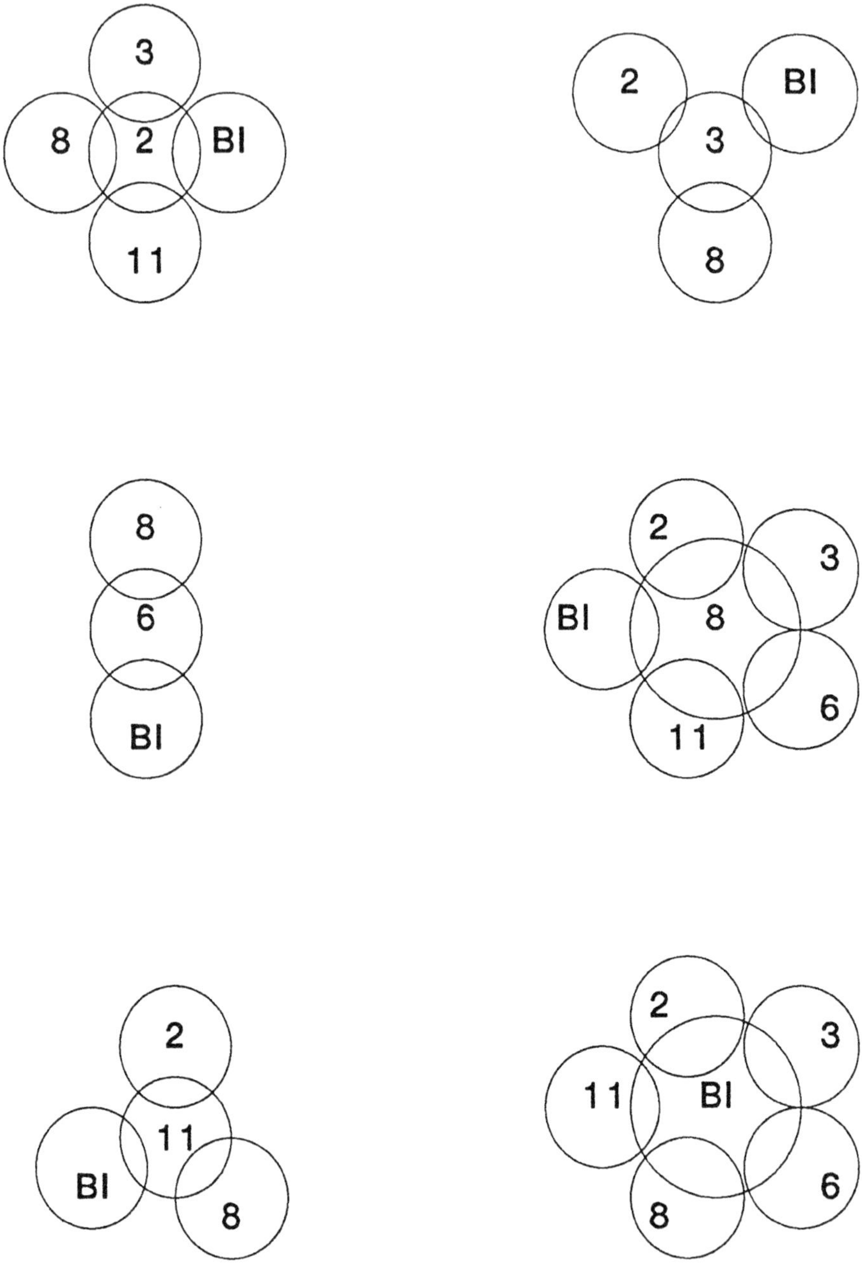

FIGURE 1. Bridging relationships of the six anastomosis groups (AG) that contain at least some isolates that produce hyphal anastomosis reactions with isolates from other AGs. This figure is not intended to illustrate intensity of relationship or relative degree of relationship, but only to indicate where inter-AG anastomosis relationships exist.

VIII. Vegetatively compatible populations (VCP)

In addition to being the basis for dividing *R. solani* into AGs, hyphal anastomosis reactions also can be used to divide AGs into smaller, more homogeneous subunits. This is possible because the C3 anastomosis reaction indicates a much closer level of relationship than does either C1 or C2. All three of these reaction types are used to place an isolate into an AG but only C3 can be used to place isolates into subunits of an AG. Ogoshi and Ui (1983) called these subunits "clones" and were first to use this methodology in an evaluation of relationships within populations of AGs 1, 2, and 3 collected from the field. With the C3 anastomosis reaction they identified "clonal" relationships among isolates within each of the three AGs. They observed that isolates from the same plant were more likely to be of the same clone than isolates from different plants and that isolates from the same field were more likely to be of the same clone than isolates from different fields. They concluded that the probability of clonal relationships increased as the distance between collection points decreased.

Believing this technique would work with other AGs as well, MacNish *et al.* (1993) used it to study diversity within populations of isolates of AG 8 collected from areas affected by bare patch disease. Bare patch disease of cereals is caused only by AG 8 and from areas within individual bare patches only members of a single zymogram group (ZG) of AG 8 can be found (MacNish and Sweetingham, 1993). Using anastomosis technique, MacNish *et al.* (1993) found that all isolates collected from discrete bare patches gave C3 reactions with one another illustrating that discrete bare patches resulted from infection by a single clone, and probably an individual, of AG 8. They also observed that bare patches immediately adjacent to one another often were caused by the same clone whereas patches separated by greater distances often were caused by different clones.

The term "clone", as it is being use in this context by Ogoshi and Ui (1983) and MacNish *et al.* (1993), implies a degree of sameness among isolates that cannot be assured based solely on the quality of an anastomosis reaction. Recognizing this, MacNish and Carling (1995) have proposed that the phrase vegetatively compatible population (VCP) be adopted to describe collections of isolates within an AG that produce C3 anastomosis reactions among themselves. VCP precisely and correctly describes a compatible relationship (MacNish *et al.*, 1996) defined on the basis of a vegetative reaction (the C3 hyphal anastomosis reaction) without claiming the sameness implied by the word clone. It is hoped that the term VCP will prove helpful as we continue sorting out relationships among subgroups of *R. solani*, rather than becoming another in the "bewildering array" (Vilgalys and Cubeta, 1994) of terms used to describe diversity in plant pathogenic fungi (Brasier and Rayner, 1987).

IX. Formal taxonomic status of the anastomosis group

The formal taxonomic status of the AGs has been the subject of much discussion. Some have suggested that the AGs should be accorded species status, but species in *Rhizoctonia* are not well understood because of the absence of required anatomical features (Vilgalys and Cubeta, 1994) or because of the lack of information about mating and mating relationships (Adams, 1988; Vilgalys and Cubeta, 1994). Many studies have supported the belief that AGs are genetically isolated populations within *R. solani* (Adams, 1988; Vilgalys and Cubeta, 1994) so adequate amounts of the required evidence probably exists to permit taxonomic formalization of some sort. It remains to be seen how this will be done, but regardless of its formal taxonomic status, the AG has and will continue to be a method of characterization that is useful to mycologists and plant pathologists who study *R. solani*.

X. SUMMARY

There is much work to be done. Today there are eight more AGs in *R. solani* than were known 25 years ago but in spite of this, there still are isolates being collected that do not fit into any of the known AGs. Doubts about the meaning and usefulness of grouping based on hyphal anastomosis reactions continue to be raised and discussed. Although it frequently does, an example being the correlation between AGs and groupings based on pectic zymograms (MacNish *et al.* 1994), grouping by hyphal anastomosis reaction still does not always correspond with other traditional and the more contemporary molecular methods of grouping. It continues to be true that little is known about the process of anastomosis and genetic factors that control it. Also, the involvement of hyphal anastomosis reactions (those used to define AG) in mating processes are not understood. In spite of this, grouping based on hyphal anastomosis reaction continues to be a useful and important part of the process of characterizing isolates of *R. solani*. Simpler, more definitive, more consistent and more useful methods are always being sought, but so far none has been found.

XI. REFERENCES

Adams GC Jr (1988) *Thanatephorus cucumeris (Rhizoctonia solani)* a species complex of wide host range. Adv. Plant Pathol. 6:535-52.

Anderson NA (1982) The genetics and pathology of *Rhizoctonia solani*. Annu. Rev. Phytopathology 20:329-347.

Baker KF (1970) Types of Rhizoctonia diseases and their occurrence. In: Parmeter JR Jr (ed.) *Rhizoctonia solani*: Biology and Pathology. (pp 125-148) University of California Press, Berkeley.

Brasier CM & Rayner ADM (1987) Whither terminology below the species level? In: Rayner ADM, Brasier CM & Moore D (eds.) Evolutionary Biology of the Fungi. (pp 379-88) Cambridge University Press, Cambridge.

Carling DE & Sumner DR (1992) *Rhizoctonia* In: Singleton LL, Mihail JD & Rush CM (eds.) Methods for Research on Soilborne Phytopathogenic Fungi. (pp 157-165) APS Press, St. Paul, Minnesota.

Carling DE, Kuninaga S & Leiner RH (1988) Relatedness within and among intraspecific groups of *Rhizoctonia solani*: A comparison of grouping by anastomosis and by DNA hybridization. Phytoparasitica 16:209-210.

Carling DE, Leiner RH & Kebler KM (1987) Characterization of a new anastomosis group (AG 9) of *Rhizoctonia solani*. Phytopathology 77:1609-1612.

Carling DE, Rothrock CS, MacNish GC, Sweetingham MW, Brainard KA & Winter SA (1994) Characterization of anastomosis group-11 (AG 11) of *Rhizoctonia solani*. Phytopathology 84:1387-1393.

Castro C (1982) Methods for the quantitative estimation of *Rhizoctonia solani* in soil. Ph.D thesis. (68 pp) University of Idaho, Moscow.

Castro C, Davis JR & Wiese MV (1988) Quantitative estimation of *Rhizoctonia solani* AG 3 in soil. Phytopathology 78:1287-1292.

Flentje NT & Stretton HM (1964) Mechanisms of variation in *Thanatephorus cucumeris* and *T. praticolus*. Aust. J. Biol. Sci. 17:686-704.

Homma Y, Yamashita Y & Ishii M (1983) A new anastomosis group (AG 7) of *Rhizoctonia solani* Kühn from Japanese radish fields. Ann. Phytopathol. Soc. Jpn. 49:184-190.

Hyakumachi M & Ui T (1987) Non-self-anastomosing isolates of *Rhizoctonia solani* obtained from fields of sugarbeet monoculture. Trans. Br. Mycol. Soc. 89:155-159.

Kronland WC & Stanghellini ME (1988) Clean slide technique for the observation of anastomosis and nuclear condition of *Rhizoctonia solani*. Phytopathology 78:820-822.

Kuninaga S (1980) Hyphal anastomosis behavior of single basidiospore isolates of *Thanatephorus cucumeris* (Frank) Donk. Higashi Nippon Gakuen J. of Liberal Arts and Sci. 6:95-106.

Kuninaga S & Yokosawa R (1984b) DNA base sequence homology in *Rhizoctonia solani* Kühn. V. Genetic relatedness within AG 6. Ann. Phytopathol. Soc. Jpn. 50:346-352.

Kuninaga S, Yokosawa R & Ogoshi A (1979) Some properties of anastomosis group 6 and BI in *Rhizoctonia solani* Kühn. Ann. Phytopathol. Soc. Jpn. 45:207-217.

MacNish GC & Carling DE (1995) Use of vegetatively compatible populations (VCP) to determine field distribution of *R. solani* AG 8. Phytopathology 85:1040.

MacNish GC & Sweetingham MW (1993) Evidence that each Rhizoctonia bare patch is dominated by an individual zymogram group (ZG) of *Rhizoctonia solani* AG 8. Aust. J. Agric. Res. 44:1175-1194.

MacNish GC, Carling DE & Brainard KA (1993) Characterization of *Rhizoctonia solani* AG 8 from bare patches by pectic isozyme (zymogram) and anastomosis techniques. Phytopathology 83:922-927.

MacNish GC, Carling DE & Brainard KA (1996) Relationship of microscopic and macroscopic vegetative reactions in *Rhizoctionia solani* and the occurence of vegetatively compatible populations (VCPs) in AG 8. Mycl. Res. 100:(in press).

MacNish GC, Carling DE, Sweetingham MW & Brainard KA (1994) Anastomosis group (AG) affinity of pectic isozyme (Zymogram) groups (ZG) of *Rhizoctonia solani* from the Western Australian cereal belt. Mycol. Res. 98:1368-1375.

MacNish GC, Carling DE, Sweetingham MW, Ogoshi A & Brainard KA (1995) Characterization of anastomosis group-10 (AG 10) of *Rhizoctonia solani*. Australasian J. Plant. Pathol. 25:252-260.

Matsumoto T (1921) Studies in the physiology of the fungi. XII. Physiological specialization in *Rhizoctonia solani* Kühn. Ann. Missouri Botan. Garden 8:1-62.

Matsumoto T, Yamamoto W & Hirane S (1932) Physiology and parasitology of the fungi generally referred to as *Hypochnus sasakii* Shirai. I. Differentiation of the strains by means of hyphal fusion and culture in differential media. J. Soc. Trop. Agric. 4:370-388.

Neate SM & Warcup JH (1985) Anastomosis grouping of some isolates of *Thanatephorus cucumeris* from agricultural soils in South Australia. Trans. Br. Mycol. Soc. 85:615-620.

Ogoshi A (1976) Studies on the grouping of *Rhizoctonia solani* Kühn with hyphal anastomosis, and on the perfect stage of groups. Bull. Natl. Inst. Agric. Sci. Ser. C. No. 30:1-63.

Ogoshi A (1987) Ecology and pathogenicity of anastomosis and intraspecific groups of *Rhizoctonia solani* Kühn. Annu. Rev. Phytopathol. 25:125-143.

Ogoshi A, Oniki M, Araki T & Ui T (1983) Studies on the anastomosis groups of binucleate *Rhizoctonia* and their perfect states. J. Fac. Agr. Hokkaido Univ. 61:244-260.

Ogoshi A & Ui T (1983) Diversity of clones within an anastomosis group of *Rhizoctonia solani* Kühn in a field. Ann. Phytopathol. Soc. Jpn. 49:239-245.

Oniki M, Ogoshi A, Araki T, Sakai, R & Tanaka S. (1985) The perfect stage of *Rhizoctonia oryzae* and *R. zeae* and the anastomosis groups of *Waitea circinata*. Trans. Mycol. Soc. Jpn. 26:189-198.

Parmeter JR Jr & Whitney HS (1970) Taxonomy and nomenclature of the imperfect stage. In: Parmeter JR Jr (ed.) *Rhizoctonia solani*: Biology and Pathology (pp 7-19) University of California Press, Berkeley.

Parmeter JR Jr, Sherwood RT & Platt WD (1969) Anastomosis grouping among isolates of *Thanatephorus cucumeris*. Phytopathology 59:1270-1278.

Richter, H & Schneider R (1953) Untersuchungen zur morphologischen und biologischen differenzierung von *Rhizoctonia solani* K. Phytopathol. Z. 20:167-226.

Rovira AD, Ogoshi A & McDonald HJ (1986) Characterization of isolates of *Rhizoctonia solani* from cereal roots in South Australia and New South Wales. Phytopathology 76:1245-1248.

Schultz H (1936) Vergleichende Untersuchungen zur Okologic, Morphologie und Systematik des Vermchrungspilzes. Abr. Biol. Reichsanst. Land-u Forstwirtsch., Berl. 22:1-41.

Stevens-Johnk J, Jones RK, Shew HD & Carling DE (1993) Characterization of populations of *Rhizoctonia solani* AG 3 from potato and tobacco. Phytopathology 83:854-858.

Tu CC & Kimbrough JW (1978) Systematics and phylogeny of fungi inthe *Rhizoctonia* complex. Bot. Gaz. 139:454-466.

Vilgalys R & Cubeta MA (1994) Molecular systematics and population biology of Rhizoctonia. Annu. Rev. Phytopathol. 32:135-155.

Yokoyama K & Ogoshi A (1986) Studies on hyphal anastomosis of *Rhizoctonia solani*. IV. Observation of imperfect fusion by light and electron microscopy. Trans. Mycol. Soc. Jpn. 27:399-413.

Yokoyama K & Ogoshi A (1988) Studies on hyphal anastomosis of *Rhizoctonia solani*. V. Nutritional conditions for anastomosis. Trans. Mycol. Soc. Jpn. 29:125-32.

Yokoyama K, Ogoshi A & Ui T (1983) Studies on hyphal anastomosis of *Rhizoctonia solani*. I. Observation of perfect fusion with light microscopy. Trans. Mycol. Soc. Jpn. 24:329-40.

Yokoyama K, Ogoshi A & Ui T (1985) Studies on hyphal anastomosis of *Rhizoctonia solani*. II. The ultrastructural changes of hyphal cells during perfect fusion. Trans. Mycol. Soc. Jpn. 26:199-207.

I.A3 A SYNOPSIS OF THE TAXONOMY OF TELEOMORPHS CONNECTED WITH *RHIZOCTONIA* S.L.

JOOST A. STALPERS[1] AND TORBEN F. ANDERSEN[2]

[1]*CBS, Oosterstraat 1, 3742 SK Baarn, THE NETHERLANDS,* [2]*Botanical Laboratory, University of Copenhagen, Gothersgade 140, DK-1123 Copenhagen K, DENMARK*

I. INTRODUCTION

The genus *Rhizoctonia* s.l., was originally characterized by the presence of sclerotia different from those of *Sclerotium* and the ability to infect living plant tissue (DeCandolle, 1815). This has already served to exclude the saprophytic ascomycetous genera *Ascorhizoctonia* (teleomorph: *Tricharina*, Yang and Korf, 1985; Moore, 1987) and *Oedocephalum* p.p. (teleomorph: *Iodophanus*, Gamundì and Ranalli, 1964; Stalpers, 1974) and to restrict it to the basidiomycetes. Duggar (1915) and Parmeter and Whitney (1970) listed additional characteristics to delimit *Rhizoctonia* s.l., among them hyphal constrictions at septa, hyphal branching at right angles just before septa, and production of monilioid cells. Unfortunately, these criteria are not sufficiently diagnostic to delimit a homogeneous taxon, and the teleomorphs of *Rhizoctonia* s.l. (as far as known) are now classified in various basidiomycete orders. A natural concept, in which anamorph taxonomy reflects teleomorph taxonomy, is now also being adopted in *Rhizoctonia* (Moore, 1987).

The type species of *Rhizoctonia* is *Rhizoctonia crocorum*, but as this typification would necessitate the introduction of the name *Moniliopsis napaeae* for the well known *Rhizoctonia solani* (Andersen and Stalpers, 1994), a proposal has been made to change the typification of *Rhizoctonia* by conservation and to adopt *R. solani* as the type. This proposal was unanimously supported at ISR '95, the International Symposium on *Rhizoctonia* in Noordwijkerhout, The Netherlands; its acceptance by the 1999 Botanical Congress is anticipated. As a consequence, *Thanatophytum* Nees, a former obligate synonym of *Rhizoctonia* would become the correct name for the anamorph of *Helicobasidium*.

The scope of this book does not allow a comprehensive study of teleomorph taxonomy, but a short analysis of the characteristics considered important in both teleomorph and anamorph classification is provided, the position of the teleomorph genera in the system is discussed, and keys to genera and species are given. The teleomorphs of *Rhizoctonia* s.l. mainly occur in the Ceratobasidiaceae and Tulasnellaceae; the key to the teleomorphs contains all genera for which a *Rhizoctonia* state is known, and some that incorrectly have been connected with *Rhizoctonia*. For many species of the genera concerned no cultural characteristics are known, and more representatives are expected to display *Rhizoctonia* characteristics. Hence for *Tulasnella* and the generic complex *Ceratobasidium* - *Thanatephorus*, condensed keys to all species have been compiled.

B. Sneh et al. (eds.),
Rhizoctonia Species: Taxonomy, Molecular Biology, Ecology, Pathology and Disease Constrol, 49–63.

II. Induction of basidiome production

Most of the work on *Rhizoctonia* spp. is performed with cultures (Warcup and Talbot 1966; 1967; 1971), which are often isolated from sources not displaying the teleomorph. Techniques to induce sporulation are particularly useful and outlined below.

A. Transfering strains from high nutrient agar to low nutrient agar

Flentje (1956) obtained basidiomes by growing the cultures on PDMA (potato dextrose 'marmite' agar, 15 g potato, 40 g Marmite, 7.5 g dextrose per liter) for 7-8 days, and then transferred them to weak nutrient agar (soil extract agar). Murray (1982b) found that the initial medium was important and recommended PDMA and DMA (dextrose 'marmite' agar, 25 g marmite, 12.5 g dextrose per liter). Murray (1984) used the following method successfully for many species of *Rhizoctonia* s.l. Strains were pregrown on DMA prepared with distilled deionized water. Inocula were taken from margins of 3-day-old colonies and transferred to dishes with WA (water agar) or CMA (cornmeal agar) in Petri dishes with or without studs, thus with varying aeration. The time needed for the production of basidia was up to 34 days for many *Ceratobasidium* strains.

B. Soil-over-culture

(Tu *et al.*, 1969; Tu and Kimbrough 1975a). The strain was grown on PDA + 5% yeast extract (PYDA = potato yeast dextrose agar). Mycelial disks were cut from the periphery of 3 day old colonies and incubated at 28°C in the dark on PYDA, which contained 0.75% yeast extract. After 3-5 days the lid was removed and the culture covered with 1 cm sterilized soil (about 90g). The soil was watered 1-3 times a day to keep the relative humidity within acceptable limits. Sporulation generally occurred after 3-14 days. Important parameters were the age of the colony before covering it with soil (sporulation occurred only when mycelium of 3-7 days old was used for inoculation), the amount of yeast extract (sporulation occurred only in the range of 0.25-0.9%), the presence of dextrose and the choice of the soil (silty clay, Urrbray loam, sandy loam and even coarse river sand), the pH of the soil (between 6-10) and possibly light. The relative humidity seems to be less critical, but 40-60% is considered optimal. Stretton *et al.* (1964) and Warcup and Talbot (1965) also obtained good results after grinding cultures grown for 7-10 days on PDMA with sterile sand in a mortar, after which they were incorporated into soil.

C. Infection of plants

(Flentje, 1956). Unsterilized soil was inoculated with 25% (v/v) of a culture grown on maizenchal-sand. Seeds of wheat or tomato or sprouted potato cutlings were planted in 7cm diam, 9cm high glass jars in the soil, and placed in a glass house. Origin of inoculum as well as soil type was important for induction of sporulation.

III. Characteristics analysis

Vegetative characteristics such as mycelial coloration, hyphal diameter, number of nuclei, length of cells, shape and size of monilioid cells, and sclerotial size, have generally been used in the characterization of *Rhizoctonia* s.l. (e.g. Duggar, 1915: Parmeter and Whitney, 1970). Andersen (1990) challenged this, arguing that, as these characteristics are mostly studied under *in vitro* conditions, care is needed in the evaluation, as most characteristics vary considerably with temperature, light and composition of the medium. In view of an integrated anamorph-teleomorph system (Moore, 1987), some of these characteristics may be useful in the description of the holomorphs along with studies of the septal pore ultrastructure, which is now considered a basic character of ordinal rank.

The diameter of hyphae is quite variable under culture conditions (Andersen, 1990), and have been reported to swell to eight times their original diameter in a concentrated substrate (Burgeff, 1936). The diameter may vary considerably within each genus, e.g., 0.5-1.5 µm in *Tulasnella permacra* (Roberts, 1993) to 3-7 µm in *T. brinkmannii* (Roberts, 1994). Only a few guidelines can therefore be suggested. *Tulasnella* and *Sebacina* spp. tend to have the narrowest hyphae, and if the hyphal diameter is less than 4 µm, the strain probably belongs to one of these genera. *Thanatephorus* and *Waitea* spp. possess the widest hyphae, and the occurrence of hyphae wider than 9-10 µm indicates that the strain probably belongs there. The cell length in the runner hyphae was once thought to be of diagnostic value, but Andersen (1990) demonstrated that such measurements are of no value in characterizing taxa in *Rhizoctonia* s.l.

Clamp connections have been reported for some *Rhizoctonia* strains, e.g. *Tulasnella* `*calospora*' (Warcup and Talbot, 1967), *T. allantospora* (Warcup and Talbot, 1971), and `*Ceratobasidium fibulatum*' (Tu and Kimbrough, 1978); among teleomorphic genera with a *Rhizoctonia*-like anamorph, only *Tulasnella* has a considerable number of clamped species (Roberts, 1992, 1993, 1994, and see the present key). The occurrence of clamped septa in the basidiome need not be constant; in some species clamps are restricted to the subhymenium (Warcup and Talbot, 1967), in others they only occur at some septa of the subicular hyphae and then not necessarily in all strains of the species concerned (Warcup and Talbot, 1971). Clamp connections are here considered valuable at the species level, but not as a main character in the delimitation of genera.

Vegetative cells can be uni-, bi- and/or multinucleate. This condition is rarely absolute: for example, a small percentage of cells of a binucleate strain may show one or three nuclei, and multinucleate species may display a large variation in number (e.g., 4-15), often correlated with hyphal width. In *Tulasnella* the number of nuclei per cell is only known for a few species which are generally binucleate, in one case multinucleate (Boidin and Lanquetin, 1984; Andersen, 1996). Also *Sebacina* shows varying numbers of nuclei, from uninucleate strains of *S. umbrina* and *S. grisea* (Andersen, 1996) to the multinucleate *S. vermifera* (Williams and Thilo, 1989). In anamorphic *Rhizoctonia*-like fungi the number of nuclei per cell has been used in the description and delimitation of genera (e.g., the binucleate *Ceratorhiza* and the multinucleate *Rhizoctonia (Moniliopsis)*. *Thanatephorus* is considered to be multinucleate (Duggar, 1915; Parmeter and Whitney, 1970), but in fact the nuclear condition is only known for about 50% of the species; *Ceratobasidium* is generally binucleate (Parmeter *et al.*, 1967, Burpee *et al.*, 1980), but *C. koleroga* has been reported to be multinucleate (Boidin and Lanquetin, 1984). Although the number of nuclei is occasionally different from the expected, it is valuable as a diagnostic character in combination with other features in some genera, but not in others. However, recent investigations suggest that bi-multinucleate relations should be used with great care in taxonomy, as for example *Ceratobasidium* and *Thanatephorus* are closer related to each other than to *Waitea* (Andersen, 1996), although the two latter are multinucleate, with anamorphs until now classified in *Moniliopsis*, and the former is binucleate and consequently placed in *Ceratorhiza* (Moore, 1987); the taxonomy of aphyllophoraceous species presents several examples of similar situations (e.g., *Gloeocystidiellum*, *Aleurodiscus*, Boidin and Lanquetin, 1984).

Monilioid cells are inflated cells which occur as more or less branched chains that remain intact (Duggar, 1915; Saksena and Vaartaja, 1961). Ontogenetical studies have shown that in *Rhizoctonia* these cells can be considered as precursors of sclerotia (Tu and Kimbrough, 1975b), and as such their presence has been taken as indicative of a relationship to this genus (Bernard, 1909; Burgeff, 1936; Saksena and Vaartaja, 1960, 1961). However, the shape and size of the monilioid cells is extremely variable, and also medium dependent, and their morphology is considered to be of limited taxonomical value (Butler and Bracker, 1970; Andersen, 1990). Certain taxa may possess characteristic monilioid cells, e.g., *Epulorhiza anaticula* (Currah *et al.*, 1987) and *Ceratorhiza pernacatena* (Zelmer and Currah, 1996), and the morphology of monilioid cells may turn out to be characteristic for more anamorph taxa in *Rhizoctonia* s.l. at the species level.

Sclerotial production has been considered as a very important character not only for distinguishing the form genus (DeCandolle, 1815; Duggar, 1915; Parmeter and Whitney, 1970), but also for the distinction of both anamorph and teleomorph taxa at the species level (e.g., Bernard, 1909; Matz, 1921; Tu and Kimbrough, 1975b). Later investigations have shown that taxa belonging to *Ceratobasidium* and *Thanatephorus* can produce sclerotia that are indistinguishable from each other (Andersen, 1996). At least in these genera the distinction of species based on sclerotial structure seems unwarranted. However, it suggests a close relation between these two genera, and certainly helps in delimiting them from other genera like *Waitea*. In teleomorphic *Tulasnella* no records of sclerotia could be traced in the literature, but they have been described in cultural studies of sclerotial anamorphs referable to this genus on account of their dolipore structure (Bernard, 1909, Currah *et al.*, 1987; Zelmer and Currah, 1996; Andersen, 1996). Nevertheless, research on cultures derived from teleomorphs is highly needed, and may show some difference in sclerotial morphology within *Tulasnella*. Sclerotia have not been reported for *Sebacina*, but in culture monilioid cells are formed.

The basidia of the Ceratobasidiaceae and Tulasnellaceae, where most of the *Rhizoctonia* species belong, display much variation. As these families are in several respects intermediates between the homo- and heterobasidiomycetes, this is very informative about possible lines of development, as described e.g., by Talbot (1965) and Donk (1958b). Fig. 1 illustrates the developmental stages in four species that are or have been associated with *Rhizoctonia*: the holo- and homobasidiomycetous *Athelia* (1), the holo- and heterobasidiomycetous *Thanatephorus* (2), the heterobasidiomycetous *Tulasnella*, which is holobasidiomycetous, but has been considered to be phragmobasidiomycetous because of false septa produced under the sterigmata (3), and the phragmobasidiomycetous and heterobasidiomycetous *Sebacina* (4). The terminology is after Donk (1954a, 1956) and Talbot (1973). The development starts with a terminal cell, that can be anything between globose and cylindrical and is called a protobasidium in the Heterobasidiomycetes and basidiole or young basidium in the Homobasidiomycetes. Then meiosis and subsequent mitosis occurs and the nuclei migrate towards the apex. Now various things may happen: protosterigmata (Donk 1954a) may be formed as small or larger semiglobose to globose protuberances. In *Athelia* and *Thanatephorus* the protosterigmata are small and form only a brief stage in a continuous development, but in *Tulasnella* they are large and there is often a pause before further development; thus this stage is more obvious and the protuberance has been called epibasidium (Martin, 1938) and the remaining main body hypobasidium. In *Sebacina* there are no external protosterigmata; here the development is internal, and the development of two longitudinal walls or septa is considered as homologous to the protosterigmata. It should be recalled that these walls or septa have a variable length: they may be short and only apical, leaving most of the basidium undivided as in *Oliveonia* (Talbot, 1965) or (almost) to the very base of the basidium (nearly) dividing it into four cells. The third developmental stage includes the development of a secondary protosterigma (Donk 1954b, 1956), which is a hyphoid elongation of the protosterigma. The stage may be completely absent (in *Athelia*), very short and sometimes difficult to distinguish (in species of *Ceratobasidium*) or quite distinct (*Tulasnella*, especially the species formerly classified in *Gloeotulasnella*) to very long (in *Sebacina*). The main function of this structure is probably to reach the air by growing through the gelatinous layer. In the absence of a gelatinous layer (or a high humidity) secondary protosterigmata are short or absent. Moreover the ability to produce a secondary protosterigma seems to be correlated with the ability to produce secondary basidiospores. The fourth and final stage is the formation of the spiculum, the apical apparatus that produces the ballistospore.

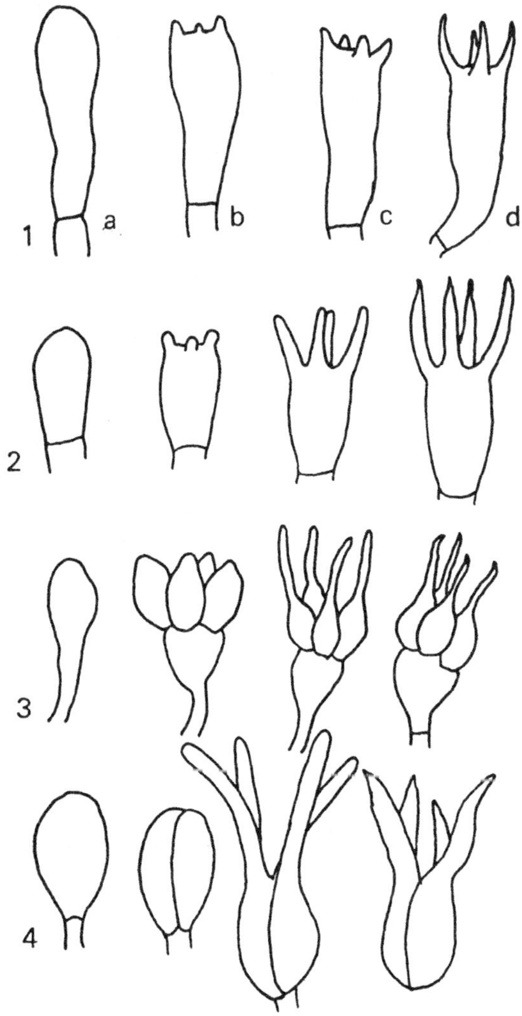

FIGURE 1. Various types of basidia found in teleomorph genera associated with *Rhizoctonia*-like anamorphs. 1: *Athelia*; 2: *Thanatephorus*; 3: *Tulasnella;* 4: *Exidia*. a: protobasidium, b: basidium with protosterigmata; c: basidium with secondary protosterigmata; d: mature basidium, sterigmata with spiculum.

The shape and septation of the basidia determines the generic classification of most *Rhizoctonia* teleomorphs. However, many teleomorphs have only been described from cultures, where some characteristics vary more widely than in nature. Humidity, medium composition, light and changes in orientation may cause aberrations, as for example many lateral basidia, larger variation in sterigmatal and basidial size (cf Murray, 1982a) etc.

Oberwinkler (1985) and Bauer and Oberwinkler (1986) launched an interesting theory on the ontogeny of basidia and basidiospores, based on observations in heterobasidiomycetous species. They interpreted the basidial cell as the final ontogenetic state in the meiotic process, thereby suggesting that all basidiospores were actually secondary spores, i.e. asexual spores or conidia.

Basidiospores in *Rhizoctonia* teleomorphs are hyaline, smooth, thin-walled, not amyloid and typically uninucleate. The shape varies from globose to ovoid, ellipsoid, cylindrical, allantoid or sigmoid and from citriform to fusiform. All species are known to be able to produce secondary spores, except *Waitea* and some species of *Thanatephorus*. The intraspecific variation increases with increasing length-width ratio, while the overall variation is larger than in homobasidiomycetous species because of the potential to produce secondary spores, which are generally slightly smaller, but otherwise indistinguishable.

IV. TAXONOMY

The genera with *Rhizoctonia* s.l. anamorphs are classified in three orders, the Platygloeales (Platygloeaceae), and the related Exidiales (Sebacinaceae) and Tulasnellales (Ceratobasidiaceae and Tulasnellaceae); the Ceratobasidiaceae are often placed in a separate order Ceratobasidiales. Sclerotium-producing homobasidiomycetous species in the genus *Athelia* have sometimes been mentioned in connection with *Rhizoctonia* s.l., while *Sistotrema* has not, although e.g., *S. brinkmannii* may produce both sclerotia and monilioid cells. However, in culture all these species may produce clamps (although sometimes rarely). Homobasidiomycetous species have not been considered here further.

KEY TO THE TELEOMORPH FAMILIES CONTAINING RHIZOCTONIA-LIKE ANAMORPHS

1a. Basidia (ustidia) transversely septate (*Helicobasidium*) **Platygloeaceae**
1b. Basidia not transversely septate **2**

2a. Basidia longitudinally septate (*Sebacina*) **Exidiaceae**
2b. Basidia not septate **3**

3a. Basidia producing rather large, globose protosterigmata, which grow out to form pyriform sterigmata (*Tulasnella*) **Tulasnellaceae**
3b. Basidia without pyriform sterigmata **4**

4a. Spores capable of producing secondary spores and/or basidia with long sterigmata, often blunt or hyphoid (*Cejpomyces, Ceratobasidium, Tofispora, Metabourdotia, Oliveonia, Scotomyces, Thanatephorus, Uthatobasidium*) **Ceratobasidiaceae**
4b. Spores never producing secondary spores, basidia with (slightly) curved, rather short sterigmata (*Waitea* and similar fungi, for example *Athelia, Botryobasidium*

Ceratobasidiaceae

Basidiome annual, effused, arachnoid, hypochnoid, pellicular, (sub)membranaceous or ceraceous. Hymenial surface smooth, sometimes discontinuous. Hyphal system monomitic. Subicular hyphae hyaline or brownish, thin- to thick-walled, often branching at right angles, often contricted near branching point, generally wider than subhymenial hyphae. Parenthesomes of dolipores with 3-4 large perforations. Subhymenial hyphae typically hyaline, relatively short-celled, thin-walled. Basidia hyaline, thin-walled, ovoid, sphaeropedunculate or barrel-shaped to subcylindrical to broadly clavate, (1-)2-4(-7) sterigmata. Sterigmata typically stout, with relatively broad base. Spores hyaline to pale yellowish, thin-walled, smooth, subglobose to ellipsoid or cylindrical, not amyloid, typically producing secondary spores, sometimes becoming secondarily 1-2-septate. Sclerotia may be present. Substrate: soil-inhabiting, mycorrhizal, parasitic or saprophytic on plants and wood.

The differences used to distinguish *Thanatephorus* from *Ceratobasidium* are gradual. The basidial shape, which is typically ovoid or sphaeropedundulate, may become short-cylindrical to broadly clavate in e.g., *C. anceps*. The basidiome of *Ceratobasidium* is typically thin, ceraceous, but it is pellicular to arachnoid in e.g., *C. anceps, C. koleroga* and *C. gramineum*. The rather narrow (up to 6 µm) and scant subicular hyphae in typical *Ceratobasidium* spp. may become rather wide (up to 10 µm in *C. koleroga*), brown sclerotia are also known in *Ceratobasidium*. Also the two-spored species (of which the non-ceraceous species have been classified in *Ypsilonidium*) also contain intermediate taxa. The ceraceous ones with an ovoid to sphaeropedunculate basidium have always been referred to *Ceratobasidium*, but *T. ovalisporus* is pellicular, forms hyphal strands and has sphaeropedunculate basidia. *Y. anomalum* shows besides septate basidiospores intermediate basidia and relatively narrow subicular hyphae; the number of nuclei is unknown. To conclude, there is a continuum of species displaying various combinations of characteristics, making the separation of the typical ceraceous, saprophytic *Ceratobasidium* with ovoid to sphaeropedunculate basidia and the typical hypochnoid, parasitic *Thanatephorus* with barrel-shaped basidia vague. *Waitea* is generally considered to belong to the Ceratobasidiaceae, despite the lack of secondary spore production and the homobasidiomycete-like basidia; the spores become secondarily septate.

Ceratobasidium D.P. Rogers (syn.: *Koleroga* Donk; type species: *Ceratobasidium calosporum* D.P. Rogers; anamorph: *Ceratorhiza* R.T. Moore)

Basidiome pellicular to ceraceous. Subicular hyphae without clamps, typically binucleate. Cystidia and other hymenial elements lacking. Basidia hyaline, thin-walled, subglobose to pyriform or sphaeropedunculate, about twice as wide as the supporting hypha. Sterigmata rather stout to stout (especially the two-spored species).

Thanatephorus Donk (syn.: *Ypsilonidium* Donk, *Aquathanatephorus* Tu and Kimbrough; incl.: *Cejpomyces* Svrcek and Pouzar, *Uthatobasidium* Donk; type species: *Hypochnus solani* Prillieux and Delacroix 1891; anamorph: *Rhizoctonia* DC.:Fr.

Basidiome arachnoid, hypochnoid, pellicular or (sub)membranaceous. Subicular hyphae often constricted near branching point, relatively wide (up to 17 µm), without clamps, multinucleate. Cystidia and other sterile hymenial elements lacking. Basidia hyaline, thin-walled, barrel-shaped to subcylindrical or broadly clavate, about the same width or slightly wider than the supporting hyphae, with (1-)2-4(-6) sterigmata. Sterigmata stout, with relatively broad base.

Talbot (1970) treated *T. cucumeris* extensively. Tu and Kimbrough (1978) distinguished *Aquathanatephorus* from *Thanatephorus* because of the globose metabasidium and swollen protosterigmata and the number of nuclei, despite the fact that they identified the anamorph as *Rhizoctonia solani*. The fungus normally grows at nearly 100% humidity, which is known to cause irregularities in size and shape of various hyphal structures including the basidia; the number of nuclei does not differ significantly from that of some strains of *R. solani* (Andersen, 1996). *Ypsilonidium* has been separated from *Thanatephorus* on the base of its basidia with two large sterigmata. However, a similar phenomenon is known for e.g., *Ceratobasidium bicorne* and *C. oryzae-sativae* and we see no reason to value this character higher in *Thanatephorus*. *Cejpomyces* was synonymized with *Thanatephorus* by Langer (1994), but the number of nuclei was not mentioned. In a number of other cases no data of the number of nuclei per cell was found; when at least the subicular hyphae turn out to be binucleate in one or more species, the position of these species may have to be reconsidered. For this very reason *Uthatobasidium* is not considered synonymous, although its species are keyed out here for convenience. The key below treats the members of the Ceratobasidiaceae as characterized in the key to the families, thus exclusive *Metabourdotia* (not treated) and *Waitea* (see below). When only a generic name is given, the genus is not known to contain *Rhizoctonia*-like species.

KEY TO THE SPECIES (C = *Ceratobasidium*, T = *Thanatephorus*, U = *Uthatobasidium*, Y = *Ypsilonidium*)

1a.	Cystidia or gloeocystidia present	***Oliveonia*** Donk
1b.	Cystidia and gloeocystidia absent	**2**
2a.	Clamps present	***Scotomyces*** Jülich
2b.	Clamps absent	**3**
3a.	Spores ornamented, citriform. Basidia with 4-6 sterigmata	***Tofispora*** G. Langer
3b.	Spores smooth. Basidia only incidentally with more than 4 sterigmata	**4**
4a.	Spores filiform-flexuous to subsigmoid, 23-36 x 3-3.6 μm	***C. calosporum*** D.P. Rogers
4b.	Spores globose, ovoid, ellipsoid or cylindrical.	**5**
5a.	Basidia with (1-)2(-3) stout sterigmata.	**6**
5b.	Basidia with (2-)4(-5) sterigmata	**12**
6a.	Spores globose to subglobose or broadly ellipsoid, at least 10 μm wide	**7**
6b.	Spores ovoid, ellipsoid or cylindrical, less than 10 μm wide	**8**
7a.	Spores (8-)9-17(-21) x (8-)9-16(-19) μm (anam.: *Rhizoctonia oryzae-sativae* (Sawada) Mordue)	***C. oryzae-sativae*** P.S. Gunnell & R.K. Webster
7b.	Spores 11-15 x 10-12 μm (anam.: *Rhizoctonia fumigata* (Nakata ex Hara) P.S. Gunnell and R.K. Webster)	***C. setiariae*** (Sawada) P.S. Gunnell & R.K. Webster
8a.	Spores ovoid to ellipsoid, 8-15(-17) x (5-)6-9(-10.5) μm	**9**
8b.	Spores ellipsoid to cylindrical, 10-17 (-20) x 5-8(-9) μm	**10**
9a.	Hyphal strands present. Subicular hyphae hyaline	***T. ovalisporus*** Cizek & Pouzar
9b.	Hyphal strands absent. Subicular hyphae brownish	***T. langlei-regis*** D.Reid
10a.	Subicular hyphae up to 6 μm wide	***C. bicorne*** J. Erikss. & Ryv.
10b.	Subicular hyphae wider. *T. sterigmaticus* complex	**11**
11a.	Basidiospores becoming 1-2-septate when old	***Y. anomalum*** Warcup & Talbot
11b.	Basidiospores remaining aseptate	***T. sterigmaticus*** (Bourdot) Talbot

12a. Spores cylindrical to slightly fusiform, (7-)9-16 x (2.5-)3-4.5(-5) µm **13**

12b. Spores globose to subcylindrical, Q (length:width) < 2.5 **14**

13a. Saprophytic. Basidiome ceraceous, adnate. Sclerotia absent. Incl. *C. angustisporum* ***C. pseudocornigerum*** M.P. Christ.

13b. Parasitic on leaves. Basidiome pellicular, separable. Sclerotia present (*Pellicularia koleroga* complex pp, see also note under 27b) ***C. stevensii*** (Burt)

14a. Subicular hyphae brownish, often some wider than 10 µm. Brown sclerotia present; hyphae multinucleate (*Thanatephorus* s. str.) **15**

14b. Subicular hyphae hyaline or pale yellowish, up to 10 µm wide (in *U. fusisporum* wider) **20**

15a. Spores globose to broadly ovoid **16**

15b. Spores ovoid to ellipsoid **17**

16a. Spores globose, hyaline, 6-7 µm diam. ***T. pennatus*** Currah

16b. Spores broadly ovoid, pale brown, 9-12 x 7-9.5 µm ***T. orchidicola*** Warcup & Talbot

17a. Spores ovoid to ellipsoid, 4-6.5 x 3-4 µm. Protosterigmata swollen. Aquatic. (Basionym: *Aquathanatephorus pendulus* Tu and Kimbrough 1978, Bot. Gaz. 139: 459) ***T. pendulus*** (Tu & Kimbrough) comb. nov.

17b. Spores larger. Protosterigmata not swollen. Terrestrial. *T. cucumeris* complex **18**

18a. Spores ovoid, 5-12.5 x 3-6.3 µm ***T. corchori*** Tu *et al.*

18b. Spores ellipsoid to ovoid **19**

19a. Basidia barrel-shaped to subcylindrical, (10-)12-20(-23) x8-12(-13) µm. Spores (7-)8-13.5 x (4-)5-7 µm (incl.: *Hypochnus sasakii* Shirai and *Corticium microsclerotium* Weber; anam.: *Rhizoctonia solani* Kühn) ***T. cucumeris*** (Frank) Donk

19b. Basidia cylindrical to suburniform or subclavate, 20-37 long, 10-13 µm wide at the apex and 8-10 µm near the base. Spores 9.5-13.5 x 5.5-6.5 µm ***T. gardneri*** Warcup

20a. Spores globose to broadly ovoid **21**

20b. Spores ovoid to ellipsoid, cylindrical or slightly fusiform **23**

21a. Basidiospores broadly ovoid or broadly ellipsoid, 7.5-10 x 6-8 µm ***C. obscurum*** D.P. Rogers

21b. Spores globose to subglobose **22**

22a. Spores 6.5-9 µm diam. ***C. sphaerosporum*** Warcup & Talbot

22b. Spores 10-12(-14) µm diam. ***C. globosisporum*** Warcup & Talbot

23a. Spores ovoid or citriform, Q (length:width) < 2 **24**

23b. Spores narrowly ellipsoid or subcylindrical to slightly fusiform, Q$\geq$ 2 **27**

24a. Spores citriform, biapiculate, 9-16(-18) x 6-10 µm. Subicular hyphae 8-14(-20) µm wide ***U. fusisporum*** (J. Schroeter) Donk

24b. Spores not biapiculate. Subicular hyphae up to 10 µm wide **25**

25a. Basidiome ochraceous. Spores 8-10 x 5-6.5 µm ***U. ochraceum*** (Massee) Donk

25b. Basidiome whitish or pale greyish. Spores narrowly ovoid, 5.2-10.5 x 3.5-7 µm **26**

26a. Cells with (2-)3-5(-10) nuclei. Brown sclerotia present ***T. corchori*** Tu *et al.*

26b. Cells binucleate. Sclerotia absent (incl. *T. papillatum*, which may be an aberrant form) ***C. ramicola*** Tu *et al.*

27a. Spores 6-7.5 x 3-3.5 µm ***C. stridii*** J. Erikss. & Ryv.

27b. Spores larger, at least on average **28**

28a. Basidia short cylindrical, 20-30 x 8-10 µm. Spores not repetitive
Cejpomyces terrigenus (Bres.) Svrcek & Pouzar

28b. Basidia obovoid, up to 15(-20) µm long. Spores producing secondary spores **29**

29a. Saprophytic on wood. Basidiome ceraceous, adnate. Spores narrowly ellipsoid, 6.5-10.5(-11.5) x 3.5-6(-7) µm ***C. cornigerum*** (Bourdot) D.P. Rogers

29b. Parasitic on leaves, causing web blights. Basidiome often pellicular, separable **30**

30a. On Gramineae. Hyphal strands absent. Spores ellipsoid, 6.5-12.7 x 3.2-6.3 µm. (syn.: *C. cereale* Murray and Burpee; anam.: *Cerathorhiza cerealis* (v.d. Hoeven) R.T. Moore)
C. gramineum (Ikata & T. Matsuura) Oniki *et al.*

30b. On leaves of dicotyledonous trees. Hyphal strands present. Spores ellipsoid to slightly fusiform, 9-15(-18) x 4.5-7(-8) µm. ***C. anceps*** (Bres. & Syd.) H.S. Jacks.

Note: there is still considerable confusion concerning the *C. anceps* - *Pellicularia koleroga* complex. Höhnel (1910) described the spores of the type *P. koleroga* as länglich-keulig, etwa gebogen', 10-12 x 4-4.2 (Q = 2.5-2.9), which is narrower than the description of Stevens and Hall (1909; 10.5-11.6 x 4.7-5.8, Q = 2-2.3), on its turn narrower than the spores of *Hypochnus ochroleucus* Noack in Saccardo and Sydow (1902), described as ovoid, 8.5-10 x 5.5-6 (Q = 1.5-1.7). Burt (1918), in his description of *Corticium stevensii* (proposed as nom. nov. for *H. ochroleucus*!) gave 8-11 x 3-4 (Q=2.6-2.8), which agrees with *P. koleroga*. There is no overlap between these taxa, which definitely need further study; we hesitate to declare any of them synonymous.

Waitea Warcup and Talbot (type species: *Waitea circinata* Warcup and Talbot)
Anamorph: *Chrysorhiza* T.F. Andersen and Stalpers gen. nov.
Hyphae multinucleatae, septa dolipori praedita. Parenthesoma 3-4-perforatum. Sclerotia globosa vel aggregata et irregularia, glabra, aurantiaca vel salmonea, brunnescentia, saepe tecta strato gelatinoso. Typus: *C. zeae* (Voorhees) T.F. Andersen and Stalpers comb. nov. Basionym: *Rhizoctonia zeae* (Voorhees 1934).

Basidiome hypochnoid to ceraceous, separable. Subicular hyphae multinucleate. Cystidia absent. Basidia subcylindrical to urniform, not much wider than supporting hyphae, with 4 curved sterigmata, up to 6 µm long. Spores hyaline, thin-walled, smooth, ellipsoid, pear-shaped or cylindrical, sometimes slightly curved, not amyloid, becoming 1-2-septate. Sclerotia globose to irregular, ochraceous to salmon, becoming brownish, often covered with a gelatinous layer.

The anamorph of *Waitea* has so far been accommodated in *Moniliopsis* (now *Rhizoctonia* s.str.) together with the anamorphs of *Thanatephorus*. However, the latter genus is more closely related to *Ceratobasidium*, because of basidia with typical elongating sterigmata and sclerotial morphology. The sclerotia of *Chrysorhiza* are smooth, globose or aggregating into irregular structures, often covered with a gelatinous layer, contrary to those of *Thanatephorus*, which are irregular, brown, without a smooth, delimited surface and never covered with a gelatinous substance.

A possible other member of the genus is the anamorph of *Waitea nuda* Clémençon, which may be identical with *Rhizoctonia oryzae* (nom. inval.).

KEY TO THE SPECIES

1a. Basidiome hypochnoid. Some subhymenial cells irregular to contorted
W. circinata Warcup& Talbot

1b. Basidiome ceraceous. Irregular, contorted subhymenial cells absent ***W. nuda*** Clémençon

Platygloeoaceae

The only genus containing *Rhizoctonia* like species is *Helicobasidium*, which has until recently been considered to belong to the Auriculariaceae in the Auriculariales (Jülich, 1981), mainly because it was interpreted to have transversely septate basidia and is not an obligate parasite. However, the genus differs from *Auricularia* in a number of characteristics that are considered of some taxonomic importance. The so-called basidiospores (sporidia) germinate by budding and grow in a yeast-like manner. Bourett and McLaughlin (1986) described the pores of *Helicobasidium mompa* Tanaka as simple, i.e. lacking dolipores and a conspicuous parenthesome. The pores were sometimes occluded by globose bodies of electron-dense material, and the presence of microbodies close to the pore was observed. The spindle pole body (SPB) resembles that of the Uredinales, although differing in the number of layers (four in rusts, five in *Helicobasidium*) and the relationship of the discs to the nuclear envelope from the late interphase through telophase, while *Auricularia fuscosuccinea* has a spheroid, monoglobular SPB like in the other basidiomycetes (Bourett and McLaughlin, 1986). Additional support came from the study of 5S-rRNA, where Gottschalk and Blanz (1985) found a strong divergence between *Rhizoctonia crocorum* and *Hirneola auricula-judae* (as close relative of *Auricularia*). *Helicobasidium* is currently classified in the Ustomycota, Platygloeales (Moore, 1990).

Helicobasidium Pat. (Type species: *Hypochnus purpureus* Tul.; anamorph: *Thanatophytum* Nees)

Basidiome annual, resupinate, arachnoid or pellicular to membranaceous or felty. Hymenial surface even, whitish to pink or purplish. Hyphal system monomitic. Generative hyphae hyaline, thin- to thick-walled, without clamps, binucleate. Simple septa without parenthesome. Cystidia absent. Basidia hyaline, thin-walled, transversally septate, cylindrical, often curved or spirally. with 2-4 hyphoid sterigmata. Spores hyaline, thin-walled, smooth, ellipsoid to cylindrical or fusiform, not amyloid.

Substrate: soil-inhabiting, saprophytic or parasitic on plants.

Only the species of *Helicobasidium* producing thick, reddish-purplish basidiomes, which are known to produce *Rhizoctonia*-like anamorphs, are keyed out.

KEY TO THE SPECIES

1a. Spores ovoid to ellipsoid, 10-13.5 x 6-8(-9) µm — ***H. purpureum*** Pat.

1b. Spores fusiform, sometimes slightly curved, (10-)13-24(-29) x 4-7 µm. (= *H. mompa* auctt.) — ***H. compactum*** (Boedijn) Boedijn

Sebacinaceae with only a single genus containing *Rhizoctonia*-like species:

Sebacina Tulasne (syn.: *Cristella* Pat.; type species: *Thelephora incrustans* Fr.:Fr.; anamorph: *Opadorhiza* R.T. Moore)

Basidiome annual, resupinate or encrusting, ceraceous to gelatinous. Hymenial surface even to warted or irregular, whitish, cream, brownish, reddish or grey. Hyphal system monomitic or (seemingly) dimitic. Generative hyphae hyaline, thin- to thick-walled, with or without clamps, binucleate or multinucleate. Dolipore septa with imperforate parenthesome, or with a single, asymmetrically positioned perforation. Thin-walled (gloeo)cystidia rare. Basidia hyaline, thin-walled, longitudinally septate, globose to subpyriform, much wider than supporting hyphae, with 2-4 hyphoid sterigmata. Spores hyaline, thin-walled, smooth, globose, ovoid, ellipsoid, cylindrical, allantoid or fusiform, not amyloid.

Because a *Rhizoctonia* anamorph has only been described for *S. vermifera* (Warcup and Talbot, 1967) and an unidentified species (Warcup and Talbot, 1966) and because our taxonomic knowledge does not permit the construction of a reliable key to the species, this is not attempted.

Tulasnellaceae with only a single genus containing *Rhizoctonia*-like species:

Tulasnella J. Schroeter (syn.: *Pachysterigma* J. Olsen 1888; *Prototremella* Patouillard 1888; *Muciporus* Juel 1897; *Gloeotulasnella* Höhnel and Litschauer 1906; type species: *Hypochnus violeus* Quélet; anamorph: *Epulorhiza* R.T. Moore)

Basidiome annual, effused, membranaceous to ceraceous or subgelatinous. Hymenial surface smooth to warted. Hyphal system monomitic. Subicular hyphae hyaline, thin- to thick-walled, not typically branching at right angles, with or without clamps, predominantly binucleate. Subhymenial hyphae hyaline, thin-walled, often somewhat irregular. Dolipores with imperforate parenthesomes, which are often recurved at the margin. Gloeocystidia sometimes present. Basidia hyaline, thin-walled, subglobose to broadly clavate, pyriform or sphaeropedunculate, often twice as wide as the supporting hypha. Sterigmata (or epibasidia) subglobose to broadly ellipsoid at first, then producing a tubular apex, broadly fusiform to pyriform. Spores hyaline, thin-walled, smooth, subglobose, cylindrical or fusiform to sigmoid, not amyloid, typically producing secondary spores. Sclerotia may be present.

KEY TO THE SPECIES

1a. Spores fusiform, allantoid, sickle-shaped or helicoid, Q (length:width) >2. **2**
1b. Spores globose, ellipsoid, ovoid, reniform or short-fusiform, Q ≤2. **19**

2a. Spores sickle-shaped to helicoid **3**
2b. Spores cylindrical, fusiform, allantoid or reniform **6**

3a. Clamps present **4**
3b. Clamps absent **5**

4a. Spores spirally coiled, 13.2-30 x 1.8-3.6 µm ***T. anguifera*** P. Roberts
4b. Spores sickle-shaped, 12-30 x 1.2-1.8 µm ***T. interrogans*** P. Roberts

5a. Gloeocystidia present, 24-54 x 6 µm. Spores sickle-shaped, 11.4 x 16.8 x 1.8-3 µm ***T. falcifera*** P. Roberts
5b. Gloeocystidia absent. Spores sickle-shaped, 20-30 x 1.8-3.6 µm ***T. helicospora*** Raunk.

6a. Spores on average larger than 20 µm **7**
6b. Spores up to 18 µm long **9**

7a. Spores narrowly curved-fusiform to sinuous, 25-50 x 2-3(-4) µm ***T. deliquescens*** (Juel) Juel
7b. Spores fusiform to curved-fusiform, up to 30 µm long **8**

8a. Hyphae up to 2 µm wide. Spores curved-fusiform with a swelling near the apiculus, 18-25 (-28) x 3-4 µm ***T. quasiflorens*** P. Roberts
8b. Hyphae up to 7 µm wide. Spores fusiform to slightly curved-fusiform, 16-30 x (3-)3.5-5 µm ***T. calospora*** (Boud.) D.P. Rogers

9a. Clamps present **10**
9b. Clamps absent **12**

10a. Spores allantoid, 7-14 x 2.5-4 µm, on average at least 8 µm long (incl. *T. caroliniana* (L.S. Olive) L.S. Olive and *T. curvispora* Donk) ***T. allantospora*** Wakef. & Pearson
10b. Spores up to 8(-8.5) µm long, on average shorter **11**

11a. Spores allantoid, 5-8(-8.5) x 2-2.5(-3) µm — ***T. permacra*** P. Roberts
11b. Spores cylindrical to slightly allantoid or slightly fusiform, 5-6.5 x 2-2.5 µm — ***T. bifrons*** Bourdot & Galzin

12a. Spores up to 9(-10) µm long — **13**
12b. Spores 9-18 µm long, on average at least 10 µm — **15**

13a. Spores 4-7(-8) x 2.5-3.5(-4) µm, reniform — **28**
13b. Spores 6-9(-10) µm long, allantoid — **14**

14a. Gloeocystidia present — ***T. aggregata*** (L.S. Olive) L.S. Olive
14b. Gloeocystidia absent (incl. *T. saveloides* P. Roberts) — ***T. rubropallens*** Bourdot & Galzin

15a. Spores strongly allantoid, (7-)9-15(-17.5) x (2.5-)3-4 µm. — ***T. danica*** Hauersl.
15b. Spores cylindrical to fusiform, sometimes slightly allantoid. (*T. fuscoviolacea* complex) — **16**

16a. Spores cylindrical, (8-)9-14 µm long — **17**
16b. Spores cylindrical to fusiform, (10-)12-18(-20) µm long — **18**

17a. Capitate structures present, capitae becoming thick-walled chlamydospores, 13.5-17 µm diam. Spores 11-14 x 4.5-5 µm — ***T. irregularis*** Warcup & Talbot
17b. Capitate structures and chlamydospores absent. Spores (8-)9-13 x 3-4.5 µm (incl. *T. pallidocremea* Jülich) — ***T. fuscoviolacea*** Bres.

18a. Spores cylindrical, slightly curved, 12-18 x 3.5-4.5 µm. Sterigmata widely spaced, nearly lateral — ***T. cruciata*** Warcup & Talbot
18b. Spores fusiform, more rarely cylindrical, 10-20 x 3.5-6 µm. Sterigmata normally apical, not widely spaced (*T. violacea* auctt.) — ***T. brinkmannii*** Bres.

19a. Gloeocystidia present. Spores 4.5-6(-6.5) x 3.5-5(-5.5) µm — **20**
19b. Gloeocystidia absent — **23**

20a. Clamps present — **21**
20b. Clamps absent — **22**

21a. Gloeocystidia 9-35 x 5.5-10 µm (incl. *T. guttulata* L.S. Olive) — ***Gloeocystidia metachroa*** Bourdot & Galzin
21b. Gloeocystidia 50-100(-200) x 5-16 µm (*T. traumatica* auctt.) — ***T. opalea*** D.P. Rogers

22a. Gloeocystidia 22-162 x 6.8-10.5 µm. Chlamydospores may be present — ***T. pacifica*** L.S. Olive
22b. Gloeocystidia 10-75 x 6-12 µm. Chlamydospores absent — **23**

23a. Gloeocystidia hyaline. Basidia 7-11 x 4-6 µm. On gymnosperms — ***T. hyalina*** Höhn. & Litsch.
23b. Gloeocystidia yellowish. Basidia 12-18 x 8-10 µm. On angiosperms — ***T. cystidiophora*** Höhn. & Litsch.

24a. Clamps present — **25**
24b. Clamps absent — **27**

25a. Spores subglobose to ellipsoid, 12.2-14.5 x 9-10 µm — ***T. conidiata*** L.S. Olive
25b. Spores 6-8.5 µm long — **26**

26a. Spores ellipsoid, 6-8.5 x 3.5-4.5 µm — ***T. cremea*** Jülich
26b. Spores subglobose to ovoid or broadly ellipsoid, 6-7.5 x 4.8-5.8 µm. (incl. *T. bourdotii* Jülich) — ***T. inclusa*** M.P. Christ.

27a. Spores reniform, 4-7(-8) x 2.5-3.5(-4) — **28**
27b. Spores globose to ellipsoid, not reniform — **29**

28a. Sterigmata widely spaced. Hyphae 1-2(-3) µm wide — ***T. dissitispora*** P. Roberts
28b. Sterigmata closely adnate. Hyphae 2.5-3.5 µm wide — ***T. tomaculum*** P. Roberts

29a. Spores up to 6 µm long. (incl.: *T. lactea* Bourdot & Galzin, *T. obscura* Bourdot & Galzin and *T. microspora* Wakef. & Pearson. These 'species' are mainly based on colour variants and slight differences in the size of spores and basidia) — ***T. eichleriana*** Bres.
29b. Spores 6-12 µm long — **30**

30a. Spores globose, (8-)10-12 μm diam. ***T. sphaerospora*** Martin

30b. Spores ovoid to ellipsoid, up to 6 μm wide. (*T. pinicola* complex = *T. violea* auctt.) **31**

31a. Spores narrowly ellipsoid, 6-8(-9) x (3.2-)3.5-4.5 μm. (incl. *T. araneosa* Bourdot & Galzin, *T. papillata* (L.S. Olive) L.S. Olive and *T. asymmetrica* Warcup & Talbot) ***T. pruinosa*** Bourdot & Galzin

31b. Spores ovoid to ellipsoid, 6.5-9(-10) x (4-)4.5-5.5(-6) μm **32**

32a. Hymenial surface whitish to pale violaceous. Basidiome up to 80 μm thick. Associated with angiosperms? (incl. *T. albida* Bourdot and Galzin) ***T. albolilacea*** Bourdot & Galzin

32b. Hymenial surface pink to purplish, plumbeous to blackish when dry. Basidiome gelatinous-plicate. Associated with gymnosperms? (incl. *T. sordida* Bourdot & Galzin and *T. tremelloidea* Wakef. & Pearson) ***T. pinicola*** Bres.

V. References.

Andersen TF (1990) A study of hyphal morphology in the form genus *Rhizoctonia*. Mycotaxon 37: 25-46.

Andersen TF (1996) A comparative taxonomical study of *Rhizoctonia* s.l. employing morphological, ultrastructural and molecular methods. Mycol. Res. (in press).

Andersen TF & Stalpers JA (1994) A check-list of *Rhizoctonia* epithets. Mycotaxon 51:437-457.

Bauer R & Oberwinkler F (1986) Experimentell-ontogenetische Untersuchungen an Phragmobasidien. Z. Mykol. 52:259-269.

Bernard N (1909) L'évolution dans la symbiose. Les orchidées et leur champignons commençeaux. Ann. Sci. Nat. Bot. Sér. IX, 9:1-196.

Boidin J & Lanquetin P (1984) Répertoire des données utiles pour effectuer les tests d'intercompatibilité chez les Basidiomycètes III - Aphyllophorales non porées. Crypt. Mycol. 5: 93-245.

Bourett TM & McLaughlin DJ (1986) Mitosis and septum formation in the basidiomycete *Helicobasidium mompa*. Can. J. Bot. 64:130-145.

Burgeff H (1936) Samenkeimung der Orchideen. Gustav Fischer, Jena.

Burpee LL, Sanders PL, Cole H Jr & Sherwood RT (1980) Anastomosis groups among isolates of *Ceratobasidium cornigerum* and related fungi. Mycologia 72:689-701.

Burt EA (1918) The Thelephoraceae of North America XV. Ann. Mo. Bot. Gdn 13: 173-354.

Butler EE & Bracker CE (1970) Morphology and cytology of *Rhizoctonia solani*. In: Parmeter JR Jr (ed.) *Rhizoctonia solani*: Biology and Pathology (pp 51) University of California Press, Berkeley.

Currah RS, Sigler L & Hambleton S (1987) New records and new taxa of fungi from the mycorrhizae of terrestrial orchids of Alberta. Can. J. Bot. 65:2473-2482.

DeCandolle AP (1815) Flore Française 3. ed. Vol. V. Desray, France.

Donk MA (1954a) A note on sterigmata in general. Bothalia 6:301-302.

Donk MA (1954b) Notes on resupinate hymenomycetes - I. On *Pellicularia* Cooke. Reinwardtia 2:425-434.

Donk MA (1956) Notes on resupinate hymenomycetes - II. The tulasnelloid fungi. Reinwardtia 3:363-379.

Donk MA (1958) Notes on the basidium. Blumea Suppl. IV:96-105.

Duggar BM (1915) *Rhizoctonia crocorum* (Pers.) DC. and *R. solani* Kühn (Corticium vagum B. & C.) with notes on other species. Ann. Mo. Bot. Gdn 2:403-458.

Flentje NT (1956) Studies on *Pellicularia filamentosa* (Pat.) D.P. Rogers - I. Formation of the perfect stage. Trans. Brit. Mycol. Soc. 39:343-356.

Gamundí IJ & Ranalli ME (1964) Estudio systemático y biológico de las Ascoboláceae de Argentina I. Nova Hedw. 7:517-533.

Gottschalk M & Blanz PA (1985) Untersuchungen an 5S ribosomalen Ribonucleinsäuren als Beitrag zur Klärung von Systematik und Phylogenie der Basidiomyceten. Z. Mykol. 51:205-243.

Höhnel F von (1910) Fragmente zur Mykologie X. Sber. Akad. Wiss. Wien, Math.-nat. Kl. 119:393-473.

Jülich W (1981) Higher taxa of Basidiomycetes. J. Cramer Verlag, Vaduz.

Langer G (1994) Die Gattung *Botryobasidium* Donk (Corticiaceae, Basidiomycetes). Bibl. Mycol. 158, J. Cramer Verlag, Berlin.

Matz J (1921) The rhizoctonias of Porto Rico. J. Dep. Agric., Porto Rico 5: 1-30.

Moore RT (1987) The genera of *Rhizoctonia*-like fungi: *Ascorhizoctonia, Ceratorhiza* gen. nov., *Epulorhiza* gen.nov., *Moniliopsis*, and *Rhizoctonia*. Mycotaxon 29:91-99.

Moore RT (1990) Order Platygloeales ord. nov. Mycotaxon 39:245-248.

Murray DIL (1982a) Unusual basidiospore isolates of *Thanatephorus cucumeris*. Trans. Br. mycol. Soc. 78: 63-566.

Murray DIL (1982b) A modified procedure for fruiting *Rhizoctonia solani* on agar. Trans. Br. mycol. Soc. 79:129-135.

Murray DIL (1984) Cultural conditions influencing basidium formation in the Ceratobasidiaceae. Aust. J. Bot. 32:101-108.

Oberwinkler F (1985) Anmerkungen zur Evolution und Systematik der Basidiomyceten. Bot. Jahrb. Syst. 107:541-580.

Parmeter JR Jr & Whitney HS (1970) Taxonomy and nomenclature of the imperfect state. In: Parmeter JR Jr (ed.) *Rhizoctonia solani*: Biology and Pathology (pp 7-19) University of California Press, Berkeley.

Parmeter JR Jr, Whitney HS & Platt WD (1967) Affinities of some *Rhizoctonia* species that resemble mycelium of *Thanatephorus cucumeris*. Phytopathology 57:218-223.

Roberts P (1992) Spiral-spored *Tulasnella* species from Devon and the New Forest. Mycol. Res. 96: 233-236.

Roberts P (1993) Allantoid-spored *Tulasnella* species from Devon. Mycol Res. 97:213-220.

Roberts P (1994) Long-spored *Tulasnella* species from Devon, with additional notes on allantoid-spored species. Mycol Res. 98:1235-1244.

Saccardo PA & Sydow P (1902) Sylloge Fungorum 16. Patavii.

Saksena HK & Vaartaja O (1960) Descriptions of new species of *Rhizoctonia*. Can. J. Bot. 38:931-943.

Saksena HK & Vaartaja O (1961) Taxonomy, morphology, and pathogenicity of *Rhizoctonia* species from forest nurseries. Can. J. Bot. 39:627-647.

Stalpers JA (1974) Revision of the genus *Oedocephalum* (Fungi imperfecti). Proc. K. Ned. Akad. Wet. C, 77:383-401.

Stevens FL & Hall JG (1909) Hypochnose of pomaceous fruits. Ann. mycol. 7:49-59.

Stretton HM, McKenzie AR, Baker KF & Fletje NT (1964) Formation of the basidial stage of some isolates of *Rhizoctonia*. Phytopathology 54: 093-1095.

Talbot PHB (1965) Studies on '*Pellicularia*' and associated genera of hymenomycetes. Persoonia 3:371-406.

Talbot PHB (1970) Taxonomy and nomenclature of the perfect state. In: Parmeter JR Jr (ed.) *Rhizoctonia solani*: Biology and Pathology (pp 20-31) University of California Press, Berkeley.

Talbot PHB (1973) Towards uniformity in basidial terminology. Trans. Brit. mycol. Soc. 61:497-512.

Tu CC & Kimbrough JW (1975a) A modified soil-over culture method for inducing basidia in *Thanatephorus cucumeris*. Phytopathology 65:730-731.

Tu CC & Kimbrough JW (1975b) Morphology, development, and cytochemistry of the hyphae and sclerotia of species in the *Rhizoctonia* complex. Can. J. Bot. 53:2282-2296.

Tu CC & Kimbrough JW (1978) Systematics and phylogeny of fungi in the *Rhizoctonia* complex. Bot. Gaz. 139 454-466.

Tu CC, Roberts DA & Kimbrough JW (1969) Hyphal fusion, nuclear condition, and perfect stages of three species of *Rhizoctonia*. Mycologia 61:775-783.

Voorhees, R.K. (1934) Sclerotial rot of corn by *Rhizoctonia zeae* n.sp. Phytopathology 24:1290-1303.

Warcup JH & Talbot PHB (1966) Perfect states of some Rhizoctonias. Trans. Brit. mycol. Soc. 49:427-435.

Warcup JH & Talbot PHB (1967) Perfect states of Rhizoctonias associated with orchids. New Phytol. 66:631-641.

Warcup JH & Talbot PHB (1971) Perfect states of Rhizoctonias associated with orchids II. New Phytol. 70:35-40.

Warcup JH & Talbot PHB (1980) Perfect states of Rhizoctonias associated with orchids III. New Phytol. 86:267-272.

Williams PG & Thilo E (1989) Ultrastructural evidence for the identity of some multinucleate rhizoctonias. New Phytol. 112:513-518.

Yang CS & Korf RP (1985) *Ascorhizoctonia* gen. nov. and *Complexipes* emend., two new genera of anamorphs assigned to species of *Tricharina* (Discomycetes). Mycotaxon 23:457-481.

Zelmer C & Currah RS (1996) *Ceratorhiza pernacatena* and *Epulorhiza calendulina* spp. nov. mycorrhizal fungi of terrestrial orchids. Can. J. Bot. (in press).

I.B1 BIOCHEMICAL METHODS

SUHA JABAJI-HARE

Department of Plant Science, Macdonald Campus of McGill University, Ste-Anne-de-Bellevue, Quebec, CANADA, H9X 3V9

I. INTRODUCTION.

In spite of the economic importance of *Rhizoctonia* species, very little is known about the genetic variation and systematic relationships within and among this species complex. Identification of fungal species using conventional taxonomic criteria is difficult to apply within the *Rhizoctonia* complex. The lack of a useful fossil record, general absence of the teleomorphic stage, high phenotypic variation and their pathological diversity have been major impediments to progress in the field of systematics. The present system of identification of *Rhizoctonia* isolates belonging to the genera *Thanatephorus* and *Ceratobasidium* is based on morphology, pathology and the ability of their hyphae to anastomose when grown on solid media (Ogoshi *et al.*, 1979; Ogoshi, 1987; Burpee *et al.*, 1980). On the basis of different anastomosis reactions (Carling and Kuninaga, 1990), most investigators have resorted to classifying their isolates into distinct anastomosis groups (AGs) and subgroups. To date, 12 recognized AGs (designated as AG 1 through AG 11 and AG BI) and 12 subgroups (designated as AGs 1-IA; 1-IB; 1-IC; 2-1, 2-2IIB, 2-2IV, 4HG-1, 4HG-II, 6HG-1, 6GV, 9TP and 9TX) in *R. solani,* and 22 AGs have been identified in binucleate *Rhizoctonia* species (Sneh *et al.*, 1991; Ogoshi *et al.*, 1991). Although the AG and subgroup concepts represent the most successful method to date for classification, they provide little insight into the genetic diversity of these fungi.

That is the reason why during recent years, there has been an increasing interest in new approaches utilizing methods of molecular biology (see also chapters I.B2 and I.B3), biochemistry, and serology that would allow us to circumvent these obstacles. This review examines the development and use of biochemical and serological techniques for systematic and genetic studies of *Rhizoctonia* spp. and will attempt to illustrate the advantages and limitations of each of these techniques.

II. BIOCHEMICAL METHODS.

A. Soluble proteins

Electrophoretic polypeptide patterns of total soluble proteins have been found to be valuable in fungal taxonomy, because they express the genetic constitution of the cell. Several studies have been made since 1960, when Chang and coworkers (1962) performed their fundamental studies on *Neurospora* spp. Since then, the utility of this technique has been useful in identifying many plant pathogenic fungi including *Rhizoctonia solani* (Clare *et al.*, 1967; Reynolds *et al.*, 1983; Kuninaga, 1986). Clare and his colleagues were the first ones to demonstrate that total soluble protein patterns revealed on starch gel, were variable within isolates of *R. solani* from crucifers and cereals, and that isolates could be grouped based on their similar protein patterns. Reynolds and coworkers (1983) were able to distinguish 5 different AGs (AGs 1 through 5) from each other based on distinct protein patterns. Using the same

B. Sneh et al. (eds.),
Rhizoctonia Species: Taxonomy, Molecular Biology, Ecology, Pathology and Disease Constrol, 65–71.

technique on AGs 1 through 7 and BI, Kuninaga (1986) claimed that total protein patterns were not only different among AGs but distinct patterns were recognizable among homologous groups within an AG. Total protein electrophoresis may be a useful method in distinguishing AGs but not the subgroups. The method is simple and easy, however environmental factors can affect the synthesis of proteins and standardized growth.

B. Isoenzymes

Apart from general proteins, multilocus enzyme electrophoresis, which has long been a standard method for eukaryotic population genetics (Rattazzi *et al.*, 1983; Selander and Whittam, 1983) and systematics (Oxford and Rollinson, 1983), has been useful in identifying genetic variability within and between fungal species (Micales *et al.*, 1986, Micales and Stipes, 1987; Stasz *et al.*, 1989; Hanson and Wells, 1991; Chen *et al.*, 1991). Isoenzymes are defined as multiple molecular forms of a given enzyme. They are usually coded by different alleles or separate genetic loci and appear on a gel as multiple bands due to their different electrophoretic mobilities (Micales *et al.*, 1986). The banding patterns in the majority of enzymes are predictable and can be interpreted as alleles or loci depending on the quaternary structure of the enzyme. Taxonomic and genetic information can be summarized by scoring the alleles according to the method of May and co-workers (1979). The frequency of shared alleles among isolates is used to calculate the coefficient of similarity among organisms using commercial computer programs and the final information can be summarized in the form of a dendrogram which is constructed using a variety of clustering procedures (Sneath and Sokal, 1973; Ferguson, 1980).

Only a few researchers have taken advantage of studying phylogenetic relationships of *R. solani* using isoenzymes (Laroche *et al.,* 1993; Liu and Sinclair, 1992; 1993). In these studies, electrophoretic patterns of isoenzymes provided a good indication of the genetic diversity among AGs and within the subgroups, and reconfirmed the genetic basis of the AG concept. Based on isoenzyme profiles derived from 10 enzyme systems, several genetically distinct subgroups (6 within AG 1 and 5 within AG 2) were distinguished and some isozyme alleles and loci were identified (Liu and Sinclair, 1993; Liu *et al.*, 1990; Liu and Sinclair, 1992). Laroche *et al.* (1992) demonstrated that isozyme profiles can discriminate between 48 isolates from potato fields belonging to AG 3 and AG 9. Based on variation in isozyme patterns of seven enzymes, three subgroups within AG 3 were identified indicating that AG 3 is a heterogeneous group. These results demonstrate that isozyme electrophoresis is a promising means for studying relationships among groups. More isozyme studies on other AGs are still needed. Recently Kaufman and Rothrock (1995) exploited this technique by studying isozyme polymorphisms within AG 11. They were able to distinguish between AG 11 isolates from Australia and Arkansas, USA.

Group identification and the study of genetic relationships within binucleate *Rhizoctonia* (*Ceratobasidium* spp.) by means of electrophoresis were investigated in only one study (Damaj *et al.*, 1993). Damaj and colleagues (1993) differentiated 17 anastomosis groups and confirmed the identity of several cross-anastomosing groups using isozyme electrophoresis. Patterns of similarity among isozyme phenotypes supported the separation of 50 isolates of binucleate *Rhizoctonia* into 4 genetically distinct groups that corresponded well with anastomosis grouping concept and teleomorph affinity. These groups are congruent with five out of seven rDNA groups identified by rDNA RFLP patterns (Cubeta *et al.*, 1991; see also Chapter I.B3).

Researchers working on the genus *Rhizoctonia* are encouraged to exploit isozyme electrophoresis since it has special merits, and is well suited to the study of genetic diversity because a) isozymes normally show high levels of variation within species, b) they are encoded by numerous loci scattered over the genome and some of these loci could provide valuable genetic markers for population studies, and c) the analyses are relatively easy to apply to large numbers of individuals.

C. Pectic zymograms

Another technique which allows the detection of isozymes of specific enzymes such as pectic enzymes, has been found useful in identifying groups within *Rhizoctonia* spp. This method is based on the detection of various pectinases (e.g., pectin esterases and polygalacturonases) in pectic acid acrylamide gels by staining with ruthenium red (Cruickshank and Wade, 1980). Sweetingham and colleagues (1986) separated 140 isolates of *Rhizoctonia* from the grainbelt region of Western Australia into 11 distinct pectic zymogram-pattern groups (ZGs). They further demonstrated that individual pectic zymograms contained isolates whose cultural characteristics and pathogenic capabilities are similar. Using the same technique, five distinct ZGs within AG 8 have been established (Neate *et al.*, 1988; Sweetingham, 1990; MacNish and Sweetingham, 1993) that also corresponded with grouping based on random amplified polymorphic DNA (RAPD) markers (Duncan *et al.*, 1993). Recently, Schneider *et al.* (1994) demonstrated the utility of pectic zymogram patterns for discriminating between AG 2 isolates pathogenic to tulips from AG 2-1 isolates, while Balali *et al.* (1994) separated field isolates of *R. solani* AG 3 into three distinct pectic zymograms, and suggested that zymogram group variation is related to the morphology of tuber sclerotia and disease symptoms.

Group identification within binucleate *Rhizoctonia* (*Ceratobasidium* spp) by means of pectic zymograms were investigated by various Australian researchers (Sweetingham *et al.*, 1986; Neate and Cruickshank, 1988). Based on pectic zymogram patterns, these authors distinguished 5 binucleate zymogram groups (CZGs) within *Ceratobasidium cornigerum* from 65 isolates of binucleate *Rhizoctonia* isolated from soil and cereals.

This electrophoretic method is simple, reliable and easy to carry out allowing researchers to screen large number of isolates at the same time. Besides its taxonomic application, it can be used to predict pathogenicity of Rhizoctonia isoaltes based on the presence and relative amounts of induced pectinases.

D. Lectins

The use of Rhizoctonia-specific lectins in serological assays to differentiate among and within AGs has a promising future. However, there is a great deal of effort and time invested in lectin extraction and purification prior to their application as a chemical marker.

Another good biochemical marker which has not been fully explored, is the application of carbohydrate-binding proteins known as lectins in serological assays. These specific proteins are thought to be involved in cell-cell adhesion and recognition in fungi (Manocha, 1990). The recent discovery of cell surface (Barak *et al.*, 1986) and intracellular lectins (Vranken *et al.*, 1987) in *Rhizoctonia solani* prompted Kellens and Peumans (1991) to isolate lectins from twenty five isolates representing six AGs (1, 2, 3, 4, 5, and 7) and five subgroups (AGs 1-IA, IB, IC; and AGs 2-1, 2-2), and to compare them in serological assays. Their results clearly demonstrated that lectins are AG specific and isolates within a particular AG can be distinguished from other AG isolates based on their serological reactions.

E. Fatty acids

Another potential biochemical method that can be applied to study genetic relationships between and among isolates of *Rhizoctonia* spp. is whole fatty acid profiles (Steven-Johnk and Jones, 1992; Stevens-Johnk *et al.*, 1993). Fatty acid methyl esters (FAME) have been extensively employed to characterize and differentiate closely related strains of bacteria (Giatis and Beaver, 1990) and to a lesser extent in fungi (Jabaji-Hare, 1988). The technique is based on the analysis and identification of FAME derivatives using gas liquid chromatography (Miller and Berger, 1985) coupled with the computer software package Microbial Identification System (MIS, Microbial ID Inc., Newark, DE).

Cluster and principal component analyses of cellular fatty acids of *R. solani* populations representing AG 2-2IIIB (mat rush type) and AG 2 (isolated from corn and turf) showed that each population was genetically distinct (Stevens-Johnk and Jones, 1992). In a subsequent study, Stevens-Johnk *et al.* (1993) used fatty acid analysis to identify populations of *R. solani* causing tobacco target spot, which were previously identified as a genetic distinct new AG 2E (Liu and Sinclair, 1992), as a new AG 3 subgroup.

III. Serological methods

Serological studies using both polyclonal and monoclonal antibodies raised against mycelial proteins, secreted proteins or lectins have been used successfully to immunologically distinguish between AGs of *R. solani*. Depending on whether the antibody is polyclonal or monoclonal, detection of a specific AG and the study of phylogenetic relationships are some of the applications.

A. Polyclonal antibodies

Adams and Butler (1979) were the first to develop specific polyclonal antisera against mycelial extracts of tester isolates belonging to six AGs (1, 2-1, 2-2, 3, 4 and 5). The antibodies were successful in separating the isolates into serological groups that corresponded well to all AGs, except for AG 2-1 and AG 2-2 which were serologically indistinguishable. More recently, Mathiew and Brooker (1991) developed polyclonal antibodies against secreted proteins from AG 8, and assayed them against isolates belonging to AGs 2-1, 3, 4, and 8. Their results demonstrated that immunoblot patterns were distinct for each AG and varied within a single AG. Although the polyclonal antisera will be of little use to detect a specific AG due to their cross-reactivity with several AGs, their immunoblot patterns may be applied in a similar manner to isozyme and RFLP patterns as a means to establish inter-and intra-groups relationships.

B. Monoclonal antibodies

While polyclonal antibodies may not have the required specificity to detect a particular AG; in such cases monoclonal antibodies become the reagent of choice. The first monoclonal antibodies raised against *R. solani* were used to identify members belonging to AG 8 (Mathiew and Brooker, 1991). Cross reaction with other AGs and unrelated soil-borne fungi was minimal, however the authors stressed the need of testing a large number of related and unrelated fungi before discounting the possibility of cross-reaction with fungi other than *R. solani*. Since then, several antibodies have been developed showing potential specificity. Dusunceli and Fox (1992) developed monoclonal antibodies that detected a range of AGs (AGs 1, 2-1, 2-2, 3, 4, 5, 6 and 8) of *Thanatephorus cucumeris* in soils as well as in plant samples. Miller and colleagues (1992) developed monoclonal antibodies that are suitable for the detection and monitoring of Rhizoctonia brown patch populations (AG 2-2 IIIB) in turf. However, these antibodies were also shown to recognize populations of *Rhizoctonia cerealis* [AG-D (=CAG 1)] the causal agent of yellow patch of turf. Recently, Thornton *et al.* (1994) produced monoclonal antibodies and developed immunological assays that qualitatively and quantitatively detected propagules of *R. solani* AG 4. Their species-specific antibody did not cross react with other related and unrelated soil fungi.

VI. CONCLUSIONS

The recent application of biochemical studies have definitely played an important role in the genetics of *Rhizoctonia* spp. These methods along with molecular methods had clearly indicated that the AG and subgroup concepts are genetically distinct and are far more complexed than what we have previously assumed. Already, some long-held notion of homogeneity of certain AGs based on frequency of hyphal anastomosis are being challenged, raising questions about the genetic variability within an AG or subgroup. However, with these new techniques, our knowledge of genetic composition of *Rhizoctonia* spp. is still comparatively limited. Most comprehensive studies using biochemical methods have been carried out on few selected AGs and subgroups and the same can be said for studies using molecular markers and DNA base sequence complementarity (Cubeta *et al.* 1995; Kuninaga, 1995; see also chapter I.B2). Similar detailed and comprehensive studies are urgently needed for other AGs. If all AGs and subgroups could be compared through their proteins and nucleic acids and placed on a single phylogenetic tree, it may provide a more accurate method to replace the conventional AG concept.

V. ACKNOWLEDGMENTS

I gratefully acknowledge L Hare, PA O'Brien and T Paulitz for critically reviewing this manuscript and the Natural Sciences and Engineering Research Council of Canada (Research Grant) for supporting this work.

VI. REFERENCES

Adams GC & Butler EE (1979) Serological relationships among anastomosis groups of *Rhizoctonia solani*. Phytopathology 69:629-633.

Balali GR, Neate SM, Scott ES & Whisson DL (1995) Heterogeneity in pectic zymogram phenotypes of field isolates of *Rhizoctonia solani* AG 3. International symposium on *Rhizoctonia*, June 27-30, Noordwijkerhout, The Netherlands. (P-1-5).

Barak R, Elad Y & Chet I (1986) The properties of L-fucose-binding agglutinin associated with the cell wall of *Rhizoctonia solani*. Arch. Microbiol. 144:346-349.

Burpee LL, Sanders PL, Cole H, & Sherwood RT (1980) Anastomosis groups among isolates of *Ceratobasidium cornigerum* and related fungi. Mycologia 72:689-701.

Carling DE & Kuninaga S (1990) DNA base sequence homology in *Rhizoctonia solani* Kühn:inter- and intra-group relatedness of anastomosis group 9. Phytopathology 80:1362-1364.

Chang LO, Srb AM & Steward FC (1962) Electrophoretic separation of the soluble proteins of *Neurospora*. Nature, London 193:756-759.

Chen W, Hoy JW & Schneider RW (1991) Comparison of soluble protein and isozyme for seven *Pythium* species and application of the biochemical data to *Pythium* systematics. Mycol. Res. 95:48 555.

Clare BG, Flentje NT, & Atkinson MR (1967) Electrophoretic patterns of oxidoreductases and other proteins as criteria in fungal taxonomy. Aust. J. Biol. Sci. 21:275-295.

Cruickshank RH & Wade GC (1980) Detection of pectic enzymes in pectin-acrylamide gels. Anal. Biochem. 197:177-181.

Cubeta M, Echandi E, Abernethy T & Vilgalys R (1991) Characterization of anastomosis groups of binucleate *Rhizoctonia* spp. using restriction analysis of an amplified ribosomal RNA gene. Phytopathology 81:1395-1400.

Damaj M, Jabaji-Hare SH, & Charest PM (1993) Isozyme variation and genetic relatedness in binucleate *Rhizoctonia* species. Phytopathology 83:864-871.

Duncan S, Barton JE & O'Brien PA (1993) Analysis of variation in isolates of *Rhizoctonia solani* by random amplified polymorphic DNA assay. Mycol. Res. 97:1075-1082.

Dusunceli F & Fox RTV (1992) Accuracy of methods for estimating the size of *Thanatephorus cucumeris* populations in soil. Soil Use and Management 8:21-26.

Ferguson A (1980) Biochemical Systematics and Evolution. Wiley and Sons, New York, 194 pp.

Giatis RD & Beaver RW (1990) Characterization of fatty acid methyl ester content of *Clavibacter michiganensis* subsp. *michiganensis*. Phytopathology 80:318-321.

Hanson LC & Wells K (1991) Characterization of three *Tremella* species by isozyme analysis. Mycologia 83:446-454.

Jabaji-Hare SH (1988) Lipid and fatty acid profiles of some vesicular-arbuscular mycorrhizal fungi:contribution to taxonomy. Mycologia 80:622-629.

Kaufman P & Rothrock CS (1995) Evaluation of isolate diversity of *Rhizoctonia solani* subgroup11. Phytopathology 85:1125.

Kellens JT & Peumans WJ (1991) Biochemical and serological comparison of lectins from different anastomosis groups of *Rhizoctonia solani*. Mycol. Res. 95:1235-1241.

Kuninaga S (1986) Grouping of *Rhizoctonia solani* Kühn with physicochemical properties. Plant Prot. Jpn. 40:35-40.

Laroche JP, Jabaji-Hare SH & Charest PM (1993) Differentiation of two anastomosis groups of *Rhizoctonia solani* by isozyme analysis. Phytopathology 82:1387-1393.

Liu ZL & Sinclair JB (1992) Genetic diversity of *Rhizoctonia solani* anastomosis group 2. Phytopathology 82:778-787.

Liu ZL & Sinclair JB (1993) Differentiation of intraspecific groups within anastomosis group 1 of *Rhizoctonia solani* using ribosomal DNA internal transcribed spacer and isozyme comparisons. Can. J. Plant Pathol. 15:272-280.

Liu ZL, Nickrent DL, & Sinclair JB (1990) Genetic relationships among isolates of *Rhizoctonia solani* anastomosis group-2 based on isozyme analysis. Can. J. Plant Pathol. 12:376-382.

MacNish GC & Sweetingham MW (1993) Evidence of stability of pectic zymogram groups within *Rhizoctonia solani* AG 8. Mycol. Res. 97:1056-1058.

Manocha, MS (1990) Cell-cell interaction in fungi. J. Plant Disease and Protection 97:655-669.

Mathiew JS & Brooker JD (1991) The isolation and characterization of polyclonal and monoclonal antibodies to anastomosis group 8 of *Rhizoctonia solani*. Plant Pathol. 40:67-77.

May B, Roberts DW & Soper RS (1979) Intraspecific genetic variability in laboratory strains of *Entomophthora* as determined by enzyme electrophoresis. Exp. Mycol. 6:289-297.

Micales JA & Stipes RJ (1987) On the conspecificity of *Endothia eugeniae* and *Cryphonectria cubensis*. Mycologia 79:707-720.

Micales JA, Bonde MR & Peterson, GL (1986) The use of isozyme analysis in fungal taxonomy and genetics. Mycotaxon 27:405-449.

Miller L & Berger T (1985) Bacteria identification by gas chromatography of whole cell fatty acids. Hewlett-Packard Application Note 228-42, Hewlett Packard, Avondale PA. 8 pp.

Miller SA, Rittenburg JH, Petersen FP & Grothaus GD (1992) From the research bench to the market place:development of commercial diagnostic kits. In: Duncan JM & Torrence R (eds.) Techniques for The Rapid Detection of Plant Pathogens. (pp 208-221) Blackwell Scientific, Oxford.

Neate SM & Cruickshank RH (1988) Pectic enzyme patterns of *Ceratobasidium* spp. and *Rhizoctonia* spp. associated with sharp eyespot-like lesions on cereals in South Australia. Trans. Brit. Mycol. Soc. 91:267-272.

Neate SM, Cruickshank RH & Rovira AD (1988) Pectic enzyme patterns of *Rhizoctonia solani* isolates from agricultural soils in South Australia. Trans. Brit. Mycol. Soc. 90:37-42.

Ogoshi A (1987) Ecology and pathogenicity of anastomosis and intraspecific groups of *Rhizoctonia solani* Kühn. Annu Rev. Phytopathol. 25:125-143.

Ogoshi A, Cook RJ & Bassett EN (1991) *Rhizoctonia* species and anastomosis groups causing root rot of wheat and barely in the Pacific Northwest. Phytopathology 81:784-788.

Ogoshi A, Oniki M, Sakai R & Ui T (1979) Anastomosis grouping among isolates of binucleate *Rhizoctonia*. Trans. Mycol. Soc. Jpn. 20:33-39.

Oxford GS & Rollinson D. (eds.) (1983) Protein Polymorphism:Adaptive and Taxonomic Significance. The Systematics Association Special Volume no.24. Academic Press, Inc. London.

Rattazzi MC, Scandalios JG, & Whitt GS (eds.) (1983). Isozymes:Current Topics in Biological and Medical Research, vol. 10. Genetics and Evolution. Alan R. Liss, Inc. New York.

Reynolds M, Weinhold AR & Morris TJ (1983) Comparison of anastomosis groups of *Rhizoctonia solani* by polyacrylamide gel electrophoresis of soluble proteins. Phytopathology 73:903-906.

Schneider JHM, Schilder MT, Dijst G & Keijer J (1995) Anastomosis behavior, pathogenicity, and pectic zymograms of *Rhizoctonia solani* AG 2 in tulip. International symposium on *Rhizoctonia*, June 27-30, Norrwikerhout, The Netherlands. (P-1-8).

Selander RK & Whittam TS (1983) Protein polymorphisms and the genetic structure of populations. In: Nei M & Koehn RK (eds.) Evolution of Genes and Proteins (pp 89-114), Sinauer Associates, Sunderland, Mass.

Sneath PHA & Sokal RR (eds.) (1973) Numerical Taxonomy. The Principals and Practice of Numerical Classification. WH Freeman and Co., San Francisco. 573 pp.

Sneh B, Burpee LL & Ogoshi A (1991) Identification of *Rhizoctonia* species. APS Press, St. Paul, Minnesota 133 pp.

Stasz TE, Nixon K, Harman GE & Weeden NF (1989) Evaluation of genetic species and phylogenetic relationships in the genus *Trichoderma* by cladistic analysis of isozyme polymorphism. Mycologia 81:577-580.

Stevens-Johnk J & Jones RK (1993) Differentiation of populations of AG 2-2 of *Rhizoctonia solani* of cellular fatty acids. Phytopathology 83:278-293.

Stevens-Johnk J, Jones RK, Shew, HD & Carling DE (1993) Characterization of populations of *Rhizoctonia solani* AG 3 from potato and tobacco. Phytopathology 83:854-858.

Sweetingham MW (1990) Coping with brown spot and root rots of lupines. J. Agric. Western Australia 31:5-13

Sweetingham MW, Cruickshnak RH & Wong DH (1986) Pectic zymograms and pathogenicity of the *Ceratobasidiaceae*. Trans. Brit. Mycol. Soc. 86:305-311.

Thornton CR, Dewey FM & Gilligan CA (1994) Development of monoclonal antibody-based immunological assays for the detection of live propagules of *Rhizoctonia solani* in the soil. In: Schots A, Dewey FM & Oliver R (eds.) Modern Assays for Plant Pathogenic Fungi:Identification, Detection and quantification. (pp 29-35) CAB International Oxford, UK.

Vranken AM, Van Damme EJM, Allen AK & Peumans WJ (1987) Purification and properties of an N-acetylgalactosamine specific lectin from the plant pathogenic fungus *Rhizoctonia solani*. FEB Letters 216:67-72.

I.B2 DNA BASE SEQUENCE COMPLEMENTARY ANALYSES

SHIRO KUNINAGA

Health Science University of Hokkaido, Ishikari-Tohbetu, Hokkaido, 061-02, JAPAN

I. Introduction

Rhizoctonia solani Kühn [teleomorph = *Thanatephorus cucumeris* (Frank) Donk] is a collective species, and a taxonomic entity composed of divergent populations. Numerous attempts have been made to recognize subspecific groups within the *R. solani* complex, with primary emphasis on the anastomosis group (AG) system (Parmeter *et al.*, 1969; Anderson, 1982; Ogoshi, 1987; Sneh *et al.*, 1991). The characterization and grouping of isolates using the AG system represents a valuable approach for understanding variations within the species as well as identifying characteristics of the various groups inciting plant diseases. The AG concept however, offers no information on the genetic variation and taxonomic relationships within and between AGs in the *R. solani* complex.

With the development of molecular biology techniques, fungal taxonomy is now in the midst of a much broader spectrum of acceptable criteria (Bruns *et al.*,1991). The formal taxonomic significance of the AG concept has been proven correct by the molecular systematic evidence which indicates that isolates within the same AG are genetically related (Sneh *et al.*, 1991; Vilgalys and Cubeta, 1994).

The aim of this article is to assess the degree of genetic resolution among and within AGs revealed through DNA comparisons of *R. solani*, DNA base composition and DNA base sequence relatedness. In addition, the taxonomic implications of the AG concept based on evidence from these molecular studies will be discussed. Recent DNA-DNA hybridization studies in other groups of fungi have significantly contributed to better understand the correlation of DNA base sequence relatedness with the definition of a fungal species. In this review, no attempts were made to characterize the recently assigned new groups of AG 2-3 (Naito and Kanematsu, 1994), AG 10 (Ogoshi,1987) and AG 11 (Carling *et al.*, 1994) in *R. solani*, since very little is known about their genetic relatedness based on DNA-DNA hybridization.

The term homology, which was used in many of the earlier DNA-DNA hybridization studies, is replaced in this review by the term complementarity or relatedness. The term homology has a rather restricted definition which specifically indicates identity through descent from a common evolutionary origin (Hillis, 1994).

II. DNA Hybridization Methods

A. DNA base composition

Knowledge of DNA base composition, expressed as the molar percentage (mole %) of guanine + cytosine (GC) content, is very useful and is established as one of the most easily applied approach for bacterial taxonomy. In fungi however, this method has scarcely been used for taxonomical purposes. Storck and Alexopoulos (1970) analyzed the DNAs of 322 fungal species and found a wide range in their GC content

B. Sneh et al. (eds.),
Rhizoctonia Species: Taxonomy, Molecular Biology, Ecology, Pathology and Disease Constrol, 73–80.

(approximately 27.5 to 64.5 mole %). This initial survey has provided molecular biologists with sufficient data to discourage using this approach in taxonomic and phylogenetic strategies in fungi. The GC content of fungal species within a genus is frequently within the range of 10 mole %.

In *R solani*, the first comparison of DNA base composition was reported by Chang and Wu (1973), using a spectrophotometric method. They reported that the majority of the isolates examined, revealed a GC content of 40 to 44 mole %. However, they did not address the AG affiliation of the isolates.

Kuninaga and Yokosawa (1980; 1982a,b; 1983; 1984a,b; 1985a,b) have compared the DNA base compositions of *R. solani* and examined the genetic relatedness within and among AGs, based on the thermal melting profile (Tm). The GC contents of the isolates within each AG were very similar (Table 1). However, the *R. solani* complex was found to have a wide variation in GC content, ranging from 43.2 to 49.5 mole %.

TABLE 1. DNA Base Composition of the various AnastomosisGroups (AGs) of *Rhizoctonia solani*

AG (subgroup)	GC mole %[a]
1 (-IA)	49.1 - 49.5
1 (-IB)	46.1 - 47.6
2-1	43.7 - 45.9
2-2 (IIIB)	44.4 - 44.6
2-2 (IV)	44.9 - 45.4
3	47.1 - 48.3
4	46.1 - 47.9
5	46.1 - 47.6
6	43.2 - 46.1
7	46.1 - 47.1
8	44.4 - 46.4
BI	43.2 - 45.9
Range	43.2 - 49.5

[a] *Escherichia coli* DNA (GC mole % = 50) was used as standard. The GC content of *R. solani* DNA was calculated from the modified equation: GC mole % = 50 + (Tm *R. solani* DNA - Tm *E. coli* DNA/ 0.41) (Kuninaga and Yokosawa,1982a,b; 1983; 1984a,b; 1985a,b).

The taxonomic uses of GC content values are mainly exclusionary. All of the investigations have indicated that closely related species have similar GC contents, but having a similar GC content does not necessarily confirm conspecificity. On the other hand, strains may be expected to differ in their GC content if they belong to different organisms. Kurtzman (1987) has reported that fungal strains showing a difference in their GC contents in excess of 2.0-2.5 mole % (based on the thermal denaturation method) probably belong to different species. The wide range of GC content values in *R. solani* suggests that AGs may actually represent distinctly separate species, although more definitive separations are realized when DNA-DNA hybridization techniques are used.

B. DNA base sequence complementarity

Molecular studies on DNA base sequence complementarity have provided convincing evidence for the possibility to demonstrate genetic relatedness in microbial species, by measuring the degree of reassociation (hybridization) between dual DNA molecules. DNA reassociation may be carried out either in a free solution or when one of the DNA species immobilized on a filter. The reassociation of DNA molecules is monitored spectrophotometrically by measuring the kinetics of duplex formation, or by using isotope-labeled preparations. DNA relatedness examined by both techniques was reported to yield similar results (Kurtzman *et al.*, 1980).

A series of DNA-DNA hybridization studies conducted by Kuninaga and Yokosawa (1982a,b; 1983; 1984a,b; 1985a,b), and Carling and Kuninaga (1990) on various AGs of *R. solani* demonstrated that DNA-DNA hybridization groups support previous groupings based on anastomosis reactions. In addition, these studies presented further evidence that isolates of different AGs are actually distantly related (Fig. 1). Similar studies by Vilgalys (1988), in which genetic relationships between different AGs of *R. solani* were evaluated with heterologous DNA-DNA hybridization techniques, were in exact accordance with the previous studies of Kuninaga and Yokosawa (1982a,b; 1983; 1984a,b; 1985a,b).

DNA-DNA hybridization values obtained for members within most AGs were at a very high level (≥90%). In AGs 2-1, 3, 5, 7, 8 and BI, isolates hybridized with other members within their respective AGs at a rate of ≥91%, indicating a high degree of genetic relatedness among isolates within these groups. The hybridization studies also demonstrated considerable genetic differences between subgroups within certain AGs: 1, 2-2 and 9. The relatively low DNA relatedness (47-87%) between certain isolates within each of these AGs corresponded well with previously defined subgroups based on cultural morphology, pathogenicity and vitamin requirement.

DNA-DNA hybridization studies also revealed previously unknown heterogeneity within AG 4 (Kuninaga and Yokosawa, 1984a; Vilgalys, 1988), and within AG 6 (Kuninaga and Yokosawa, 1984b). Within AG 4, isolates of subgroups HG-1 and HG-II hybridized at 88% or higher with other members of their groups, but at the low values of 30 to 47% between the groups. Within AG 6, isolates of subgroup HG-I hybridized with other members of the group at ≥91%. Isolates of the heterogeneous subgroup GV hybridized with other members of the group at 55 to 66%. The extent of DNA relatedness between isolates of HG-I and GV ranged from 47 to 62%.

C. Taxonomic implications of DNA base sequence complementarity

The review has focused on interpretations of DNA base sequence relatedness as well as on taxonomic implications of the AG concept in *R. solani*. DNA relatedness depends highly on the kind of DNA used and reaction conditions determining hybridization. Considering these reservations however, DNA-DNA hybridization techniques have been very useful for defining species and assessing current taxonomical criteria for bacteria and yeasts. In extensive DNA comparisons among species of bacteria and yeasts, strains showing ≥60-70% DNA relatedness are considered to be members of the same species (Price *et al.*,1978; Kurtzman, 1987; Wayne *et al.*, 1987). In yeast however, a few cases are known where low DNA relatedness (20%) did not preclude interfertility (Kurtzman, 1987). At present, these exceptional results were only reported in yeasts.

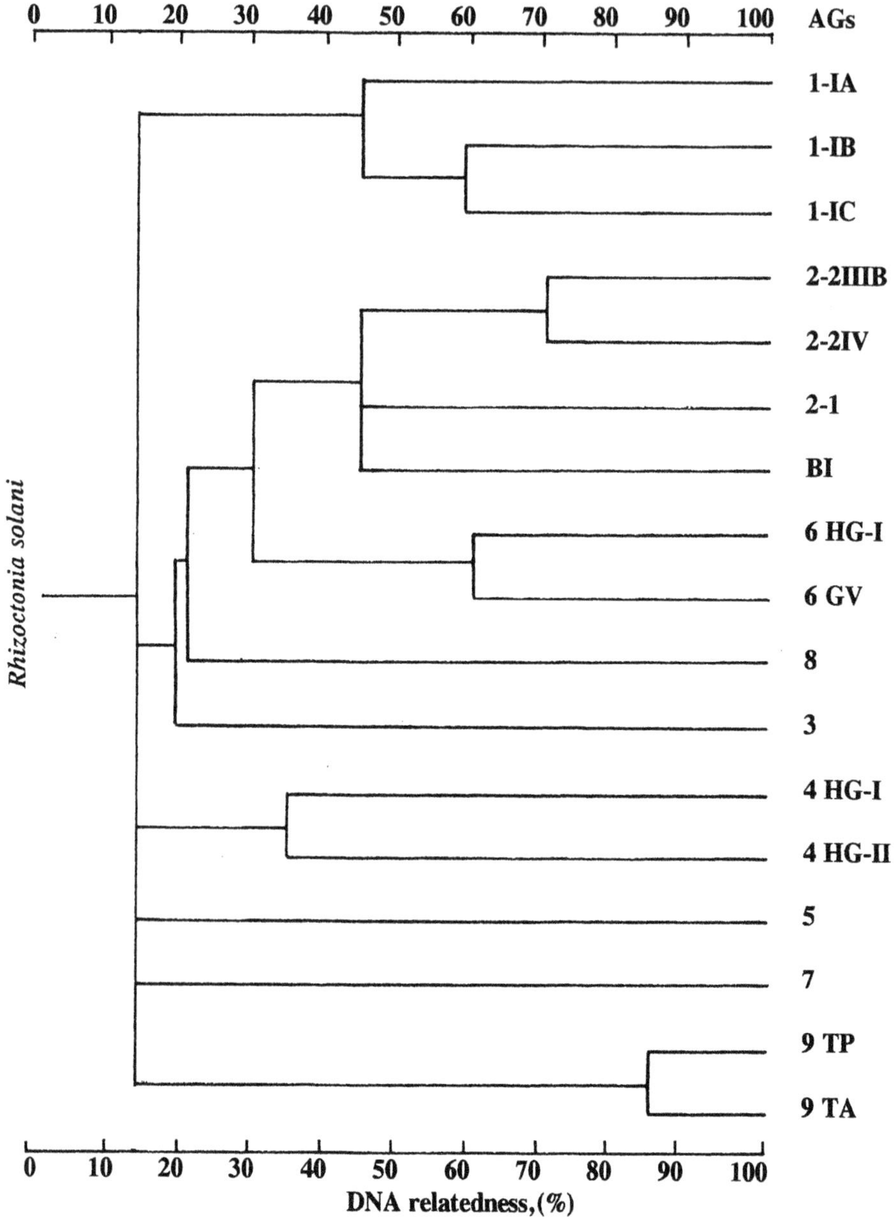

FIGURE. 1. Dendrogram of *Rhizoctonia solani* anastomosis groups (AGs) based on DNA relatedness among its AGs. The dendrogram was constructed according to the unweighed pair-group method with arithmetic averaging (UPGMA) from DNA-DNA hybridization values (%) (Kuninaga and Yokosawa, 1982a,b; 1983; 1984a,b; 1985a,b; Carling and Kuninaga, 1990).

In fungi other than yeasts, evaluation of genetic divergence with DNA-DNA hybridization has been demonstrated during the past 10 years (Fig. 2). These comparisons have also provided insight into the correlation between the DNA relatedness and the fungal species concept.

DNA-DNA hybridization values >85% represent strains of the same taxon, while values of <60-65% represent strains of different species. DNA-DNA hybridization values of 60 to 84% within the same species indicate strains of different groups (e.g. varieties, pathotype or formae speciales).

Consequently, the interpretation of DNA relatedness in the *R. solani* complex suggests that different AGs may apparently represent distinct evolutionary units, corresponding to separate species. Each population of AGs 1-IA, 1-IB, 1-IC, 2-1, 2-2IIIB, 2-2IV, 3, 4HG-I, 4HG-II, 5, 6HG-I, 6GV, 7, 8, 9TP, 9TX and BI is regarded as having a formal taxonomic status. The subgroups IA and IB within AG 1 have been previously classified into different species based on their sclerotial form. However, AGs in *R. solani* can not be distinguished by teleomorph morphology and taxonomically meaningful characteristics. These findings do not allow yet to classify these groups into different species. Therefore, the groups may be recognized according to their DNA relatedness as "genomic species."

III. CONCLUSIONS

Kinetic analyses of whole-cell DNA reassociation estimate a haploid genome size of *R. solani* to be around 2.8×10^7 nucleotide base pairs. The fungal genome consists of approximately 8-17% repeated sequences (Kuninaga, unpublished.). The characteristics of such a fungal genome has made it possible to answer basic questions regarding their taxonomic relatedness by DNA-DNA hybridization methods. In fungi however, the lower limit of DNA relatedness values suggesting species delimitation is not yet well defined. Further studies are required to determine whether the correlation of DNA relatedness with the species concept are also applicable to other groups of fungi.

The percentage of DNA that hybridizes among most AGs is typically ≤15%. Thus, the limited resolution afforded by DNA-DNA hybridization of the whole genome does not provide sufficient understanding of the evolutionary relationships in *R. solani*. At present, some molecular techniques are available to enable resolving the degree of sequence variation of fungal DNA. Further studies on phylogenetic relationships in the *R. solani* complex comparing their ribosomal DNA sequences could provide useful information to determine whether most AGs are in fact representing independent evolutionary units.

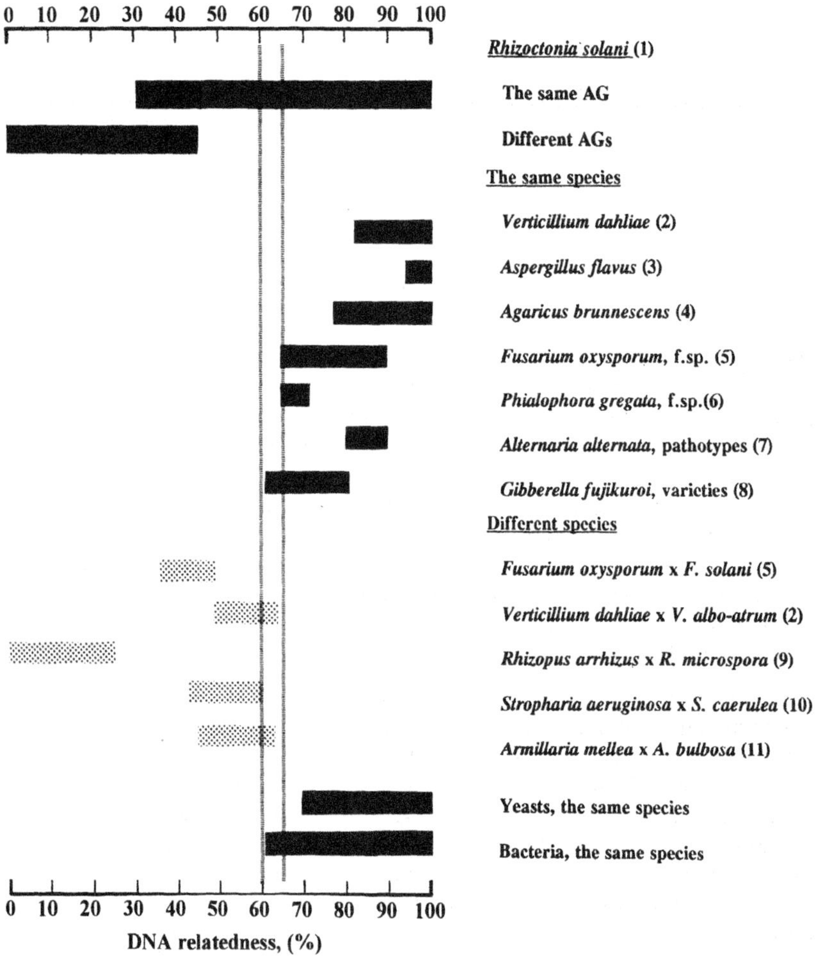

FIGURE 2. Genetic relatedness among various fungi including *Rhizoctonia solani* based on DNA-DNA hybridization. ■ *R. solani*; ■ The same species; ░ Different species. The vertical lines represent the lower limit of DNA-DNA hybridization values (60 to 65%) suggesting species delimitation. (1) Kuninaga and Yokosawa (1982a,b; 1983; 1984a,b; 1985a,b); Carling and Kuninaga (1990), (2) Kuninaga and Yokosawa (unpublished), (3) Yamamoto *et al.* (1992), (4) Kurtzman *et al.* (1986), (5) Horgen *et al.* (1984), (6) Kuninaga and Yokosawa (1989), (7) Kuninaga and Yokosawa (unpublished), (8) Ellis (1988), (9) Ellis (1985), (10) Jahnke and Bahweg (1986), (11) Jahnke *et al.* (1987), (12) Kurtzman (1987), (13) Wayne *et al.* (1987).

IV. ACKNOWLEDGMENTS

The author thanks D Carling, MA Cubeta and R Yokosawa for their valuable comments and suggestions on this manuscript.

V. REFERENCES

Anderson NA (1982) The genetic and pathology of *Rhizoctonia solani*. Annu. Rev. Phytopathol. 20:329-347.

Bruns TD, White TJ & Taylor JW (1991) Fungal molecular systematics. Annu. Rev. Ecol. Syst. 22:525-564.

Carling DE & Kuninaga S (1990) DNA base sequence homology in *Rhizoctonia solani* Kühn: inter- and intragroup relatedness of anastomosis group-9. Phytopathology 80:1362-1364.

Carling DE, Rothrock CS, MacNish GC, Sweetingham MW, Brainard KA & Winters SW (1994) Characterization of anastomosis group 11 (AG 11) of *Rhizoctonia solani*. Phytopathology 84:1387-1393.

Chang SC & Wu LC (1973) Base composition of *Rhizoctonia solani* Kühn. Memories of the College of Agriculture, Natl. Taiwan Univ. 14: 22-26.

Ellis JJ (1985) Species and varieties in the *Rhizopus arrhizus-Rhizopus oryzae* group as indicated by their DNA complementarity. Mycologia 77:243-247.

Ellis JJ (1988) Section *Liseola* of *Fusarium*. Mycologia 80:255-258.

Hillis DM (1994) Homology in Molecular Biology. In: Hall BK (ed.) Homology.(pp 339-368) Academic Press, California.

Horgen PA, Arthur R, Davy O, Moum F, Herr N, Straus N & Anderson J (1984) The nucleotide sequence homologies of unique DNA's of some cultivated and wild mushrooms. Can. J. Microbiol. 30:587-593.

Jahnke KD & Bahweg G (1986) Assessing natural relationships in the Basidiomycetes by DNA analysis. Trans. Br. mycol. Soc. 87:175-191.

Jahnke KD, Bahweg G & Worrall JJ (1987) Species delimitation in the *Armillaria mellea* complex by analysis of nuclear and mitochondrial DNAs. Trans. Br. mycol. Soc. 88:572-575.

Kuninaga S & Yokosawa R (1980) A comparison of DNA base compositions among anastomosis groups in *Rhizoctonia solani*. Ann. Phytopathol. Soc. Japan 46:150-158.

Kuninaga S & Yokosawa R (1982a) DNA base sequence homology in *Rhizoctonia solani* Kühn. I. Genetic relatedness within anastomosis groups 1. Ann. Phytopathol. Soc. Japan 48:659-667.

Kuninaga S & Yokosawa R (1982b) DNA base sequence homology in *Rhizoctonia solani* Kühn. II. Genetic relatedness within anastomosis groups 2. Ann. Phytopathol. Soc. Japan 48:668-673.

Kuninaga S & Yokosawa R (1983) DNA base sequence homology in *Rhizoctonia solani* Kühn. III. Genetic relatedness within AG 3, AG 5, AG 7 and AG BI. Ann. Phytopathol. Soc. Japan 49:647-652.

Kuninaga S & Yokosawa R (1984a) DNA base sequence homology in *Rhizoctonia solani* Kühn. IV. Genetic relatedness within AG 4. Ann. Phytopathol. Soc. Japan 50:322-330.

Kuninaga S & Yokosawa R (1984b) DNA base sequence homology in *Rhizoctonia solani* Kühn. V. Genetic relatedness within AG 6. Ann. Phytopathol. Soc. Japan 50:346-352.

Kuninaga S & Yokosawa R (1985a) DNA base sequence homology in *Rhizoctonia solani* Kühn. VI. Genetic relatedness among seven anastomosis groups. Ann. Phytopathol. Soc. Japan 51:127-132.

Kuninaga S & Yokosawa R (1985b) DNA base sequence homology in *Rhizoctonia solani* Kühn. VII. Genetic relatedness between AG BI and other anastomosis groups. Ann. Phytopathol. Soc. Japan 51:133-138.

Kuninaga S & Yokosawa R (1989) Genetic relatedness within and between formae speciales of *Fusarium oxysporum* as measured by DNA-DNA reassociation kinetics. Ann. Phytopathol. Soc. Japan 55:216-223.

Kurtzman CP, Smiley MJ, Johnson CJ, Wickerham LJ & Fuson GB (1980) Two new and closely related heterothallic species. *Pichia amylophilia* and *Pichia mississippiensis*: Characterization by hybridization and deoxyribonucleic acid reassociation. Int. J. Syst. Bacteriol. 30:208-216.

Kurtzman CP, Smiley MJ, Robnett CJ & Wicklow DT (1986) DNA relatedness among wild and domesticated species in the *Aspergillus flavus* group. Mycologia 78:955-959.

Kurtzman CP (1987) Impact of nucleic acid comparison on the classification of fungi. Proc. Indian Acad. Sci. 97:185-201.

Naito S & Kanematsu S (1994) Characterization and pathogenicity of a new anastomosis subgroup AG 2-3 of *Rhizoctonia solani* Kühn isolated from leaves of soybeans. Ann. Phytopathol. Soc. Japan 60:681-690.

Ogoshi A (1987) Ecology and pathogenicity of anastomosis and intraspecific groups of *Rhizoctonia solani*. Annu. Rev. Phytopathol. 25:125-143.

Parmeter JR Jr, Sherwood RT & Platt WD (1969) Anastomosis grouping among isolates of *Thanatephorus cucumeris*. Phytopathology 59:1270-1278.

Price CW, Fuson GB & Phaff HJ (1978) Genome comparison in yeast systematics: delimitation of species within the genera *Schwanniomyces*, *Saccharomyces*, *Debaryomycetes*, and *Pichia*. Microbiol. Rev. 42:161-193.

Sneh B, Burpee L & Ogoshi A (1991) Identification of *Rhizoctonia* Species. APS press. St. Paul, Minnesota 133 pp.

Storck R & Alexopoulos CJ (1970) Deoxyribonucleic acid of fungi. Bacteriol. Rev. 34:126-154.

Vilgalys R (1988) Genetic relatedness among anastomosis groups in *Rhizoctonia solani* as measured by DNA/DNA hybridization. Phytopathology 78:698-702.

Vilgalys R & Cubeta MA (1994) Molecular systematic and population biology of *Rhizoctonia*. Annu. Rev. Phytopathol. 32:132-155.

Wayne LG, Brenner DJ, Colwell RR, Grimont PAD, Kandler O, Krichevsky MI, Moore WEC, Murray RGE, Stackebrandt E, Starr MP & Truper HG (1987) Report of the ad hoc committee on the reconciliation of approaches to bacterial systematic. Int. J. Syst. Bacteriol. 37:463-464.

Yamamoto H, Kobayashi K & Ogoshi A (1992) Genetic relatedness between *Phialophora gregata* f. sp. *adzukicola* and *sojae*. Trans. Mycol. Soc. Japan 33:461-465.

I.B3 MOLECULAR ANALYSIS OF RIBOSOMAL RNA GENES IN *RHIZOCTONIA* FUNGI

MARC A. CUBETA[1], RYTAS VILGALYS[2] AND DOLORES GONZALEZ[3].

[1]*Department of Plant Pathology, North Carolina State University, Vernon G. James Research and Extension Center, 207 Research Station Road, Plymouth, NC 27962, USA;*
[2]*Department of Botany, Duke University, Durham, NC 27708, USA;*
[3]*Instituto de Ecologia, A.C., 91000 Xalapa, Vera Cruz, MEXICO*

I. Introduction

A. Ribosomal RNA genes

The DNA sequences that encode ribosomal RNAs have been useful for understanding phylogenetic and taxonomic relationships and patterns of genetic variation in fungi (Bruns *et al.*, 1992). In fungi, ribosomal genes (rDNA) are located within mitochondria or nuclei and contain numerous regions of high sequence conservation and variability over their entire length (White *et al.*, 1990). The nuclear rDNA genes of fungi are arranged as a series of tandemly repeated units that include genes for the 18S (small), 5.8S and 28S (large) rRNA subunits. Vilgalys and Gonzalez (1990a) determined that *Thanatephorus praticola* (anamorph = *Rhizoctonia solani* AG 4) had approximately 59 copies (repeated units) per haploid genome. The 28S and 18S rRNA subunits are separated by an intergenic spacer (IGS) region that usually contains a 5S rRNA subunit and the 5.8S subunit is flanked by two internal transcribed spacer (ITS1 and ITS2) regions. In contrast to nuclear rDNA genes, mitochondrial (mt) rDNA genes have a lower guanine and cytosine content (25-40%), possess more methylated nucleotides and lack a 5.8S subunit (White *et al.*, 1990). Except for analysis of the small mt rRNA subunit (Liu and Sinclair, 1992), mt rDNA genes have not been critically examined in *Rhizoctonia* fungi.

B. Molecular Methods for Examining rDNA Genes

A variety of molecular methods have been employed to examine rDNA genes in *Rhizoctonia* fungi to infer taxonomic and phylogenetic relationships and develop diagnostic protocols. Genetic differences between anastomosis groups (AGs) and subgroups of *Rhizoctonia solani* and binucleate *Rhizoctonia* spp. have been demonstrated with restriction fragment length polymorphism (RFLP) analysis of rDNA genes using Southern blotting or polymerase chain reaction (PCR) methods (Vilgalys and Cubeta, 1994). RFLP analysis, using Southern blotting methods, require more DNA than PCR-based methods (usually $<1\mu g$ vs. $<$ng of total genomic DNA, respectively), radioactive or bioluminescent capabilities and are often more labor intensive. In contrast, PCR-generated RFLPs are directly observable after electrophoresis and substantially reduce poor digestion of methylated DNA associated with certain restriction endonucleases (Vilgalys and Gonzalez, 1990b). An abundance of well-defined oligonucleotide primers are also available to amplify nuclear and

B. Sneh et al. (eds.),
Rhizoctonia Species: Taxonomy, Molecular Biology, Ecology, Pathology and Disease Constrol, 81–86.

mitochondrial rDNA regions (Vilgalys and Hester, 1990; White *et al.* 1990; Gardes and Bruns, 1993).

Restriction site data (i.e., presence and absence of restriction fragments) obtained from digestion of organellar or PCR-amplified DNA can be used to estimate genetic distance and provide characters for phylogenetic analysis (Swofford and Olsen, 1990; Bruns *et al.*, 1991). A similarity matrix can be constructed with restriction site data, from which a dendrogram showing genetic relationships can be made. These data can also be used to construct phylogenetic trees from pairwise distance estimates or using character-based methods (i.e., parsimony or maximum likelihood). RFLP data generated using Southern blotting or PCR-based methods are also useful for developing DNA restriction maps and providing information about rDNA gene gene duplication and length mutation.

II. RFLP ANALYSIS OF *RHIZOCTONIA* FUNGI

Two studies of RFLP variation using Southern blotting methods have examined nuclear rDNA. Vilgalys and Gonzalez (1990b) and Jabaji-Hare *et al.* (1990) examined 87 and 37 *R. solani* strains representing 10 AGs from various geographic origins, respectively. In each study, *R. solani* genomic DNA was digested with different restriction endonucleases and analyzed by Southern blotting with radiolabeled rDNA probes. Vilgalys and Gonzalez (1990b) revealed 55 unique RFLP patterns from 87 strains after Southern blotting of digested DNA with a plasmid probe (pTP1) containing the entire 8.8 kilobase (kb) rDNA repeat from a *T. practicola* AG 4 strain. Each AG was characterized by one or more unique RFLP patterns not found in any other AG. Strains within AGs: 3, 4, 7, 8 and BI possessed a single invariant RFLP pattern, whereas strains within AGs: 1, 2, 5 and 6 possessed multiple RFLP patterns. In addition, strains belonging to each of three subgroups in AG 1 and AG 2 could not be differentiated based on RFLP patterns and exhibited considerable genetic variability, even within strains of the same subgroup.

In the studies by Jabaji-Hare *et al.* (1990), 38 *R. solani* strains were examined by Southern blotting of digested DNA with a plasmid probe (pAm2) containing a 6.8 kb rDNA repeat from *Armillaria ostoyae* (Romagn.) Herink (Anderson *et al.*, 1989). Their results indicate that strains belonging to AGs: 1, 2, 3 and 6 possessed one or more unique RFLP patterns not found in any other group. Strains belonging to AGs: 4, 5, 7, 8, 9 and BI produced a variety of RFLP patterns, some of which were shared among them. In contrast to the results of Vilgalys and Gonzalez (1990b), some RFLP patterns of rDNA of certain AGs were not unique but were shared with other AGs, while AG 4 strains were characterized by two distinct RFLP patterns and each subgroup of AG 1 had identical and indistinguishable RFLP patterns. Although these studies provide additional evidence for rDNA variation among *R. solani* AGs and subgroups, differences among results may be attributed to restriction endonucleases, probes and strains used in each study.

Detailed restriction analyses of PCR-amplified rDNA regions of *R. solani* and binucleate *Rhizoctonia* spp. have been conducted by several research groups (Vilgalys and Cubeta, 1994). Liu and Sinclair (1992; 1993) examined the ITS1, ITS2 and 5.8S rDNA regions in several *R. solani* subgroups using PCR and restriction analyses. Based on restriction analysis of ITS1, ITS2 and 5.8S sequences from 131 field strains representing AG 1 and AG 2, patterns of rDNA variation revealed at least six subgroups within AG 1 (designated 1A-F) and five subgroups within AG 2 (designated 2A-E). Evidence for the five subgroups within AG 2 was supported with isozyme data (Liu *et al.*, 1990), however subgroups within AG 1 could not be distinguished alone based on isozyme analysis (Liu and Sinclair, 1993). Although isozyme phenotypes alone could not be used to define subgroups within AG 1, several distinct phenotypes were identified. The genetically distinct AG 2-E strains from tobacco in North Carolina examined by Liu and Sinclair (1992), were subsequently identified as belonging to AG 3 based on anastomosis behavior, vitamin requirement and fatty acid analysis (Johnk *et al.*, 1993). A recent genetic characterization of *R. solani* AG 2-3, a newly recognized foliar pathogen of soybean, using RFLP analysis of nuclear rDNA ITS1, ITS2 and 5.8S resions, has provided additional evidence for another genetically distinct subgroup within AG 2 (Kanematsu and Naito, 1995).

Liu *et al.* (1993) subsequently used restriction analysis of ITS1, ITS2 and 5.8S rDNA regions to examine 30 *R. solani* strains representing nine different AGs. Patterns of RFLP variation revealed two subgroups in AG 3, two in AG 4, three in AG 5, and two in AG 10. Two strains of AG 6, one of AG 7, one of AG 8, six of AG 9 and two of AG-BI each appeared to represent a single subgroup. The length of this rDNA region ranged from 630-740 base pairs (bp). Collectively, Liu and colleagues have designated 25 different subgroups within the *R. solani* anastomosis group classification system based on PCR amplification and restriction analysis of ITS1, ITS2 and 5.8S rDNA regions. Caution should be exercised in directly equating rDNA diversity with the subgroup concept, particularly in the absence of additional biochemical, mating, morphological and molecular criteria. Recently, Liu *et al.* (1995) constructed rDNA restriction maps for these 25 *R. solani* subgroups based on amplification and digestion of the 18S and ITS1, ITS2 and 5.8S rDNA regions. Four types of DNA restriction maps were identified in the 18S region, however, most *R. solani* subgroups remain largely unresolved, except for the two distinct subgroups of AG 10.

The genetic relationships among AGs of binucleate *Rhizoctonia* spp. have also been investigated using RFLP analysis of PCR-amplified rDNA (Cubeta *et al.*, 1991). In this study, 13 of 21 AGs of binucleate *Rhizoctonia* spp. were distinguished using restriction analysis of a portion of the 28S rDNA region. Patterns of rDNA variation revealed seven major groupings that also corresponded well with anastomosis designation, host range and teleomorph affinity. For example, strains previously classified as AG-D (=CAG-1) (*Ceratobasidium cereale*) formed a distinct genetic group from other binucleate AGs. Similar genetic groupings representing AG-A (=CAG-2) and AG-B(o) corresponded with (*C. cornigerum*). Similar patterns of isozyme variation revealed at least five different groups of binucleate *Rhizoctonia* spp. that corresponded well with the rDNA groups (Damaj *et al.*, 1993).

III. rDNA SEQUENCE ANALYSIS OF *RHIZOCTONIA* FUNGI

Comparative analysis of rDNA genes have been extensively used to examine taxonomic and phylogenetic relationships of fungi (Anderson *et al.*, 1989; Bruns *et al.*, 1991; Hibbett, 1992). With the advent of PCR and automated sequencing technology, it is possible to amplify and study DNA sequence variation within the rDNA region of *Rhizoctonia* fungi at almost any desired taxonomic level. DNA sequence data can provide an unlimited source of information and greater resolution of genetic differences within and among *Rhizoctonia* fungi. However, phylogenetic studies of this taxonomically and genetically diverse group of fungi are just emerging.

Recently, Boysen *et al.* (1995) conducted a phylogenetic analysis of the ITS1, ITS2 and 5.8S rDNA regions of nine *R. solani* AG 4 strains using asymmetric PCR sequencing techniques. The phylogenetic analysis revealed three distinct subgroups within AG 4 that correlated with ecological habitat and virulence. Two hypovirulent AG 4 strains had similar rDNA deletions (50-60 bp) that were not present in seven virulent AG 4 strains. These seven virulent strains were subsequently separated based on ecological habitat. Four and three of the seven strains were obtained from hot, arid areas and cooler, higher elevation regions in Spain, respectively.

In a separate study, a phylogenetic analysis of the 28S rDNA region indicated little sequence variation among 10 AGs of *R. solani* (approximately 2%), however, several evolutionary groupings were identified (Gonzalez and Vilgalys, 1991b). For example, strains representing different subgroups of AG 1 had a common evolutionary origin as did different subgroups within AG 4 and AG 7. Gonzalez (1991a) also identified several *R. solani* AGs that possessed 400 bp insertions in the 28S rRNA subunit which complicated phylogenetic analysis and inferences. These insertions may represent Group I introns that have been previously identified in nuclear rDNA genes of other organisms (Burke, 1988). Although these studies provide preliminary information suggesting that AGs and subgroups are independent evolutionary groups, a larger sample of *R. solani* strains is needed to determine whether AGs and subgroups represent independent evolutionary lineage's.

IV. CONCLUSIONS

In general, the analysis of rRNA genes presented in this manuscript supports previous hyphal anastomosis (Burpee *et al.*, 1980; Anderson, 1982; Ogoshi *et al.*, 1983; Ogoshi, 1987; Sneh *et al.*, 1991), protein (Reynolds *et al.*, 1983; Adams, 1988) and DNA/DNA hybridization (Kuninaga and Yokosawa, 1984, 1985a, 1985b; Vilgalys, 1988; Carling and Kuninaga, 1990) studies indicating that AGs of *R. solani* and binucleate *Rhizoctonia* spp. are genetically distinct and that considerable genetic variability can also exist within an AG. However, several questions still need to be resolved. What are the biological, ecological and evolutionary implications of this genetic variability? How much rDNA variation exists within an AG and what additional criteria are required to designate additional subgroups in *Rhizoctonia*? Are subgroups independent evolutionary units and do subgroups represent biological species? If subgroups represent species, then what are populations and other units of organization in *Rhizoctonia*? In the future, a fairly large and comprehensive rDNA sequence data base should be available to provide conceptual advances in our knowledge of evolutionary relationships and species concepts (Brasier, 1987; Mayer, 1984) in *Rhizoctonia*. rDNA sequence data should also facilitate the development of subgroup-specific PCR primer sets and oligonucleotide probes for diagnostic and experimental purposes.

V. ACKNOWLEDGMENTS

We gratefully acknowledge Suha Jabaji-Hare, Louise Lewis and Pamela Puryear for reviewing this manuscript and the United States Department of Agriculture (88-01007 and 91-00853) for supporting this work.

VI. REFERENCES

Adams GC Jr (1988) *Thanatephorus cucumeris* (*Rhizoctonia solani*), a species complex of wide host range. Adv. Pl. Pathol. 6:535-552.

Anderson NA (1982) The genetics and pathology of *Rhizoctonia solani*. Annu. Rev. Phytopathol. 20:329-347.

Anderson JB, Bailey SS & Pukkilla PJ (1989) Variation in ribosomal DNA among biological species of *Armillaria mellea*, a genus of root-infecting fungi. Evolution 43:1652-1662.

Boysen M, Borja M, del Moral C, Salazar O & Rubio V (1995) Identification at strain level of *Rhizoctonia solani* AG 4 isolates by direct sequence of asymmetric PCR products of the ITS region. Curr. Genetics (In press).

Brasier CM (1987) The dynamics of fungal speciation. In: Rayner ADM, Brasier CM and Moore D (eds.) Evolutionary Biology of the Fungi. (pp 231-283) Cambridge University Press, Cambridge, England.

Bruns TD, White TJ & Taylor JW (1991) Fungal molecular systematics. Annu. Rev. Ecol. Syst. 22:525-564.

Bruns TD, Vilgalys R, Barns SM, Gonzalez D, Hibbett DS, Lane DJ, Simon L, Stickel S, Szaro TM, Weisburg WG & Sogin ML (1992) Evolutionary relationships within fungi: analysis of nuclear small subunit rRNA sequences. Mol. Phylog. Evol. 1:231-241.

Burke JM (1988) Molecular genetics of group I introns: RNA structures and protein factors required for splicing-a review. Gene 73:273-294.

Burpee LL, Sanders PL, Cole H Jr & Sherwood RT (1980) Anastomosis groups among isolates of *Ceratobasidium cornigerum* and related fungi. Mycologia 72:689-701.

Carling DE & Kuninaga S (1990) DNA base sequence homology in *Rhizoctonia solani* Kühn: inter-and intragroup relatedness of anastomosis group-9. Phytopathology 80:1362-1364.

Cubeta MA, Echandi E, Abernethy T & Vilgalys R (1991) Characterization of anastomosis groups of binucleate *Rhizoctonia* spp. using restriction analysis of an amplified ribosomal RNA gene. Phytopathology 81:1395-1400.

Damaj M, Jabaji-Hare SH & Charest P-M (1993) Isozyme variation and genetic relatedness in binucleate *Rhizoctonia* species. Phytopathology 83:864-871.

Gardes M & Bruns TD (1993) ITS primers with enhanced specificity for basidiomycete application to the identification of mycorrhizae and rusts. Mol. Ecol. 2:113-118.

Gonzalez D (1991a) Molecular systematics of the *Rhizoctonia solani* complex based on evidence from ribosomal DNA. M.Sc. Thesis, Duke University, 1991.

Gonzalez D & Vilgalys R. (1991b) A molecular approach to classification within the *Rhizoctonia solani* complex. Mycol. Soc. Amer. Newsl. 42:15 (Abst.).

Hibbett DS (1992) Ribosomal RNA and fungal systematics. Trans. Mycol. Soc. Jpn. 33:533-556.

Jabaji-Hare SH, Meller Y, Gill S & Charest PM (1990) Investigation of genetic relatedness among anastomosis groups of *Rhizoctonia solani* using cloned DNA probes. Can. J. Plant Pathol. 12:393-404.

Stivens-Johnk J, Jones RK & Shew HD (1993) Characterization of populations of *Rhizoctonia solani* AG 3 from potato and tobacco. Phytopathology 83:854-858.

Kanematsu S & Naito S (1995) Genetic characterization of *Rhizoctonia solani* AG 2-3 by analyzing restriction fragment length polymorphism of nuclear ribosomal DNA internal transcribed spacers. Ann. Phytopathol. Soc. Jpn. 61:18-21.

Kuninaga S & Yokosawa R (1984) DNA base sequence homology in *Rhizoctonia solani* Kühn. IV. Genetic relatedness within AG 4. Ann. Phytopathol. Soc. Jpn. 50:322-330.

Kuninaga S & Yokosawa R (1985a) DNA base sequence homology in *Rhizoctonia solani* Kühn. VI. Genetic relatedness among seven anastomosis groups. Ann. Phytopathol. Soc. Jpn. 51:127-132.

Kuninaga S & Yokosawa R (1985b) DNA base sequence homology in *Rhizoctonia solani* Kühn. VII. Genetic relatedness between AG BI and other anastomosis groups. Ann. Phytopathol. Soc. Jpn. 51:133-138.

Liu ZL, Nickrent DL & Sinclair JB (1990) Genetic relationships among isolates of *Rhizoctonia solani* anastomosis group 2 based on isozyme analysis. Can. J. Plant Pathol. 12:376-382.

Liu ZL & Sinclair JB (1992) Genetic diversity of *Rhizoctonia solani* anastomosis group 2. Phytopathology 82:778-787.

Liu ZL & Sinclair JB (1993) Differentiation of intraspecific groups within anastomosis group I of *Rhizoctonia solani* using ribosomal DNA internal transcribed spacer and isozyme comparison. Can. J. Plant Pathol. 15:272-280.

Liu ZL, Domier LL & Sinclair JB (1993) Subgroup-specific ribosomal DNA polymorphism of the *Rhizoctonia solani* species complex. Mycologia 85:795-800.

Liu ZL, Domier LL & Sinclair JB (1995) Polymorphism of genes coding for nuclear 18S rRNA indicates genetic distinctiveness of anastomosis group 10 from other groups in the *Rhizoctonia* species complex. Appl. Envir. Microbiol. 61:2659-2664.

Mayr E (1984) Species concepts and their application. In: Sober E (ed.) Conceptual Issues in Evolutionary Biology, (pp 531-540). Cambridge, MA: MIT press.

Ogoshi A (1987) Ecology and pathogenicity of anastomosis and intraspecific groups of *Rhizoctonia solani* Kühn. Annu. Rev. Phytopathol. 25:125-143

Ogoshi A, Oniki M, Araki T & Ui T (1983) Anastomosis groups of binucleate *Rhizoctonia* in Japan and North America and their perfect states. Trans. Mycol. Soc. Jpn.24:79-87.

Reynolds M, Weinhold AR & Morris TJ (1983) Comparison of anastomosis groups of *Rhizoctonia solani* by polyacrylamide gel electrophoresis of soluble proteins. Phytopathology 73:903-906.

Sneh B, Burpee LL & Ogoshi A (1991) Identification of *Rhizoctonia* Species., APS Press, St. Paul, Minnesota. 133 pp.

Stivens-Johnk J, Jones RK & Shew HD (1993) Characterization of populations of *Rhizoctonia solani* AG 3 from potato and tobacco. Phytopathology 83:854-858.

Swofford DL & Olsen GJ (1990) Phylogenetic reconstruction. In: Hillis DM & Moritz C (eds.) Molecular Systematics. (pp 411-501) Sinauer Assoc Inc. Sunderland MA.

Vilgalys R (1988) Genetic relatedness among anastomosis groups in *Rhizoctonia* as measured by DNA/DNA hybridization. Phytopathology 78:698-702.

Vilgalys R & Gonzalez D (1990a) Organization of ribosomal DNA in the basidiomycete *Thanatephorus practicola*. Curr. Genet. 18:277-280.

Vilgalys R & Gonzalez D (1990b) Ribosomal DNA restriction fragment length polymorphisms in *Rhizoctonia solani*. Phytopathology 80:151-158.

Vilgalys R & Hester M (1990) Rapid identification and mapping of enzymatically amplified ribosomal DNA from several *Cryptococcus* species. J. Bacteriol. 172:4238-4246.

Vilgalys R & Cubeta MA (1994) Molecular systematics and population biology of *Rhizoctonia*. Annu. Rev. Phytopathol. 32:132-155.

White TJ, Bruns TD, Lee S & Taylor JW (1990) Amplification and direct sequencing of fungal ribosomal RNA genes for phylogenetics. In: Gelfand MAIDH, Sninsky JJ and White TJ (eds.) PCR Protocols: A Guide to Methods and Applications. (pp 315-322), Academic Press, New York.

I.B4 INTEGRATED BIOCHEMICAL, CULTURAL AND NUMERICAL METHODS

J. ELIZABETH M. MORDUE, SABINE BANNIZA, PAUL D. BRIDGE, MICHAEL A. RUTHERFORD AND MARK HOLDERNESS

International Mycological Institute, Bakeham Lane, Egham, Surrey TW20 9TY, UNITED KINGDOM

I. INTRODUCTION

In recent years there has been a considerable expansion in the types of characters available for characterising and grouping filamentous fungi. In addition to the more established morphological and anastomosis groupings and pathological behaviour, recent studies within the genus *Rhizoctonia* have included sequence analysis of the rDNA gene cluster, RFLPS, RAPDs, whole cell fatty acid and isoenzyme analyses (Sweetingham *et al.*, 1986; Vilgalys and Gonzales 1990; Laroche *et al.*, 1992; Liu and Sinclair, 1992; Duncan *et al.*, 1993; Johnk and Jones, 1993; O'Brien, 1994; Vilgalys and Cubeta, 1994; Tommerup *et al.*, 1995). Although not recently emphasised within *Rhizoctonia*, other biochemical linkages such as enzyme production, secondary metabolite production and carbon and nitrogen source utilisation have been used with other fungal groups (Bridge and Hawksworth, 1985; Manczinger and Polner, 1987; Frisvad, 1992).

Most of the methods used to group *Rhizoctonia* isolates can be separated into two categories based on functional or genetic properties (Bridge *et al.*, 1995). Functional properties are those that can be directly related to some feature or activity of the organism, such as enzyme production; while genetic properties are those that reflect some aspect of the genetic make-up or relatedness of the organisms, such as anastomosis groups (AGs) or DNA sequence analysis. Some methods such as isoenzyme analysis or the sequencing of coding regions of the DNA may have qualities relating to both categories. In multidisciplinary investigations, similarity groupings defined by one approach will not necessarily be supported by those from another. A number of recent studies have integrated phenotypic and genotypic characters and these studies have illustrated some of the complex situations that may occur. The ability of different techniques to reflect relationships at different levels has been illustrated in the genus *Ganoderma*, where palm pathogens form a single homogeneous group by pectinase isozyme analysis. However, within that functional group molecular methods, such as RFLPs of mtDNA, have revealed considerable variation (Miller *et al.*, 1995). Within the vascular wilt pathogen *Fusarium oxysporum* some host-specific special forms, when investigated by molecular methods, are fairly homogeneous and consist of very similar races and a small number of vegetative compatibility groups (e.g., special form *vasinfectum*), whereas others (e.g., special form *cubense*) consist of many vegetative compatibility groups and well separated races (Bridge *et al.*, 1993; Thomas *et al.*, 1994).

B. Sneh et al. (eds.),
Rhizoctonia Species: Taxonomy, Molecular Biology, Ecology, Pathology and Disease Constrol, 87–98.

In most studies on *Rhizoctonia*, characteristics from one approach have either been used in isolation or, more commonly related to AGs, which have become accepted as a "baseline" classification. With the increase in characteristics available for the differentiation of fungal populations it has become increasingly important to consider both the type of information provided, and the relationship of this information to the ecology and ancestry of the organism.

As a result of the complexity of the character sets available, numerical methods are widely used to group organisms on the basis of their common characteristics. Several numerical approaches have been used with filamentous fungi with morphological, biochemical and molecular characters and with combinations of these characters. The most common techniques used are those that result in a dendrogram, although non-hierarchic ordination methods have also been considered (Dabinett and Wellman, 1978; Bridge *et al.*, 1989; Frisvad, 1994). Numerical methods are also used extensively with molecular characters to produce phylogenetic trees that may represent evolutionary relationships. These trees are commonly based on DNA sequence data and estimates of the variance of the distribution, in the form of bootstrap values, are applied to each major branch point (Swofford and Olsen, 1990).

II. MOLECULAR, BIOCHEMICAL, CULTURAL AND MORPHOLOGICAL CONSIDERATIONS

A. Cultural methods and morphology

The influence of culture media, incubation temperature, light and other physical environmental conditions on growth and morphology of *Rhizoctonia solani* is implicit in every paper that includes information on these topics in its methods section. The morphological and cytological characteristics that are common to all isolates of *R. solani* (Parmeter, 1970) and similar characters in other *Rhizoctonia* species have a genetic basis. Nevertheless, many of these characters will have functional significance and there is much general evidence of variability within and between isolates (Sneh *et al.*, 1991). Central to any integrated study is assessment of morphological components of that variability.

Cell dimensions, hyphal branch/septum distances and general morphological characters of hyphal growth pattern have been analysed frequently (Watanabe and Shiyome, 1975; Armentrout *et al.*, 1986; Andersen, 1990), and are reproducible within the limits of the systems studied. Analysis of structure and development of sclerotia has been less common (Tu and Kimbrough, 1975; Yang *et al.*, 1989; Kousik and Snow, 1990), although sclerotium frequency, shape and size are recorded as significant colony characters (Sneh *et al.*, 1991; D'yakov, 1993). Failure to correlate morphology with strictly defined cultural/environmental parameters hinders comparison, as Andersen (1990) indicated. Currah and Zelmer (1992) largely avoided use of cell dimensions and branching as identification key characters, although they did include hyphal width with other characters in species descriptions. Mordue *et al.* (1989) found morphometric characters more distinctive between than within *Rhizoctonia* species. They derived quantitative morphological characters from assimilative tests, as did Tu and Kimbrough (1975) for cytochemical studies. Papers including morphological attributes frequently emphasize the value of combining several characters to differentiate strains (Kim, 1993; Johnk and Jones, 1993).

B. Biochemical methods

Biochemical methods include a wide variety of techniques; many are recorded by quantitative parameters and are functionally expressed. Included are long established observations such as the influence of CO_2 on strain distribution (Durbin, 1959) and thiamine auxotrophy/prototrophy (Parmeter, 1970; Carling and Kuninaga, 1990; Johnk and Jones, 1993). Tu and Kimbrough's (1975) histological observations are biochemically based. Some comparatively simple assimilative and physiological tests have been used to differentiate isolates. Mordue *et al.* (1989) included growth on five basal media with additional adaptations and tests for lipolytic activity, gelatinase, tyrosinase, ß-galacturonase, phenol oxidases, peroxidase and RNAase (of which tyrosinase, phenol oxidases and lipolytic activity were particularly distinctive). Although biochemical/morphological correlation can undoubtedly be developed further, current biochemical emphasis is on electrophoretic profiles of proteins and isoenzymes, fatty acid analysis and serological assays as discussed by Jabaji-Hare (chapter I.B1).

One form of functional biochemical test that has rarely been considered in integrated studies is differential fungicide sensitivity. Many fungicides, like propiconazole (Eeckhout *et al.*, 1990) show effects on morphology and ultrastructure of *R. solani in vitro* and therefore have potential to provide biochemical information for strain differentiation. Appreciable variability in sensitivity of AGs to fungicides, including pencycuron, has also been observed (Kataria *et al.*, 1991; Ueyama *et al.*, 1993). Chen *et al.* (1990) successfully used PDA -containing carbendazim and PDA plus the antibiotic jingangmycin as differential media for AGs. Like Mordue *et al.*, they recorded their results in morphological terms of growth, zonation and production of sclerotia and pigments. However, test strains are often not fully characterized in laboratory-based fungicide testing.

C. Molecular methods

The earliest techniques that are generally considered to have a genetic basis are isoenzyme studies, and these have been widely implemented within *Rhizoctonia*. The most generally reported technique is analysis of patterns produced from electrophoresis of pectin-degrading enzymes; this was used to designate eleven "zymogram groups" (ZGs) within *Rhizoctonia* (Sweetingham *et al.*, 1986). The pectic zymogram technique detects different extracellular enzymes and different isoforms of individual extracellular enzymes, while other isoenzyme methods have also been employed that show different forms of individual intracellular enzymes.

Most molecular methods involve the analysis of some region of DNA, and several different approaches have been used within *Rhizoctonia*. Probably the earliest forms of DNA analysis were the determination of mol % G+C contents and DNA/DNA hybridization values (Kuninaga and Yokosawa, 1980; 1985). Although these techniques were of value in individual cases, they did not always provide clear correlation with AG. Other early DNA based studies included sequencing ribosomal DNA genes for determining phylogenetic relationships between *Rhizoctonia* and other genera (Blanz and Gottschalk, 1984). More recently less specific molecular techniques, including restriction fragment length polymorphisms (RFLPs) from random genomic probes and random amplification of polymorphic DNA (RAPD), have been used to delineate populations within anastomosis groups (Liu and Sinclair, 1992; Duncan *et al.*, 1993).

III. The pathogen: integration with pathology

R. solani is widespread throughout the world, in both uncultivated and cultivated soils. Although capable of saprophytic or mycorrhizal growth (Currah and Zelmer, 1992), the species is primarily recognised as one of the most widespread and economically-important pathogens known to man. The species can infect a wide range of crops, causing seed decay, root rots, damping-off, stem cankers, fruit decay and leaf, sheath and web blight.

Frequently, the results of *in vitro* characterization are reflected functionally by differences found in pathogenicity. Conversely, differences in pathogenicity can suggest functional properties for characterization. However, isolates belonging to one AG can vary substantially in aggressiveness and host range whilst isolates belonging to different AGs may produce similar disease expressions (e.g., infection cushions, Kousik *et al.*, 1995) on related hosts (Sneh *et al.*, 1991; Chase, 1991). Isolates lacking host specificity may still differ markedly in virulence to a specific host (Bolkan and Ribeiro, 1985; Roh and Kim, 1987). Early work by Schulz (1936) and Richter and Schneider (1953) indicated that only one group of isolates (AG 3) was highly host specific and pathogenic on potato, but it is now known that isolates from AGs 1, 2, 4, 5 and 9 also can be potato pathogens (Abe and Tsuboki, 1987; Chand and Logan, 1983; Bains and Bisht, 1995) and the host specificity must be integrated with other characters for diagnostic purposes.

Involvement of culturally and biochemically defined characters in plant infection processes may be direct or indirect. There is no direct evidence that the similarity of lesions on wheat and lupins after infection with isolates in the same ZG (Sweetingham *et al.*, 1986) was due to pectinase activity. Keon *et al.* (1987) debate whether pectinases determine pathogenicity or merely modify symptom expression. Sclerotial production, much studied *in vitro*, is highly variable but in some instances correlates with AGs and/or field pathogenicity (Chand and Logan, 1983). The differential ability of isolates to produce sclerotia as seed and soil-borne inoculum and saprophytic survival agents is of major importance in the subsequent development of disease, reductions in crop yield and possibilities for disease control (Hide *et al.*, 1973; Hide and Cayley, 1982; Little and Cooke, 1988; Damicone *et al.*, 1993) and also provides analytical characters, as does host varietal resistance (Rashid and Bernier, 1993).

IV. Data analysis

A. General considerations

1. Methods of analysis and character coding. Many recent studies on *Rhizoctonia* have contained some element of numerical data analysis such as dendrograms, ordinations or phylogenetic trees. These methods provide convenient ways to represent numerical data obtained from the electrophoresis of proteins and DNA fragments. The various techniques are well established and have been used widely in biological taxonomy to estimate both inter- and intra-group relationships. Traditional taxonomic groupings are hierarchical, in that a genus consists of species that can be further divided into subspecies or other groups. Cluster analysis is also an hierarchic approach and will always tend to group organisms. The validity of cluster analysis therefore needs to be carefully considered, as it may be more appropriate to use non-hierarchic methods in which commonality of characters may not imply true relatedness (Bridge *et al.*, 1996).

2. Integrated approaches. The integration of data from different approaches is not a new concept with filamentous fungi, and biochemical and physiological characters were being considered in addition to morphology as long ago as 1956 (Abe, 1956). There have however been few extensive and integrated studies utilizing cultural and numerical data within *Rhizoctonia*, even though a number of authors have linked molecular studies and pectinase isoenzyme studies with AG and pathogenicity data. Mordue *et al.* (1989) combined physiological and simple biochemical characters with morphology in a comparison of orchid mycorrhizal isolates with representative AG isolates, and used cluster analysis and an ordination technique based on taxonomic distances with the mixed data. Their conclusions enabled separation of orchid endophytes from *R. solani* and also indicated certain within-group affinities.

Liu and Sinclair (1992) studied genetic diversity by pectinase isoenzyme polymorphism and restriction analysis of the rDNA ITS regions within a characterized set of isolates of AG 2. Cluster analyses for each approach generated dendrograms each consisting of five distinct groups comparable to AGs and pathogenicity data. Similar approaches have been undertaken by Duncan *et al.* (1993) who, for strains of eight AGs, constructed a similarity matrix from RAPD-PCR band patterns and identified groups highly correlated with anastomosis groups and pectinase isoenzyme groups. O'Brien (1994) compared rDNA RFLPs to anastomosis and pectin isoenzyme groups. These techniques address different facets of the genome and so it is important to consider what information is contained in the different approaches when comparisons are made between them. While character weighting schemes can be developed to represent the different numbers of genes involved in pathways giving phenotypic characters, the loss or absence of an individual character may not imply that all of the corresponding genes have been lost.

B. Application and interpretation of numerical methods

Recently, Laroche *et al.* (1992) investigating AG 3 and AG 9 isolates by electrophoresis of seven intracellular isoenzymes with dendrograms and Principal Coordinate ordinations, generated two groups conforming directly with AGs. Although numerical techniques are often used to group individual organisms, they can also be used to group sets of organisms. This approach was used by Johnk and Jones (1993) who analysed whole cell fatty acids in populations of AG 2-2, creating an isolate library to compare isolates within and between populations, and a population library to compare populations. Cluster analysis of both libraries generated dendrograms showing three distinct groups, a grouping also obtained by Principal Coordinate analysis of fatty acid percentage composition. Later (Johnk and Jones, 1994), isolates of AG1 were analysed and cluster analysis of isolate and population libraries showed that fatty acid composition of AG 1-IA and IB was very similar, while AG 1-IC was distinct.

Banniza *et al.* (1994) considered groupings within isolates of *R. solani* from rice and associated plants. AG 1 isolates predominated, and various different approaches have been used to show further discrimination within the AG. The results obtained from morphological, biochemical and molecular approaches can be analysed separately by numerical techniques, and variations on these techniques used to compare and combine findings from the different data sets. As an example, Fig. 1 shows the dendrogram (obtained from 93 bands from a series of PCR amplifications with four primers based on simple repeated motifs) of a representative set of *R. solani* isolates. This set of 23 isolates consisted of 20 that could be assigned to AG 1 and three that did not anastomose with tester isolates of AGs 1, 2 or 4. Similarities were calculated with the Percent similarity coefficient and the resulting data clustered by UPGMA. The majority of the isolates form a single large group and at the bottom of the dendrogram are five ungrouped isolates comprising four from non-rice hosts (IMI 359622, 360298, 360336 and 360016) and one from rice (360314).

Similarly, Fig. 2 shows an example dendrogram obtained from 30 morphological characters for the same isolates. In this dendrogram most rice isolates form a single group which also includes one of the previously ungrouped isolates (359622) and the rice isolate which was distinct by PCR (360314). Again three isolates from non-rice hosts (360298, 360336 and 360016) are ungrouped, as is one further isolate (361171), possibly because it failed to produce sclerotia in culture.

Data from different approaches may be combined in a single dendrogram. Fig. 3 shows the dendrogram obtained from an unweighted analysis of the PCR dataset with the morphological characters. The morphologically atypical rice isolate (361171) is now included in the main group although the other three ungrouped isolates remain distinct. In this instance the arrangements are supported by pectinase zymograms as two ungrouped isolates (360298 and 360336) also showed a different pectinase isoenzyme pattern. Combining characters is therefore reinforcing the validity of separating isolates which have several distinct genotypic and phenotypic characters. An overall placement is attained and particular subgroups within the dendrogram reflecting particular character sets may be identified.

IV. Discussion

R. solani is a fungus of great importance as a plant pathogen. It is also an extremely good experimental organism and most of the methods now in use for the genus *Rhizoctonia* are integrated with or adapted from those developed on this species complex. The number of taxa involved in the diseases ascribed to *Rhizoctonia* remains controversial (Andersen and Stalpers, 1994; Ogoshi, 1987; Moore, 1987; see also chapters I.A1, I.A3, and V.14) and the precise relationships of biological species, anastomosis groups, their subgroups and zymogram groups to taxonomic species, special forms, pathogenicity or ecological concepts such as that of mycelial individualism (Rayner, 1991) remains undefined. Integration of AG data with classical taxonomy and nomenclature presents problems because the groups do not have any nomenclatural status: formal diagnoses have not been published nor permanently preserved reference material deposited. Tester isolates recently recommended (Sneh *et al.*, 1991) may differ in subsidiary characters from those used in early work; for instance an American isolate is recommended for AG 8, originally characterized from Australia (Neate and Warcup, 1985). During experimental work, each isolate should be treated as an independent operational taxonomic unit, to be integrated by a combination of its characters with other units and thence into a systematic classification.

A practical problem with use of AG as a "baseline" classification is that the necessary tester isolates and expertize may be inaccessible, particularly in developing countries where Rhizoctonia diseases are often extremely important. The technique cannot be applied retrospectively unless living isolates have been maintained. The number of such isolates preserved as living genetic resources is small in comparison with the number of records of *Rhizoctonia* diseases. The total amount of preserved material in biosystematic reference collections worldwide is, however, much greater (over 1300 specimens of the *R. solani* complex in IMI herbarium for example, excluding those associated with current projects). Although there are restrictions, morphological, biochemical and molecular data can be extracted from dead material (Paterson and Hawksworth, 1985; Lee and Taylor, 1990). An important aim of integrated studies must be identification of marker characters for use with specimens in all available forms, thus facilitating coordination of investigations separated by space, time and technique.

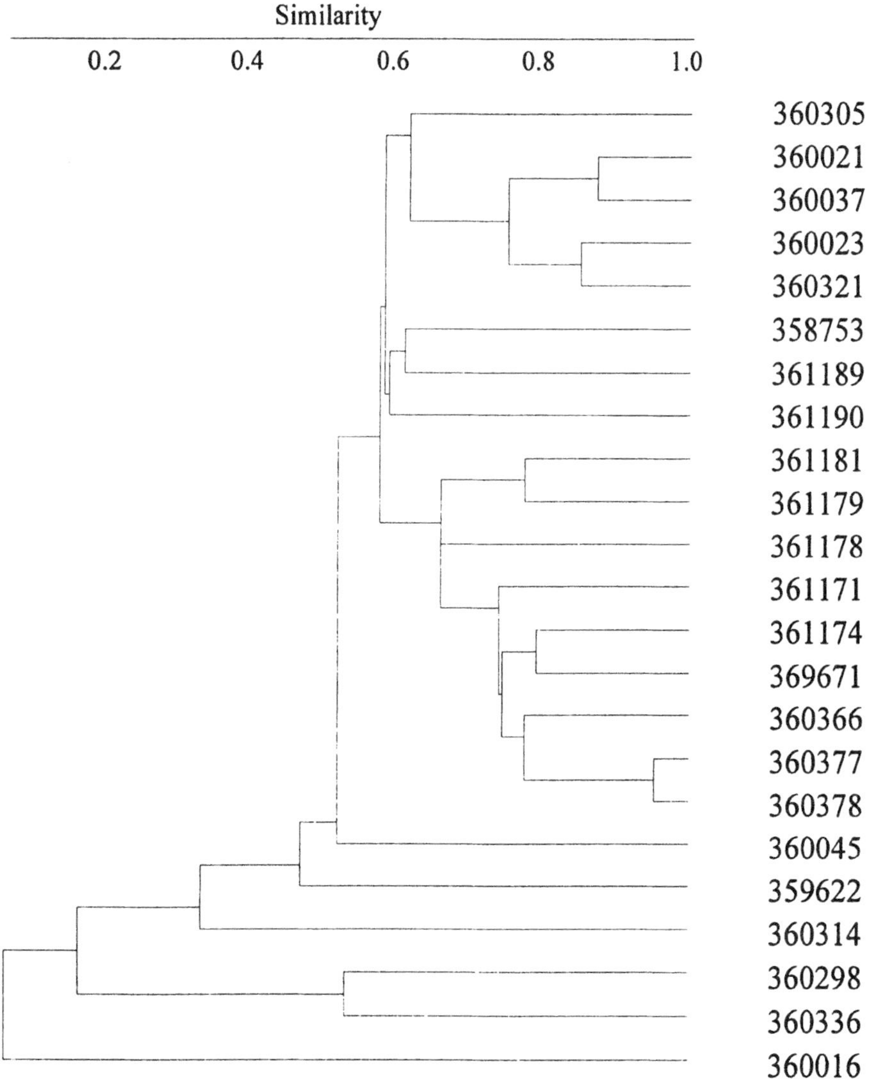

FIGURE 1. Dendrogram obtained from Percent similarity coefficient after UPGMA clustering of band data generated by PCR amplification of *Rhizoctonia solani* DNA by four different primers.

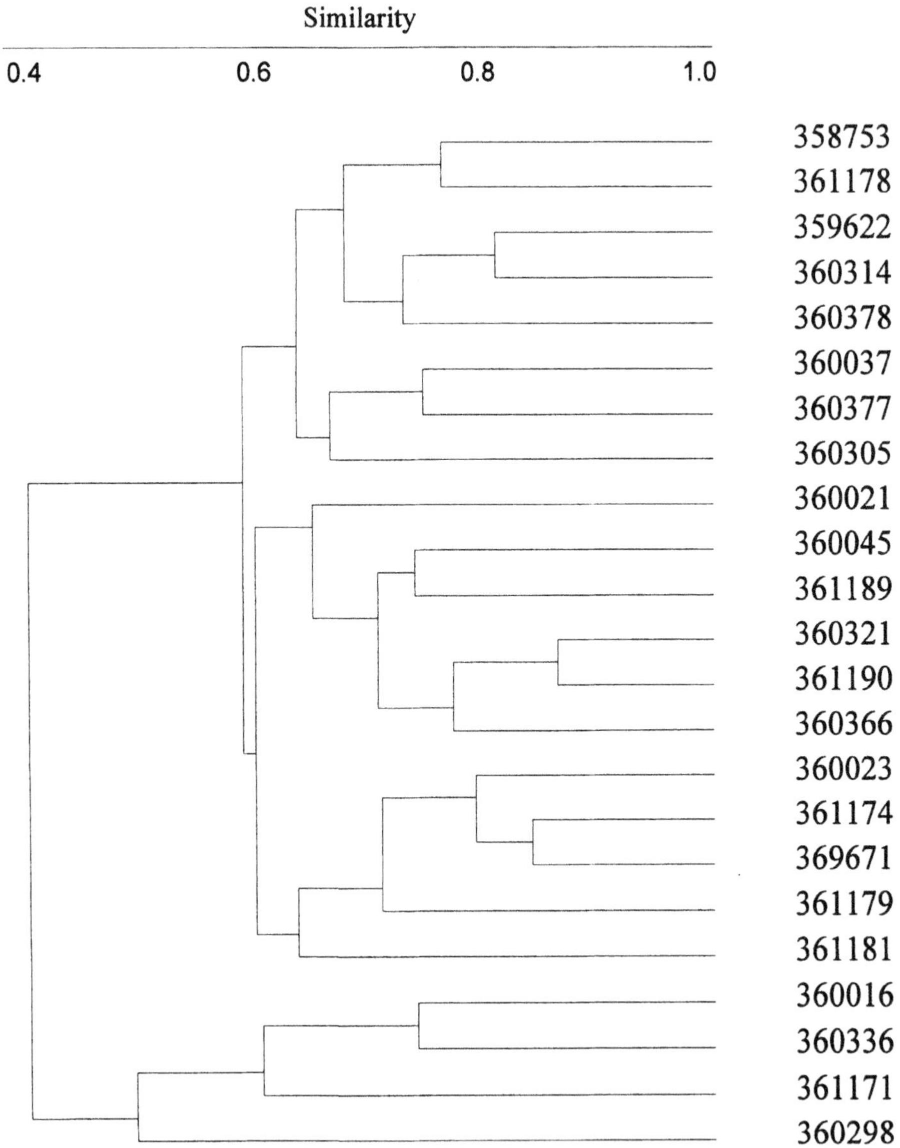

FIGURE 2. Dendrogram obtained from Percent similarity coefficient after UPGMA clustering of data for 30 morphological characters exhibited by *Rhizoctonia solani*.

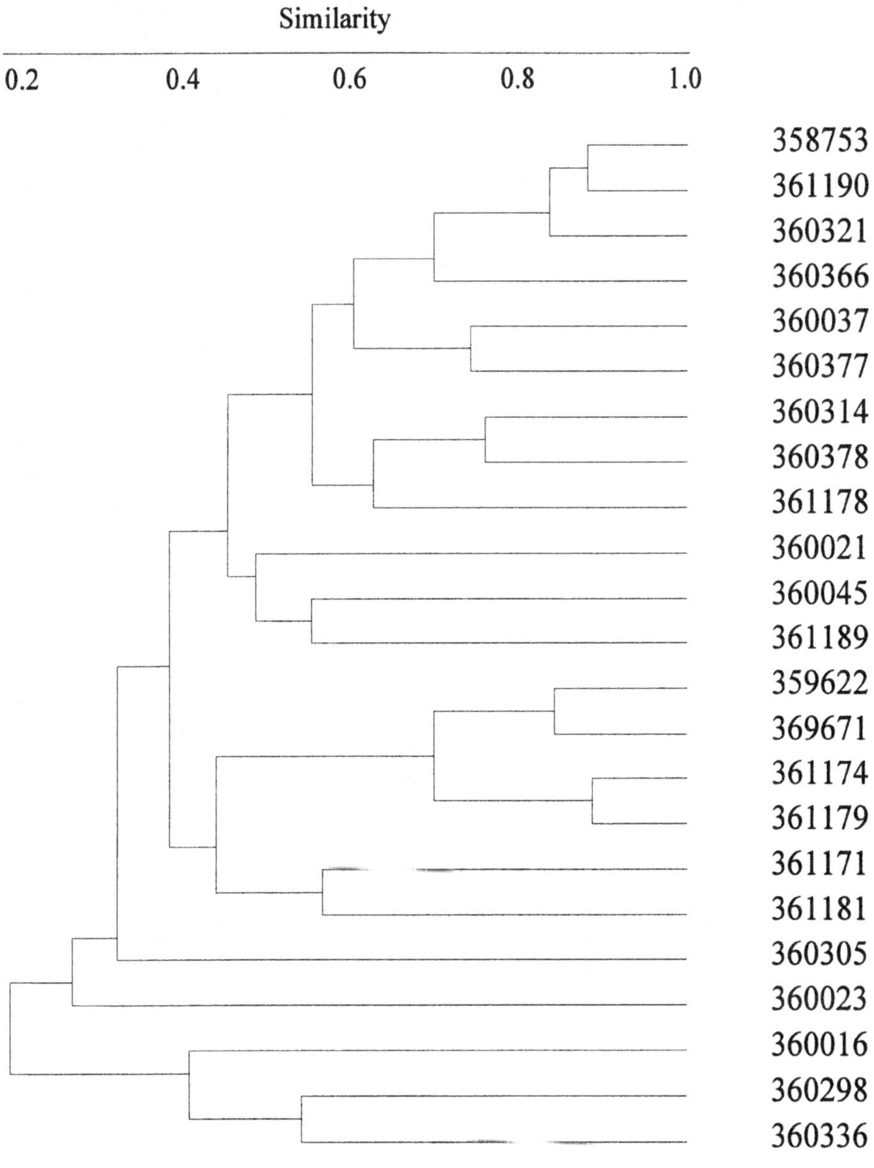

FIGURE 3. Dendrogram obtained from Percent similarity coefficient after UPGMA clustering of combined morphological and PCR generated DNA band data for *Rhizoctonia solani*.

Since the early molecular phylogenetic studies in which *Rhizoctonia* was included (Blanz and Gottschalk, 1984), interest in phylogeny has expanded greatly. Vilgalys and Cubeta (1994), Vilgalys and Gonzalez (1990) and Cubeta (chapter I.B3), use molecular data only. Tehler (1995) expressed concern at the tendency for morphological and molecular data to be analysed separately and similar arguments can be advanced for biochemical and other functional characters. Some of the problems of integration of data sets for phylogenetic estimates are highlighted by Lutzoni and Vilgalys (1995), who noted the possibility of different phylogenetic histories between organism and molecule. Integrated studies cannot be expected to overcome all the theoretical and applied problems in *Rhizoctonia* research. Nevertheless, they can expand the predictive capacity of results from many investigations.

V. Acknowledgements

The authors would like to acknowledge the UK Overseas Development Administration for providing financial support for research project EMC X0228, 'Characterization of plant pathogenic strains of *R. solani*', undertaken at the International Mycological Institute.

VI. References

Abe H & Tsuboki (1987) Anastomosis groups of *Rhizoctonia solani* Kühn from potatoes. Bull. Hokkaido Prefect. Agric. Exp. Stn 40:61-70.

Abe S (1956) Studies on the classification of the penicillia. J. Gen. Appl. Microbiol. 2:1-344.

Andersen TF (1990) A study of hyphal morphology in the form genus *Rhizoctonia*. Mycotaxon 37:26-46.

Andersen TF & Stalpers JA (1994) A check-list of *Rhizoctonia* epithets. Mycotaxon 51:437-457.

Armentrout VN, Downer AJ & Nameth ST (1986) A simplified branching assay for *Rhizoctonia solani*. Mycologia 78:657-663.

Bains PS & Bisht VS (1995) Anastomosis group identity and virulence of *Rhizoctonia solani* isolates collected from potato plants in Alberta, Canada. Plant Dis. 79:241-242.

Banniza S, Holderness M, Rutherford MA, Bridge PD, Mordue JEM, Johanson A, Simons SA & Paterson RRM (1994) Characterization of *Rhizoctonia* strains in the rice-based cropping systems of southeast Asia and west Africa. In: Rajan A and Ibrahim Y (eds.). Proceedings, 4th Int. conf. on Plant Protection in the Tropics (pp. 419-421). Malaysian Plant Protection Society, Kuala Lumpur.

Blanz PA & Gottschalk M (1984) A comparison of 5S ribosomal RNA nucleotide sequences from smut fungi. Syst. Appl. Microbiol. 5:518-526.

Bolkan HA & Ribeiro WRC (1985) Anastomosis groups and pathogenicity of *Rhizoctonia solani* isolates from Brazil. Plant Dis. 69:599-601.

Bridge PD & Hawksworth DL (1985) Biochemical tests as an aid to the identification of *Monoascus* species. Lett. Appl. Microbiol. 1:25-29.

Bridge PD, Hawksworth DL, Kozakiewicz Z, Onions AHS, Paterson RRM, Sackin MJ & Sneath PHA (1989) A reappraisal of the terverticillate penicillia using biochemical, physiological and morphological features I. Numerical taxonomy. J. Gen. Microbiol. 135:2941-2966.

Bridge PD, Ismail MA & Rutherford MA (1993) An assessment of aesculin hydrolysis, vegetative compatibility and DNA polymorphism as criteria for characterizing pathogenic races within *Fusarium oxysporum f. sp. vasinfectum*. Plant Pathol. 42:264-269.

Bridge PD, Holderness M, Paterson RRM & Rutherford MA (1995) Multidisciplinary characterization of fungal plant pathogens. Bull. OEPP 25:125-131.

Bridge PD, Prior C, Sagbohan J, Lomer CJ, Carey M & Buddie A (1996) Molecular characterization of isolates of *Metarhizium* from locusts and grasshoppers. Biodivers. Conservat. 5 (in press).

Carling DE & Kuninaga S (1990) DNA base sequence homology in *Rhizoctonia solani* Kühn: inter- and intragroup relatedness of anastomosis group 9. Phytopathology 80:1362-1364.

Chand T & Logan C (1983) Cultural and pathogenic variation in potato isolates of *Rhizoctonia solani* in Northern Ireland. Trans. Brit. Mycol. Soc. 81:585-589.

Chase AR (1991) Characterization of *Rhizoctonia* species isolated from ornamentals in Florida. Plant Dis. 75:234-238.

Chen JS, Ge QX & Zhang BX (1990) [BPDA and JPDA media for rapid differentiation of anastomosis groups of *Rhizoctonia solani*]. Acta Agric. Univ. Zhejiang. 16:213-218.

Currah RS & Zelmer C (1992) A key and notes for the genera of fungi mycorrhizal with orchids and a new species in the genus *Epulorhiza*. Rep. Tottori Mycol. Inst. 30:43-59.

Dabinett PE & Wellman, AM (1978) Numerical taxonomy of certain genera of Fungi Imperfecti and Ascomycotina. Can. J. Bot. 56:2031-2049.

Damicone JP, Patel MV & Moore WF (1993) Density of sclerotia of *Rhizoctonia solani* and incidence of sheath blight in rice fields in Mississippi. Plant Dis. 77:257-260.

Duncan S, Barton JE & O'Brien P (1993) Analysis of variation in isolates of *Rhizoctonia solani* by random amplified polymorphic DNA assay. Mycol. Res. 97:1075-1082.

D'yakov Yu T (1993) [The biological species criteria of fungi (with a review of the taxonomic structure of *Rhizoctonia*-like fungi)] Mikol. Fitopatol. 27:68-82.

Durbin RD (1959) Factors affecting the vertical distribution of *Rhizoctonia solani* with reference to CO_2 concentration. Amer. J. Bot. 46:22-25.

Eeckhout E van, Blackwell M, Rush MC (1990) In vitro effects of propiconozole on morphology and ultrastructure of *Rhizoctonia solani*. Mycol. Soc. Am. Newsl. 41:42.

Frisvad JC (1992) Chemometrics and chemotaxonomy: a comparison of multivariate statistical methods for the evaluation of binary fungal secondary metabolite data. Chemometrics Intelligent Lab. Syst. 14:253-269.

Frisvad JC (1994) Correspondence, principal coordinate and redundancy analysis used on mixed chemotaxonomical qualitative and quantitative data. Chemometrics Intelligent Lab. Syst. 23:213-229.

Hide GA & Cayley GR (1982) Chemical techniques for control of stem canker and black scurf (*Rhizoctonia solani*) diseases of potatoes. Ann. appl. Biol. 100:105-116.

Hide GA, Hirst JM & Stedman OJ (1973) Effects of black scurf (*Rhizoctonia solani*) on potatoes. Ann. appl. Biol. 74:139-148.

Johnk JS & Jones RK (1993) Differentiation of populations of AG 2-2 of *Rhizoctonia solani* by analysis of cellular fatty acids. Phytopathology 83:278-283.

Johnk JS & Jones RK (1994) Comparison of whole-cell fatty acid compositions in intra-specific groups of *Rhizoctonia solani* AG 1. Phytopathology 84:271-275.

Kataria HR, Verma PR & Gisi U. (1991) Variability in the sensitivity of *Rhizoctonia solani* anastomosis groups to fungicides. J. Phytopath. (Berl.) 133:121-133.

Keon JPR, Byrde RJW & Cooper RM (1987) Some aspects of fungal enzymes that degrade plant cell walls. In: Pegg GF and Ayres PG (eds.) The fungal infection of plants (pp 133-157) Cambridge University Press, Cambridge.

Kim WG (1993) Morphological and cultural characteristics of fungi causing rice sclerotial diseases. Korean J. Mycol. 21:16-22.

Kousik CS & Snow JP (1990) Comparison of sclerotial development among anastomosis groups of *Rhizoctonia solani*. Phytopathology 80:1042 (abstr.).

Kousik CS, Snow JP, Berggren GT & Harville BG (1995) Effect of temperature on virulence of *Rhizoctonia solani* on soybean leaves and seedlings. Plant Pathol. 44:580-586.

Kuninaga S & Yokosawa R (1980) DNA base sequence homology in *Rhizoctonia solani* Kühn. Ann. Phytopathol. Soc. Japan 46:150-158.

Kuninaga S & Yokosawa R (1985) DNA base sequence homology in *Rhizoctonia solani* Kühn. VI. Genetic relatedness among seven anastomosis groups. Ann. Phytopathol. Soc. Japan 51:127-132.

Laroche JP, Jabaji-Hare SH & Charest PM (1992) Differentiation of two anastomosis groups of *Rhizoctonia solani* by isozyme analysis. Phytopathology 82:1387-1393.

Lee SB & Taylor JW (1990) Isolation of DNA from fungal mycelia and single spores. In: Innis MA, Gelfand DH, Sninsky JJ & White TJ (eds.) PCR protocols. A guide to methods and applications (pp 282-287) Academic Press, San Diego

Little G & Cooke LR (1988) The effects of fungicides applied to seed potato tubers on *Rhizoctonia solani* stem canker and black scurf. Tests of Agrochemicals and cultivars Ann. Appl. Biol. 112:34-35.

Liu ZL & Sinclair JB (1992) Genetic diversity of *Rhizoctonia solani* anastomosis group 2. Phytopathology 82:778-787.

Lutzoni F & Vilgalys R (1995) Integration of morphological and molecular data sets in estimating fungal phylogenies. Can. J. Bot. 73:649-659.

Manczinger L & Polner G (1987) Cluster analysis of carbon source utilization patterns of *Trichoderma* isolates. Syst. Appl. Microbiol. 9:214-217.

Miller RNG, Holderness M, Bridge PD, Paterson RRM, Meon S, Hussin MZ & Hilsley EJ (1995) A multidisciplinary approach to the characterization of *Ganoderma* in oil-palm cropping systems. In:Buchanan PK, Hseuh RS & Moncalvo P (eds.) *Ganoderma*:systematics, phytopathology and pharmacology (pp 57-66) National Taiwan University, Taipei.

Moore RT (1987) The genera of *Rhizoctonia*-like fungi: *Ascorhizoctonia, Ceratorhiza* gen. nov., *Epulorhiza* gen. nov., *Moniliopsis*, and *Rhizoctonia*. Mycotaxon 29:91-99.

Mordue JEM, Currah RS & Bridge PD (1989) An integrated approach to *Rhizoctonia* taxonomy:cultural, biochemical, and numerical techniques. Mycol. Res. 92:78-90.

Neate SM & Warcup JH (1985) Anastomosis grouping of some isolates of *Thanatephorus cucumeris* from agricultural soils in South Australia. Trans. Brit. Mycol. Soc. 85:615-620.

O'Brien PA (1994) Molecular markers in Australian isolates of *Rhizoctonia solani*. Mycol. Res. 98:665-671.

Ogoshi A (1987) Ecology and pathogenicity of anastomosis and intraspecific groups of *Rhizoctonia solani* Kühn. Annu. Rev. Phytopathol. 25:125-143.

Parmeter JR Jr (ed.)(1970) Biology and Pathology of *Rhizoctonia solani*. University of California Press, Berkeley.

Paterson RRM & Hawksworth DL (1985) Detection of secondary metabolites in dried cultures of *Penicillium* preserved in herbaria. Trans. Brit. Mycol. Soc. 85:95-100.

Rashid KY & Bernier CC (1993) Genetic diversity among isolates of *Rhizoctonia solani* and sources of resistance in Vicia faba. Can. J. Plant Pathol. 15:23-28.

Rayner ADM (1991) The phytopathological significance of mycelial individualism. Annu. Rev. Phytopathol. 29:305-323.

Richter H & Schneider R (1953) Untersuchungen zur morphologischen und biologischen differenzierung von *Rhizoctonia solani* Kühn. Phytopathol. Z. 20:165-226.

Roh MJ & Kim HK (1987) Pathogenicity of *Rhizoctonia* isolates from southern horticultural area in Korea. Korean J. Plant Pathol. 3:217-222.

Schulz H (1936) Vergleichende Untersuchungen zur Okologie, Morphologie und Systematic des `Vermehrungspilzes'. Arb. Biol. BundAnst. Land-u. Forstw. 22(1):1-41.

Sneh B, Burpee L & Ogoshi A (1991) Identification of *Rhizoctonia* Species. APS Press, St. Paul, Minnesota. 133 pp.

Sweetingham MW, Cruickshank RH & Wong DH (1986) Pectic zymograms and taxonomy and pathogenicity of the *Ceratobasidiaceae*. Trans. Brit. Mycol. Soc. 86:305-311.

Swofford DL & Olsen GJ (1990) Phylogeny reconstruction. In: Hillis DM & Moritz C (eds.) Molecular Systematics (pp 411-501) Sinauer Associates, Sunderland, Massachusetts.

Tehler A (1995) Morphological data, molecular data, and total evidence in phylogenetic analysis. Can. J. Bot. 73:667-676.

Thomas V, Rutherford MA & Bridge PD (1994) Molecular differentiation of two races of *Fusarium oxysporum* special form *cubense*. Lett. Appl. Microbiol. 18:193-196.

Tommerup IC, Barton JE & O'Brien PA (1995) Reliability of RAPD fingerprinting of three basidiomycete fungi, *Laccaria, Hydnangium* and *Rhizoctonia*. Mycol. Res. 99:179-186.

Tu CC & Kimbrough JW (1975) Morphology, development, and cytochemistry of the hyphae and sclerotia of species in the *Rhizoctonia* complex. Can. J. Bot. 53:2282-2296.

Ueyama I, Araki Y, Kurogochi S & Yamaguchi I (1993) Metabolism of the phenylurea fungicide, pencycuron, in sensitive and tolerant strains of *Rhizoctonia solani*. J. Pestic. Sci. 18:109-117.

Vilgalys R & Cubeta, MA (1994) Molecular systematics and population biology of *Rhizoctonia*. Annu. Rev. Phytopathol. 32:135-155.

Vilgalys R & Gonzales D (1990) Ribosomal DNA restriction length polymorphisms in *Rhizoctonia solani*. Phytopathology 80:151-158.

Watanabe B & Shiyomi M (1975) Hyphal morphology of *Rhizoctonia solani* Kühn and related fungi isolated from sugarcane in Taiwan. Trans. Mycol. Soc. Japan 16:253-263.

Yang XB, Snow JP & Berggen GT (1989) Morphogenesis oof microsclerotia and sassaki-type sclerotia in *Rhizoctonia solani*, anastomosis groups iA and 1B. Mycol. Res. 93:429-434.

Section II.

GENETICS AND PATHOGENICITY OF *RHIZOCTONIA* SPP.

II.1 GENETICS OF *RHIZOCTONIA* SPECIES

GERARD C. ADAMS

Department of Botany and Plant Pathology, Michigan State University, East Lansing, Michigan 48824 USA.

I. INTRODUCTION

The genetics of *Rhizoctonia solani* and related species is complex and remains poorly understood. When *R. solani* was first studied it was apparent that it was a ubiquitous species, that it could parasitize an enormous host range, and that it occurred in a great variety of cultural forms. The scientific community actively sought an understanding of what mechanisms underlay the organisms successful and broad adaptations as a pathogen. But an understanding did not come easily and I believe that in large part this was due to our confusion in whether to view the organism as a functionally sexual or functionally asexual (non-sexual) species.

At the time that Flentje and Stretton (1964) first approached the question of the mechanism of genetic variation, they already understood that much of the variation encountered in the taxon was due to the presence of the different anastomosis groups. From the outset, the anastomosis groups were viewed intuitively as genetically isolated groups, or likely biological species (Anderson, 1982). Today we understand this to be true but also recognize that further genetic isolation occurs in subgroups, within the anastomosis groups (AGs). The subgroups are probably subspecies of each AG. The subspecies are sympatric and they are the unit of evolution. The total number of biological species in the taxon *Thanatephorus cucumeris* is still being uncovered.

The question of the mechanism of variation remained to be asked for each biological species, as it was found that an anastomosis group may exhibit most of the full range of variation seen in the taxon, including the capability to exploit many different hosts and habitats. However, studies were limited primarily to representative isolates of AGs 1-IC, 4, and 2 that were amenable to reproducing sexually under laboratory conditions (Adams, 1988) until recently when homokaryons of AG 8 were produced through generation of uninucleate protoplasts (Yang *et al.*, 1992).

The direction of scientific inquiry shifts depending on whether we view a biological species of *T. cucumeris* as functionally sexual or asexual. When considering an asexual species we turn our attention to heterokaryosis as the mechanism having the most potential for permitting genetic recombination without sex (meiosis). When considering a sexual species we examine outbreeding, and again we turn towards heterokaryosis and the constraints genetically controlling its function. Heterokaryosis and the concomitant heterokaryon is crucial because it can bring together two nuclei with different genotypes in one cytoplasm. Field isolates have been found to be heterokaryotic (Bolkan and Butler, 1974; Anderson *et al.*, 1972; Garza-Chapa and Anderson, 1966; Whitney and Parmeter, 1963).

II. THE NON-SEXUAL SPECIES VIEW

The apparent absence of the sexual fruiting in nature and the difficulty producing it in the laboratory induced most plant pathologists to view *R. solani* as a non-sexual fungus. This inspired reflections of the potential of the heterokaryotic condition for asexual recombination, during the 1960s and 1970s. Heterokaryosis confers on a haploid fungus the advantages of heterozygosity of the diploid (Caten and Jinks,

B. Sneh et al. (eds.),
Rhizoctonia Species: Taxonomy, Molecular Biology, Ecology, Pathology and Disease Constrol, 101–116.

1966), such as exploiting the genetic variability of the interacting homokaryotic nuclei and, perhaps, mitochondria. It was speculated that recombination of alleles was a theoretical potential in a heterokaryon environment in nature. Without rigid control of nuclear pairing and conjugate division, the genetic balance of the heterokaryon might vary by simple change in the proportions of nuclei. The ratio of the genetically different nuclei might shift in response to selective pressures on favored alleles in sectors of growth exploiting new habitat. Shifting ratios of nuclei in the unbalanced heterokaryon might also be complemented by the independent shifting of the genetically different mitochondria providing for somatic flexibility that might continually adjust to changing environment (Burnett, 1983; Buxton, 1954; Jinks, 1952). Further reflection on this idealized non-sexual organism considers the potential that more than two nuclei with different genotypes might cohabit in the coenocytic cell (Bolkan and Butler, 1974). Combined with the phenomena of self and non-self anastomosis (Stenlid and Rayner, 1991), the fungal colony might unite with others to become a genetically plastic, vigorously communicating, cross-feeding, cytoplasmic-streaming super-mycelium that encompasses hectacres. Throughout the soma heterokaryotic - monokaryotic pairings of different germplasms might occur along with rare instances of mutation, parasexual recombination, and mitochondrial recombination, providing significant genetic recombination of alleles in the super-mycelium.

The integration of the above behavioral components that would permit such a non-sexual organism has only in part been rigorously rejected in experimental research on *R. solani*. However, much is inferred from research on other fungi and those fungi have not revealed this level of genetic plasticity (Esser and Baich, 1994; Beeching *et al.*, 1989). In the past plant pathologists seized upon the parasexual cycle as the principal mechanism of variability in asexual fungi. Nevertheless, since the 1980s few mycologists have considered parasexuality to likely have significant influence in the genetics of *Rhizoctonia* because of the recognition of the presence of somatic (heterogenic) compatibility systems, and the absence of proof that such asexual events of recombination occur in basidiomycetes that have served as genetic models (Parmeter *et al.*, 1963; Rayner, 1991). As Parmeter *et al.* (1963) pointed out, research in *R. solani* has been hampered by a lack of satisfactory criteria for determining when compatible heterokaryotic association of nuclei has occurred. Genetically controlled systems of somatic incompatibility are widespread and have usually been found in species able to form heterokaryons (Esser and Baich, 1994; Rayner *et al.*, 1984). Incompatibility between non-mating homokaryons is dependent on genotypic dissimilarity as homokaryons must be identical at certain loci before they will form balanced heterokaryons. Today we believe the significance of heterokaryosis as an asexual system of variation in fungi is greatly reduced because heterokaryon formation is restricted to pairs of strains with more or less identical genotypes (Esser and Baich, 1994; Ainsworth *et al.*, 1990; Rayner *et al.*, 1984). In evaluating the importance of heterokaryosis in the wild, it is necessary to consider the extent of the differences between nuclei that are capable of co-existing in heterokaryotic associations (Gordon, 1993; Caten and Jinks, 1966), and perhaps also differences in mitochondria (May, 1988). The hypothesis that parasexuality in heterokaryosis is the mechanism of genetic variation has given way to newer hypotheses of chromosome rearrangements and horizontally inherited genetic elements, discussed below and in other chapters.

III. Somatic incompatibility

Somatic incompatibility (SI) is an asexual genetic system, the genetic basis of which is poorly understood. It is generally a complex polygenic and heterogenic system in basidiomycetes (Esser and Baich, 1994; Rayner, 1991). Control of the heterogenic system is primarily by nuclear genes but perhaps also by as yet unrecognized mitochondrial genes (May, 1988). SI is tested by pairing heterokaryons and the responses effectively delineate individual genotypes in a population, not interspecific or intraspecific groups within a taxon. The latter are differentiated in some basidiomycetes also by heterogenic somatic incompatibility system called the intersterility system which is studied via pairings of homokaryons. The two systems are often confused in the literature (Esser and Baich, 1994).

The SI system appears to control the survival of anastomosing cells in basidiomycetes (Rayner, 1991). The SI incompatible reaction should be recognizable in *R. solani* as the Category 2 (C2) type anastomosis reaction, i.e., cell-wall fusion and cytoplasm collapse in the joined cell (Carling *et al.*, 1988), particularly when it occurs among isolates within a well-defined genetically isolated population such as a subgroup or zymogram group (ZG). The compatible reaction is Category 3 (C3), cell-wall fusion with cytoplasm of the joined cell remaining alive. Therefore, SI also restricts horizontal gene flow, but little is known of this except that dsRNA presumably cannot be transmitted outside an SI group. Cubeta *et al.* (1993) described two different incompatible responses, a clear zone or a white line, delimiting the two interacting colonies of a heterokaryon and one of its parental homokaryons. Syminis *et al.* (1987) reported on the macroscopic appearance of hyphal interactions in SI pairings of *T. cucumeris* and correlated them with cytoplasmic exchange. Field isolates (presumed heterokaryons) of AG 8 have been studied using an AG-specific method of SI characterization based on the anastomosis group's unusual formation of tufts in heterokaryon x heterokaryon pairings (He-He pairings) between different individuals, and merging reactions between somatically compatible individuals (Yang *et al.*, 1993b). A genetically and morphologically well-characterized SI system in bipolar basidiomycetes is that of *Heterobasidion annosum* (Hansen *et al.*, 1993a,b). Hansen *et al.* (1993a) have found each nuclear type in a heterokaryon acts independently and each has 3-4 loci that control compatibility. A single allelic difference at any of the loci, some of which are multiallelic, is sufficient to cause heterogenic incompatibility (epistatic control). However, they did not find cell death following anastomosis of incompatible heterokaryons and attributed this to dedikaryotization and mating occurring in the gap between the incompatible heterokaryons (Hansen *et al.*, 1993b).

Although SI reactions can distinguish individuals, additional markers are necessary to estimate the relatedness between the individuals. Somatic incompatibility grouping (SIG) in combination with nuclear and mitochondrial markers are being used in population genetic studies of basidiomycetes, notably *Armillaria* , by identifying the population structure of genets (Rizzo *et al.*, 1995; Smith *et al.*, 1994; Harrington *et al.*, 1992). Genets are genetic individuals and all descendants (ramets) whether asexually propagated by hyphal proliferation/fractionation or otherwise (Rayner, 1991). SI grouping has contributed significantly to estimating basidiospore success in establishing colonies, to determining the size and longevity of genets, and to revealing the contributions of sib-matings, outcrossing, and disturbances in habitat to population structure (Rizzo *et al.*, 1995; Smith *et al.*, 1992; 1994; Worrall, 1994; Kay and Vilgalys, 1992).

The ecology of *Armillaria* species is different than that of *Rhizoctonia* species, but it is noteworthy that in the sexually reproducing mushroom opportunities for basidiospore colonization and expansion of new genets was revealed to be restricted and remarkably limited (Rizzo *et al.*, 1995). In the midwestern pine forest there was little evidence of random mating, only small inbreeding populations, and only local distribution of basidiospores (Rizzo *et al.*, 1995). Similar population structure, but with the occurrence of more random mating, was evident in a second forest niche (Smith *et al.*, 1994).

Studies of *Rhizoctonia* ecology and population genetics should exploit SIG methodology. The frequent mechanical disturbance common in agricultural sites however, would be expected to enhance the colonization success of the rarely produced basidiospores of *T. cucumeris* and magnify the contribution of sexual reproduction to population structure and dynamics, compared to *Armillaria* spores. Basidiospore infection and colonization of leaves of sugarbeet followed by senescence and incorporation into the soil, for example, provides an additional strategy for magnifying the contribution of sexuality to population structure and genetics.

Alternatively, if SIG and DNA markers show non-random associations between nuclear and extranuclear genetic markers in studies of *R. solani*, that would be evidence of clonality or of an asexually propagating population (Gordon, 1993).

IV. INTERSTERILITY GROUPS

The genetic systems of intersterility (IS) prevent or limit sexual mating between individuals from reproductively isolated groups. Therefore the IS system is superimposed on the mating system. IS groups in the basidiomycetes are interpreted as biological species and many taxa are composed of IS groups (Esser and Baich, 1994; Burnett, 1983). The AGs and subgroups of *T. cucumeris* are probably intersterility groups, presumably, there will be a system of genetic incompatibility alleles superimposed on the regulation exerted by the H-factors, that restricts or makes impossible vertical gene exchange among the subgroups. Confirming that subgroups are IS groups will require the capability of inducing sporulation, accurately identifying homokaryons and synthesized heterokaryons, and scoring numerous markers for genetic analysis. Potentially, preparation of homokaryons from protoplasts of heterokaryons, and synthesis of heterokaryons without forcing could overcome a portion of the obstacles (Phillips, 1993; Yang *et al.*, 1992; 1993a). Molecular DNA technology should provide sufficient markers. Suitable models for study might include the interactions of homokaryons among the AG 8 ZGs, among the subgroups within AG 2-2, between AG 2-2 and AG 2-1, or between AG 1-IA and AG 1-IB subgroups. Chase and Ullrich (1990a) have shown that in *H. annosum* at least five genes (loci), each with two alternate alleles designated (+) and (-), regulate intersterility. For mating to be successful between groups the individuals must be homoallelic for the (+) allele of at least one of the five loci (Chase and Ullrich, 1990b).

Unraveling the IS genetic system should provide the key to our understanding of speciation in *T. cucumeris* as IS groups are believed to represent newly emerged or emerging species (Chase and Ullrich, 1990a). Thorough genetic analysis of IS should help us define, in generally accepted biological categories such as 'sympatric subspecies', the zymogram groups, the subgroups, and subsequently the anastomosis groups.

V. CHROMOSOME POLYMORPHISM

Other aspects in the exploration of the mechanisms of variation in *T. cucumeris* include determination of chromosome length and number. These are now determinable by pulse-field electrophoresis. Frequent quantitative and qualitative changes in chromosomes appear to be normal in both Ascomycetes and Basidiomycetes (Mills and McCluskey, 1990). Chromosome length polymorphisms are believed to initially arise through gross chromosomal rearrangements such as insertions of transposons, deletions, duplications, and translocations. Translocations and the occurrence of fragile sites on chromosomes will segregate in progeny and produce populations with great variability in chromosome size (McCluskey *et al.*, 1994). In *Ustilago hordei*, heat shock treatment produced a strain with a mutant phenotype and a 50kb deletion near the terminus of one arm of a 940kb chromosome (McCluskey *et al.*, 1994). The mutant phenotype and the deletion cosegregated in ordered tetrads. McCluskey *et al.* (1994) hypothesized that deletions alone may account for the magnitude of karyotype variation observed in *U. hordei*..

The contribution of chromosome polymorphism to generating genotypic and phenotypic variation in a population is not yet understood, except when these changes are accompanied by changes in ploidy or the unstable inheritance of smaller chromosomes. There is a high tolerance for aneuploidy in *Saccharomyces cerevisiae* (Bakalinsky and Snow, 1990) and an unexpectantly high occurrence of deletions in chromosome fragments in *Septoria tritici* (McDonald and Martinez, 1990, 1991). Small chromosomes characterized by eccentric construction, unstable inheritance, and dispensable coding functions are called B-chromosomes, first described in Angiosperms. *Ustilago hordei*, *Nectria haematococca, Cochliobolus heterostrophus*, *Magnaporthe grisea* and perhaps *Tilletia indica* appear to have supernumerary B-chromosomes (Tzeng *et al.*, 1992; Miao *et al.*, 1991; McCluskey and Mills, 1990). In *U. maydis* the smallest chromosome was surmised to have arisen by interchromosomal recombination (Kinscherf and Leong, 1988) and in *U. hordei* a 2:0 segregation was noted for a chromosome approximately 1000kb long (McCluskey and Mills, 1990). Differences have been observed in karyotypes of teliospore and monosporidial lines in *T. indica*, indicating that chromosomal alterations, loss and/or gain, may occur through meiosis (Russell and Mills, 1994; Tooley *et al*, 1995). Frequent gross rearrangements and translocation, and unstable inheritance, of chromosomes provide mechanisms of variation with profound implications for genetic evolution in *T. cucumeris*. Keijer *et al.* (in press) have found karyotype variation within an AG. More significantly, they observed polymorphism in chromosome size in homokaryotic progeny compared with parent strains. Meiosis apparently can generate genetic variation in *T. cucumeris* by chromosome alterations. Kistler and Miao (1992) suggest that rapid genetic changes in fungi may perhaps be explained by: a) frequent transpositional events, b) sorting of cytoplasmic elements, and c) loss of B-chromosomes. They also suggest that past reports of parasexuality could perhaps be explained today by mechanisms involving chromosome instability or duplication. Cytoplasmic elements are discussed by others in chapter II.3.

VI. THE SEXUAL SPECIES VIEW

Should not we consider *T. cucumeris* a functionally sexual species? Formation of the sexual state has been reported for all the anastomosis groups (Anderson, 1982; Ogoshi, 1987; Sneh *et al.*, 1991; Carling *et al.*, 1994). An unbiased view of the sexual morphology of the *Rhizoctonia*-like fungi reveals that it is excellently adapted for function. There is no evidence of evolutionary reduction of this function in the structure of the hymenium, for example, loss of forcible spore discharge, loss of repetitive spore germination, or loss of primary or repetitive sterigmata. The broad and long sterigmata of the Tulasnellales efficiently lift the basidiospore and orient them to escape the boundary layer, which they do, with robust forcible discharge. The sterigmata of these fungi are probably more resistant to desiccation, and discharge spores with greater force, than those of common agarics. The hymenium is adaptable to forming on many living plant parts. But in addition to these standard dispersal mechanisms, the Tulasnellales have repetitive forcible spore discharge that permits a spore a second means of disseminations away from an unsuitable habitat. The hymenium is excellently designed for dissemination of the species in a pre-agricultural world where dissemination of asexual propagules by machinery, wind-blown crop debris, and transport on vegetative plant propagation materials would be absent.

Dissemination by spores throughout the temperate and tropical world might have occurred repeatedly before agriculture.

T. cucumeris continues to be frequently observed disseminating by sexual spores today. For example, sporulation is commonly observed with AG 3 on potato stems (Butler, 1993). Sporulation of AG 2-2 isolates as web blights in the United States and elsewhere is common (Naito and Sugimoto, 1980; Anderson, 1982; Herr, 1982; Olaya and Abawi, 1994; see also chapter IV.3). In the southeast and humid regions of Texas, AG 1-IA and AG 1-IB sporulate occasionally as sheath blights on rice and AG 1-IA and AG 1-IB as web blights on corn, soybean, and sorghum (Jones and Belmar, 1989), as they do on various hosts in other parts of the world (Tu *et al.*, 1977; Galindo *et al.*, 1983).

An additional reason to consider *T. cucumeris* a functionally sexual species includes the successful fruiting of strains in the laboratory that have not been observed fruiting in nature. In instances in which cultures sporulate in the laboratory but are not seen to do so in nature, I believe we are dealing with sexual populations, with the possible exception of AG 8 in western Australia (Yang *et al.*, 1994b). Evidence that outbreeding is the norm in nature is revealed by collections of single-basidiospore cultures with macroscopically visible morphological variability among progeny of a parent. Sexual sporulation evidently is common but unobserved and therefore should be the most important influence on genetic variation in *T. cucumeris*. Methods to induce laboratory sporulation have become more standardized and controlled (Adams and Butler, 1983a,b) and new approaches are encouraging. Recent studies have suggested that an occasional isolate might be stimulated by a cold pretreatment of two days at 8C, and by photoperiod (Noel *et al.*, 1995), or by fungistasis (Kangatharalingam and Carson, 1988). A technique that I believe will show greater potential success is liquid culture in the presence of the living host plant and diurnal temperature gradients (Hietala *et al.*, 1994). This technique might be particularly effective for isolates that inhabit roots of a plant throughout the growing season fruiting in autumn, such as AG 3 on potato.

Genetically *T. cucumeris, Ceratobasidium cornigerum* and the other teleomorphs of *Rhizoctonia*-like fungi appear to be typical sexual basidiomycetes. Experimental research has demonstrated that in several AG of *T. cucumeris* there exists at least remnants and at most a complete standard system of sexual recombination typical of more thoroughly studied basidiomycetes (Puhalla and Carter, 1976; Adams and Butler, 1982). The heterokaryon appears to be controlled by the mating type genes that maintain a balanced ratio of paired nuclei in basidiomycetes through physical pairing and conjugate division (Raper, 1983). Thus natural selection would not be capable of directly acting on the balance of the two nuclei in an advancing hypha. However, an intriguing exception appears to be isolates of AG 8 (Yang *et al.*, 1992). It is difficult to understand the nature of sexuality in this AG because the characteristic test for heterokaryosis involving tuft reactions differs from AG 1-IC and AG 4. Evidence indicates AG 8 isolates might be frequently trikaryons (Yang *et al.*, 1994a,b). Additionally, heterokaryons apparently are less restricted in their participation in the Buller Phenomenon as tuft formation, presumably representing compatible matings, occurs between a heterokaryon and a homokaryon (He-Ho pairing) as well as between two heterokaryons (He-He pairings; Yang *et al.*, 1993b). It appears that the Buller phenomenon is so universal and predictable among AG 8 isolates that it can be used for identifying individual genotypes (somatically compatible strains) by rare instances when new heterokaryotic tufts are not generated in He-He pairings (Yang *et al.*, 1993b). Studies of the mating system in AG 8 involving sufficient DNA markers are needed. Unraveling the nature of heterokaryons, heterokaryotization, and the mating system in AG 8 will likely revolutionize our understanding of diversity in the genetics of *Rhizoctonia*.

VII. THE ROLE OF THE HOMOKARYON

Considering the amount of research effort expended on *T. cucumeris* worldwide, it is sobering to view the deficiencies in our fundamental knowledge of the role of basidiospores, homokaryotic colonies, and sporulation in its life history. *T. cucumeris*, a complex taxon, likely varies in the importance for the role of homokaryons in separate populations and habitats. Single-basidiospores have been shown to be pathogenic in numerous studies (Butler, 1993). Flentje and Stretton (1964) and Papavizas (1964) reported poor to good survival and competitive saprophytic activity in sterilized and unsterilized soil, respectively, and found great variation, 0.1-1.0 mm h^{-1}, in growth rate of single-spore cultures. Puhalla and Carter (1976) and Anderson (1982) observed that AG 2-2 homokaryons were difficult to grow on agar medium, and apparently so are AG 3 homokaryons (Julian *et al.*, 1995). Vilgalys and Cubeta (1994) are reinvestigating survival and competitiveness in soil of homokaryons of AG 4 and suggestions are that survival may be poor. Homokaryons of AG 4 and AG 1-1C grow vigorously and are pathogenic, but measurable though slight decreases in growth rate are evident in AG 4 homokaryons compared to heterokaryons in our research. Mycelial growth rate of homokaryons is variable and polygenically controlled in other basidiomycetes (Ullrich and Raper, 1977). Homokaryons appear to be less virulent than their parental heterokaryons, and the genetic factors controlling virulence and pathogenicity in heterokaryons appear to be dominant (Yang *et al.*, 1994a; Butler, 1993).

Understanding survival and competitiveness of homokaryons is important for understanding the impact of homothallism, dedikaryotization, and He-Ho pairing as mechanisms of genetic variation. He-Ho pairing appears to be a newly recognized possibility in *T. cucumeris* AG 4 (Cubeta *et al.*, 1993) and AG 8 (Yang *et al.*, 1993b). This Buller Phenomenon would increase outbreeding effectively to 1 (i.e., 100%). Once *T. cucumeris* becomes viewed as a functionally sexual fungus the mechanism of variation becomes standard meiotic recombination. Most mysteries are clarified but some that remain are the role of homothallism, of He-Ho pairing, and of inheritance of the mitochondrial genome.

VIII. NUCLEAR AND MITOCHONDRIAL MIGRATION DURING HETEROKARYOSIS

Nuclear migration in basidiomycetes is common following mating but its biological significance and selective advantage is not clear. The potential for mitochondria to concomitantly migrate, or form a mosaic has been questioned. Recent investigations in other genera have shown that although nuclear migration is usually bidirectional, mitochondria do not migrate with the nuclei and this creates a dikaryon that is a mosaic of the two mitochondrial types of the homokaryons (Rizzo and May, 1994; Smith *et al.*, 1990; May and Taylor, 1988; Hintz *et al.*, 1988; Casselton and Economou, 1985; Baptista-Ferreira *et al.*, 1983). In *Coprinus cinereus* such mitochondrial mosaics formed antagonistic barrage zones (incompatible SIG reactions) between themselves and thus acted as two individuals (May, 1988), whereas in *A. bulbosa* the mitochondrial mosaics behave as one individual (Smith *et al.*, 1990). A narrow zone of hyphae exists at the junction of the mating monokaryons/homokaryons where mitochondria of different genotypes do cohabit in heteroplasmic fused cells. It has been suggested that recombination of mitochondrial genomes can occur in the cells of the zone (Baptista-Ferreira *et al.*, 1983), but evidence of this is absent in nature (Carvalho *et al.*, 1995).

Nuclear migration was not believed to occur in matings of *T. cucumeris* because homokaryotic pairings produced the vigorous heterokaryotic tuft only at the narrow junction of hyphal conjugation. However, recently Cubeta *et al.* (1993) have observed unilateral and bilateral nuclear migration in AG 4 matings with the use of isolates that produced colored (brown) heterokaryons. They also observed a 'bow-tie' interaction in Ho-Ho pairings that was similar to well-studied nuclear migration responses in *Stereum hirsutum* which forms clamp connections (Rayner, 1991). Studies using somatic incompatibility reactions, phenotypic and genotypic markers indicated slow and limited nuclear migration occurred in some but not all successful matings in *T. cucumeris* (Cubeta *et al.*, 1993). Also some instability in recently synthesized heterokaryons was observed which they interpreted as occurring during a transitional mosaic stage present preceeding the time when the two genetically different nuclei become balanced in a 1:1 ratio. Further studies are needed to clarify the nature of the presumptive mosaic heterokaryon and of mosaics between Ho-Ho pairings with similar mating alleles. Behavior of mitochondria is not yet known in *T. cucumeris*. So the heterokaryon might be a mosaic, or perhaps transitionally mosaic, then one mitochondrial type predominating.

The best recent studies of the existence and survival of mitochondrial mosaics in nature have concerned the diploid of *A. bulbosa*, *A. gallica* and *A. ostoyae* (Carvalho *et al.*, 1995; Rizzo *et al.*, 1995; Smith *et al.*, 1990). In these species bidirectional inheritance of mitochondria occurs in laboratory matings. In all instances, the mitochondrial haplotype of the heterokaryotized individual was that of the resident haploid, whether nuclear donor or recipient. Surprisingly, only one mitochondrial haplotype is found in an individual in field populations and this poses a perplexing situation as it argues for either unidirectional inheritance or rapid selection for exclusively one mitochondrial haplotype in a particular nuclear background (Carvalho *et al.*, 1995; Rizzo *et al.*, 1995; Rizzo and May, 1994). Carvalho *et al.* (1995) suggested that mitochondrial heteroallelism, resulting from inheritance of two genetically different mitochondria, might exist only in a transient state followed by rapid segregation of mitochondrial types. Evidently, there is a cost to maintaining multiple mitochondrial genotypes within an individual (Smith *et al.*, 1990).

The fungi within *Rhizoctonia* have some representatives with unusual nuclear conditions. For example, a stable and fertile uninucleate *Ceratobasidium* species is common on conifer seedlings and likely represents a diploid (Hietala *et al.*, 1994). Additionally, some fertile *Ceratobasidium* populations may exhibit a stable complement of three nuclei per hyphal compartment (trikaryotic condition? Uchida *et al.*, 1986) and the heterokaryotic condition of AG 8 suggests a trikaryon (Yang *et al.*, 1994, 1992). These fungi will provide us with challenging research subjects, particularly in exploring Ho-Ho mating and He-Ho mating. Recent research in *A. gallica* and *A. ostoyae* is revealing unique mechanisms of heterokaryotization between diploids and haploids including the phenomenon of nuclear replacement, recombination without meiosis, and stable triploidy (Carvalho *et al.*, 1995; Rizzo and May, 1994). Additionally, unusual matings between non-outcrossing and outcrossing strains of *S. hirsutum* (Ainsworth *et al.*, 1990) are revealing unexpected mechanisms of genetic variation also involving nuclear replacement. Similar genetic systems of variation might be occasionally operating in *Rhizoctonia* species.

IX. THE ROLE OF HOMOTHALLISM

From the standpoint of sexuality of a homothallic basidiomycete, if homokaryons are not vigorous the fungus becomes functionally heterothallic. If we extrapolate from laboratory studies, homothallic homokaryons range from non-existent to rare in AG 3 and some AG 2 and AG 4, to common in AG 1-IC and some Australian AG 2. Determining the percentage of homokaryons that exhibit self-fertility and determining the fecundity of the self-fertile homokaryons has been necessary to the understanding of sexuality in *T. cucumeris*. Early studies of homothallism were based on the examination of nuclear number in basidiospores to separate primary homothallism from secondary (Stretton and Flentje, 1967; Anderson *et al.*, 1972). Counts of nuclear numbers in basidiospores are imprecise because it is difficult to determine if a spore began as binucleate or became binucleate. We observe nuclear division occurring immediately in spores swelling upon contact with moist substrate, and poison agars often do not act swiftly enough to stop swelling and the first mitotic division. We found that AG 1-IC progeny possessing one mating type allele averaged 44% self-fertile whereas AG 4 averaged 8%, under rigidly controlled laboratory conditions (Adams and Butler, 1982). Controlled studies revealed significantly lower fecundity in AG 4 self-fertile homokaryons ranging from 100-fold to 10^6 fold lower fecundity compared to heterokaryons. Even if 100% self-fertility occurs, this may have no influence on genetic variation if competitiveness and survival are insignificant. For this reason we have viewed the occurrence of self-fertility in *T. cucumeris* as an accidental feature of a heterothallic fungus (Adams and Butler, 1982), but the alternative view is that self-sterility is the result of a homothallic fungus carrying sterility factors which block basidial formation (Stretton and Flentje, 1967). In nature, there is potential for rapid selection for homothallism in *T. cucumeris* if it permits increased competitiveness in a niche. Homothallism, therefore, could provide for increased plasticity for genetic variation in specific niches.

X. THE MATING SYSTEMS

Many mycologists consider heterothallism and the A and B mating type loci as definitive characteristics of the ancestral basidiomycete (Ullrich and Raper, 1977). The condition of homothallism and unifactorial systems are examples of specialized de-evolution/evolution. *T cucumeris* AG 1-IC and AG 4 have an obvious "bipolar" mating system typical of many basidiomycetes (Puhalla and Carter, 1976; Adams and Butler, 1982) and referring to the taxon as homothallic confuses the understanding that it is a mated and outbreeding organism, with a complex system to favor mating and outbreeding (Adams, 1988). New research may uncover obligate outbreeding in AG 3 and some AG 2, if homokaryons are proven unable to form a vigorous mycelium in nature. I await the demonstration by molecular genetics that the heterokaryon incompatibility H-factor of *T. cucumeris* is the A locus mating type factor of the Basidiomycete.

XI. THE MOLECULAR MODEL OF THE MATING SYSTEM

Our understanding of the sexual system in *T. cucumeris* is based, primarily, on research on other Basidiomycetes. The H-factor can be conceptualized as the A locus mating type. The A locus is a complex of genes which are separated into two subcomplexes, known classically as the α and β loci (Day, 1960). The Aα and Aβ factors are closely linked. The genes encode a homeobox-containing family of proteins (Casselton and Kües, 1994; Stankis *et al.*, 1992). A homeodomain can be defined as a coding region for a protein with a DNA binding site. Homeodomains are regulatory and activate transcription, committing the cell to a particular developmental pathway. We know, for example, that heterokaryon formation initiates a developmental sequence of events including nuclear pairing, conjugate division, clamp formation, nuclear migration, and mitochondrial replacement, in typical basidiomycetes (Raper, 1983). Likely all these events are triggered when the homeodomain proteins bind to specific DNA sequences. In the α locus, the α A is described as coding for homeodomain proteins Y and Z, and the Aβ factor codes also for Y and Z proteins but apparently is non-functional. The Z protein has acidic regions that activate transcription. The Y protein has serine-rich and basic regions that resemble those of H1 histone proteins (25% identity; Stankis *et al.*, 1992). Z and Y may form a heterodimer that blocks the function of haploid-specific genes, similar to the homeodomain proteins of *S. cerevisiae* (Yang *et al.*, 1995). The ability of homeodomain proteins to interact together appears to be a common theme in mating-type regulation in fungi and possibly in developmental regulation in other eucaryotes.

Attached to each homeodomain are the regions that code for the different "alleles" of the mating type locus. The alleles are actually ideomorphs, regions of dissimilar DNA sequence that occur in one location (locus) on the mating type factor. Research in *Schizophyllum* has revealed that Z and Y proteins of mating type allele A3 are only 42-52% homologous to those of allele A4 (Stankis *et al.*, 1992; Yang *et al.*, 1995). The revelation that one allele will not share much homology with the sequence of another allele is a revolutionary concept for mycologists. Anderson *et al.* (1972) and Puhalla and Carter (1976) noted rare occurrences of new alleles (nonparental H-factors) and hypothesized that they arose from cross-over events during meiosis. It is interesting to review this and other reports of interactions among the basidiomycete alleles of the A locus and its Aα and Aβ subfactors now that we can conceptualize the molecular model. Recombination between Y and Z from different alleles might occur during meiosis and create recombinants in which Aα -regulated development occurs. Rephrased, this implies that recombination might create primary homothallism. The likelihood of recombination between Y and Z would be very low, however, because the DNA encoding these genes is heterogeneous (Kües *et al.*, 1994).

The second locus of tetrapolar basidiomycetes, the B mating type locus has recently been cloned in *S. commune* (Casselton and Kües, 1994). A 5kb fragment containing the B locus encodes components of a pheromone response pathway including pheromones and pheromone receptors (Gooday, 1994; Banuett, 1991 as the A locus). Thus a cell-signaling and recognition mechanism seems to be an integral part of Basididiomycete mating. The pheromone response pathway in the bipolar *T. cucumeris* might be nonfunctional, functionally "on", or might be triggered environmentally instead of by the sexual system. I would suggest that this genetic factor could be involved in the recognition response of hyphae preceeding anastomosis. Research seeking remnants of the complex B locus could uncover fascinating new biochemical explanations for the bipolar condition for primarily homothallism, and for the species recognition response in *T. cucumeris*.

XII. A MOLECULAR MODEL OF HOMOTHALLISM

Conceptualizing the origination of primary homothallism by alteration of the functional homeodomain and ideomorph sub-complex is a challenge. One molecular model of a mutation that created a derived homothallism has been described for *C. cinereus* (Kües *et al.*, 1994). It involved the fused combination of the Y and Z homeodomain proteins into a single chimeric and functional protein that created self-compatibility and fertility. The self-compatible mutation described occurred within the A6 allele of the A locus and constitutively activated sexual development.

This model of primary homothallism, however, is not sufficient to describe the type of homothallism occurring in *T. cucumeris* where self-fertility of progeny does not approach 100%, is not vigorous in fecundity, and mating persists with control by mating type alleles. Uncovering the molecular model of homothallism in *T. cucumeris* will be a demanding challenge. The approach to understanding would likely be through cloning of the H-factor (A locus) and any remnants of the B locus. Site-specific transformation with whole and partial portions of the factors, gene disruption, and/or sequencing could uncover an understandable biochemical model for homokaryotic fruiting in *T. cucumeris*. Recently, numerous mutations that confer self-compatibility (homothallism) have been generated in the B mating locus of tetrapolar basidiomycetes (Casselton and Kües, 1994). Perhaps studies of remnants of the B locus in *T. cucumeris* will reveal a corresponding mutation that directed evolution to a unifactorial and "homothallic" system. To complete our understanding of sexuality in

T. cucumeris we will also need to discover the molecular and physiological characteristics that define the heterokaryon, especially those developmental events initiated by mating and triggered by the homeodomain proteins.

XIII. TRANSFORMATION, TRANSFECTION AND TRANSPOSABLE ELEMENTS

Many linear plasmids have been found in the species-complex *R. solani*, AGs 1-IC, 1-IB, 2-2, 3, 4, 5 and 6 (Miyasaka *et al.*, 1990; Jabaji-Hare *et al*, 1994). The plasmids, many of which are invertrons (Sakaguchi, 1990), have nucleotide sequences at the 5' termini that form inverted repeats, plasmid ends protected by either covalently bound terminal proteins, or hairpin loops (without bound proteins) and protected from exonuclease digestion by their peculiar DNA structures (Jabaji-Hare *et al*, 1994; Miyashita *et al.*, 1990). Most of the plasmids are associated with mitochondria (Samac and Leong, 1989a) but at least one has nucleus-derived sequence (Wako *et al.*, 1991). The biological function of the plasmids is unknown but they occur in unusually high copy numbers in the host cytoplasm and have presumptive origins of replication that are terminal inverted repeat sequences (TIRs). These linear plasmids will be useful for constructing transformation vectors with autonomous replication (Samac and Leong, 1989a,b) because the TIRs have the ability to function as replicons. Linear plasmids can be used to construct vectors that are covalently closed circular DNA or linearized DNA. Autonomously replicating sequences on transformation vectors of *S. cerevisiae* are maintained at high copy numbers and have high transformation frequencies compared to integrative vectors. The TIR of *N. haematococca* pFSC1 conferred autonomous replication when spliced into an integrative transformation vector (Samac and Leong, 1989b). Construction of such a vector for *R. solani* would facilitate the development of molecular genetic techniques, simplify the isolation and cloning of genes, facilitate transformation of mitochondria, and likely the vector will function in heterologous hosts, for example, other *Rhizoctonia* teleomorphs.

The most thoroughly studied linear plasmid from *R. solani* is pRS64. This plasmid apparently is stably integrated into chromosomal DNA and acts as a Transposable element (Hongo *et al.*, 1994; Wako *et al.*, 1991). Study of the sequence of this and other plasmids will help identify their integrated forms in the fungal DNA and reveal how integration over the eons has impacted evolution of the mitochondrial and nuclear genetics of *Rhizoctonia*.

Advances in molecular genetic techniques are beginning to accelerate our understanding of the genetic environment influencing *Rhizoctonia*. The construction and use of complementary DNA (cDNA) has allowed the identification and expression of the protein product encoded by the pRS64 linear plasmid (Hongo *et al.*, 1994). cDNA will be the principal tool in the study of dsRNA genomes and hypovirulence (Chen *et al.*, 1994) potentially permitting the construction of synthetic infectious hypovirus transcripts for transfection and biological control of *R. solani* (see chapter II.3).

XIV. POPULATION GENETICS

While I have mentioned that little progress has been made in our understanding of the specific genetics of *T. cucumeris*, one area of study has been most fruitful. The rapid progress in our understanding of the characterization of the evolutionary unit, i.e., the genetically isolated, intrabreeding populations of a biological species. *T. cucumeris.* has now been divided by numerous criteria into what is believed to be genetically distinct groups. For example, there are 12 known AGs including 6 bridging groups (Carling *et al.*, 1994), at least 14 recognizable groups defined by polymorphisms in the internal transcribed spacers of the nuclear ribosomal DNA (Liu *et al.*, 1993), at least 18 subgroups defined primarily by DNA/DNA hybridization

(Carling and Kuninaga, 1990; Vilgalys, 1988) but also by isozymes (Laroche *et al.*, 1992; Liu and Sinclair, 1992) and cellular fatty acids (Johnk and Jones, 1993), and 5 zymogram groupings in AG 8 alone (MacNish *et al.*, 1994). While it is first necessary to define a subgroup by the above and other methods prior to population genetic studies, only two methods have been demonstrated to recognize allele frequencies within a population, isozyme polymorphisms (Liu and Sinclair, 1992) and pectic enzyme polymorphisms (MacNish *et al.*, 1994). The crucial criterion of demonstrating that enzyme polymorphisms act as alleles of loci by segregating in progeny has been accomplished only with the pectic enzymes (Yang *et al.*, 1994b). Outsiders frequently complain about our preoccupation with anastomosis and subgroup grouping, and isozyme and DNA subgrouping of strains but these studies have matured and our concept of the evolutionary unit, the subgroup, is coalescing in perfect synchrony for beginning studies of population genetics.

The recent arrival of DNA fingerprinting technology should allow us to directly approach and answer the most cogent and fundamental questions about genetic variation of the biological species of *T. cucumeris*. DNA technology can provide the abundant and neutral single copy and multicopy markers needed for formal studies of populations in a locale; once the population has been carefully defined to the subgroup. Two fingerprinting probes, the oligonucleotides $(TCC)_5$ and $(CAT)_5$, have been useful in detecting variable dispersed repetitive DNA in fungi (DeScenzo and Harrington, 1994) and have been effective as markers for population genetic studies of *A. ostoyae* in pine forests (Rizzo *et al.*, 1995). These probes or similar repetitive oligonucleotides could be useful for studies of population genetics of a subgroup of *T. cucumeris* in field studies, particularly when combined with somatic incompatibility grouping (Cubeta *et al.*, 1993), mitochondrial DNA restriction fragment length polymorphisms (RFLP), cloned DNA probes, random amplified polymorphic DNA polymerase chain reaction products (RAPD), and pectic enzyme polymorphisms. Methodology for producing reliable RAPD fingerprinting of *R. solani* has been characterized (Tommerup *et al.*, 1995). Equations for mathematical analysis of population genetic structure using RAPD markers (Lynch and Mulligan, 1994) and repetitive oligonucleotide probes (Rizzo *et al.*, 1995; Rosewich and McDonald, 1994) have been described.

The equations of population genetics will permit direct calculation of outbreeding, inbreeding, population structure, spatial orientations and genetic variation. As long as we utilize sufficient numbers of markers and have an understanding of how they are inherited (Rosewich and McDonald, 1994), population genetic research of this nature will answer most of the original questions asked about genetic variation in *Rhizoctonia*. This work has already begun in Australia where some fingerprinting probes have been cloned and utilized in combination with other biochemical and genetic markers (Balali *et al.*, 1995). Similar studies will be needed in many geographical regions, in many unique habitats, and for distinct subgroups. Our seemingly endless struggle with the species concept of *T. cucumeris* has unexpectantly led to the key to our ultimate understanding of the organisms' genetics, ecology, epidemiology, and biology.

XV. References

Adams GC (1988) *Thanatephorus cucumeris* (*Rhizoctonia solani*) a species of wide host range. In: Sidhu GS (ed.) Advances in Plant Pathology. Volume 6. Genetics of Plant Pathogenic Fungi. (pp 535-552). Academic Press, New York.

Adams GC & Butler EE (1982) A re-interpretation of the sexuality of *Thanatephorus cucumeris* anastomosis group four. Mycologia 74:793-800.

Adams GC & Butler EE (1983a) Influence of nutrition on the formation of the formation of basidia and basidiospores of *Thanatephorus cucumeris* (anamorph:*Rhizoctonia solani*). Phytopathology 73:147-151.

Adams GC & Butler EE (1983b) Environmental factors influencing the formation of basidia and basidiospores of *Thanatephorus cucumeris* (anamorph: *Rhizoctonia solani*). Phytopathology 73:152-155.

Ainsworth AM, Rayner ADM, Broxholme SJ & Beeching JR (1990) Occurrence of unilateral genetic transfer and genomic replacement between strains of *Stereum hirsutum* from non-outbreeding and outbreeding populations. The New Phytologist 115:99-128.

Anderson NA (1982) The genetics and pathology of *Rhizoctonia solani*. Annu. Rev. Phytopathol. 20:329-347.

Anderson NA, Stretton HM, Groth JV & Flentje NT (1972) Genetics of heterokaryosis in *Thanatephorus cucumeris*. Phytopathology 62:1057-1065.

Bakalinsky AT & Snow R (1990) The chromosomal constitution of wine strains of *Saccharomyces cerevisiae*. Yeast 6:367-382.

Balali GR, Whisson DL, Scott ES & Neate SM (1995) DNA fingerprinting and pectic zymogram phenotypes of field isolates of *Rhizoctonia solani* AG 3. Mycological Res. 100: (in press).

Banuett F (1991) Life cycle determinants of the plant pathogen *Ustilago maydis*. In: Bennett JW & Lasure LL (eds.) More Gene Manipulations in Fungi. (pp 217-233). Academic Press, New York.

Baptista-Ferreira JLC, Economou A & Casselton LA (1983) Mitochondrial genetics of *Coprinus* :recombination of mitochondrial genomes. Curr. Genet. 7:405-407.

Beeching JR, Ainsworth AM, Broxholme SJ, Pryke JA & Rayner ADM (1989) Investigation of genetic transfer between strains of the basidiomycete, *Stereum hirsutum*, using molecular and morphological criteria. New Phytol. 113:505-512.

Bolkan HA & Butler EE (1974) Studies of heterokaryosis and virulence of *Rhizoctonia solani*. Phytopathology 64:513-522.

Burnett JH (1983) Speciation in fungi. Trans. Brit. Mycol. Soc. 81:1-14.

Butler EE (1993) *Rhizoctonia solani*. In: Lyda S & Kenerley CM (eds.) Biology of Sclerotial-Forming Fungi. (pp 87-112). Texas Agric. Exp. Stn. Press, College Station, Texas.

Buxton EW (1954) Heterokaryosis and variability in *Fusarium oxysporum* f. sp. *gladioli* (Snyder & Hansen). J. Gen. Microbiol. 10:71-84.

Carling DE, Rothrock CS, MacNish GC, Sweetingham MW, Brainard KA & Winters SW (1994) Characterization of anastomosis group 11 (AG 11) of *Rhizoctonia solani*. Phytopathology 84:1387-1393.

Carling DE & Kuninaga S (1990) DNA base sequence homology in *Rhizoctonia solani* Kühn:inter- and intragroup relatedness of anastomosis groups. Phytopathology 80:1362-1364.

Carling DE, Kuninaga S & Leiner RH (1988) Relatedness within and among intraspecific groups of *Rhizoctonia solani* :a comparison of grouping by anastomosis and DNA hybridization. Phytoparasitica 16:209-210.

Carvalho DB, Smith ML & Anderson JB (1995) Genetic exchange between diploid and haploid mycelia of *Armillaria gallica*. Mycol. Res. 99:641-647.

Casselton LA & Economou A (1985) Dikaryon formation. In: Moore D, Casselton LA Wood DA & Frankland J (eds.) Developmental Biology of Higher Fungi. (pp 213-229). Cambridge University Press, Cambridge.

Casselton LA & Kües U (1994) Mating-type genes in Homobasidiomycetes. In: Wessels JGH & Meinhardt F (eds.) The Mycota. (pp 307-321). Springer-Verlag, New York.

Caten CE & Jinks JL (1966) Heterokaryosis:its significance in wild homothallic ascomycetes and fungi imperfecti. Trans. Brit. Mycol. Soc. 49:81-93.

Chase TE & Ullrich RC (1990a) Genetic basis of biological species in *Heterobasidion annosum*:Mendelian determinants. Mycologia 82:67-72.

Chase TE & Ullrich RC (1990b) Five genes determining intersterility in *Heterobasidion annosum*. Mycologia 82:73-81.

Chen B, Choi G & Nuss DL (1994) Attenuation of fungal virulence by synthetic infectious hypovirus transcripts. Science 264:1762-1763.

Cubeta MA, Briones-Ortega R & Vilgalys R (1993) Reassessment of heterokaryon formation in *Rhizoctonia solani* anastomosis group 4. Mycologia 85:777-787.

Day PR (1960) The structure of the A mating type locus in *Coprinus lagopus*. Genetics 45:641-650.

DeScenzo, RA & Harrington, TC (1994) Use of $(CAT)_5$ as a DNA fingerprinting probe for fungi. Phytopathology 84:534-540.

Esser K (1965) Heterogenic incompatibility. In: Esser K & Raper JR (eds.) Incompatibility in Fungi. (pp 6-12). Springer-Verlag, New York.

Esser K & Baich R (1994) Heterogenic incompatibility. In: Wessels JGH & Meinhardt F (eds.) The Mycota. (pp 211-232). Springer-Verlag, New York.

Flentje NT & Stretton HM (1964) Mechanisms of variation in *Thanatephorus cucumeris* and *T. practicolus*. Aust. J. Biol. Sci. 17:686-704.

Galindo JJ, Abawi GS, Thurston HD & Galvez G (1983) Sources of inoculum and development of bean web blight in Costa Rica. Plant Dis. 67:1016-1021.

Garza-Chapa R & Anderson NA (1966) Behavior of single-basidiodpore isolates and heterokaryons of *Rhizoctonia solani* from flax. Phytopathology 56:1260-1268.

Gooday GW (1994) Hormones in mycelial fungi. In: Wessels JGH, & Meinhardt F (eds.) The Mycota (pp 401-411). Springer-Verlag, New York.

Gordon TR (1993) Genetic variation and adaptive potential in an asexual soilborne fungus. In: Reynolds DR & Taylor JW (eds.) The Fungal Holomorph: Mitotic, Meiotic and Pleomorphic Speciation in Fungal Systematics. (pp 217-224). CAB International, London.

Hansen EM, Stenlid J & Johansson M (1993a) Somatic incompatibility and nuclear assortment in *Heterobasidion annosum*. Mycol. Res. 97:1223-1228.

Hansen EM, Stenlid J & Johansson M (1993b) Genetic control of incompatibility in the root rotting basidiomycete *Heterobasidion annosum*. Mycol. Res. 97:1229-1233.

Harrington TC, Worrall JJ & Baker FA (1992) *Armillaria*. In: Singleton LL, Mihail JD & Rush C (eds.) Methods for Research on Soilborne Phytopathogenic Fungi. (pp 81-85). APS Press, St. Paul, Minnesota.

Herr LJ (1982) Characteristics of hymenial isolates of *Thanatephorus cucumeris* on sugar beets in Ohio. Plant Dis. 66:246-249.

Hietala AM, Sen R & Lilja A (1994) Anamorphis and teleomorphic characteristics of a uninucleate *Rhizoctonia* sp. isolated from roots of nursery grown conifer seedlings. Mycol. Res. 98:1044-1050.

Hintz WE Anderson JB & Horgen PA (1988) Nuclear migration and mitochondrial inheritance in the mushroom *Agaricus bitorquis*. Genetics 119:35-41.

Hongo M, Miyasaka A, Suzuki F & Hashiba T (1994) Expression of the linear DNA plasmid pRS64 in the plant pathogenic fungus *Rhizoctonia solani*. Molec. Gen. Genet. 245:265-271.

Jinks JL (1952) Heterokaryosis:a system of adaptation in wild fungi. Proc. Royal Soc. B 140:105-145.

Jabaji-Hare SH, Burger G, Forget L & Lang BF (1994) Extrachromosomal plasmids in the plant pathogenic fungus *Rhizoctonia solani*. Curr. Genet. 25:423-431.

Johnk JS & Jones RK (1993) Differentiation of populations of AG 2-2 of *Rhizoctonia solani* by analysis of cellular fatty acids. Phytopathology 83:278-283.

Jones RK & Belmar SB (1989) Characterization and pathogenicity of *Rhizoctonia* spp. isolated from rice, soybean, and other crops grown in rotation with rice in Texas. Plant Dis. 73:1004-1010.

Julian M, Dullemans A, Houteerman P & Keijer J (1995) Studies on the genetics of *Rhizoctonia solani* anastomosis groups 2 and 3. International Symposium on *Rhizoctonia*, ISR'95. p. 35 (abstract). Misset Publ., Doetinchem, The Netherlands.

Kangatharalingam N & Carson ML (1988) Technique to induce sporulation in *Thanatephorus cucumeris*. Plant Dis. 72:146-148.

Kay E & Vilgalys R (1992) Spatial distribution and genetic relationships among individuals in a natural population of the oyster mushroom *Pleurotus ostreatus*. Mycologia 84:173-182.

Keijer J, Houterman PM, Dullemans AM & Korsman MG (1996) Heterogeneity in electrophoretic karyotype within and between anastomosis groups of *Rhizoctonia solani*. Mycol. Res. 100: (in press)

Kinscherf TC & Leong SA (1988) Molecular analysis of the karyotype of *Ustilago maydis*. Chromosoma (Berlin) 96:427-433.

Kistler HC & Miao VP (1992) New modes of genetic change in filamentous fungi. Annu. Rev. Phytopathol. 30:131-152.

Kües U, Göttgens B, Stratmann R, Richardson WV, O'Shea SF & Casselton LA (1994) A chimeric homeodomain protein causes self-compatibility and constitutive sexual development in the mushroom *Coprinus cinereus*. EMBO J. 13:4054-4059.

Laroche JP, Jabaji-Hare SH & Charest P-M (1992) Differentiation of two anastomosis groups of *Rhizoctonia solani* by isozyme analysis. Phytopathology 82:1387-1393.

Liu ZL, Domier LL & Sinclair JB (1993) ISG-specific ribosomal DNA polymorphism of the *Rhizoctonia solani* species complex. Mycologia 85:795-800.

Liu ZL, & Sinclair JB (1992) Genetic diversity of *Rhizoctonia solani* anastomosis group 2. Phytopathology 82:778-787.

Lynch M & Milligan BG (1994) Analysis of population genetic structure with RAPD markers. Molecular Ecology 3:91-99.

MacNish GC, Carling DE, Sweetingham MW & Brainard KA (1994) Anastomosis group (AG) affinity of pectic enzyme (zymogram) groups (ZG) of *Rhizoctonia solani* from the western Australian cereal belt. Mycol. Res. 98:1369-1375.

May G (1988) Somatic incompatibility and individualism in the coprophilous basidiomycete *Coprinus cinereus*. Trans. Birt. Mycol. Soc. 91:443-451.

McCluskey K, Agnan J & Mills D (1994) Characterization of genomic plasticity in *Ustilago hordei*. Curr. Genet. 26:486-493.

McCluskey K & Mills D (1990) Identification and characterization of chromosome length variation among strains representing fourteen races of *Ustilago hordei*.. MPMI. 3:366-373.

McDonald BA & Martinez JP (1990) Restriction fragment length polymorphisms in *Septoria tritici* occur at a high frequency. Curr. Genet. 17:133-138.

McDonald BA & Martinez JP (1991) Chromosome length polymorphisms in a *Septoria tritici* population. Curr. Genet. 19:265-271.

Miao VP, Covert SF & Van Etten HD (1991) A fungal gene for antibiotic resistance on a dispensable ("B") chromosome. Science 254:1773-1776.

Mills D & McCluskey K (1990) Electrophoretic karyotypes of fungi:the new cytology. MPMI 3:351-357.

Miyasaka A, Chen CL & Hashiba T (1990) Detection and properties of plasmid-like DNA in isolates from nine anastomosis and intraspecific groups of *Rhizoctonia solani*. J. Gen. Microbiol. 136:1791-1798.

Miyashita S-I, Hirochika H, Ikoda J-E & Hashiba T (1990) Linear plasmid DNAs of the plant pathogenic fungus *Rhizoctonia solani* with unique terminal structures. Molec. Gen. Genet. 220:165-171.

Naito S & Sugimoto T (1980) Relationship between basidiospore dispersal of *Thanatephorus cucumeris* (Frank) Donk and development of foliage blight of sugar beets. Ann. Phytopath. Soc. Japan 46:216-223.

Noel T, Axiotis S & Labarère J (1995) Study of nutritional and environmental factors affecting the fruiting competence of *Rhizoctonia solani* AG 1. J. Phytopathol. 143:173-177.

Ogoshi A (1987) Ecology and pathogenicity of anastomosis and intraspecific groups of *Rhizoctonia solani* Kühn. Annu. Rev. Phytopathol. 25:125-143.

Olaya G & Abawi GS (1994) Influence of inoculum type and moisture on development of *Rhizoctonia solani* on foliage of table beets. Plant Dis. 78:805-810.

Papavizas GC (1964) Survival of single-basidiospore isolates of *Rhizoctonia practicola* and *Rhizoctonia solani*. Can. J. Microbiol. 10:739-745.

Parmeter JR Jr, Snyder WC & Reikle RE (1963) Heterokaryosis and variability in plant-pathogenic fungi. Annu. Rev. Phytopathol. 1:51-76.

Phillips AJL (1993) The use of protoplasts for the preparation of homokaryons from heterokaryotic isolates of *Rhizoctonia solani*. Mycol. Res. 97:456-460.

Puhalla JE & Carter, WW (1976) The role of the H locus in heterokaryosis in *Rhizoctonia solani*. Phytopathology 66:1348-1353.

Raper CA (1983) Controls for development and differentiation of the dikaryon in basidiomycetes. In: Bennett JW & Ciegler A (eds.) Secondary Metabolism and Differentiation in Fungi. (pp 195-238). Marcel Dekker, New York.

Rayner ADM (1991) The challenge of the individualistic mycelium. Mycologia 83:48-71.

Rayner ADM, Coates D, Ainsworth AM, Adams TJH, Williams END & Todd NK (1984) The biological consequences of the individualistic mycelium. In: Jennings DH & Rayner ADM (eds.) The Ecology and Physiology of the Fungal Mycelium. (pp 509-540). Cambridge University Press, Cambridge

Rizzo DM & May G (1994) Nuclear replacement during mating in *Armillaria ostoyae* (Basidiomycotina). Microbiol.140:2115-2124.

Rizzo DM, Blanchette RA & May G (1995) Distribution of *Armillaria ostoyae* genets in a *Pinus resinosa-Pinus banksiana* forest. Can. J. Bot. 73:776-787.

Rosewich UL & McDonald B (1994) DNA fingerprinting in fungi. Methods Molec. Cell. Biol. 5:41-48.

Russell BW & Mills D (1994) Morphological, physiological, ad genetic evidence in support of a conspecific status for *Tilletia caries, T. controversa*, and *T. foetida*. Phytopathology 84:576-582.

Sakaguchi,K (1990) Invertrons, a class of structurally and functionally related genetic elements that include linear plasmids, transposable elements, and genomes of Adeno-type viruses. Microbiol. Rev. 54:66-74.

Samac DA & Leong SA (1989a) Mitochondrial plasmids of filamentous fungi:characteristics and use in transformation vectors. MPMI. 2:155-159.

Samac DA & Leong SA (1989b) Characterization of the termini of the linear plasmids from *Nectria haematococca* and their use in construction of an autonomously replicating transformation vector. Curr. Genet. 16:187-194.

Smith ML, Bruhn JN & Anderson, JB (1994) Relatedness and spatial distribution of *Armillaria* genets infecting red pine seedlings. Phtopathology 84:822-829.

Smith ML, Bruhn JN & Anderson JB (1992) The fungus *Armillaria bulbosa* is among the largest and oldest living organisms. Nature (London) 356:428-431.

Smith ML, Duchesne LC, Bruhn JN & Anderson JB (1990) Mitochondrial genetics in a natural population of the plant pathogen *Armillaria*. Genetics 126:575-582.

Sneh B, Burpee L & Ogoshi A (1991) Identification of *Rhizoctonia* Species. APS Press, St. Paul, Minnesota. 135 pp.

Stankis MM, Specht CA, Yang H, Glasson L, Ullrich RC & Novotny CP (1992) The Aα mating locus of *Schizophyllum commune* encodes two dissimilar multiallelic homeodomain proteins. Proc. Natl. Acad. Sci. USA 89:7169-7173.

Stenlid J & Rayner ADM (1991) Patterns of nuclear migration and heterokaryosis in pairings between sibling homokaryons of *Heterobasidion annosum*. Mycol. Res. 95:1275-1283.

Stretton HM, Flentje NT & McKenzie AR (1967) Homothallism in *Thanatephorus cucumeris*. Aust. J. Biol. Sci. 20:113-120.

Syminis T, Bandy BP & Tavantzis SM (1987) Vegetative incompatibility in *Rhizoctonia solani*. Phytopathology 77:1619-1620.

Tommerup IC, Barton JE & O'Brien PA (1995) Reliability of RAPD fingerprinting of the basidiomycete fungi, *Laccaria*, *Hydrangium* and *Rhizoctonia*. Mycol. Res. 99:179-186.

Tooley PW, Carras MM, Beck R, Peterson G & Bonde MR (1995) Separation of *Tilletia indica* chromosomes using CHEF gel electrophoresis. Mycologia 87:61-67.

Tu CC, Chang YH & Schenck NC (1977) Leafspot caused by *Thanatephorus cucumeris* on jute, and survival of single basidiospore isolates in soil. Plant Dis. Rep. 61:80-84.

Tzeng T, Lyngholm LK, Ford CF & Bronson CR (1992) A restriction fragment polymorphism map and electrophoretic karyotype of the maize pathogen *Cochliobolus heterostrophus*. Genetics 130:81-96.

Uchida JY, Aragaki M & Yahata PS (1986) Basidiospore formation by *Ceratobasidium* sp. on agar. Mycologia 78:587-592.

Ullrich RC & Raper JR (1977) Evolution of genetic mechanisms in fungi. Taxon 26:169-179.

Vilgalys R & Cubeta MA (1994) Molecular systematics and population biology of *Rhizoctonia*. Annu. Rev. Phytopathol. 32:135-155.

Wako T, Ishikawa T & Hashiba T (1991) Unique DNA plasmid pRS64 associated with chromosomal DNAs of the plant pathogenic fungus *Rhizoctonia solani*. J. Gen. Microbiol. 137:2817-2821.

Whitney HS & Parmeter JR Jr (1963) Synthesis of heterokaryons in *Rhizoctonia solani* Kühn. Can. J. Bot. 41:879-886.

Worrall JJ (1994) Population structure of *Armillaria* species in several forest types. Mycologia 86:401-407.

Yang H Shen GP, Park DC, Novotny CP & Ullrich RC (1995) The Aα mating-type transcripts of *Schizophyllum commune*. Exp. Mycol. 19:16-25.

Yang HA, Zhou J, Sivasithamparam K & O'Brien PA (1994a) Variation in cultural morphology and pathogenicity among protoplast-regenerated strains of *Rhizoctonia solani*. FEMS Microbiol. Lett. 115:83-86.

Yang HA, Zhou J, Sivasithamparam K, Tommerup IC, Barton JE & O'Brien PA (1994b) Genetic variability in pectic enzymes of *Rhizoctonia solani* isolates causing bare-patch diseases of cereals. J. Phytopathol. 141:259-266.

Yang HA, Sivasithamparam K & O'Brien PA (1993a) Improved method for protoplast regeneration on *Rhizoctonia solani*. Soil Biol. Biochem. 25:633-636.

Yang HA, Sivasithamparam K & O'Brien PA (1993b) Mycelial interactions and the potential use of tuft formation in characterizing *Rhizoctonia solani* isolates infecting cereals. Aust. J. Bot. 41:253-262.

Yang HA, Tommerup IC, Sivasithamparam K, & O'Brien PA (1992) Heterokaryon formation with homokaryons derived from protoplasts of *Rhizoctonia solani* anastomosis group eight. Exp. Mycol. 16:268-278.

II.2 TRANSLOCATION AND TRANSFER IN *RHIZOCTONIA*: MECHANISMS AND SIGNIFICANCE

JAMES W DEACON

Institute of Cell & Molecular Biology, University of Edinburgh, Mayfield Road, Edinburgh EH9 3JH, UNITED KINGDOM

I. Introduction

Many important aspects of the biology of *Rhizoctonia* depend on translocation in the hyphae. For example, *R. solani* can grow considerable distances through soil by translocating nutrients from a food base. The infection process depends on mobilization of resources to infection cushions or other pre-penetration structures (Dodman, 1970). Nutrient translocation enables *Rhizoctonia* spp. to serve as mycorrhizal symbionts of orchids. Translocation also mediates the remobilization of mycelial components into developing sclerotia, for survival in soil. In addition, the control of nuclear distribution and migration is central to the population genetics of *Rhizoctonia* (Vilgalys and Cubeta, 1994). And an understanding of the mobilization and transfer of extrachromosomal genetic elements in *Rhizoctonia* could have major implications for disease management. The control of some of these processes could be within grasp, because the mycoparasite *Verticillium biguttatum* can suppress sclerotium production by *R. solani* AG 3, possibly by altering the normal patterns of translocation (Van den Boogert and Deacon, 1994).

Despite these points, the processes of nutrient translocation and organellar transfer in *Rhizoctonia* have received little study. This review assembles the available information and attempts to place it in the context of *Rhizoctonia* ecology and pathology.

II. Translocation and the growth of *Rhizoctonia* through soil

The early studies by Blair (1943) demonstrated that *R. solani* behaves differently from most other soilborne pathogens, because it seems to have a natural role as a saprotroph in soils. *R. solani* grew extensively from a food base into natural soil, and the growth, once initiated, continued even after the food base was removed. Yet, the growth of *R. solani* was suppressed if the soil was supplemented with finely ground organic matter such as dried grass or wheat straw, which promoted the activities of other microorganisms. So, *R. solani* seems to subsist on the more refractory organic components such as cellulose that are not exploited by competing saprotrophs. Many subsequent studies have confirmed these points (Garrett, 1970), but there has been little comparative study of the cellulolytic activity or growth through soil by different anastomosis groups and species of *Rhizoctonia*. A monoclonal antibody probe might be useful for such studies (Thornton *et al.*, 1993); it recognized antigens of *R. solani* AGs 1, 3, 4, 5 and 7, but not of AG 2-2 or *R. cerealis*. A major unresolved issue is the degree to which different forms of *Rhizoctonia* depend on nutrient translocation compared with nutrient capture for their growth in soil.

B. Sneh et al. (eds.),
Rhizoctonia Species: Taxonomy, Molecular Biology, Ecology, Pathology and Disease Constrol, 117–125.

A role for nutrient translocation was suggested by the work of Henis and Ben-Yephet (1970). Using inoculum produced *in vitro*, they found that *R. solani* did not infect bean seedlings from single propagules (400-500µm) placed 2 cm away from the host in soil, but it infected from clusters of eight or more propagules, probably because the mycelial resources were pooled by anastomosis, enabling growth through soil. Studies with natural inoculum units retrieved from field soils have given similar results (Naiki, 1985): the larger inoculum units are the most infective.

For the Rhizoctonia patch diseases of cereals, which are common in non-ploughed sites, an intact mycelial network seems to be more important than discrete inoculum units. *R. solani* AG 8 is the main cause of these diseases, but *R. oryzae* also has been implicated in Britain (Burton *et al.*, 1988) and a distinct strain of *R. solani* causes crater disease of wheat in South Africa (Deacon and Scott, 1985). Individual disease patches of 100m or more can represent a single clone of AG 8, as evidenced by somatic compatibility and pectic zymogram patterns (MacNish *et al.*, 1993). Many recent reports confirm the early finding by Hynes (1937) that these diseases are effectively controlled by mouldboard ploughing. A major effect of ploughing would be to fragment mycelial networks, the effect of which was shown experimentally (McDonald and Rovira, 1985). However, there are additional factors in field sites. Deacon and Scott (1985) noted that Rhizoctonia patch diseases occur in soils that potentially impede root growth. Root impedance is known to enhance nutrient-exudation, and nutrient exudates favor infection of many plants by *R. solani* (Hayman, 1970). In crater disease the infection of roots is always initiated at the junction between the loose surface soil and the undisturbed soil - a black clay. The Rhizoctonia patch diseases more commonly occur on light soils, but these can be as impeditive as clays (Harrod, 1975). In several parts of Australia (Henry and Brown, 1995) these diseases have been alleviated by using seed drills that cut a slit below the seed. This seems unlikely to cause major disturbance of mycelial networks, but it would reduce root impedance. So, these diseases certainly require an intact mycelial network, but this may have obscured the role of an equally important factor-a major degree of root impedance which predisposes the plants to attack.

III. TRANSLOCATION AND SCLEROTIUM FORMATION

R. solani responds rapidly to nutrient-stress by remobilizing the mycelial resources into sclerotia. Christias and Lockwood (1973) quantified this by placing mycelial mats on a bed of sterile glass beads leached with water. Sclerotia were produced within 4 days, and accounted for 39% of the original mycelial dry weight, 26% of the original carbohydrate content and 50% of the original nitrogen content. Three other sclerotial fungi behaved similarly in converting the mycelial resources into newly formed sclerotia. However, *R. solani* was unique when the mycelial mats were placed on natural soil rather than glass beads, because the dry weight and nitrogen content of newly formed sclerotia were almost double those of the original mycelia, and the carbohydrate content was similar to the original mycelia. In response to nutrient shift-down *R. solani* not only remobilized its mycelial nutrients into sclerotia but also incorporated further nutrients from soil.

R. solani AG 3 produces sclerotia (black scurf) on potato tubers, reducing their marketability. Field trials in the Netherlands have shown that the mycoparasite *Verticillium biguttatum* can reduce these symptoms. Consistent with this, Van den Boogert and Deacon (1994) showed that *V. biguttatum* is a biotrophic parasite, and when established on only part of a *R. solani* AG 3 colony *in vitro*, it suppressed sclerotium production over the whole colony. In practice this could mean that *Verticillium* need be established on only part of a tuber surface to achieve control. When two somatically incompatible strains of *R. solani* AG 3 were paired on agar plates, then *V. biguttatum* suppressed sclerotium production only on the colony on which it was growing; but it suppressed sclerotia over the whole plate if two compatible strains were paired and formed a continuous network. In field conditions *Verticillium* might thus be most effective if, as might often be true, individual plants or tubers are infected by a single pathogen strain.

V. biguttatum was suggested to exert its effect by generating a nutrient sink in the host mycelium, equivalent to the behavior of some biotrophic plant pathogens. To explore this, D. Scott and J. W. Deacon (unpublished) used video microscopy to follow cytoplasmic movements in *R. solani* hyphae growing through a band of *Verticillium* spores on water agar films. Initially, *V. biguttatum* penetrated the hyphae and formed haustoria, but the infected hyphae continued to grow, with normal cytoplasmic streaming. There was steady movement of cytosol from the older regions to the *Rhizoctonia* tips, but major organelles did not pass through the dolipore septa. After 24h of interaction some of the initially infected host cells had died, and the parasite had formed haustoria in other host cells. *Rhizoctonia* cytoplasm continued to stream into the infected zone, even if all the host tips were dead. At this stage, also, any *Rhizoctonia* tips that had grown through and beyond the infection zone stopped growing, and cytoplasmic flow was reversed, moving back to the infection zone. Thus, *V. biguttatum* generated at least a localized sink in the host network. After 48h of interaction, the older regions of *Rhizoctonia* hyphae showed continuous, extremely rapid cytosolic streaming, 8-11μm/sec, towards the infection zone. These hyphae were highly vacuolated, but the vacuoles were peripheral in the hyphae and did not move. The cytosol streamed through the centers of the hyphae, and through the septa which seemed to be partly degraded. Non-parasitized colonies of equivalent age did not show rapid cytosolic streaming. Based on these and other videotaped observations (see later) Fig. 1 shows an interpretation of the sequential changes in cytoplasmic organization in hyphae of *R. solani*.

IV. Experimental Study of Nutrient Translocation

Thrower and Thrower (1968) used a "double dish" method to study nutrient translocation in *R. solani*. When the fungus was inoculated in a central dish of nutrient-rich agar it grew completely across an outer dish of nutrient-free agar, translocating nutrients in the hyphae. Both $^{32}PO_4$ and ^{14}C-labelled metabolites, supplied in the inner dish, were carried to the colony margin in the outer dish. The transport of both types of label was much lower if the outer dish contained phosphate or sugars, and also lower if label was applied when the colony had reached the edge of the outer dish than if the hyphal tips were still growing. So nutrient translocation was an "adaptive" process, occurring maximally when needed. Consistent with this, Littlefield *et al.* (1967) showed that $^{32}PO_4$ supplied locally to *R. solani* colonies accumulated maximally in sclerotia, which are major nutrient sinks.

More recent studies with radio labelling have focused on bi-directional translocation, but they did not involve *Rhizoctonia*. For example, Olsson and Jennings (1991) showed that $^{32}PO_4$ and ^{14}C-labelled metabolites could be translocated simultaneously in opposite directions in the hyphae of *Rhizopus nigricans*. The rates

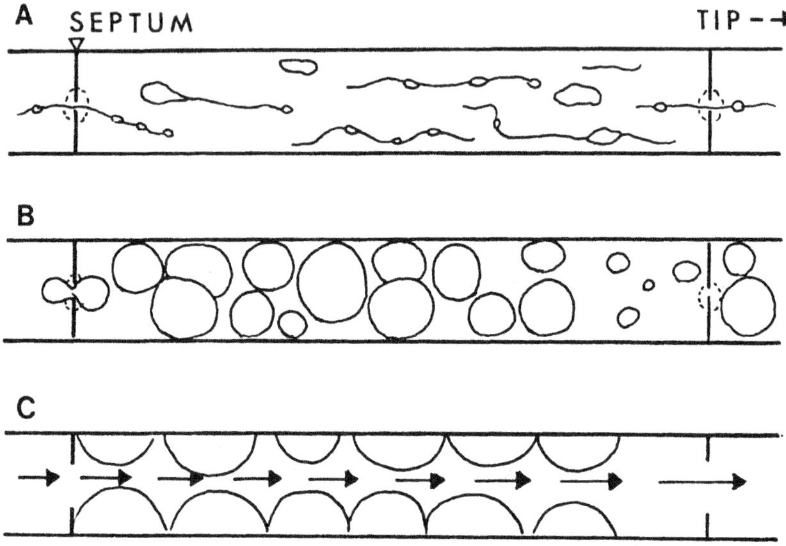

FIGURE 1. Progressive changes in the cytoplasmic organization of hyphae of *Rhizoctonia solani*, based on video microscopy of growing colonies. (A) Young hyphae and tip cells: slow, progressive protoplasmic streaming to the tips; vacuoles small, dynamic, frequently attached to tubules that extend and retract, with small swellings passing along the tubules and through the dolipore septa. (B) Older hyphal regions: more rapid protoplasmic streaming towards the tips; large, discrete vacuoles, less numerous towards the front of hyphal compartments; occasionally seen to constrict and pass through septa. (C) Oldest hyphal regions, or younger regions supplying a major nutrient sink: vacuoles occupy the periphery of the hypha and seldom move; rapid cytosolic flow (8-12 $\mu m\ sec^{-1}$) carries small particles through center of the hyphae and through septa which offer no resistance and seem to be partly dissolved.

of movement agreed closely with theoretical models for diffusion, so it was suggested that diffusion occurred both outside the hyphae (in the substratum) and within the hyphae. However, these workers emphasized the potential efficiency of diffusion as a translocation mechanism within hyphae, because it would be essentially one-dimensional (linear) rather than three-dimensional in an external environment. In further work, Olsson (1995) used narrow rectangular dishes filled to half their length with glucose agar and half with mineral nutrient agar. Fungi were inoculated at the mid-point and their growth densities towards each end of the dish were assessed by image analysis. Fig. 2 shows the mycelial density profile of *R. solani* in these conditions and in control (non-gradient) dishes of glucose-mineral agar. On the gradient dish there was a high mycelial density (similar to the control) when *R. solani* grew into the glucose-deficient zone, which was interpreted as evidence of active translocation of carbon. But the mycelial density was reduced progressively as *R. solani* grew into the mineral-deficient zone, suggesting that one or more critical mineral nutrients moved by diffusion. *Botrytis cinerea, Mucor hiemalis* and *Trichoderma koningii*, among others, showed similar profiles to *R. solani*. In contrast, *Geotrichum candidum* seemed to depend on diffusion of both carbon and mineral nutrients, *Mortierella echinulata* showed active translocation of minerals but diffusion of carbon, and several wood-rotting or litter-decomposing basidiomycetes showed active translocation of both carbon and minerals, which might be expected for fungi that grow in substrata of very low nitrogen content. In any case, the findings place *R. solani* within a category of translocation behavior. They indicate that *R. solani* can mobilize carbon actively from a food base across carbon-deficient environments, but that biomass production in carbon-rich zones could be limited if the fungus depends on import of mineral nutrients to these zones.

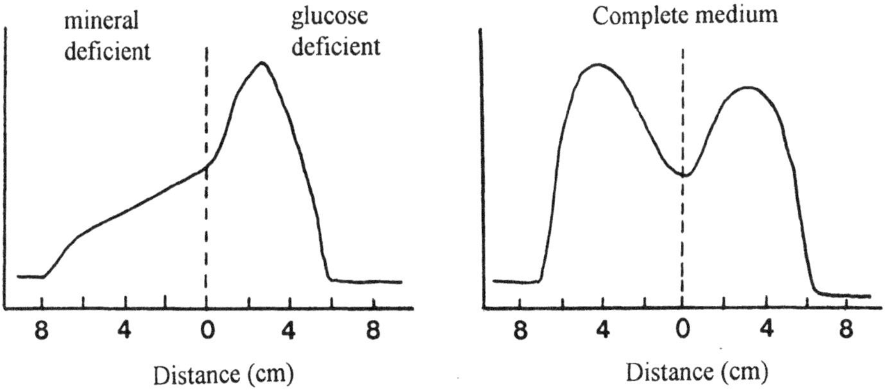

FIGURE 2. Mycelial density profiles of colonies of *Rhizoctonia solani* on agar, assessed by image analysis at different distances (cm) from an inoculation point (0); vertical axis is linear scale of pixel values. One colony was grown on a gradient strip of agar deficient in glucose on one side and deficient in mineral nutrients on the other side of the inoculation point. One colony was on a non-gradient strip containing both glucose and mineral nutrients. Redrawn and simplified from Olsson (1995).

V. Translocation Mechanisms

Many workers have suggested that the flow of cytoplasm to growing hyphal tips can serve for nutrient translocation. Consistent with this, Stewart and Deacon (1995) showed that several vital fluorochromes are absorbed by fungal spores or hyphae and then transported into new growth. However, cytoplasmic flow to hyphal tips would counteract diffusion in the opposite direction unless there is an alternative route for this. One possible route involves a motile tubular vacuole system, demonstrated recently for all the major taxonomic groups of mycelial fungi (Rees *et al.*, 1994). Hyphae were treated with carboxyfluorescein diacetate (CFDA), a membrane-permeable compound that is absorbed and accumulates rapidly in the vacuoles, where it is cleaved enzymatically to yield the fluorescent, membrane-impermeant compound, carboxyfluorescein (CF) which remains in the vacuolar system. When the hyphae were observed by fluorescence microscopy, CF was seen within a system of narrow vacuolar tubules that dilated and contracted as swellings passed along them in a peristaltic manner. The tubules also formed connections with larger vacuoles, apparently serving as a transport system. Movement occurred in all directions, independent of cytoplasmic streaming, and in the mycorrhizal basidiomycete *Pisolithus tinctorius* the tubules were shown to extend through the dolipore septa, apparently transporting fluorochrome through the septal pores (Shepherd *et al.*, 1993).

Polyphosphate is a common vacuolar component, usually seen as granules in electron micrographs. However, Orlovich and Ashford (1993) showed that these granules are an artifact of chemical fixation methods, which damage the vacuolar membrane, allowing calcium to diffuse in and precipitate phosphates. When hyphae of *Pisolithus* were prepared for electron microscopy by freeze-substitution, then the vacuoles did not contain calcium and the polyphosphate was in soluble form. Thus, it seems that hyphae have a specialized transport system that could serve in bi-directional translocation of phosphate and other substances that enter the vacuoles. It is difficult to show this for *Rhizoctonia* because the pigmentation of hyphae interferes with fluorescence. However, a dynamic tubular vacuole system exactly like that described for *Pisolithus* has been recorded by video microscopy of living non-stained hyphae of *R. solani* AG 3 infected by *V. biguttatum* (Fig. 1; J. W. Deacon, videotape demonstration, International Symposium on *Rhizoctonia*, July 1995).

VI. MOVEMENT OF ORGANELLES AND GENETIC ELEMENTS

A. Nuclear migration

Nuclei cannot pass through dolipore septa, so there is limited opportunity for nuclear exchange and migration in *Rhizoctonia*. However, Cubeta *et al.* (1993) obtained good evidence of nuclear migration when homokaryons of *R. solani* AG 4 were paired on agar and hyphal explants were taken from 4 cm behind the interaction zone after 15 days. Colonies obtained from the explants were tested for somatic compatibility with the pure homokaryons, and killing reactions were used as evidence of putative heterokaryons. This indicated that in 12 out of 30 successful homokaryon pairings the nuclei had migrated 4 cm back from the interaction zone in both directions; in a further 3 cases the nuclei had migrated in one direction. The use of RAPD markers seemed to confirm this, although the possible involvement of mitochondrial genes or growth of heterokaryotic hyphae into the pre-existing colonies could not be excluded. If nuclear migration did occur, as seems likely, it would have involved the partial dissolution of dolipore septa by β-glucanases, as reported in other basidiomycetes.

The two nuclear types in a heterokaryon often show faithful conjugate division so that there is a 1:1 ratio in the tip cells (Vilgalys and Cubeta, 1994). However, there are several reports of heterokaryon instability in *R. solani* AGs 1 and 4, leading to mosaic cultures of presumed heterokaryotic and homokaryotic regions (Anderson, 1982; Cubeta *et al.*, 1993). Further complexity has been revealed by excision and culturing of single hyphal tips from colonies of AG 4 (Fig 3; P. McCabe, D. Scott and J. W. Deacon, unpublished). From one strain a total of 78 hyphal tip explants yielded 8 daughter colonies that were somatically incompatible with the parent, and that showed various compatibility reactions with one another as evidenced by killing zones in pairing tests. It was not possible to regenerate a parent-compatible strain by pairing the daughter strains in any combination. Similar results were obtained by hyphal tip culture of a second AG 4 strain, but the frequency of parent-incompatible daughters was less than 1%. We could not obtain parent-incompatible strains by hyphal tip culture of *R. solani* AG 3 strains. The relevance of these findings to field conditions is unknown, but perhaps the fragmentation of inoculum or regrowth from a few surviving cells in a pocket of inoculum could lead to changes in the genetic structure of a population.

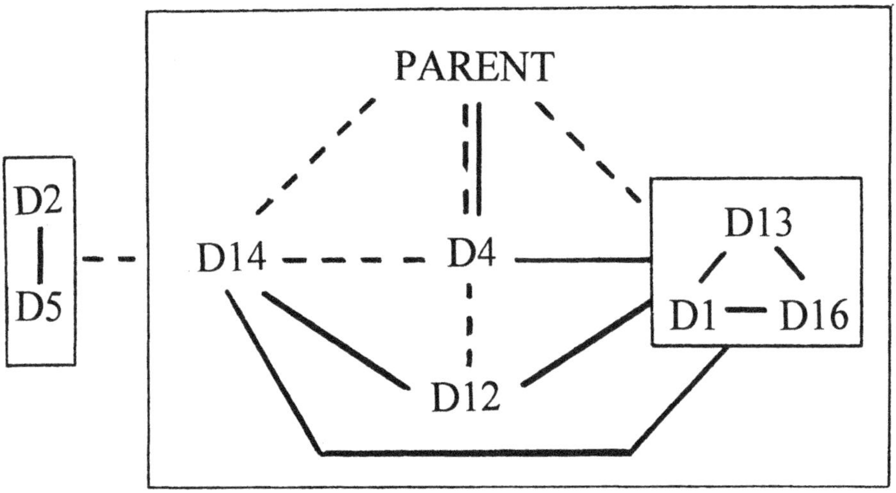

FIGURE 3. Somatic compatibility matrix, based on presence or absence of killing reactions in colony pairings, for a parental strain of *R. solani* AG 4 and daughter strains (D1, D2, etc.) derived from the parent by hyphal-tip subculture. Strains linked by a single solid line are somatically compatible; strains linked by a single broken line are incompatible; strains linked by double line (Parent and D4) are weakly compatible. Strains D1 and D2 are mutually compatible but incompatible with all other strains. (P. McCabe, D. Scott and J.W.Deacon, unpublished).

B. Transmission of double-stranded RNA and plasmids

Double-stranded RNA is common in field strains of *R. solani* (Zanzinger *et al.*, 1984) but its effects on virulence are still unclear (Castanho *et al.*, 1978; Finkler *et al.*, 1985; Kousik *et al.*, 1994). There is little direct evidence on the cellular location of dsRNA in *Rhizoctonia*, but *R. solani* can be "cured" of dsRNA by hyphal tip subculture approximately 1% curing frequency was reported by Finkler *et al.* (1985) indicating that dsRNA is not mobilized or maintained efficiently in the tips. This is compatible with electron microscopical evidence for many fungi, where virus-like particles are uncommon in tip cells. However, Newhouse *et al.* (1983) found them in the tips of *Cryphonectria parasitica* and suggested that this is related to the slow growth of hypovirulent (dsRNA-containing) strains of this pathogen.

Castanho *et al.* (1978) reported the transmission of a "diseased" phenotype of *R. solani* by hyphal anastomosis, and equated this with transmission of dsRNA. Finkler *et al.* (1985) also reported the transfer of dsRNA during hyphal anastomosis, but it was associated with transmission of virulence, although not confirmed to be conferred by the dsRNA. These workers had used two compatible strains of *R. solani* AG 4 with differential fungicide resistance (nuclear-encoded) and they subcultured hyphae from the interaction zone after 24h or 48h. Two out of 25 isolations after 48h yielded colonies with the nuclear marker and colony morphology of the "recipient" (dsRNA-free) strain, indicating that nuclei had not been transferred, but they had acquired dsRNA from the opposing strain; none of the 24h explants showed evidence of transmission. So, it seems that transmission of dsRNA is relatively slow but it can occur in the absence of nuclear transfer.

Linear DNA plasmids also are common in *R. solani* (Miyasaka *et al.*, 1990). In most fungi they are associated with mitochondria, but attempts to establish this for *R. solani* have been inconclusive. Unlike dsRNA, these plasmids seem to be maintained stably in tip cells of *R. solani*, because in this laboratory it was impossible to cure *R. solani* AG 4 of its linear DNA plasmid by repeated excision of hyphal tips, whereas the strain was readily cured of dsRNA by this method (P. McCabe and J. W. Deacon, unpublished). Much remains to be learned about the locations, mobilities and phenotypic effects of these extrachromosomal genetic elements in *Rhizoctonia*. From the comments earlier, it would seem fruitful to study their mobilization in response to mycoparasitic infection or to the development of sclerotia.

VII. Acknowledgments

I am grateful to Dr P. McCabe and Mr D. Scott for the supply of unpublished information.

VIII. References

Anderson NA (1982) The genetics and pathology of *Rhizoctonia solani*. Annu. Rev. Phytopathol. 20:329-344.

Blair ID (1943) Behaviour of the fungus *Rhizoctonia solani* Kühn in soil. Ann. appl. Biol. 30:118-127.

Burton RJ, Coley-Smith JR & Wareing PW (1988) *Rhizoctonia oryzae* and *R. solani* associated with barley stunt disease in the United Kingdom. Trans. Br. mycol. Soc. 91:409-417.

Castanho B, Butler EE & Shepherd RJ (1978) The association of double-stranded RNA with *Rhizoctonia* decline. Phytopathology 68:1515-1519.

Christias C & Lockwood JL (1973) Conservation of mycelial constituents in four sclerotium-forming fungi in nutrient-deprived conditions. Phytopathology 63:602-605.

Cubeta MA, Briones-Ortega R & Vilgalys R (1993) Reassessment of heterokaryon formation in *Rhizoctonia solani* anastomosis group 4. Mycologia 85:777-787.

Deacon JW & Scott DB (1985) *Rhizoctonia solani* associated with crater disease (stunting) of wheat in South Africa. Trans. Br. mycol. Soc. 85:319-327.

Dodman RL (1970) Factors affecting the prepenetration phase of infection by *Rhizoctonia solani*. In: Toussoun TA, Bega RV & Nelson PE (eds.) Root Diseases and Soil-borne Pathogens (pp 116-121) University of California Press, Berkeley.

Finkler A, Koltin Y, Barash I, Sneh B & Pozniak D (1985) Isolation of a virus from virulent strains of *Rhizoctonia solani*. J. gen. Virol. 66:1221-1232.

Garrett SD (1970) Pathogenic Root-infecting Fungi. Cambridge University Press, Cambridge.

Harrod MF (1975) Field behavior of light soils. In:(Anon) Soil Physical Conditions and Crop Production. M.A.F.F. Technical Bulletin 29. H.M.S.O., London.

Hayman DS (1970) The influence of cottonseed exudate on seedling infection by *Rhizoctonia solani*. In: Toussoun TA, Bega RV & Nelson PE (eds.) Root Diseases and Soil-borne Pathogens (pp 99-102) University of California Press, Berkeley.

Henis Y & Ben-Yephet Y (1970) Effect of propagule size of *Rhizoctonia solani* on saprophytic growth, infectivity, and virulence on bean seedlings. Phytopathology 60:1351-1356.

Henry FJ & Brown JS (1995) Control of Rhizoctonia root rot with different sowing systems. Abstracts, International Symposium on *Rhizoctonia*, Noordwijkerhout, Netherlands, 1995, P. 109.

Hynes HJ (1937) Studies on Rhizoctonia root-rot of wheat and oats. Sci. Bull. Dept. Agric. New South Wales, No. 58.

Kousik CS, Snow JP & Valverde RA (1994) Comparison of double-stranded RNA components and virulence among isolates of *Rhizoctonia solani* AG 1-IA and AG 1-IB. Phytopathology 84:44-49.

Littlefield LJ, Wilcoxson RD & Sudia TW (1965) Translocation of phosphorus-32 in *Rhizoctonia solani*. Phytopathology 55:536-542.

McDonald HJ & Rovira AD (1985) Development of inoculation technique for *Rhizoctonia solani* and its application to screening cereal cultivars for resistance. In: Parker CA, Rovira AD, Moore KJ & Wong PTW (eds.) Ecology and Management of Soilborne Plant Pathogens (pp 174-176). APS Press, St Paul. Minnesota

MacNish GC, Carling DE & Brainard KA (1993) Characterization of *Rhizoctonia solani* AG 8 from bare patches by pectic isozyme (zymogram) and anastomosis techniques. Phytopathology 83:922-927.

Miyasaka A, Chen CL & Hashiba T (1990) Detection and properties of plasmid-like DNA in isolates from nine anastomosis and intraspecific groups of *Rhizoctonia solani*. J. Gen. Microbiol. 136:1791-1798.

Naiki T (1985) Population and survival of sclerotia of *Rhizoctonia solani* in soil. In: Parker CA, Rovira AD, Moore KJ & Wong PTW (eds.) Ecology and Management of Soilborne Plant Pathogens (pp 51-53). APS Press, St Paul. Minnesota

Newhouse JR, Hoch HC & MacDonald WL (1983) The ultrastructure of *Endothia parasitica*. Comparison of a virulent with a hypovirulent isolate. Can. J. Bot. 61:389-399.

Olsson S (1995) Mycelial density profiles of fungi on heterogeneous media and interpretation in terms of nutrient reallocation patterns. Mycol. Res. 99:143-153.

Olsson S & Jennings DH (1991) Evidence for diffusion being the mechanism of translocation in the hyphae of three molds. Exper. Mycol. 15:302-309.

Orlovich DA & Ashford AE (1993) Polyphosphate granules are an artifact of specimen preparation in the ectomycorrhizal fungus *Pisolithus tinctorius*. Protoplasma 173:91-102.

Rees B, Shepherd VA & Ashford AE (1994) Presence of a motile tubular vacuole system in different phyla of fungi. Mycol. Res. 98:985-992.

Shepherd VA, Orlovich DA & Ashford AE (1993) Cell-to-cell transport via motile tubules in growing hyphae of a fungus. J. Cell Sci. 105:1173-1178.

Stewart A & Deacon JW (1995) Vital fluorochromes as tracers for fungal growth studies. Biotechnic and Histochem. 70:57-65.

Thornton CR, Dewey FM & Gilligan CA (1993) Development of monoclonal antibody-based assays for the detection of live propagules of *Rhizoctonia solani* in soil. Plant Pathol. 42:763-773.

Thrower KB & Thrower SL (1968) Movement of nutrients in fungi. I. The mycelium. Aust. J. Bot. 16:71-80.

Van den Boogert PHFJ & Deacon JW (1994) Biotrophic mycoparasitism by *Verticillium biguttatum* on *Rhizoctonia solani*. Eur. J. Plant Pathol. 100:137-156.

Vilgalys R & Cubeta MA (1994) Molecular systematics and population biology of *Rhizoctonia*. Annu. Rev. Phytopathol. 32:135-155.

Zanzinger DH, Bandy BP & Tavantzis SM (1984) High frequency of finding double-stranded RNA in naturally occurring strains of *Rhizoctonia solani*. J. Gen. Virol. 65:1601-1605.

II.3 EXTRACHROMOSOMAL ELEMENTS AND DEGREE OF PATHOGENICITY IN *RHIZOCTONIA SOLANI*

VICTOR RUBIO[1] STELLOS M. TAVANTZIS[2]
AND DILIP K. LAKSHMAN[2]

[1]*Centro Nacional de Biotecnologia (CSIC-UAM). Campus Cantoblanco, Universidad Autonoma de Madrid. 28049 Madrid, SPAIN.*
[2]*University of Maine, Plant Biology and Pathology Department, Orono, Maine 04469, USA*

I. INTRODUCTION

The advent of molecular biology has played a pivotal role in advancing our understanding of the genetics and biology of filamentous fungi such as *Neurospora crassa* and *Aspergillus nidulans*. There exists a great need for similar advances in the molecular genetics of other fungi, especially those causing economically important plant diseases. *Rhizoctonia solani* Kühn [teleomorph, *Thanatephorus cucumeris* (Frank) Donk], is an important soilborne pathogen attacking numerous crops, worldwide. Understanding the genetic structure of field soil populations, the functions of the genes involved in virulence expression and genetic recombination is necessary for the design of Rhizoctonia disease management that will be both effective and environmentally sound.

The genetics of *R. solani* is very complex. It is accepted today that *R. solani* is a genetically heterogeneous species with many genetically isolated subgroups (Sneh *et al.*, 1991; Carling *et al.*, 1994; see chapter I.A2).

To understand the pathogenic behavior of *R. solani*, it is important to study the genetic factors involved in the expression of virulence in this organism. Although knowledge regarding the genetics of *R. solani* is essentially non-existent, several research groups have been studying extrachromosomal genetic elements such as encapsidated or unencapsidated double-stranded RNA (dsRNA) (Castanho *et al.*, 1978; Zanzinger *et al.*, 1984; Finkler *et al.*, 1985; Hyakumachi *et al.*, 1985; Kousik *et al.*, 1994) and DNA plasmids (Hashiba *et al.*, 1984; Miyasaka *et al.*, 1990; Miyashita *et al.*, 1990; Hongo *et al.*, 1994; Jabaji-Hare *et al.*, 1994). In this chapter, we present an overview on the nature of these elements, and their potential role(s) in regulating virulence (e.g., degree of pathogenicity) in *R. solani*.

II. EXTRACHROMOSOMAL ELEMENTS IN FUNGI

Linear DNA plasmids having polypeptides covalently bound to their 5'-ends or hairpin loops at both ends constitute a group of extrachromosomal elements called invertrons (Sakaguchi, 1990). They are found in fungi, usually in mitochondria, and bacteria such as *Streptomyces spp*. Invertron genome organization is similar to that of viruses, such as adenovirus or bacteriophage ϕ29 (Sakaguchi, 1990).

Knowledge concerning plasmids found in filamentous fungi is not as advanced as that on prokaryotic plasmids, particularly at the functional level. Plasmids in fungi are found mainly in mitochondria. Some of these plasmids share common sequences with

B. Sneh et al. (eds.),
Rhizoctonia Species: Taxonomy, Molecular Biology, Ecology, Pathology and Disease Constrol, 127–138.

the mitochondrial genome, while others are not related to mitochondrial DNA (mtDNA). The latter group has been described in different species of the genera *Agaricus, Ceratocystis, Claviceps, Fusarium, Gaeumannomyces* and *Neurospora* (for review see Samac and Leong 1989).

Double-stranded RNAs (dsRNA) comprise the other class of fungal extrachromosomal elements. Most dsRNAs are presumed to be of viral nature (mycoviruses), since they have RNA-dependent RNA polymerase (RDRP) activity associated with the encapsidated virions (Finkler *et al.*, 1988; Tavantzis and Bandy, 1988), or their sequence contains an RDRP-related open reading frame (ORF) (Shapira *et al.*, 1991a; Lakshman and Tavantzis, 1994, 1995). It is quite possible, however, that some dsRNA elements are not viral, in the traditional sense of the term, since they carry essential genes involved in cellular maintenance functions (Jiang, Lakshman, and Tavantzis, unpublished).

Mycoviruses, like reoviruses, may possess more than one dsRNA component. If the largest segment contains all the essential genetic information for virion replication, the respective virus is grouped as a Totivirus. Mycoviruses in which the capsid protein ORF is located on a dsRNA segment different from that of the RNA polymerase-containing segment are classified as Partiviruses (Ghabrial, 1994); the smaller segments are considered satellite dsRNAs, even if they are encapsidated. In one particular case, a virus from *Cryphonectria parasitica,* due to the similarity in genetic organization and expression strategy to other plant viruses and the fact that it is non-encapsidated, a new family with a single genus, Hypovirus, was approved (Ghabrial, 1994).

Infection of fungal spheroplasts with cell-free mycovirus preparations has been an extremely difficult task (Ghabrial, 1994). It is not clear whether the problem lies in the fact that the wrong type of inoculum has been employed, e.g., encapsidated dsRNA as opposed to single-stranded RNA, or virion infectivity is inactivated upon purification. In *C. parasitica* it has been possible to routinely infect spheroplasts using an *in vitro* synthesized transcript of a full-length cDNA clone of a hypovirulence-associated dsRNA Hypovirus (Chen *et al*, 1994).

The expression of dsRNA molecules in filamentous fungi is well documented at the molecular level mainly in two cases: The production of killer toxins in *Ustilago maydis* (Koltin and Day, 1976; Koltin, 1989), and the dsRNA associated to hypovirulence in *C. parasitica* (Nuss, 1992).

The coding capacity of dsRNA found in a Hypovirus was documented first in the case of the *C. parasitica* (Shapira *et al.*, 1991a). One of the encoded polypeptides is a cysteine-type proteinase, phylogenetically related to the single-stranded RNA Potyvirus-encoded proteinase HC-Pro. Two ORF's encode two autocatalytic polyproteins. These proteins are found only in hypovirulent strains containing the Hypovirus dsRNA (Choi *et al.*, 1991; Hillman *et al.*, 1989; Shapira *et al.*, 1991b). The development of a DNA-mediated transformation system, in *C. parasitica* (Hillman *et al.*, 1989, Churchill *et al.*, 1990) has been used to demonstrate the role of the Hypovirus in causing hypovirulence in *C. parasitica* (for review see Nuss, 1992). Furthermore, *in vitro* synthesized, full-length transcript of the Hypovirus L-dsRNA was introduced by electroporation into other virulent, dsRNA-free *Cryphonectria* and *Endothia* species. Transformants were converted to hypovirulent, thus expanding the potential uses of dsRNA-mediated hypovirulence to other species and genera (Chen *et al.*, 1994).

Ongoing work has shown that a heterotrimeric GTP-binding protein is down-regulated in hypovirulent strains of *C. parasitica.* This observation is in accordance with data obtained by using cultures transfected with cloned forms of the Hypovirus (transgenic co-suppression) (Choi *et al.*, 1995). Differential display analysis reveals that many genes are regulated in an identical manner in both Hypovirus-infected and transgenic CPG-1 suppressed strains (*C. parasitica* G-1 protein (CPG-1), is the product of the gene *cpg-1*, and it is a G-protein α subunit). The work of Choi *et al.*, 1995 is the first report showing the involvement of a G protein-linked signal transduction pathway in fungal pathogenesis, and suggesting a molecular basis for virus-mediated hypovirulence. Heterotrimeric guanine nucleotide-binding proteins (G proteins) are a family of regulatory proteins that play an essential role in the response of eukaryotic cells to a variety of environmental stimuli (Gilman, 1987).

Other plant pathogenic filamentous fungi in which dsRNA may be associated with up- or down- regulation of virulence are *Ophiostoma ulmi* (Buisman) Nannf. (Rogers *et al.*, 1987), *Gaeumannomyces graminis var. tritici* Walker (Lapierre *et al.*, 1970; Lemaire *et al.*, 1970), *Helminthosporium victoriae* Meehan & Murphy (Ghabrial *et al.*, 1979), and *Rhizoctonia solani* Kühn (Castanho and Butler, 1978a; Finkler *et al.*, 1985; Barathan and Tavantzis, 1990; Patiño *et al.*, 1992; Lakshman and Tavantzis, 1994).

III. EXTRACHROMOSOMAL ELEMENTS IN *R. SOLANI*

A. DNA Plasmids of *R. solani*

A linear DNA 2.7kb plasmid, designated pRS64, found in isolates of *R. solani* anastomosis group 4 (AG 4) was described by Hashiba *et al.* (1984). They reported that isolates harboring the plasmid were hypovirulent, whereas virulent isolates had no detectable plasmid. Subsequently, it was shown that the 2.7kb band of pRS64 was comprised of three different plasmids of the same size, PRS64-1, -2, and -3, which all possessed hairpin loops and homologous regions at both termini (Miyashita *et al.*, 1990). When a larger number of isolates was examined, it was apparent that the correlation between plasmid and hypovirulence or with Rhizoctonia decline was not valid (Miyashita *et al.*, 1990). A 0.5kb poly(A)$^-$ RNA, cross-hybridizing with pRS64, was found in cultures possessing the plasmids but not in plasmidless cultures. The cDNA from the above RNA has a 100%, 73% and 84% homology to pRS64-1, -2 and -3, respectively. The cDNA clone has a putative 68 amino acid ORF (ORF1-1). Antiserum raised against the product of ORF1-1, produced in *E. coli*, gave a positive reaction with mycelial extracts from cultures containing the plasmids but not with those from plasmidless cultures (Hongo *et al.*, 1994). Furthermore, it appears that plasmid-hybridizing DNA sequences occur on nuclear DNA of AG 4 isolate (RI-64), from which the 3 plasmids originated (Wako *et al.*, 1991). Radiolabeled probes, specific to pRS64-1, -2, and -3, respectively, hybridized to a 1.1Mb chromosomal DNA band separated from other chromosomal and smaller size DNA bands by pulse field electrophoresis.

In another study forty eight of 114 field isolates of *R. solani* were found to possess plasmids in a study that included cultures from 9 AGs (Miyasaka *et al.*, 1990). The size of these plasmids varied from 2.2kb to 7.0kb. All 8 of the 18 AG 4 isolates tested possessed a 2.7kb plasmid, whereas all 4 of the 16 AG 5 isolates examined had a

3.2kb plasmid. Cultures belonging to other AGs had plasmids of varying sizes. A high degree of genetic similarity was found among plasmids from isolates within the same AG, but no relationship was detected among plasmids from isolates belonging to different AGs.

A second study, conducted in Canada, included 54 isolates from 5 AGs (1 to 5), and revealed a 2.4kb plasmid (pRS 104) found in AG 4 isolates, and a 3.6kb plasmid (pRS 188) found in AG 5 isolates (Jabaji-Hare *et al.*, 1994). Both plasmids were partially sequenced. In addition, this work showed that: a) the two plasmids have the same structure as that of the plasmids described in the first study (Miyashita *et al.*, 1990); b) there is no correlation between plasmid presence with virulence alteration; c) plasmids are associated with mitochondria in cellular fractionation experiments; d) no significant sequence homology occurs between the two *R. solani* plasmids and sequences available in data bases; and e) no hybridization was detected between plasmid and nuclear or mitochondrial DNA. This finding contradicts the results of the above mentioned Japanese study on AG 4 (isolate RI-64) (Wako *et al.*, 1991).

A third study of DNA plasmids in *R. solani* was carried out by the senior author's group in Spain. The survey included 15 AG 4 isolates, 13 from Spain and 2 from Israel. Thirteen of these isolates were virulent and possessed a 2.6kb DNA plasmid, and 2 were hypovirulent and contained no plasmids (Rubio *et al.*, unpublished). Plasmid bands from isolate Me 8-2, obtained from *Cucumis melo* in Southern Spain (CECT 20171), were gel-purified, treated with S1 nuclease to disrupt potential secondary structures at the termini, and cloned into pUC19. The restriction maps of the inserts confirmed that they corresponded to the 2.6-plasmid band. The data suggested the presence of two plasmids, pMe 8-2.J1 (2.65Kb), and pMe 8-2.J2 (2.55Kb), the latter being more abundant than the former. The above plasmids have different but related restriction maps (Fig. 1). They also differ from the three pRS64 plasmids found in AG 4 isolates from Japan (Miyashita *et al.*, 1990): enzymes that restrict the pRS64 plasmids (*Bam*HI, *Kpn*I and *Bgl*II) do not digest pMe 8-2.J1 or pMe 8-2.J2. Furthermore, an oligonucleotide probe, representing a conserved sequence from the termini of the pRS64 plasmids did not hybridize with either pMe 8-2 plasmid (Rubio *et al.*, unpublished). Sequence analysis revealed that whereas the two pME 8-2 plasmids are identical at their termini, they possess unique middle regions. Moreover, the sequences of these plasmids did not align with the sequences available in GenBank.

Southern blot hybridization showed that unlike the pRS64 plasmids, the pME 8-2 plasmids do not share common nucleotide sequences with genomic DNA of *R. solani*. However, they have considerable sequence homology with plasmids of similar size from isolates of the same AG, but not with those from different AGs stressing the fact that AGs are genetically distinct. So, as expected, there was no sequence homology between the pMe 8-2 plasmids (AG 4) and the pRS 188 plasmids (AG 5, Canada). These findings are in agreement with previously published results obtained with Japanese isolates (Miyasaka *et al.*, 1990).

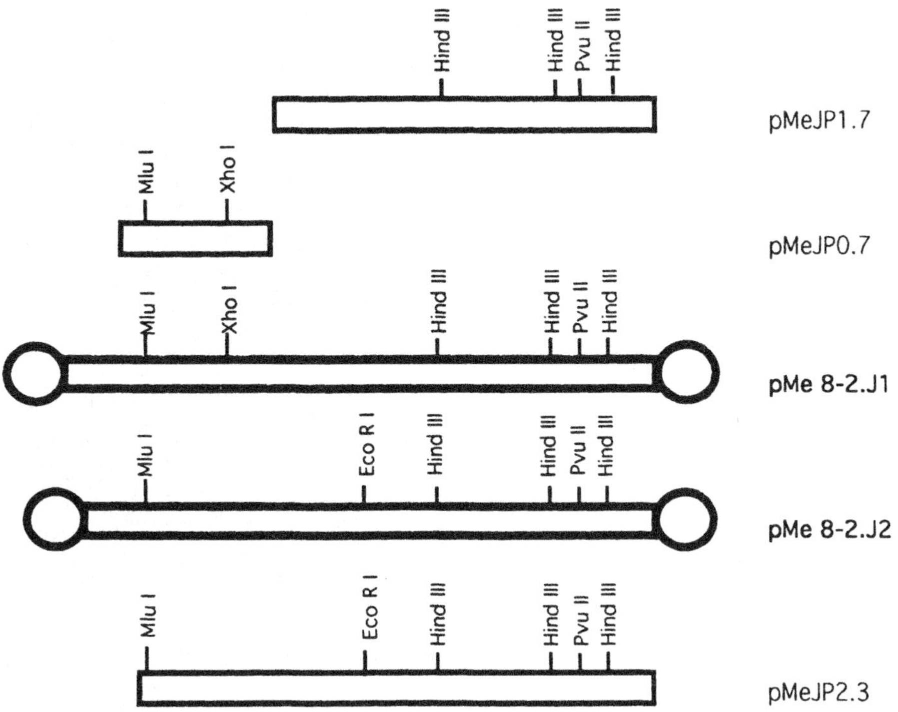

FIGURE 1. Two plasmids of about the same size, in strain Me 8-2 (pMe 8-2.J1 and pMe 8-2.J2) *R. solani* AG 4. pMeJP1.7, pMeJP0.7 and pMeJP2.3, represent tree pUC19 cloned segments of the plasmids after S-1 nuclease treatment. 1.7, 0.7 and 2.3 on the names of the clones represent the sizes in kb respectively. This clones have been completely sequenced. pMe 8-2.J1 has a unique *Xho*I site not present in pMe 8-2.J2, while pMe 8-2.J2 contains a unique *Eco*RI site not observed in pMe 8-2.J1. Other sites *Hind* III, *Pvu*II and *Mlu*I are equivalent in both plasmids. pMe 8-2.J1 is 100bp longer than pMe 8-2.J2 in the 5' end (top part in the figure). The circles at the end of the two plasmids represent areas of strong secondary structure.

B. dsRNA elements

1. Biological roles: a complex system. Castanho *et al.* (1978), reported on a cytoplasmically controlled degenerative disease of *R. solani* referred to as Rhizoctonia decline. This condition was characterized by a reduced rate of growth and sclerotia production, loss of mycelial pigmentation, presence of double-stranded (dsRNA) and hypovirulence similar to that of *C. parasitica*. Upon hyphal-tip isolation, a healthy, virulent culture (189HT5), lacking detectable dsRNA, was recovered from a diseased culture (189a). Hypovirulence and dsRNA were co-transmitted from 189a to 189HT5.

The cured isolate (189HT5) caused severe damping-off of sugar beet seedlings. In treatments including both 189HT5 and the hypovirulent 189a, no significant damping-off occurred, suggesting biological control (Castanho and Butler, 1978b). Furthermore, of the 16 isolates tested, 13 lacked dsRNA and were virulent, whereas the remaining 3 contained dsRNA and were hypovirulent. These results motivated the initiation of three independent projects in Maine, USA (Zanzinger *et al.*, 1984), Japan

(Hyakumachi *et al.*, 1985) and Israel (Finkler *et al.*, 1985). The findings of the first two studies were in agreement: dsRNA was ubiquitous in natural populations of *R. solani*, and there was no apparent correlation between the mere presence of dsRNA and degree of pathogenicity. A recent report by Kousik and coworkers, carried out in Louisiana, USA (1994) confirmed these findings. The above studies included *R. solani* isolates from five AGs (1-5), and several host plant species per AG. In some cases (e.g., AG 3), however, only a single host species is known. In contrast, the results of the third study (Finkler *et al.*, 1985) suggested that dsRNA was associated with virulence. Recent work in Spain, also suggests that dsRNA is associated with virulence (Patiño *et al.*, 1992). Of the 15 AG 4 isolates used, 13 from Spain and 2 from Israel, 13 possessed dsRNA and were virulent and 2 lacked detectable dsRNA and were nonpathogenic. The degree of pathogenicity of these isolates was assessed on 11 different hosts. Thus, indirect experimental evidence suggests that *R. solani* cultures covering a wide spectrum of virulence (nonpathogenic to highly virulent) possess dsRNA. We believe that the controversy as to whether dsRNA is associated with high degree of virulence or hypovirulence in *R. solani* is due to the genetic diversity of dsRNA occurring in natural populations of *R. solani*, the complex genetics of this fungus (comprised of at least 12 genetically isolated AGs) and the host genotype. For a detailed recent review the reader is referred to Tavantzis and Lakshman (1995). Bharathan and Tavantzis (1990, 1991) addressed the question of genetic variability of dsRNAs found in *R. solani*. Fifty one isolates from AGs 1 to 5 were included in the study that entailed analysis of over 45 Northern blots. The data showed that there exists a relatively high degree of genetic heterogeneity among dsRNAs from the same isolate. These results were expected, since fungal dsRNA viruses are multipartite (Buck, 1986). More importantly, a high degree of genetic heterogeneity exists among dsRNAs from cultures of the same AG, and of the same or different geographic location (Baranthan and Tavantzis, 1990; 1991). There was no cross-hybridization among dsRNAs from isolates of different AGs. This is consistent with the fact that the different AGs are genetically isolated. In one case, however, strong cross-hybridization occurred among 3 dsRNAs of varying sizes from 3 isolates belonging to three different AGs (Baranthan and Tavantzis, 1990). The 3 cultures originated from distant locations and different crops. More importantly, all three are hypovirulent. Actually, they are the only naturally occurring hypovirulent cultures among several dozen isolates tested by the Tavantzis group over a period of several years. These facts suggest that the dsRNA sequence shared by the above 3 isolates might be responsible for their hypovirulent condition, and might originate from an ancestral genotype that gave rise to the three respective AGs (AGs 2, 3, and 4). A detailed overview of several lines of indirect evidence suggesting that particular dsRNA segments are associated with virulence or hypovirulence is presented in Tavantzis and Lakshman (1995). The findings on the genetic diversity of dsRNAs provide at least a partial explanation for the conflicting results on the association of dsRNA with virulence (or hypovirulence) in *R. solani*. At least three different AGs were used in the three major studies that examined whether or not such a correlation exists (Castanho and Butler, 1978a; Zanzinger *et al.*, 1984; Finkler *et al.*, 1985; Lakshman and Tavantzis, 1994).

2. Characterization of dsRNA. Some of the dsRNAs found in *R. solani* are components of encapsidated icosahedral virus particles (Finkler *et al.*, 1985; Tavantzis and Bandy, 1988) or occur unencapsidated (Finkler *et al.*, 1988; Tavantzis *et al*, unpublished). Tavantzis and Bandy (1988) characterized the RDRP properties from a

33nm virus found in the AG 2 isolate Rhs 717, and showed that the two dsRNA genomic components of the virus encode two major translation products of 71kDa and 77kDa. Both polypeptides were specifically immuno-precipitated by antibodies to purified virions. The two genomic dsRNA components of the virus Rhs 717 have been sequenced to completion (E Strauss, *et al.*, unpublished).

Current work is focusing on generating knowledge which will allow us to formulate meaningful questions to be addressed, at least in part, by the use of a transformation system. The Tavantzis group has been working on the following genetic model, which is expected to provide some answers concerning the origin, nature and biological functions of the dsRNAs under study. Three hypovirulent cultures, Rhs 1A1, Rhs 1A2, and Rhs 1A3, originated as spontaneous sectors of the virulent AG 3 culture Rhs 1AP, which was isolated from a field-grown potato tuber. Rhs 1AP has 2 dsRNAs of 23kb and 6.5kb (Fig. 2). The hypovirulent Rhs 1A1

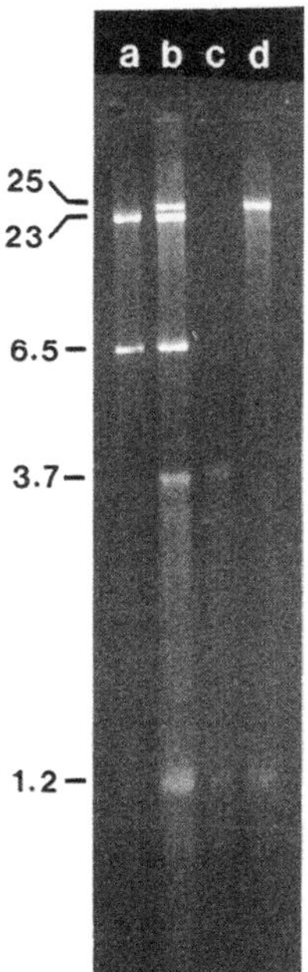

FIGURE 2. (left) Fractionation of double-stranded RNAs (dsRNA) from AG 3 strains Rhs 1AP (lane a), Rhs 1A1 (lane b), Rhs 1A2 (lane c), and Rhs 1A3 (lane d) on an 1% agarose gel stained with ethidium bromide. Numbers on the left indicate approximate sizes (kb) of the respective dsRNA elements (explanation on the text) (From Lakshman and Tavantzis, 1994).

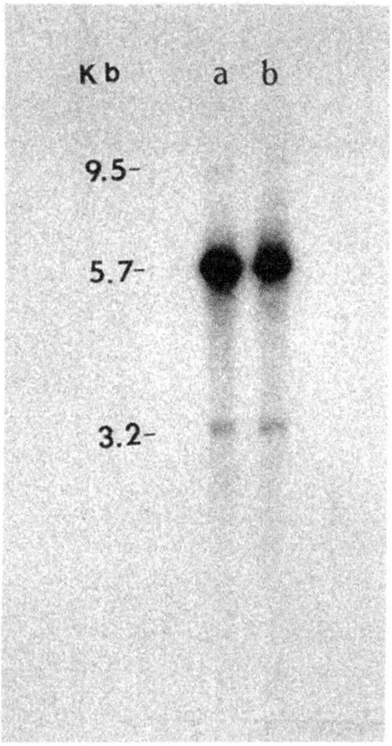

FIGURE 3. (right) Southern blot of total (a) and nuclear (b) DNA extracted from AG 3 isolate Rhs 1A1. DNA was digested with *Bam*H1 and hybridized with ^{32}P-labelled cDNA probe specific to M_2 dsRNA. The blots demonstrate that dsRNA sequences are present in very small amounts as DNA copies. Size estimation of the major band (5.7 Kb) and two weak bands (9.5kb and 3.2Kb) was based on parallel migration of the 1Kb ladder DNA (BRL, Gaithersburg, MD, USA).

contains the above 2 plus 3 novel dsRNAs of 25 kb, 3.6 kb, and 1.2 kb. The 5 dsRNAs are genetically unrelated (Lakshman and Tavantzis, 1994). Rhs 1A2 possesses the 3.6-kb and 1.2-kb, while Rhs 1A3 has the 25-kb and 1.2-kb dsRNAs. The phenotype, pathotype, and dsRNA content of the 4 isolates have remained unchanged for several years. Thus, the onset of hypovirulence in the 3 spontaneous mutant cultures coincided with the appearance of new, or loss of existing, dsRNAs. Polymerase chain reaction (PCR) experiments, and Southern blot hybridization analyses (Fig. 3) have shown that, although not detectable by ethidium bromide staining or Northern blot hybridization, sequences representing these dsRNAs are present in very small amounts in all 4 cultures either in DNA or RNA form (Lakshman and Tavantzis, 1994, 1995). To our knowledge, this is the first account of fungal dsRNA having DNA copies.

cDNA libraries for the above dsRNA elements have been prepared (Lakshman and Tavantzis, 1994), and the 3.6-kb dsRNA (M2) has been sequenced to completion (Lakshman and Tavantzis, 1995). One of the open reading frames (ORF) of M2 has all 4 motifs of a dsRNA viral RNA-dependent RNA polymerase (RDRP) and is phylogenetically related to the RDRP of a hypovirulence-associated, mitochondrial dsRNA from *Cryphonectria parasitica,* which, in turn, is related to the T and W dsRNAs from yeast (Polashock and Hillman, 1994) (Fig. 4). The 6.5-kb (M1) and 23-

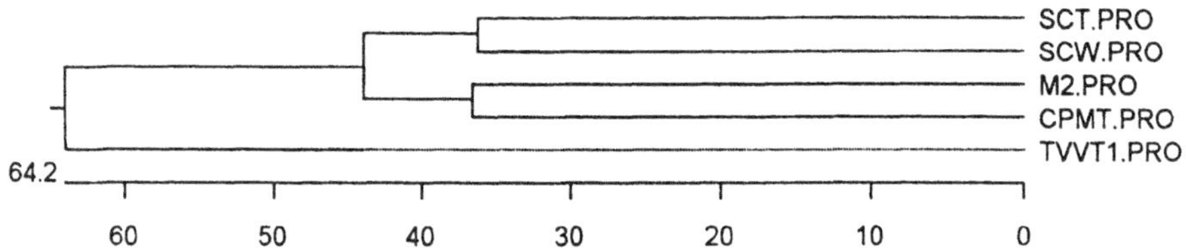

FIGURE 4. Phylogenetic distance dendrogram of the putative RNA-dependent RNA polymerase (RDRP) of four fungal dsRNAs namely, *Saccharomyces cerevisiae* T (SCT.PRO), and W (SCW.PRO) dsRNAs, M_2 dsRNA of *Rhizoctonia solani* AG 3 isolate Rhs 1A1 (M_2.PRO), and the *Cryphonectria parasitica* hypovirulence-associated mitochondrial dsRNA NB631 (CPMT.PRO). The M_2.PRO RDRP is a putative fusion of two ORFs at a potential ribosomal frameshift site in the M_2 sequence. The putative RDRP of a *Trichomonas vaginalis* virus TVv (TVVT1.PRO) was included as an outgroup in this analysis. The analysis shows that M_2 dsRNA is more closely related to NB631 dsRNA than the other two *S. cerevisiae* dsRNAs. Phylogenetic analysis is based on Clustal V method with a PAM250 residue weight table included in the 1994 version of the LASERGENE package (DNASTAR Inc., Madison, WI, USA).

kb (L1) dsRNAs also possess RDRP genes (Tavantzis *et al.* unpublished). The L1 RDRP is phylogenetically related to RDRP genes from the rice enigmatic dsRNA and a number of plant viruses (Lakshman and Tavantzis, 1995).

3. Biocontrol of *R. solani* using hypovirulent isolates. The interest generated by cytoplasmically-transmitted hypovirulence in fungi is due to the hope that the phenomenon can be adopted and used effectively under commercial conditions in a sustainable agricultural scheme. Thus, in addition to the biocontrol study conducted by Castanho and Butler (1978b), Bandy and Tavantzis (1990) carried out greenhouse and field experiments to determine if Rhs 1A1 was capable of providing biocontrol against other AG 3 isolates. Potato seedpieces (tubers) were disinfected with formaldehyde to eliminate interference by tuber-borne sclerotia of the fungus. In plants inoculated with both Rhs 1A1 and virulent isolate Rhs 27, the area of infected stem tissue was significantly (P=0.05) reduced by 56% as compared with that of plots inoculated with the virulent Rhs 27 alone (Bandy and Tavantzis, 1990). Although the mechanism responsible for the reduced disease severity is unknown, it is hoped that upon optimization, cytoplasmic hypovirulence could become an effective Rhizoctonia disease management tool in a futuristic, environment-friendly world.

Interestingly, potato plants inoculated with Rhs 1A1 alone exhibited a greater (P=0.05) growth response, expressed as a 4-fold increase in the dry weight of stolons (yield predictor), an 1.7-fold increase in the dry weight of stems (samples included foliage), and reached full bloom, which coincides with onset of tuberization, one week earlier than the respective control plants (Tavantzis, 1988; Bandy and Tavantzis, 1990).

The increased growth response of potato to the hypovirulent isolate Rhs 1A1 appeared to be induced by a phytohormone. *In vitro* cultures of *R. solani* produce phenylacetic acid (PAA) which is capable of causing the same disease symptoms on potato as the pathogen itself (Frank and Francis, 1976). PAA application mimics the Rhizoctonia disease; symptoms develop in the absence of physical contact between the host and the pathogen (Boosalis, 1950; Sherwood and Lindberg, 1962; Wyllie, 1962; Orellana and Mandava, 1983). PAA and its metabolites from plant sources and *R. solani* act as plant growth regulators (auxins) at low concentrations, whereas at high concentrations they act as toxins (Chamberlain and Wain, 1971; Milborrow *et al.*, 1975; Frank and Francis, 1976). We have shown that the amount of PAA (μg/g dry weight of mycelium) produced by the hypovirulent isolate Rhs 1A1 was lower than that produced by virulent AG 3 isolates by a factor of 10 (Tavantzis, 1988; Tavantzis *et al.*, 1989; Tavantzis and Lakshman, 1995).

IV. CURRENT WORK AND FUTURE OUTLOOK

The experimental evidence reviewed in previous sections of this chapter may suggest, but does not unequivocally prove, that dsRNA elements have important functions in *R. solani*. The most direct experimental approach for attributing specific functions to particular dsRNA elements is transfection of selected *R. solani* cultures with cDNA, single-stranded RNA or viral forms of the respective dsRNA element. The Rubio, Tavantzis and Keijer groups, have began a collaborating effort to develop a transformation system for *R. solani* that will allow the scientific community to address important questions concerning the roles of dsRNA and other genetic factors in this complex organism.

At the present time, in order to obtain further experimental evidence as to the function of the above 5 dsRNAs, Tavantzis group constructed a series of isogenic lines which differ by a single element [e.g., the 6.5-kb (M1) or the 3.6-kb dsRNA (M2)]. Pathogenicity data suggest that the 6.5-kb dsRNA enhances virulence, whereas the 3.6-dsRNA moderates virulence (Jiang *et al.*, 1995; ibid, unpublished). These isogenic lines are being analyzed to determine whether or not genetic changes other than the replication of the particular dsRNA are responsible for the shift in phenotype (and pathotype).

V. ACKNOWLEDGMENTS

This work was supported by grants BIO93 0923 C02 02 from the Commission Interministerial de Ciencia y Tecnologia (CICYT), Spain and BIOT/CT91030 and BIO2/CT94-3011 from the European Union to VR and grants 93-37311-9444 and 95-37303-2410 from the US Dept. of Agriculture, OGPS-NRI to SMT. We acknowledge Donald Nuss, Gerard Adams, Suha Jabaji-Hare and Baruch Sneh for critical reading of the manuscript and valuable suggestion for this review.

VI. REFERENCES

Bandy BP & Tavantzis SM (1990) Effect of hypovirulent *Rhizoctonia solani* on Rhizoctonia disease, growth, and development of potato. Am. Potato J. 67:189-199.

Bharathan N & Tavantzis SM (1990) Genetic diversity of double-stranded RNAs in *Rhizoctonia solani*. Phytopathology 80:631-635.

Bharathan N & Tavantzis SM (1991) Assessment of genetic relatedness among double-stranded RNA components from *Rhizoctonia solani* isolates of diverse geographic origin. Phytopathology 81: 411-415.

Boosalis MG (1950) Studies on the parasitism of *Rhizoctonia solani* Kühn on soybeans. Phytopathology 40:820-831.

Buck KW (1986) Fungal virology- an overview. In: Buck KW (ed.) Fungal Virology, (pp 1-84) CRC Press, Boca Raton, FL

Carling DE, Rothrock CS, MacNish GC, Sweetingham MW, Brainard KA & Winters SW (1994) Characterization of anastomosis group 11 (AG 11) of *Rhizoctonia solani*. Phytopathology 84:1387-1393.

Castanho B & Butler EE (1978a) Rhizoctonia decline: A degenerative disease of *Rhizoctonia solani*. Phytopathology 68: 1505-1510.

Castanho B & Butler EE (1978b) Rhizoctonia decline: studies on hypovirulence and potential use in biological control. Phytopathology 68:1511-1514.

Castanho B, Butler EE & Shepherd RJ (1978) The association of double-stranded RNA with Rhizoctonia decline. Phytopathology 68: 1515-1518.

Chamberlain VK & Wain RL (1971) Studies on plant growth-regulating substances-the influence of ring substituents on the plant growth regulating activity of phenylacetic acid. Ann. Appl. Biol. 69:65-72.

Chen B, Choi GH & Nuss DL (1994) Attenuation of fungal virulence by synthetic infectious Hypovirus transcripts. Science 264:1762-1764.

Choi GH, Shapira R & Nuss DL (1991) Cotranslational autoproteolysis involved in gene expression from a double-stranded RNA genetic element associated with hypovirulence of the chestnut blight fungus. Proc. Natl. Acad. Sci. USA 88:1167-1171.

Choi GH, Chen B & Nuss DL (1995) Virus-mediated or transgenic suppression of a G-protein **a** subunit and attenuation of fungal virulence. Proc. Natl. Acad. Sci. USA 92:305-309.

Churchill ACL, Ciuffetti LM, Hansen DR, Van Etten HD & Van Alfen NK (1990) Transformation of the fungal pathogen *Cryphonectria parasitica* with a variety of heterologous plasmids. Curr. Genet. 17:25-31.

Finkler A, Koltin Y, Barash I, Sneh B & Pozniak D (1985) Isolation of a virus from virulent strains of *Rhizoctonia solani*. J. Gen. Virol. 66:1221-1232.

Finkler A, Ben Zvi BS, Koltin Y & Barash I (1988) Transcription and in vitro translation of the dsRNA virus isolated from *Rhizoctonia solani*. Virus Genes 1:205-219.

Frank JA & Francis SK (1976) The effect of a *Rhizoctonia solani* phytotoxin on potatoes. Can. J. Bot. 54:2536-2540.

Ghabrial SA, Sanderlin AS & Calvert LA (1979) Morphology and virus-like particle content of *Helminthosporium victoriae* colonies regenerated from protoplasts of normal and diseased isolates. Phytopathology 69:312-315.

Ghabrial SA (1994) New developments in fungal virology. Advances in Virus Research 43:303-388.

Gilman AG (1987) G proteins: Transducers of receptor-generated signals. Annu. Rev. Biochem. 56: 615-649.

Hashiba T, Homma Y, Hyakumachi M & Matsuda J (1984) Isolation of a DNA plasmid in the fungus *Rhizoctonia solani*. J. Gen. Microbiol. 130:2067-2070.

Hillman BI, Rae B, Tartaglia J & Nuss DL (1989) Elucidating the structure and function of double-stranded RNA genetic elements associated with biological control of chestnut blight. In: Staskawicz B, Ahlquist P, and Yoder O (eds.) Molecular Biology of Plant-Pathogen Interaction. Vol. 101 (pp 59-70) Alan R. Liss, Inc., New York

Hongo M, Miyasaka A, Suzuki F & Hashiba T (1994) Expression of the linear DNA plasmid pRS64 in the plant pathogenic fungus *Rhizoctonia solani*. Mol. Gen. Genet. 245:265-271.

Hyakumachi M, Sumino A, Ueda I & Shikada E (1985) Relationship between the presence of dsRNA in *Rhizoctonia solani* and pathogenicity. Ann. Phytopathol. Soc. Jpn. 51:372-373.

Jabaji-Hare SH, Burger G, Forget L & Lang BF (1994) Extrachromosomal plasmids in the plant pathogenic fungus *Rhizoctonia solani*. Curr. Genet. 25:423-431.

Jiang J, Lakshman DK & Tavantzis SM (1995) Coincidence of transmission of double-stranded RNA (dsRNA) from *Rhizoctonia solani*. Phytopathology 85:1211 (Abst.).

Koltin Y (1989) The *Ustilago maydis* virus encoded toxin and its cellular effects. In: Nevalainen H and Penttilä M (eds.) Proceedings of the EMBO-Alko workshop on Molecular Biology of Filamentous Fungi, Vol.6 (pp 149-157) Foundation for Biotechnical and Industrial Fermentation Research, Helsinki

Koltin Y & Day PR (1976) Inheritance of killer phenotypes and double-stranded RNA in *Ustilago maydis*. Proc. Natl. Acad. Sci. USA. 73:594-598.

Kousik CS, Snow JP & Valverde RA (1994) Comparison of double-stranded RNA components and virulence among isolates of *Rhizoctonia solani* AG 1-IA and AG 1-IB. Phytopathology 84:44-49.

Lakshman DK & Tavantzis SM (1994) Spontaneous appearance of genetically distinct double-stranded RNA elements in *Rhizoctonia solani*. Phytopathology 84:633-639.

Lakshman DK & Tavantzis SM (1995) Sequence analysis of a hypovirulence-associated double-stranded RNA (dsRNA) from *Rhizoctonia solani*. Phytopathology 85:1211 (Abst.).

Lapierre HJ, Lemaire JM, Juan B & Molin G (1970) Mise en évidence des particules virales associées à une perte de pathogénicit chez le piétin-échaudage des céréales, conséquences agronomiques prévisibles. C. R. Acad. Agr. (Fr.) 56:134-138.

Lemaire JM, Lepierre H, Jouan B & Bertrand J (1970) Découverte de particules virales chez certaines souches d'*Ophiobolus graminis*, agent du Pietin-echaudage des céréales: Conséquences agronomique prévisibles. C. R. Hebd. Seanc. Acad. Agric. (Fr.) 56:1134-1138.

Milborrow BV, Purse JG & Wightman F (1975) On the auxin activity of phenylacetic acid. Ann. Bot. 39:1143-1146.

Miyasaka A, Chen CL & Hashiba T (1990) Detection and properties of plasmid-like DNA in isolates from nine anastomosis and intraspecific groups of *Rhizoctonia solani*. J. Gen. Microbiol. 136:1791-1798.

Miyashita S, Hirochika H, Ikeda J-E & Hashiba T (1990) Linear plasmid DNAs of the plant pathogenic fungus *Rhizoctonia solani* with unique terminal structures. Mol. Gen. Genet. 220:165-171.

Nuss, D.L. (1992) Biological control of chestnut blight: an example of virus-mediated attenuation of fungal pathogenesis. Microbiological Reviews 56:561-576.

Orellana RG & Mandava NB (1983) m-Hydroxyphenylacetic and m-Methoxyphenylacetic acids of *Rhizoctonia solani*: Their effect on specific root-nodule activity and histopathology in Soybean. Phytopath. Z. 107:159-167.

Patiño C, Morales S, del Moral C, Propín J, Colombo V, Kalo-Koenova EY, Canosa I, Goldstein J & V. Rubio (1992) Viruses and Plasmids in Fungi: Studies in *Rhizoctonia solani*, *Ustilago maydis* and *Penicillium chrysogenum*. In: Villa TG and Abalde J (eds.) Profiles on Biotechnology, (pp 143-154) .Servicio de Publicaciones. Universidad de Santiago, Santiago de Compostela.

Polashock JJ & Hillman BI (1994) A small mitochondrial double-stranded (ds) RNA element associated with a hypovirulent strain of the chestnut blight fungus and ancestrally related to yeast cytoplasmic T and W dsRNAs. Proc. Natl. Acad. Sci. 91:8680-8684.

Rogers HJ, Buck KW & Brasier CM (1987) A mitochondrial target for double-stranded RNA in diseased isolates of the fungus that causes Dutch elm disease. Nature 329:558-60.

Sakaguchi K. (1990) Invertrons, a class of structurally and functionally related genetic elements that includes linear DNA plasmids, transposable elements, and genomes of Adeno-type viruses. Microbiological Reviews 54:66-74.

Samac DA & Leong SA (1989) Mitochondrial plasmids of filamentous fungi: Characteristics and use in transformation vectors. MPMI 2:155-159.

Shapira R, Choi GH & Nuss DL (1991a) Virus-like genetic organization and expression strategy for a double-stranded RNA genetic element associated with biological control of chestnut blight. EMBO J. 10:731-739.

Shapira R, Choi GH, Hillman BI & Nuss DL (1991b) The contribution of defective RNAs to the complexity of viral-encoded double-stranded RNA populations present in hypovirulent strains of the chestnut blight fungus *Cryphonectria parasitica*. EMBO J. 10:741-746.

Sherwood RT & Lindberg CG (1962) Production of a phytotoxin by *Rhizoctonia solani*. Phytopathology 52:586-587.

Sneh B, Burpee L & Ogoshi A (1991) Identification of *Rhizoctonia* Species. APS Press, Minnesota, 133 pp.

Tavantzis SM (1988) Biological control of Rhizoctonia disease and increased growth response of potato effected by hypovirulent *R. solani*. In: Engelstad S, Coli WM & Carlson JL (eds.) Innovations in Pest Management.(pp 44-46) U. Mass. Press.

Tavantzis SM & Bandy BP (1988) Properties of a mycovirus from *Rhizoctonia solani* and its virion-associated RNA polymerase. J. Gen. Virol. 69:1465-1477.

Tavantzis SM & Lakshman DK (1996) Virus-like double-stranded RNA elements and hypovirulence in phytopathogenic fungi. In: Kohmoto K, Singh RP and Singh US (eds.) Pathogenesis and Host-Parasite Specificity in plant Diseases: Histopathological, Biochemical, Genetic and Molecular Basis. Vol. III (pp 249-267) Elsevier (Pergamon) Press, Oxford, UK.

Tavantzis SM, Perkins BL, Bushway RJ & Bandy BP (1989) Correlation between *in vitro* synthesis of phenylacetic acid and virulence in *Rhizoctonia solani*. Phytopathology 79:1199 (Abst.).

Wako T, Ishikawa T & Hashiba T (1991) Unique DNA plasmid pRS64 associated with chromosomal DNAs of the plant pathogenic fungus *Rhizoctonia solani*. J. Gen. Microbiol. 137:2817-2821.

Wyllie TD (1962) Effect of metabolic by-products of *Rhizoctonia solani* on the roots of chippewa soybean seedlings. Phytopathology 52:202-206.

Zanzinger DH, Bandy BP & Tavantzis SM (1984) High frequency of finding double-stranded RNA in naturally occurring isolates of *Rhizoctonia solani*. J. Gen. Virol. 65:1601-1605.

II.4 CULTURE PRESERVATION METHODS FOR MAINTANING THE GENETIC INTEGRITY OF *RHIZOCTONIA* SPP. ISOLATES

BARUCH SNEH[1] AND GERARD C. ADAMS[2]

[1]Department of Botany and Institute for Nature Conservation Research, The George S. Wise Faculty of Life Sciences, Tel Aviv University, Ramat Aviv, ISRAEL 69978. [2]Department of Botany and Plant Pathology, Michigan State University, East Lansing, Michigan 48824 USA

I. INTRODUCTION

Keeping the genetic integrity of *Rhizoctonia* spp. isolates is of great importance. When cultures are stored at inappropriate conditions and/or transferred many times on agar, they may undergo profound genetic changes and lose important characteristics such as: vigor, virulence (Butler, 1980), ability of certain non pathogenic isolates to increase plant growth or protect against pathogenic *Rhizoctonia* (Sneh, unpublished), cultural characteristics etc. Since the rates of mitotic recombination and mutation are likely to correspond to those of cell division and metabolic activity, an ideal system for germplasm preservation would be storage at a temperature where cultures remain viable, but are in a metabolic inactive state (Shannon *et al.*, 1975). This may be accomplished by storage below -139°C, at which there is no growth of ice crystals and the rates of other biophysical processes are too slow to affect cell survival. *Rhizoctonia* spp. isolates lose their viability when their cultures on PDA are kept in the refrigerator (4°C). Information on various preservation methods and data about the survival of isolates preserved by these methods that were kindly provided to us by *Rhizoctonia* researchers together with reports from the literature are assembled in this chapter.

II. PRESERVATION ON AGAR

A. Storage on PDA

For short-term storage at 4°C *Rhizoctonia* isolates are kept on PDA amended with vitamins (1.0mg/l thiamin, 1.0mg/l niacin, 20µg/l biotin, 0.5mg/l Ca-panthothenate, 0.5mg/l pyridoxine, 0.1mg/ml p-aminobenzoic acid). After a period of growth on water-agar the strains are grown at 23°C in the dark on a defined nutrient medium (DNM), modified after Steinberg. DNM consists of 15g/l sucrose, 2g/l asparagine, 0.6g/l KH_2PO_4, 0.8g/l K_2HPO_4, 1g/l $MgSO_4$, 10mg/l $CaCl_2$, 2.5mg/l $ZnSO_4.7H_2O$, 2.5mg/l H_3BO_4, 0.5mg/l $MnSO_4.H_2O$, 1.7mg/l NaFeEDTA, 0.3mg/l $CuSO_4.5H_2O$ 0.1mg/l $(NH_4)_6Mo_7O_{24}.4H_2O$ and 20g/l agar amended with vitamins (V. Rubio, Personal communication).

Where suitable equipment and liquid nitrogen are not available, isolates of *Rhizoctonia* spp. may be conserved for long periods in 10% glycerol solution at -80°C (Anderson, personal communication). Radial strips of cultures on PDA are cut and placed in vials containing 4-5 ml of 10% glycerol. The vials are placed in a

B. Sneh et al. (eds.),
Rhizoctonia Species: Taxonomy, Molecular Biology, Ecology, Pathology and Disease Constrol, 139–145.

refrigerator at 4°C overnight, transferred to a freezer at -15 to -20°C and after a few hours, stored in a deep freezer at -80°C .

G. Adams (unpublished) stored 138 isolates of *Rhizoctonia* in 10% glycerol at -80°C according to the directions of Anderson (in Sneh *et al.,* 1991). This method involved placing strips of PDA agar culture into 10% glycerol and refrigerating (5°C) for a day, freezing at -20°C for a week, and then storing at -80°C for 3 years. The 138 isolates included 21 AG 1A, 3 AG 1C, 11 AG 2-1, 4 AG 2-2, 19 AG 3, 50 AG 4, 3 AG 5, 8 *R. oryzae-sativae*, 4 *R. oryzae*, 6 binucleate *Rhizoctonia* and 9 *R. solani* of unknown grouping. Adams reports that survival was negligible except where visible sclerotia were included in the glycerol vials. Survival was 100% for sclerotia. Sclerotia were primarily those of AG 1A, AG 1C, and *R. oryzae-sativae.*

B. Storage on Malt Peptone Agar (MPA)

Cultures of *Rhizoctonia* spp. isolates can be stored for several years on Malt peptone agar (MPA) at 4°C. Isolates of AG 3, AG 2-1 and AG 2-t maintained their viability and pathogenicity under these conditions for at least 5 years (Dijst, 1988; Van den Boogert and Jager, 1984). However, if these isolates were kept on PDA under these conditions for 5 years, the cultures only still produced white loose sclerotia and white mycelium, instead of the brown mycelium and large firm brown sclerotia. In a subsequent study some isolates of other AGs were not well preserved on MPA under these conditions (JHM Schneider, Personal communication).

C. Storage under oil

Covering cultures on agar slants (30° to the horizontal) in 30 ml universal bottles with mineral oil prevents dehydration and slows down the metabolic activity and growth through reduced oxygen tension. This method was first extensively used by Buell & Weston (1947), and subsequent reports have indicated its wide application and success (Dade, 1960; Fennell, 1960; Smith *et al.* 1970; Onions, 1971, 1977). Mature healthy cultures are covered by 10 mm of sterile mineral oil (liquid paraffin or medicinal paraffin specific gravity 0.830-0.890). If the oil is deeper than 10 mm the fungus may not receive sufficient oxygen and may die, while if the depth is less, exposed mycelium or agar on the sides of the container may allow moisture to evaporate and the culture to dry out.

Retrieval from a culture under oil is achieved by removal of a small amount of the fungus colony on a mounted needle and draining away as much oil as possible. The fungus is then streaked onto a suitable agar medium. Inoculating an agar slope centrally sometimes has better results as excess oil can drain down the slope allowing the fungus to grow more typically towards its top. More than one subculture may be necessary after retrieval as the growth rate can very often remain reduced due to adhering oil. The fungal mycelium can normally recover when it is re-isolated from the agar plate on to fresh media. There is an added risk of personal contamination by spattering of oil containing fungus when sterilizing inoculation needles in the flame (Fennell, 1960).

In the International Mycological Institute (IMI), 90 cultures of *Rhizoctonia* spp. have been recovered after periods of 4-46 years from cultures under oil. Although many of them could be recovered after such long periods, this method allowed deterioration of cultures. Of the recovered cultures after long storage, 11 isolates had a typical growth, 59 were slightly deteriorated, 30 isolates were poorly recovered (D. Smith, personal communication).

D. Freeze drying

In the IMI A spin freeze drying method was used, in which the material is cooled at a rate of 10°C /min down to -30°C to reduce ice formation during the cooling process. Freshly isolated cultures are better preserved than older cultures. However, Hyphae of *Rhizoctonia* are damaged even in this method due to ice crystals formation. Except for one culture of *R. tuliparum*, which grew normally after 3 years of storage, the other cultures of *Rhizoctonia* spp. (*T. cucumeris*, *T praticola*, *R. fumigata*, *R. tuliparum*, and *C. cornigerum*) were restricted in their growth after being stored for 7-20 years (D. Smith, Personal communication). Too severe lyophilization may kill the isolates and absorption of moisture during preservation may also affect viability (B. Sneh, unpublished).

E. Preservation in Silica gel

Silica gel is used successfully for dry storage of various fungal spores. This method has not been proven good for *R. solani* cultures in the IMI (D. Smith, Personal communication).

III. PRESERVATION IN SAND CULTURES

A mixture of: river sand (middle size) -100g, heavy clay, -100g and potting soil - 200g is air dried, sieved and mixed thoroughly with 20ml water. The moisture is placed in test tubes (up to 6cm high) and 2ml water are added/tube. The tubes are closed with cotton plugs and autoclaved for one hour twice (in two consecutive days). An The tubes are inoculated and incubated for a couple of weeks at 25-27°C. The tubes are then kept at 8-10°C for 5 years. The isolates are recovered by transfer to sterilized oat grain to re-nurture them and the colonized grain are used to inoculate new test tube which can be kept for the next 5 years) Cultures preserved in this way were viable for at least 20 years (Schneider, 1958; Doornik, 1981).

IV. PRESERVATION ON SOIL WHEAT-BRAN

This procedure was developed by Butler (1980). Fine dry sandy loam soil is mixed with 4% (w/w) wheat bran. Six ml samples of the mixture are placed in screw-cap vials, 3 ml of water containing 500µg/ml chloramphenicol (or novobiocin) are added, the vials are covered with cotton plugs and autoclaved for one hour at 121°C on each of three consecutive days. A mycelial disk from the margin of an actively growing culture of *Rhizoctonia* is placed in a vial and incubated at 23-27°C for several weeks, until the soil is dry. The cotton plugs are then replaced with sterile screw caps and tightened, sealed with parafilm, and stored at -20°C. This method has received further evaluation including the examination of virulence following the publication by Butler (1980). Stored cultures were examined for viability in 1989 and 1995 by inoculating plates of PDA amended with yeast extract (0.5g/l), malt extract (0.5g/l), peptone (0.5g/l), and novobiocin (100µg/l). Butler reports that in the 1989 study cultures had been stored a minimum of 11 years at -20°C. Of 57 AG 1 isolates 97% were still viable, of 31 AG 2 isolates 91% remained viable, of 26 AG 3 only 38% were viable, and of 87 AG 4 isolates 71% remained viable. Additionally, 24 *R. oryzae* and 3 *R. zeae* isolates 100% remained viable. The 1989 evaluation showed excellent survival of heterokaryons and homokaryons. The heterokaryons included field isolates as well as those synthesized by pairing two or three different homokaryons. Isolates of binucleate *Rhizoctonia* and other species also survived well. Most cultures formed sclerotia on the amended PDA.

In the 1995 evaluation of the soil wheat-bran storage method Butler reports that isolates stored for a minimum of 17 years were tested for viability and pathogenicity. Survival was more effectively evaluated on PDA amended with 1g/l yeast extract and 100mg/l tetracycline than on unamended PDA. Very high levels of pathogenicity on germinating cotton seedlings were evident for all the AG 4 isolates tested. A few AG 2-1 and AG 2-2 isolates were also highly pathogenic. However, the AG 2 cultures appeared to have poor development of pigmentation on the amended PDA, although sclerotia were formed. A test of 24 homokaryons of AG 1-IC showed that 23 isolates survived after 30 years in soil wheat-bran storage (Butler, personal communication).

V. PRESERVATION IN PLANT MATERIAL

A. Colonizing seedlings

Cultures of *Rhizoctonia* spp. isolates may be preserved on colonized plant material. For example, one week old conifer germlings were placed on sterile distilled water, inoculated with (uni- or bi-nucleate) *Rhizoctonia* spp. isolates and incubated at 25°C for 3 days. The isolates colonizing the seedlings kept at 4°C remained viable and pathogenic for at least 2 years (R. Sen, personal communication).

B. Colonizing cereal grain

Rhizoctonia spp. isolates are well preserved in colonized cereal grain, such as wheat, barley, oat or rice (Sneh *et al.*, 1986; Naito *et al.*, 1993; Kuznia and Windels, 1994; LL Burpee; B Nelson, personal communication). Cereal grain are immersed in water in erlenmeyer flasks supplemented with 250µg/ml chloramphenicol, heated until boiling and left to settle for 10 min. The water is decanted and the grain are autoclaved at 121°C for an hour on two consecutive days. The grain are distributed in screw-cap vials and autoclaved again. Mycelial disks from the margins of actively growing cultures of *Rhizoctonia* spp. are placed in the vials and incubated at 23-27°C for 7-10 days (modification of Sneh *et al.*, 1986). After incubation, the vials containing the colonized wheat seeds may be placed in a desiccator to ensure complete dryness before placing them in cold storage. It is important to keep the infested grain air dry for good survival of the isolates. The caps are then tightened, sealed with parafilm, and stored at -25°C, -80°C or -195°C (liquid nitrogen).

Naito *et al.* (1993) reported on preservation of *Rhizoctonia* spp. isolates colonizing barley grain (also on rice grain) after air drying were kept at for -20°C for at least 10 years. Isolates of AGs 1-6, BI and binucleate *Rhizoctonia* spp. AGs - B, C, D, E and K (52 isolates) recovered after such a long period under these conditions were viable and pathogenic. However, some isolates of AG 6 and BI have lost their viability. Additional reports on preservation of *Rhizoctonia* spp. isolates.

Kuznia and Windels (1994) added autoclaved oat and barley grain to cultures of *R. solani* and incubated them for 14-21 days. The colonized grain were then placed directly into cryovials, cooled to 4°C overnight, -15°C for 24 hours, and then stored at -80°C for 21-26 months. All the isolates tested survived and these included 9 AG 1, 5 AG 2-1, 51 AG 2-2, 2 AG 3, 59 AG 4, and 30 AG 5.

VI. THE CRYOPRESERVATION METHOD.

The cryogenic storage temperatures commonly used are those of liquid nitrogen (-196°C) and liquid nitrogen vapor (-150°C and below). Cultures are preserved cryogenically in culture collections centers such as: the American Type Culture

Collection (ATCC) USA, International Mycological Institute (IMI) UK and the Gene Bank Culture Collection Tsukuba, Japan.

Description of the apparatus and the procedures for cryogenic storage at the ATCC have been reported by Jong (1978), Jong and Atkins (1985), and Jong and Davis (1987). Three agar disks, 5 mm in diameter, from the advancing edge of a culture of *Rhizoctonia* are removed with a sterile cork borer and placed in 2.0 ml screw-cap polypropylene vials (Vanguard International, Neptune, NJ) or in borosilicate ampules (T.C. Wheaton Co., Minville, NJ). A volume of 0.4 ml of 10% (v/v) glycerol in distilled water, or 5% dimethyl sulfoxide (DMSO) is added as a cryoprotective. The vials or ampules are sealed prior to a precooling process.

The vials or ampules are placed in pre-labelled aluminum canes (Nasco #A545, Fort Atkinson, WI) in metal canisters which are then placed in the freezing chamber of a programmed freezer (Cryo-med, 49659 Leona Dr., Mt. Clemens, MI). The initial cooling is carried out at a rate of -1°C per min from ambient temperature to -40°C. Subsequent cooling to below -100°C is rapid and uncontrolled. The vials or ampules are immediately transferred to storage in a liquid nitrogen refrigerator, either in liquid nitrogen at -196°C or in liquid nitrogen vapor at approximately -150°C.

To recover the cultures, it is important to thaw the vials or ampules fast. Thawing is achieved in a water bath at 37°C, with rapid agitation for 1-2 min. The cultures are then transferred to PDA.

In the IMI the following method for cryopreservation is used: 0.5ml aliquots of fungal suspension in 10% (v/v) glycerol are dispensed into 1ml borosilicate glass ampoules which are heat sealed and precooled to 7°C. Alternatively agar plugs are cut from growing colonies and 10 placed in each of six 2ml plastic screw cap cryotubes. These are then cooled at 1°C/min to -50°C in a Planer programmable freezer. The ampoules are then plunged into liquid nitrogen and cooled rapidly to -196°C. Thawing is carried out rapidly in a water bath at +37°C until all the ice crystals have melted. The ampoules are opened and the contents streaked onto a suitable growth medium.

Smith and Onions (1983; 1994) reported that cultures of *T. cucumeris* (among other fungi) survived under the cryogenic method for at least 12 years. In the IMI Most of the following isolates of *Rhizoctonia* spp. have been stored for periods of 3-25 years using this method. *R. solani* (*T. cucumeris*) (31 isolates), *T. praticola* (1), *Rhizoctonia* sp. (4), *R. fragariae* (2), *R. numerati* (1), *R. oryzae sativae* (2), *C. cereale* (1) *C. cornigerum* (2) *C. ramicola* (1). The isolates stored in this method were well preserved and exhibited a typical growth when recovered,. except for one culture of *R. solani* and one culture of *R. oryzae* sativae which were poorly recovered.

In addition to cultures, protoplasts of *R. solani* may be preserved in liquid nitrogen (Hashiba *et al.*, 1985). Protoplasts are suspended in 0.6 M sucrose in combination with 5% polyethylene glycol (PEG) or 5% ethylene glycol (EG) as cryoprotectants (5% DMSO is also a satisfactory cryoprotectant). Cooling and thawing procedures are the same as described above. Viability of protoplasts after cryopreservation was 95%.

B Nelson (personal communication) studied cryogenic preservation of *Rhizoctonia* at -80°C. *R. solani* (AG 1-IA, 1-IB, 1-IC, 2-1, 2-2, 3 to 9 and BI) and binucleate *Rhizoctonia* (AGs- A, Ba, Bb, Bo, C to L, O, R and S) were grown on PDA for 15 days and small pieces (5 x 20 mm) of agar plus mycelium were placed in 1 ml cryogenic vials with sterile 10% glycerol, water or no cryopreservation agent. The

AGs were also grown on autoclaved barley grain for a similar time and colonized grain were placed in vials with no cryopreservation agent. Vials were either placed directly into -80°C or into -15°C for a 24 hour cool down before freezing at -80°C. Survival of isolates of different AGs was tested by placing thawed agar pieces or frozen barley grain directly onto PDA. After four and five years storage, survival of isolates of all AGs of *R. solani* and binucleate *Rhizoctonia* in barley grain (with or without the 24 hr cool down period) was excellent. Survival varied in the other cryogenic treatments, but was generally higher for *R. solani* than for the binucleate *Rhizoctonia*. Even in the absence of a cryopreservation agent, survival of *R. solani* on PDA was good for all AGs. Some binucleate *Rhizoctonia* such as AGs E, F, G, H, L, O, R and S, did not survive or had poor survival. Cryogenic preservation of *Rhizoctonia* spp. on barley grain at -80°C is an efficient and convenient method for storing cultures, but isolates of some AGs may store well on PDA in 10% glycerol or water.

VII. CONCLUSIONS

Storing cultures of *Rhizoctonia* spp. by the cryopreservation method (liquid nitrogen at -196°C) provides the best protection of the genetic integrity of the isolates during long term storage periods. This method is used in the centers of the culture collections such as the ATCC (USA), IMI (UK), Tsukuba (Japan) and others, as well as by *Rhizoctonia* researchers. Special equipment for such a procedure are not available in most of the labs of *Rhizoctonia* researchers. To get good preservation, a special care has to be taken to get gradual slow cooling of the cultures. It is necessary to avoid internal ice formation at faster rates of cooling and shrinkage at the slower rates to ensure viability and retention of characters. *Rhizoctonia* spp. usually fail to survive spin freeze-drying unless the strain produces structures that are resistant to desiccation, such as sclerotia. However, if controlled cooling on a shelf freeze-drier is used (Kolkowski & Smith, 1995), greater success would be expected. Storage under a layer of mineral oil retains viability for long periods. Up to 47 years has been recorded at IMI, but it allows deterioration, even after short-term storage of 3 years.

Storing cultures of *Rhizoctonia* spp. colonizing cereal grain is widely used by *Rhizoctonia* researchers. This method does not require special expensive equipment and is therefore, more handy to use in all labs. The dry colonized grain can be stored for long periods at 4, -20, -80 and -196°C and the recovered cultures keep their genetic integrity as well. Another advantage of this method is that the isolates are accompanied with a rich food base which helps keeping the vigor and virulence upon recovery of the isolates.

When recovered isolates are inoculated on germinating host seeds and reisolated on PDA, their vigor, virulence and capability of the non pathogenic isolates among them to protect and increase plant growth are strengthened.

VIII. ACKNOWLEDGMENTS

We would like to express our sincere thanks to all of the following colleagues who provided us the information regarding the various methods that they use for preservation of *Rhizoctonia* spp. cultures: Lee L Burpee, Edward E Butler, Gerda Dijst, Jaap Keijer, Shigeo Naito, Berlin Nelson, Victor Rubio, Johannes HM Schneider, Robin Sen, and David Smith.

IX. REFERENCES

Buell CB & Weston WH (1947) Application of the mineral oil conservation method to maintaining collections of fungus cultures. Amer. J. Bot. 34:555-561.

Butler EE (1980) A method for long-time culture storage of *Rhizoctonia solani*. Phytopathology 70:820-821.

Dade H A (1960) Laboratory methods in use in the culture collection, IMI. in: Herb IMI Handbook, pp 78-83. Kew: IMI.

Dijst G (1988) Formation of sclerotia by *Rhizoctonia solani* on artificial media and potato tubers. Neth. J. Plant Pathol. 94:233-242.

Doornik AW (1981) Temperature dependence of the pathogenicity of *Rhizoctonia solani* in some bulb crops as an intrinsic property of the isolate. Neth. J. .Plant Pathol. 87: 139-147.

Fennell DI (1960) Conservation of fungus cultures. Bot. Rev. 26:79-141.

Hashiba T, Tetsuka Y & Suzui T (1985) Technique for long-term preservation of protoplasts of fungi in liquid nitrogen. Japan J. Freez. Dry. 31:63-67.

Jong SC & Atkins WBA (1985) Conservation, collection and distribution of cultures. In: Howard DH & Howard LF (eds.) Fungi Pathogenic for Humans and Animals, Part B. Pathogenicity and Detection II. (pp 153-193). Marcel Dekker, New York.

Jong SC & Davis EE (1987) Germplasm preservation of edible fungi in culture through cryogenic storage. In: Proc. Int. Symp. Sci. Tech. Cult. Edible Mushrooms. (pp 213-225). Elsvier Sci. Publ., Amsterdam.

Jong SC (1978) Conservation of the cultures. In: Chang ST & Hay WH (eds.). The Biology and Cultivation of Edible Mushrooms. (pp 119-135) Academic Press, New York.

Kolkowski JA & Smith D (1995) Cryopreservation and freeze-drying of fungi. In: Day JG & MR McLellan (eds.). Methods in Molecular Biology 38, Cryopreservation and freeze-drying protocols (pp 49-61) Humana Press. Tolowa, New Jersey.

Kuznia RA & Windels CE (1994) Preservation of *Pythium* species and *Rhizoctonia solani* at -80°C. Phytopathology 84:1149 (Abstr.).

Naito S, Yamaguchi T, Sugimoto T & Homma Y (1993) A simple method for the long-time cuture storage of *Rhizoctonia* spp. on barley grain. Ann. Rept. Plant Prot. North Japan 44:20-23.

Onions AHS (1971) Preservation of fungi. In: Methods in Microbiology 4 (pp113-151). Academic Press. London and New York.

Onions AHS (1977) Storage of fungi by mineral oil and silica gel for use in the collection with limited resources. In: Proceedings of the Second International Conference on Culture Collections. University of Queensland, Brisbane: World Federation for Culture Collections.

Schneider R (1958) Untersuchungen u"ber Variabilita"t und Taxonomie von *Fusarium avenaceum* (Tr) Sacc. Phytopath. Z. 32: 95-126.

Shannon JF, Ghama RL & Jong SC (1975) The role of liquid nitrogen refrigeration at the American Type Culture Collection. In: Rinfre AR & LaSalla B (eds.) (pp 1-18) National Academy of Sciences, Washington, D.C.

Smith DH, Lewis FH & Fergus CL (1970) Long term preservation of Botryosphaeria ribis and Dibotryon morbosum. Plant Dis. Rep. 54:217-218.

Smith D & Onions, AHS (1983) The preservation and maintenance of living fungi. Commonwealth Mycological Institute. Page Bros. Ltd., Norfolk, U.K.

Smith D & Onions AHS (1994) The Preservation and Maintenance of Living Fungi. Second edition. IMI Technical Handbooks (No. 2, pp 122), Wallingford, UK: CAB International..

Sneh B, Burpee LL & Ogoshi A (1991) Identification of *Rhizoctonia* Species. APS Press, St. Paul Minnesota. 133pp.

Sneh B, Zeidan M, Ichielevich-Auster M, Barash I & Koltin Y (1986) Increased growth responses induced by a nonpathogenic isolate of *Rhizoctonia solani*. Can. J. Bot. 64:2372-2378.

Van den Boogert PHJF & Jager G (1984) Biological control of *Rhizoctonia solani* on potatoes by antagonists. 3. Inoculation of seed potatoes with different fungi. Neth. J. Plant Pathol. 90:117-126.

Section III.

PLANT - PATHOGEN INTERACTIONS OF *RHIZOCOTNIA* SPP.

III.1 THE INITIAL STEPS OF THE INFECTION PROCESS IN *RHIZOCTONIA SOLANI*

JAAP KEIJER

DLO-Research Institute for Plant Protection (IPO-DLO), Wageningen, The Netherlands and DLO-State Institute for Quality Control of Agricultural Products (RIKILT-DLO). P.O. Box 230, 6700 AE Wageningen, THE NETHERLANDS

I. INTRODUCTION

This chapter aims to review the initial steps of the infection process and host plant specialization of *Rhizoctonia solani*. A subject that has previously been reviewed by Dodman and Flentje (1970). In this chapter the infection process is defined as the steps in the disease cycle between attachment and tissue colonization (Fig. 1).

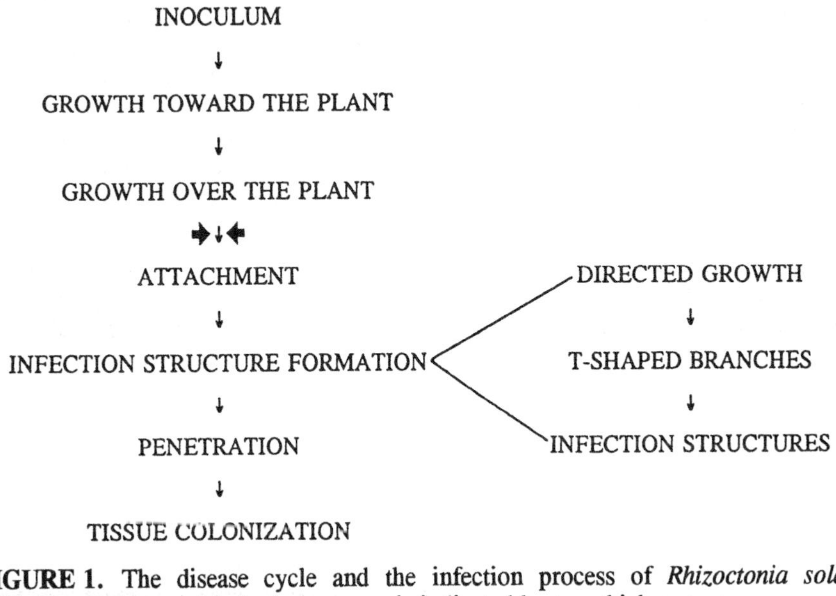

FIGURE 1. The disease cycle and the infection process of *Rhizoctonia solani*. Initiation of the actual infection process is indicated by two thick arrows.

II. INOCULUM

A plant pathogen has to be present in order to cause disease. The numbers and form in which it is present as well as its nutritional state will influence the outcome of disease. *R. solani*, which does not make vegetative spores, can be present as basidiospore, mycelium or sclerotium. Basidiospores, responsible for a wide range of *Rhizoctonia* leaf diseases (Baker, 1970; see also chapter IV.3), are very fragile and are therefore not suited for long term survival. However, they are an important source of

B. Sneh et al. (eds.),
Rhizoctonia Species: Taxonomy, Molecular Biology, Ecology, Pathology and Disease Constrol, 149–162.

genetic variation (Papavizas and Ayers, 1965; Flentje *et al.*, 1970; Anderson and Stretton, 1978; Adams, 1988; Julian *et al.*, 1996; Keijer *et al.*, 1996a) and a means of long distance dispersal (see chapter IV.3). Sclerotia, undifferentiated aggregations of thick-walled melanized cells, are the primary survival structures and therefore an important source of inoculum (see chapter IV.4). Sclerotia are formed in soil or on plant residues and have been reported to survive for several years (Sherwood, 1970). Both basidiospores and sclerotia will germinate and form mycelium prior to infection. Besides infection, mycelium also has a role in dispersal and survival; it can rapidly grow through and colonize soil (Papavizas, 1970) and is relatively persistent, especially on colonized debris (Papavizas and Davey, 1962).

III. MYCELIAL GROWTH

Upon germination, that is moisture and temperature dependent (Sherwood, 1970), sclerotia form mycelial threads that can grow toward the plant. When cellophane bags are placed around the roots of plants, hyphal aggregates are formed on the cellophane surface (Kerr, 1956), indicating the attraction of the hyphae by the roots. Similar aggregates are formed on cellophane or membranes of mixed esters of cellulose that are placed over seedlings (Fig. 2). These aggregates should not be confused with infection cushions, because they lack specific characteristics of infection cushion formation such as directed growth and the formation of T-shaped branches. Under various experimental conditions it has been shown that mycelial growth is stimulated by plant exudates (Kerr and Flentje, 1957; Flentje *et al.*, 1963; De Silva and Wood, 1964). The growth stimulation rate correlates with higher amino acids, carbohydrates, phenols and organic acids concentrations present in the exudates of younger plants as compared to older plants (Nour El Dein and Sharkas, 1964; Martinson, 1965; Reddy, 1980). This saprophytic growth stimulation will increase the inoculum density and may consequently influence disease formation. On the other hand, in the interaction between *R. solani* and cotton it has been found that larger hyphal aggregates are induced by root exudates of resistant plants than by susceptible plants (El Samra *et al.*, 1981). As in other cases it is not resolved whether the growth enhancement is due to enhanced nutrient availability or to the presence of specific inducing components.

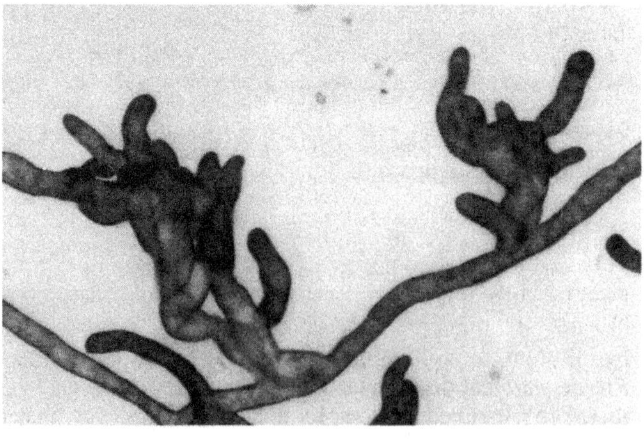

FIGURE 2. *Rhizoctonia solani* AG 5 hyphal aggregates on a HATF (Millipore) membrane that was placed over total egg-plant seedlings. The hyphae over the stem are stained with trypan blue.

IV. THE INFECTION PROCESS

When *R. solani* hyphae come in contact with the plant, they start to grow over the plant surface. These hyphae are round and not attached to the host plant. At this point the actual infection process can initiate. The distinct stages that can be observed are listed in Fig. 1. After 10 to 12h hyphae that are flattened and closely and firmly attached to the plant surface can be observed (Christou, 1962; Armentrout and Downer 1987). In the next stage, described as directed growth (Dodman and Flentje, 1970; Marshall and Rush 1980a; Armentrout and Downer, 1987; Armentrout *et al.*, 1987; Kousik *et al.*, 1990), the attached hyphae start to follow the anticlinal walls of the contiguous epidermal cells. Also some appressed hyphae that change growing direction, grow in an indiscriminate direction over the epidermal cells. Three to five hours later (t=15 h) side branches are formed at right angels (Armentrout and Downer, 1987). These so-called T-shaped branches are a determining characteristic for infection structures of *R. solani*. At this point the process can deviate in two manners. Either the side branches give rise to short swollen hyphae or appressorial structures or to repetitive formation of T-shaped branches from one or several parent hyphae result in the formation of complex infection structures (t=18h; Fig. 3). In their most extreme form, the complex infection structures can be seen as massive dome shaped structures, infection cushions. Like appressoria, complex infection structures are firmly attached to the epidermis by the hyphal branches and their swollen tips. Next, several of the swollen tips simultaneously form infection pegs (Flentje, 1957; Fukutomi and Takada, 1979). The pegs will penetrate the cuticle and the epidermal cell wall. Inside the plant, the hypha resume normal diameter, giving rise to the so-called primary hypha. Subsequently, hyphae rapidly invade the epidermal tissue and the outer layers of the cortex, growing intracellularly as well as in the intercellular spaces, depending on the strain (Christou, 1962; Ruppel, 1973; Weinhold and Motta, 1973; Fukutomi and Takada, 1979; Kenning and Hanchey, 1980; Marshall and Rush, 1980a; Yang *et al.*, 1992). In seedlings the epidermal tissue will collapse resulting in seedling collapse.

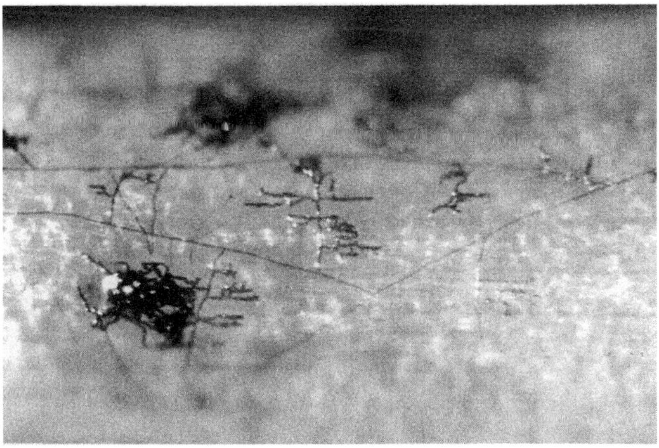

FIGURE 3. Infection cushion formation of a *Rhizoctonia solani* AG 2-1 isolate on the stem of a cauliflower seedling. Directed growth, T-shaped branches and an infection cushion can be seen. The hyphae are stained with trypan blue.

A. Attachment

Attachment of hyphae can clearly be seen with interference contrast microscopy. Round hyphae on the plant surface become flattened and appear closely appressed to epidermal cells. Where the round hyphae are easily washed off the stems of seedlings, the flattended hyphae stick. Staining and scanning electron microscopy of the attached hyphae reveals the presence of a mucilaginous sheath around the hyphae (Flentje, 1957; Fukutomi and Takada, 1979; Matsuura, 1986). This sheath is not present around the hyphae behind the point of attachment (Flentje, 1957). Attachment seem to be a prerequisite for infection to proceed. This is supported by the observation that an AG 2 mutant (ropy) that is apparently disturbed in adhesion is also not infectious (Flentje *et al.*, 1967). Factors governing the attachment are critical for understanding the regulation of the infection process.

B. Directed growth

Directed growth (Dodman and Flentje, 1970; Marshall and Rush 1980a; Armentrout and Downer, 1987; Armentrout *et al.*, 1987), which is characterized by growth of the hyphae along the anticlinal epidermal cell walls, is illustrated in Fig. 4. Also adhering hyphae growing on the plant surface in an apparent indeterminate manner are observed. Directed growth is best explained by a response of *R. solani* to enhanced leakage of nutrients or specific stimulatory components near the contact points between the epidermal cells, but also the topology of the surface is suggested to play a role (Dodman *et al.*, 1968b; Armentrout *et al.*, 1987). Either way, adhesion is a prerequisit for directed growth. Apparently the fungus is only able to sense either the surface or specific chemicals after the onset of adhesion.

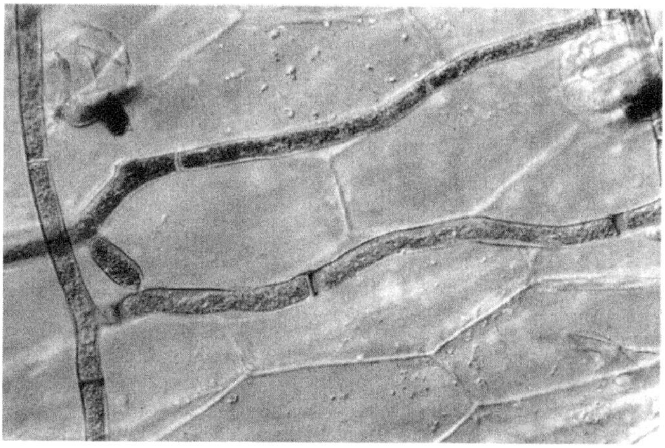

FIGURE 4. Directed growth of a *Rhizoctonia solani* AG 3 isolate on the stem of tomato

C. T-shaped branches and the formation of infection structures

In the sequence of the infection process, the formation of T-shaped branches are a determining characteristic. In this phase, hyphal internodes become shorter and short side branches are formed at right angles, which branch again at right angles forming the so-called feet (Armentrout and Downer, 1987). This is illustrated in Fig. 5. In this way simple branched appressoria-like structures with swollen hyphal tips can be formed (Fig. 6a). In other cases the parental hyphae continue to form side branches at short distances and the side branches show repetitive formation of feet. This results in an hyphal aggregate that is formed by one, but more often by several parental hyphae.

Finally branching can result in a tightly packed dome-shaped infection cushion (Fig. 6b). Also here, the hyphal tips are swollen and tightly adhered to the surface. (Dodman *et al.*, 1968b; Fukutomi and Takada, 1979, Matsuura, 1986).

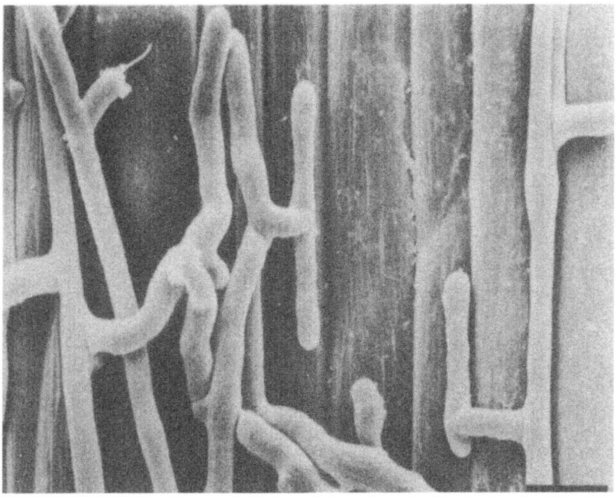

FIGURE 5. Scanning electron micrograph of a *Rhizoctonia solani* AG 4 isolate on the stem of a cotton seedling showing T-shaped branches and feet. (Reprinted with permission from Armentrout and Downer, 1987)

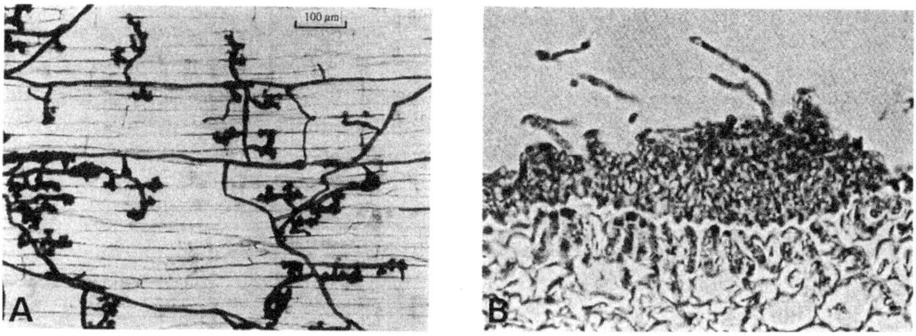

FIGURE 6. Panel A shows simple appressoria-like infection structures of an AG 5 isolate on potato (Reprinted with permission from Murray, 1982). Panel B shows a cross section of a dome shaped infection cushion (Reprinted with permission from Dodman *et al.*, 1968b)

D. Penetration

Penetration occurs typically by formation of thin infection pegs from tightly adhered swollen hyphal tips (Fig. 7). Generally, infection pegs are formed simultaneously from several hyphal tips (Flentje, 1957; Fukutomi and Takada, 1979). This may have a function in increasing the chance to find weak sites to penetrate or may have a role in overcoming local cellular defense reactions. The formation of a thin peg allows the fungus to exert maximal hydrostatic pressure on the plant surface. That hydrostatic pressure may be of importance in penetration is indicated by the observation that a hypovirulent strain is also melanin deficient (Sneh *et al.*, 1989). In

Magnaporte grisae and *Colletotrichum species* melanin has been shown to play an essential role in strengthening the cell wall allowing the fungus to generate sufficient hydrostatic pressure to penetrate (Howard and Ferrari, 1989; Kubo and Furusawa, 1991). Counter pressure has to be generated by adherence to the plant surface (Howard and Ferrari, 1989). By the formation of the complex infection structures *R. solani* optimizes adherence by attaching over a large surface area.

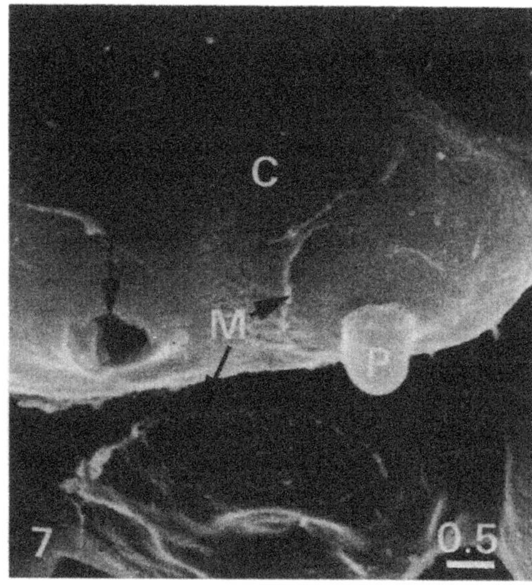

FIGURE 7. Scanning electron micrograph of a *Rhizoctonia solani* AG 1 isolate shows a penetration peg (P) from the flattened cell (C) with adhesive mucilage material (M) at the base of an infection cushion. (Reprinted with permission from Matsuura, 1986)

Although *R. solani* seems maximally specialized toward mechanical penetration of intact plant tissue, enzymes are probably also involved in preparation of the infection court (Trail and Köller, 1990) and in penetration (see chapter III.2). Based on hyphal tip morphology, enzymes are probably excreted locally from the infection peg tip (Fukutomi and Takada, 1979). Their role is therefore difficult to establish without the availability of mutants in which the gene of interest is specifically inactivated.

V. Entry routes and infection structures

To study the infection process, we developed an *in vitro* assay method that is schematically depicted in Fig. 8 (Keijer *et al.*, 1996b). Seedlings are grown under controlled aseptic conditions, with the roots adjacent to the agar and the stems off the agar. *R. solani* is introduced as small $3mm^3$ agar plugs with growing mycelium. The assay permits study of the behaviour of both the plant and the pathogen without disturbing the interaction. In the assay, the infection process of 32 isolates belonging to AG 1 through AG 6 and AG BI were studied on cauliflower, arabidopsis, eggplant, tomato and potato.

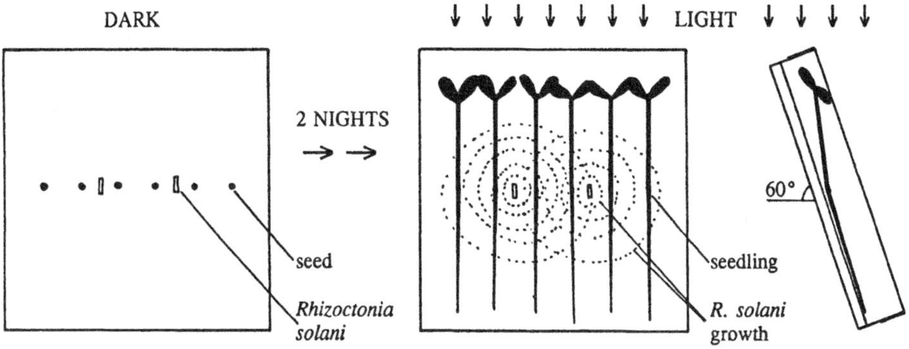

FIGURE 8. Schematic representation of the bioassay used for infection studies. Briefly: Surface sterilized seeds are germinated in the dark at 20°C on a layer of Gamborgs B-5 salts agar in a square petri-dish that are placed under a 60 degree angle. After 2 days the plates are transferred to a 12 hr day-night regime. *Rhizoctonia solani* is introduced as 3 mm^3 agar plugs from a 1 week old plate at day 0 or day 7.

The usual way of penetration of *R. solani* seems through intact tissue by the formation of infection structures, although it has been reported that *R. solani* can enter the plant through wounds (for example: Ullstrup, 1936; Nakayama, 1940; Khadga *et al.*, 1963.), lenticells (Ramsey, 1917) and stomata (Ullstrup, 1936; Christou, 1962; Dodman *et al.*, 1968a; Murray, 1982). Two cases of stomatal penetration were examined in more detail (J Keijer *et al.*, unpublished). In the interaction between cauliflower and AG BI isolate, localized necrotic lesions were formed on the stem of the seedling (Fig. 9a). Microscopic observation showed that all these lesions involved a stomata with a *R. solani* hypha attempting to enter. The observed necrosis was limited to the guard cells and few surrounding cells and did not spread. Fluorescence microscopy showed in each case a single dead hypha (Fig. 9b). Presumably, the attempt of *R. solani* to enter through the stomata resulted in a hypersensitive response, killing the stomatal cells as well as the entering hypha. A similar observation was reported by Christou (1962). In the other case, hyphae were observed entering the stomata without the occurrence of any symptoms. Further microscopic examination revealed hyphae that seemed to enter and then grow back, leaving the stomata (Fig. 10). Dodman *et al.* (1968b) mention that no consistent attraction of hyphal tips to stomata was observed, which agrees with our observations. Together, the observations indicate that stomatal penetration is not usual. Furthermore, entrance of hyphae in stomata should be interpreted with care, especially since we and others (Khadga *et al.*, 1963) did not find any evidence for stomatal penetration resulting in disease. Although Ullstrup (1936) and Murray (1982) reported progressive disease resulting from stomatal penetration, we only observed disease upon direct penetration after the formation of simple or complex infection structures.

The type of infection cushion formed may differ for different isolates. In our studies, we observed similarity of infection cushion morphology within AGs, independent of the plant species on which the cushions were formed. AG 2-1 isolates formed large compact dome shape infection cushions, with few hyphae connecting the separate cushions (Fig. 11a). Like AG 2-1, AG 4 isolates formed dome shaped infection cushions with few interconnecting hyphae. Generally the AG 4 cushions were larger in width, length and height (multilayered) than those of AG 2-1. AG 3 isolates formed relatively small infection cushions as condensed areas in a network of interconnecting hyphae (Fig. 11d). In addition to complex infection cushions simple appressoria-like infection structures were also generally encountered. Our observations agree with previous descriptions of the appearance of infection cushions of an AG 2-1 isolate on canola (Yang *et al.*, 1992), of an AG 3 isolate on the stems of potato (Hofman and Jongebloed, 1988), of AG 4 on bean and on radish (Dodman *et al.*, 1968b) and on cotton seedlings (Weinhold and Motta, 1973; Armentrout *et al.*, 1987). The infection cushions of AG 1 (Matsuura, 1986; Hirooka *et al.*, 1989) resemble those of AG 3 in that the cushions are formed in a net of interconnecting hyphae, but in contrast to AG 3 the cushions seem less complex, consisting of loosely intertwined hyphae.

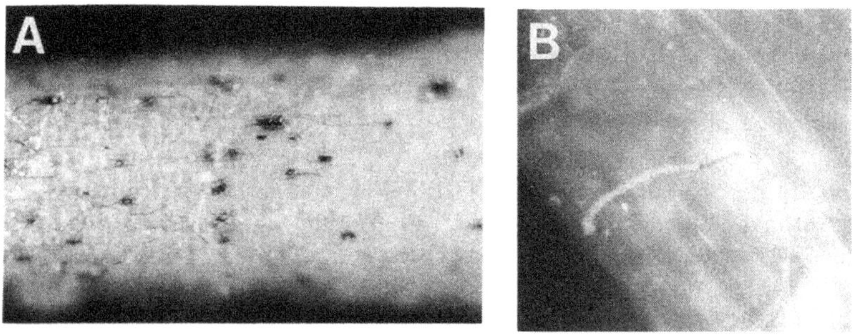

FIGURE 9. Panel A shows the hypersensitive response of cauliflower as the result of the attempts to stomatal penetration of a *Rhizoctonia solani* bridging isolate. Panel B shows the resulting killing of the penetrating hypha.

In our studies, the amount of disease that is caused by a particular isolate shows a linear correlation with the number and size of infection cushions that were formed by this isolate (Keijer *et al.*, 1996b). Similarly, Marshal and Rush (1980a) observed that the amount of lesions caused by an AG 1 isolate on different rice cultivars correlated linearly with the number of infection cushions. Others also observed that fewer infection cushions are formed on resistant versus susceptible cultivars of rice (Groth and Nowick, 1992) and soybean (Kousik *et al.*, 1994) and on resistant mustard versus susceptible canola (Yang *et al.*, 1992).

In their studies on rice, Marshall and Rush (1980a) also found that on cultivars with a low disease severity, resistant cultivars, only appressoria were formed. On susceptible cultivars increasing numbers of infection cushions were formed with the number of appressoria remaining constant. A similar observation was made by Stockwell and Hanchey (1984) for an AG 4 isolate on bean. These observations indicate that complex infection structures are more efficient, perhaps due to allowing simultaneous penetration over a larger area allowing for escape from defense responses. It remains unclear why enhanced formation of complex infection structures occurs on susceptible cultivars as compared to resistant cultivars. It may be caused by increased amounts of specific or non specific inducing compounds or to decreased amounts of inhibitory compounds. If specific compounds are involved and could be identified, these could be used for selection in resistance breeding programs.

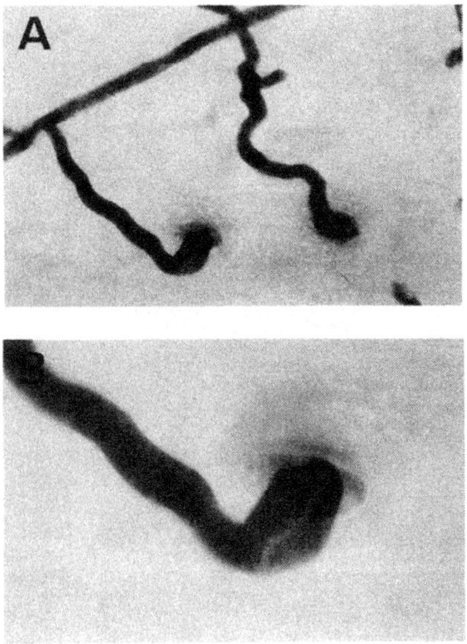

FIGURE 10. Hyphae of a *Rhizoctonia solani* AG 3 isolate entering and leaving stomata of a cauliflower seedling. Panel B is a 2.5 fold magnification over panel A. The hyphae are stained with trypan blue.

VI. HOST PLANT SPECIALIZATION

The infection process of *R. solani* suits its ectotrophic behaviour: The mode of entry is predominantly through intact tissue with apparent emphasis on mechanical penetration. This allows the fungus to potentially infect a wide range of hosts. Nevertheless, host plant specialization occurs within the *R. solani* species complex. For example, AG 3 isolates are predominantly recovered from diseased potatoes (Bandy *et al.*, 1988; Carling and Leiner, 1990) and AG 2-1 isolates are predominantly recovered from diseased cauliflower (Dijst, pers. comm.). In our experimental set-up (Fig. 8) a preference of AG 3 isolates for *Solanaceae* (potato, and tomato) was observed, while the isolates from the other AGs that were tested (predominantly AG 2) preferred *Cruciferae* (cauliflower and arabidopsis; Keijer *et al.*, 1996b). This agrees with observations in more complex field experiments (Richter and Schneider, 1953).

For a more detailed analysis of host plant specialization, two isolates with reciprocal pathogenicity were selected: Isolate 21R6 (AG 2-1), that causes disease to cauliflower seedlings but not at all to tomato seedlings, and isolate 3R8 (AG 3) that 3R8 is highly pathogenic to potato and tomato, but hardly to cauliflower (Keijer *et al.*, 1996b). Microscopic observation of the stems showed that in the pathogenic combinations (21R6 on cauliflower and 3R8 on tomato seedlings) the isolates go through the characteristic stages of the infection process, ultimately resulting in tissue colonization and seedling collapse. The hyphae attach, grow aligned to the epidermal cells, form T-shaped branches, infection cushions and penetrate. In the non-pathogenic combinations (21R6 on tomato and 3R8 on cauliflower), the hyphae grow to the plant and grow over the plant, but do not adhere nor proceed to further steps of the infection process (Fig. 11). Together these observations clearly demonstrate that host plant specialization is, at least in part, determined in the initial steps of the infection process.

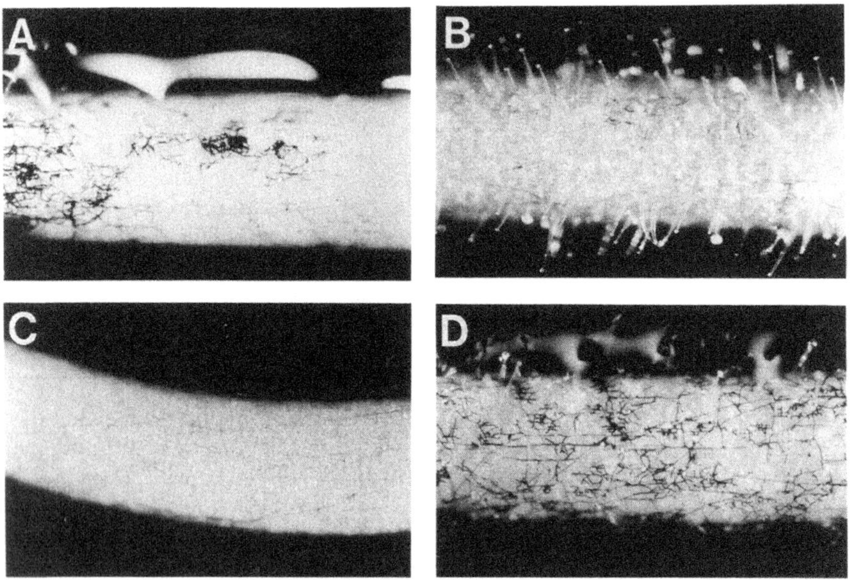

FIGURE 11. Reciprocal infection cushion formation and hyphal growth of *Rhizoctonia solani* isolates 21R6 (panels A and B) and 3R8 (panel C and D) on the stems of cauliflower (panels A and C) and tomato (panels B and D) seedlings. The hyphae are stained with trypan-blue.

VII. CHEMICAL OR SURFACE INDUCTION

It is not clear how host plant specialization is regulated. Initiation of the infection process as a result of thigmo-differentiation has been suggested. De Silva and Wood (1964) observed that infection cushions of two *R. solani* isolates could be host plant dependently induced by epidermal strips of the stem, but not by collected root exudates. Armentrout *et al.* (1987) studied infection cushion formation on polystyrene replicas of cotton seedlings stems. They observed that infection cushion formation was dependent on the presence of nutrients. No evidence for a specific chemical stimulus was found, since whole-plant exudates could be replaced by sucrose and, like infection cushion formation on whole plants (Weinhold and Bowman, 1974), could be repressed by methyl-glucose. Stimulation of internode shortening and branching in the presence of nutrients by replicas as compared to flat surfaces resulted in the conclusion that infection cushion formation is initiated by a topological signal.

In contrast, neither Marshall and Rush (1980b) nor Kousik *et al.* (1994) found infection cushion formation on polystyrene replicas of rice sheaths or soybean leaves, respectively. Both groups observed induction of infection cushions on the collodion membranes that were placed over susceptible plant parts, thereby providing support for induction of infection cushions by a chemical stimulus, a hypothesis originally put forward by Flentje (1957). In later papers Flentje and co-workers were able to show induction of infection cushions of AG 2-1 isolates on collodion membranes over agar blocks with radish stem exudates (Flentje *et al.*, 1963). Agar blocks without exudates did not induce infection cushions (Dodman and Flentje, 1970). Regrettably, Flentje *et al.* (1963) did not include agar blocks with exudates from other, susceptible, resistant or non-host plants in their experiments. Marshall and Rush (1980b) found for an AG 1 isolate that both complex and simple, appressoria-like infection structures were induced on collodion membranes that were placed over the stems of susceptible rice cultivars. On collodion membranes placed over stem pieces of resistant cultivars, only the appressoria-like infection structures were induced, but not the complex infection cushions. Similarly, using AG 1-IA , AG 1-IB and AG 1-IC isolates, Kousik *et al.* (1994) found that significantly more infection cushions were induced on collodion membranes placed over leaves of susceptible cultivars of soybean than on collodion membranes placed over resistant cultivars of soybean. In contrast to the AG 1 isolates, AG 4 and AG 5 isolates that formed infection cushions on soybean leaves did not form infection cushions on collodion membranes that were placed over soybean leaves (Kousik *et al.*, 1994). In our own experiments, we observed induction of infection cushions of isolate 21R6 (AG 2-1) on collodion membranes placed over cauliflower seedlings (Fig. 12), but not on collodion membranes placed over tomato seedlings (Keijer *et al.*, unpublished). Included as a positive control, isolate 3R8 (AG 3) formed infection cushions on collodion membranes placed over tomato seedlings. Together these observations firmly support the hypothesis that a chemical stimulus is the principal determinant in the induction of infection cushions. It should be noted that induction of infection structures was only observed on collodion membranes, but not on a number of other membrane surfaces (for example: cellophane, polyvinilidene difluoride (hydrophyllic and hydrophobic), polycarbonate and mixed cellulose esters

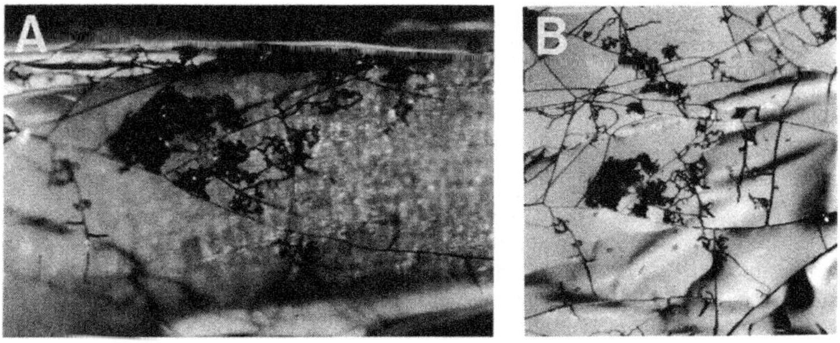

FIGURE 12. *Rhizoctonia solani* isolate 21R6 infection cushion formation on a collodion membrane that was placed over cauliflower seedlings. Panel A shows the membrane on the stem of the seedling. Panel B shows the same membrane after removal from the seedling. The hyphae are stained with trypan blue.

(cellulose acetate and cellulose nitrate, Keijer *et al.*, unpublished), indicating that the surface constitution is an important, but probably not determining, factor in host-plant specific infection structure formation.

Although a diffusible component seems the principal determinant for the induction of adherence and the subsequent steps of the infection process, nothing is known about the nature or mechanism of action of such component(s). Although induction by activation has been suggested, no proof is given and in theory a stimulus could also function by inhibition of constitutive repression of infection structure formation. With respect to regulation, no information is available whether a stimulus has to be present continuously and during all stages of the infection process, or that one or more transient stimuli are sufficient for induction and progression of the infection process.

VIII. CONCLUDING REMARKS

Over a number of years insight has been obtained in the morphology of the infection process of *R. solani*, a process with characteristics that fit the ectotrophic behaviour of this fungus. The initial steps of the infection process are characterized by adhesion of hyphae and an altered growth pattern: Directed growth and the formation of T-shaped branches. The simple and complex infection structures that are subsequently formed allow the fungus to penetrate intact plant tissue.

Differences in susceptibility and host plant specialization are, at least in part, determined in the initial steps of the infection process (Marshall and Rush, 1989b; Kousik *et al.*, 1994; Keijer *et al.*, 1996b). By using molecular approaches to screen for differentially induced genes, it should be possible to obtain insights into the components and mechanisms that regulate the infection process, and in particular those factors that govern its induction. Understanding of the regulation of infection will open new avenues to the development of control measures for *R. solani*, based on resistance breeding and rotation management.

IX. ACKNOWLEDGEMENTS

The contribution of Annette Dullemans, Petra Houterman and Marja Korsman to the research described in this chapter is gratefully acknowledged.

X. REFERENCES

Adams GC (1988) *Thanatephorus cucumeris* (*Rhizoctonia solani*), a species complex of wide host range. Adv. Plant Pathol. 6:535-552.

Anderson NA & Stretton HM (1978) Genetic control of pathogenicity in *Rhizoctonia solani*. Phytopathology. 68:1314-1317.

Armentrout VN & Downer AJ (1987) Infection cushion development by *Rhizoctonia solani* on cotton. Phytopathology 77:619-623.

Armentrout VN, Downer AJ, Grasmick DL & Weinhold AR (1987) Factors affecting infection cushion development by *Rhizoctonia solani* on cotton. Phytopathology 77:623-630.

Baker KF (1970) Types of *Rhizoctonia* disease and their occurence. In: Parmeter JR (ed.) *Rhizoctonia solani*, biology and pathology (pp 125-148). University of California Press, Berkeley.

Bandy BP, Leach SS & Tavanzis SM (1988) Anastomosis group 3 is the major cause of *Rhizoctonia* disease of potato in Maine. Plant Dis. 72:596-598.

Carling DE & Leiner RH (1990) Virulence of isolates of *Rhizoctonia solani* AG 3 collected from potato plant organs and soil. Plant Dis. 74:901-903.

Christou T (1962) Penetration and host-plant relationships of *Rhizoctonia solani* in the bean plant. Phytopathology 52:291-300tionships of

De Silva RL & Wood RKS (1964) Infection of plants by *Corticium solani* and *C. praticola* - effect of plant exudates. Trans. Brit. Mycol. Soc. 47:15-24.

Dodman RL, Barker KR & Walker JC (1968a) Modes of penetration by different isolates of *Rhizoctonia solani*. Phytopathology 58:31-33.

Dodman RL, Barker KR & Walker JC (1968b) A detailed study of the different modes of penetration by *Rhizoctonia solani*. Phytopathology 58:1271-1276.

Dodman RL & Flentje NT (1970) The mechanism and physiology of plant penetration by *Rhizoctonia solani*. In: Parmeter JR Jr (ed.) *Rhizoctonia solani*, Biology and Pathology (pp 149-160). University of California Press, Berkeley.

El Samra IA, El Faham YM & Kamara AM (1981) Selective induction of infection cushions by *Rhizoctonia solani* in relation to host responses. Phytopath. Z. 102:122-126.

Flentje NT (1957) Studies on *Pellicularia filamentosa* (Pat.) Rogers III. Host penetration and resistance and strain specialization. Trans. Brit. Mycol. Soc. 40:322-336.

Flentje NT, Dodman RL & Kerr A (1963) The mechanism of host penetration by *Thanatephorus cucumeris*. Aust. J. Biol. Sci. 16:784-799.

Flentje NT, Stretton HM & McKenzie AR (1967) Mutation in *Thanatephorus cucumeris*. Aust. J. Bot. Sci. 20:1173-1180.

Flentje NT, Stretton HM & AR McKenzie (1970) Mechanisms of variation in *Rhizoctonia solani*. In: Parmeter JR (ed.) *Rhizoctonia solani*, Biology and Pathology. (pp 52-65). University of California Press, Berkeley.

Fukutomi M & Takada H (1979) Ultrastructure of infection process of *Rhizoctonia solani* Kühn in cucumber hypocotyls. Ann. Phytopath. Soc. Jpn 45:453-462.

Groth DE & Nowick EM (1992) Selection for resistance to rice sheat blight through number of infection cushions and lesion type. Plant Dis. 76:721-723.

Hofman TW & Jongebloed, PHJ (1988) Infection process of *Rhizoctonia solani* on *Solanum tuberosum* and effects of granular nematocides. Neth. J. Plant Pathol. 94:234-252.

Hirooka T, Miyagi Y, Araki F & Kuno H (1989) Biological mode of action of flutolanil in its systemic control of rice sheath blight. Phytopathology 79:1091-1094.

Howard RJ & Ferrari MA (1989) Role of melanin in appressorium function. Exp. Mycol. 13:403-418.

Julian MC, Debets F & Keijer J (1996) Independence of sexual and vegetative incompatibility mechanisms of *Thanatephorus cucumeris* (*Rhizoctonia solani*) anastomosis group 1. Phytopathology (in press).

Keijer J, Houterman PM, Dullemans AM & Korsman MG (1996a) Heterogeneity in electrophoretic karyotype within and between anastomosis groups of *Rhizoctonia solani*. Mycol. Res. (in press).

Keijer J, Dullemans AM, Houterman PM & Korsman MG (1996b) Analysis of *Rhizoctonia solani* host plant specialization in a novel *in vitro* pathogenicity assay. (Submitted).

Kenning LA & Hanchey P (1980) Ultrastructure of lesion formation in *Rhizoctonia* infected beans hypocotyls. Phytopathology: 70:998-1004.

Kerr A (1956) Some interactions between plant roots and pathogenic soil fungi. Aust. J. Biol. Sci. 9:45-52.

Kerr A & Flentje NT (1957) Host infection of *Pellicularia filamentosa* controlled by chemical stimuli. Nature 179:204-205.

Khadga BB, Sinclair JB & Exner BB (1963) Infection of seedling cotton hypocotyl by an isolate of *Rhizoctonia solani*. Phytopathology 53:1331-1336.

Kousik CS, Snow JP & Berggren GT (1990) Infection cushions development by *Rhizoctonia solani* Kühn on soybean leaves and factors affecting it. Phytopathology 80:968 (abst).

Kousik CS, Snow JP & Berggren GT (1994) Factors affecting infection cushion development by *Rhizoctonia solani* Ag 1-IA and IB on soybean leaves. Plant Pathol. 43:237-244.

Kubo Y & Furusawa I (1991) Melanin biosynthesis: prerequisite for successful invasion of the plant host by appressoria of *Colletotrichum* and *Pyricularia*. In: Cole CT and Hoch HC (eds.) The Fungal Spore and Disease Initiation in Plants and Animals. (pp 205-208). Plenum Press, New York.

Marshal DS & Rush MC (1980a) Relation between infection by *Rhizoctonia solani* and *R. oryzae* and disease severity in rice. Phytopathology 70:941-946.

Marshall DS & Rush MC (1980b) Infection cushion formation on rice sheats by *Rhizoctonia solani*. Phytopathology 70:947-950.

Martinson CA (1965) Formation of infection cushions by *Rhizoctonia solani* on synthetic films in soils. Phytopathology 55:129 (abst.)

Matsuura K (1986) Scanning electron microscopy of the infection process of *Rhizoctonia solani* in leaf sheats of rice plants. Phytopathology 76:811-814.

Murray DIL (1982) Penetration of barley root and coleoptile surfaces by *Rhizoctonia solani*. Trans. Brit. Mycol. Soc. 79:354-360.

Nakayama T (1940) A study on the infection of cotton seedlings by *Rhizoctonia solani*. Ann. Phytopathol. Soc. Japan 10:93-103.

Nour El Dein MS & Sharkas MS (1964) The pathogenicity of *Rhizoctonia solani* in relation to different tomato root exudates. Phytopathol. Z. 51:285-290.

Papavizas GC (1970) Colonization and growth of *Rhizoctonia solani* in soil. In: Parmeter JR Jr (ed.) *Rhizoctonia solani*, Biology and Pathology (pp 108-122). University of California Press, Berkeley.

Papavizas GC & Ayers WA (1965) Virulence, host-range and pectolytic enzymes of single basidiospore isolates of *Rhizoctonia praticola* and *Rhizoctonia solani*. Phytopathology 55:111-116

Papavizas GC & Davey CB (1962) Activity of *Rhizoctonia* in soil as affected by carbon dioxide. Phytopathology 52:759-766.

Ramsey GB (1917) Form of potato disease produced by *Rhizoctonia solani*. J. Agric. Res. 9:421-426.

Reddy MN (1980) Studies on groundnut hypocotyl exudates and the behaviour of *Rhizoctonia solani* in influencing the disease. Plant Soil 55:445-454.

Richter H & Schneider R (1953) Untersuchungen zur morphologischen und biologischen differenzierung van *Rhizoctonia solani* Kühn. Phytopathol. Z. 20:165-226.

Ruppel EG (1973) Histopathology of resistant and susceptible sugar beet roots inoculated with *Rhizoctonia solani*. Phytopathology 63:123-126.

Sherwood RT (1970) Physiology of *Rhizoctonia solani*. In: Parmeter JR Jr (ed.) *Rhizoctonia solani*, Biology and Pathology (pp69-92). University of California Press, Berkeley.

Sneh B, Ichielevich-Auster M & Shomer I (1989) Comparative anatomy of colonization of cotton hypocotyls and roots by virulent and hypovirulent isolates of *Rhizoctonia solani*. Can. J. Bot. 67:2142-2149.

Stockwell V & Hanchey P (1984) The role of the cuticle in resistance of beans to *Rhizoctonia solani*. Phytopathology 74:1640-1642.

Trail F & Köller W (1990) Diversity of cutinases from plant pathogenic fungi: Evidence for relationship between enzyme properties and tissue specificity. Physiol. Mol. Plant Pathol. 36:495-508.

Ullstrup AJ (1936) Leaf blight of aster caused by *Rhizoctonia solani*. Phytopathology 26:981-990.

Weinhold AR & Bowman T (1974) Repression of virulence in *Rhizoctonia solani* by glucose and 3-O-methyl-glucose. Phytopathology 64:985-990.

Weinhold AR & Motta J (1973) Initial host responses in cotton to infection by *Rhizoctonia solani*. Phytopathology 63:157-162.

Yang J, Verma PR & Tewari JP (1992) Histopathology of resistant mustard and susceptible canola hypocotyls infected by *Rhizoctonia solani*. Mycol. Res. 96:171-179.

III.2 *RHIZOCTONIA SOLANI*: PENETRATION, COLONIZATION, AND HOST RESPONSE

ALBERT R. WEINHOLD[1] AND JAMES B SINCLAIR[2]

[1]*University of California, 151 Hilgard Hall, Berkeley, CA 94720 USA,*
[2]*University of Illinois, 1101 W. Peabody, Urbana, IL 61801-4723 USA.*

I. INTRODUCTION

Studies on penetration, colonization and host response have been done with representative isolates of AGs 1, 2, 3 and 4. Most of the information available, however, is for AG 4, although sufficient work has been done with the other groups to provide a basis for comparison and contrast.

It is necessary to decide what constitutes the initial stage in the infection process. In almost all cases host penetration occurs from some type of infection structure. The infection structure is also where substances are exchanged between the pathogen and the host plant. These include materials such as extracellular enzymes from *Rhizoctonia* and exudates from the host. Thus, to provide continuity, a brief discussion of the formation of these structures will be included in this presentation. A detailed treatment of the early stages in the infection process is presented in chapter III.1.

II. INFECTION STRUCTURE FORMATION

It is clear that host penetration by *Rhizoctonia solani* usually involves some type of infection structure. Infection through stomata has been reported but is limited to isolates of AGs that infect leaves. There is considerable variation in infection structure morphology. Infection structures generally are described as either lobate appressoria or dome-shaped infection cushions. Dodman *et al.* (1968a) studied the modes of penetration by different isolates of *R. solani*. They found that isolates from stems and roots of plants usually penetrated from infection cushions, whereas isolates from foliage commonly penetrated from lobate appressoria. This correlation held for 72% of the isolates they studied. As more work has been done, involving different anastomosis groups and different hosts, it has become apparent that there is no clear distinction between appressoria and cushions.

The initial response of *R. solani* to contact with the surface of a compatible host is typically the alignment of hyphae along the junction between epidermal cell walls. This is followed by lateral branching, hyphal swelling and coiling, and additional right angle branching to form the so-called feet. At this stage the structure could be considered a lobate appressoria. If the branching process continues, a more complex structure is formed and this may proceed until a dome-shaped infection cushion of several cell layers is produced. The point at which an infection structure becomes sufficiently complex to be considered an infection cushion depends, in part, on the perception of the investigator.

Despite some confusion in terminology the morphological complexity of the infection structure is characteristic of each AG. Although there can be variation

B. Sneh et al. (eds.),
Rhizoctonia Species: Taxonomy, Molecular Biology, Ecology, Pathology and Disease Constrol, 163–174.

among isolates within a group some generalizations can be made. Infection structures formed by members of AG 1 are discrete areas of hyphal branching and coiling but they are a single layer of hyphae which are loosely intertwined. A good scanning electron photomicrograph and description is presented by Matsuura (1986). Similar pictures and descriptions have been published by other investigators (Hirooka *et al.*, 1989).

In contrast to the relatively simple infection structures formed by AG 1, the AG 2 cushions are composed of tightly coiled hyphae several cell layers thick. The formation of infection cushions by AG 2-1 on the susceptible canola cultivar Westar is described by Yang *et al.* (1992). They observed a single unbranched hyphae growing on the surface of the hypocotyl along the groove between anticlinal walls of epidermal cells 24 h after inoculation. Hyphae grew tightly and regularly on the hypocotyl surface and branched almost at right angles. The lateral hyphae often branched again bidirectionally when the hyphal tip reached another groove. The resulting hyphae grew parallel to the long axis of the hypocotyl and continued to form further branches. The branches of the hyphae tended to coil and overlap to form a tight infection cushion.

The infection structures formed by AG 3 on potato appear to be intermediate in complexity between those of AG 1 and AG 2-1 formed on their respective hosts. Their formation is described by Hofman and Jongebloed (1988). Initially, runner hyphae grew mainly longitudinally along the sprout. The hyphae tended to follow the junctions of the epidermal cells and at certain sites primary branches formed. They consisted of straight cells but shorter than those of the longitudinal hyphae. From primary branches secondary branches developed which consisted of short swollen cells. These often branched several times to form infection cushions i.e. dense masses of swollen cells.

Of all the anastomosis groups, AG 4 has been studied most extensively and the formation of infection structures has been described by several investigators (Armentrout and Downer, 1987; Dodman *et al.*, 1968b; Fukutomi and Takada, 1987; Weinhold and Motta 1973). The first stages of infection structure formation are very similar for all of the anastomosis groups that have been studied. Hyphae grow longitudinally along the host surface with a tendency to follow the junction between epidermal cells. At certain sites lateral branches form and the infection structure is initiated. With AG 4 however, the branching, curling and aggregation of hyphae continues which leads to the formation of a tightly compacted dome-shaped infection cushion that is many cell layers thick. The cushions are tightly appressed to the cuticle and the hyphal cells in contact with the host often have swollen, flattened tips.

III. Role of pathogen nutrition in the infection process

Energy is required for *R. solani* to form infection structures and penetrate the host plant. The influence of nutrition was shown in a study in which the amount of carbon or nitrogen in the inoculum was varied (Weinhold *et al.*, 1969). The nutritional status of *R. solani* mycelium used to inoculate cotton hypocotyls was influenced by growing the pathogen on a liquid medium containing various concentrations of either glucose or asparagine. There was a direct relationship between the concentration of carbon or nitrogen in the medium and disease severity as determined by the area of macerated tissue. When inoculum was deficient in nutrients, mycelium would grow vegetatively on the host surface but infection structures did not develop.

In further studies it was shown that exogenous sources of nutrients can have a marked effect on the pathogenic activities of *R. solani* (Weinhold *et al.*, 1972). Mycelium growing from inoculum deficient in nitrogen rapidly absorbed nitrogen containing compounds from the external environment and utilized them in pathogenesis. The potential importance of soil as a source of nutrients for *R. solani* was shown by the fact that in the presence of soil, disease severity on cotton resulting from nitrogen deficient inoculum was increased. At all levels of inoculum nutrition seed inoculation resulted in greater disease severity than did hypocotyl inoculation. This provides additional evidence that seed exudates are an important source of nutrients available to seed and seedling pathogens.

IV. THE INFECTION PROCESS

Following infection structure formation the infection process proceeds to penetration, colonization, and subsequently host response leading to disease. As with infection structure formation there are differences among AGs in penetration, colonization, and host response. Therefore, each step in the process will be considered separately and the four AGs for which information is available will be compared and contrasted.

A. Penetration

A description of penetration into leaf sheaths of rice by AG 1 has been provided by Matsuura (1986). Infection cushions were closely appressed to the epidermis of the leaf sheaths and it appeared that individual hyphae became attached to the epidermis by means of a mucilage-like material. It was commonly observed that numerous fine penetration pegs developed from the flattened cells at the base of the infection structures. Penetration was effected by these pegs, which were 1-2 microns in diameter. Once the infection pegs reached the lumen of epidermal cells they resumed normal diameter giving rise to infection hyphae. Penetration through stomata was infrequently observed. Hirooka *et al.*, (1989) studied the penetration of rice plant leaf sheaths by AG 1-IA. Their description of the process was very similar to that presented by Matsuura.

The penetration of potato sprout tissue by AG 3 was described by Hofman and Jongebloed (1988). Under the infection cushions, many hyphae penetrated the epidermal cells. Scanning electron photomicrographs clearly show the formation of penetration pegs. This process may be both mechanical and enzymatic. The scanning electron microscope showed light-colored tissue around the penetration sites. Such a halo indicates the cell walls are locally less thick. This is likely due to enzymatic action.

Isolates of AG 2-1 formed infection pegs and penetrated directly through the cuticle and epidermis of canola (Yang *et al.*, 1992). Mucilage-like materials were not observed to form during the penetration process. Ruppel (1973) compared penetration of resistant and susceptible sugar beet roots by *R. solani*. Roots were penetrated directly by individual hyphae or by hyphae from infection cushions. After penetration necrosis and some tissue degradation preceded advancing hyphae.

Penetration of host tissue by AG 4 has been studied far more extensively than any of the other AGs (Christou, 1962; Dodman *et at.*, 1968a and 1968b; Fukutomi and Takada, 1979; Kenning and Hanchey, 1980; Lisker *et al.*, 1976; Weinhold and Motta, 1973). The observations by the different investigators who studied the process on a variety of hosts, including bean, radish, cucumber, and cotton are generally consistent. Penetration is directly through the cuticle by means of fine infection pegs of approximately 1-2 microns in diameter. The primary purpose of the infection pegs appears to be to penetrate the cuticle. In general the cuticle is not degraded and can often be seen separated from the epidermal cell walls. After penetrating the cuticle the hyphae resume their normal diameter. In some cases infection pegs continue through the epidermal cell wall before expanding to normal size. Penetration may, at times, be accomplished by hyphae without formation of infection pegs. Weinhold and Motta (1973) stated that there was no evidence that infection pegs were involved in penetration of cotton hypocotyls. Dodman *et al.*, (1968b) observed that infection pegs were involved in penetration of bean but not radish. They suggested this may be a function of cuticle strength. Stockwell and Hanchey (1984) with bean and Marshall and Rush (1980) with AG 1 on rice suggested that cuticle may be a factor in host resistance. Several investigators have observed the formation of mucilage-like material that held hyphae together in the infection cushion and served to adhere the cushion to the host surface during penetration (Armentrout and Downer, 1987; Fukutomi and Takada, 1987). Whether this is a widespread phenomenon with AG 4 is unclear. Kenning and Hanchey (1980) reported little or no mucilage at the edge of cushions. In other studies (Dodman *et al.*, 1968b; Weinhold and Motta, 1973) there was no mention of mucilage. This could have been because it was absent or because the method of tissue preparation or type of observation would not detect its presence.

R. solani has been shown to produce cutinase (Baker and Bateman, 1978; Kolattukudy, 1985) and it is suspected that penetration of the cuticle involves, at least in part, enzymatic action. Trail and Koller (1990) presented evidence indicating a relationship between cutinase properties and the expression of tissue specificity by directly penetrating fungi. The role of cutinases in plant-pathogen interactions is the subject of chapter III.1. Direct evidence is available for the involvement of enzymes in the penetration of the cell wall. Weinhold and Motta (1973) studied the initial host response in cotton to infection by *R. solani* AG 4. Fresh sections approximately 50 microns thick were cut with a Vibratome so there were no artifacts created by tissue fixation. They detected the loss of pectic substances in epidermal cell walls during the early, extensive branching stage of pathogen development (Fig. 1A). As cushions began to develop many cells beneath the pathogen exhibited loss of pectic substances (Fig. 1B). The initial loss of these materials was not uniformly distributed but occurred in discrete localized areas. A few hours after cushion formation, approximately 18 h after inoculation, the protoplast of the cells beneath the structures became slightly discolored. At this stage removal of pectic substances and loss of cell wall structure was readily apparent (Fig. 1C).

Observations of host tissue sections with polarized light revealed considerable change in the crystalline structure of cell wall cellulose (Fig. 2). Throughout the initial process of lesion formation there was no penetration of host tissue by the pathogen. Sections stained with Sudan IV show this clearly, with the cuticle well defined beneath the infection cushion (Fig. 1E). Only after cell discoloration and development of a cavity beneath the infection cushion did the fungus grow into the host tissue (Figs. 1D

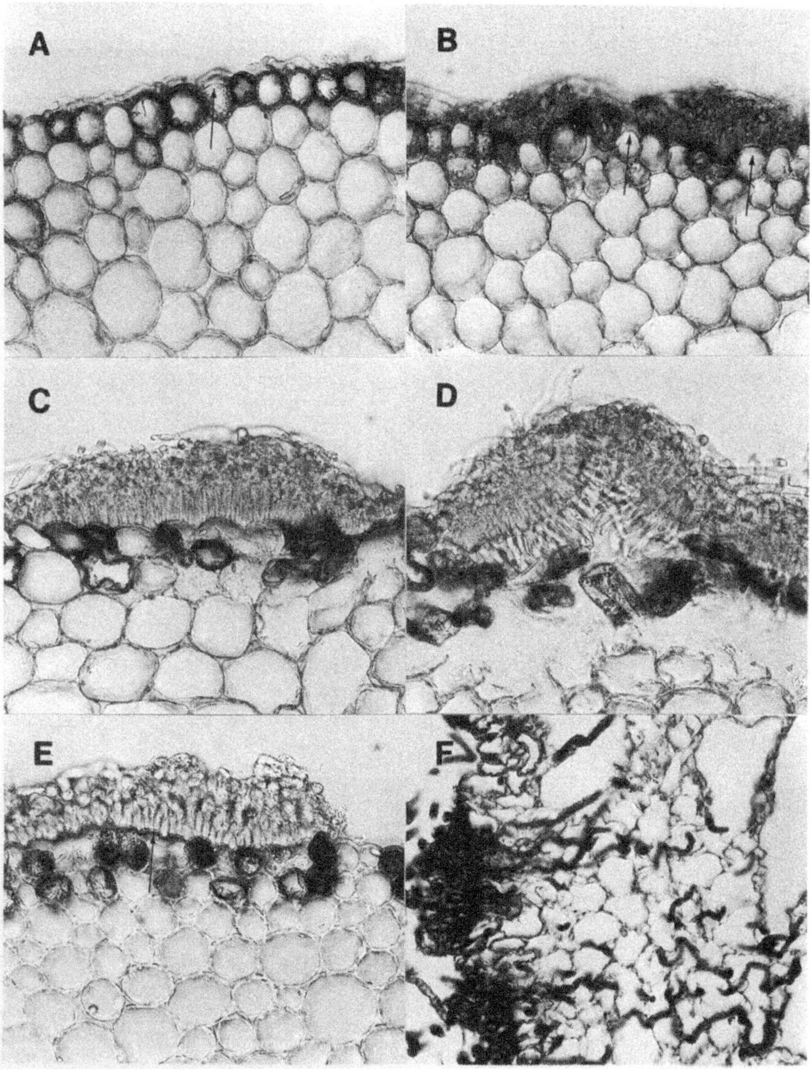

FIGURE 1. Progressive steps in attack of cotton hypocotyl by *Rhizoctonia solani*. Sections A-D stained with ruthenium red. A) Extensive hyphal branching but infection cushions have not formed. Some loss of ruthenium red positive materials (arrow) in epidermal cell walls can be detected. B) Initial infection cushion formation with loss of ruthenium positive materials (arrow) more pronounced. C) Infection cushions were well formed and loss of cell wall structure was apparent. D) Extensive damage to host cells with pathogen growing into cavity formed below infection cushion. E) Pathogen development comparable to C but section stained with Sudan IV. Cuticle (arrow) is intact and tightly appressed against base of infection cushion. Extensive cell wall damage but no penetration. F) Section through edge of macerated cavity after penetration showing hyphae in relatively intact cells.

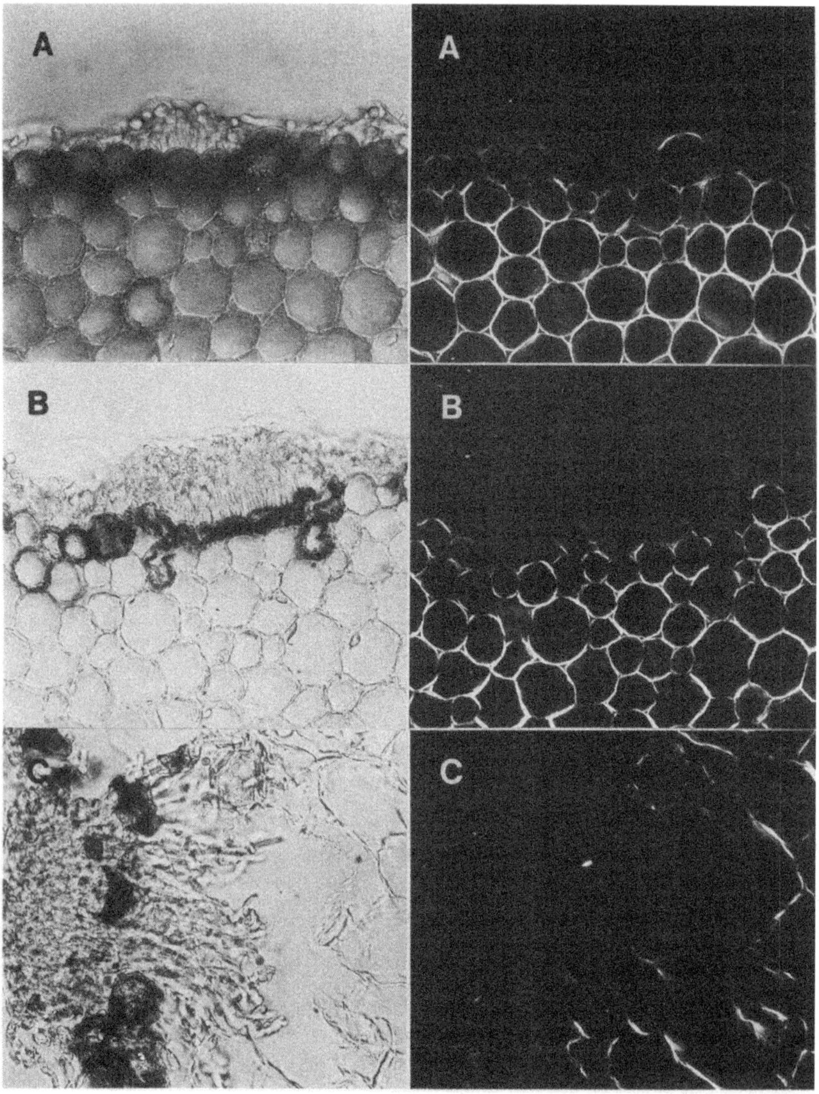

FIGURE. 2. Progressive steps in attack of cotton hypocotyl by *Rhizoctonia solani*. Paired plates are identical. Those on right were taken with polarized light. Sections were not stained. A) early infection cushion formation. Note loss of birefringence in cell walls beneath the infection cushion. B) Cell wall damage extending into second layer of epidermal cells. C) Pathogen growing into macerated cavity. Extensive cell wall damage is occurring in advance of the pathogen.

and 1F). The infection process in bean and radish was observed and found to be similar to that in cotton. These observations are consistent with the conclusion that *R. solani* AG 4 produces pectinolytic and cellulolytic enzymes during the early stages of infection and that these enzymes play a crucial role in host penetration. Marcus *et al.*, (1986) suggested that endopectinlyase may be involved in tissue degradation in the later stages of infection. They also reported the interesting observation that endopectinlyase was present in virulent but not in hypovirulent isolates of *R. solani*. Also, *R. solani* AG 2-2IIIB, the causal agent of soybean seedling root rot (Liu and Sinclair, 1991), has been shown to produce at least 10 distinct extracellular enzymes (Bertagnolli *et al.*, 1996). The reported extracellular enzymes secreted by *R. solani* includes pectinase, pectin lyase, pectin methylesterase, cellulase, and phosphatase (Bateman and Basham, 1976; Ko and Hora, 1970; Lister *et al.*, 1975; Sherwood, 1966; Tseng and Bateman, 1968) Another line of evidence for the involvement of pectin degrading enzymes is the report that an exogenous supply of glucose markedly reduced lesion development on cotton (Weinhold and Bowman, 1974). A solution of 28 mM glucose prevented the production of pectinase by *R. solani in-vitro*. This led to the conclusion that the exogenous supply of glucose interfered with disease development through catabolyte repression of pectinolytic enzyme production and thus prevented tissue maceration. Consistent with the early production of cell wall degrading enzymes was the demonstration by Brookhouser and Weinhold (1979) that cotton seed and hypocotyl exudates induced the production of pectinase.

B. Colonization

Penetration of the epidermal cell wall by *R. solani* is followed by branching in the cell lumen and growth into adjacent cells. Studies on AG 1 have been limited to penetration but colonization of host tissue by AGs 2-1, 3, and 4 has been well documented. With these three AGs colonization is characterized by cell wall degradation and cell damage in advance of the hyphae.

Colonization of potato tissue by AG 3 was initially restricted to one or two cell layers underneath the infection cushion (Hofman and Jongebloed, 1988). Lesions developed only under infection cushions and with a margin that exceeded the infection cushion by about 2mm. Finally, lesions were up to 12 cell layers deep, reaching the vascular bundle of the stem. Girdling of a sprout by this type of lesion caused death, which occurred mostly on young sprouts. Except during initial colonization of epidermal cells the mycelium was never observed in healthy tissue. On older lesions the mycelial mass of the infection had increased and straight-walled brown hyphae were present on noninfected parts of the sprout. Even at this stage, colonization of plant tissue was restricted to the part directly under the infection cushion. Therefore AG 3 does not colonize potato tissue progressively (Hofman and Jongebloed 1988).

With both AG 2-1 (Yang *et al.*, 1992) and AG 4 (Fukutomi and Takada, 1979; Kenning and Hanchey, 1980; Van Etten *et al.*, 1967; Weinhold and Motta, 1973) colonization is more extensive than with AG 3. In the susceptible canola cultivar Westar the pathogen (AG 2-1) penetrated the epidermis and rapidly invaded the host tissue up to the fourth layer of cells. Hyphae extended in all directions within the outer layer of cortex, both inter and intra-cellularly. The advancing hyphae grew rapidly in the cells and branched extensively. Then they grew deeper into the cortex and endodermis, and finally the vascular tissue. Eighty- four hours after inoculation the entire cortex was colonized. The invaded cells were collapsed and tissues were filled with mycelium. Cells away from the advancing hyphae, however, were apparently normal. In severely infected hypocotyls the entire cortex was completely disintegrated. The endodermis, phloem, and xylem parenchyma cells were affected but not the xylem elements. It appeared that the fungus was not able to colonize cells with lignified walls (Yang *et al.*, 1992).

Colonization of host tissue by AG 4 is similar to that described for AG 2-1 except that the process is somewhat more rapid with more distinct changes in the cells in advance of growing hyphae. A detailed description of the colonization of bean is provided by Kenning and Hanchey (1980). After 24 hours the pathogen had penetrated to the second layer of cortical cells, after 36 hours to the sixth cortical layer and after 48 hours to the ninth cortical layer. Mycelial growth occurred both inter and intracellularly.

Extensive cell swelling was found in all regions adjacent to and up to two cell layers away from the pathogen. The cell walls appeared to offer little resistance to penetration since hyphal constrictions were not common. Extensively degraded host cell walls consisted of a herringbone pattern of fibrils and smaller densely-stained granules. The pattern of digestion of the cell walls supports the hypothesis that the granules represent degraded pectic materials.

In addition to the extracellular enzymes known to be produced by *R. solani* evidence for the secretion of DNAse, RNAse, alpha-amylase, chitinase, beta-glucanase, xylanase, protease, and urease has been reported (Bertagnolli *et al.*, 1996). Additional studies using HPLC chromatography and SDS-polyacrylamide gel electrophoresis suggest the presence of at least seven, as yet unidentified, extracellular proteins produced by *R. solani* (Bertagnolli and Sinclair Unpublished). The role of these proteins in the penetration and colonization process has not been determined.

C. Host Response

Detailed descriptions are available for the reaction of canola to infection by AG 2-1 (Yang *et al.*, 1992) and of bean by AG 4 (Kenning and Hanchey, 1980). The reaction of two hosts to isolates of two different anastomosis groups appears to be very similar. Therefore, the reaction of canola to AG 2-1 will serve as an example. Infection hyphae penetrated directly through the cuticle and frequently through the junction between the anticlinal walls of epidermal cells. The cytoplasm of the invaded cells was collapsed. In adjacent epidermal cells plasmolysis and destruction of cellular organelles occurred. Plasmolysis was also observed in parenchyma cells of the second layer of the cortex. A granular appearance and vacuolation of the cytoplasm was observed in most cells of infected tissues. Cell wall swellings were observed in all infected cells, in adjacent cells, and those two or three cells away from the hyphae. A change in the fine structure of the cellulose layers of the cell wall was easily recognized. The nucleus showed little change in its form in the early stages of infection. In cells with severe plasmolysis the nuclear envelope was partially disintegrated. In severely infected tissue the mitochondria were completely disorganized. In infected tissue the plastids were swollen, the plastid membrane became indiscernible and partially disintegrated but the grana structure could still be seen. Additionally, the endoplasmic reticulum in the infected cells became dilated and swollen. In general, the degeneration of organelles in infected cells proceeded rapidly. The entire degeneration of organelles and cytoplasm could be observed as early as 36h after inoculation. The early plasmolysis and cytoplasm damage is consistent with the finding of Lai and Weinhold (1968) that increases in cell wall membrane permeability could be detected in mung bean hypocotyls 14 to 18 h after inoculation with AG 4. These results suggest that increased cell membrane permeability is an initial host response to infection by *R. solani*.

V. Host Resistance

The following discussion considers resistance primarily from the perspective of host reaction. The availability of germplasm for resistance and strategies for molecular resistance breeding are reviewed in chapters VI.C1 and VI.C2 respectively. Host resistance to *R. solani* is typically an incremental reduction in disease severity. This is probably a reflection of the rather nonspecific cell damage that precedes colonization by the pathogen. Yang *et al.* (1992) compared the infection process of AG 2-1 in resistant mustard and susceptible canola. Infection cushions formed on both hosts but were appreciably less frequent on the resistant host. Bassi *et al.* (1979) studied the histopathology of resistant and susceptible tomato fruit infected by *R. solani*. They found that the susceptible cultivar was penetrated by numerous infection pegs emanating from infection cushions whereas in the resistant fruit penetration was by individual hyphae. Advancement of the pathogen in all tissues was slower in the resistant than in the susceptible host. Enzyme inhibition has also been implicated in resistance to *Rhizoctonia*. Bugbee (1993) found that a protein from sugar beet roots inhibited pectin lyase isolated from *R. solani*. Pectin lyase inhibitory activity was greater in cell wall extracts from rotted and adjacent tissue than from healthy tissue. Inhibitor concentration was greater in root rot resistant germ plasm than a susceptible cultivar. It was also higher in root than hypocotyl or crown tissue.

Cuticle thickness has been implicated as a factor in resistance in two host /pathogen systems. Marshall and Rush (1980) in their study of infection cushion formation on rice sheaths by AG 1 observed abundant wax deposits on the outer sheath of resistant but not on susceptible cultivars. Cultivars intermediate in resistance had varying amounts of wax on their outer sheaths. Removal of wax deposits from resistant cultivars with chloroform, followed by inoculation, resulted in a susceptible reaction. *R. solani*, AG 4, is a pathogen of juvenile tissue and as plants age they become resistant. Stockwell and Hanchey (1984) investigated two factors suggested to be important in this resistance, calcification of cell walls and increased cuticle thickness. They altered the cuticle of older plants in several ways and found that the fungus formed infection cushions and lesions on the treated older plants similar to those formed on young plants. They suggested that cuticle thickness plays a more important role than calcification of cell walls in the resistance of older plants to *R. solani*.

The limitation in lesion size that occurs in beans infected with AG 4 is another host reaction that has received considerable attention. Calcium accumulation at the lesion site (Bateman, 1964; Bateman and Lumsden, 1965) and accumulation of phytoalexins to levels inhibitory to the pathogen (Smith *et al.*, 1975) have been implicated as factors in limiting lesion size.

A recent and exciting development is the creation of transgenic plants with increased levels of endochitenase and the demonstration that such plants are more resistant to *R. solani* AG 4 (Broglie *et al.*, 1991). Benhamou *et al.* 1993) compared the ultrastructural features of infected tissues of wild-type and transgenic plants expressing the bean chitinase gene. They provided evidence that the increased resistance of the transgenic canola seedlings to infection by *R. solani* was associated with a decrease in the amount of fungal colonization and pathogen viability. Benhamou *et al.* in 1992 (1993a) demonstrated that purified bean chitinase could lyse the hyphal walls of *R. solani in vitro*. They attributed this to the large amount of chitin exposed on the surface of the hyphae. A cytological study (Benhamou, 1993b) indicated that a similar phenomenon occurred *in planta*. The authors state, however, that it is likely that other hydrolytic enzymes such as B-1,3 glucanase are also involved in hyphal degradation. Vierheilig *et al.* (1993) confirmed that transgenic plants expressing a chitinase gene show increased resistance to *R. solani*. They formed transgenic *Nicotiana sylvestris* plants constitutively expressing vacuolal chitinase of *N. tabacum*. These plants exhibited less colonization by *R. solani* than wild-type plants. The transgenic plants, however, were still normally colonized by the VA mycorrhizal fungus *Glomus mosseae*. The authors conclude that it appears possible to make use of the antifungal properties of chitinase in transgenic plants without interfering with mycorrhizal development.

The resistance conferred by enhanced levels of constitutive chitinase has the characteristics of quantitative resistance. The severity of diseases caused by *R. solani* is influenced by many factors, including inoculum level, quantity and quality of nutrients available to the pathogen and environmental conditions, primarily temperature, which influences the relative growth rates of both pathogen and host. Consequently any plant characteristics that would reduce the rate of host colonization could have a significant impact on disease severity.

VI. SUMMARY

Although considered separately penetration, colonization and host reaction are all part of one, interrelated, process. In examining data available for the anastomosis groups that have been investigated it is apparent that there are more similarities than differences. Therefore, a generalized description of the infection process can be developed.

When hyphae contact the surface of a susceptible host a recognition phenomenon occurs resulting in repeated branching and the formation of an infection structure. There is evidence to suggest that both chemo (Flentje *et al.*, 1963) and contact (Armentrout, 1987) tropisms are involved in host recognition but the precise mechanism remains unknown. It is interesting to note that an exogenous supply of 3-O-methyl glucose inhibited pathogenesis by AG 1 (Kousik *et al.*, 1994) and AG 4 (Weinhold and Bowman, 1974). In both cases it was reported that methyl glucose interfered with host recognition and infection cushion formation. Armentrout *et al.* (1987) suggested that the methyl glucose effect may be attributed to interference with mucilage formation and adherence so that physical contact with the hypocotyl is inadequate to stimulate cushion formation. Whatever the basis for the effect the fact that methyl glucose inhibits infection structure formation by two anastomosis groups indicates the existence of a common host recognition mechanism.

Following infection structure formation, penetration is directly through the cuticle and epidermal cell wall, although AG 1 will, on occasion, penetrate through stomata. Fine penetration pegs, 1-2 microns in diameter, are often involved in penetration. Pegs appear to be common but penetration without pegs has been reported. Thinning of the cuticle at the point of penetration, the smooth sides of the penetration pores, and the fact that *R. solani* is known to produce cutinase provide evidence of enzymatic involvement in cuticle penetration. If cutinase is involved the amount produced is probably small because in general the cuticle remains intact, except in the areas of penetration. Penetration without pegs may be a function of cuticle strength, as suggested by Dodman *et al.*, (1968), or it may be because certain isolates produce sufficient cutinase that penetration can occur without reduction in hyphal diameter.

Colonization of host tissue is accomplished by the production of hydrolytic enzymes that degrade cell walls one or more cell layers beyond the advancing hyphae. With AG 4 there is evidence of cell wall degrading enzyme activity prior to penetration (Weinhold and Motta, 1973). Damage to cell membranes and changes in cytoplasm can also be detected in cortical cells prior to colonization. In addition to swelling and degradation of cell walls the cytological changes in infected plants include, the formation of reaction zones, plasmolysis, and collapsing of cytoplasm. Pathogenesis by *R. solani* is characterized by severely damaging or killing host cells prior to, or immediately following, penetration and colonization is primarily a process of hyphal growth into moribund tissue.

VII. REFERENCES

Armentrout VN & Donner AJ (1987) Infection cushion development by *Rhizoctonia solani* on cotton. Phytopathology 77: 619-623.

Armentrout VA, Downer AJ, Grasmick DL & Weinhold AR (1987) Factors affecting infection cushion development by *Rhizoctonia solani* on cotton. Phytopathology 77:623-630.

Baker CJ & Bateman DF (1978) Cutin degradation by plant pathogenic fungi. Phytopathology 68: 1577-1584.

Bassi A Jr., Moore EL & Batson WE Jr. (1979) Histopathology of resistant and susceptible tomato fruit infected with *Rhizoctonia solani*. Phytopathology 69: 56-559.

Bateman DF (1964) An induced mechanism of tissue resistance to polygalacturonase in Rhizoctonia-infected hypocotyls of bean. Phytopathology 54:438-445.

Bateman DF & Lumsden RD (1965) Relation of calcium content and nature of the pectic substances in bean hypocotyls of different ages to susceptibility to an isolate of *Rhizoctonia solani*. Phytopathology 55:734-738.

Bateman DF & Basham HG (1976) Degradation of plant cell walls and membranes by microbial enzymes. In: Heitefuss R and Williams P H (eds.) Encyclopedia Plant Physiology. Vol 4 (pp 316-355) Springer-Verlag, New York.

Benhamou N, Broglie K, Broglie R & Chet I (1993a) Antifungal effect of bean endochitinase on *Rhizoctonia solani*: ultrastructural changes and cytochemical aspects of chitin breakdown. Can. J. Microbiol.39:318-328.

Benhamou N, Broglie K, Chet I & Broglie R (1993b) Cytology of infection of 35S -bean chitinase transgenic canola plants by *Rhizoctonia solani* : Cytochemical aspects of chitin breakdown *in vivo*. Plant J. 4 :295-305.

Bertagnolli BL, Dal Soglio FK & Sinclair JB (1996) Extracellular enzyme profiles of the fungal pathogen *Rhizoctonia solani* isolate 2B-12 and of two antagonists, *Bacillus megaterium* strain B153-2-2 and *Trichoderma harzianum* isolate Th008. I. Possible correlations with inhibition of growth and biocontrol. Physiol. Mol. Plant Pathol. 45: (in press)

Broglie K, Chet I, Holliday M, Cressman R, Biddle P, Knowlton S, Mauvais CJ & Broglie R (1991) Transgenic plants with enhanced resistance to the fungal pathogen *Rhizoctonia solani*. Science 254:1194-1197.

Brookhouser LW & Weinhold AR (1979) Induction of polygalacturonase from *Rhizoctonia solani* by cotton seed and hypocotyl exudates. Phytopathology 69:599-602.

Bugbee WM (1993) An inhibitor of pectin lyase from cell walls of sugar beet. Phytopathology 83:63-68.

Christou T (1962) Penetration and host-parasite relationships of *Rhizoctonia solani* in the bean plant. Phytopathology 52:291-380.

Dodman RL, Barker KR & Walker JC (1968a) Modes of penetration by different isolates of *Rhizoctonia solani*. Phytopathology 58: 31-33.

Dodman RL, Barker KR & Walker JC (1968b) A detailed study of the different modes of penetration by *Rhizoctonia solani*. Phytopathology 58:1271-1276.

Fukutomi M & Takada H (1979) Ultrastructure of infection process of *Rhizoctonia solani* Kühn in cucumber hypocotyls. Ann. Phytopathol. Soc. Jpn. 45:453-462.

Hirooka T, Mayagi Y, Araki F & Kunoh H (1989) Biological mode of action of flutolanil in its systemic control of rice sheath blight. Phytopathology 79:1091-1094.

Hofman TW & Jongebloed PHJ (1988) Infection process of *Rhizoctonia solani* on *Solanum tuberosum* and effects of granular nematocides. Neth. J. Plant Pathol. 94:243-252.

Kenning LA & Hanchey P (1980) Ultrastructure of lesion formation in Rhizoctonia-infected bean hypocotyls. Phytopathology 70: 998-1004.

Ko WH. & Hora FA (1970) Production of phospholipases by soil microorganisms. Soil Sci. 110:355-358.

Kolattukudy PE (1985) Enzymatic penetration of the plant cuticle by fungal pathogens. Ann. Rev. Phytopathol. 23:233-250.

Kousik CS, Snow JP & Berggren GT (1994) Factors affecting infection cushion development by *Rhizoctonia solani* AG 1-IA and 1-IB on soybean leaves. Plant Pathol. 43:237-244.

Lai M-T & Weinhold AR (1968) Permeability changes in *Phaseolus aureus* associated with infection by *Rhizoctonia solani*. Phytopathology 58: 40-245.

Lister N, Katan J & Henis Y (1975) Sequential production of polygalacturonase, cellulase, and pectin lyase by *Rhizoctonia solani*. Can. J. Microbiol. 21: 298-1304.

Lisker N, Katan J & Henis Y (1976) Scanning electron microscopy of the infection of beans by *Rhizoctonia solani* propagules. Ann. Bot. 40:625-629.

Liu ZL & Sinclair JB (1991) Isolates of *Rhizoctonia solani* AG 2-2 pathogenic to soybeans. Plant Dis. 75:682-687.

Marcus L, Barash I, Sneh B, Koltin Y & Finkler A (1986) Purification and characterization of pectinolyic enzymes produced by virulent and hypovirulent isolates of *Rhizoctonia solani* Kühn. Physiol. Mol. Plant Pathol. 29:325-336.

Marshall DS & Rush MC (1980) Infection cushion formation on rice sheaths by *Rhizoctonia solani*. Phytopathology 70:947-950.

Matsuura K (1986) Scanning electron microscopy of the infection process of *Rhizoctonia solani* in leaf sheaths of rice plants. Phytopathology 76:811-814.

Ruppel EG (1973) Histopathology of resistant and susceptible sugar beet roots inoculated with *Rhizoctonia solani*. Phytopathology 63:123-126.

Sherwood RT (1966) Pectin lyase and polygalacturonase production by *Rhizoctonia solani* and other fungi. Phytopathology 56:279-286.

Smith DA, Van Etten HD & Bateman DF (1975) Accumulation of phytoalexins in *Phaseolus vulgaris* hypocotyls following infection by *Rhizoctonia solani*. Physiol. Plant Pathol. 5: 1-64.

Sneh B, Ichielevich-Auster M & Shomer I (1989) Comparative anatomy of cotton hypocotyls and roots by virulent and hypovirulent isolates of *Rhizoctonia solani*. Can. J. Bot. 67:2142-2149.

Stockwell V & Hanchey P (1984) The role of the cuticle in resistance of bean to *Rhizoctonia solani*. Phytopathology 74: 1640-1642.

Trail F & Koller W (1990) Diversity of cutinases from plant pathogenic fungi: Evidence for a relationship between enzyme properties and tissue specificity. Physiol. Mol. Plant Pathol. 36:495-508.

Tseng TC & Bateman DF (1968) Production of phosphatidases by phytopathogens. Phytopathology 58:1437-1438.

Van Etten HD, Maxwell DP & Bateman DF (1967) Lesion maturation, fungal development, and distribution of endopolygalacturonase and cellulase in Rhizoctonia-infected bean hypocotyl tissues. Phytopathology 57:121-126.

Vierheilig H, Alt M, Neuhaus J-M, Boller T & Wiemken A (1993) Colonization of transgenic *Nicotiana sylvestris* plants, expressing different forms of *Nicotiana tabacum* chitinase, by the root pathogen *Rhizoctonia solani* and by the mycorrhizal symbiont *Glomus mosseae*. Mol. Plant-Microbe Interact. 6:261-264.

Weinhold AR & Bowman T (1974) Repression of virulence in *Rhizoctonia solani* by glucose and 3-O-methyl glucose. Phytopathology 64:985-990.

Weinhold AR & Motta J (1973) Initial host responses in cotton to infection by *Rhizoctonia solani*. Phytopathology 63:157-162.

Weinhold AR, Bowman T & Dodman RL (1969) Virulence of *Rhizoctonia solani* as affected by the nutrition of the pathogen. Phytopathology 59: 1601-1605.

Weinhold AR, Dodman RL & Bowman T (1972) Influence of exogenous nutrition on virulence of *Rhizoctonia solani*. Phytopathology 62:278-281.

Yang J, Verma PR & Tewart JP (1992) Histopathology of resistant mustard and susceptible canola hypocotyls infected by *Rhizoctonia solani*. Mycol. Res. 96:171-179.

Section IV.

ECOLOGY OF *RHIZOCTONIA* SPP., POPULATION AND DISEASE DYNAMICS.

IV.1 IDENTIFICATION AND DETECTION OF *RHIZOCTONIA SOLANI* USING SEROLOGICAL AND DNA MARKER TECHNIQUES.

PHILIP A. O'BRIEN

School of Biological and Environmental Sciences, Murdoch University, Murdoch WA 6150, AUSTRALIA; Western Australian State Agricultural Biotechnology Center, Murdoch University, Murdoch WA 6150, AUSTRALIA.

I. INTRODUCTION

Techniques for identification of fungal pathogens are necessary for understanding the epidemiology and etiology of the organism and for assessing the effects of control strategies. Detection techniques also allow the level of the pathogen to be monitored so that disease outbreaks can be predicted and control strategies implemented at an early stage. In recent years considerable progress has been made in the development of identification methods based on the use of antibodies, or detection of DNA markers. Such tests are simple to use and have a high degree of specificity. This chapter reviews progress in the development of such tests for identification of phytopathogenic fungi in general, with special reference to *R. solani*.

II. IDENTIFICATION OF *R. SOLANI*

A. RFLP analysis

In this technique, DNA extracted from isolates is digested with a restriction enzyme which cuts the DNA only at specific sequences. The digest is electrophoresed on an agarose gel, and blotted onto a nylon membrane. The membrane is hybridized to labeled cloned fragments of DNA (Bruns *et al.*, 1991). The banding pattern obtained can be used to estimate the relationship between the isolates, and the groups they are derived from. Closely related anastomosis subgroups will have very similar patterns, whilst more distantly related subgroups will have a greater number of differences between them. The differences are caused by mutations which create and destroy restriction enzyme recognition sites. New patterns are generated by using a different probe. The probe fragment can be chosen at random from a library, although many studies focus on the rRNA genes because of their high variability and high copy number.

Analysis of RFLP's in isolates of *R. solani*. revealed that there is considerable genetic variability within a subgroup although it did not correlate with other biological characteristics (Jabaji-Hare *et al.*, 1990; O'Brien, 1994; Vilgalys and Gonzalez, 1990). However all studies have indicated that there were significantly greater differences between groups, than within groups, supporting the division of isolates into AG/subgroups . Of the AG assayed, AG 2-1 was found to have the highest level of intragroup variation. (Liu *et al.*, 1993) using restriction digestion of PCR amplified ribosomal DNA were able to identify unique patterns for 13 out of 14 *R. solani* subgroups analyzed. Only subgroups 5A and 8 could not be differentiated.

B. Sneh et al. (eds.),
Rhizoctonia Species: Taxonomy, Molecular Biology, Ecology, Pathology and Disease Constrol, 177–183.

Although RFLP analysis is a very useful technique, it suffers from the disadvantages that it requires relatively large amounts of DNA (5-10μg/analysis), and is a complex procedure which takes several days. The availability of DNA thus becomes a limiting factor on the number of analyses that can be carried out. Moreover, the activity of restriction enzymes is affected by other substances (such as polysaccharides) in the DNA extract. Because of these difficulties RFLP analysis has been largely superseded by the simpler technique of RAPD-PCR.

B. Random amplification of polymorphic DNA by polymerase chain reaction. (RAPD-PCR)

This is a variation of the PCR reaction (see below) in which short oligonucleotide primers bind to complementary, or partially complementary sites on genomic DNA, thus allowing regions of the genome to be amplified by PCR (Weising *et al.*, 1995). Unlike the PCR reaction, the temperature at which the primers anneal to the DNA in RAPD-PCR is low enough to permit annealing of the primers to sites that are not completely complementary. Electrophoresis of the amplification products on an agarose or polyacrylamide gel generates a band pattern which is subgroup specific (Duncan *et al.*, 1993). The primers used for RAPD-PCR consist of arbitrarily selected sequences, with the only criterion that they have greater than 50% G+C content. No prior knowledge is required about the regions to be amplified. The technique is simple to carry out, it requires very little DNA (1-10ng), the equipment is readily available, and since different primers give different fingerprints a large number of genetic loci can be included in the analysis. Variations of the RAPD-PCR reaction include, inter-SSR PCR (Zietkiewicz *et al.*, 1994), DAF (Caetano-Anolles *et al.*, 1993), AP-PCR (Welsh and McClelland, 1990). These techniques differ with respect to the length of the primer.

(Duncan *et al.*, 1993) used RAPD-PCR to analyze isolates from different zymogram groups (ZG). Some of the primers used gave highly complex patterns with multiple bands. These patterns enabled differentiation of all groups analyzed. Isolates from the same AG, but from different ZG gave different patterns with two of the four primers tested. Other primers gave a very simple pattern which nonetheless was useful for differentiating groups. Some primers gave no pattern with *R. solani*, although they gave patterns with other fungal species. There are no rules for predicting whether a primer will give a complex or simple pattern, or no pattern. This has to be determined for each primer and species of fungus. (Tommerup *et al.*, 1995) investigated factors affecting the reproducibility of RAPD-PCR, and described conditions for obtaining reproducible RAPD-PCR fingerprints for *Rhizoctonia*, *Laccaria*, and *Hydnangium*. (Cubeta *et al.*, 1993) used RAPD-PCR to demonstrate that tuft formation between opposing AG 4 isolates did not correlate with heterokaryon formation.

C. Pulsed field gel electrophoresis (PFGE)

PFGE is a method of electrophoretically separating chromosome size DNA molecules (≤12Mbp) by electrophoresis in agarose gels using alternating electric fields (Hardwood, 1993). The DNA is electrophoresed directly from protoplasts embedded and lysed within blocks of agarose. The technique enables intact fungal chromosomes to be separated and viewed. Thus an electrophoretic karyotype can be constructed for each strain/species. Karyotypes have been constructed for a number of fungal species including *Agaricus bisporus*, (Lodder *et al.*, 1993), *Cochliobolus heterostrophus* (Tzeng *et al.*, 1992), and *Aspergillus nidulans* (Brody and Carbon, 1989). In each case the number of chromosomes observed correlated well with the number of linkage groups identified by genetic analysis. However, a surprising degree of chromosome length polymorphism within species was also observed PFGE is particularly useful in genetic mapping experiments as it enables RFLP probes to be assigned to linkage groups. There is only one report of PFGE analysis of *R. solani*. Wako *et al.* (1991) analyzed the DNA from two AG 4 isolates by OFAGE. The DNA resolved into six chromosomal bands ranging from 0.8 to 3.8Mb in size.

D. DNA sequencing

Genetic variability at the DNA level can be analyzed by DNA sequencing. The region to be sequenced is first amplified by PCR, and the amplification products then sequenced. The region most commonly used for sequencing is the ribosomal RNA genes. This is because they contain both highly conserved regions which can be used for the design of PCR primers, and variable regions which are useful for sequence information (Bruns *et al.*, 1991). The technique is easy to carry out, requires very little DNA, and yields a lot of information. An additional advantage of this approach is that the sequence information can be used to design PCR primers specific for an AG/subgroup. The application of this method to analysis of *R. solani* is described in the previous chapter.

E. Accessory genetic elements

These consist of small molecular weight DNA (plasmids) or double stranded RNA (dsRNA) molecules. Both types are widespread in phytopathogenic fungi (Nuss and Koltin, 1990; Samac and Leong, 1989). Plasmids may exist either in the nucleus or the mitochondrion, and be circular or linear, whilst dsRNA exist in the cytoplasm. These elements are detected as small bands which migrate ahead of the chromosomal DNA during agarose gels electrophoresis. The two types can be differentiated by digestion of the DNA with RNAase.

Both plasmids and dsRNA elements are widespread in *R. solani*. Accessory genetic elements in *R. solani* are discussed in chapter II.3. Approximately 0-60% of isolates contain either single or multiple size classes of plasmids and/or dsRNA (Bharathan and Tavantzis, 1991; Kousik *et al.*, 1994; Miyasaka *et al.*, 1990; Yang *et al.*, 1994). Analysis of plasmids in isolates from various locations in Japan showed that plasmids from the same AG/subgroup although of different sizes, were homologous to each other, but not to plasmids from different AG/subgroup (Miyasaka *et al.*, 1990). (Bharathan and Tavantzis, 1991) similarly analyzed dsRNA elements in isolates from AGs 1 to 5 from diverse geographic locations. Quantitative cross-hybridization between dsRNA's showed that the sequence homology of dsRNA's from isolates of the same AG, but different locations ranged from 32-80%. However, homology among dsRNA's of common origin was higher than among those of distant origins.

Limitations to the use of accessory elements as genetic markers for the identification of AG/subgroup includes their absence from many isolates, and their variation in size. They can possibly be used as an adjunct to other techniques to confirm the identity of an AG/subgroup. However, their variability may be an advantage, enabling the tracking of specific isolates from one location to another, leading to identification of the mechanisms by which this spread occurs.

III. DETECTION OF *R. SOLANI*

A. Serological methods

Antibodies which bind to fungal mycelium in a species specific manner can be used to detect the presence of fungal pathogens. The binding is detected most commonly by an ELISA reaction, in which the antibody is added to the wells of a microtitre tray which has been coated with antigen. Specific binding of the antibody to the antigen, is detected by the addition of a second enzyme linked antiimmunoglobulin antibody which binds to the antigen- antibody complex. The amount of enzyme activity trapped in the wells is proportional to the amount of antigen present. Antibody detection tests have been developed for a number of fungal species including *R. solani* AG 3 (Thornton *et al.*, 1993), *Spongospora subterranea* (Harrison *et al.*, 1994), *Verticillium* (van de Koppel and Schots, 1994), *Heterobasidion annosum* (Galbraith and Palfreyman, 1994), and *Phytophthora cinnamomi* (Cahill and Hardham, 1994).

A disadvantage of antibody tests may be a lack of specificity of the antibody. The antibodies are not always sufficiently specific to differentiate pathotypes of the same species (Nicholson and Rezanoor, 1994). (Robb *et al.*, 1994), or even closely related species (Burns *et al.*, 1994).

B. DNA probes

DNA probes consist of cloned DNA sequences that are specific to the organism, being detected. The probe may be labeled either radioactively or chemically, and hybridized to DNA extracted from the sample being analyzed. Hybridization results in retention of the labeled probe on a membrane containing the extracted DNA. Probe detection tests are highly specific, and very sensitive. Moreover they are not affected by environmental conditions or stage of the life cycle. Probe sequences are isolated relatively easily by screening RAPD-PCR products (Dobrowolski and O'Brien, 1993) or by screening randomly cloned fragments from a plasmid library (Jabaji-Hare *et al.*, 1990; O'Brien, unpublished). Probes have now been isolated for a number of species of fungal pathogens including; *Phytophthora cinnamomi* (Dobrowolski and O'Brien, 1993), *Pseudocercospora herpotrichoides* (Nicholson and Rezanoor, 1994), *R. solani* AG 8 (Matthew, 1992), *R. solani* AG 11 (O'Brien unpublished), *R. solani* AG 3 (Jabaji-Hare *et al.*, 1990).

C. Polymerase chain reaction (PCR)

In PCR specific regions of a genome are amplified by the binding of oligonucleotide primers to sites flanking the region (Henson and French, 1993). The products are detected by electrophoresis on agarose, and photography, although alternative detection and quantification methods have been developed (Kemp *et al.*, 1990; Khudyakov *et al.*, 1994). PCR tests are very sensitive, and highly specific. Tests for detection of subgroups such as pathotypes are easily developed. The greatest advantage of PCR is that multiple samples can be analyzed simultaneously. The reactions can be carried out in microtitre trays (96 reactions), and the amplification products analyzed automatically using a plate reader connected to a computer. A disadvantage is that amplification is sensitive to the presence of contaminants which co-extract with the DNA (O'Brien, unpublished). To detect false negatives, an internal standard fragment which is co-amplified with the target sequence should be added to the PCR reaction (Uberla *et al.*, 1992).

Fungi for which PCR detection tests have been developed include: *Stagonospora nodorum* and *Septoria tritici* (Beck and Lignon, 1995), *Verticillium* (Beck and Lignon, 1995; Carder *et al.*, 1994; Robb *et al.*, 1994), *Phytophthora parasitica* and *Phytophthora citrophthora* (Ersek *et al.*, 1994), *Phytophthora cinnamomi*, (O'Brien and Farbey, 1995), *Gaumannomyces graminis* (Schesser *et al.*, 1991). To date, a PCR detection test for *R. solani* has not been developed, although a test is being developed for detection of AG 8 strains (Whisson personal communication).

IV. Quantification

For a detection test to be useful, it should allow quantification of the pathogen in the sample so that the threat of a disease outbreak, and the effect of control strategies on the fungal population can be more accurately assessed.

In ELISA tests a colored product is generated by the activity of the enzyme (e.g., horseradish peroxidase) coupled to the secondary antibody. The amount of enzyme activity bound to the antigen antibody complex is proportional to the amount of antigen present in the sample. The relationship between the amount of enzyme activity and antigen can be determined using a calibration curve.

(Whisson *et al.*, 1995) described a method for quantification of detection of *R. solani* AG 8 using a DNA probe. Known amounts of a standard DNA fragment are included in parallel on the slot-blot with the test samples. The intensity of signals from the standards provides a calibration curve from which the amount of DNA in the test sample can be determined.

A similar approach can be used to quantify PCR reactions. When the amount of product from the target and the internal standard are equivalent, it is assumed that equivalent amounts were initially present in the sample. (Uberla *et al.*, 1992). A number of replicate PCR reactions containing the same amount of the target, but different amounts of the standard fragment are set up and amplified, and the ratios of the products measured. For this method to be valid it is necessary: (a) that both sequences are amplified by the same primers; and (b) that the sequences are only slightly different in size so that they can be separated by electrophoresis. If using a non electrophoretic method of analysis, the products may be the same size, but the method must be able to differentiate the products of the target and standard

V. Prospects for the Future

Diagnostic tests should be simple, rapid, specific, sensitive, and quantifiable. At the present state of development serological methods fulfill the criteria of simplicity and rapidity to a greater extent than PCR based tests. However, developments in PCR technology have reduced the complexity of the procedure to the point where they will soon rival that of serological methods. An additional advantage is that new PCR tests are more easily developed than serological tests since subgroup specific DNA sequences are easily isolated (Dobrowolski and O'Brien, 1993; Jabaji-Hare *et al.*, 1990). However this difficulty may be overcome with the use of combinatorial antibody libraries in phage display vectors from which pathogen specific antibodies can be isolated by affinity chromatography (Barbas *et al.*, 1991).

VI. Acknowledgment

I gratefully acknowledge S. Neate for critically reviewing this manuscript, and the Western Australian Grains Council for supporting this work.

VII. REFERENCES

Barbas CF, Kang AS, Lerner RA & Benkovic SJ (1991) Assembly of combinatorial antibody libraries on phage surfaces: The gene III site. Proc. Nat. Acad. Sci. USA 88:7978-7982.

Beck JJ & Lignon JM (1995) Polymerase chain reaction assays for the detection of *Stagonospora nodorum* and *Septoria tritici* in wheat. Phytopathology 85:319-324.

Bharathan N & Tavantzis SM (1991) Assessment of genetic relatedness among double stranded RNAs from isolates of *Rhizoctonia solani* from diverse geographic origins. Phytopathology 81:411-415.

Brody H & Carbon J (1989) Electrophoretic karyotype of *Aspergillus nidulans*. Proc. Nat. Acad. Sci. USA 86:6260-6263.

Bruns TD, White TJ & Taylor JW (1991) Fungal molecular systematics. Ann Rev. Ecol. Syst. 22:525-564.

Burns R, Vernon ML & George EL (1994) Monoclonal antibodies for the detection of *Pyrenophora graminea*. In: Schots A, Dewey F M & Oliver R (eds.), Modern Assays for Plant Pathogenic Fungi: Identification, Detection, and Quantification, (pp 199-203), CAB International, Wallingford.

Caetano-Anolles G, Bassam BJ & Greshoff PM (1993) Enhanced detection of polymorphic DNA by multiple arbitrary amplicon profiling of endonuclease digested DNA: identification of markers tightly linked to the supernodulation locus in soybean. Mol. Gen. Genet. 241:57-64.

Cahill DM & Hardham AR (1994) Exploitation of zoospore taxis in the development of a novel dipstick immunoassay for the specific detection of *Phytophthora cinnamomi*. Phytopathol. 84:193-200.

Carder JH, Morton A, Tabrett AM & Barbara DJ (1994) Detection and differentiation by PCR of subspecific groups within two *Verticillium* species causing vascular wilts in herbaceous hosts. In: Schots A, Dewey F M & Oliver R (eds.), Modern Assays for Plant Pathogenic Fungi: Identification, Detection, and Quantification, (pp 91-97), CAB International, Wallingford.

Cubeta MA, Briones-Ortega R & Vilgalys R (1993) Reassessment of heterokaryon formation in *Rhizoctonia solani* anastomosis group 4. Mycologia 85:777-787.

Dobrowolski M & O'Brien PA (1993) Isolation of a species specific DNA probe for *Phytophthora cinnamomi* by RAPD-PCR. FEMS Microbiol. Letts. 113:43-47.

Duncan S, Barton JE & O'Brien PA (1993) Analysis of variation in isolates of *Rhizoctonia solani* by random amplified polymorphic DNA assay. Mycological Res 97:1075-1082.

Ersek T, Shoelz JE & English JT (1994) PCR amplification of species specific DNA sequences can distinguish among *Phytophthora* species. Appl. & Envir. Microbiol 60:2616-2621.

Galbraith DN, & Palfreyman JW (1994) Detection of *Heterobasidion annosum* using monoclonal antibodies. In: Schots A, Dewey F M & Oliver R (eds.), Modern Assays for Plant Pathogenic Fungi: Identification, Detection, and Quantification, (pp 105-110), CAB International, Wallingford.

Hardwood A (1993) Protocols for Gene Analysis Humana Press, Totowa, NJ.

Harrison JG, Lowe R, Wallace A & Williams NA (1994) Detection of *Spongospora subterranea* by ELISA using monoclonal antibodies. In: Schots A, Dewey F M & Oliver R (eds.), Modern Assays for Plant Pathogenic Fungi: Identification, Detection, and Quantification, (pp 23-27), CAB International, Wallingford.

Henson JM & French R (1993) The polymerase chain reaction and plant disease diagnosis. Ann Rev Phytopathol 31:81-109.

Jabaji-Hare SH, Meller Y, Gill S & Charest PM (1990) Investigation of genetic relatedness among anastomosis groups of *Rhizoctonia solani* using cloned DNA probes. Can. .J. Plant Pathol. 12:393-404.

Kemp DJ, M.Churchill, Smith DB, Biggs BA, Foote SJ, Person MG, Samaras N, Deacon NJ & Doherty R (1990) Simplified colorimetric analysis of polymerase chain reactions: detection of HIV sequences in AIDS patients. Gene 94:223-228.

Khudyakov YE, Gaur L, Singh J, Patel R & Fields HA (1994) Primer specific solid phase detection of PCR products. Nucleic Acids Res. 22:1320-1321.

Kousik CS, Snow JP & Valverde RA (1994) Comparison of double-stranded RNA components and virulence among isolates of *Rhizoctonia solani* AG 1-1A and AG 1-1B. Phytopathology 84:44-49.

Liu ZL, Domier LL & Sinclair JB (1993) subgroup -Specific ribosomal DNA polymorphisms of the *Rhizoctonia solani* species complex. Mycologia 85:795-800.

Lodder S, Gull K & Wood D (1993) An electrophoretic karyotype of the cultivated mushroom *Agaricus bisporus*. Current Genetics 24:496-499.

Matthew JS (1992). Molecular diversity between anastomosis groups of *Rhizoctonia solani* (PhD Thesis, Adelaide University).

Miyasaka A, Chen CL & Hashiba T (1990) Detection and properties of plasmid like DNA in isolates from nine anastomosis and intraspecific groups of *Rhizoctonia solani*. J. Gen. Microbiol. 136: 1791-1798.

Nicholson P & Rezanoor HN (1994) DNA probe for the R-type of eyespot disease of cereals *Pseudocercosporella hypotrichoides*. In: Schots A, Dewey F M & Oliver R (eds.), Modern Assays for Plant Pathogenic Fungi: Identification, Detection, and Quantification, (pp 17-22), CAB International, Wallingford.

Nuss DL & Koltin Y (1990) Significance of dsRNA elements in plant pathogenic fungi. Annu. Rev. Phytopathol 28:37-58.

O'Brien PA (1994) Molecular markers in Australian isolates of *Rhizoctonia solani*. Mycological Res 98:665-671.

Robb J, Hu X, Platt H & Nazar R (1994) PCR based assays for the detection and quantification of *Verticillium* species in potato. In: Schots A, Dewey F M & Oliver R (eds.), Modern Assays for Plant Pathogenic Fungi: Identification, Detection, and Quantification, (pp 83-90). CAB International, Wallingford.

Samac DA & Leong SA (1989) Mitochondrial plasmids of filamentous fungi: characteristics and use in transformation vectors. MPMI 2:155-159.

Schesser K, LUder A, & Henson JM (1991) Use of polymerase chain reaction to detect take all fungus *Gaeumannomyces graminis* in infected wheat plants. Appl. Env. Microbiol. 57:553-556.

Thornton CR, Dewey FM & Gilligan CA (1993) Development of monoclonal antibody based immunological assays for the detection of live propagules of *Rhizoctonia solani* in soil. Plant Pathol. 42:763-773.

Tommerup IC, Barton JE & O'Brien PA (1995) Reliability of RAPD fingerprinting of three basidiomycete fungi, *Laccaria, Hydnangium*, and *Rhizoctonia*. Mycol. Res. 99:179-186.

Tzeng T, Lyngholm LK, Ford CF & Bronson CR (1992) A restriction fragment length polymorphism map and electrophoretic karyotype of the fungal maize pathogen *Cochliobolus heterostrophus*. Genetics 130:81-96.

Uberla K, Platzer C, Diamantstein T & Blankenstein. T **(1992)** Generation of competitor DNA fragments for quantitative PCR. PCR Meth. Appl. 1:136-139.

van de Koppel MM & Schots A (1994) A double (monoclonal) antibody sandwich for the detection of *Verticillium* species in roses. In: Schots A, Dewey F M & Oliver R (eds.), Modern Assays for Plant Pathogenic Fungi: Identification, Detection, and Quantification, (pp 99-104), CAB International, Wallingford.

Vilgalys R & Gonzalez D (1990) Ribosomal DNA restriction fragment length polymorphisms in *Rhizoctonia solani*. Phytopathology 80:151-158.

Wako T, Ishikawa T & Hashiba T (1991) Unique DNA plasmid pRS64 associated with chromosomal DNAs of the plant pathogenic fungus *Rhizoctonia solani*. J. Gen. Microbiol. 137:2817-2821.

Weising K, Nybom H, Wolff K & Meyer W (1995) DNA Fingerprinting in Plants and Fungi (London: CRC Press).

Welsh J & McClelland M (1990) Fingerprinting genomes using PCR with arbitrary primers. Nucleic Acids Res 18:7213-7218.

Whisson DL, Herdina & Francis L (1995) Detection of *Rhizoctonia solani* AG 8 in soil using a specific DNA probe. Mycol. Res. (in press).

Yang HA, Sivasithamparam K, Alemohammad J, Barton JE & O'Brien PA (1994) Association of *Rhizoctonia* strains with bare patch disease of wheat in Western Australia. J. Plant Pathol. 43:878-884.

Zietkiewicz E, Rafalski A & Labuda D (1994) Genome fingerprinting by simple sequence repeat (SSR)-anchored polymerase chain reaction amplification. Genomics 20:176-183.

IV.2

SAMPLING AND QUANTIFICATION OF *RHIZOCTONIA SOLANI* IN SOIL

STEPHEN M. NEATE[1] AND JOHANNES H. M. SCHNEIDER[2]

[1]*CSIRO Division of Soils/CRC for Soil and Land Management, P.B. No 2 Glen Osmond, 5064, AUSTRALIA.* [2]*IPO-DLO, P.O. Box 9060, 6700 GW Wageningen, THE NETHERLANDS*

I. INTRODUCTION

In this chapter we consider sampling and quantification of the disease and of *Rhizoctonia* in soil and on plants. We assess critically, both techniques that are most commonly used and techniques that are new or that have promise but have been little used.

II. STATISTICAL CONSIDERATION OF SAMPLING

A. Sampling strategies

Sampling strategies depend on the aim to be fulfilled and the spatial distribution of the disease or the pathogen. For a discussion of the statistical principles of various traditional sampling strategies see Cochran (1977) and Thompson (1992). Most sampling strategies applied in *Rhizoctonia* research have been used to establish a collection of isolates rather than to describe the spatial and temporal characteristics of the disease and pathogen. Arbitrary sampling with selected sampling sites and units rather than random sampling in which each sample unit has an equal probability of being selected, was often used. Sampling designs used have been systematic sampling (Clark *et al.*, 1978), sampling in a systematic pattern (Damicone *et al.*, 1993), arbitrary sampling from contiguous quadrats (Campbell and Pennypacker, 1980) and transects (Naiki and Ui, 1977). Many of these sampling strategies are not well suited to highly patchy disease distributions which is typical for many Rhizoctonia diseases.

Problems occur with size of the sampling unit on patchily distributed characters, as the size of the unit increases, the statistical distribution moves from irregular to regular. Furthermore for some characters to be measured (e.g., numbers of colony forming units or length of hyphae/g soil), only a small amount of material can be sampled, leading to data reliability being dependant on how representative the samples are.

B. Spatial pattern analysis

Rhizoctonia spp. show different spatial patterns in the field. AG 8 occurs in distinct patches (MacNish, 1985), whereas AG 3 has a tendency to patchiness (Jager and Velvis, 1995). Few attempts have been made to statistically analyse spatial distribution (Table 1). Methods for study of spatial pattern analysis are: a) probability distributions, b) indices of aggregation, c) mapping, d) spatial autocorrelation, and e) geostatistics. For general discussion of spatial pattern analysis see Upton and Fingleton (1985), and Journel and Huijbreghts, (1978) and for phytopathology references see Campbell and Noe (1985), Gilligan (1988) and Madden and Hughes (1995).

B. Sneh et al. (eds.),
Rhizoctonia Species: Taxonomy, Molecular Biology, Ecology, Pathology and Disease Constrol, 185–195.

TABLE 1. Spatial pattern analysis in *Rhizoctonia* research: Anastomosis groups and host displayed in maps, the aim of the study and the spatial statistic applied.

AG	Host	Aim	Spatial statistic	Reference
1 2-2 3	rice sugar beet potato	distribution of clones within fields	none	Ogoshi and Ui, 1983
1	soybean	within season development	LIP, paired quadrat variance	Yang *et al.*, 1990
2-2	sugar beet	decline		Hyakumachi, 1983
2-2	sugar beet		Chi-square for distribution fitting	Hyakumachi & Ui, 1982
2-t	tulip	reduce sampling	geostatistics	Schneider Kocks, 1995
3	potato	search for mechanisms	none	Jager & Velvis, 1995
8	cereals	change of patches between seasons	none	MacNish & Lewis 1985: MacNish, 1985
8	cereals		none	MacNish & Sweetingham, 1993
8	barley	assessment of yield loss	none	Cotterill, 1991
	groundnut	spatial distribution	geostatistics	Lannou & Savary, 1991

1. Probability distributions Frequency distribution fitting can be used for testing randomness against clustering. The chi-square-statistic is used to test goodness of fit of observed data against theoretical frequency distributions, for example, a Poisson distribution indicates randomness, whereas a negative binomial distribution indicates clustering. Hyakumachi and Ui (1982) mapped the decline of Rhizoctonia root rot in two fields during nine years. Using a chi-square-test for the spatial distribution of diseased sugar beet plants at two sites, they found randomness in some years and clustering in others. Nicot *et al.* (1984) discuss conflicting results using probability distributions and aggregation indices.

2. Indices of aggregation Indices of aggregation also provide information on the spatial pattern and additionally can be used to evaluate the effects of management and ecological variables (Campbell and Noe, 1985). Most common used indices of aggregation are a) variance/mean ratio, b) the k-value from the negative binomial distribution, c) Lloyd's index of mean crowding, d) Lloyd's index of patchiness (LIP), e) Morisita's index, and f) Taylor's index b.

Yang *et al.* (1991) found that LIP decreased between two ratings in the season for five out of six soybean fields with Rhizoctonia foliar blight. They suggested that early in the season the disease is highly aggregated because it is soil-borne and that aggregation decreases because of foliar spread during the season.

3. Mapping Mapping is a convenient, simple technique for displaying spatial patterns and has been applied in *Rhizoctonia* research with different aims (Table 1). Mapping lacks any basis for statistical analysis, but may be more informative than distribution fitting and indices of aggregation (Gilligan, 1988).

4. Spatial autocorrelation techniques Spatial autocorrelation analysis tests whether the observed value of a variable at one site is significantly dependant on values of the variable at neighbouring sites. Methods most commonly used are Joint-Count statistics, Moran I statistics, and correlogram analysis. Correlograms are graphs (or tables) which show how autocorrelation changes with distance (Upton and Fingleton, 1985). Corellogram analyses have been successfully used to detect detailed differences in disease patterns and inoculum distribution in several pathosystems (Shew *et al.*, 1984; Reynolds *et al.*, 1988; Campbell and van der Gaag, 1993).

5. Geostatistics Geostatistics are based on the theory of regionalized variables, that is, the value of a variable depends on its spatial position (Matheron, 1963). The description of the spatial dependence between regionalized variables as a function of the distance between sample sites provides the necessary statistical tools for optimal spatial interpolation (Journel and Huijbregts, 1978).

Geostatistics has recently become a popular tool for analysing disease epidemics (Chellemi *et al.*, 1988; Stein *et al.*, 1994; Larkin *et al.*, 1995). Lannou and Savary (1991) applied geostatistics for mapping and analysing disease incidence of *Rhizoctonia* blight in groundnut plots. Schneider and Kocks (1995) used geostatistics for mapping and reducing sampling intensity of Rhizoctonia disease in flower bulb fields. They concluded that regular grid sampling of as little as 10% of the field provided good information on the spatial distribution of *Rhizoctonia* affected plants.

III. Sampling and quantification in soil

When selecting a technique to measure a character of *Rhizoctonia* it is important to recognise that the suitability of a method will be determined by the ecological behaviour of the *Rhizoctonia* anastomosis group under study and the type of disease it produces. Consideration should be given to: a) all the soil isolation methods that involve mechanical disturbance will influence the size and viability of the mycelial fragments which are extracted, and therefore will not give information about absolute amounts of the fungus, b) absolute quantities of the fungus in natural soil are unlikely be estimated if the methods have only been calibrated against artificially infested soil, c) baiting techniques give only information about the organisms that are in close enough proximity to the bait to be attracted to it and capable of growing on it, d) agar techniques only give information about the organisms that are capable of surviving the soil extraction process and then growing on the medium under the selected conditions. An assessment of the limitations of the techniques discusssed below is presented in Table 2.

TABLE 2. Some potential limitations of techniques useful for quantification of *Rhizoctonia* in soil or on plants.

	Experience in identifying hyphae necessary	Only small quantity of soil assayed	Only detects fungi capable of growth in the test	Relatively tedious technique	Influenced by competition between organisms	Fungus must be close to bait	Does not differentiate dead from live hyphae	Only detects hyphae actively growing in soil	Disturbance caused by technique may affect result	More qualitative than quantitative	d by soil type *	Does not indicate size of CFU	Does not indicate capacity to cause disease
Whole soil / pellet plating	x	x	x	x	x				x	?		x	x
Agar film/membrane	x	x					?		?				x
Soil fraction plating	x		x	x	x				x	x	?	x	x
Baiting with plants	x		x			x		x		?		x	x
Baiting with other material	x		x			x		x		?		x	
Solid surfaces in soil	?		x					x	x	x			x
Direct observation	x	x		x			x			x	?		x
Plating plants on agar media	x		x		x					?		x	x
Molecular							?				?		x
Antibodies							?	?			?		x
Biochemical analysis		?					?			?	?	x	x
Vital stains	x	x	x	x				x	x		?		x
Radio isotopes		?							x		x		x

*May be due to signal quenching in the case of radio isotopes and fluorescent dyes

Note that these are the subjective opinions of the authors and will change as the techniques develop.

A. Whole Soil

1. Soil pellet sampler The soil pellet sampler (Henis *et al.*, 1978) is a whole soil plating method which quantifies the volume of soil used and favors isolation of *Rhizoctonia* by use of Ko and Hora's (1971) *Rhizoctonia* selective medium. It has been suggested that the soil-pellet sampling technique accurately measures the protected population of *R. solani* in soil, i.e., the fungus inside particulate organic matter or sclerotia, but may not be sensitive enough to detect hyphae in soil (Neate and Benger, 1995). The soil pellet sampler is generally calibrated using artificially inoculated soil and this may not reflect the results obtained with natural fungal units.

2. Agar film and membrane filter In the agar film method (Jones and Mollison, 1948) a soil is ground in water, molten agar added allowed to dry, stained and observed. In contrast, in the membrane filter method, soil dilutions are stained and the organisms collected by vacuum filtration on a membrane filter for observation (Sundman and Sivela, 1978). Wiseman (1996) has successfully combined the membrane filter method with vital stains to assess *R. solani* AG 8 in inoculated field soil in pots. Both the agar film and membrane technique may be improved by the use of fluorescent labelled monoclonal antibodies or vital stains to stain the specific fungus. In both methods the intensity of grinding or mixing influences the size and viability of the mycelial fragments but should not influence the total length of hyphae recorded.

B. Soil fraction plating

Boosalis and Scharen (1959) described a method where soil is dispersed in water, the supernatant is poured through a sieve and the debris on the sieve collected and plated on agar. Variations on the technique are: a) dry sieving the soil followed by differential sedimentation, plating on agar and baiting the sediment (Papavizas, 1968), b) wet sieving soil and dispersing the retained debris through cooled molten agar (Weinhold, 1977), c) blending soil with H_2O_2, sieving, mixing retained particles with further H_2O_2 and retaining the particles in suspension for counting or plating (Ui *et al.*, 1976), d) using a semi-automatic elutriator to separate the soil into different density and size fractions and dispersing the retained fractions into cooled molten agar (Clark *et al.*, 1978). *R. solani*, binucleate Rhizoctonias and *Waitea circinata* may be isolated by these methods. These methods isolate both pathogenic and non-pathogenic Rhizoctonias.

C. Baiting with plant material

Many different organic materials have been used for baiting *Rhizoctonia* from soil. They include for *R. solani,* untreated plant stems (Papavizas and Davey, 1962; Sneh *et al.*, 1966; Huang and Kuhlman, 1989), plant seeds (Papavizas *et al.*, 1972) or other plant tissues (Sneh *et al.*, 1966), for binucleate Rhizoctonias, untreated plant stems (Huang and Kuhlman, 1989) and for *R. zeae,* fungicide soaked plant stems (Windham and Lucas, 1987). Disadvantages with baiting are that: a) the best bait may vary between Rhizoctonias and soil types and needs to be determined empirically, b) the pathogen has to be attracted to the bait and grow toward it, which may result in an underestimation of the quiescent *Rhizoctonia* in soil, c) the pathogen has to be re-isolated from the bait and d) the number of baits required for an accurate estimation varies with the inoculum density.

D. Solid Surfaces in Soil

Neate and Benger (1995) estimated the three-dimensional growth of *R. solani* in soil by measurement of growth of *R. solani* on nylon netting. Buried cellulose nitrate filter membranes have also been used to measure two-dimensional mycelial growth of *R. solani* (Glenn and Sivasithamparam, 1990). The use of biologically inert fabrics for recovery of mycelia from soils provide a relatively discontinuous surface (i.e., large holes) that permits free movement of gases and water and allows penetration by roots and soil organisms (Waid and Woodman, 1957), however the technique is most suited to situations where soil is inoculated with *Rhizoctonia,* and due to the uncertainty of its effect on pathogen growth it is unlikely to give an absolute measurement of growth in soil.

E. Direct observation of soil

An advantage of this method is that relationships between *Rhizoctonia* and soil components or other soil organisms can be determined. Both resin (Nicholas *et al.*, 1965) and gelatine embedding (Minderman, 1956) have been used to study groups of organisms, but no studies have attempted to quantify Rhizoctonias using the system. However, the development of specific antibodies with fluorescent labelling (Thornton *et al.*, 1993) may allow calculation of fungal densities or hyphal length at the species or anastomosis group level.

IV. SAMPLING AND QUANTIFICATION ON/OF PLANTS

A. Plating on agar media

Rhizoctonia is readily isolated from newly infected plant material. The sections may be left untreated, surface sterilized or washed in water and then incubated on either non-selective or *Rhizoctonia* selective media. Commonly used selective media are given in Sneh *et al.* (1991). Selective media need to be calibrated for the pathosystem being studied. Warcup (personal communication) found that modifying incubation temperatures (e.g.,. for AG 8 incubating at 10^0C rather than 25^0C), examining plates at frequent intervals, use of different isolation media and even avoiding detergents when washing glassware could increase the frequency of isolation of some Rhizoctonias. It has been noted that with AG 8 the isolation frequency declines with the age of the infection (Samuel and Garrett, 1932) so that time of sampling will influence isolation frequency.

B. Visual assessment

Typical measurements are either incidence or severity of disease and may be expressed as numbers of lesions per plant or plant organ, percentage of plants or plant organs infected or dead, or some combination of these to give a rating scale. Incidence can give information about disease loss providing a relationship between the two has been determined. However, unlike diseases where incidence on a plant part results in total loss, many Rhizoctonia diseases result in increased loss with increased severity. The relationship between degree of severity and yield loss is complicated in diseases such as bare patch of wheat, as plants can recover or die depending on the seasonal conditions.

An alternative approach is to look at the number or length of fungal structures on plant organs. This has not been used regularly for *R. solani,* perhaps because of the type of soft rot damage caused to the tissues makes extraction of plant roots difficult, however, Wyss *et al.* (1990) used a grid line intersection method to estimate percentage root infection by *R. solani*.

V. PLANT AND SOIL METHODS

A. Molecular techniques

The advantages of molecular techniques are that the degree of specificity can be chosen and the techniques can be quick. Recently, Herdina and Whisson (1995) have developed a quantitative diagnostic test to determine levels of *Rhizoctonia* AG 8 in soil. The current technique uses a radioactive probe and has an estimated detection level of 100 hyphal cells per gram of soil. A high specificity, high copy number molecular probe has been developed for AG 3 (Balali *et al.*, 1996) and development of similar probes for other anastomosis groups will allow widespread use of molecular techniques.

B. Monoclonal and polyclonal antibodies

With both monoclonal and polyclonal antibodies, problems may occur if the antigen is not constitutive, is only expressed at a particular stage of the organisms' life cycle or detects dead fungal tissue.

A summary of the characteristics of antibodies that have been developed for *Rhizoctonia* is presented in Table 3. Monoclonal antibody *Rhizoctonia* detection kits developed by Agri-Diagnostics Associates are currently available commercially. The field and laboratory kits available detect *Rhizoctonia* to genus level. Due to a lack of specificity of the antibodies used in the test, the current system is of most use where it is known that there is only a single anastomosis group present or that one group dominates, it will be less useful in more complex systems. Using fungal protein from pure culture, it is estimated that, the minimum quantity of fungus that is detected with the kits is about $0.1 \mu g$ and there is discrimination of protein quantities of an order of magnitude (Miller *et al.*, 1992). Monoclonal antibodies produced by Agri-Diagnostics have been used in an unreplicated experiment to detect *T. cucumeris* in soil and could distinguish 0, 0.1 and 1.0 $\mu g/g$ of the fungus in soil (Dusunceli and Fox, 1992).

Thornton *et al.* (1993) have developed Diagnostic-ELIZA and DIP-STICK assays for *Rhizoctonia* using more specific antibodies, however they acknowledge difficulties with sensitivity, high background signals and detection of dead cells which they overcame by a biological amplification step followed by centrifugation of the soil extracts and heating. No indication of the current levels of sensitivity were provided. They also describe a soil colony staining technique using a fluorescent labelled antibody. The method uses the biological amplification step followed by drying of the semi-selective agar medium and then staining.

C. Biochemical analysis

There are a number of references to use of these types of techniques with fungi other than *Rhizoctonia* (Whipps *et al.*, 1982). Two soil enzymes that have been successfully used to monitor *Rhizoctonia* in sterile soil are β-galactosidase and phosphotase (Beam *et al.*, 1977). Chitin and mannan are two cell wall components that have been used successfully to assay for Basidiomycotina in various substrates and may be useful for *Rhizoctonia*. Whipps *et al.* (1982) discuss a number of different potential biochemicals and their advantages and disadvantages, however the most obvious problem is lack of specificity so that experiments need to be run under sterile conditions.

VI. General methods

A. Aerial photography

Remote sensing can be used to measure both disease severity and incidence, as stressed plants differ from healthy plants in the wavelength of light they reflect. Another alternative is measurement of temperature differences between stressed and healthy plants using infrared thermometry (Pinter, 1979). Disease incidence is easier to measure than severity, as severity requires that a relationship between image color and severity and yield have been previously determined. Assessment should be made when the differences between diseased and healthy plants are greatest, which requires either a knowledge of disease development or multiple sampling over time.

TABLE 3. Summary of published work on the development and use of antibodies for quantification of *Rhizoctonia*

Authors	Antibody type	Antigen type	Antigen location	Specificity	Use
Matthew & Brooker, 1991	monoclonal MAb.2D1-IgM MAb.1D6-IgG	pectolytic enzymes?	extracellular induced	MAb.2D1 AG 8, 2-1, 3, 4 MAb.1D6 AG 8, 3	pure culture
Miller *et al.*, 1988	monoclonal	not stated	not stated	*R. solani* *R. cerealis* *R. oryzae*	plant samples
Dusunceli & Fox, 1992	ex Miller *et al.*, 1988			AG 1-6, 8	soil extracts
Thornton *et al.*, 1993	monoclonal EH2[3] - IgA EE1[2], HD10[2], IFIO[1], DB3[1] -IgM	EH2[3] -protein EE1[2], HD10[2], IFIO[1], DB3[1] - carbohydrate All water soluble	extracellular	AG 4, 3, *T. cucumeris*	soil extracts, colony staining

Area of *Rhizoctonia* patches in wheat was recorded with true colour photography (MacNish and Lewis 1985). In many pathosystems, *R. solani* causes a reduction in plant growth rather than total loss, so that estimates of losses would require calibration of the colour changes or patch area with yield loss for the particular location and season.

B. Vital stains

Use of vital stains in the study of *Rhizoctonia* has potential for both soil and plant assessments. Although *Rhizoctonia* was not tested, more than 50 fungi have been tested either *in-vitro* or in soil for their ability to take up eight different fluorescent vital stains or five different non-fluorescent vital stains (Söderström, 1977; Cohen, 1984; Stewart and Deacon, 1995). The stains differed in their ability to stain different fungal structures and in different environments and therefore need to be tested in the system being investigated. The stains themselves are not specific so other means of identifying the hyphae is required.

C. Radio isotopes

The most commonly used radionuclides for study of mycorrhizas are ^{32}P, ^{14}C and ^{35}S and all have potential for use in studies involving *Rhizoctonia*. Grossbard and Stranks (1959) showed that after several weeks ^{60}Co was only translocated 2 cm in *Rhizoctonia* hyphae growing in soil, however Skinner and Bowen (1974) showed translocation of ^{32}P in *Rhizopogon* up to 12 cm. Radio isotopes may be useful for quantification of growth in soil or on plants artificially infested with *Rhizoctonia*.

VII. Conclusions and Challenges

Little attention has been devoted to understanding the spatial patterns of the various Rhizoctonia diseases so that sound sampling strategies can be used in the study of the fungus or disease. No doubt this is in part due to a lack of suitable existing statistical tools, or because they were being used largely in non-biological disciplines. Areas of sampling which need to be addressed are to develop sampling strategies for highly patchily distributed distributions and to test their validity over time.

In the area of quantification, we need to prove the usefulness of the molecular techniques being developed, their sensitivity and their robustness in different soil types. An area which appears to be immediately profitable is combining the antibody techniques with traditional techniques such as soil embedding or agar film or soil membrane techniques. Many of the quantification techniques have been tested or calibrated by adding and then detecting a known number of artificial inoculum units to soil. We need to test the validity of the assumption that these artificial fungal units behave similarly to natural fungal units. The new molecular techniques may provide the means to do this.

VIII. Acknowledgments

SMN greatfully acknowledges that part of this work was written while he was a guest at IPO-DLO Wageningen, The Netherlands.

IX. References

Balali GR, Whisson DL, Scott ES & Neate SM (1996) DNA fingerprinting probe specific to isolates of *Rhizoctonia solani* AG 3. Mycol. Res.(in press).

Beam EW, Curl EA & Rodriguez-Kabana R (1977) Effects of the herbicides fluometron and prometryn on *Rhizoctonia solani* in soil cultures. Can. J. Microbiol. 23:617-623.

Boosalis MG & Scharen AL (1959) Methods for microscopic detection of *Aphanomyces euteiches* and *Rhizoctonia solani* associated with plant debris. Phytopathology 49:192-198.

Campbell CL & Pennypacker SP (1980) Distribution of hypocotyl rot caused in snap bean by *Rhizoctonia solani*. Phytopathology 70:521-525.

Campbell CL & Noe JP (1985) The spatial analysis of soilborne pathogens and root diseases. Annu. Rev. Phytopathol. 23:129-148.

Campbell CL & van der Gaag DJ (1993) Temporal and spatial dynamics of microsclerotia of *Macrophomina phaseolina* in three fields in North Carolina over four to five years. Phytopathology 83: 1434-1440.

Chellemi DO, Rohrbach KG, Yost RS & Sonoda RM (1988) Analysis of the spatial pattern of plant pathogens and diseased plants using geostatistics. Phytopathology 78:221-226.

Clark CA, Sasser JN & Barker KR (1978) Elutriation procedures for quantitative assay of soils for *Rhizoctonia solani*. Phytopathology 68:1234-1236.

Cochran WC (1977) Sampling techniques. (3rd ed.) Wiley, New York.

Cohen SD (1984) Detection of mycelium and oospores of *Phytophthora megasperma* forma specialis *glycinea* by vital stains in soil. Mycologia 76:34-39.

Cotterill PJ (1990) Assessment of yield loss caused by Rhizoctonia root rot in a barley crop sown following cultivation at Nhill, north-west Victoria. Australas. Plant Pathol. 19:77-78.

Damicone JP, Patel MV & Moore WF (1993) Density of sclerotia of *Rhizoctonia solani* and incidence of sheath blight in rice fields in mississippi. Plant. Dis. 77:257-260.

Dusunceli F & Fox RTV (1992) Accuracy of methods for estimating the size of *Thanatephorus cucumeris* populations in soil. Soil Use Manage. 8:21-26.

Gilligan C (1988) Analysis of the spatial pattern of soilborne plant pathogens. In: Kranz J & Rotem J (eds.) Experimental Techniques in Plant Disease Epidemiology. (pp 85-98) Springer, Berlin.

Glenn OF & Sivasithamparam K (1990) The effect of soil compaction on the saprophytic growth of *Rhizoctonia solani*. Plant Soil 121:282-286.

Grossbard E & Stranks D (1959) Translocation of cobalt-60 and cesium-137 by fungi in agar and soil cultures. Nature 184:310-314.

Henis Y, Ghaffar A, Baker R & Gillespie SL (1978) A new pellet soil-sampler and its use for the study of population dynamics of *Rhizoctonia solani* in soil. Phytopathology 68:371-376.

Herdina & Whisson DL (1995) Detection of *Rhizoctonia solani* AG 8 in soil using a specific DNA probe. Mycol. Res. 99:1299-1302.

Huang JW & Kuhlman EG (1989) Recovery and pathogenicity of *Rhizoctonia solani* and binucleate *Rhizoctonia*-like fungi in forest nurseries. Plant Dis. 73:968-872.

Hyakumachi, M. (1983) Disease decline phenomenon of sugar beet root rot. Shokubutsu Boeki (Plant Protection) 37:532-537.

Hyakumachi, M. & Ui T. (1982) Decline of Rhizoctonia root rot during sugar beet monoculture. Mem. Fac. Agric. Hokkaido Univ. 13:445-454.

Jager G & Velvis H (1994) Geostatistical analysis of the spatio-temporal development of downy mildew epidemics in cabbage. Phytopathology 84:1227-1239.

Journel AG & Huijbregts CJ (1978) Mining geostatistics. Academic Press. New York.

Ko W & Hora FK (1971) A selective medium for the quantitative determination of *Rhizoctonia solani* in soil. Phytopathology 61:707-710.

Lannou C & Savary S (1991) The spatial structure of spontaneous epidemics of different diseases in a groundnut plot. Neth. J. Plant Pathol. 97:355-368.

Larkin RP, Gumpertz ML & Ristaino JB (1995) Geostatistical analysis of *Phytophthora* epidemic development in commercial bell pepper fields. Phytopathology 85:191-203.

MacNish GC (1985) Mapping Rhizoctonia patch in consecutive cereal crops in Western Australia. Plant Pathol. 34:165-174.

MacNish GC & Lewis S (1985) Methods of measuring Rhizoctonia patch of cereals in Western Australia. Plant Pathol. 34:159-164.

MacNish GC & Sweetingham MW (1993) Evidence that each Rhizoctonia bare patch is dominated by an individual zymogram group (ZG) of *Rhizoctonia solani* AG 8. Aust. J. Agric. Res. 44:1175-1194.

Madden LV & Hughes G (1995) Plant disease incidence -distribution, heterogeneity, and temporal analysis. Annu. Rev. Phytopathol. 33:529-564.

Matheron G (1963) Principles of geostatistics. Econ. Geol. 58:1246-1266.

Matthew JS & Brooker JD (1991) The isolation and characterization of polyclonal and monoclonal antibodies to anastomosis group 8 of *Rhizoctonia solani*. Plant Pathol. 40:67-77.

Miller SA, Rittenburg JH, Petersen FP & Grothaus GD (1992) From the research bench to the market place: development of commercial diagnostic kits. In: Duncan JM Torrance L (ed.) Techniques for the rapid detection of plant pathogens. (pp 208-221) Blackwell Scientific Publications, Oxford.

Minderman G (1956) The preparation of microtome sections of unaltered soil for the study of soil organisms in situ. Plant Soil 8:42-48.

Naiki T & Ui T (1977) Population and distribution of sclerotia of *Rhizoctonia solani* Kuehn in sugarbeet field soil. Soil Biol. Biochem. 9:377-381.

Neate SM & Benger AM (1995) A nylon netting technique for the study of the three dimensional growth of *Rhizoctonia solani* in soil. Soil Biol. Biochem. 27:1617-1621.

Nicholas DP, Parkinson D & Burgess NA (1965) Studies of fungi in a podzol. II. Application of the soil-sectioning technique to the study of amounts of fungal mycelium in the soil. J. Soil Sci. 16:258-269.

Nicot PC, Rouse DI & Yandell BS (1984) Comparison of statistical methods for studying spatial patterns of soilborne plant pathogens in the field. Phytopathology 74:1399-1402.

Ogoshi A & Ui T (1983) Diversity of clones within an anastomosis group of *Rhizoctonia solani* Kühn in a field. Ann. Phytopathol. Soc. Jpn 49:239-245.

Papavizas GC (1968) Survival of root-infecting fungi in soil. VIII. Distribution of *Rhizoctonia solani* in various physical fractions of naturally and artificially infested soils. Phytopathology 58:746-751.

Papavizas GC & Davey CB (1962) Isolation of pathogenicity of Rhizoctonia saprophytically existing in soil. Phytopathology 52:834-840.

Pinter P, Jr., Stanghellini ME, Reginato RJ, Idso SB, Jenkins AD & Jackson RD (1979) Remote detection of biological stresses in plants with infrared thermometry. Science, USA 205:585-587.

Reynolds KM, Madden LV & Ellis MA (1988) Spatio-temporal analysis of epidemic development of leather rot of strawberry. Phytopathology 78:246-252.

Samuel G & Garrett SD (1932) *Rhizoctonia solani* on cereals in South Australia. Phytopathology 22:827-836.

Schneider JHM & Kocks GC (1995) Mapping Rhizoctonia disease using geostatistics to reduce sampling. Abstract No. P-5-5. International Symposium on *Rhizoctonia*. Noorwijkerhout, The Netherlands.

Shew BB, Beute MK & Campbell CL (1984) Spatial pattern of southern stem rot caused by *Sclerotium rolfsii* in six North Carolina peanut fields. Phytopathology 74:730-735.

Skinner MF & Bowen GD (1974) The uptake and translocation of phosphate by mycelial strands of pine mycorrhizas. Soil Biol. Biochem. 6:53-56.

Sneh B, Katan J, Henis Y & Wahl I (1966) Methods for evaluating inoculum density of *Rhizoctonia* in naturally infested soil. Phytopathology 56:74-78.

Sneh B, Burpee L & Ogoshi A (1991) Identification of *Rhizoctonia* species. The American Phytopathological Society, St Paul, Minnesota, 133 pp.

Söderström BE (1977) Vital staining of fungi in pure cultures and in soil with fluorescein diacetate. Soil Biol. Biochem. 9:59-63.

Stein A, Kocks GC, Zadocks JC, Frinking HD, Russein MA & Myers DE (1994) A geostatistical analysis of the spatio-temporal development of downy mildew epidemics in cabbage. Phytopathology 84:1227-1239.

Stewart A & Deacon JW (1995) Vital fluorochromes as tracers for fungal growth studies. Biotechnic & Histochem. 70:57-65.

Sundman V & Sivela S (1978) A comment on the membrane filter technique for estimation of length of fungal hyphae in soil. Soil Biol. Biochem. 10:399-401.

Thompson SK (1992) Sampling. Wiley, New York Wiley series in probablility and mathematical statistics

Thornton CR, Dewey FM & Gilligan CA (1993) Development of monoclonal antibody-based immunological assays for the detection of live propagules of *Rhizoctonia solani* in soil. Plant Pathol. 42:763-773.

Ui T, Naiki T & Akimoto M (1976) A sieving-flotation technique using hydrogen peroxide solution for determination of sclerotial population of *Rhizoctonia solani* Kuehn in soil. Ann. Phytopathol. Soc. Jpn 42:46-48.

Upton GJG & Fingleton B (1985) Spatial data analysis by example. Vol. 1. Point pattern and Quantitative Data. Wiley, Chichester.

Weinhold AR (1977) Population of *Rhizoctonia solani* in agricultural soils determined by a screening procedure. Phytopathology 67:566-569.

Whipps JM, Haselwandter K, McGee EEM & Lewis DH (1982) Use of biochemical markers to determine growth, development and biomass of fungi in infected tissues, with particular reference to antagonistic and mutualistic biotrophs. Trans. Brit. Mycol. Soc. 79:385-400.

Windham AS & Lucas LT (1987) A qualitative baiting technique for selective isolation of *Rhizoctonia zeae* from soil. Phytopathology 77:712-714.

Wiseman BM (1996) Characterisation of Rhizoctonia bare patch decline. PhD Thesis, University of Adelaide.

Wyss P, Boller T & Wiemken A (1990) Glyceollin production in soybean during the process of infection by *Glomus mosseae* and *Rhizoctonia solani*. Agric. Ecosyst. & Environ. 29, 451-456.

Yang XB, Snow JP & Berggren GT (1990) Patterns of Rhizoctonia foliar blight on soybean and effect of aggregation on disease development. Phytopathology 81:287-293.

IV.3 BASIDIOSPORE DISPERSAL AND SURVIVAL

SHIGEO NAITO

Tohoku National Agricultural Experiment Station, Akahira 4, Shimo-kuriyagawa, Morioka, 020-01 JAPAN

I. INTRODUCTION

Rhizoctonia solani Kühn is well known as a root-infecting fungus. Under certain conditions such as high relative humidity, it attacks aerial parts of a wide variety of plants. These aerial diseases occur in countries having tropical, temperate and subpolar climates. The common names given to these diseases are based on the particular symptoms occurring on various hosts and include: sheath blight, web blight, aerial blight, foliage or leaf blight, target spot, leaf spot, brown spot, and others (Baker, 1970; Sneh *et al.*, 1991). Symptoms of these diseases can occur on plants from ground level to more than 100 cm above ground in tall plants. Under natural field conditions, hymenia of the teleomorph state of *R. solani* [*Thanatephorus cucumeris* (Frank) Donk] are formed by isolates of certain anastomosis groups (AGs). Hymenia are formed on a wide variety of plant hosts, either on the stem bases, or on other plant and soil surfaces, which are protected under the plants' canopies. Airborne basidiospores can be disseminated rapidly over long distances, reaching the upper plant parts, and eventually, causing severe aerial epidemics under suitable moist conditions. Basidiospore infections have been reported in several host plants: citrus (Stahel, 1940), sugar beet (Kotila, 1947; Naito, 1984, 1990), soybean (Naito *et al.*, 1995), bean (Echandi, 1965; Galindo *et al.*, 1983), jute (Tu *et al.*, 1977), ground nut (Dubey and Dwivedi, 1992a), Peruvian barbasco (Crandall, 1950), rubber tree (Carpenter, 1951); cotton (Luke *et al.*, 1974), tobacco (Vargas, 1973; Shew and Main, 1985; 1990), sesame (Matsuura and Takahashi, 1953), rice (Fukazu *et al.*, 1960), *Eupatorium cannabinum* (Dwivedi and Dubey, 1985), tomato (Kodama *et al.*, 1982), sugarcane (Duvey and Dwivedi, 1992b), cabbage (Hoshi *et al.*, 1995), night shade (Shew and Main, 1985), *Sida rhombifolia* (Galindo *et al.*, 1983), *Amaranthus lividus* and *Chenopodium album* (Naito *et al.*, 1995). When infections are caused by hyphae or sclerotia originating from soil, the disease spreads considerably slower than when epidemics originate from airborne propagules. The rapid spread of aerial epidemics occurs primarily due to the presence of basidiospores as primary and secondary inocula. Nevertheless, the severity of field epidemics is decisively influenced by micro- and macro environmental as well as cultural conditions.

The present review focuses on symptoms of aerial diseases originating from *T. cucumeris* basidiospore infections, disease development, hymenia formation, leaf infection with basidiospores, their dispersal and survival, the role of basidiospores in spreading of the aerial disease epidemics and factors affecting these epidemics.

II. FOLIAR DISEASE SYMPTOMS ORIGINATING FROM BASIDIOSPORE INFECTIONS

Colors of foliar disease symptoms originating from basidiospore infections vary from dark green to brown. Lesion size and shape also vary. Two distinct types of lesions can be distinguished. Primary lesions are small, 1-3 mm or < 1 mm diameter,

B. Sneh et al. (eds.),
Rhizoctonia Species: Taxonomy, Molecular Biology, Ecology, Pathology and Disease Constrol, 197–205.

circular necrotic spots surrounded by a brown border, mostly developing on young or upper leaves, but rarely occurring on mature lower leaves. Typical foliar symptoms may be observed on sugar beets (Kotila, 1947; Naito, 1984, 1990), soybean (Naito *et al.*, 1995), sesame (Matsuura and Takahashi, 1953), tobacco (Vargus, 1973; Shew and Main, 1990) and citrus (Stahel, 1940). Appearance of the small spots on the upper leaves indicates that the infection source was airborne basidiospores. The small primary lesions do not extend much and can be distinguished from other enlarged lesions (Naito, 1984; Shew and Main, 1990). When temperature or humidity decrease, "shot" holes are formed in the center of the primary lesions. When humidity remains high, primary lesions are accompanied by enlarged lesions.

Secondary infections originating from hyphal infections form secondary lesions, which appear as dark green to brown or reddish brown with circular, zonate, star-, target- or irregular-shaped necrosis, 3-5 mm in diameter, up to more than half of the leaf. Their shapes change during lesion development and may vary in different hosts. In sugar beet, the healthy green areas are generally scattered in between the enlarged lesions (Naito, 1984; 1990). Secondary lesions appear on the healthy tissue adjacent to the primary lesion, while some secondary lesions may appear on unfolded leaves or on lower leaves of outer whorls, where no primary lesions are present.

III. HYMENIA FORMATION IN THE FIELD

A. Fructification

Grayish-white hyphae and powdery-white or -pale brown hymenia develop initially on the soil or on the stem surface. Then, they develop on the lower side of the infected leaves, which are protected from direct sunlight under the plants' canopies. Fruiting bodies were recorded on a wide variety of crops and on soils around their plants: sugar beet (Kotila, 1943, 1947; Coons, 1953; Herr, 1982; Naito, 1984; Windels *et al*, 1994), bean (Echandi, 1965; Galindo *et al.*, 1983), tobacco (Vargas, 1973; Shew and Main, 1990; Shew and Melton, 1995), Jute (Tu *et al.*, 1977), cotton (Luke *et al*, 1974), egg plant (Kodama *et al.*, 1982), tomato (Date *et al.*, 1988), rubber tree (Lorenz, 1948; Carpenter, 1949), rice (Sawada, 1919; Fukazu *et al.*, 1960; Jones and Belmar, 1989), soybean (Atkins and Lewis, 1954; Naito, 1984; Naito and Kanematsu, 1994; Jones and Belmar, 1989; Yang *et al.*, 1990), potato (Baker, 1970; Naito, 1984), corn (Naito, 1984; Jones and Belmar, 1989), groundnut (Duvey and Dwivedi, 1992a), camphor tree, pea, peanut (Sawada, 1919), sesame (Matsuura and Takahasi, 1953), cabbage (Hoshi *et al*, 1995), citrus (Stahel, 1940), sugarcane (Duvey and Dwivedi, 1992b), sorghum (Jones and Belmar, 1989), *Eupatorium cannabium* (Dwivedi and Duvey, 1985), *Robinia pseudoacacia* (Ito and Kontani, 1952), black gram (Saksena, 1973), barbasco (Crandall, 1950), kenaf and roselle (Crandall, 1948), wheat, oat, adzuki bean, rape, Japanese dock, dandelion, spinach, turnip, radish, barley, great burdock, barnyard grass and low bamboo (Naito, 1984). In reports published before 1970, anastomosis groups have not been determined. Anastomosis groups of basidiospore isolates obtained in northern Japan are as follows: *R. solani* AGs 1-IA, 2-1, 2-2IV, 2-3, 3, 4, 5 and binucleate *Rhizoctonia* spp. (Naito, 1984; Naito *et al.*, 1995). Of these groups, AG 5 and binucleate *Rhizoctonia* spp. were found to develop hymenia on stems of a wide variety of hosts and surrounding soil surfaces. Relationship between AGs and fructification may depend on the nature of the soilborne or the saprophytic phase and on cultural and climatic factors.

B. Effect of climatic factors and cultural practices on fructification.

Climatic factors, such as temperature, relative humidity, wind and aeration, determine the fructification of *T. cucumeris*. Generally, the fungus forms abundant hymenia at relatively high temperature (≥20°C) and high moisture (≥90% RH) conditions in fields (Fukazu *et al*, 1960; Ito and Kontani, 1952; Dubey and Dwivedi, 1992a; Naito, 1984; Shew and Main, 1985). Isolates of AG 2-2 IV in sugar beet fields usually start to produce hymenia abundantly, 2 or 3 days after heavy rainfall (≥5-10mm/day), in which air temperatures are above 20°C (Naito, 1984). When the cover-soil method (Ogoshi, 1976) was applied to cultures in a growth chamber, it took 10, 4, 3 and 2 days for an isolate of AG 2-2 IV to develop hymenia at temperatures of 15.4, 21.5, 24.0 and 28.9°C, respectively (Naito, 1984). However, the times for fruiting differed among isolates and AGs (Ogoshi, 1976).

Usually, hymenia appear after the inter row space is covered by the plants' canopies, which keep relative humidity high and protect the leaves from drying by the wind. Therefore, planting densities or broad leaf plants strongly influence hymenia production (Fukazu *et al.*, 1960; Naito, 1984, 1995). In sugar beets, wheat, potato, soybean and corn fields artificially infested with isolates of AG 2-2 IV, relative sizes of leaf areas covered with hymenia were in the following order: potato> wheat> sugar beet> corn> soybean (Naito, 1984). When soybeans were inter-sown with winter wheat, conditions were favorable for fructification and foliar blight with basidiospore infection by *T. cucumeris* (AG 2-3) (Naito *et al.*, 1995). Delay in wheat harvest time frequently results in an increase in disease severity (Naito, *et al.*, 1995).

Hymenia are not usually formed during rainy days. However, they develop abundantly, several days after rainfall. Good aeration is essential for fruiting of *T. cucumeris* (Kotila, 1929; Adams and Butler, 1983b). Isolates of AGs 1 and 4 produced many fruiting bodies on spore-inducing agar media at an air flow rate of 45L/h, whereas both never fruited without such aeration (Adams and Butler, 1983b).

Cultural practices may influence hymenia formation through changes in micro environmental conditions. Isolates of AG 2-2 IV produced many fruiting bodies in inter tilled or molded fields, but less in non tilled ones. This may be attributed to active hyphal growth promoted by cultivation. Fruiting bodies were rarely observed at the same spot after a single tillage, but when soil was inter tilled again, fruiting bodies developed again at the same location (Naito, 1984). Application of barnyard manure to rice paddy fields promoted plant growth, and hymenia were consequently formed (Fukazu *et al.*, 1960). Application of high concentrations of sodium nitrate to a spore-inducing medium promoted hymenia development (Adams and Butler, 1983a).

IV. Basidiospore dispersal

A. Basidiospore dispersal and disease incidence

Although quite a few diseases of aerial plant organs are caused by basidiospore infections (as outlined above), little information is available regarding the importance of basidiospores as inocula in the epidemiology of these diseases. Naito (1984) studied seasonal changes of aerial dispersal of basidiospores and the consequent spread of sugar beet foliage blight in fields artificially infested with AG 2-2 IV colonized barley grains. Data recorded from spore traps indicated that foliage blight incidence did not occur before basidiospores were discharged. The first trapped basidiospores originated from basidia which were formed on the soil surface around infected sugar beet with root rot symptoms (around mid July). First initial symptoms consisting of primary and secondary lesions appeared a few days later around sugar beet root rot patches. At the beginning of August, secondary lesions extended and many fruiting bodies were formed on the lower side of the infected leaves. Subsequently, repeated basidiospore infections resulted in an increase in disease severity, followed by rapid spread all over the field. Such epidemics cause a uniform disease pattern in the field, which is in contrast to patches formed by pathogen spread through soil. In certain years, when climatic and cultural conditions are favorable for foliar blight, such epidemics may spread all over sugar beet fields in certain regions. Basidiospore densities in the air are closely related to the number of leaves harboring secondary lesion symptoms and with disease severity. Number of primary lesions per leaf may reach more than 150. During each September, when temperatures decrease, spore dispersal and disease incidence do not advance and subsequently fruiting bodies disappear. It is therefore concluded that the rapid spread of epidemics incited by basidiospore dissemination plays an important role in epidemiology of foliage blight in sugar beets.

B. Vertical and horizontal dispersal of basidiospore

Vertical dissemination of basidiospores in sugar beet fields was evaluated by placing water agar traps at a range of height above the soil surface (Naito, 1984). Frequency of basidiospore trapping correlated with their densities in the air and with height, as summarized in Fig. 1. During one night, up to 180 basidiospores were trapped on a 18 x 18 mm surface at a height of 185 cm. Similar trends were recorded in a rice paddy field infected by AG 1-IA showing sheath blight symptoms (Fukazu *et al*, 1960). Because basidia were trapped abundantly at a height of 185 cm (Naito, 1984), it is quite possible that basidiospores can be disseminated for long distances (more than 300 m), as suggested by Kotila (1947). In a bean field located 70 m from a sugar beet field with severe foliage blight, 10 basidiospores were trapped at a height of 83 cm on a water agar trap exposed over night, although no hymenia could be observed on the bean plants. Wind velocity and direction are known to influence the distance and direction of *Pyricularia oryzae* conidiospore dispersal (Suzuki, 1975). Similar effects on basidiospores would be anticipated.

Additional and more accurate studies using new methods such as vegetative compatibility and DNA markers are required for more accurate modeling of horizontal basidiospore dispersal and their consequent influence on foliar disease epidemics.

C. Diurnal changes in basidiospore release

Spore dispersal occurs almost exclusively during the night hours (Carpenter, 1949; Echandi, 1965; Fukazu *et al.*, 1960; Kotila, 1947; Luke; 1974; Naito, 1984; Tu *et al.*, 1977). However, little information is available regarding the effect of diurnal changes on basidiospore discharge in the field. Basidiospores were trapped overnight on agar plate traps above an infected sugar beet field beginning in the early evening. The number of trapped spores rapidly increased with time from approximately 9:00 p.m., reaching maximum numbers from midnight to dawn (Fig. 2). Time and duration of basidiospore discharge correlated with suitable conditions of dew deposition and high relatively humidity necessary for basidiospore germination and germ tube penetration. After sunrise, numbers of spores trapped decreased rapidly and no spores were trapped during the daytime. Basidiospore discharge occurs periodically and appears to be diurnal phenomena regulated by light and dark periods (Carpenter, 1949; Baker & Martinson, 1970).

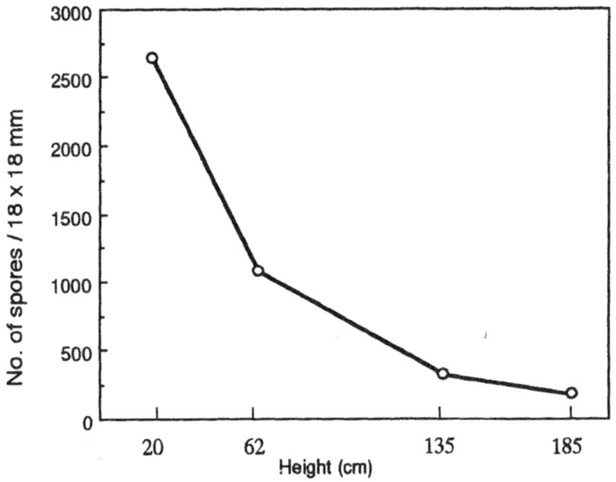

FIGURE 1. Vertical distribution of *Thanatephorus cucumeris* basidiospores in a sugar beet field (2.3ha). Basidiospores were trapped by agar plate from 5:30 pm to 8:30 am of the following morning. (Aug. 11 - 12, 1978)

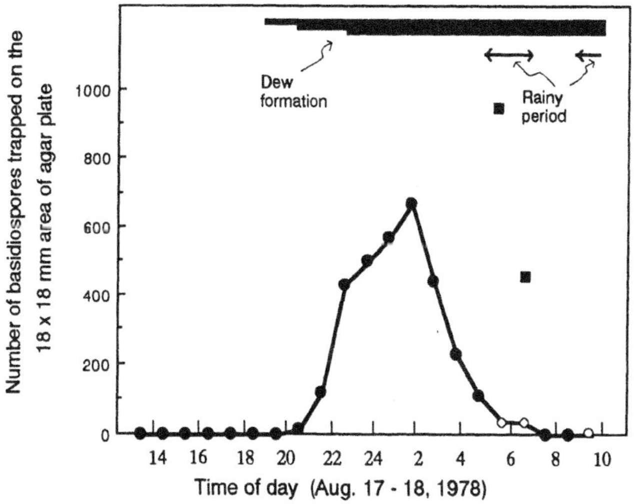

FIGURE 2. Diurnal change of basidiospores discharged from basidial mats on the ventral side of sugar beet leaves in a field. Basidiospores were trapped during clear or cloudy periods (•), during a rainy period (o), under the canopy of an umbrella in a rainy period (▪),

V. Germination and Infection of Basidiospores Deposited on Leaves

A. Effects of relative humidity and temperature on basidiospore germination

Earlier studies have emphasized the essential importance of relative humidity and temperature on basidiospore germination *in vitro*, (reviewed by Baker and Martinson, 1970). AG 2-2 IV basidiospores germinated on agar media at 15 - 32°C, with an optimum of 28°C, and developed typical vegetative hyphae. On sugar beet leaves, basidiospores germinated with frequencies of 19.8, 36.2 and 51.2 % at 16, 20 and 28°C, respectively. At these temperatures, infection rates were 1.2, 15.8 and 19.1 %, respectively (Naito, 1984).

Basidiospores can germinate only under high relative humidity conditions (>99% RH) (Baker and Martinson, 1970; Naito, 1984), however, they are exposed frequently to low relative humidity in fields. When basidiospores placed on cellophane membranes were exposed to lower relative humidities (81-32 % RH), they kept their ability to germinate for 18 days (Naito, 1984). However, it is unclear whether those viable spores maintained their ability to infect the host.

Among the basidiospores deposited on the leaf surface at night, some will be exposed to direct sunlight during day time. Basidiospores placed on water agar were exposed to the direct sunlight (for 30-180 min between 12:00 15:00pm in the summer). Many spores germinated after 30min exposure, but lost their ability to germinate after 60 min exposure (Naito, 1984). Similar results were obtained in winter tests (<0°C), made to avoid the increase in temperature of the agar surface that occurred due to exposure to sunlight in the summer. Control basidiospores kept in the shade outside, germinated well.

B. Infection and lesion development

The basidiospore infection process and subsequent foliar lesion development were similar on sugar beet infected by AG 2-2 IV (Naito, 1984), tobacco by AG 3 (Shew and Main, 1990) and soybean by AG 2-3 (Naito *et al.*, 1995). The infection process on sugar beet leaves with basidiospores of AG 2-2 IV is as follows. The basidiospores germinated on the leaves; formed appressoria within a few hours, and penetrated directly into the epidermal cells. The invading hyphae produced a stroma-like structure within the epidermal cells or the upper layer cells of the mesophile. Subsequently, hyphae grew out radially from this structure and formed small (≤1mm diameter), primary lesions after 5 or 6 days' incubation. The stroma-like structures develop only from basidiospore infections, and are never produced from hyphal infections. They are essential for the development of primary lesions (Naito, 1984). These structures have been found also in tobacco (Shew and Main, 1990), soybean (Naito *et al.*, 1995) and the weeds *Amaranthus lividus* and *Chenopodium album* var. *centrorubrum* (Naito *et al.*, 1995), which were infected with basidiospores of AGs 3, 2-3, and 2-2IV, respectively. The function of the stroma-like structure is not clear, but it is probably different from ordinary sclerotia which were formed during the later stages of lesion development (Naito and Sugimoto, 1981). Further studies are needed to find out whether basidiospores of other AGs also produce the stroma-like structure prior to primary lesion formation. Germinating basidiospores did not develop any vegetative hyphae on the host surface (Naito, 1984).

In moist weather, the hyphae growing out from the primary lesions spread on the leaf surface, and the hyphal tips enter the leaf through stomata, causing one or two secondary lesions, 3 mm in diameter, around the primary lesion. Then, hyphae growing out from the first secondary lesion enter the leaf repeatedly in the same manner and form irregular-shaped coalesced lesions. Eventually, most of the leaf area is covered by a large lesion. Fruiting bodies are formed on these infected leaves as described above. Secondary lesions may also originate from hyphae growing from on the soil surface onto the plant.

VI. BASIDIOSPORES SURVIVAL IN SOIL

A. Survival of basidiospores which land on soil

A considerable number of *R. solani* clones inhabit sugar beet field soils (Ogoshi and Ui, 1981). It is not clear yet whether basidiospores deposited on the soil surface can germinate and colonize fresh plant residues, moribund host or develop fungal clones via plant infections. Naito (1984) investigated the survival of basidiospores landing on natural soil by modifying the direct assay method of soils for fungistasis described by Lingappa and Lockwood (1963). Basidiospores could germinate on autoclaved soil, but not on natural soil. On mixtures of different proportions of autoclaved soils with natural soil, basidiospore germination decreased with increasing the proportion of the natural soil. It can therefore be concluded that basidiospores landing on natural field soil will not survive unless they are deposited on host plant parts, probably due to the nutrient sink-induced soil fungistasis.

B. Survival of fungal clones originating from basidiospores, via plant infection

Isolates colonizing infected plants originated from basidiospores are likely to infest and survive in soil. However, more studies are needed to confirm this assumption. When sugar beet leaves, on which sclerotia occasionally are formed, were buried in soil and over wintered in the field, many isolates could be recovered from the colonized debris until April or May, but soon after this period their viability decreased rapidly (Naito, 1984). Few of the hyphal and sclerotial propagules, which survived until the early summer, could infect sugar beet and subsequently form fruiting bodies on soil and foliage. When diseased leaves recovered from soil in April, May and June were placed on moist filter paper and covered with soil, they formed basidia on the soil surface.

VII. CONCLUSIONS

One complete cycle from basidiospore germination and infection, through secondary basidiospore dispersal, as described above, may take around 2 weeks in favorable field conditions. Therefore, such cycles are likely to be repeated 2-3 times when foliar epidemics spread rapidly all over the field. It may be concluded that *T. cucumeris* basidiospores play an important role in the epidemiology of aerial diseases in certain hosts when environmental conditions are favorable for hymenial formation, basidiospore dispersal, infections and lesion development. Meteorological analysis may provide a good tool for epidemiological modeling of aerial diseases caused by *T. cucumeris* basidiospores, similar to the simulation models of rice sheath blight reviewed by Hashiba (see chapter V.7).

VIII. References

Adams GC Jr & Butler EE (1983a) Influence of nutrition on the formation of basidia and basidiospores in *Thanatephorus cucumeris*. Phytopathology. 73: 147-151.

Adams GC Jr & Butler EE (1983b) Environmental factors influencing the formation of basidia and basidiospores in *Thanatephorus cucumeris*. Phytopathology 73:152-155.

Atkins JG Jr & Lewis W D (1954) *Rhizoctonia* aerial blight of soybeans in Louisiana. Phytopathology 44:215-218.

Baker KF (1970) Types of *Rhizoctonia* diseases and their occurrence. In: Parmeter Jr JR (ed.) Biology and Pathology of *Rhizoctonia solani* (pp 125-148) Univ. of California Press, Berkeley.

Baker R & Martinson CA (1970) Epidemiology of diseases caused by *Rhizoctonia solani*. In: Parmeter JR Jr (ed.) Biology and Pathology of *Rhizoctonia solani* (pp 172-188) :Univ. of California Press, Berkeley.

Carpenter JB (1949) Production and discharge of basidiospores by *Pellicularia filamentosa* (Pat.) Rogers on Hevea rubber. Phytopathology 39:980-985.

Carpenter JB (1951) Target leaf spot of the Hevea rubber tree in relation to host development, infection, defoliation, and control. U.S. Dept. Agric. Tech. Bull. 1028:1-34.

Coons GH (1953) Some problems in growing sugar beets. In: Stefferud A (ed.) Plant Disease, Yearbook of Agriculture (pp 509-524) USDA, Washington DC .

Crandall BS (1948) *Pellicularia* target spot leaf disease of kenaf and roselle. Phytopathology 38: 503-505.

Crandall BS (1950) Leaf diseases of Peruvian barbasco, *Lonchocarpus ultilis*. Phytopathology 40:34-43.

Date H, Yagi S, Okamoto Y & Oniki M (1984) Leaf blight of tomato caused by *Thanatephorus cucumeris* (Frank) Donk. Kinki Chugoku Agric. Res. 76: 12-16. (Japanese with English summary)

Dubey SC & Dwivedi RP (1992a) Survival of *Thanatephorus cucumeris* causing web blight of groundnut. Indian J. Mycol. Plant Pathol. 22:49-53.

Dubey SC & Dwivedi RP (1992b) Banded blight of sugarcane caused by *Thanatephorus cucumeris* in Uttar Pradesh. Agric. Sci. Dig. (Karnal) 12:90-92.

Dwivedi RP & Dubey SC (1985) Web blight of *Eupatorium cannabinum* L. caused by *Thanatephorus cucumeris*. Indian J. Mycol. Plant Pathol. 16: 309.

Echandi E (1965) Basidiospore infection by *Pellicularia filamentosa* (=*Corticium micro-sclerotia*), the incitant of web blight of common bean. Phytopathology 55:698-699.

Fukazu R, Kashizaki T & Hirayama S (1960) Studies on secondary spread of rice sheath blight caused by basidiospores of *Pellicularia filamentosa*. Bull. Kohchi Pre. Agric. Exp. Stn. 2: 26-38. (In Japanese).

Galindo JJ, Abawi GS, Thurston HD & Galvez G (1983) Source of inoculum and development of bean web blight in Costa Rica. Plant Dis. 67: 1016-1021.

Herr LJ (1982) Characteristics of hymenial isolates of *Thanatephorus cucumeris* on sugar beets in Ohio. Plant Dis. 66:246-249.

Hoshi H, Horie H & Iijima T (1995) Occurrence of head rot of cabbage caused by *Thanatephorus cucumeris* in Tokyo. Ann. Phytopathol. Soc. Jpn. 61:218-219 (Abstract in Japanese).

Ito K & Kontani S (1952) The causal fungus of web-blight of leguminous woody plants. Bull. Gov. For. Exp. Stn. 54: 45-78 (Japanese with English summary).

Jones RK and Belmar SB (1989) Characterization and pathogenicity of *Rhizoctonia* spp.isolated from rice, soybean, and other crops grown in rotation with rice in Texas. Plant Dis. 73:1004-1010.

Kotila JE (1929) A study of the biology of a new spore forming *Rhizoctonia, Corticium praticola*. Phytopathology 19:1059-1100.

Kotila JE (1943) A new sugar beet leaf blight caused by a strain of *Corticium solani*. Phytopathology 33:6-7.

Kotila JE (1947) *Rhizoctonia* foliage blight of sugar beets. J. Agric. Res. 74:289-314.

Kodama T, Horimoto K & Ogoshi A (1982) On the brown spot of eggplants caused by *Thanatephorus cucumeris* (Frank) Donk (*Rhizoctonia solani*) AG 3. Ann. Phytopathol. Soc. Jpn. 48:356-357. (Abstract in Japanese).

Lingappa BT & Lockwood JL (1963) Direct assay of soils for fungistasis. Phytopathology 53: 529-531.

Lorenz RC (1948) A new leaf disease of Hevea in Peru. J. For. 46: 27-30.

Luke WJ, Pinckard JA & Wang SC (1974) Basidiospore infection of cotton bolls by *Thanatephorus cucumeris*. Phytopathology 64:107-111.

Matsuura Y & Takahashi K (1953) Studies on the disease of *Sesamum indicum* L. in Japan. 1. On a new shot-hole leaf disease caused by a strain of *Corticium vagum* B. et. C. Sci. Rep. Fac. Agric. Ibaraki Univ. 1:1-10. (Japanese with English summary)

Naito S (1984) Studies on foliage blight of sugar beets. Res. Bull. Hokkaido Natl. Agric. Exp. Stn. 139:145-188. (Japanese with English summary)

Naito S (1990) Ecological role of basidiospores of *Thanatephorus cucumeris* (Frank) Donk in the incidence of foliage blight of sugar beets in Japan. JARQ 23:268-275.

Naito S & Kanematsu S (1994) Characterization and pathogenicity of a new anastomosis subgroup AG 2-3 of *Rhizoctonia solani* Kühn isolated from leaves of soybean. Ann. Phytopathol. Soc. Jpn. 60:681-690.

Naito S, Mochida H, Nakajima T & Ohto Y (1995) Infection with basidiospores of *Thanatephorus cucumeris* (AG 2-3 of *Rhizoctonia solani*) and development of soybean foliar blight lesions. Ann. Phytopathol. Soc. Jpn. 61: 362-368.

Naito S & Sugimoto T (1981) Histopathological observation on root rot of sugar beets by different anastomosis groups of *Rhizoctonia solani* Kühn. Res. Bull. Hokkaido Natl. Agric. Exp. Stn. 131:95-110 (Japanese with English summary).

Ogoshi A (1976) Studies on the grouping of *Rhizoctonia solani* Kühn with hyphal anastomosis and on the perfect stages of groups. Bull. Natl. Inst. Agric. Sci. Ser. C. No. 30: 1-63 (Japanese with English summary).

Ogoshi A & Ui T (1983) Diversity of clones within an anastomosis group of *Rhizoctonia solani* Kühn in a field. Ann. Phytopathol. Soc. Jpn 49:239-245. (Japanese with English summary).

Saksena HK & Dwivedi RP (1973) Web blight of black gram caused by *Thanatephorus cucumeris*. Indian. J. Farm. Sci. 1: 58-61.

Sawada K (1919) Descriptive Catalogue of the Formosan Fungi, Part 1, Agric. Exp. Stn. Gov. Formosa. Special Bull. 19: 471-493 (Japanese).

Shew HD & Main CE (1985) *Rhizoctonia* leaf spot of flue-cured tobacco in North Carolina. Plant Dis. 69:901-903.

Shew HD & Main CE (1990) Infection and development of target spot of flue-cured tobacco caused by *Thanatephorus cucumeris*. Plant Dis. 74:1009-1013.

Shew HD & Melton TA (1995) Target spot of tobacco. Plant Dis. 79: 6-11.

Sneh B, Burpee L and Ogoshi A (1991) Identification of *Rhizoctonia* species. APS Press, St. Paul. Minnesota 133 pp.

Suzuki H (1975) Meteorological factors in the epidemiology of rice blast. Ann. Rev. Phytopathol. 13:239-256.

Stahel G (1940) *Corticium areolatum*, the cause of the areolate leaf spot of citrus. Phytopathology 30: 119-130.

Tu CC, Cheng YH & Schenck NC (1977) Leaf spot caused by basidiospores of *Thanatephorus cucumeris* on jute, and survival of single-basidiospore isolates in soil. Plant Dis. Rep. 61:80-84.

Vargas GE (1973) Infeccion por basidiosporas de *Thanatephorus cucumeris* causant de una enfermedad foliar en tabaco. Turrialba 23:357-359.

Windels CE, Kuznia RA & Call J (1994) First report of *Thanatephorus cucumeris* (=*Rhizoctonia solani* AG 3 and AG 5) on sugar beet. Phytopathology 84:1161.

Yang XB, Berggren GT & Snow JP (1990) Types of *Rhizoctonia* foliar blight on soybean in Louisiana. Plant Dis. 74:501-504.

IV.4 SCLEROTIA FORMATION BY *RHIZOCTONIA* SPECIES AND THEIR SURVIVAL

DONALD R. SUMNER

Department of Plant Pathology, University of Georgia, Coastal Plain Experiment Station, Tifton, GA 31793-0748, USA

I. SCLEROTIA FORMATION

A. Historical perspectives

Rhizoctonia has a "loose type" of sclerotium that is not organized into a rind, cortex, and medulla, and only *Rhizoctonia* spp. have sclerotia of this type (Townsend and Willetts, 1954). Isolates of AG 1-IA that have "sasakii type" sclerotia are exceptional and have dark outer cells and have been referred to as a "primitive rind type" (Tu and Kimbrough, 1975). Sclerotia are formed primarily from compact masses of monilioid cells, but they may be formed from undifferentiated hyphae. The barrel-shaped cells average 20-22μm wide, and 30-35μm long, and are thicker than the ordinary hyphae. The central area of the sclerotium may have a pseudo-parenchymatous appearance. The sclerotia enlarge by repeated branching and formation of new cells (Townsend and Willetts, 1954). Sclerotial cells have the same number of nuclei as undifferentiated hyphae (Tu and Kimbrough, 1975).

In culture, sclerotia may range in size from barely detectable up to 6 mm in diameter, and may coalesce into crusts of a few centimeters in diameter. They may be irregular, globose, subglobose, irregular globose and pitted, flat, flattened on the bottom and rounded on the top, or formed into crusts. They may be light tan to dark brown (Butler and Bracker, 1970). Aeration and high humidity or condensed moisture apparently are essential for sclerotia formation. In culture, they may be small and infrequent on nutritionally poor media, but large and frequent on rich media. High numbers of sclerotia are formed and mature rapidly on media with high glucose and peptone (Sherwood, 1970) and on green bean pods (van Bruggen and Arneson, 1985).

Light influences sclerotial formation and morphology, and some isolates produce sclerotia in light but not in dark. In constant light, sclerotia were more appressed and compact, and occurred mostly in the center of Petri dishes compared with cultures grown in constant darkness (Durbin, 1959b).

B. Recent Research

Adenosine 3',5'-cyclic monophosphate (cyclic AMP) induced formation of sclerotia in an AG 1 isolate that did not usually form sclerotia, but did not induce sclerotial formation in several other mycelial isolates (Hashiba and Ishikawa, 1978). In culture AG 1-IA produced few sclerotia without phosphate, and Mg increased the number and weight of sclerotia. On a P-free medium the branching internodes were longer than when P was present. Phosphate induced the interweaving of hyphae and the formation of sclerotia. P was not essential in the early stages of initiation of sclerotial initials, but was esscntial for the development, enlargement, and maturation of sclerotia. In contrast, sclerotia were initiated and matured without Mg, but the mature weight of sclerotia was lower if Mg was deficient (Moromizato *et al.*, 1991).

B. Sneh et al. (eds.),
Rhizoctonia Species: Taxonomy, Molecular Biology, Ecology, Pathology and Disease Constrol, 207–215.

The stages of sclerotial formation in AG 1-IA were hyphal branching, hyphal aggregation and network formation, initials formation, the formation of whitish immature sclerotia, and maturation with pigmentation (Moromizato *et al.*, 1980). During maturation of sclerotia there is rapid RNA synthesis, a decline in polyA (+)-RNA, and a decrease in permeability of the sclerotia, followed by increased protease and ribonuclease and a rapid increase in ribosomes (Hashiba and Staples, 1976). Sclerotia of AG 1-IA formed on rice plants have a high proportion of `empty' cells in their outer layer, which play an important role in enabling them to float on the water surface of the rice paddies (Hashiba and Yamada, 1981).

In AG 1-IB, growing points from which lateral branches originated developed into 'trunk' hyphae lateral to the main hyphae. The `trunk' hyphae and lateral branches made up the skeleton of the microsclerotium (Yang *et al.*, 1989). Amino acids containing sulfur inhibited the formation of sclerotia of AG 1-IA. Lateral branching was limited, but aggregation of the hyphae was not affected. Sclerotial formation probably was inhibited because of the reduction in interweaving hyphae caused by reduced lateral branching, and the inhibition of cell enlargement as sclerotia matured (Moromizato *et al.*, 1980).

In experiments with soil in a laboratory, AGs 1 and 2-1 produced sclerotia that were firm and irregularly globose to globose, with a smooth or pitted surface. In contrast, sclerotia of AGs 2-2, 3, 4, and 5 were soft with an indefinite shape and irregular surface. AG 1 had three well defined layers, AG 2 had two well defined layers, and the other AG's had only loosely arranged brown cells (Naiki and Ui, 1978).

II. SURVIVAL OF SCLEROTIA

A. In plant debris

Most *Rhizoctonia* spp. probably survive in soil between annual crops in colonized plant debris. The crown-rot pathogen of sugar beet produced sclerotia on plant debris fragments that ranged from $<10\mu m$ to $>500\mu m$ long, or thick-walled, irregularly branched hyphae of monilioid chain-like aggregations of cells within fragments. In a temperate climate, both the sclerotia and the thick-walled, monilioid hyphae survived 7 months and caused seedling blight and damping-off of sugar beet. In contrast, hyphae on the surface of plant debris were devoid of protoplasm and did not germinate after 7 months (Boosalis and Scharen, 1959). The pathogens studied were probably AG 2-2 and AG 4. Wet-screening soil through a 0.35 mm mesh sieve removed particles that contained essentially all propagules of AG 4 that induced seedling disease in cotton (Weinhold, 1977). Large debris particles (> 6 mm long) caught on a 4-mesh sieve may serve as additional reservoirs of inoculum (Roberts and Herr, 1979a). By wet-screening topsoil (15 cm deep) for plant debris *R. solani* was detected in all of the peanut-growing areas of Texas and New Mexico, USA. Isolates of AG 4 were primarily recovered from plant debris, but a few isolates of binucleate *Rhizoctonia* were identified (Woodard and Jones, 1983).

In Australia, isolates of AG 8 were recovered from plant debris particles in soil 0-5, 5-10, and 10-15 cm deep in direct-drilled wheat. The fungus was found in particles ranging in size from 250μm to >2.0mm. When *Rhizoctonia* was recovered, the particles usually contained embedded monilioid cells. Dark-colored, partly decomposed plant debris was an important source of inoculum. *Rhizoctonia* rarely grew from sclerotia in soil (Neate, 1987).

Rhizoctonia spp. may survive in contaminated irrigation water in ponds containing debris and soil eroded from fields. However, *R. solani* was found infrequently in surface irrigation ponds in Georgia, USA compared with *Pythium* spp. and other pathogens. Constructing ponds to prevent eroded soil and debris from entering the water, and the use of floating inlet pipes to minimize the intake of bottom sediment and plant debris may reduce contamination in irrigation water (Shokes and McCarter, 1979).

B. Vertical distribution of sclerotia in soil

R. solani was isolated rarely from subsoil (Papavizas *et al.*, 1975, Sumner *et al.*, 1981). This may be because of the low tolerance of *R. solani* to high concentrations of CO_2. At 20 % CO_2, growth of *Rhizoctonia* clones of aerial types, surface types, and subterranean types were inhibited 87-73, 62-40, and 43-20 %, respectively, and Durbin (1959a) suggested that tolerance to a microenvironment of high CO_2-low O_2 may be a characteristic of isolates that are found commonly in soil. In snap bean culture, AG 4 was found primarily in the upper 5 cm of soil, and rarely below 10 cm deep (Papavizas *et al.*, 1975). In a multiple-cropping system of corn-snap bean or corn-lima bean AG 4 and other *Rhizoctonia* spp. were found more commonly in the upper 15 cm of soil when soil was disk-harrowed 10-15 cm deep than when soil was deep-turned with a moldboard plow 25-30 cm deep (Sumner *et al.*, 1986).

At the Coastal Plain Experiment Station on loamy sand and sand soils, population densities of *R. solani* (primarily AG 4) were determined with a multiple-pellet soil sampler (Henis *et al.*, 1978) following different tillage practices. From 1978-1983 more than 40 experiments were conducted comparing conventional tillage (moldboard plowing) with the conservation tillage treatments of disk-harrowing (10-15 cm deep), subsoiling under the row (25-35 cm deep) and planting, subsoiling under the row and bedding or ridging soil behind the subsoiler, or chiseling (15-20 cm deep). Soil was assayed after tillage and planting in a variety of cropping systems including vegetable and agronomic crops (Sumner *et al.*, 1981, Sumner *et al.*, 1986).

Population densities of AG 4 were variable following different crops and averaged 6, 11, 12, and 19 colony-forming units (CFU)/100 g of oven-dried soil after moldboard plowing, disk-harrowing, subsoiling and planting, chiseling, and subsoiling and bedding, respectively. Moldboard plowing reduced populations of AG 4 significantly in topsoil ($P=0.05$) in 20, 19, and 43 % of the experiments compared with disk-harrowing, subsoiling and planting, or subsoiling and bedding, respectively. In contrast, moldboard plowing was not significantly different from chisel plowing. In one experiment, a double-crop of sweet corn-soybean was followed by a double-crop of turnip-peanut. Soil was collected in 10-cm increments to a depth of 40-cm 10 days after planting peanut. Tillage practices were moldboard plowing, in-row subsoiling and planting, chisel-plowing, and disk-harrowing. The inoculum of *R. solani* was confined primarily to the upper 10 cm of soil in the conservation tillage treatments, but was distributed throughout the upper 30 cm of soil following moldboard plowing, and much less inoculum was in the upper 10 cm of soil with moldboard plowing than with the other treatments. *R. solani* was rarely found 30-40 cm deep (Sumner *et al.*, 1981).

In a separate investigation in cotton in the Piedmont region of Georgia, disk-harrowing followed by subsoiling and bedding was compared with disk-harrowing and bedding. Population densities of *R. solani* were low, primarily confined to the upper 18 cm of soil, and not influenced by subsoiling (Hussey and Roncadori, 1977).

In India, soil was collected at 2.5 cm intervals to a 35 cm depth in two soils, a sandy loam and a black cotton soil, and soil pellets were incubated on selective media (Ko and Hora, 1971). *Rhizoctonia* grew from 80-98 % of the pellets in the upper 15 cm of soil, from 40-60 % of the pellets 15-25 cm deep, and from none of the pellets below 25-30 cm deep. When fenugreek was cultivated continuously *Rhizoctonia* survived to a depth of 30 cm (Hiremath and Prasad, 1985).

C. Sclerotia of different anastomosis groups

1. AG 1. The sclerotia of the rice sheath blight pathogen (AG 1-IA) are dense and sink in water, but within 15 days cell contents in the outer layers disappear and the cells are empty. The sclerotia then become buoyant and float on water (Hashiba and Mogi, 1975). Both buoyant and nonbuoyant sclerotia were found in soil in rice fields in late winter in Arkansas, USA, 0-0.6, 0.6-3.8, and 3.8-7.6 cm deep, but numbers and viability decreased with depth. Nonbuoyant sclerotia, when dried 24-48 hrs, became buoyant and then nonbuoyant (Lee, 1980). In Mississippi, USA, fewer sclerotia were found in the top 7.5 cm of fields where rice was planted following 2 or 3 years of soybean compared with rice alternated with soybean (Damicone *et al.*, 1993). In Taiwan, the number of sclerotia per kg of dry soil averaged 36.1, 3.2, 3.3, 2.6, and 0.9 in the 0-1, 1-5, 5-10, 10-15, and 15-20 cm depth, respectively, after harvest of the first rice crop. Germination averaged 45.4, 19.0, 14.8, 8.5, and 5.5 %, respectively, from the same depths. Most were buoyant, and germination of sclerotia floating in nylon screen bags in a paddy 0-120 days after transplanting averaged 36.4 % (Leu and Yang, 1985).

In India, sclerotia buried 10 cm deep in low land or wet land field soils submerged in water survived only 3-4 mo, but sclerotia buried in upland soil survived 11 mo. Soil temperatures ranged from 20-30 C. Roy (1986) concluded that sclerotia in the temperate rice lands in the USA and Japan may survive longer than in the wetlands of the tropics. In Kerela, India, the pathogen lost its viability in infected rice debris buried 5 cm deep in flooded or dry wetland or garden soils in 3-4 wks, but survived 6 months in dry rice straw in a laboratory (Lakshmanan and Nair,1986). In March in Korea, *R. solani* was isolated most frequently from sclerotia recovered from the top 3 cm of soil in paddy fields. The percentage of sclerotia > 2.0 mm, 1-2 mm, and < 1.0 mm in diameter were 15, 38, and 47, respectively, and the larger sclerotia maintained the viability longer than the smaller sclerotia. The germination of overwintered sclerotia was similar to the germination of sclerotia collected from paddy fields after harvest (Kim and Kim, 1988).

In Nigeria, the web-blight pathogen of cowpea survived the dry season in naturally infested leaves on the soil surface, but not in leaves buried 5 cm deep in soil (Onesirosan and Sagay, 1975). In Costa Rica, sclerotia and mycelium free in soil or in colonized plant debris were the primary sources of inoculum of bean web blight (Galindo *et al.*, 1983). The web blight pathogen of peanut survived 11 months in plant debris and 21 months as sclerotia buried in soil in pots in India (Dubey and Dwivedi, 1992). In Ohio, AG 1-IB and AG 1-IC survived 5 months during the winter in infested barley grains on the soil surface or buried 15 cm deep and induced bottom rot in lettuce (Herr, 1993).

2. AG 2. The sugarbeet root-rot pathogen (AG 2-2) survives in infested plant tissues and as sclerotia (Herr, 1976, Roberts and Herr, 1979a, Hyakumachi and Ui, 1982). In Japan, sclerotia are considered to be more important than infested plant debris in the overwintering survival of the pathogen (Hyakumachi and Ui, 1982,1984). In the spring, more sclerotia germinated and had 5 or more emerged hyphae in areas of fields with a high disease incidence (DI) than in areas with a low DI (Hyakumachi *et al.*, 1983). Non-self-anastomosising isolates occur, but they do not form sclerotia, or survive in soil as hyphae or in colonized organic matter (Hyakumachi and Ui, 1987). In the USA, the pathogen survives in infested plant debris in surface soil and 10-20 cm deep (Ruppel, 1991).

Crown and brace root rot in corn is caused by AG 2-2 IIIB, and the fungus apparently overwinters as sclerotia or in colonized plant debris in the southeastern USA. The pathogen survived at low inoculum densities for 10 months in fallow soil (Bell and Sumner, 1987), and for 38 months in a rotation of a corn-snap bean double crop and a peanut monocrop (Sumner and Bell, 1994).

3. AG 3. In the Netherlands tubers had fewer sclerotia when they were produced from seed potatoes grown in an acidic sand soil compared with seed potatoes produced in a near neutral marine sandy loam (Jager and Velvis, 1983). The direct effect of pH may have caused the differences. However, soil antagonists may have been involved, and that will be covered in another chapter. Volatile tuber exudates may increase or decrease sclerotia production in potato, and black scurf increases after haulm destruction. Dijst (1990) showed that inhibitory volatiles decreased during maturation, and suggested that quick separation of plant residues from the tubers after haulm killing may reduce the formation of sclerotia.

4. AG 4. As indicated previously, survival of AG 4 is affected by tillage practices that influence vertical distribution in soil. The fungus survived 2 years in peanut shells of pods on the soil surface, or buried 7.6 or 25.4 cm deep, in Georgia. However, the inoculum density was greatly reduced after 6 mo, especially in buried pods (Baird *et al.*, 1993). Soil was infested with cornmeal-sand cultures, placed in clay pots, and buried in a field. Average populations 10 cm deep in fallow soil were 563, 296, 44, and 13 cfu/100 g of oven-dry soil 0, 86, 211, and 283 days after infestation. In field microplots of loamy sand infested with AG 4 and planted for 3 years in a peanut monocrop, corn-snapbean double-crop rotation, populations of AG 4 were 23, 9, 48, and 18 cfu/100 g of oven-dried soil 3, 10, 19, and 32 months after infestation, respectively (Sumner and Bell, 1994). In a 3-years rotation of a corn-bean double crop on a sand soil, indigenous populations of AG 4 in topsoil in conservation tillage ranged from 2 to 33 cfu/100 g following legumes and from 1 to 8 cfu/100g following corn (Sumner *et al.*, 1986).

5. Binucleate *Rhizoctonia* spp. In Georgia, some binucleate *Rhizoctonia* spp. survived up to 40 weeks in fallow field soils in buried clay pots at temperature ranges of -22 to 40°C, but some isolates did not survive 12 weeks (Bell and Sumner, 1987). *Ceratobasidium* spp. AG R (=CAG 5; Burpee *et al.*, 1980) survived in soil for 19 months in a corn-snapbean double-crop in rotation with a peanut monocrop. AGs -A, -E and -F (=CAGs 2, 3 and 5 respectively) colonize peanut pods, and shells and seeds left in soil at harvest may provide a means of survival in soil for many months (Sumner and Bell, 1994). In North Carolina AG G (*R. fragariae*) survived 10 months in precolonized snapbean stems and 4-8 months in precolonized oat kernels buried at

1, 15, and 30 cm depths in soil, respectively (Cubeta *et al.*, 1991). When sclerotia of AG D (CAG 1, *R. cerealis*) were buried in field soil in a greenhouse, 79 % survived for 6 mo. In contrast, the pathogen was reisolated from only 14 % of naturally infected buried wheat straws after 110 days (Pitt, 1964).

6. *R. zeae*. In artificially infested soil in pots in a fallow field, populations declined very slowly from an average of 616 to 166 cfu/g of soil from 0 to 283 days after infestation, respectively (Bell and Sumner, 1987). The fungus commonly reinvades fumigated soil when corn is planted, and residual populations range from 6-43 cfu/g of soil (Sumner and Bell, 1994). In fields in the Georgia coastal plain, populations vary from <1 to 30-40 cfu/g of soil.

D. Influence of environmental conditions on survival of sclerotia

Several environmental factors may influence survival of propagules of *Rhizoctonia*. Edaphic factors include soil texture, compaction, and water potential. Other factors are temperature, soil and plant nutrition, soil amendments (including composts), other soil microbes, and soil fauna. Viable propagules of *Rhizoctonia* spp. were found more frequently in coarse sand than in fine sand, silt, or clay (Papavizas, 1968). Adding clay to sandy loam soils reduced survival (Lewis, 1979). The AG's occurring in fine textured soils are different from those occurring in coarse textured soils (Roberts and Herr, 1979b, Herr and Roberts, 1980). Soil compaction increased the density of hyphae/g of soil of AG 2-1 compared with noncompacted soil (Glenn and Sivasithamparam, 1990). Survival of *R. solani* in moist soil is lower than in dry soil (Benson and Baker, 1974; Ploetz and Mitchell, 1985). *R. oryzae-sativae* sclerotia survived longer at 50 % water-holding capacity (WHC) than at either 3 % or 100 % WHC (Pan and Islam, 1991).

Soil temperatures of 10-15°C are the most favorable for survival of many *R. solani* isolates in temperate zones (Benson and Baker, 1974; Papavizas *et al.*, 1975). Sclerotia of *R. solani* from rice survived for 18 months in wet soil at 20 or 30°C, and in dry soil at 20°C, but only 5 % survived 6 months at 30°C in dry soil (Manian and Manibhushanrao, 1990). In soil fertilized with N, P, and K, alone and in all combinations, survival of the rice sheath blight fungus was reduced by all treatments compared to nonfertilized soil, but K and PK especially reduced survival. Reductions in survival may have been related to the effects of the nutrients on other soil microflora (Kannaiyan and Prasad, 1983).

Sclerotia of *R. solani* AG 6 (Kobayashi *et al.*, 1984) and sclerotia and mycelium of some *Rhizoctonia* spp. (Burton and Coley-Smith, 1985) contain antibiotics. The antibiotics may be beneficial in the survival of some *Rhizoctonia* spp., but antibiotics were not found in sclerotia of AGs 1, 2, 3, 4 and 5, or BI (Kobayashi *et al.*, 1984).

Soil amendments with plant materials, composts, sewage, and manures; and the soil microflora and microfauna may have an influence on the survival of *Rhizoctonia* spp. These topics will be covered in detail in other chapters.

III. References

Baird RE, Bell DK, Sumner, DR, Mullinix BG & Culbreath AK (1993) Survival of *Rhizoctonia solani* AG 4 in residual peanut shells in soil. Plant Dis. 77:973-975.

Bell DK & Sumner DR (1987) Survival of *Rhizoctonia solani* and other soilborne basidiomycetes in fallow soil. Plant Dis. 71:911-915.

Benson DM & Baker R (1974) Epidemiology of *Rhizoctonia solani* preemergence damping-off of radish: Survival. Phytopathology 64:1163-1168.

Boosalis MG & Scharen AL (1959) Methods for microscopic detection of *Aphanomyces euteiches* and *Rhizoctonia solani* and for isolation of *Rhizoctonia solani* associated with plant debris. Phytopathology 49:192-198.

Burpee LL, Sanders PL, Cole H Jr & Sherwood RT (1980) Anastomosis groups among isolates of *Ceratobasidium cornigerum* and related fungi. Mycologia 72:689-701.

Burton RJ & Coley-Smith JR (1985) Antibiotics in sclerotia and mycelium of Rhizoctonia species. Trans. Br. Mycol. Soc. 85:447-453.

Butler EE & Bracker C (1970) Morphology and cytology of *Rhizoctonia solani*. In: Parmenter JR Jr (ed.) *Rhizoctonia solani*: Biology and Pathology (pp 32-51) Univ. California Press, Berkeley.

Cubeta MA, Echandi E & Gumpertz ML (1991) Survival of binucleate *Rhizoctonia* species, biological control agents, in soil and plant debris under field conditions. Biol. Control 1:218-226.

Damicone JP, Patel MV & Moore WF (1993) Density of sclerotia of *Rhizoctonia solani* and incidence of sheath blight in rice fields in Mississippi. Plant Dis. 77:257-260.

Dijst G (1990) Effect of volatile and unstable exudates from underground potato plant parts on sclerotium formation by *Rhizoctonia solani* AG 3 before and after haulm destruction. Neth. J. Plant Pathol. 96:155-170.

Dubey SC & Dwivedi RP (1992) Survival of *Thanatephorus cucumeris* causing web blight of groundnut. Ind. J. Mycol. & Plant Pathol. 22:49-53.

Durbin RD (1959a) Factors affecting the vertical distribution of *Rhizoctonia solani*, with special reference to CO2 concentration. Amer. J. Bot. 46:22-25.

Durbin RD (1959b) Some effects of light on the growth and morphology of *Rhizoctonia solani*. Phytopathology 49:59-60.

Galindo JJ, Abawi GS, Thurston HD & Gálvez G (1983) Source of inoculum and development of bean web blight in Costa Rica. Plant Dis. 67:1016-1021.

Glenn OF & Sivasithamparam K (1990) The effect of soil compaction on the saprophytic growth of *Rhizoctonia solani*. Plant Soil 121:282-286.

Hashiba T & Ishikawa T (1978) Effect of adenosine 3'5'-cyclic monophosphate on induction of sclerotia in *Rhizoctonia solani*. Phytopathology 68:1723-1727.

Hashiba T & Mogi S (1975) Developmental changes in sclerotia of the rice sheath blight fungus. Phytopathology 65:159-162.

Hashiba T & Staples RC (1976) Nucleic acid metabolism in developing sclerotia of the rice sheath blight fungus (*Rhizoctonia solani* Kühn). J. Gen. Microbiol. 96:239-245.

Hashiba T & Yamada M (1981) Viability of sclerotia formed on rice plants and on culture media by *Rhizoctonia solani*. Ann. Phytopathol. Soc. Jpn. 47:464-471.

Henis Y, Ghaffar A, Baker R & Gillespie, SL (1978) A new pellet soil-sampler and its use for the study of population dynamics of *Rhizoctonia solani* in soil. Phytopathology 68:371-376.

Herr LJ (1976) In field survival of *Rhizoctonia solani* in soil and in diseased sugarbeets. Can. J. Microbiol. 22:983-988.

Herr LJ (1993) Host sources, virulence and overwinter survival of *Rhizoctonia solani* anastomosis groups isolated from field lettuce with bottom rot symptoms. Crop Prot. 12:521-526.

Herr LJ & Roberts DL (1980) Characterization of Rhizoctonia populations obtained from sugarbeet fields with differing soil textures. Phytopathology 70:476-480.

Hiremath PC & Prasad CKPS (1985) Vertical distribution of *Rhizoctonia solani* in soil and effect of culture filtrate on the rhizosphere mycoflora of fenugreek. J. Soil Biol. & Ecol. 5:126-128.

Hussey RS & Roncadori RW (1977) Vertical distribution of soil microorganisms following subsoiling in a cotton management system. Phytopathology 67:783-786.

Hyakumachi M & Ui T (1982) The role of the overwintered plant debris and sclerotia as inoculum in the field occurred with sugarbeet root rot. Ann. Phytopathol. Soc. Jpn. 48:628-633.

Hyakumachi M & Ui T (1984) Comparison of soil-pellet and sieving-flotation methods for enumeration of sugarbeet-root rot fungus (*Rhizoctonia solani* AG 2-2). Ann. Phytopathol. Soc. Jpn. 50: 551-553.

Hyakumachi M & Ui T (1987) Non-self-anastomosing isolates of *Rhizoctonia solani* obtained from fields of sugarbeet monoculture. Trans. Br. Mycol. Soc. 89:155-159.

Hyakumachi M, Yamamoto Y & Ui T (1983) Survival and pathogenicity of sclerotia produced by sugarbeet root rot fungus (*Rhizoctonia solani* AG 2 type 2) in the areas with different degree of disease incidence. Ann. Phytopathol. Soc. Jpn. 49:18-21.

Jager G & Velvis H (1983) Suppression of *Rhizoctonia solani* in potato fields. II. Effect of origin and degree of infection with *Rhizoctonia solani* of seed potatoes on subsequent infestation and on formation of sclerotia. Neth. J. Pl. Pathol. 89:141-152.

Kannaiyan S & Prasad NN (1983) Effect of nitrogen, phosphorus and potassium on the survival and saprophytic activity of *Rhizoctonia solani*. Madras Agric. J. 70:506-510.

Kim WG & Kim CK (1988) Density of overwintered sclerotia in paddy fields in Korea, viability of the sclerotia and pathogenicity of the sclerotial fungi. Korean J. Plant Pathol. 4:207-217.

Ko W & Hora FK (1971) A selective medium for the quantitative determination of *Rhizoctonia solani* in soil. Phytopathology 61:707-710.

Kobayashi K, Inami S, Nakajima T & Ui T (1984) Antibacterial antibiotics detected in sclerotia of *Rhizoctonia solani* AG 6. Ann. Phytopathol. Soc. Jpn. 50:263-269.

Lakshmanan P & Nair MC (1986) Survival of *Rhizoctonia solani* causing sheath blight of rice under various conditions. Madras Agric. J. 73:325-328.

Lee FN (1980) Number, viability, and buoyancy of *Rhizoctonia solani* sclerotia in Arkansas rice fields. Plant Dis. 64:298-300.

Leu LS & Yang HC (1985) Distribution and survival of sclerotia of rice sheath blight fungus, *Thanatephorus cucumeris*, in Taiwan. Ann. Phytopathol Soc. Jpn. 51:1-7.

Lewis JA (1979) Influence of soil texture on survival and saprophytic activity of *Rhizoctonia solani* in soils. Can. J. Microbiol. 25:1310-1314.

Manian S & Manibhushanrao K (1990) Influence of some factors on the survival of *Rhizoctonia solani* in soil. Trop. Agric. 67:207-208.

Moromizato Z, Ishizaki F, Takara, K & Tamori M (1991) The effects of phosphorous and magnesium on sclerotium formation in *Rhizoctonia solani* Kühn. Ann. Phytopathol. Soc. Jpn. 57:649-656.

Moromizato Z, Matsuyama N & Wakimoto S (1980) The effect of amino acids on sclerotium formation of *Rhizoctonia solani* Kühn (AG 1). II. Developmental process of sclerotium and its inhibition with several amino acids. Ann. Phytopathol. Soc. Jpn. 46:21-25.

Naiki T & Ui T (1978) Ecology and morphological characteristics of the sclerotia of *Rhizoctonia solani* Kühn produced in soil. Soil Biol. Bioch. 10:471-478.

Neate SM (1987) Plant debris as a source of inoculum of Rhizoctonia in wheat. Trans. Br. Mycol. Soc. 88:157-162.

Onesirosan PT & Sagay SO (1975) Survival of two pathogens on cowpea over the dry season. Plant Dis. Rep. 59:1018-1020.

Pan S & Islam SJ (1991) Effect of soil moisture and soil amendment on survival of sclerotia of *Rhizoctonia orzae-sativae*. J. Mycopathol. Res. 29:77-82.

Papavizas GC (1968) Survival of root-infecting fungi in soil. VIII. Distribution of *Rhizoctonia solani* in various physical fractions of naturally and artificially infested soils. Phytopathology 58:746-751.

Papavizas GC, Adams PB, Lumsden, RD, Lewis, JA, Dow, RL, Ayers WA & Kantzes, JG (1975) Ecology and epidemiology of *Rhizoctonia solani* in field soil. Phytopathology 65:871-877.

Pitt D (1964) Studies on sharp eyespot disease of cereals. II. Viability of sclerotia: persistence of the causal fungus, *Rhizoctonia solani* Kühn. Ann. Appl. Biol. 54:231-240.

Ploetz RC & Mitchell DJ (1985) Influence of water potential on the survival and saprophytic activity of *Rhizoctonia solani* AG4 in natural soil. Can. J. Bot. 63:2364-2368.

Roberts DL & Herr LJ (1979a) Superiority of a soil debris isolation method over a beet seed colonization method for assay of *Rhizoctonia solani* at high soil inoculum densities. Can. J. Microbiol. 110-113.

Roberts DL & Herr LJ (1979b) Soil populations of *Rhizoctonia solani* from areas of healthy and diseased beets within four sugarbeet fields differing in soil texture. Can. J. Microbiol. 25:902-910.

Roy AK (1986) Survival of sclerotia of *Rhizoctonia solani* f. sp. *sasakii* in relation to moisture regime of soil. Indian Phytopathol. 39:259-263.

Ruppel EG (1991) Survival of *Rhizoctonia solani* in fallow field soil and buried sugarbeet roots at three depths. J. Sugar Beet Res. 28:141-153.

Sherwood RT (1970) Physiology of *Rhizoctonia solani*. In: Parmenter JR Jr (ed.) *Rhizoctonia solani*: Biology and Physiology (pp 69-92) Univ. California Press, Berkeley.

Shokes FM & McCarter SM (1979) Occurrence, dissemination, and survival of plant pathogens in surface irrigation ponds in southern Georgia. Phytopathology 69:510-516.

Sumner DR & Bell DK (1994) Survival of *Rhizoctonia* spp. and root diseases in a rotation of corn, snap bean, and peanut in microplots. Phytopathology 84:113-118.

Sumner DR, Doupnik B Jr & Boosalis MG (1981) Effects of reduced tillage and multiple cropping on plant diseases. Annu. Rev. Phytopathol. 19:167-187.

Sumner DR, Smittle DA, Threadgill, ED, Johnson, AW & Chalfant RB (1986) Interactions of tillage and soil fertility with root diseases in snap bean and lima bean in irrigated multiple-cropping systems. Plant Dis. 70:730-735.

Townsend BB & Willetts HJ (1954) The development of sclerotia of certain fungi. Trans. Brit. Mycol. Soc. 37:213-221.

Tu CC & Kimbrough JW (1975) Morphology, development, and cytochemistry of the hyphae and sclerotia of species in the *Rhizoctonia* complex. Can. J. Bot. 53:2282-2296.

Van Bruggen AHC & Arneson PA (1985) A quantifiable type of inoculum of *Rhizoctonia solani*. Plant Dis. 69:966-969.

Weinhold AR (1977) Population of *Rhizoctonia solani* in agricultural soils determined by a screening procedure.. Phytopathology 67:566-569.

Woodard KE & Jones BL (1983) Soil populations and anastomosis groups of *Rhizoctonia solani* associated with peanut in Texas and New Mexico. Plant Dis. 67:385-387.

Yang XB, Snow JP & Berggren GT (1989) Morphogenesis of microsclerotia and sasakii-type sclerotia in *Rhizoctonia solani*, anastomosis group 1, intraspecific groups IA and IB. Mycol. Res. 93:429-423.

IV.5 PATCH DYNAMICS AND BARE PATCH

GORDON C. MACNISH

School of Biological & Environmental Sciences, Murdoch University, Murdoch, WA, 6150, AUSTRALIA

I. Introduction

Rhizoctonia solani Kühn [*Thanatephorus cucumeris* (Frank) Donk] causes a number of patch diseases. These include Rhizoctonia bare patch of cereals and other crops (*R. solani* AG 8), crater disease of cereals (Deacon and Scott, 1985; Smith and Wehner, 1986, 1988; Opperman and Barnard 1992; AG unknown), patch of tulips (AG 2-1-tulip; see Chapter V.4) and several turf patches (AG 1-1A, AG 2-2; Green *et al.*, 1994; Zarlengo *et al.*, 1994; Burpee, 1995). Of these, Rhizoctonia bare patch will be used as a model for the study of patch dynamics.

Dynamics imply movement or change over time and will be taken to include patch initiation, patch expansion, patch demise and decline, patch coalescing and changes in patch configuration between seasons.

II. Rhizoctonia bare patch disease

A. Background

Rhizoctonia bare patch disease was first reported and identified in Australia (Samuel 1928; Neate and Warcup 1985). It has subsequently been reported in England (Dillon Weston and Garrett, 1943), Canada (Benedict and Mountain, 1956), Scotland (McKelvie, 1978; Murray, 1981) and the USA (Weller *et al.*, 1986). In Australia it is reported from southern cereal growing regions (Murray and Brown, 1987) and has a wide host range including cereals, legumes, rapeseed, grasses and other pasture species (MacNish, 1983).

The most striking symptom of bare patch is the distinct edge to the patch with an abrupt change from stunted plants inside to apparently healthy plants immediately outside. Plants within the patch have stunted roots with brown pinched-off 'spear tip' ends, while plants outside the patch that are apparently healthy, may also have some affected roots.

Bare patch is reduced, but not eliminated, by tillage (Neate, 1984; MacNish, 1985b; Rovira and Venn, 1985; Weller *et al.*, 1986; Jarvis and Brennan, 1986) and use of nitrogen fertiliser (MacNish, 1985b). The disease can affect cultivated crops in a mild form and will usually reappear in a severe form with a return to reduced tillage (MacNish, 1985b).

Until recently, no satisfactory explanation was available for the formation of patches, the distribution of patches in the field and apparently sudden changes in size and number of patches between seasons.

B. Methods used to study Rhizoctonia bare patch

1. Detection and isolation of *R. solani* AG 8 using undisturbed soil cores The use of wheat seedlings grown in undisturbed soil cores removed from patches combined

B. Sneh et al. (eds.),
Rhizoctonia Species: Taxonomy, Molecular Biology, Ecology, Pathology and Disease Constrol, 217–226.

with the use of a semi-selective medium and elimination of root surface sterilization has allowed the isolation of the *R. solani* AG 8 throughout the year (Dubé, 1971; MacNish, 1984; MacNish and Sweetingham, 1993a).

2. Identification of *R. solani* isolates using pectic isozymes The use of pectic isozymes to produce zymogram patterns (Sweetingham *et al.*, 1986) has so far lead to the detection of only five zymogram groups (ZG1-1 to ZG1-5) within isolates of *R. solani* AG 8. These five groups appear to be very stable (MacNish and Sweetingham, 1993b) and some have been isolated from all states in Australia where the disease is found. Two ZG1s (1-1 and 1-2) have also been detected in the USA (MacNish *et al.*, 1993a).

3. Anastomosis among isolates of *R. solani* AG 8 There are four categories of anastomosis reaction between isolates of *R. solani* (Carling *et al.*, 1988). Of these, Categories 2 (C2) and 3 (C3) have been employed in the study of patch dynamics involving *R. solani* AG 8. An anastomosis test between isolates from different ZG1s always gives a C2 anastomosis reaction (MacNish *et al.,* 1995). An anastomosis test of isolates from within the same ZG1 can either give a C2 or a C3 reaction.

The C2 anastomosis reaction is a vegetative incompatibility reaction while a C3 reaction is a vegetative compatibility reaction. MacNish *et al.* (1995) propose that isolates from within a ZG1 that give a C3 reaction can be allocated to a Vegetatively Compatible Population (VCP) and that all isolates belonging to a VCP are closely related. Thus, the C3 reaction can be used to show relationships of isolates from within and between patches (MacNish *et al.*, 1993a).

III. FACTORS INVOLVED IN PATCH FORMATION

One explanation for patch formation assumes that *R. solani* AG 8 is widely distributed in the soil and patches are caused by localized environmental factors influencing the pathogen. These factors could make the plant susceptible to attack by the pathogen or make the pathogen more virulent. Another explanation is that patches are caused by discrete colonies of *R. solani* AG 8. It has also been proposed that pathogens other than *R. solani* may be involved in the bare patch disease or that *R. solani* AG 8 is a primary colonizer that is rapidly replaced by a range of common root rotting fungi which exacerbate the root rot and cause the development of patches (Benedict and Mountain, 1956; Meagher and Chambers, 1971; Meagher *et al.*, 1978; Patel, 1983; Harris and Moen, 1985; Roberts and Sivasithamparam, 1986, 1987). The latter two proposals have not received strong support.

It has been reported (Ogoshi *et al.* 1990) that both *R. solani* AG 8 and *R. oryzae* Ryker are involved in Rhizoctonia root rot of cereals in the USA. Isolates, however, were obtained from crops with no obvious evidence of root rot (or patches) and no claim was made that *R. oryzae* was a cause of bare patch. MacNish and Sweetingham (1993) reported the isolation of *Waitea* spp. from both inside and outside patches and could show no correlation between distribution of *Waitea* spp. and patch formation. In a study of 12 patches in Western Australia, Roberts and Sivasithamparam (1986) recorded an isolation frequency of 10% for avirulent binucleate isolates of *Ceratobasidium* sp. from within patches and found an increase in relative frequency of isolation towards the edge of the patch. They proposed that this increase in frequency may be related to the abrupt edging of patches. From a larger study, MacNish and Sweetingham (1993) found frequency of isolation of binuculate *Rhizoctonia* spp.

ranged from 7% to over 90%, but they were unable to show any correlation between distribution or frequency of isolation and patch formation.

A. Physical factors and patch formation

There is a possibility that changes in soil properties could influence soil organisms and lead to changes in population densities and expression of disease. Rovira and Simon (1985) measured chemical properties including $CaCO_3$, pH, total nitrogen, total organic carbon, total phosphorus, bicarbonate-extractable phosphorus, chloride and conductivity of soil from inside or outside patches at three sites. Except for chloride concentration at one site, they found no differences in the chemical properties. MacNish (unpublished) examined soil properties of nine patches with a range of cropping histories and measured bulk density, soil pH, water penetration, NH_4-nitrogen, NO_3-nitrogen, total nitrogen and organic carbon content. No differences were detected between inside and outside patches.

As crater disease in South Africa involves root impedance in heavy clay soil, Deacon and Scott (1985) proposed that Rhizoctonia bare patch on sandy soils could be associated with root impedance caused by the packing together of sand particles. Studies on sandy soils in Scotland by McKelvie (1978) have failed to support this hypothesis. In soil compaction studies in Australia in a sandy soil field, MacNish (1989) achieved soil strengths ranging from 0.17 to 0.55 MPa at 6 cm to 0.26 to 1.99 MPa at 14 cm but failed to detect any effect on bare patch or root rot. The above results and the finding by Rovira and Simon (1985) that fumigation eliminates bare patch suggest that patches are of a biological nature.

B. Biological factors and patch formation

Daniels (1963) proposed that patches developed where there is a localized increase in virulence of *R. solani,* while Baker and Cook (1974) suggested patches formed where antagonistic populations receded, rather than an advance of the pathogen colonizing the soil. These hypotheses have not been confirmed. Others have proposed that localized stimuli cause a localized increase in the population of widely distributed pathogenic strains of *R solani*, leading to the development of patches (Kerr, 1955; De Beer, 1965).

MacNish (1985a) proposed that small changes linked to seasonal or localized factors could change the conduciveness of soil around foci of inoculum leading to the development (or disappearance) of patches. Under this hypothesis, if the conditions were highly suppressive there would be no patches; if less suppressive, a few moderate patches; if conducive, a cluster of moderate to severe patches would develop. If conditions were very conducive, there might be just one large severe patch incorporating all the patches in the cluster plus all the area between these patches. However, if as MacNish (1985a) proposed, an environmental stimulus could cause many foci to combine into a single patch, it seems likely that within some patches there would be foci from a number of different ZG1s. If by chance all the foci were from the same ZG1, it seems likely that in at least some cases there would be foci from a number of different VCPs.

To test the hypothesis that each patch is colonized by a single isolate of *R. solani* AG 8 and will therefore not contain a mixture of ZG1s, MacNish and Sweetingham (1993a) conducted extensive temporal and spatial sampling of 28 patches. Except for one patch were two samples out of 108 were from a different ZG1, there was only one ZG1 isolated from each patch. The exceptions were probably contaminants. Multiple isolates recovered from different locations in the same patch or recovered over a period of as much as 10 months, always gave the same VCP (MacNish *et al.*, 1993a). These results support the single isolate hypothesis and suggest that it is unlikely that some physical or biological factor causes many different foci to combine to form a patch.

IV. PATCH DYNAMICS

A. Patch initiation

If each patch is the results of the soil being colonized by an individual isolate of *R. solani* AG 8, the patch will, except for coalescing patches, have been initiated from an infection focus. A number of attempts to initiate patches have been made. Flentje and Saksena (1957) added hyphal mats of the 'root attacking' strain (AG 8) to soil in the field and found the isolate failed to persist after the first year. In laboratory experiments, Baker *et al.* (1967) found that all or most of the hyphae added to soil from areas other than where the culture had originated, had died after 3 months.

MacNish (unpublished) attempted to initiate patches in September 1991 and in May 1992 in an area of pasture where *R. solani* AG 8 could not be detected. The forms of inoculum used were undisturbed soil cores removed from natural field patches, river-bed sand artificially colonized with *R. solani* AG 8 and millet seed colonized with *R. solani* AG 8. The first two were added to the soil as intact cores and the third as a bulk sample placed in a hole (2.5 diameter x 9.0 cm deep) in the soil. The test area was sown to barley in 1993, 1994 and 1995. Patches were not observed in the 1993 crop but were evident in 1994 and 1995. Although this experiment suggests a lag phase exists before patches will become evident, it does demonstrate that patches can be initiated from colonized soil or organic matter.

B. Patch expansion

It has been shown that patches expand at different rates (Kerr, 1955; MacNish and Sweetingham, 1993a). There are also different rates of expansion around the circumference of individual patches (MacNish *et al.*, 1993b) and this would account for many patches showing varying shapes (MacNish, 1985a). Ludbrook *et al.* (1953) proposed that patches expand between seasons, but the work of MacNish *et al.* (1993b) demonstrated that patch expansion takes place during the growing season. Confusion about patch expansion probably occurred because at the beginning of the growing season the young plant is susceptible to stunting, while later in the season although the soil is colonized by the pathogen, the maturing plant is not affected by stunting. The actual size of the patch (colonized soil) will not be evident until the following season when susceptible plants will again be affected (MacNish and Sweetingham, 1993a). Thus, large changes can occur in patch size without being noticed, giving the impression of expansion between seasons.

Tillage appears to have major effects on patch expansion and there is considerable evidence that patches tend to be elongated in the direction of sowing (Ludbrook *et al.*, 1953; MacNish, 1985a). These observations suggest that the pathogen is spread more rapidly by tillage (presumably by movement of soil) than by growth through the soil, but there is also the possibility that the pathogen could be spreading along the drill row by using the host as a bridge.

C. Patch demise and decline

It is well established that patches can disappear between growing seasons (Kerr, 1955; MacNish, 1985a). Just as the pathogen can colonize the soil during the growing season, it can also die out during the growing season (MacNish *et al.*, 1993b). Because the plants usually remain stunted, this gradual death of the pathogen will not be obvious, but the isolation of the pathogen from the soil becomes increasingly difficult during the growing season. MacNish *et al.* (1993b) showed changes in the configuration of two coalesced patches over a six year period, were partly due to differential growth rates, but the most dramatic changes were due to the death of the pathogen in large parts of the combined patch. Plants growing in these areas in the following season showed no evidence of patch disease.

MacNish (1988) reported a decline in Rhizoctonia root rot in two field experiments where continuous wheat was sown following one or two cultivations. Root rot peaked in the seventh or ninth year, and then declined to negligible levels over the next three years. Lucas *et al.* (1993) found in the glasshouse that a decline in root rot could be induced by successive plantings of wheat. Roget (1995) reported an experiment where there was a build-up in both patch area and root rot to peak in the fourth or fifth year followed by a decline over the next five to seven seasons. This decline was demonstrated in plots continuously cropped to wheat or in plots sown to wheat in alternate year rotations with various break crops. The decline occurred in both cultivated plots and plots direct drilled with minimum tillage. This decline appears to be due to the establishment of a general microbial suppressiveness (Roget, 1995; Wiseman *et al.*, 1996).

D. Patch coalescing

A six year study of two patches that coalesced demonstrated major changes in configuration of the patches over that period (MacNish *et al.*, 1993b). In a situation where there were many patches, coalescing of patches could be expected to cause dramatic patch patterns to appear (MacNish, 1985a).

E. Changes in patch configuration between seasons

Out of 74 patches mapped in one season by Kerr (1955), 22 had disappeared by the following season and remaining 52 were in the same position in both seasons but with increases in diameter that ranged from 0 to 50 cm. In the second season 28 new patches appeared. In a study over four consecutive seasons, MacNish (1985a) mapped bare patches in an area (0.225 ha) of crop sown with minimum tillage (Fig. 1). His study showed that there could be dramatic changes in shape, size and area of patch between seasons (Fig. 1). The area of crop that was patch in 1979 to 1982 was 27%, 28%, 10% and 21% respectively. By using overlays to determine differences in patch distribution between 1980 and 1981, MacNish (unpublished) showed some new patches appeared in 1981 and some patches from 1980 had disappeared. Some patches enlarged while in other patches, some parts enlarged and other parts disappeared. Other patches remained unchanged between seasons and for some only vestiges of the original patch remained.

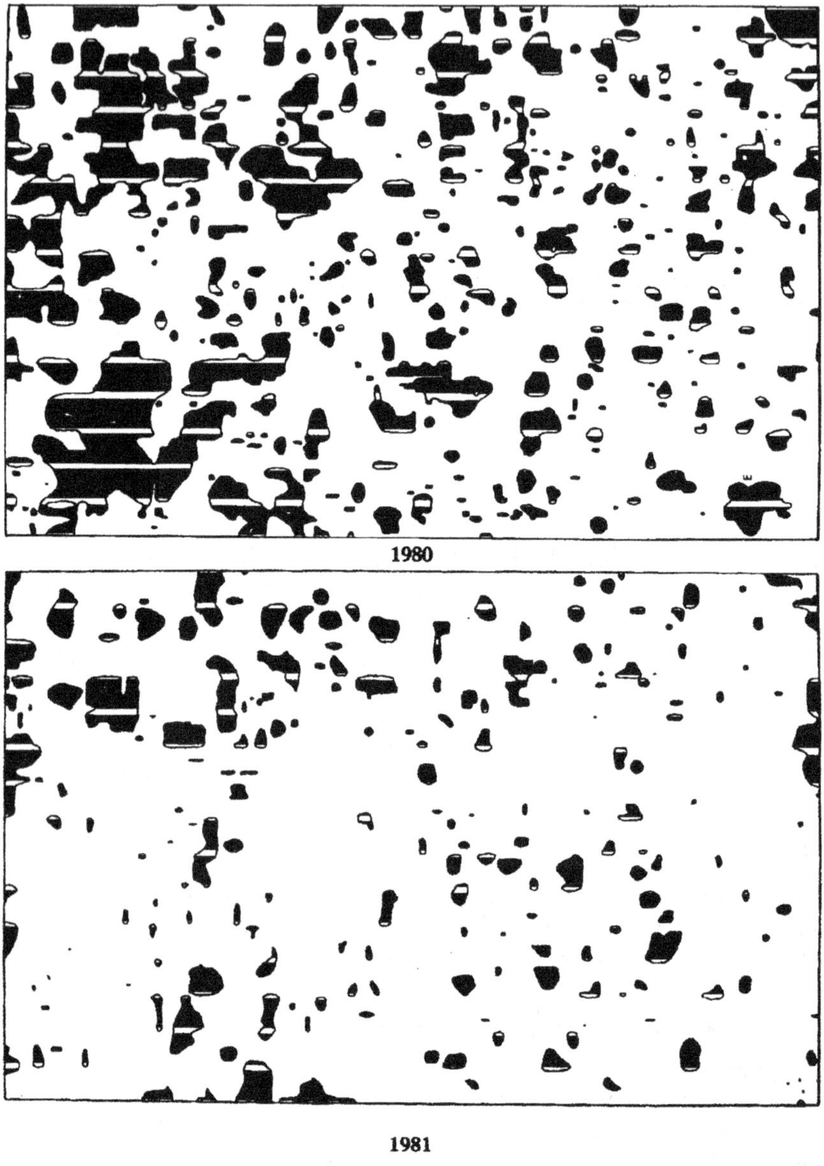

FIGURE 1. Maps of Rhizoctonia bare patch (black areas) showing the contrast between two seasons. The area (60 x 37.5 m, composed of 15 plots 60 m long) was sown to crop with minimum tillage in both 1980 and 1981. White bands across patches are unsown gaps (50 cm wide) between drill-widths. (From MacNish, 1985a)

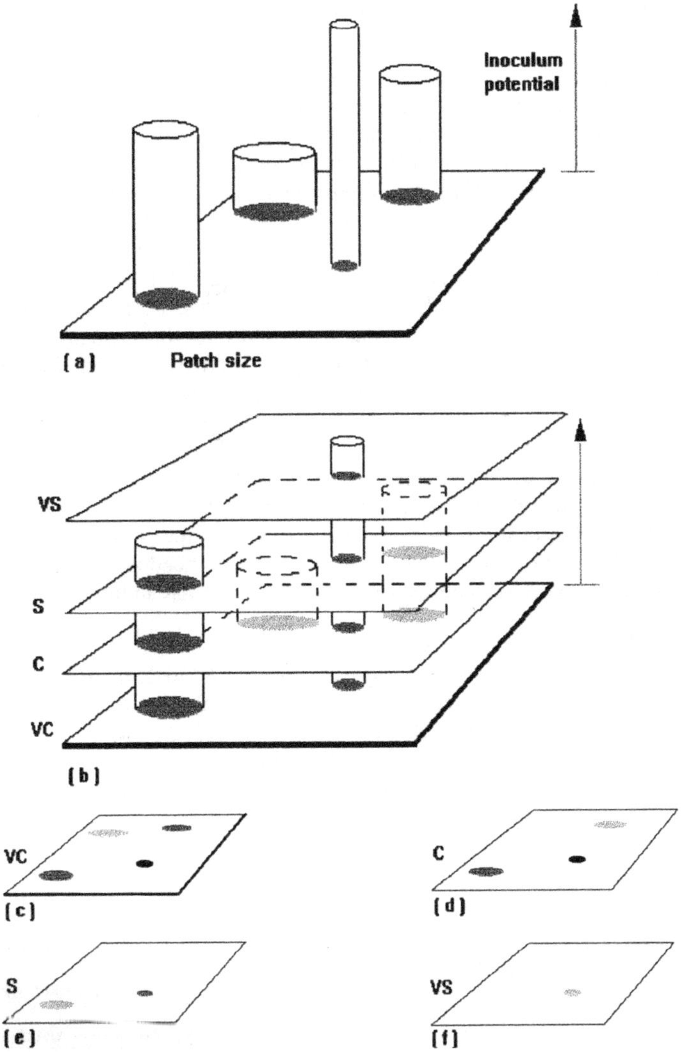

FIGURE 2. Stylized diagram showing proposed explanation for some inter-seasonal changes in patch configuration based on an interaction between seasonal suppressiveness and inoculum potential. (a) An area with four patches that vary in size and inoculum potential. (b) Same area with four different levels of seasonal suppressiveness superimposed (VC = very conducive season, C = conducive season, S = suppressive season and VS = very suppressive season). Parts of (b) separated to give: (c) very conducive season with one severe patch (black), two moderate patches (dark gray) and one mild patch (light gray); (d) conducive season with one severe patch, one moderate patch and one mild patch; (e) suppressive season with one moderate patch and one mild patch and (f) very suppressive season with one mild patch.

V. A PROPOSED EXPLANATION FOR INTER-SEASON DIFFERENCES IN PATCH CONFIGURATION

The above elements of patch dynamics can explain much but not all of the changes in patch configuration between seasons. I propose that some of the inter-season differences in patch configuration can be explained by differences in seasonal suppressiveness. Under this proposal, changing level of suppressiveness between seasons interact with the level of inoculum potential of the pathogen in each patch of colonized soil to cause changes in the number and severity of patches in the crop (or other host) (Fig. 2). In a very suppressive season, those patches of soil colonized by a pathogen with a low inoculum potential will be hidden and will not be expressed as patches in the crop. Only those patches of soil colonized by a pathogen with a very high inoculum potential will be evident as patches in the crop and these may be expressed as mild patches (Fig. 2f). In a very conducive season, however, all the patches will be seen in the crop with a range from mild through to severe patches (Fig. 2c). Thus in one season there may be only a few mild patches and in the next season there may be many patches with a range of severities without any major changes in the area of soil colonized by the pathogen. I envisage that this changing suppressiveness is related to seasonal changes and is not a general microbial suppressiveness that appears to be responsible for a decline in Rhizoctonia bare patch (Roget, 1995; Wiseman *et al.*, 1996). Further work is needed to test this hypothesis and to determine the possible causes for changes in seasonal suppressiveness.

VI. CONCLUSIONS

It now seems clear that patches are caused by *R. solani* AG 8 and evidence indicates that each patch is caused by a single isolate colonizing the soil from a single infection focus. Apparent radical changes in patch configuration between seasons can be explained by differential growth rates, tillage effects, coalescing of patches, decline and demise of patches and possibly, changes in seasonal suppressiveness interacting with differences in the inoculum potential of the pathogen in different patches.

VII. REFERENCES

Baker KF & Cook RJ (1974) Biological Control of Plant Pathogens. W.H. Freeman and Company, San Francisco 433 pp.

Baker KF, Flentje NT, Olsen CM & Stretton HM (1967) Effect of antagonists on growth and survival of *Rhizoctonia solani* in soil. Phytopathology 57:591-597.

Benedict WG & Mountain WB (1956) Studies on the etiology of a root rot of winter wheat in southwestern Ontario. Can. J. Bot. 34:159-174.

Burpee LL (1995) Interactions among mowing height, nitrogen fertility, and cultivar affect the severity of Rhizoctonia blight of tall fescue. Plant Dis. 79:721-726.

Carling DE, Kuninaga S & Leiner RH (1988) Relatedness within and among intraspecific groups of *Rhizoctonia solani*: A comparison of grouping by anastomosis and by DNA hybridization. Phytoparasitica 16:209-210.

Daniels J (1963) Saprophytic and parasitic activities of some isolates of *Corticium solani*. Trans. Br. Mycol. Soc. 46:485-501.

Deacon JW & Scott DB (1985) *Rhizoctonia solani* associated with crater disease (stunting) of wheat in South Africa. Trans. Br. Mycol. Soc. 85:319-327.

De Beer JF (1965) Studies on the ecology of *Rhizoctonia solani*, Kühn. Ph.D. Thesis, University of Adelaide.

Dillon Weston WAR & Garrett SD (1943) *Rhizoctonia solani* associated with a root rot of cereals in Norfolk. Ann. Appl. Biol. 30:79.

Dubé AJ (1971) Studies on the growth and survival of *Rhizoctonia solani*. Ph.D. Thesis, University of Adelaide.

Flentje NT & Saksena H (1957) Studies on *Pellicularia filamentosa* (Pat.) Rogers. II. Occurrence and distribution of pathogenic strains. Trans. Br. Mycol. Soc. 40:95-108.

Green DE, Fry JD, Pair JC & Tisserat NA (1994) Influence of management practices on Rhizoctonia large patch disease in zoysiagrass. HortSci. 29:186-189.

Harris JR & Moen R (1985) Replacement of *Rhizoctonia solani* on wheat seedlings by a succession of root-rot fungi. Trans. Br. Mycol. Soc. 84:11-20.

Jarvis RJ & Brennan RF (1986) Timing and intensity of surface cultivation and depth of cultivation affect Rhizoctonia patch and wheat yield. Aust. J. Exp. Agric.. 26:703-708.

Kerr A (1955) Studies on the parasitic and saprophytic activities of *Pellicularia filamentosea* (Pat.) Rogers and *Sclerotinia homecarpa* Bennett. Ph.D. Thesis, University of Adelaide.

Lucas P, Smiley RW & Collins HP (1993) Decline of Rhizoctonia root rot on wheat in soil infested with *Rhizoctonia solani* AG 8. Phytopathology 83:260-265.

Ludbrook WV, Brockwell J & Riceman DS (1953) Bare-patch disease and associated problems in subterranean clover pastures in South Australia. Aust. J. Agric. Res. 4:403-414.

MacNish GC (1983) Rhizoctonia patch in Western Australian grain belt. Australas Plant Pathol. 12:49-50.

MacNish GC (1984) The use of undisturbed soil cores to study methods of controlling Rhizoctonia patch of cereals. Plant Pathol. 33:355-359.

MacNish GC (1985a) Mapping Rhizoctonia patch in consecutive cereal crops in Western Australia. Plant Pathol. 34:165-174.

MacNish GC (1985b) Methods of reducing Rhizoctonia patch of cereals in Western Australia. Plant Pathol. 34:175-181.

MacNish GC (1988) Changes in take-all (*Gaeumannomyces graminis* var. *tritici*), Rhizoctonia root rot (*Rhizoctonia solani*) and soil pH in continuous wheat with annual applications of nitrogenous fertilizer in Western Australia. Aust. J. Exp. Agric. 28:333-341.

MacNish GC (1989) Compaction, cultivation and Rhizoctonia bare patch. Abstracts of papers, Seventh Conference of the Australas Plant Pathol. Soc., Brisbane, Queensland, 1989, No. 17.

MacNish GC & Sweetingham MW (1993a) Evidence that each Rhizoctonia bare patch is dominated by an individual zymogram group (ZG) of *Rhizoctonia solani* AG 8. Aust. J. Agric. Res. 44:1175-1194.

MacNish GC & Sweetingham MW (1993b) Evidence of stability of pectic zymogram groups within *Rhizoctonia solani* AG 8. Mycol. Res. 97:1056-1058.

MacNish GC, Carling DE & Brainard KA (1995) Relationships of microscopic and macroscopic vegetative reactions in Rhizoctonia solani and the occurrence of vegetatively compatible populations (VCPs) in AG 8. Mycol. Res. 100: (in press).

MacNish GC, Carling DE & Brianard KA (1993a) Characterization of *Rhizoctonia solani* AG 8 from bare patches by zymogram and anastomosis techniques. Phytopathology 83:922-927.

MacNish GC, McLernon CK & Wood DA (1993b) The use of zymogram and anastomosis techniques to follow the expansion and demise of two coalescing bare patches caused by *Rhizoctonia solani* AG 8. Aust. J. Agric. Res. 44:1161-1173.

McKelvie A (1978) Barley stunt disorder. North Scot. Coll. Agric. Digest 1:45-50.

Meagher JW & Chambers SC (1971) Pathogenic effects of *Heterodera avenae* and *Rhizoctonia solani* and their interaction on wheat. Aust. J. Agric. Res. 22:189-194.

Meagher JW, Brown RH & Rovira AD (1978) The effects of cereal cyst nematode (*Heterodera avenae)* and *Rhizoctonia solani* on the growth and yield of wheat. Aust. J. Agric. Res. 29:1127-1137.

Murray DIL (1981) *Rhizoctonia solani* causing barley stunt disorder. Trans. Br. Mycol. Soc. 76:383-395.

Murray GM & Brown JF (1987) The incidence and relative importance of wheat diseases in Australia. Australas Plant Pathol. 16:34-37.

Neate SM (1984) Minimum cultivation and root diseases of wheat. Ph.D. Thesis, University of Adelaide.

Neate SM & Warcup JH (1985) Anastomosis grouping of some isolates of *Thanatephorus cucumeris* from agricultural soils in South Australia. Trans. Br. Mycol. Soc. 85:615-620.

Ogoshi A, Cook RJ & Bassett EN (1990) *Rhizoctonia* species and anastomosis groups causing root rot of wheat and barley in the Pacific Northwest. Phytopathology 80:784-788.

Opperman L & Barnard RO (1992) Improved growth and grain yield of wheat in deep mouldboard ploughed crater disease and disease-free soil on the Springbok Flats. Phytophylactica 24:335-338.

Patel BK (1983) Interrelationships between soil-borne pathogens on *Triticum aestivum*. Ph.D. Thesis, University of Adelaide.

Roberts FA & Sivasithamparam K (1986) Identity and pathogenicity of *Rhizoctonia* spp. associated with bare patch disease of cereals at a field site in Western Australia. Neth. J. Plant Pathol. 92:185-195.

Roberts FA & Sivasithamparam K (1987) Effect of interaction of *Rhizoctonia* spp. with other fungi from cereal bare patches on root rot of wheat. Trans. Br. Mycol. Soc. 89:256-259.

Roget DK (1995) Decline in root rot (*Rhizoctonia solani* AG 8) in wheat in a tillage and rotation experiment at Avon, South Australia. Aust. J. Exp. Agric. 35: 1009-1013.

Rovira AD & Simon A (1985) Growth, nutrition and yield of wheat in calcareous sandy loams of South Australia: Effects of soil fumigation, fungicide, nematicide and nitrogen fertilizers. Soil Biol. Biochem. 17:279-284.

Rovira AD & Venn NR (1985) Effect of rotation and tillage on take-all and Rhizoctonia root rot in wheat. In: Parker CA, Rovira AD, Moore KJ, Wong PTW & Kollmorgen JF (ed.) Ecology and Management of Soilborne Plant Pathogens. (pp 255-258) APS Press, St. Paul. Minnesota.

Samuel G (1928) Two "stunting" diseases of wheat and oats. J. Agric., South Australia 32: 40-43.

Smith EM & Wehner FC (1986) Induction of crater disease of wheat in the field by artificial infestation of the soil with *Rhizoctonia solani*. Phytophylactica 18:127-130.

Smith EM & Wehner FC (1988) Effect of ripping and ploughing of soil on crater disease of wheat caused by *Rhizoctonia solani*. Phytophylactica 20:345-347.

Sweetingham MW, Cruickshank RH & Wong DH (1986) Pectic zymograms and taxonomy and pathogenicity of the *Ceratobasidiaceae*. Trans. Br. Mycol. Soc. 86:305-311.

Weller DM, Cook RJ, MacNish GC, Bassett EN, Powelson RL & Paterson RR (1986) Rhizoctonia root rot of small grains favored by reduced tillage in the Pacific Northwest. Plant Dis. 70:70-73.

Wiseman BM, Neate SM, Ophel-Keller K & Smith SE (1996) Suppression of *Rhizoctonia solani* anastomosis group 8 in Australia and its biological nature. Soil Biol. Biochem. 28: (In Press).

Zarlengo PJ, Rothrock CS & King JW (1994) Influence of shading on the response of tall fescue cultivars to *Rhizoctonia solani*. AG 1-IA. Plant Dis. 78:126-129.

IV.6 MECHANISMS INVOLVED IN DISEASE DECLINE

MITSURO HYAKUMACHI

Faculty of Agriculture, Gifu University,
Yanagido 1-1 Gifu 501-11, JAPAN

I. INTRODUCTION

Disease decline phenomena in incidence or severity of diseases caused by certain soil-borne plant pathogenic fungi or induction of soil suppressiveness have been reported in fields under monoculture (King, 1923; Menzies, 1959; Itoi *et al.*, 1964; Weinhold *et al.*, 1964; Glynne, 1965; Mauer and Baker, 1965; Araki, 1967; Gerlagh, 1968; Shipton; 1975; 1977; Hornby, 1979; Hyakumachi, 1983a; 1983b; Hyakumachi and Ui, 1982a; Hyakumachi *et al.*, 1990). As the disease decline phenomena are examples of naturally occurring biological control, accurate evaluation of the factors involved in the decline may lead to develop practical biological control strategies. Although the phenomenon has been reported for many soil-borne fungal pathogens, except for the take-all decline (TAD), not many studies have been conducted to elucidate the mechanism(s) involved in the decline phenomenon in the field. One of the important reasons might be that it takes long time (10 years or more) to carry out such studies. Disease decline patterns may be different in various edaphic factors, geographical locations, and seasonal variations. Furthermore, reproduction of disease decline is rather inconsistent. On the other hand, many studies regarding mechanisms involved in disease decline or induced soil suppressiveness in the laboratory have been reported (Anwar, 1949; Gelragh, 1968; Henis *et al.*, 1978; 1979; Wijetunga and Baker, 1979; Liu and Baker, 1980; Zogg, 1976; Hyakumachi and Ui, 1982b; Davik and Sundheim, 1984; Chern and Ko, 1989).

As for the decline of diseases caused by *Rhizoctonia* spp., root rot declines in sugarbeet and wheat monoculture are the only few cases reported from field studies (Hyakumachi and Ui, 1982a; Hyakumachi, 1983a; 1983b; Hyakumachi *et al.*, 1990; MacNish, 1988; Roget, 1995). The decline of turfgrass patch disease caused by *Rhizoctonia* spp. is known empirically, but a detailed study on this phenomenon has not yet been done. However, Rhizoctonia disease decline (RDD) in the laboratory was studied in more detail (Henis *et al.*, 1978; 1979; Wijetunga and Baker, 1979; Chet and Baker, 1980; Liu and Baker, 1980; Hyakumachi and Ui, 1982b; Davik and Sundheim, 1984; Chern and Ko, 1989; Noda *et al.*, 1993; Tsuji *et al.*, 1994).

II. DEMONSTRATION OF RHIZOCTONIA DISEASE DECLINE (RDD)

Rhizoctonia disease decline (RDD) or induction of soil suppressiveness by monoculture have been reported in naturally or in artificially infested fields, as well as artificially infested soil in the greenhouse.

A. Occurrence of RDD in naturally or artificially infested fields

The nature of disease decline due to monoculture in sugar beet root rot in naturally infested fields with *R. solani* (AG 2-2IV) in Hokkaido, Japan (with some variability among fields), was manifested by severe disease during the first several years, while a

B. Sneh et al. (eds.),
Rhizoctonia Species: Taxonomy, Molecular Biology, Ecology, Pathology and Disease Constrol, 227–235.

gradual decline was observed in the following years with a most considerable decline during the 5th to the 8th year (Hyakumachi and Ui, 1982a; Hyakumachi *et al.*, 1990), and thereafter followed by less severe outbreaks. However, in artificially infested fields the severe disease was only evident in the first year of monoculture and declined from the second year onwards. While disease severity in the first year was positively correlated with inoculum concentration, the declines in the next years were similar in the field plots infested with different inoculum levels (Hyakumachi, 1983a; 1983b; Hyakumachi *et al.*, 1990).

Disease decline was also observed in a wheat monoculture field in Australia (MacNish, 1988; Roget, 1995). Severity of root rot caused by *R. solani* (AG 8) reached its peak on the fourth year and subsequently declined gradually, reaching negligible disease severity levels on the 11th year (Roget, 1995).

B. Occurrence of RDD in greenhouse pot experiments

Relatively more studies regarding RDD were carried out in the greenhouse, mainly on damping-off of various crops such as: radish, cucumber, sugarbeet, alfalfa, and cauliflower, caused by AGs: 1-1B, 1-1C, 2-1, 2-2 or 4 (Henis *et al.*, 1979; Liu and Baker, 1980; Hyakumachi and Ui, 1982b; Davik and Sundheim, 1984; Chern and Ko, 1989; Noda *et al.*, 1993). In these studies, new sets of seeds were replanted every 1-3 weeks in the soil initially infested with the pathogen and usually a remarkable rapid disease decline could be demonstrated. Usually, disease incidence increased during the first 1-5 replantings and after a peak of disease incidence or severity, it declined. The peak of disease incidence or severity varied in the different reports. Maximum disease incidence occurred at the first planting (Hyakumachi and Ui, 1982b; Noda *et al.*, 1993) or at the 2nd to the 5th replantings (Henis *et al.*, 1978; 1979; Chet and Baker, 1980; Liu and Baker, 1980; Chern and Ko, 1989). However, disease decline occurred 4 to 5 weeks after inoculation, regardless of the crop growing period of replanting. Root rot decline in sugarbeet or wheat was delayed for 8 to 24 weeks, respectively (Hyakumachi, 1983a; Lucas *et al.*, 1993). When soil was initially infested with the pathogen and replanted with sugarbeet or wheat every 1 or 3 months, the peak of root rot appeared at the first cropping and disease decline occurred during the second or third replanting (2 or 6 months, respectively). The unique event of 'no peak' of disease incidence due to artificial infestation in the greenhouse was similar to that found in the artificially infested field.

III. CHARACTERIZATION OF RHIZOCTONIA DISEASE DECLINE

A. Disease decline differences between field and pot experiments

There are several differences in disease decline phenomena between field and pot experiments, but the most definite difference is the period of planting. In the field, crops are usually planted only once a year, but in pot experiments, the plantings period are very short (1-4 weeks). It is well known that Rhizoctonia diseases occur in field patches and are not evenly spread all over the fields. Hence, when the disease declines, the part of the field in which the decline occurred has to be carefully followed and examined, particularly in the evaluation of the suppressive nature of the soil, in which disease decline was evident. It does not mean that the disease declined all over the field. On the other hand, in pot experiments, disease decline occurs evenly. The difference is that the whole soil in the pot uniformly expressed the suppressiveness. However, inconsistency in disease decline occurs even among replicates in pot experiments (Chern and Ko, 1989).

Although disease decline in pot experiments may be reflected similarly to that in the field, it may not be right to simply relate to pot experiments as actually mimicking field conditions. This has already been emphasized for the take-all decline (Walker, 1975). Therefore, final conclusions regarding disease decline phenomena should be derived from field studies.

B. Can successive croppings of host crops generate disease decline or induce soil suppressiveness?

Disease decline could not always be generated by repeated replantings of host crops, even in soil infested with the pathogen. Thus, successive plantings of sugarbeet, alfalfa and wheat did not generate disease decline (Liu and Baker, 1980). However, Chet and Baker (1980) demonstrated disease decline even in successive cropping of sugarbeet and alfalfa using the same soil after adjustment of soil pH from 8.2 to 6.5 or 5.7, respectively. In case of wheat, the pathogen *R. solani* AG 4 that presumably lacked virulence on this cultivar (i.e. AG 4 caused only superficial brown discoloration of cortical tissue and did not kill plants) did not induce disease decline (Liu and Baker, 1980).

C. Effect of inoculum type on disease decline

Different types of inoculum preparations such as: sclerotia, colonized cereal grains, living mycelial fragments or chopped potato-soil (CPS) were used to demonstrate the disease decline phenomenon in *Rhizoctonia solani* (Chet and Baker, 1980; Liu and Baker, 1980; Hyakumachi and Ui, 1982b; Davik and Sundheim, 1984; Chern and Ko, 1989). Propagule size of the inoculum may greatly influence the rate of development of soil suppressiveness or disease decline after a series of replantings. Disease decline was generated when small propagules of mycelial fragments (<250mm), but not when large propagules (>590mm), were used as inoculum (Wijetunga and Baker, 1979).

D. Anastomosis groups of *Rhizoctonia* spp. and disease decline

So far, disease decline was reported for AGs 2-2 and 8 in sugarbeet and wheat root rots in the field (Hyakumachi *et al.*, 1990; Roget, 1995). In pot experiments, a decline of damping-off caused by AGs 1-1B, 1-1C, 2-1, 2-2, and 4 of radish, cucumber, sugarbeet, and cauliflower were reported (Chet and Baker, 1980; Hyakumachi and Ui, 1982b; Davik and Sundheim, 1984; Chern and Ko, 1989; *et al.*, 1993a; Tsuji *et al..*, 1994); and of root rot by AG 2-2 in sugarbeet and AG 8 in wheat (Hyakumachi, 1983b; Lucas *et al.*, 1993). Other AGs have not been studied for their ability to cause disease declinc.

IV. MECHANISMS OF DISEASE DECLINE

Disease decline may result from abiotic factors such as physico-chemical factors in soil or climatic factors and/or biotic factors such as changes within or outside the pathogen population, and antagonism by other microorganisms during the monoculture (Shipton, 1975; Hornby, 1979). The physico-chemical soil characteristics, air temperature, precipitation, or duration of exposure to sun light have not been shown to affect Rhizoctonia disease decline except in the findings of Chet and Baker (1980) on the effect of soil pH in a pot experiment. While biotic factors are mainly responsible for disease decline (Henis *et al.*, 1978; 1979; Wijetunga and Baker, 1979; Liu and Baker, 1980; Hyakumachi and Ui, 1982a; Hyakumachi, 1983a; Davik and Sundheim, 1984; Chern and Ko, 1989; Hyakumachi *et al.*, 1990).

A. Decline in Propagule density

Rhizoctonia disease decline is associated with qualitative and quantitative changes in the pathogen's propagules in soil. Qualitative changes within the pathogen may include the loss of pathogenicity or loss of the ability to produce progeny propagules. Among the isolates of *R. solani*, some of AG 2-2, which were obtained from monoculture fields, were of low pathogenicity, low ability to produce sclerotia and abnormally slow growing, or they did not anastomose with themselves (Hyakumachi and Ui, 1984; 1987). These isolates could also be obtained from artificially infested soils, or by continuous transfer to growth media or by single spore isolation. The occurrence of such variations in isolates may be considered as one of the contributing factors to disease decline. However, information regarding qualitative mechanisms involved in disease decline within the pathogen are still scarce and insufficient. Presence of double stranded RNA (dsRNA) or plasmid DNA in *R. solani* has been reported to be related to hypovirulence of the pathogen (Castanho *et al.*, 1978; Hashiba *et al.*, 1984). However, recent studies have shown that there is no relationship between the presence of dsRNA or plasmid DNA with virulence (Finkler *et al.*, 1985; Hyakumachi *et al.*, 1986; Miyasaka *et al.*, 1988).

Viability of *Rhizoctonia* spp. propagules in soil after several cropping of monoculture, where the disease decline phenomenon was evident, was considerably reduced. Reduction in propagule viability was found in naturally infested, as well as in artificially infested fields. In fields where disease occurred naturally, the number of viable sclerotia in soil around diseased roots varied considerably from year to year during monoculture (Hyakumachi and Ui, 1982a; Hyakumachi *et al.*, 1990). While few, or no sclerotia were recorded in some years, many sclerotia were recorded in other years, but very few of those were viable. A close correlation was found between inoculum density in the previous year and disease severity in the next year (Hyakumachi and Ui, 1982a; Hyakumachi *et al.*, 1990). The build up in population levels of antagonistic microorganisms may be responsible for the fluctuation in propagule density of *Rhizoctonia* spp. in soil (Hyakumachi, 1983a; Hyakumachi *et al.*, 1990). The mycophagous gnat, *Pnyxia scabiei*, may be an additional causal agent affecting propagule viability decline (Naito and Sugimoto, 1988; Naito *et al.*, 1988; Naito and Makino, 1995). In an artificially infested field, number of *R. solani* (AG 2-2) sclerotia remained constant until the following spring (Hyakumachi, 1983a; Hyakumachi *et al.*, 1990), but their viability dropped markedly. A negative correlation between disease severity and sclerotial germination was observed, i.e. sclerotia around roots with a low disease severity showed high frequency of germination, while sclerotia around roots with a high disease severity showed low frequency of germination. Sclerotia around severely diseased roots did not germinate because of the greater abundance of antagonistic organisms or mycophagous gnat (Hyakumachi, 1983a; Naito *et al.*, 1988; Hyakumachi *et al.*, 1990; Naito and Makino, 1995). Reduced inoculum density results in disease decline (Hyakumachi and Ui, 1982a, Hyakumachi, 1983a, 1983b; Hyakumachi *et al.*, 1990). This finding differs from that of Shipton (1975) in take-all decline soils, where the inoculum density of the pathogen did not decline significantly, while its ability to cause disease did.

B. Suppressiveness in soil with disease decline

Soil suppressiveness was induced by replanting with a susceptible crop to the following soil-borne pathogens; *R. solani* (Henis *et al.*, 1979; Chet and Baker, 1980;

Hyakumachi *et al.*, 1990), *Pythium ultimum* (Hancock, 1967; Lifshitz *et al.*, 1984) and *Gaeumannomyces graminis* (Hornby, 1975; Cook and Rovira, 1976), or with resistant cultivars to *Fusarium oxysporum* (Sneh *et al.*, 1987). Presence or absence, degree, or persistence of disease suppressiveness in soil where disease decline was shown were assessed by inoculation with *R. solani*, planting host crops and evaluating the disease incidence. In an artificially infested sugarbeet field, Rhizoctonia root rot was severe in the first year of infestation, but declined rapidly in subsequent years. The incidence of Rhizoctonia root rot varied according to the amount of inoculum. It was very low when low inoculum levels were applied and very high when high inoculum levels were used. When equal inoculum levels were added to these plots in the subsequent year, surprisingly, disease was severe in the plots where incidence had been very low in the previous year, but was lower in plots where disease incidence had been very high in the previous year (Hyakumachi, 1983a). Moreover, the suppressive effect persisted in soil where root rot severity had declined. But this trend did not extend over 3 years (Hyakumachi, 1983a), suggesting that the suppressive effect of soil following disease decline in sugarbeet root rot did not persist for a very long time. On the other hand, when the field plot was infested with the pathogen but not planted with sugarbeet, soil suppressiveness did not develop (Hyakumachi, 1983b). Soils obtained from fields where plots were not infested during continuous croppings, from the area where root rot had already declined during more than 3 previous years, and from an area where root rot incidence was severe, were collected from the naturally infested field, and reinfested to evaluate the disease suppressiveness level.

Disease suppressiveness developed significantly in soils where crops were severely diseased, but not in two areas where disease severity was low, or the decline has been observed in several previous years (Hyakumachi, 1983a; 1983b; Hyakumachi *et al.*, 1990). These results indicate that the presence of the pathogen, the suitable host and the antagonist(s), are required to develop soil suppressiveness to *Rhizoctonia* spp. Absence of any of these three components prevents the build up of soil suppressiveness in the field. Naito and Sugimoto (1988) obtained a similar result in sugarbeet monoculture fields artificially infested with *R. solani*. These results suggest that high level of disease severity, or presence of diseased root debris in soil are essential for building up of disease suppressiveness. A similar indication was recently reported by Roget (1995) for development of suppression in wheat root rot decline. Fahima and Henis (1990) suggested that disease suppressiveness develops as a consequence of damage caused by the pathogen to infected roots. Increased root exudation provide the antagonists with an important food base. Thus, enhances the aggressiveness of the antagonist on the pathogen's germinating propagules around the infected root. In addition to protection of plants against their respective pathogens, antagonistic microorganisms also multiply on roots of diseased plants. This may explain the preferred multiplication of *Trichoderma hamatum* in infected roots, as well as the development of suppressiveness following replanting of radish in *R. solani*-infested soil. Rovira and Wildermuth (1981) suggested a similar model, by generating specific suppression against take-all on infected roots. In sugarbeet, disease suppressiveness level was correlated with disease severity of the previous year. When field soil was repeatedly reinfested disease incidence declined in the first two years, but disease severity was high in the following year. It declined in the subsequent two years (Hyakumachi, 1983b). Disease suppressiveness could not be transferred from one soil to the other. However, it could be transferred from a field with a decline of bare patch disease of wheat by adding 10% (w/w) of the suppressive soil to an autoclaved or pasteurized soil (Wiseman *et al.*, 1995).

In soil initially infested with *R. solani*, disease suppressiveness developed during successive plantings of the host in pot experiments (Chet and Baker, 1980; Hyakumachi and Ui, 1982b; Davik and Sundheim, 1984; Chern and Ko, 1989; Lucas, 1993; Noda *et al.*, 1993; Tsuji *et al.*, 1994). Once disease suppressiveness was established, its effect continued for as long as 1 year or more, when soil kept at 4°C (Noda *et al.*, 1993). When disease suppressiveness decreased proportionally with dilution of suppressive soil in a conducive soil, the suppressive effect could not be transferred. Disease suppressiveness could easily be established in soil in pots, upon initial infestation with *R. solani*, after a few replantings of host plants (at 1 to 5 weeks intervals). Disease suppressiveness level could be increased by repeated infestation and replanting (Lucas *et al.*, 1993; Noda *et al.*, 1993; Tsuji *et al.*, 1994). Disease incidence could thus be rapidly reduced after 2 or 4 replantings. Introduction of nonpathogenic *R. solani*, could not generate soil suppressiveness (Hyakumachi, unpublished data). However, disease suppressiveness could also be established by soil infestation with the pathogen in the absence of a host in pot experiments (Hyakumachi and Ui, 1982b; Van Den Boogert and Jager, 1983). Induction of disease suppressiveness was also observed with other pathogens such as *Gaeumannomyces graminis* var. *tritici* (Zogg and Amiet, 1980; Zogg and Yaggi, 1974; Wildermuth, 1982a, 1982b).

V. ROLE OF MICROORGANISMS IN DISEASE SUPPRESSIVENESS

Several investigators have indicated that soil microorganisms are responsible for the development of soil suppressiveness (Gerlagh, 1968; Baker and Cook, 1974; Cook and Rovira, 1976; Cook and Baker, 1983; Hornby, 1979; Rovira and Wildermuth, 1981). It has also been postulated that the soil suppressive effect against Rhizoctonia diseases is directly related to population levels of the antagonistic microorganisms, which suppress hyphal growth (Hyakumachi, 1983b; Hyakumachi *et al.*, 1990), parasitize and lyse the pathogen's hyphae (Chet and Baker, 1980; Liu and Baker, 1980) and kill its sclerotia (Hyakumachi, 1983b; Hyakumachi *et al.*, 1990). Population levels of several different microorganisms were higher in soils where the disease decline phenomenon was obtained than where disease was not apparent.

The nature of disease suppressiveness in soils where disease decline occurred is different in different experimental conditions. The suppressive effect of some soils was nullified when subjected to about 60°C for 30 minutes (Cook and Rovira, 1976), but 80°C or 100°C were required in some other soils (Hyakumachi, 1983b; Hyakumachi *et al.*, 1990; Wiseman *et al.*, 1995). Accordingly, the suppressive microbial agent(s) may differ in different soils. *Bacillus* spp. were reported to be the primarily responsible agents for disease suppressiveness in sugarbeet root rot decline field soil (Hyakumachi, 1983b; Hyakumachi *et al.*, 1990). *Trichoderma* spp. were confirmed to be the agents causing disease suppressiveness in various Rhizoctonia disease decline soils in pot experiments (Henis *et al.*, 1979; Wijetunga and Baker, 1979; Chet and Baker, 1980; Liu and Baker, 1980; Noda *et al.*, 1993; Tsuji *et al.*, 1994) and in natural suppressive soil (Chet and Baker, 1981). *Bacillus* spp. and *Trichoderma* spp. were shown to generate the soil suppressiveness when introduced into conducive soils (Wijetunga and Baker, 1979; Chet and Baker, 1980; Liu and Baker, 1980; Hyakumachi, 1983b; Hyakumachi *et al.*, 1990; Noda *et al.*, 1993).

Gerlagh (1968) classified the suppressive effects of soils against take-all as: general suppression and specific suppression. Cook and Rovira (1976) defined *Bacillus* spp. and *Trichoderma* spp. to be the responsible agents for general, as well as specific suppression. Hence, the cause of Rhizoctonia disease decline could be both general and/or specific in nature.

VI. Conclusions

The take-all decline phenomenon was intensively studied and yielded many publications, while relatively few studies have been reported for Rhizoctonia disease decline, especially in field conditions. More studies were conducted in pot experiments. Rhizoctonia disease decline could be explained as follows: In the disease suppressive soil, following severe disease incidence, hyphal growth of the pathogen was strongly suppressed, hyphae were quickly lysed and the pathogen died earlier due to increase in the antagonistic activity of the soil microbial populations. These lead to a reduction in inoculum density and inoculum potential and consequently to disease decline. Although the decline phenomena reported for field and pot experiments seem to be similar, the results indicate that they were different in various aspects and could not be directly correlated. For better understanding of the decline phenomena, it would be better to separately consider their occurrence in the field and in the greenhouse, because suppression under these two different conditions are quite different. So far, two Rhizoctonia diseases; root rot of sugarbeet caused by *R. solani* AG 2-2 and root rot of wheat caused by *R. solani* AG 8, were reported to decline in continued monoculture fields. They show the important role(s) of the accumulated microbial population levels in the diseased plant debris during successive plantings on disease suppression in soil following disease decline. In both cases soil suppressiveness was not nullified by heat treatment at 60°C but did at 80°C or 100°C (Hyakumachi, 1983b; Wiseman *et al.*, 1995), and in the case of sugarbeet root rot decline, *Bacillus* spp. were considered as the major responsible agent(s) for disease suppressiveness (Hyakumachi, 1983b). While in pot experiments, many reports have shown the role of *Trichoderma* spp. on disease suppressiveness (Henis *et al.*, 1979; Wijetunga and Baker, 1979; Chet and Baker, 1980; Liu and Baker, 1980; Noda *et al.*, 1993; Tsuji *et al.*, 1994). Eventually, the mechanism(s) of disease decline under field and pot conditions are also quite different. When the aim of the study of disease decline is to understand and elucidate the mechanism of disease decline occurring in the field, it may be better to conduct the studies for clarification of the mechanism(s) only in the field. Instead, when the aim is to study disease decline in order to find the effective microorganisms for controlling Rhizoctonia diseases, it might be useful to mimic the decline phenomenon in the greenhouse (pot experiments). Artificially, disease suppression could be obtained but the mechanisms involving disease suppressiveness of soil where disease declined in these experiments are different from those in naturally occurring suppressiveness in disease decline soil. A special care should be taken, not to confuse these two aims.

VII. References

Anwar AA (1949) Factors affecting the survival of *Helminthosporium sativum* and *Fusarium lini* in soil. Phytopathology 39:1005-1019.

Araki T (1967) Soil conditions and the violet and white root rot diseases of fruit trees. Bull. Nat. Inst. Agric. Sci. Series C, No. 21:1-109.

Baker KF & Cook RJ (1974) Biological Control of Plant Pathogens. WH Freeman, San Francisco

Castanho B, Butler EE & Shepherd RJ (1978) The association of double stranded RNA with decline. Phytopathology 68:1515-1519.

Chern LL & Ko WH (1989) Characteristic of inhibition of suppressive soil created by monoculture with radish in the presence of *Rhizoctonia solani*. J. Phytopathol. 126:237-245.

Chet I & Baker R (1980) Induction of suppressiveness to *Rhizoctonia solani* in soil. Phytopathology 70:994-998.

Chet I & Baker R (1981) Isolation and biocontrol potential of *Trichoderma hamatum* from soil naturally suppressive to *Rhizoctonia solani*. Phytopathology 71:286-290.

Cook RJ & Baker KF (1983) The Nature and Practice of Biological Control of Plant Pathogens. APS Press, Minnesota.

Cook RJ & Rovira AD (1976) The role of bacteria in the biological control of *Gaeumannomyces graminis* by suppressive soils. Soil Biol. Biochem. 8:269-273.

Davik JE & Sundheim L (1984) Induced suppression of *Rhizoctonia solani* in Norwegian soils. Meld. Nor. Landbrukshoesk 63:1-12.

Fahima T & Henis Y (1990) Interactions between pathogen, host and biocontrol agent: multiplication of *Trichoderma hamatum* and *Talaromyces flavus* on roots of diseased and healthy hosts. In: Hornby D, Cook RJ, Henis Y, Ko WH, Rovira AD, Schippers B & Scott PR (eds.) Biological Control of Soil-Borne Plant Pathogens. (pp 165-180) C.A.B. International, Wallingford.

Finkler A, Koltin Y, Barash I, Sneh B & Pozniak D (1985) Isolation of a virus from virulent strains of *Rhizoctonia solani*. J. gen. Virol. 65:1221-1232.

Gerlagh M (1968) Introduction of *Ophiobolus graminis* into new polders and its decline. Neth. J. Plant Pathol. 74, (Suppl. 2):1-97.

Glynne MD (1965) Crop sequence in relation to soil-borne pathogens. In: Baker KF & Snyder WC (eds.) Ecology of Soil-borne Plant Pathogens. (pp 423-425) University of California Press, Berkeley.

Hancock JG (1979) Occurrence of soils suppressive to *Pythium ultimum*. In: Schippers B & Gams W (eds.) Soil-Borne Plant Pathogens. (pp 183-189) Academic Press, London, New York, San Francisco.

Hashiba T, Homma Y, Hyakumachi M & Matsuda I (1984) Isolation of a DNA plasmid in the fungus *Rhizoctonia solani*. J. gen. Microbiol. 130:2067-2070.

Henis Y, Ghaffar A & Baker R (1978) Integrated control of *Rhizoctonia solani* damping-off of radish: Effect of successive plantings, PCNB, *Trichoderma harzianum* on pathogen and disease. Phytopathology 68: 900-907.

Henis Y, Ghaffar A & Baker R (1979) Factors affecting suppressiveness to *Rhizoctonia solani*. Phytopathology 69:1164-1169.

Hornby D (1979) Take-all decline: a theorist's paradise. In: Schippers B & Gams W (eds.) Soil-Borne Plant Pathogens. (pp 133-156) Academic Press, London, New York, San Francisco.

Hyakumachi M (1983a) Decline phenomenon of sugarbeet root rot. Shokubutsu Boeki (Plant Protection) 37:532-537.

Hyakumachi M (1983b) Studies on the sugarbeet root rot decline. Ph.D thesis, Hokkaido University.

Hyakumachi M & Ui T (1982a) Decline of Rhizoctonia root rot during sugarbeet monoculture. Mem. Fac. Agric. Hokkaido Univ. 13:445-454.

Hyakumachi M & Ui T (1982b) Decline of damping-off of sugarbeet seedlings caused by *Rhizoctonia solani* AG 2-2. Ann. Phytopathol. Soc. Japan. 48:600-606.

Hyakumachi M & Ui T (1984) Properties of diseased isolates of *Rhizoctonia solani* isolated from soil. Ann. Phytopathol. Soc. Japan 50:255-262.

Hyakumachi M & Ui T (1987) Non-self-anastomosing isolates of *Rhizoctonia solani* obtained from fields of sugarbeet monoculture. Trans. Br. mycol. Soc. 89:155-159.

Hyakumachi M, Kanzawa. K & Ui T (1990) Rhizoctonia root rot decline in sugarbeet monoculture. In: Hornby D, Cook RJ, Henis Y, Ko WH, Rovira AD, Schippers B & Scott PR (eds.) Biological Control of Soil-Borne Plant Pathogens. (pp 227-247) CAB International, Wallingford.

Hyakumachi M, Sumino A, Uyeda I & Shikata E (1986) Relationship between dsRNA of *Rhizoctonia solani* and hypovirulence. International Workshop on Viruses of Fungi and Lower Eukaryots. European Molecular Biology Organization. Jerusalem, Israel. p 34.

Itoi, S, Hinata, T, Hayasaka, T, Kubomura, Y & Nakayama K (1964) Studies on white and violet root rots of mulberry trees. II. Relationship between the type of root infection with *Rosellinia necatrix* and some physical properties of soils, and survival period of this fungus in infected root tissue. Bull. of the Sericultural Exp. Sta. 18:529-543.

King CJ (1923) Habits of the cotton root rot fungus. J. Agric. Res. 26:405-418.

Lifshitz R, Dupler M, Elad Y & Baker R (1984) Hyphal interactions between *Pythium nunn* and several soil fungi. Can. J. Microbiol. 30:1482-1487.

Liu S & Baker R (1980) Mechanism of biological control in soil suppressive to *Rhizoctonia solani*. Phytopathology 70:404-412.

Lucas P, Smiley RW & Collins HP (1993) Decline of Rhizoctonia root rot on wheat in soils infested with *Rhizoctonia solani* AG 8. Phytopathology 83:260-265.

MacNish GC (1988) Changes in take-all (*Gaeumannomyces graminis* var. *tritici*), Rhizoctonia root rot (*Rhizoctonia solani*) and soil pH in continuous wheat with annual applications of nitrogenous fertilizer in Western Australia. Aust. J. Exp. Agric. 28:33-341.

Mauer CL & Baker R (1965) Ecology of plant pathogens. II. Influence of glucose, cellulose, and inorganic nitrogen amendments on development of bean root rot. Phytopathology 55:69-72.

Menzies JD (1959) Occurrence and transfer of a biological factor in soil that suppress potato scab. Phytopathology 49: 648-652.

Miyasaka A, Hashiba T & Ehara T (1990) Detection and properties of plasmid-like DNA in isolates from nine anastomosis and intraspecific groups of *Rhizoctonia solani*. J. gen. Microbiol. 136:1791-1798.

Naito S & Makino S (1995) Control of sclerotia of *Rhizoctonia solani* by a sciarid fly, *Pnyxia scabiei*, in soil. JARQ 29:31-37.

Naito S & Sugimoto T (1988) Control of sugarbeet damping-off (*Rhizoctonia solani* Kühn) by mycophagous *Pnyxia scabiei* (Hopkins). Ann. Phytopath. Soc. Jpn. 54:317-318.

Naito S, Makino S & Sugimoto T (1988) *Pnyxia scabiei* (Hopkins) feeding on sclerotia of *Rhizoctonia solani* Kühn and its population changes in sugarbeet root rot field. Ann. Phytopath. Soc. Jpn. 54:52-59.

Noda T, Honda H, Kageyama K & Hyakumachi M (1993) Role of microbes for soil suppression induced by repeated infestation of *Rhizoctonia solani* and *Sclerotium rolfsii*. Ann. Phytopath. Soc. Jpn. 59: 48-749 (Abstr.).

Roget DK (1995) Decline in root rot of wheat caused by *Rhizoctonia solani* AG 8 in a tillage and rotation trial at Avon, South Australia. International symposium on *Rhizoctonia*, Facts and Cahllenges in Pathology, Taxonomy, Ecology and Disease Control. June 27-30, Noordwijkerhout, The Netherlands (Abstr. No. P-7-6)

Rovira AD & Wildermuth GB (1981) The nature and mechanism of suppression. In: Asher MJC & Shipton PJ (eds.) Biology and Control of Take-all. (pp 385-415) Academic Press, London.

Shipton PJ (1975) Take-all decline during cereal monoculture. In: Bruehl GW (ed.) Biology and Control of Soil-Borne Plant Pathogens. (pp137-144). APS Press, St. Paul, Minnesota.

Shipton PJ (1977) Monoculture and soilborne plant pathogens. Ann. Rev. Phytopath. 15:387-407.

Sneh B, Pozniak D & Salomon D (1987) Soil suppressiveness to Fusarium wilt of melon, induced by repeated cropping of resistant varieties of melon. J. Phytopathology 120:347-354.

Tsuji T, Noda T, Kageyama K & Hyakumachi M (1994) Induction of disease suppressiveness by repeated infestation of soil with soil-borne pathogens in radish system and their factors. Ann. Phytopath. Soc. Jpn. 60:781 (Abst.).

Van Den Boogert PHJF & Jager G (1983) Accumulation of hyperparasites of *Rhizoctonia solani* by addition of live mycelium of *R. solani* to soil. Neth. J. Pl. Path. 89:223-228.

Walker J (1975) Take-all diseases of Gramineae: a review of recent work. Rev. Pl. Path. 54:113-144.

Weinhold AR, Oswald JW, Bowman T, Bishop J & Wright D (1964) Influence of green manures and crop rotation on common scab of potato. Am. Potato J. 41: 265-273.

Wiseman BM, Neate SM, Ophel-Keller K & Smith SE (1995) Suppression of *Rhizoctonia solani* anastomosis group 8 in Australia and its biological nature. International symposium on *Rhizoctonia,* Facts and Cahllenges in Pathology, Taxonomy, Ecology and Disease Control. June 27-30, Noordwijkerhout, The Netherlands (Abstr. No. P-7-7)

Wijetunga C & Baker R (1979) Modelling of phenomena associated with soil suppressive to *Rhizoctonia solani* Phytopathology 69:1287-1293.

Wildermuth GB (1982a) Soils suppressive to *Gaeumannomyces graminis* var. *tritici*: Effect on other fungi. Soil Biol. Biochem. 14:561-567.

Wildermuth GB (1982b) Soils suppressive to *Gaeumannomyces graminis* var. *tritici*: Induction by other fungi. Soil Biol. Biochem. 14:569-573.

Zogg H (1976) Problems of biological soil disinfection. Soil Biol. Biochem. 8:299-303.

Zogg H & Amiet JA (1980) Laboratory studies on decline with different foot and root rot pathogens of wheat. Phytopath. Z. 97:193-213.

Zogg H & Jaggi W (1974) Studies on the biological soil disinfection. Phytopath. Z. 81:160-169.

IV.7 ANALYZING CROP LOSSES DUE TO *RHIZOCTONIA SOLANI*: RICE SHEATH BLIGHT, A CASE STUDY

S. SAVARY AND T. W. MEW

Division of Entomology & Plant Pathology, International Rice Research Institute, P.O. Box 933, 1099 Manila, PHILIPPINES

I. Introduction

The host range of *Rhizoctonia solani* is so wide, and the damage caused to specific crops is so variable, that a case study may provide an appropriate approach to the general issue of measuring crop losses at the field level. Rice sheath blight provides a good example because the symptoms of the disease are fairly representative of symptoms caused by *R. solani* to a wide range of both monocots and dicots.

However, it needs to be stressed that representatives of AG 1 of *R. solani* do not attack only rice, but also other crop species such as maize, sorghum, soybean, mungbean, and cowpea. In tropical Asia. These crops are often used in very diverse rice-based cropping systems. Crop losses due to a certain pathogen such as AG 1 of *R. solani* in such complex systems have not been documented. The relationships between disease intensity and the resulting crop loss in a particular pathosystem may be affected by climatic environmental conditions and, for instance, this relationship may be different for rice in temperate or in tropical environments. This chapter focuses on rice sheath blight only in a tropical environment, as an example to highlight methodological issues, rather than specific results.

II. Disease symptoms and injury mechanisms

A. Injury, damage and loss

In this section we use the concepts of (Zadoks, 1985) injury (the visible result of host-pathogen interaction, and the measurable effects of the pathogen on the host in the field), damage (which is equivalent to the older term 'crop loss', i.e., the consequence of injury on yield), and loss (the economic consequences of damage). The relationship between damage and injury is materialized by a damage function; the relationship between loss and damage is materialized by a loss function (Zadoks, 1985). Relations among these concepts are shown in Fig. 1 (Savary, 1991).

B. Rice sheath blight injury: disease symptoms

The way the pathogen affects the standing crops has been described in several reference texts (Kozaka, 1961; Ou, 1987) and by Hashiba (chapter V.9). The main symptoms of the disease are lesions on the sheaths, at the base of the plants, or close to the water line in irrigated crops. The size, shape, and color of the lesions may vary considerably. Under favorable, especially humid, conditions, typical lesions have a brown margin with a grayish center. These lesions can expand extremely fast on the sheath, causing the corresponding leaf blade to wilt and die within a few days. Favorable conditions also allow the pathogen to spread as mycelial strands from the lower part of the plant (sheaths) to the upper part of the plant (leaves), and also from a tiller to another in one plant, or between neighboring plants. These mycelial strands therefore provide the means for the disease to spread in the canopy, both vertically

B. Sneh et al. (eds.),
Rhizoctonia Species: Taxonomy, Molecular Biology, Ecology, Pathology and Disease Constrol, 237–246.

and horizontally (Kozaka, 1961). When contact is established between a mycelial strand and a healthy sheath or leaf area, a new lesion may establish. The rate of spread of these strands, and therefore the expansion of the disease, can be very fast (the horizontal spread of the disease has been estimated as up to an average of 20cm/day under field conditions) (Savary *et al.*, 1995). Usually, sclerotia are formed on tissues that have been colonized by the pathogen, but it is not unusual to see sclerotia formed on the margin of lesions, very close to, or even on, tissues that are apparently healthy. In most cases, the sclerotia are easily detached from their substrate. One typical feature of the disease is the very fast wilting it induces on tissues (sheath or leaves) that are colonized by the pathogen. As a result, typical sheath blight symptoms are transient, and measurable for a very short period of time only. When inspecting a rice crop affected by sheath blight, several symptoms are observed, that are more or less characteristic of the causal pathogen: typical, water-soaked lesions on the sheaths or the leaves; dried lesions that no longer exhibit the typical characteristics of the disease; green tissues that are covered by mycelial strands, but do not appear colonized; dead leaves and sheaths, where drying lesions may be distinguished, but can hardly be attributed to *R. solani* with absolute certainty; in some cases, plants where most leaves and sheaths are dead, bearing the previous signs.

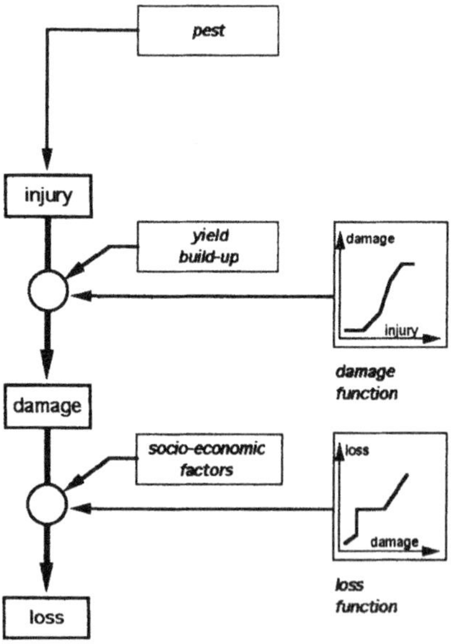

FIGURE 1. Diagrammatic relationships between the concepts of injury, damage, and loss (Source: Savary, 1991).

C. Damage mechanisms

This rapid overview of disease symptoms indicates that infection of rice plants by *R. solani* triggers a number of damage mechanisms. The following mechanisms may be considered:

a. A loss of phothosynthetically active leaf area due to the establishment of lesions on both the leaves and the sheaths;

b. An alteration of photosynthetic and respiratory processes in the green tissues due to the activities of enzymes released by the colonizing pathogen;

c. A disruption of water and carbohydrate flows within a tiller caused by lesions established at its base, and altering the photosynthesis of apparently unaffected leaves;

d. An accelerated senescence of leaves due to the wide array of lytic enzymes produced by such an efficient necrotrophic pathogen;

e. An accelerated senescence of tillers that may lead to the destruction of potentially fertile tillers (i.e., destruction of sinks where photosynthates accumulate), or reduce the compensation ability of the crop to other environmental stresses;

f. An alteration of the resistance of culm, leading to premature lodging.

There has been little quantitative research addressing damage mechanisms in rice sheath blight. A reduction in chlorophyll content together with a reduced (net) photosynthesis of infected tissues was demonstrated by Damodaram Nadu *et al.* (1981). An increase of the respiration rate and a decrease in the transpiration rate was reported by Roy (1982). Gangopadyay and Chakrabarti (1982) also reported an inhibition of formation of secondary roots. Recently, efforts have been initiated to incorporate damage mechanisms in a deterministic simulation model of rice development and growth (Ellings, 1994). Only two components are so far considered: a reduction of maximal photosynthesis, and an increased respiration. This modeling approach is still undergoing further testing, and additional mechanisms certainly need to be considered. Surprisingly, one of the most striking effects of sheath blight in the field, the accelerated senescence of infected leaves and sheaths, does not seem to have been thoroughly documented. Analysis of field experiments in the Philippines indicated that the relative rate of tiller mortality associated with sheath blight was approximately half of relative rate of spontaneous tiller senescence in healthy plants (Savary and Willocquet, unpublished results). In other words, sheath blight was responsible for around 50 % increase in tiller death in inoculated plants under field conditions.

Another area that will require attention of modellers is the spatial characteristics of the disease. Simulation models for crop growth and development currently do not handle the spatial nature of crop injury, which is a key feature for rice sheath blight epidemics. Disease aggregation leads to limited compensation, both at the plant or hill level (between tillers) and at the plant population level (between plants or hills), and may be translated in simulations to systematically underestimate damage. We shall address further the issue of disease aggregation.

III. A METHODOLOGICAL ISSUE: DISEASE MEASUREMENT

A large number of assessment methods have been used to measure disease in the field. One main reason for this is the injury inflicted by *R. solani* to the rice crop is exceedingly difficult to capture in a simple, universal, observational system. Also, this diversity of approaches makes comparisons between results difficult. Although the purpose of this section is not to present a review of methods, it is useful to present examples that illustrate the diversity of approaches used to quantify rice sheath blight injury in the field.

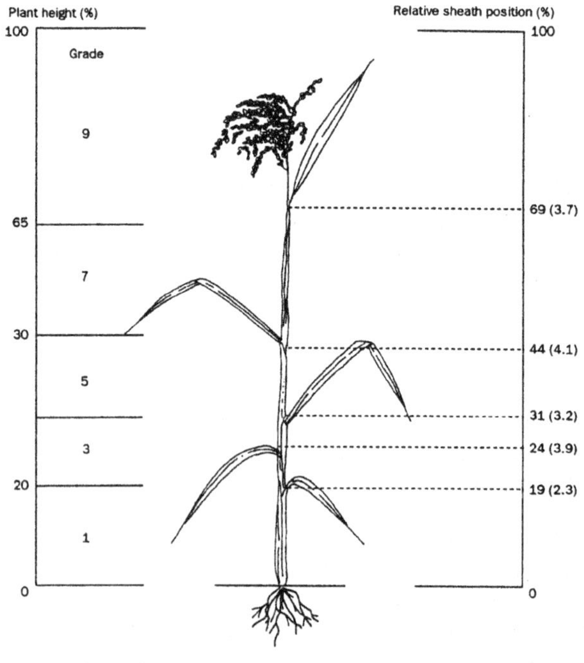

FIGURE 2. An example of assessment technique for intensity of rice sheath blight: the relative lesion height (from Ahn *et al.*, 1986).

a. A first approach consists in the development of formulas based on categorized levels of disease intensity on plant units (whole plants or tillers). For instance, Ono (1953) developed an equation based on five levels of severity at the plant level. In a similar way, Yoshimura (1954) used a formula where four categories of tillers in a plant population are defined on the basis of the number of sheaths in a tiller that are infected.

b. A second approach consists in considering the progress of the disease towards the upper part of the canopy. The concept of relative lesion height was used by Hashiba (1984) and Ahn *et al.* (1984) to measure disease intensity (Fig. 2). This approach makes sheath blight severity directly dependent on the upwards disease progress, relative to the height of the infected plant. This is linked to the empirical knowledge that (a) sheath blight can only spread in a rice field, and therefore reach high intensity, if the pathogen is capable to access the upper part of the canopy where most of the contacts among neighboring tillers and plants occur, and (b) damage is linked to the amount of injury inflicted to the upper part of the canopy, especially the flag leaves.

c. A third approach, which has seldom been followed, consists in considering the experimental plot as a whole, and assess the distribution and the intensity of the disease simultaneously. Marchetti and Bollich (1991) developed a disease index based on visual estimates of the proportion field area where the disease is rated at prescribed disease levels. This approach has definitely a strong appeal as it makes justice to the spatial distribution of the disease.

This range of methods has been used for many purposes: assessment of varietal resistance, measurement of the efficiency of control measures (predominantly, fungicides), surveys in farmer's fields, epidemiological studies on rice sheath blight, and crop loss studies. An important issue is that accurate measure of injury is crucial for an adequate representation of damage. Ideally, in rice sheath blight, the measurement of injury should take into account (a) the density of lesions on the host plants, (b) the distribution of symptoms in the vertical dimension of the canopy, and (c) the spatial extent of symptoms in the crop.

In some cases, some methods are applied for two objectives at a time. For instance, the relative lesion height technique, translated into a 9-point rating system (the SES scale, IRRI, 1988) has been widely used to compare the resistance of rice cultivars to ShB, as well as assess crop losses. Such duplication of objectives may sometimes lead to confusion. Ahn and Mew (1986) however demonstrated that this approach was indeed providing a good representation of injury, and could be closely correlated to damage.

IV. Experimental design to measure damage

Plant pathologists must consider the representativeness of experimental designs when conducting damage (crop loss) estimates (Teng, 1987). One major difficulty in studying damage caused by rice sheath blight is the aggregated nature of the disease: this can only be done in plots that are large enough, and where the disease is uniformly distributed. Another design which would ideally better address the issue consists in large plots where the experimenter would mimic the appearance of disease foci at various densities, and follow their expansion.

V. Damage-loss relationships

The importance of rice sheath blight was recently reviewed by Ou (1987), Gangopadyay and Chakrabarti (1982), and Premalatha Dath (1990). The magnitude of damage reported varies considerably from one study to another. Thus, among experimental designs and environments. It seems that a maximum relative damage of approximately 30 to 50% is a quite common observation. Whether this represents actual situations in farmers' fields is difficult to assess. In most cases, relative damage (i.e., proportion or percentage of the yield of a disease-free control) is reported, rather than absolute damage (i.e., the yield difference itself). This hides the fact that damage and attainable yield are not independent, and this may have considerable implications.

Several disease management systems, based on simulation of epidemiological processes, of injury-loss relationships, and on various management options have been developed, especially in Japan (Hashiba, 1984; Hashiba and Ijiri, 1989) and in China (Fan *et al.*, 1992). To a large extent, these systems are based on site-dependent information, and sometimes heavily rely on regression, or regression-like approaches. While these studies represent a wealth of information on epidemiological processes and on damage-loss relationships, whether they could be extrapolated to new sites remains to be shown.

Over the past years, IRRI has been conducting a series of experiments where several input factors (fertilizer level, water management, crop establishment method, potential yield of cultivars) and 'pests' (in the broadest sense: pathogens, insects, and weeds) are simultaneously manipulated. These experiments are conducted in rainy as well as in dry seasons, in farmers' fields or at IRRI's experimental farm. A typical experiment would for instance involve two input factors (say, three water regimes and two fertilizer levels), producing six main units, and three pests, each of them at two possible levels: introduced or not. The plots where no pests are introduced provide a measure of the attainable yield, i.e., the yield a crop would produce in a given production situation in the absence of injury (Savary and Zadoks, 1992). Sheath blight is among the nine pests considered so far in these experiments, along with fungal (brown spot), bacterial (bacterial blight), viral (tungro) diseases, three types of insect pests, and two types of weed species. Therefore it is possible to address sheath blight damage in the context of combination of injuries - a rule rather than an exception in farmers' fields (Savary *et al.*, 1994). In these experiments, sheath blight injury is measured as the maximum severity (% sheath and leaf area covered by lesions) observed in the course of a cropping season. Currently, results are available pertaining to a little more than 400 individual rice plots, covering a range of 2 to 11 tons/ha in attainable yield and a range of pest combinations. Results from these experiments are briefly described here, concentrating on sheath blight only.

In terms of yield reductions (Fig. 3) increased sheath blight injury corresponds to increasing damage. For example, for a 5 tons/ha attainable yield, the damage is 0.18 tons/ha at a 10% injury level, while it is 0.94 tons/ha at 50%. Initial damage (at, say, 10% injury level) also increases with increasing attainable yield: from 0.10 tons/ha at an attainable yield of 3 tons/ha, it becomes 0.34 tons/ha at 5 tons/ha. In the same way, at 30% injury level the damage increases from 0.42 to 1.62 tons/ha with attainable yield increasing from 5 to 9 tons/ha. When considering relative damage (i.e., damage expressed as % of the attainable yield), a more than proportional increase with increasing attainable yield is indicated (Fig. 4). Among the various injuries considered so far in these experiments, damage caused by sheath blight is the only one to increase in relative terms with increasing attainable yield. It contrasts strongly, for instance, with tungro injury which tends to decrease in relative terms with increasing attainable yield. These results therefore indicate that, although sheath blight injury may cause relatively little damage at low attainable yield, the same level of injury can cause considerable damage at high attainable yield.

VI. AN EMPIRICAL APPROACH TO THE FUTURE

As a conclusion, we offer here two examples where better insight in the pathosystem could help assessing damage - either current or potential. This example may suffice to explain why crop loss assessment is a key for disease management.

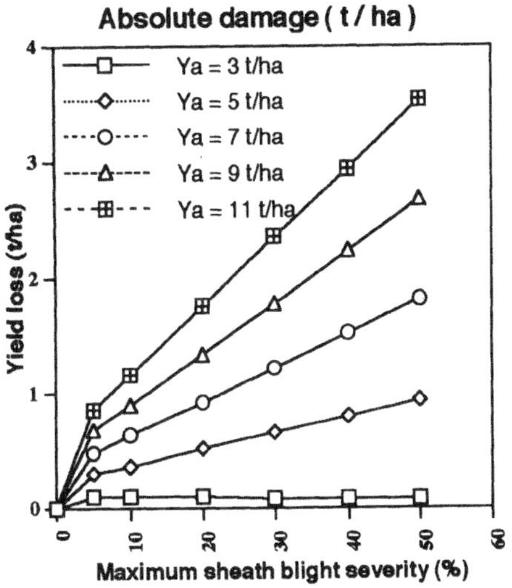

FIGURE 3. Variation in absolute damage (tons/ha, vertical axis) due to rice sheath blight injury (maximum severity in the course of a cropping season in %, horizontal axis) with varying attainable yield (Ya). Five curves are shown, with attainable yield varying from Ya = 3 tons/ha to Ya = 11 tons/ha.

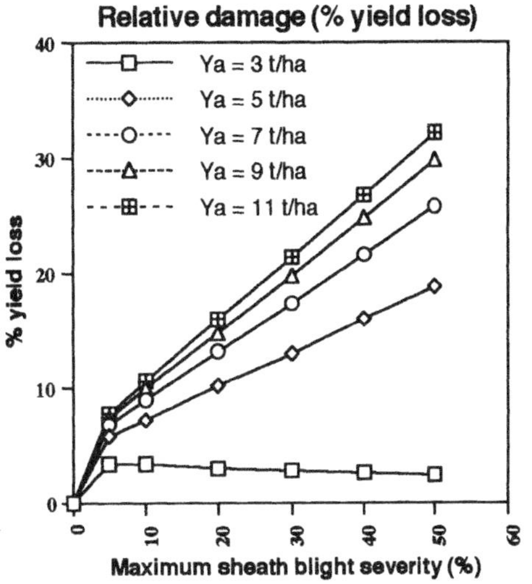

FIGURE 4. Variation in relative damage (% yield loss, vertical axis) due to rice sheath blight injury (maximum severity in the course of a cropping season in %, horizontal axis) with varying attainable yield (Ya). Five curves are shown, with attainable yield varying from Ya = 3 tons/ha to Ya = 11 tons/ha.

A. Perspective 1: a new rice plant type

Researchers are contemplating the deployment of a new rice plant type that would have fewer, but fertile, tillers; it would also have fewer, but longer, stronger, and long-lived leaves. The new type would be especially adapted for direct-seeding crop establishment with high fertilizer inputs (Vergara *et al.*, 1990). From the plant pathological viewpoint, and specifically in terms of vulnerability to sheath blight, the deployment of cultivars of this type may have consequences in terms of both injury and damage.

a. A relatively high leaf area index would be achieved with a smaller number of leaves per plant. A smaller number of leaves per field unit area would translate into fewer contacts between plants, thus in a reduced capacity of the pathogen to spread. Possibly, the crop micro-climate would be slightly different only from that of a crop planted to a 'conventional' variety. It may therefore be anticipated that the spread of the disease would be reduced, not primarily because of microclimatic reasons, but as a result of reduced ability of the pathogen to disperse. As a result, injury should not be anticipated to be higher in this crop as compared to conventional ones. One might even expect a reduced injury.

b. The number of tillers plays a key role in the yield of the new type: nearly each tiller, being potentially fertile, contributes to yield, and therefore matters. There would be little room to maneuver and little compensation capacity among tillers. This might apply, albeit to a lesser degree, to the few, efficient, long-lived leaves. As a result, the new type might be very sensitive in terms of damage to low levels of injury.

The possible outcome of the deployment of this new plant type is thus difficult to predict. It is possible that sheath blight could cause substantial damage, even at low injury levels. This example illustrates our strong need for a modeling approach that would operate at the process level, so that yet-to-come relationships in the pathosystem could be explored.

B. Perspective 2: changes in cropping intensity

Intensification in rice production is a current process. It involves, among key components: (a) an increase in cropping intensity; (b) an increase in fertilizer supply; (c) a permanent, reliable access to irrigation water; (d) short cycle, semi-dwarf, high yielding cultivars. This leads to (a) the build-up of primary inoculum; (b) increased leaf area index based on a larger number of leaves per plant at any point of time in the cropping season, i.e., increased contact frequency in the crop and increased efficiency of inoculum in a darker, more humid canopy (Savary *et al.*, 1995); (c) an easy access of the pathogen to the upper layer of the canopy, and therefore, and easier spread. From the injury view point, therefore, there is every reason to see the current increase in injuries wherever intensification takes place. On the other hand increasing attainable yield is accompanied by higher damage (Fig. 3) in a more-than-proportional pattern (Fig. 4).

The overall result is an unsustainable situation, where farmers are compelled to spray fungicides. Many Vietnamese farmers are now considering sheath blight a key constraint, with an increased reliance on chemical control (Mai *et al.* 1993). The sole reliance on a single management technique has too often proven to be unreliable (Teng *et al.*, 1993). Here again, simulation modeling should be used as a strategic tool to assess management alternatives - which are many (Mew, 1991), and minimize this dependency. Such alternatives involve a better crop husbandry, including management of weeds, and new options for control itself, including biological control (Mew and Rosales, 1986).

VII. References

Ahn SW, de la Peña R., Candole BL & Mew TW (1986) A new scale for rice sheath blight (ShB) assessment. Int. Rice Res. Newsl. 11:17.

Ahn SW & Mew TW (1986) Relation between rice sheath blight (ShB) and yield. Int. Rice Res. Newsl. 11:21-22.

Damodaram Naidu V, Srinivasa Rao B & Murty PSS (1981) Influence of sheath blight infection on the levels of chlorophyll and $^{14}CO_2$ uptake in rice. Indian Phytopathol. 34:30-33.

Ellings A (1994) Structure and development of blight, a model to simulate the effects of bacterial leaf blight and sheath blight on rice. In: Ellings A and Rubia EG (eds.) Analysis of damage mechanisms by pests and diseases and their effects on rice yield. (pp 9-30) Systems Analysis for Rice Production research proceedings.

Fan K, Peng S, Zou H, Li B, Kang X, Jian H & Xiang X (1992) A simulation model for the prediction and management of rice sheath blight caused by *Rhizoctonia solani* SIMPMRS. Chinese J. Rice Res. 6:165-172.

Gangopadyay S & Chakrabarti NK (1982) Sheath blight on rice. Rev. Plant Pathol. 61:451-460.

Hashiba T (1984) Forecasting model and estimation of yield loss by rice sheath blight disease. Jpn. Agric. Res. Q. 18:92-98.

Hashiba T & Ijiri T (1989) Estimation of yield loss and computerized forecasting system (BLIGHTAS) for rice sheath blight disease. Trop. Agric. Res. Ser. No. 22:163-172.

International Rice Research Institute (1988) Standard Evaluation System for Rice, 3rd Edition. International Rice Testing Program, IRRI, Los Baños, 54p.

International Rice Research Institute (1993) IRRI Rice Almanac 1993-1995. IRRI, Los Baños, Philippines, 142 p.

Kozaka T (1961) Ecological studies on sheath blight of rice plant caused by Pellicularia sasakii (Shirai) S. Ito, and its chemical control. Chugoku Agric. Res. 20:1-133.

Mai V, Cuc NT, Hung NQL, Chau NM, Chien HV, Escalada MM & Heong K (1993) Farmers' perceptions of rice pest problems and management tactics used in Vietnam. International-Rice-Research-Notes 18:31.

Mew TW (1991) Disease management in rice. In: CRC Handbook of Pest Management in Agriculture (pp 279-299) 2nd Edition. Vol. III; CRC Press, Inc. Boca Raton, Ann Arbor, Boston.

Mew TW & Rosales AM (1986) Bacterization of rice plants for control of sheath blight caused by *Rhizoctonia solani*. Phytopathology 76:260-1264.

Ono K (1953) Damage of rice sheath blight to rice. J. Pl. Prot. Tokyo 7:99-103.

Ou SH (1987) Rice Diseases. Second Edition. C.A.B. International, Farnham House, Farnham Royal, Slough. 380 pp.

Premalatha DA (1990) Sheath Blight Disease of Rice and its Management. Associated Publishing Company, New Delhi. 129p.

Roy AK (1982) Effect of sheath blight infection on respiration and transpiration of rice plants. Int. Rice Res. Newsl. 7:20-21.

Savary S (1991) Approches de la Pathologie des Cultures Tropicales. Exemple de l'Arachide en Afrique de l'Ouest. Editions Karthala / ORSTOM, Paris. 288 pp.

Savary S & Zadoks JC (1992) Analysis of crop loss in the multiple pathosystem groundnut-rust-leaf spot. I. Six experiments. Crop Prot. 11:9-109.

Savary S, Elazegui FA, Moody K Litsinger JA & Teng PS (1994) Characterization of rice cropping practices and multiple pest systems in the Philippines. Agric. Syst. 46:385-408.

Savary S, Castilla NP, Elazegui FA, McLaren CG, Ynalvez MA & Teng S (1995) Direct and indirect effects of nitrogen supply and disease source structure on rice sheath blight spread. Phytopathology 85:959-965

Teng PS (1987) Quantifying the relationship between disease intensity and yield loss In: PS Teng, (ed.). Crop Loss Assessment and Pest Management (pp 105-113), APS Press, St Paul Minnesota.

Teng PS, Savary S & Revilla I (1993) Systems of plant protection. In: Crop Protection and Sustainable Agriculture (pp 116-139). Ciba Foundation Symposium 177. Wiley, Chichester..

Vergara BS, Venkateswarlu B, Janoria M, Ahn JK, Kim JK & Visperas RM (1990) Rationale for a low-tillering rice plant type with high density grains. in: Direct-Seeded Flooded Rice in the Tropics. (pp 39-53), Selected papers from the International Rice Research Conference, 27-31 Aug. 1990, Seoul, Korea. IRRI P.O. Box 933, 1099 Manila, Philippines.

Yoshimura S (1954) On the scale for estimating degree of severity of sheath blight by *Hypochnus sasakii* Shirai in rice plants. Ann. Phytopathol. Soc. Jpn. 9:58-60.

Zadoks JC (1974) The role of epidemiology in modern phytopathology. Phytopathology 64:818-823.

Zadoks JC (1985) On the conceptual basis of crop loss assessment: the threshold theory. Annu. Rev. Phytopathol. 23:455-473.

Section V.

CHARACTERIZATION OF *RHIZOCTONIA* SPP. ISOLATES, DISEASE OCCURENCE AND MANAGEMENT IN VARIOUS CROPS

V.1 OILSEED RAPE AND CANOLA DISEASES INCITED BY *RHIZOCTONIA* SPECIES

PRITHWI RAJ VERMA

Agriculture and Agri-Food Canada Saskatoon Research Centre 107 Science Place Saskatoon, Saskatchewan CANADA S7N 0X2

I. INTRODUCTION

Rhizoctonia solani infects many members of the Brassicaceae (Flentje *et al.*, 1963; Akhtar, 1969; Chauhan and Saksena, 1974; Tahvonen *et al.*, 1984; Gupta, 1985; Huber *et al.*, 1992). Pre- and post emergence damping-off, and root rots of seedling or adult plants are serious diseases of oilseed rape (*Brassica napus* L.) and turnip rape (*B. rapa* L.) (rapeseed) in western Canada (Berkenkamp and Vaartnou, 1972; Rimmer and Platford, 1982; Ellis, 1983; Sippell *et al.*, 1985b). Although disease severity varies depending on the region and weather conditions at the time of spring planting, most rapeseed crops in western Canada are affected to some extent each year. In Manitoba and Saskatchewan, the incidence of root rot has generally been low (Petrie and Vanterpool, 1970; Rimmer and Platford, 1982). In Alberta, however, root rot has been recognised as a serious disease for at least 20 years. Partial to nearly complete losses of plant stands, with disease incidence as high as 80-100%, have been recorded in some fields in the Peace River region of northern Alberta (Ellis, 1983; Sippell *et al.*, 1985a, b). The predominant type of root rot in the Peace River region is called "brown girdling root rot" due to the production of large, sunken light brown lesions circumscribing the tap root and lateral roots. Girdled plants may be left with only a short stub of the tap root and are vulnerable to desiccation and lodging. Brown girdling root rot causes poor pod development, pod sterility, shrivelled seed, low seed weight, lodging and uneven ripening of plants. Surveys conducted in the Peace River region of northern Alberta in 1983 and 1984 estimated annual yield losses of 30% due to this disease (Sippell *et al.*, 1985a).

Although *Fusarium* spp., and *Pythium* spp. are also associated with rapeseed root rot, *R. solani* is the principal pathogen responsible for damping-off and seedling and adult plant root rot in western Canada (Petrie, 1973; Vanterpool, 1974; Acharya *et al.*, 1984; Kaminski and Verma, 1985; Sippell *et al.*, 1985b; Calman *et al.*, 1986; Calman and Tewari, 1987; Gugel *et al.*, 1987). This is also true in Finland, where damping-off due to *R. solani* occurred in 90% of the spring oilseed rape fields surveyed in 1981 and 1982 (Tahvonen *et al.*, 1984).

In addition to root and collar rot, *R. solani* also causes a severe leaf blight of oilseed rape and mustard [*B. juncea* (L.) Czern.] in India. The diseased leaves develop brownish patches with chlorotic margins and abscise prematurely (Chauhan and Saksena, 1974; Gupta, 1985). The leaf blight phase of *R. solani* has not been reported from temperate countries.

B. Sneh et al. (eds.),
Rhizoctonia Species: Taxonomy, Molecular Biology, Ecology, Pathology and Disease Constrol, 249–258.

II. THE PATHOGENPR VERMA

A. Characterization of *R. solani* populations

In western Canada, *R. solani* isolates recovered from field soils, and from diseased rapeseed seedlings and adult plants were characterized into anastomosis groups (AGs). Of the 83 *R. solani* isolates obtained from diseased seedlings and adult plants from Saskatchewan, 48% were assigned to AG 2-1, and 50% to AG 4, while the remaining 2% did not anastomose with any AG tester (Kaminski and Verma, 1985). Of the 69 *R. solani* isolates from west central Alberta, 41% belonged to AG 2-2, 32% to AG 2-1, and 14% to AG 4; the remaining 13% were not assignable to any tester AG (Hwang *et al.*, 1986). Gugel *et al.* (1987) found that all 68 *R. solani* isolates obtained from diseased seedlings and adult plants from the Peace River region of Alberta belonged exclusively to AG 2-1; however, isolates recovered from Peace River soils were characterized as 59% AG 2-1, 25% AG 4 and 16% other AGs. The soil-borne population of *R. solani* in Saskatchewan infested rapeseed fields belonged to AG 2-1 (36%), AG 2-2 (6%), AG 3 (4%), AG 4 (53%) and AG 5 (2%). In Saskatchewan, the AG 4 isolates were detected in soils of all textures, but AG 2-1 isolates were confined to loam, silty clay loam, and silty clay soils. In these surveys, isolates from field-infected seedlings belonged to AG 2-1 and AG 4, with frequencies of 94 and 6%, respectively. Among the isolates from field-infected adult plants, 69% were AG 2-1, 29% were AG 4 and 2% were AG 2-2 (Yitbarek *et al.*, 1987). All *R. solani* isolates from rapeseed in Indiana, USA belonged to AG 2-1 (Huber *et al.*, 1992).

B. Cultural Characteristic

Colonies of *R. solani* AG 2-1 from rapeseed plants growing on potato dextrose agar (PDA) are light brown with concentric zones of aerial mycelium; sclerotia are distributed in the zones of aerial mycelia. Colonies of AG 4 on PDA are white to brown-grey with a leathery appressed or mealy textured mycelium, and form very few sclerotia around the inoculum disc. The mean growth rates of AG 2-1 and AG 4 isolates from rapeseed were 0.34 and 0.51 mm/h, respectively, at 18°C (Hwang *et al.*, 1986; Kaminski and Verma, 1985).

C. Pathogenicity of *R. solani* isolates

R. solani isolates differ considerably in virulence. Isolates of AG 2-1 from diseased rapeseed seedlings and adult plants were generally more virulent than those of AG 2-2 and AG 4, although there is a wide range of virulence among isolates within the same AG (Gugel *et al.*, 1987; Kaminski and Verma, 1985; Yitbarek, 1987; Yitbarek *et al.*, 1987). Isolate virulence was influenced by their source. Isolates obtained from seedlings and adult plants were more virulent than isolates from soil of the same AG. Similarly, seedling isolates of AG 2-1 and AG 4 were more virulent than isolates from adult plants. In petri plate tests, soil isolates induce less severe hypocotyl and radicle infection than seedling and adult plant isolates; seedling isolates prevented seed germination more than the isolates from adult plants.

AG 2-1 isolates are mainly responsible for pre- and post-emergence damping-off and seedling root rot in all rapeseed growing areas in western Canada and in the state of Indiana in the United States. AG 4 isolates mainly attack adult plants and cause basal stem rot late in the growing season. AG 2-2 isolates, although pathogenic on rapeseed seedlings, are not as prevalent and wide-spread as AG 2-1 and AG 4 isolates in rapeseed fields (Gugel *et al.*, 1987; Kaminski and Verma, 1985; Yitbarek *et al.*, 1987).

D. Infection Process

The sequence of events in the infection process is similar in both susceptible and resistant hosts. Infection cushions are formed on both susceptible and resistant hypocotyls, but are more frequent and form earlier on susceptible than on resistant hypocotyls. In hypocotyls of the susceptible cv. Westar, hyphal growth and infection cushions were observed within 24h after inoculation, with cell maceration occurring within two days compared to resistant cv. Arda where infection cushions did not occur until 48-72h, and cell maceration required three days.. Infection develops more rapidly and penetrates deeper in susceptible than in resistant hosts. In all tissues of Arda, hyphae grew more slowly, branched less frequently, oriented irregularly, and were confined largely to the parenchymatous tissue of the cortex. Hyphae were not observed in endodermis or vascular tissues five days after inoculation (Yang, 1989; Yang *et al.*, 1992b)..

E. Factors affecting disease development

The incidence and severity of pre-and post-emergence damping-off and root rot in rapeseed fields is influenced by weather and soil conditions at the time of spring planting and by the inoculum density of the predominant anastomosis groups of *R. solani* in the field soil. Seedling infection by AG 2-1 isolates is favoured by cool weather at the time of spring planting, whereas warm weather in summer is more conducive for infection of mature plants by AG 4 isolates.

Kaminski and Verma (1985) found that AG 2-1 isolates grew at 2°C but not at 36°C with optimum growth observed at 24°C. The optimum growth temperature for AG 4 isolates was 26°C with no growth at 2°C and some growth at 36°C. Temperatures in the range of 18-30°C support luxuriant growth of both AGs, and AG 4 isolates have a higher growth rate than AG 2-1 isolates at these temperatures.

Yitbarek *et al.* (1988) in a factorial growth chamber experiment comparing virulence of AG 2-1 and AG 4 isolates, found that night/day air temperatures, isolate, inoculum density, and all possible interactions among these factors had a significant effect on the incidence and severity of pre-emergence damping-off in rapeseed. In general, the incidence of pre-emergence damping-off increased progressively with increased inoculum density, but the virulence of both isolates was governed principally by temperature. Isolate AG 2-1 caused significantly more pre-emergence damping-off at all inoculum densities (1400 through 22800 viable propagules/L soil-less potting mix) at the lower night/day temperature of 7/8 and 7/12°C, as well as at the lower inoculum densities (1400-5700 viable propagules) in the 7/18°C regime. In contrast, more pre-emergence damping-off was caused by AG 4 isolate at all inoculum densities in the 26/35°C temperature regime, and at the lower inoculum densities (1400-5700 viable propagules) under the 12/18 and 19/25°C regimes. Both AG 2-1 and AG 4 isolates induced a similar incidence of pre-emergence damping-off at the higher inoculum densities (11400-22800 viable propagules) in the 7/18, 12/18 and 19/25°C regimes.

Field experiments in Saskatchewan in 1985 and 1986 demonstrated that the difference in virulence between AG 2-1 and AG 4 isolates is primarily dependent on soil temperature. Teo *et al.* (1988) showed that early planting of rapeseed, before the soil had warmed to 15°C, was more favourable to damping-off by AG 2-1 than AG 4, while damping-off due to AG 4 increased as the soil warmed to 20°C.

The influence of soil temperature on the virulence of AG 2-1 and AG 4 isolates is also reflected in the frequency with which these AGs were isolated from diseased seedlings and mature rapeseed plants (Yitbarek *et al.*, 1987). Isolates from seedlings were almost exclusively of AG 2-1, while AG 4 were isolated almost five times more frequently from mature plants than from seedlings.

Soil moisture had no consistent effect in field tests on AG 2-1 or AG 4-induced pre-emergence damping-off in rapeseed (Teo *et al.*, 1988). Generally, the AG 2-1 isolate was significantly more virulent than the AG 4 isolate with early seeding and at high soil moisture. Surveys of brown girdling root rot conducted in 50-70 fields annually from 1983 through 1985 in the Peace River region of Alberta also reported that disease development was favoured by high rainfall and high soil moisture (Sippell *et al.*, 1985a). Disease incidence and severity were highest in 1983, intermediate in 1984, and lowest in 1985, which corresponded with declining annual rainfall from 318 mm in 1983, 288 mm in 1984, and 161 mm in 1985. These surveys also recorded higher root-rot severity in fields with low levels of nitrogen, phosphorus and potassium, and in fine-textured soils with high levels of copper.

III. Control Measures

A. Plant resistance

1. Screening plant lines for resistance. All rapeseed/canola cultivars currently grown in Canada are susceptible to *R. solani* (Acharya *et al.*, 1984; Yang, 1989; Yang and Verma, 1992). Using a mixture of 35 noncharacterized isolates of *R. solani*, Acharya *et al.* (1984) evaluated 300 *B. napus* and *B. rapa* genotypes in a growth chamber, and found 10 lines, five from each of the two species, that were significantly more resistant to *R. solani* than commercial cultivars. All 10 lines were later found to be highly susceptible to a canola isolate of AG 2-1 (Yang, 1989). Harrison (1987) also evaluated over 200 *B. rapa* and *B. napus* accessions in a naturally-infested field nursery and found significant differences in disease severity among the accessions.

Yang and Verma (1992) evaluated 122 cultivars/lines of *B. rapa*, *B. napus*, *B. juncea, B. nigra* (L.) Koch, *B. tournefortii* Gouan, *B. carinata* Braun, *B. oleracea* L. var. *capitata* L., *B. oleracea* L. var. *botrytis* L., *Sinapis alba* L., *Raphanus sativus* L., and *Camelina sativa* L. in a growth chamber for resistance against an isolate of *R. solani* AG 2-1; 26 selected genotypes were also evaluated in field plots artificially-infested with AG 2-1. None of the 122 cultivars/ lines were immune, although there were significant differences in susceptibility among and within genera and species. Of the 11 species tested, all eight lines of *S. alba* and one line of *C. sativa* exhibited elevated resistance to *R. solani*. Moderate resistance in some of the hosts appeared to be genetically controlled and could be utilized in breeding programmes. Although interspecific crosses between *S. alba* and the *Brassica* oilseed crops are very difficult to obtain, it may be possible to transfer the *R. solani* resistance from *S. alba* to other commercial *Brassica* species. The report by Van Lelivelt *et al.* (1989) of the transfer of beet cyst nematode resistance from *S. alba* to *B. napus,* and successful interspecific hybridizations between *B. juncea* and *S. alba* (Mohapatra and Bajaj, 1987), and between *B. napus* and S. alba (Ripley and Arnison, 1990) using embryo rescue strengthen this possibility.

In a growth chamber test, Kataria *et al.*, (1993) screened 16 cultivars of six *Brassica* species against an isolate of *R. solani* AG 2-1. Pre-emergence damping-off was most severe in *B. carinata, B. rapa, B. oleracea,* and *B. napus*, and least severe in *B. juncea* and *B. nigra;* lower pre-emergence damping-off in *B. juncea* and *B. nigra* may be related to faster emergence of seedlings in these two than the other four species. Although all six species suffered severe post-emergence seedling root rot, more occurred in *B. juncea* and *B. nigra* than in the four other species. Significant differences in the levels of susceptibility/resistance to pre-and post-emergence damping-off occurred among cultivars within the *Brassica* species.

2. Effect of age of host plant. Rapeseed plants can be infected by isolates of *R. solani* AG 2-1 and AG 4 at all growth stages. However, the plants acquire some resistance to infection as they age. Yang and Verma (1992) compared the disease severity ratings of nine rapeseed and mustard cultivars following inoculation at 0, 3, and 5 weeks after planting, and found that the older the plant at the time of inoculation, the lower the disease rating; differences in disease ratings between 3 and 5 weeks were highly significant. Gugel *et al.* (1987) also reported a significant decline in disease ratings for both AGs of *R. solani* with increasing plant age; this decline was particularly marked with the isolate of AG 2-1.

3. Effect of bean endochitinase gene in transgenic rapeseed. Rapeseed plants containing high constitutive levels of bean endochitinase are more resistant to infection by *R. solani* than are plants that lack the chimeric chitinase gene through stronger root development and increased survival upon infection (Broglie *et al.*, 1991). Since reduced fungal biomass, increased hyphal alteration and lysis, and chitin breakdown were common features in transgenic rapeseed plants, it was concluded that the constitutive expression of the bean endochitinase gene was partly responsible for the protection against *R. solani* (Benhamou *et al.*, 1993).

4. Antimicrobial compounds/phytoalexins. *Camelina sativa* is moderately resistant to *R. solani* AG 2-1 (Yang and Verma, 1992), possibly due to the presence of antimicrobial compounds in the roots (Conn *et al.*, 1994). Four antimicrobial compounds capable of inhibiting mycelial growth of *R. solani* were extracted and purified from *C. sativa* roots. Of the four compounds, 10-methylsulfinyl decylisothiocyanate and methyl 1-methylindole-3-carboxylate were produced in *C. sativa* roots only, while the phytoalexins, camalexin and methoxycamalexin were produced in roots, stems, leaves, and siliquae.

5. Role of the plant cuticle. The plant cuticle constitutes the first defensive barrier to pathogen penetration. A thick cuticle may increase the mechanical strength of the hypocotyl epidermal layer, reduce the speed of infection, and give the plant more time to mobilize its defenses following recognition of the pathogen on the host plant surface. Cuticle penetration by *R. solani* is most likely due to a mechanical force created by the growth of infection cushions. A thicker cuticle can be important in resisting this mechanical force and preventing pathogen penetration.

Yang *et al.* (1992a) conducted studies to determine whether differences in cuticle thickness and morphology of epicuticular wax exist between resistant and susceptible genera. As *B. napus* and *S. alba* plants develop from 1 to 3 weeks of age, they become increasingly resistant to *R. solani* AG 2-1 seedling root rot (Gugel *et al.*, 1987; Yang and Verma, 1992). One-week old seedlings of *S. alba* were invariably more resistant than those of *B. napus*. *B. napus* seedlings did not show an obvious cuticle layer, but at 3 weeks a cuticle was present with a concomitant increase in resistance to *R. solani*. Three-week-old plants of *S. alba* had a much lower disease severity rating and a significantly thicker cuticle layer than *B. napus* plants of the same age. These results suggest that the cuticle plays an important role in the resistance of *S. alba* and older plants of *B. napus* to infection by *R. solani*.

Hypocotyl surfaces from 3-week-old plants of susceptible *B. napus* cvs. Westar and Midas, and resistant *S. alba* cv. Arda, were covered with an amorphous layer of two types of wax crystals (Yang *et al.*, 1992a). There was no apparent difference among the three cultivars in the type of wax deposits present, but there appeared to be more wax deposits on susceptible *B. napus* than on resistant *S. alba* hypocotyls. Whether wax is a major cuticle component in seedling hypocotyls requires further investigation, but the finding of fewer wax deposits on *S. alba*, a resistant cultivar, suggests that wax may not be important in affording resistance to *R. solani* hypocotyl rot at the seedling stage.

6. Host calcium content and resistance. More calcium oxalate crystals were found on hypocotyls of the susceptible *B. napus* cv. Westar than on hypocotyls of resistant *S. alba* cv. Arda and *C. sativa*; one-week-old plants had more calcium oxalate crystals than 3-week-old plants (Yang *et al.*, 1993). There was no significant difference in total calcium content in non-infected hypocotyl tissues among the three plant genera, and the calcium content did not increase with increasing age of plant tissue. The production of oxalic acid in culture and the formation of calcium oxalate crystals in infected tissues suggests that calcium in plant cells is sequestered by oxalic acid produced by *R. solani*. Oxalic acid secreted by *R. solani* may play a role in pathogenesis, but calcium content of the host is not associated with resistance in *S. alba* and *C. sativa*. Some other factors, that can inhibit growth of the pathogen and prevent oxalic acid production may have a greater role in generic and age-related resistance. In resistant hosts, phytoalexins produced as a result of infection may be involved. *Rhizoctonia*-resistant crucifers produce larger amounts and different kinds of phytoalexins and antimicrobial compounds than susceptible crucifers (Conn *et al.*, 1994; Tewari *et al.*, 1988). These can inhibit the growth of the fungus and may lead to less oxalic acid production and calcium sequestration in resistant plants.

B. Chemical Control

The lack of adequate genetic resistance to *R. solani* in currently grown rapeseed and canola cultivars, and the absence of cultural methods for suppressing *R. solani* populations in the field, caused research to be directed towards investigating chemical control. A systematic screening of fungicides against *R. solani* damping-off and seedling root rot was initiated at the Agriculture and Agri-Food Canada Research Centre at Saskatoon in 1988.

1. Efficacy of fungicides as seed treatments to control *R. solani* AG 2-1-induced damping-off and seedling root rot in cultivars of various *Brassica* species. Kataria *et al.* (1993) evaluated the efficacy of Benodanil, carboxin, cyproconazole, fenpropimorph, furmecyclox, iprodione, pencycuron and tolclofos-methyl seed treatments against damping-off and seedling root rot in one cultivar from each of *B. rapa*, *B. napus, B. juncea, B. nigra, B. carinata* and *B. oleracea* in soil-less mix infested with an isolate of *R. solani* AG 2-1. All eight fungicides controlled pre-emergence damping-off, but the level of control at the lower rates of fungicides differed among different *Brassica* species. Benodanil, furmecyclox, iprodione and tolclofos-methyl provided very high levels of control against damping-off for all *Brassica* species. Carboxin, cyproconazole, pencycuron and fenpropimorph were less effective against damping-off in *B. oleracea* than in the other five species.

Cyproconazole was also less effective in controlling pre-emergence damping-off in *B. rapa*.

Furmecyclox provided very good control of post-emergence seedling root rot in all species. Benodanil, iprodione and tolclofos-methyl were most effective in *B. napus* and *B. juncea*. Cyproconazole provided only moderate control of root rot in *B. napus* and poor control in the other species tested.

Kataria *et al.* (1993) also evaluated the efficacy of one seed treatment rate of each of the above eight fungicides against *R. solani* AG 2-1-induced pre-emergence damping-off and post-emergence seedling root rot in four cultivars from each of *B. juncea, B. napus,* and *B. rapa*. Cyproconazole at 0.1 g a.i./kg seed, and each of the seven other fungicides at 3 g a.i./kg seed, provided almost complete control of pre-emergence damping-off in all cultivars of all three *Brassica* species with little or no difference in the level of control among cultivars or fungicides.

The eight fungicides showed marked differences in their relative efficacy for control of seedling root rot; efficacy also varied among cultivars within species. There was a significant cultivar x fungicide interaction with most treatments. Fermycyclox, pencycuron, tolclofos methyl, and iprodione provided excellent control of root rot in all cultivars of the three species, although there were some differences in efficacy among the three species and their cultivars. Benodanil and fenpropimorph provided only moderate control, and carboxin and cyproconazole gave poor control against seedling root rot in all cultivars.

Since the six *Brassica* species differ considerably in seed weight and the number of seeds per g, the amount of fungicide applied per seed is also different for each species (Kataria and Verma, 1990). Thus, differences in the level of disease control in different *Brassica* species may be partly due to the difference in the amount of fungicide applied per seed. It was concluded that furmecyclox, iprodione and tolclofos-methyl consistently provided the most effective control of *R. solani* AG 2-1 damping-off and seedling root rot in all *Brassica* species and cultivars tested. Since these fungicides also exhibit a strong activity against AG 4 isolates of *R. solani* (Kataria and Verma, 1990), they should give effective control of both AG 2-1 and AG 4 populations in rapeseed/canola and mustard crops.

2. Interaction of fungicide and insecticide combinations against damping-off and seedling root rot in canola. Carboxin (Vitavax) and iprodione (Rovral) are widely used as a canola seed treatments to control *R. solani* damping-off in western Canada. Carboxin and iprodione are applied to canola seed in mixtures with the insecticide gamma-HCH (Lindane) and marketed as Vitavax RS and Rovral ST, respectively.

A variety of both insecticides and fungicides are used together in canola fields to control both flea beetles (*Phyllotreta* spp.) and root rot. Kataria and Verma (1993) tested two seed dressing insecticides (cloethocarb and gamma-HCH) and two granular insecticides (carbofuran and terbufos) alone and in combination with five seed-dressing fungicides (carboxin, cyproconazole, furmecyclox, iprodione, and tolclofos-methyl) against *R. solani* AG 2-1-induced damping-off and seedling root rot in canola cv. Westar (*B. napus*) in a growth chamber experiment. Each fungicide provided almost complete control of damping-off, but furmecyclox, iprodione, and tolclofos-methyl provided far better control against seedling root rot than carboxin and cyproconazole. Cloethocarb and carbofuran increased the incidence of damping-off and, when used in combination with fungicides, lowered the level of control of damping-off and root rot. Gamma-HCH and terbufos decreased the incidence of damping-off, and terbufos also reduced seedling root rot. Combined treatments with gamma-HCH and terbufos gave better control than either gamma-HCH or terbufos alone. Disease control by fungicides was enhanced by the use of either gamma-HCH or terbufos together with fungicide treatments.

Maximum disease control with insecticide-fungicide combinations was achieved when terbufos granules were applied at the time the pots were sown with fungicide + gamma-HCH-treated seed. Adding terbufos granules to the treated seed 1-4 wk before seeding resulted in a considerable loss of efficacy of each of the five fungicide + gamma-HCH seed treatments. Storing the non-treated seed with terbufos granules also resulted in a higher incidence of pre-emergence damping-off and seedling root rot.

C. Biological

Few studies have investigated the use of biocontrol agents to control damping-off and root rot in rapeseed/canola. In *B. napus* fields in Pakistan, Iqbal, *et al.* (1977) observed that vesicular arbuscular-mycorrhizal plants suppressed root infection by *R. solani*, and Akhtar (1969) obtained substantial control of damping-off in mustard plants grown in soil amended with *Trichoderma harzianum*. Gugel *et al.* (1986) found that an indigenous population of non-pathogenic binucleate *Rhizoctonia* suppressed damping-off and root rot of rapeseed caused by *R. solani* AG 2-1 and AG 4 in growth chamber and field tests. Dahiya and Woods (1987) isolated a strain of *Pseudomonas fluorescens* that produced three antibiotics (procyanin, pyrrolnitrin and phenazine carboxamide) inhibitory to *R. solani;* treating canola seed with *P. fluorescens* controlled damping-off in infested peat-mix soil. Reddy *et al.* (1994) evaluated seven bacterial strains (5 of *Pseudomonas fluorescens;* 1 of *Coryneform* group; and 1 of *Enterobacter agglomerans)* as canola seed treatments in field tests and identified two strains of *P. fluorescens* (63-49 and U-14) that were effective and consistent in suppressing *R. solani* irrespective of the field or year tested when compared to the non-bacterization controls.

IV. SUMMARY

R. solani AG 2-1 is the principal pathogen of damping-off and seedling root rot in oilseed rape and canola in western Canada and the United States; AG 4 isolates mainly attack adult plants and cause stem rot. Seedling infection by AG 2-1 is favoured by cool weather at the time of planting, whereas warm weather late in the growing season is more conducive for infection of mature plants by AG 4 isolates. Survey data show that disease development is favoured by high soil moisture, low levels of nitrogen, phosphorus and potassium and in fine-textured soils with high levels of copper. Carboxin (Vitavax RS) and iprodione (Rovral ST) in mixtures with insecticide gamma-HCH (Lindane) are although recommended seed treatments to control damping-off and seedling root rot, they do not provide control against the mature plant root rot.

V. REFERENCES

Acharya SN, Verma PR, Dueck J & Downey RK (1984) Screening rapeseed/canola for resistance to damping-off and seedling root rot caused by *Rhizoctonia solani*. Can. J. Plant Pathol. 6:325-328.

Akhtar CM (1969) Biological control of *Rhizoctonia solani* causing seed decay and damping-off of mustard. West Pakistan J. Agric. Res. 7:29-52.

Benhamou N, Broglie K, Chet I & Broglie R (1993) Cytology of infection of 35S-bean chitinase transgenic canola plants by *Rhizoctonia solani*: cytochemical aspects of chitin breakdown *in vivo*. Plant J. 4:295-305.

Berkenkamp B & Vaartnou H (1972) Fungi associated with rape root rot in Alberta. Can. J. Plant Sci. 52: 973-976.

Broglie K, Chet I, Holliday M, Knowlton S, Cressman R, Biddle P & Broglie R (1991) Transgenic plants with enhanced resistance to the fungal pathogen, *Rhizoctonia solani*. Science 254:1194-1197.

Calman AI, Tewari JP & Mugala M (1986) *Fusarium avenaceum* as one of the causal agents of seedling blight of canola in Alberta. Plant Dis. 70:694.

Calman AI & Tewari JP (1987) Involvement of *Pythium* spp. in the pre-emergence phase of canola seedling blight. Can. J. Plant Pathol. 9:274.

Chauhan LS & Saksena HK (1974) A new *Rhizoctonia* leaf blight of rapes and mustard. Indian J. Farm Sci. 2:98-99.

Conn KL, Browne LM, Tewari JP & Ayer WA (1994) Resistance to *Rhizoctonia solani* and presence of antimicrobial compounds in *Camelina sativa* roots. J. Plant Biochem. Biotech. 3:125-130.

Dahiya JS & Woods DL (1987) Control of *Rhizoctonia solani*, causal agent of brown girdling root rot of canola/rapeseed by *Pseudomonas fluorescens*. Can. J. Plant Pathol. 9:275-276.

Ellis PJ (1983) Diseases of rapeseed in the Peace River region of Alberta. M.P.M. Thesis. Simon Fraser University, Burnaby, B.C., Canada: 89 p.

Flentje NT, Dodman RL & Kerr A (1963) The mechanisms of host penetration by *Thanatephorus cucumeris*. Aust. J. Biol. Sci. 16:784-799.

Gugel RK, Verma PR & Kaminski DA (1986) Protection of canola seedlings from *Rhizoctonia solani* infection by a nonpathogenic, binucleate *Rhizoctonia*-like isolate. Can. J. Plant Pathol. 8:349.

Gugel RK, Yitbarek SM, Verma PR, Morrall RAA & Sadasivaiah RS (1987) Etiology of the Rhizoctonia root rot complex of canola in the Peace River region of Alberta. Can. J. Plant Pathol. 9:119-128.

Gupta DK (1985) Root and collar rot of mustard in sikkim. Indian J. Mycol. Plant Pathol. 15:225.

Harrison LM (1987) Evaluation of *Brassica* accessions for resistance to *Rhizoctonia solani* and the effects of seeding date, soil temperature and moisture on disease development. M.Sc. Thesis, University of Manitoba, Winnipeg, Canada: 109 p.

Huber DM, Christmas EP, Herr LJ, McCay-Buis TS & Baird R (1992) *Rhizoctonia* crown rot of canola in Indiana. Plant Dis. 76:1251-1253.

Hwang SF, Swanson TA & Evans IR (1986) Characterization of *Rhizoctonia solani* isolates from canola in west central Alberta. Plant Dis. 70:681-683.

Iqbal SH, Kureshi KS & Ahmad JS (1977) Influence of vesicular arbuscular mycorrhiza on damping-off caused by *Rhizoctonia solani* in *Brassica napus*. Biologia 23:197-208.

Kaminski DA & Verma PR (1985) Cultural characteristics, virulence, and *in vitro* temperature effect on mycelial growth of *Rhizoctonia* isolates from rapeseed. Can. J. Plant Pathol. 7:256-261.

Kataria HR & Verma PR (1990) Efficacy of fungicidal seed treatments against pre-emergence damping-off and post-emergence seedling root rot of growth chamber grown canola caused by *Rhizoctonia solani* AG 2-1 and AG 4. Can. J. Plant Pathol. 12:409-416.

Kataria HR & Verma PR (1993) Interactions of fungicide and insecticide combinations against *Rhizoctonia* damping-off and root rot in canola. Ann. appl. Biol. 123:233-246.

Kataria HR, Verma PR & Rakow G (1993) Fungicidal control of damping-off and seedling root rot in *Brassica* species caused by *Rhizoctonia solani* in the growth chamber. Ann. appl. Biol. 123:247-256.

Mohapatra D & Bajaj YPS (1987) Interspecific hybridization in *Brassica juncea* x *Brassica hirta* using embryo rescue. Euphytica 36:321-326.

Petrie GA (1973) Diseases of *Brassica* species in Saskatchewan. 1970-72. III, Stem and root rots. Can. Plant Dis. Surv. 53:88-92.

Petrie GA & Vanterpool TC (1970) Diseases of rape and other crucifers in Saskatchewan in 1969. Can. Plant Dis. Surv. 50:106-107.

Reddy MS, Covert DC, Craig KA, Rennie RJ, McKenzie D, Verma PR & Mortensen K (1994) Suppression of pre-emergence damping-off of canola, caused by *Rhizoctonia solani*, by rhizobacteria. In: Ryder MH, Stephens PM & Bowen GD (Eds.): Improving Plant Productivity with rhizosphere bacteria. (pp 80-82) Proc. 3rd Int. Workshop on Plant growth promoting rhizobacteria, 7-11 March, Adelaide, South Australia.

Rimmer SR & Platford RG (1982) Manitoba rapeseed disease survey 1978-1980. Can. Plant Dis. Surv. 62:45-49.

Ripley VL & Arnison PG (1990) Hybridization of *sinapis alba* L. and *Brassica napus* L. via embryo rescue. Plant Breeding 104:26-33.

Sippell DW, Davidson JGN & Cox M (1985a) Factors affecting severity of root rot of canola in the Peace River region. Proc. 6th Annu. Meet. Plant Pathol. Soc. 7-8 November 1985, (p 34). Lethbridge, Alberta.

Sippell DW, Davidson JGN & Sadasivaiah RS (1985b) *Rhizoctonia* root rot of rapeseed in the Peace River region of Alberta. Can. J. Plant Pathol. 7:184-186.

Tahvonen R, Hollo J, Hannukkala & Kurppa A (1984) *Rhizoctonia solani* damping-off on spring turnip rape and spring rape *(Brassica* spp.) in Finland. J. Agric. Sci. Finland 56:143-154.

Teo BK, Yitbarek SM, Verma PR & Morrall RAA (1988) Influence of soil moisture, seeding date, and *Rhizoctonia solani* isolates (AG 2-1 and AG 4) on disease incidence and yield in canola. Can. J. Plant Pathol. 10:151-158.

Tewari JP, Conn KL, & Dahiya JS (1988) Dynamic response of rapeseed and other crucifers to *Alternaria brassicae* and *Rhizoctonia solani*. Proc. 5th Intern. Congr. Plant Pathol., 20-27 August, 1988: (p 242) Kyoto, Japan.

Van Lelivelt CL, Furderiks HJ & Van Dijk N (1989) Transfer of resistance to beet cyst nematode to Swede rape via interspecific crosses with related species. Plant Breed. Abstr. 59:59.

Vanterpool TC (1974) *Pythium polymastum* pathogenic on oilseed rape and other crucifers. Can. J. Bot. 52:1205-1208.

Yang J (1989) Studies on *Rhizoctonia solani* diseases of rapeseed: germplasm screening, histo-cytopathology, cuticle thickness and calcium content in relation to resistance. M.Sc. Thesis, University of Saskatchewan, Saskatoon, Canada: 141 p.

Yang J & Verma PR (1992) Screening genotypes for resistance to pre-emergence damping-off and post-emergence seedling root rot of oilseed rape and canola caused by *Rhizoctonia solani* AG 2-1. Crop Prot. 11:443-448.

Yang J, Verma PR & Lees GL (1992a) The role of cuticle and epidermal cell wall in resistance of rapeseed and mustard to *Rhizoctonia solani*. Plant and Soil 142:315-321.

Yang J, Verma PR & Tewari JP (1992b) Histopathology of resistant mustard and susceptible canola hypocolyls infected by *Rhizoctonia solani*. Mycol. Res. 96:171-179.

Yang J, Tewari JP & Verma PR (1993) Calcium oxalate crystal formation in *Rhizoctonia solani* AG 2-1 culture and infected crucifer tissue: relationship between host calcium and resistance. Mycol. Res. 97:1516-1522.

Yitbarek SM (1987) Studies of *Rhizoctonia* diseases of rapeseed/canola in Saskatchewan. M.Sc. Thesis, University of Saskatchewan, Saskatoon, Canada:114 p.

Yitbarek SM, Verma PR & Morrall RAA (1987) Anastomosis groups, pathogenicity, and specificity of *Rhizoctonia solani* isolates from seedling and adult rapeseed/canola plants and soils in Saskatchewan. Can. J. Plant Pathol. 9:6-13.

Yitbarek SM, Verma PR, Gugel RK & Morrall RAA (1988) Effect of soil temperature and inoculum density on pre-emergence damping-off of canola caused by *Rhizoctonia solani*. Can. J. Plant Pathol. 10:93-98.

V.2 CHARACTERIZATION OF *RHIZOCTONIA* ISOLATES, DISEASE OCCURRENCE AND MANAGEMENT IN CEREALS

MARK MAZZOLA[1], RICHARD W. SMILEY[2], ALBERT D. ROVIRA[3] AND R. JAMES COOK[4]

[1]*USDA-ARS, Tree Fruit Research Lab, 1104 N. Western Ave., Wenatchee WA 98801 USA.* [2]*Oregon State University, Columbia Basin Agricultural Research Center, P.O. Box 370, Pendleton, OR 97801 USA* [3]*CSIRO Division of Soils, Private Bag No. 2, Glen Osmond, South Australia 5064.* [4]*USDA-ARS, Root Disease & Biological Control Unit, 365 Johnson Hall, WSU, Pullman, WA 99164 USA*

I. INTRODUCTION

Cereal crops are susceptible to infection by several species and inter- and intra-specific groups of *Rhizoctonia*. Symptoms caused by these pathogens are diverse and include preemergence damping-off, root rot, foliar blight, hypocotyl rot and sheath blight. Root diseases are the most damaging to cereals, and therefore, are the primary focus of this paper. In addition, although all cereals are susceptible to infection by *Rhizoctonia* spp., diseases of wheat and barley are emphasized due to the limited number of reports on the diseases and species of *Rhizoctonia* that affect other cereal crops. Diseases of rice caused by *Rhizoctonia* spp. are discussed in the chapter by T. Hashiba.

II. SYMPTOMS AND EIOLOGY OF CEREAL DISEASES CAUSED BY *RHIZOCTONIA*

A. Root diseases

Root diseases of cereals caused by *Rhizoctonia* spp. have been named Rhizoctonia root rot (Pumphrey *et al.*, 1987), Rhizoctonia patch (MacNish, 1984), bare patch (Samuel and Garrett, 1932), stunted patch (Price, 1970), barley stunt (Murray and Nicholson, 1979), purple patch (Hynes, 1933), and crater disease (Scott *et al.*, 1979).

The most acute form of root disease caused by *Rhizoctonia* spp. is characterized above ground by the development of circular patches of stunted and/or chlorotic plants that produce little or no grain. When disease pressure is very high, the patches may coalesce to create the appearance of uniform disease development across the entire field. Low disease pressure results in individual stunted plants interspersed within rows of productive plants, or minor root rot characterized by subclinical root infections that do not cause stunting or other symptoms of reduced plant vigor. Moderate root damage can delay crop maturation up to three weeks even in the absence of patches of stunted plants (Smiley and Wilkins, 1993).

Symptoms on infected roots can vary depending on the causal agent. Infection by *R. solani* AG 8 results in development of sunken brown lesions, and as growth of the fungus progresses, roots may be girdled and severed with the remaining root stub having a "spear-tip" appearance (Wiese, 1987). *R. oryzae* also causes these symptoms at high soil temperatures (>20° C), but at lower temperatures it causes yellowing and

B. Sneh et al. (eds.),
Rhizoctonia Species: Taxonomy, Molecular Biology, Ecology, Pathology and Disease Constrol, 259–267.

necrosis of the cortex without severing the root stele (Smiley and Uddin, 1993; Mazzola *et al.*, 1995). *R. oryzae* and *R. cerealis* can also retard or prevent seminal and lateral root development, a phenomenon that is not caused by isolates of *R. solani* AG 8 (Mazzola *et al.*, 1995). Crater disease occurs on wheat grown on the Springbok Flats, Transvaal, South Africa, under conditions where root growth is inhibited by soil structure (Deacon and Scott, 1985). A unique symptom associated with crater disease is the development of bead-like swellings composed of hyphae of the causal fungus on infected roots of wheat, barley, oats, rye or corn.

B. Seed rot and seedling diseases

Rhizoctonia spp. cause seed rot and preemergence damping-off of cereals. *R. oryzae* caused preemergence damping-off of wheat (Mazzola *et al.*, 1995) and barley (Burton *et al.*, 1988), but isolates of *R. solani* AGs 4, 5 and 8 that caused severe root rot on these hosts did not affect seedling emergence (Burton *et al.*, 1988; Rush *et al.*, 1994; Mazzola *et al.*, 1995). Isolates of *R. cerealis* from wheat following a bluegrass seed crop reduced wheat emergence as much as 40% in greenhouse experiments (M. Mazzola, unpubl.). Some isolates of *R. solani* AGs 2, 4 and 5 cause minor reductions in emergence of corn and sorghum but do not cause extensive root damage (Rush *et al.*, 1994).

C. Culm and stalk diseases

Sharp eyespot is characterized by development of circular to elliptical lesions on the leaf sheath near the base of wheat and barley culms. Although the disease typically results in minimal damage, infections that extend through the pseudostem or culm tissue can result in blight of seedlings or lodging of mature plants, respectively (Wiese, 1987). Causal agents of sharp eyespot on wheat have been reported as *R. solani* AG 4 in Arkansas (Sterne and Jones 1978) and *R. cerealis* in Ohio (Lipps and Herr, 1982) and Europe (Pitt, 1964; Van der Hoeven in Boerema and Verhoeven, 1977; Hollins *et al.*, 1985). Both *R. solani* AG 4 (Ploetz *et al.*, 1985) and *R. cerealis* (Van der Hoeven and Bollen, 1980) were reported to cause sharp eyespot on rye. Hollins *et al.* (1985) reported large ranges in susceptibility of winter wheat cultivars to infection by *R. cerealis*. *R. cerealis* severely damaged winter wheat following bluegrass harvested for amenity turfgrass sod in New York state (R.W. Smiley, unpubl.), and Richardson and Cook (1985) reported that isolates of *R. cerealis* were particularly virulent when wheat followed grass in a rotation. Sharp eyespot of wheat became more prevalent as fungicides and resistant cultivars were introduced to control eyespot caused by *Pseudocercosporella herpotrichoides* (Kapoor and Hoffman, 1984; T. Murray, personal commun.).

R. solani also causes a foliar blight of grain sorghum (Bell *et al.*, 1973), and *R. zeae* causes ear and stalk rot of corn (McGee, 1988).

D. Causal agents of Rhizoctonia diseases of cereals

The species complex *R. solani* has been subdivided into anastomosis groups, with hyphal fusion occurring between isolates of the same AG but not between isolates in different AGs. Methods have recently been developed, including the establishment of pectic zymogram groups (Sweetingham *et al.*, 1986; Neate *et al.*, 1988; MacNish *et al.*, 1993) and vegetatively compatible populations (MacNish and Carling, 1995), to further characterize populations of *R. solani* AG 8 that infect wheat and barley. These techniques provide additional tools to study the ecology and population dynamics of this pathogen.

Considerable variation in the dominant causal agent of Rhizoctonia seed rot and root rot of wheat has been observed across geographic regions. *R. solani* AG 8 has generally been considered the primary causal agent of root rot of wheat and barley in the US Pacific Northwest (Ogoshi *et al.*, 1990; Smiley and Uddin, 1993), South Australia and New South Wales (Neate and Warcup, 1985; Rovira *et al.*, 1986), and the United Kingdom (Burton *et al.*, 1988). Other species and intraspecific groups reported to cause root rot of wheat and barley include: *R. solani* AG 4 and AG 5 in Texas (Rush *et al.*, 1994); *R. oryzae* in the US Pacific Northwest (Ogoshi *et al.*, 1990; Smiley and Uddin, 1993; Mazzola *et al.*, 1995) and United Kingdom (Burton *et al.*, 1988); *R. zeae* in Minnesota (Kuznia and Windels, 1994); and *Rhizoctonia* (W-AK) in Alaska (Leiner and Carling, 1994). Yang *et al.* (1994) determined that highly virulent multinucleate *Rhizoctonia* spp. were more abundant than weakly virulent isolates inside stunted patches and that the converse was true outside these patches. Roberts and Sivasithamparam (1986) suggested that in Western Australia, bare patch of wheat and barley is caused by a complex of fungi composed of multiple anastomosis groups of *Rhizoctonia* spp., including AGs 2-1, 2-2, and 8 and perhaps binucleate *Rhizoctonia* spp., as well as other associated fungi. Harris and Moen (1985) demonstrated the transient existence of *Rhizoctonia* spp. in a rapidly evolving succession of root colonists, and Roberts and Sivasithamparam (1987) showed that combinations of fungi involved in the succession caused more severe root disease than the same fungi used singly. A complex of fungi, as opposed to a single pathogenic agent, was proposed as the cause of the bare patch manifestation of this disease. This differs from the conclusions reached by MacNish and Sweetingham (1993) that each bare patch is dominated by a single zymogram group of *R. solani*.

Damage to wheat and barley caused by *Rhizoctonia* spp. is often isolate and/or temperature-dependent. A continuous distribution in severity of disease, from none to very severe, was observed for 19 isolates of *R. oryzae* tested separately on wheat seedlings (Mazzola *et al.*, 1995) and for 21 isolates of *Rhizoctonia* (W-AK) on barley seedlings (Leiner and Carling, 1994). In contrast, all four isolates of *R. solani* AG 8 pathogenic on wheat caused similar severities of seedling stunting and root rot (Mazzola *et al.*, 1995). Neate (1985) found that the virulence of isolates of *R. solani* AG 8 varied on different cereals. Some isolates of *R. oryzae* caused more root rot and damping-off of wheat at higher temperatures (>20°C) than at lower temperatures (<15°C) (Ogoshi *et al.*, 1990; Smiley and Uddin, 1993; Mazzola *et al.*, 1995), but virulence of other isolates was unaffected by temperature (Mazzola *et al.*, 1995). Virulence of *R. solani* AG 8 and *R. zeae* on wheat and barley was temperature-dependent (Ogoshi *et al.*, 1990; Smiley and Uddin, 1993; Leiner and Carling, 1994; Mazzola *et al.*, 1995), but *Rhizoctonia* (W-AK) caused similar amounts of root rot on barley seedlings at 15, 22 and 25°C (Leiner and Carling, 1994).

R. zeae and *R. solani* AG 2 and AG 4 all cause root rot of corn (Sumner and Bell, 1982). *R. zeae* was less virulent on corn than isolates of *R. solani* AG 2 or AG 4 and caused only mild root rot on primary and lateral roots. *R. solani* AG 2 caused severe crown, brace and lateral root rot when irrigated corn was grown in monoculture or in rotation with peanut or soybean, but these symptoms were rarely observed when corn was grown in non-irrigated fields (Sumner and Bell, 1982).

III. Disease management

A. Cultivation and crop residue management

Rhizoctonia root rot of wheat and barley is generally more severe when the crop is sown directly into cereal stubble ("direct drill" or "no-till") than when planted into a seedbed prepared using conventional cultivation practices (Neate *et al.*, 1982; MacNish, 1984, 1985; Rovira, 1986; Weller *et al.*, 1986; Pumphrey *et al.*, 1987; Smiley *et al.*, 1994). Conventional tillage, either with mold-board ploughing so as to invert the top 10-20 cm of soil and bury residue from the previous crop or disturbance of the 0-10 cm layer with chisel points as used in Australia, has been the primary measure used to control Rhizoctonia root rot of cereals. The incidence and severity of root infection is often inversely correlated with intensity and depth of cultivation (Jarvis and Brennan, 1986; Pumphrey *et al.*, 1987). Neate (1984) and Rovira (1986) proposed that the reduction in Rhizoctonia root rot in response to cultivation is due to fragmentation of plant debris harboring the fungus, resulting in reduced propagule size and inoculum potential through increased exposure of the pathogens to microbial antagonists. Neate (1991) demonstrated that severing of hyphae without moving or reducing the size of inoculum could be a major factor in the reduction of disease associated with cultivation. Other factors also may be involved. In Oregon, root rot on wheat after summer fallow increased directly with the amount of residue on the soil surface at the time of planting, was suppressed by burning wheat stubble before planting, and was unaffected by residue and/or tillage in a pea-wheat rotation that had much higher microbial diversity and biomass than the wheat-fallow rotation (Smiley *et al.*, 1996).

Inoculum of *Rhizoctonia* in a wheat field sown by direct drilling was concentrated in the upper 10 cm of the soil profile (Neate, 1987). Deep cultivation could alter the vertical distribution of propagules so that the plant does not encounter a high inoculum density until the seedling is well established with an extensive root system. Conversely, deep-furrow seeding, typical for winter wheat planted into fallow in the low-rainfall areas of the Pacific Northwest, could place the seed below the infested layer, such that seminal roots could escape infection. Crown roots, on the other hand, would be fully exposed to inoculum and could not escape infection. Severity of damage to the mature plant is highest for infections that sever main root axes at a point near the crown. Rhizoctonia root rot can develop into a crown rot and cause premature blight of adult plants, including "white heads," but this symptom is rare (R. J. Cook, unpubl.).

Seed drills differ greatly in the amount of soil disturbance created during direct drilling. Increasing the depth and intensity of soil disturbance under the seed placement zone while drilling can increase wheat yields in the presence or absence of *Rhizoctonia*, and may be as effective as pre-plant cultivations for reducing damage from Rhizoctonia root rot (Jarvis and Brennan, 1986; Schmidt and Belford, 1994). Roget *et al.* (1996) demonstrated that direct drilling with narrow sowing points, which disturbed the soil below the seed, reduced disease incidence in wheat. Farmers in the US Pacific Northwest have minimized the impact of Rhizoctonia root rot in direct-drilled wheat by installing either a chisel point or coulter in front of each drill opener, such that soil is loosened within and infested residue moved away from the narrow zone where seedlings will become established (R.J. Cook and R.W. Smiley, unpubl.). Smiley and Uddin (1993) demonstrated in reverse the protective value of the "tillage" band during planting; damage from *R. solani* AG 8 remained confined to the row in which the pathogen was applied with the seed and did not spread to adjacent rows (25 cm) during the entire 11-month growing season.

B. Crop rotation

Despite the broad host range of *R. solani* isolates pathogenic to wheat (Neate, 1985), crop rotation is an effective means to reduce the incidence of root rot of wheat and barley. In the US Pacific Northwest, peas and lentils are commonly grown in rotation with cereals and are susceptible to isolates of *R. solani* AG 8 that cause root rot of wheat and barley (Cook and Veseth, 1991). Reduced root rot in wheat-pea compared to wheat-summer fallow or wheat-oilseed rape rotations is thought to occur in response to greater microbial biomass and/or diversity in the wheat-legume rotation, thereby limiting the inoculum potential of *R. solani* (Smiley *et al.*, 1994, 1996). It also is possible that the fibrous roots of cereals of both the harvested crop and volunteer plants provide a larger inoculum source of *R. solani* than that supported by the susceptible, but more sparse tap root systems of peas and lentils, thereby increasing the amount of residual inoculum available for infection of the succeeding cereal crop (Cook and Veseth, 1991). In Australia, the incidence of bare patch was lower when wheat was grown in rotation with medic, peas or wheat than when planted after annual grass-medic pasture (Rovira, 1986). Although crop rotation may afford some protection against Rhizoctonia root rot by lowering the inoculum potential of the pathogen in soil, rotation can only provide complete control of this disease when wheat or barley follows a long period of plant-free fallow (Cook and Veseth, 1991).

While rotations often reduce severity of Rhizoctonia root rot in short-term experiments there is growing evidence that disease incidence and/or severity also decline under long-term monoculture in the US Pacific Northwest (Lucas *et al.*, 1993; Smiley *et al.*, 1994, 1996). A disease decline phenomenon for root rot caused by *R. solani* AG 8 (Lucas *et al.*, 1993) is considered equivalent to Rhizoctonia root rot decline in sugarbeet monoculture (Hyakumachi *et al.*, 1990) and take-all decline in cereal monoculture (Shipton, 1975). Smiley and Wilkins (1993) reported that productivity of barley in *Rhizoctonia*-infested soil increased with the number of years of no-till management. Roget (1995) has reported a decline in the incidence of Rhizoctonia root rot in wheat over 15 years in different rotations in which all plant residues were retained. Disease decline was attributed to increased biological activity in the soil resulting from the high inputs of organic matter. Wiseman *et al.* (1995) reported that this disease suppression was transferable to sterilized soil and was destroyed by gamma irradiation or pasteurization of soil. Earthworms are more active in conservation tillage systems and may contribute directly, and significantly to suppression of Rhizoctonia root rot (Stephens *et al.*, 1994) in soils that are favorable to the development of earthworm populations.

C. Management of volunteer (self-sown) cereals and weeds

Volunteer (self-sown) cereals and grass weeds serve as hosts of *R. solani* between the harvesting of one crop and the planting of the next. The incidence of root rot in direct-drilled wheat and barley was most severe if planting was done immediately (1-3 days) after herbicide treatment. Timely applications of paraquat or glyphosate herbicides well in advance of planting (e.g. at least 2-3 weeks) to control volunteer plants and weeds have resulted in a significant reduction in the incidence of Rhizoctonia root rot of direct drilled wheat and barley (Roget *et al.*, 1987; Smiley *et al.*, 1992). The longer interval between herbicide application and planting apparently allows additional time for saprophytic microorganisms to displace *Rhizoctonia* in the dying roots of treated plants, resulting in lower inoculum potential of the pathogen.

Damage from Rhizoctonia root rot has been amplified on wheat and barley in soils treated with sulfonylurea herbicides such as chlorsulfuron (Rovira and McDonald, 1986; Smiley and Wilkins, 1992). These herbicides stunt root growth which may exacerbate the damage to roots caused by *R. solani*.

D. Seed treatment

Many chemical and biological seed treatments have been evaluated for controlling Rhizoctonia seed and root rot of cereals. Triadimenol seed treatment was effective in controlling infection of wheat by *R. solani* AG 4 (Mathieson and Rush, 1991) but not *R. solani* AG 8 (Smiley *et al.*, 1990a). Tolclofos-methyl and flutolanil were fungitoxic to *R. solani* AG 8 and *R. oryzae* and reduced the incidence of Rhizoctonia root rot on winter wheat seedlings, but did not improve grain yields (Smiley *et al.*, 1990a, 1990b). Difenoconazole in combination with the biocontrol agents *Pseudomonas fluorescens* Pf-5, *Pseudomonas chlororaphis* 30-84, or *Bacillus* sp. L324 reduced damping-off and root rot of wheat seedlings caused by *Rhizoctonia* spp., but were more active against *R. oryzae* and *R. cerealis* than *R. solani* AG 8 (M. Mazzola, unpubl.). Development of consistently effective seed treatments will be difficult due to the multiple species of *Rhizoctonia* capable of causing disease on cereals, the differential sensitivity of these pathogens to fungicides (Smiley *et al.*, 1990a; Kataria *et al.*, 1991) and antibiotics produced by biological control agents (M. Mazzola, unpubl.), and the prolonged period during which winter cereals remain susceptible to damage by *Rhizoctonia* spp. Disease protection and positive yield responses to fungicide seed treatments were more prevalent for spring- than autumn-sown cereals (Smiley *et al.*, 1991), presumably because protection remains higher over a longer proportion of the plant growth cycle for spring- than autumn-sown cereals.

E. Plant nutrition

There are numerous reports of plant growth responses to application of nitrogen fertilizer in *Rhizoctonia*-infested soils. Results have varied from disease suppression (MacNish, 1985) to no effect (Pumphrey *et al.*, 1987), or increased root rot severity (Smiley *et al.*, 1990b). Inorganic sources of N (e.g. ammonium nitrate) have led to higher disease incidences than organic sources (pea vines or cow manure) in a long-term (60 yr) experiment (Smiley *et al.*, 1996). Wall *et al.* (1994) performed elaborate evaluations of wheat responses to nitrogen in soils infested or not infested with *R. solani* AG 8, and concluded that nitrogen had no effect on root mass in infested soil. Zinc deficiency reduces host resistance to *R. solani* AG 8 and increases damage to roots of wheat (Thongbai *et al.*, 1993a, 1993b) and barley (Thongbai *et al.*, 1993c).

F. Host plant resistance or tolerance

Wheat and barley cultivars resistant to Rhizoctonia root rot are not available although cultivars and breeding lines vary significantly in tolerance to root damage. For example 2% of 216 lines yielded nearly as well when grown in soil artificially infested with *R. solani* AG 8 as in adjacent plots of non-infested (check) soil at each of two locations in the US Pacific Northwest (R.W. Smiley and P.K. Zwer, unpubl.). Development of cultivars with acceptable levels of tolerance continues to be impeded by high season-to-season and field-to-field variability in cultivar performance (Neate, 1989).

IV. Conclusions

Rhizoctonia root rot of wheat has been one of the major impediments to the adoption of direct-drill sowing of cereals in the Pacific Northwest of the United States and in southern Australia. However, our improved understanding of the epidemiology of *Rhizoctonia* spp. affecting wheat and barley has led to the development of management strategies with the potential to effectively control this disease.

V. References

Bell DS, Harris H, & Wells HD (1973) Rhizoctonia blight of grain sorghum foliage. Plant Dis. Rep. 57:549-550.

Boerema GH & Verhoeven AA (1977) Check-list for scientific names of common parasitic fungi. Series 26: Fungi on field crops: cereals and grasses. Neth. J. Plant Pathol. 83:165-204.

Burton RJ, Coley-Smith, JR & Wareing PW (1988) *Rhizoctonia oryzae* and *R. solani* associated with barley stunt disease in the United Kingdom. Trans. Br. Mycol. Soc. 91:409-417.

Cook RJ & Veseth RJ (1991) Wheat Health Management. APS Press, St. Paul.

Deacon JW & Scott DB (1985) *Rhizoctonia solani* associated with crater disease (stunting) of wheat in South Africa. Trans. Br. Mycol. Soc. 85:319-327.

Harris JR & Moen R (1985) Replacement of *Rhizoctonia solani* on wheat seedlings by a succession of root-rot fungi. Trans. Br. Mycol. Soc. 84:11-20.

Hollins TW, Jellis GJ, & Scott PR (1985) Sharp eyespot of cereals and Rhizoctonia of potato. In: Parker CA, Rovira AD, Moore KJ & Wong PTW (eds.) Ecology and Management of Soilborne Plant Pathogens (pp 34-35) Amer. Phytopathol. Soc., St. Paul.

Hyakumachi M, Kanzawa K & Ui T (1990) Rhizoctonia root rot decline in sugarbeet monoculture. In: Hornby D (ed.) Biological Control of Soil-Borne Plant Pathogens (pp 227-247) C.A.B. International, Wallingford.

Hynes HJ (1933) 'Purple patch' of wheat and oats. A disease caused by the fungus *Rhizoctonia solani*. Agric. Gaz. New South Wales 44:879-883.

Jarvis R & Brennan RF (1986) Timing and intensity of surface cultivation and depth of cultivation affect Rhizoctonia patch and wheat yield. Aust. J. Exp. Agric. 26:703-708.

Kapoor IJ & Hoffman GM (1984) Antagonistic effects of soil microbes on *Rhizoctonia solani* and Rhizoctonia-like fungi (*Ceratobasidium* sp.) associated with foot rot of cereals. Z. Pflanzenkr. Pflanzenschutz 91:186-196.

Kataria HR, Hugelshofer U & Gisi U (1991) Sensitivity of *Rhizoctonia* species to different fungicides. Plant Pathol. 40:203-211.

Kuzia RA & Windels CE (1994) *Rhizoctonia zeae* pathogenic to spring wheat and sugar beet seedlings. Phytopathology 84:1159.

Leiner RH & Carling DE (1994) Characterization of *Waitea circinata (Rhizoctonia)* isolated from agricultural soils in Alaska. Plant Dis. 78:385-388.

Lipps PE & Herr LJ (1982) Etiology of *Rhizoctonia cerealis* in sharp eyespot of wheat. Phytopathology 72:1574-1577.

Lucas P, Smiley RW & Collins HP (1993) Decline of Rhizoctonia root rot on wheat in soils infested with *Rhizoctonia solani* AG 8. Phytopathology 83:260-265.

MacNish GC (1984) The use of undisturbed soil cores to study methods of controlling Rhizoctonia patch of cereals. Plant Pathol. 33:355-359.

MacNish GC (1985) Methods of reducing Rhizoctonia patch of cereals in Western Australia. Plant Pathol. 34:175-181.

MacNish GC & Carling DE (1995) Use of vegetatively compatible populations to determine field distribution of *R. solani* AG 8. Phytopathology 85:1040.

MacNish GC & Sweetingham MW (1993) Evidence that each Rhizoctonia bare patch is dominated by an individual zymogram group (ZG) of *Rhizoctonia solani* AG 8. Aust. J. Agric. Res. 44:1175-1194.

MacNish GC, Carling DE & Brainard KA (1993) Characterization of *Rhizoctonia solani* AG 8 from bare patches by pectic isozyme (zymogram) and anatomosis techniques. Phytopathology 83:922-927.

Mathieson JT & Rush CM (1991) Influence of five fungicides on Rhizoctonia root rot of hard red winter wheat. Plant Dis. 75:983-986.

Mazzola M, Wong OT & Cook RJ (1995) Virulence of *Rhizoctonia oryzae* isolates on wheat and detection of the pathogen in plant tissue using a PCR protocol. Phytopathology 85:1133.

McGee DC (1988) Maize Diseases: A Reference Source for Seed Technologists. APS Press, St. Paul.

Murray DIL & Nicholson TH (1979) Barley stunt disorder in Scotland. Plant Pathol. 28:200-201.

Neate SM (1984) Minimum cultivation and root diseases of wheat. PhD thesis. University of Adelaide.

Neate SM (1985) *Rhizoctonia* in South Australian wheat fields. In: Parker CA, Rovira AD, Moore KJ & Wong PTW (eds.) Ecology and Management of Soilborne Plant Pathogens (pp 54-56) Amer. Phytopathol. Soc., St. Paul.

Neate SM (1987) Plant debris in soil as a source of inoculum of *Rhizoctonia* in wheat. Trans. Br. Mycol. Soc. 88:157-162.

Neate SM (1989) A comparison of controlled environment and field trials for detection of resistance in cereal cultivars to root rot caused by *Rhizoctonia solani*. Plant Pathol. 38:494-501.

Neate SM (1991) Some mechanisms of action of a narrow sowing point which influences Rhizoctonia root disease and yield of cereals. Abstract of the 8th Australasian Plant Pathology Society Conference, Sydney.

Neate SM & Warcup JH (1985) Anastomosis grouping of some isolates of *Thanatephorus cucumeris* from agricultural soils in South Australia. Trans. Br. Mycol. Soc. 85:615-620.

Neate, SM, Rovira AD & Warcup JH (1982) Minimum cultivation and root disease in wheat. Biennial Report of the Waite Agricultural Research Institute, 1980-1981. 142.

Neate, SM, Cruickshank RH & Rovira AD (1988) Pectic enzyme patterns of *Rhizoctonia solani* isolates from agricultural soils in South Australia. Trans. Br. Mycol. Soc. 90:37-42.

Ogoshi A, Cook RJ & Bassett EN (1990) *Rhizoctonia* species and anastomosis groups causing root rot of wheat and barley in the Pacific Northwest. Phytopathology 80:784-788.

Pitt D (1964) Studies on sharp eyespot disease of cereals. Ann. Appl. Biol. 54:77-89.

Ploetz RC, Mitchell DJ & Gallaher RN (1985) Characterization and pathogenicity of *Rhizoctonia* species from a reduced-tillage experiment multicropped to rye and soybean in Florida. Phytopathology 75:833-839.

Price RD (1970) Stunted patches and deadheads in Victorian cereal crops. Tech Bull. 23, Dept. Agric., Victoria.

Pumphrey FV, Wilkins DE, Hane DC & Smiley RW (1987) Influence of tillage and nitrogen fertilizer of Rhizoctonia root rot (bare patch) of winter wheat. Plant Dis. 71:125-127.

Richardson MJ & Cook RJ (1985) *Rhizoctonia* on small-grain cereals in Great Britain. In: Parker CA, Rovira AD, Moore KJ & Wong PTW (eds.) Ecology and Management of Soilborne Plant Pathogens (pp 63-65) Amer. Phytopathol. Soc., St. Paul.

Roberts FA & Sivasithamparam K (1986) Identity and pathogenicity of *Rhizoctonia* spp. associated with bare patch disease of cereals at a field site in Western Australia. Neth. J. Plant Pathol. 92:185-195.

Roberts FA & Sivasithamparam K (1987) Effect of interaction of *Rhizoctonia* spp. with other fungi from cereal bare patches on root rot of wheat. Trans. Br. Mycol. Soc. 89:256-259.

Roget DK (1995) Decline in root rot (*Rhizoctonia solani* AG 8) in wheat in a tillage and rotation experiment at Avon, South Australia. Aust. J. Exp. Agric. 35:(in press).

Roget DK, Venn NR & Rovira AD (1987) Reduction of Rhizoctonia root rot of direct-drilled wheat by short-term chemical fallow. Aust. J. Exp. Agric. 27:425-430.

Roget DK, Neate SM & Rovira AD (1996) The effect of sowing point design and tillage practice on the incidence of Rhizoctonia root rot, take-all and cereal cyst nematode. Aust. J. Exp. Agric. (in press).

Rovira AD (1986) Influence of crop rotation and tillage on Rhizoctonia bare patch of wheat. Phytopathology 76:669-673.

Rovira AD & McDonald HJ (1986) Effects of the herbicide chlorsulfuron on Rhizoctonia bare patch and take-all of barley and wheat. Plant Dis. 70:879-882.

Rovira AD, Ogoshi A & McDonald HJ (1986) Characterization of isolates of *Rhizoctonia solani* from cereal roots in South Australia and New South Wales. Phytopathology 76:1245-1248.

Rush CM, Carling DE, Haverson RM & Mathieson JT (1994) Prevalence and pathogenicity of anastomosis groups of *Rhizoctonia solani* from wheat and sugar beet in Texas. Plant Dis. 18:349-352.

Samuel G & Garret SD (1932) *Rhizoctonia solani* on cereals in South Australia. Phytopathology 22:827-836.

Schmidt CP & Belford RK (1994) Increasing the depth of soil disturbance increases yields of direct drilled wheat on the sandplain soils of Western Australia. Aust. J. Exp. Agric. 34:777-781.

Scott DB, Visser CPN & Rufenacht EMC (1979) Crater disease of summer wheat in African drylands. Plant Dis. Rep. 63:836-840.

Shipton PJ (1975) Take-all decline during cereal monoculture. In: Bruehl GW (ed.) Biology and Control of Soil-Borne Plant Pathogens (pp 137-144) Amer. Phytopathol. Soc., St. Paul.

Smiley RW & Uddin W (1993) Influence of soil temperature on Rhizoctonia root rot (*R. solani* AG 8 and *R. ozyzae*) of winter wheat. Phytopathology 83:777-785.

Smiley RW & Wilkins DE (1992) Impact of sulfonylurea herbicides on Rhizoctonia root rot, growth, and yield of winter wheat. Plant Dis. 76:399-404.

Smiley RW & Wilkins DE (1993) Annual spring barley growth, yield and root rot in high- and low residue tillage systems. J. Prod. Agric. 6:270-275.

Smiley RW, Wilkins DE & Klepper EL (1990a) Impact of fungicide seed treatments on Rhizoctonia root rot, take-all, eyespot, and growth of winter wheat. Plant Dis. 74:782-787.

Smiley RW, Uddin W, Ott S & Rhinhart KEL (1990b) Influence of flutolanil and tolclofos-methyl on root and culm diseases of winter wheat. Plant Dis. 74:788-791.

Smiley RW, Wilkins DE & Case CE (1991) Barley yields as related to use of seed treatments in eastern Oregon. J. Prod. Agric. 4:400-407.

Smiley RW, Ogg AG & Cook RJ (1992) Influence of glyphosate on Rhizoctonia root rot, growth and yield of barley. Plant Dis. 76:937-942.

Smiley RW, Ingham RE, Uddin W & Cook GH (1994) Crop sequences for managing cereal cyst nematode and fungal pathogens of winter wheat. Plant Dis. 78:1142-1149.

Smiley RW, Collins HP & Rasmussen PE (1996) Diseases of wheat in long-term agronomic experiments at Pendleton, Oregon. Plant Dis. 80:(in press).

Stephens PM, Davoren CW, Ryder MH, Doube BM & Correll RL (1994) Field evidence for reduced severity of Rhizoctonia bare-patch disease of wheat, due to the presence of the earthworms *Aporrectodea rosea* and *Aporrectodea trapezoides*. Soil Biol. Bioch. 26:1495-1500.

Sterne RE & Jones JP (1978) Sharp eyespot of wheat in Arkansas caused by *Rhizoctonia solani*. Plant Dis. Rep. 62:56-60.

Sumner DR & Bell DK (1982) Root diseases induced in corn by *Rhizoctonia solani* and *Rhizoctonia zeae*. Phytopathology 72:86-91.

Sweetingham MW, Cruickshank RH & Wong DH (1986) Pectic zymograms and taxonomy and pathogenicity of the *Ceratobasidiaceae*. Trans. Br. Mycol. Soc. 86:305-311.

Thongbai P, Hannam RJ, Graham RD & Webb MJ (1993a) Interaction between zinc nutritional status of cereals and Rhizoctonia root rot severity. I. Field observations. Plant Soil 153:207-214.

Thongbai P, Graham RD, Neate SM & Webb MJ (1993b) Interaction between zinc nutritional status of cereals and Rhizoctonia root rot severity. II. Effect of Zn on disease severity of wheat under controlled conditions. Plant Soil 153:215-222.

Thongbai P, Webb MJ & Graham RD (1993c) Zinc deficiency predisposes winter cereals to Rhizoctonia root rot. In: Barrow NJ (ed.) Plant Nutrition - From Genetic Engineering to Field Practice (pp. 681-684). Kluwer Acad. Publ., Dordrecht.

Van der Hoeven EP & Bollen, GJ (1980) Effect of benomyl on soil fungi associated with rye. I. Effect on the incidence of sharp eyespot caused by *Rhizoctonia cerealis*. Neth. J. Pl. Pathol. 86:163-180.

Wall PC, Neate SM, Graham RD, Reuter DJ, & Rovira AD (1994) The effect of Rhizoctonia root disease and applied nitrogen on growth, nitrogen uptake and nutrient concentrations in spring wheat. Plant Soil 163:111-120.

Weller DM, Cook RJ, MacNish G, Bassett EN, Powelson RL & Peterson RR (1986) Rhizoctonia root rot of small grains favored by reduced tillage in the Pacific Northwest. Plant Dis. 70:70-73.

Wiese, MV (1987) Compendium of Wheat Diseases (2nd Edition). APS Press, St. Paul.

Wiseman BM, Neate SM, Ophel-Keller K and Smith, SE (1995) Suppression of *Rhizoctonia solani* anastomosis group 8 in Australia and its biological nature. Abstract of the International Symposium on Rhizoctonia, Noordwijkerhout, the Netherlands, June 27-30, 1995.

Yang HA, Sivasithamparam K, Alehohammad J, Barton JE & O'Brien PA (1994) Association of *Rhizoctonia* strains with bare patch disease of wheat in Western Australia. Plant Pathol. 43:878-884.

V.3 COTTON DISEASES INCITED BY *RHIZOCTONIA SOLANI*

CRAIG S. ROTHROCK

Department of Plant Pathology, University of Arkansas
Fayetteville AR 72701, USA.

I. INTRODUCTION

Cotton (*Gossypium* spp.) originated as a tropical/subtropical perennial plant, but is produced as an annual crop in many temperate regions around the world. The epicotyl of the cotton plant is poorly developed and thus the first true leaves may not expand for a period of several weeks. As a result of planting cotton under marginal environmental conditions (i.e., temperate regions) and the crops slow physiological development, the plant remains in the seedling stage for an extended period. Seedling diseases on cotton are a worldwide problem (Hillocks, 1992). An estimated 2.8% loss in production was associated with seedling diseases over the last ten years, 1986-1995, in the United States (Proceedings of the Beltwide Cotton Production Research Conferences, National Cotton Council). The primary pathogens in the seedling disease complex of cotton include *Rhizoctonia solani*, *Pythium* spp., *Thielaviopsis basicola*, and *Fusarium* spp. (Watkins, 1981; Ridgway *et al.*, 1984). *R. solani* is a prevalent and important pathogen in the seedling disease complex of cotton wherever cotton is grown (Ray and McLauglin, 1942; Fulton and Bollenbacher, 1959; Ranney, 1962; Ford *et al.*, 1981; Johnson *et al.*, 1969, 1978; Davis, 1975; Smirnova, 1980; Roy and Bourland, 1982; Melero-Vara and Jimenez-Diaz, 1990; Colyer *et al.*, 1991a; Lisker and Meiri, 1992; Moustafa-Mahmoud *et al.*, 1993).

R. solani has been associated with all stages of seedling disease including seed decay, preemergence damping-off (death of the seedling prior to emergence), and postemergence damping-off (death of the seedling after emergence). Huisman (1988) found that the isolation of *R. solani* was consistently associated with a high incidence of seedling tissue damage. It is very difficult to determine causal agent of seed rots and preemergence damping-off. The symptoms associated with damage due to *R. solani* are most often associated with postemergence damping-off. Symptoms include brown to reddish-brown sunken lesions on the hypocotyl near the soil line. This sunken lesion, caused by the death and collapse of cortical cells is often referred to as 'soreshin'. Often the lesion will girdle the hypocotyl of the cotton seedling creating a condition referred to as 'wirestem', from which the seedling usually dies. Superficial hyphae may be present on the surface of the lesion from which soil particles often adhere.

The cotton plant becomes increasingly resistant to disease with age. However, the disease can extend beyond the seedling stage as lesions on the stem above and below the soil line (Neal, 1942). Diseased plants have less ability to compensate for other stresses and there is often a delay in flowering and boll maturity. Brown and McCarter (1976) found that transplanted seedlings with hypocotyl lesions and healthy plants grew equally well at 28 C, but at 19 C plants with lesions had less shoot and root growth than healthy plants. In field studies over two years, transplanted healthy seedlings yielded 39% and 13% more seed cotton than plants transplanted with lesions.

B. Sneh et al. (eds.),
Rhizoctonia Species: Taxonomy, Molecular Biology, Ecology, Pathology and Disease Constrol, 269–277.

In addition to seedling disease, *R. solani* may cause other disease symptoms on cotton. *R. solani* was found to be one of the primary agents in boll rot in Louisiana (Pinckard and Luke, 1967; Luke *et al.*, 1974). Inoculum came primarily from the discharge of basidiospores from hymenia on cotton stems that formed during the latter part of the growing season (Luke *et al.*, 1974). Bolls inoculated with spores began to rot in 5 to 7 days. The fungus generally infected the bract, proceeded into the base of the boll, and entered the carpel wall or staminal column. Neal (1944) reported a leaf spot caused by *R. solani*. Even though considerable leaf damage could occur, the disease was not considered to cause any yield reduction.

II. Pathogen variation

Pathogen diversity has not been critically examined for *R. solani* on cotton. Anastomosis group (AG) 4 is the primary group responsible for seedling disease on cotton. Weinhold (1977) reported AG 4 was associated with soils having histories of cotton production. Rothrock *et al.* (1995) also found only *R. solani* AG 4 when *R. solani* was characterized from soils with a history of cotton production and cotton seedlings at two sites in Arkansas. Disease surveys in Spain also identified all *R. solani* isolates from cotton as AG 4 (Melora-Vara and Jimenez-Diaz, 1990).

Earlier work by Maier (1959) characterized 10 cultural types of *R. solani* among a group 245 isolates from cotton plants with symptoms or soil planted to cotton. He noted a large variation in virulence, but virulence of isolates was similar on *Gossypium hirsutum* and *Gossypium barbadense*. In controlled experiments, pathogenicity on cotton of other anastomosis groups of *R. solani* can readily be demonstrated (Carling *et al.*, 1994; Rothrock unpublished)

III. The infection process

The infection process of *R. solani* has been examined extensively for cotton. Hyphae of *R. solani* initially grow vertically along the hypocotyl following the anticlinal longitudinal walls (Khadga *et al.*, 1963; Armentrout and Downer, 1987; Armentrout *et al.*, 1987). Side branches of the fungus grow obliquely or at right angles across a few epidermal cells. Hyphae then align in the grooves between epidermal cells and produce lateral branches that terminate in a T-shaped branch. Tips from these branches elongate to form hyphae parallel to first. This accumulation of axial hyphae and lateral branches form the infection cushion. The dome-shaped infection cushion varies in size from 300 to 500μm in diameter rising 120-170μm above the host surface (Khadga *et al.*, 1963). Mucilaginous material was observed along the hyphae allowing the hyphae to adhere to the plant surface (Armentrout and Downer, 1987). Hyphal tips on the underside of the cushion become swollen and produce infection pegs which penetrate the cuticle and the epidermal cell wall into the cell lumen or through intercellular spaces (Khadga *et al.*, 1963). Many epidermal cells can be penetrated under an infection cushion. No infection pegs were observed by Weinhold and Motta (1973), only direct penetration by hyphal tips. Sneh *et al.* (1989) observed no infection cushion formation using a virulent isolate of AG 4. The isolate grew extensively over the surface of the seedling and penetrated via hyphal tips.

Infection cushion formation was reported to be mainly due to the topography of the hypocotyl surface (Armentrout *et al.*, 1987). Hypocotyl replicas caused internode shortening, hyphal alignment, foot formation and branching to form a cushion-like structure in the presence of sucrose or hypocotyl exudates. On hypocotyls covered with cellophane no hyphal alignment or foot formation occurred, however some hyphal aggregation was observed. Hyphal alignment, foot formation, and infection cushions also did not form on dark grown hypocotyls. Infection cushion formation did occur on dark grown hypocotyls with the addition of sucrose. Nutrients appeared to play only a stimulatory role.

The infection process and the development of disease symptoms occurs very quickly. In 10 to 12 hr after inoculation, the pathogen grew on the surface of the hypocotyl. Characteristic branching of *R. solani* occurred within 15 hr after inoculation and growth was dense by 18 hr. Infection cushions were formed by 21 hr (Armentrout and Downer, 1987; Weinhold and Motta 1973). Khadga *et al.* (1963) observed no apparent discoloration or death of cells below the infection cushion prior to penetration. The first sign of discoloration occurred as early as 24 hr and distinct lesions with tissue maceration occurred by 30 hr (Weinhold and Motta, 1973; Armentrout *et al.*, 1987). The pathogen grows intercellularly and intracellularly and eventually colonizes all cells except the xylem (Khadga *et al.*, 1963; Sneh *et al.*, 1989)

Weinhold and Motta (1973) first detected the loss of pectic materials from the cell walls as early as 18 hr. At this stage there was extensive hyphal branching and infection cushion formation. After 24 to 30 hr, an extensive cavity had formed beneath the cushions. However, unlike Khadga *et al.* (1963) no penetration had been detected. Changes in the crystalline nature of cellulose in the cell wall also were observed prior to penetration. Endopolygalacturonase (EPG) was detected in cotton seedling hypocotyl extracts 18 hr after inoculation with *R. solani* (Brookhouser and Weinhold, 1979; Brookhouser *et al.*, 1980). Evidence was found that *R. solani* produced EPG in response to seed exudates or to hypocotyl exudates (Brookhouser *et al*, 1980).

IV. Host Defense

Resistance of cotton seedlings to *R. solani* increases with age. A number of investigators have demonstrated differences in the susceptibility of seedlings between 5 and 12 days after planting (Hunter, 1974; Hunter *et al.*, 1978a). Several mechanisms have been suggested for these changes in resistance. Hunter and Guinn (1968) suggested cotton plants become more resistant to disease with age due to an increase in cuticle thickness. However, most research on host defense mechanisms has focused on biochemical defenses.

Seven different terpenoid aldehydes (desoxyhemigossypol, des-oxy-6-methoxyhemigossypol, hemigossypol, 6-methoxy-hemigossypol, gossypol, 6-methoxygossypol, and 6,6'-di-methoxygossypol) have been reported in cotton (Hunter *et al.*, 1978a). These terpenoid compounds are found in cotton seedlings in localized lysigenous glands in the stem and in the root in the epidermis and in scattered cortical parenchyma cells extending from the stem root transition to within several centimeters of the root tip (Mace *et al.*, 1974). Because of the fungitoxic properties of these compounds it has been suggested that they may act as phytoalexins or precursors of phytoalexins (Hunter *et al.*, 1978a). Concentrations of aldehydes, gossypol equivalents, were greater in seedlings challenged by *R. solani* than seedlings not inoculated with the pathogen (Hunter *et al.*, 1978b). The concentration of these terpenoid compounds also was higher in 12 day-old than in 5 day-old seedlings. However, the level of increase following infection was similar for seedlings of both ages.

Production and oxidation of catechin in response to infection by *R. solani* also has been suggested to be partially responsible for resistance in older seedlings. Catechins are the major components of the polyphenols found in cotton tissues. Following infection, catechins in the seedling are oxidized and the tannins formed can inhibit growth and polygalacturonase activity of the pathogen (Hunter, 1978). Catechin initially is found in the epidermis and endodermis and concentrations increase with seedling age. At 24 hr after inoculation, the concentration of phenolic compounds, primarily catechin, were greater in infected than noninfected seedlings (Hunter, 1974). This increase also was greater in older seedlings. Seedling extracts inhibited polygalacturonase activity in proportion to phenolic content.

Veech (1976) examined the activity of peroxidases which mediate oxidation of catechin. Within 12 hr after inoculation peroxidase activity had increased in the outer tangential walls of epidermal cells below the mycelium. As portions of the epidermis and outer cortex in the lesion became discolored, peroxidase activity was high in the cortical cell walls near this brown tissue. During the early stages of disease development the cortical cells with greatest peroxidase activity were separated from brown cells by at least one layer of cells. However, Veech found no differences in peroxidase activity between 5 day-old and 11 day-old seedlings.

V. THE ROLE OF THE ENVIRONMENT IN DISEASE

Poor stand and seedling disease is often associated with early planting (Johnson *et al.*, 1969; Colyer *et al.*, 1991a). In Tennessee, seedling disease was negatively correlated with soil temperature and positively correlated with soil water (Johnson *et al*, 1969). The role of the environment in seedling disease damage caused by *R. solani* is less clear. The frequency of isolation of *R. solani* from necrotic hypocotyls did not correlate with soil temperature or soil water (Johnson *et al.*, 1969). Based on isolation frequency, seedlings were equally susceptible when grown at a mean minimum temperature of 14-24°C. Below 13°C, *R. solani* was rarely isolated from seedlings. This is in contrast with seedling disease associated with *Pythium* spp. which was related to high soil water and low soil temperatures. Huisman (1988) also found colonization of cotton roots by *Pythium* spp. was associated with high soil water content, whereas colonization by *R. solani* remained relatively constant over a range of soil moistures.

Arndt (1957) reported *R. solani* was active over a wide range of temperatures from 18 to 33°C in controlled environmental studies. Preemergence damping-off was greatest at 21°C, whereas at temperatures higher than 21°C postemergence damping-off was of greater importance. Similarly, Hunter *et al.* (1960) found that isolates of *R. solani* had different temperature optima, but the most virulent isolate was most aggressive from 24 to 32°C, and still caused considerable disease at 20°C.

Low temperature has been found to predispose cotton to seedling disease caused by *R. solani*. Germinated seeds exposed to 10°C for three to five days and then planted in *R. solani* infested soil had reduced emergence compared to germinated seed not exposed to low temperatures (McCarter and Roncadori, 1971; Shao and Christiansen, 1982). This indicated that increased susceptibility at low temperatures may be largely a host response. Hunter and Guinn (1968) also reported that low root temperatures increased disease severity. They found that sugar content in the hypocotyl was greater at 10 and 15°C compared to 20 and 30°C. Hypocotyl extracts from seedlings grown at low temperatures supported more growth of the pathogen *in vitro* than extracts from plants grown at higher root temperatures. Hayman (1969) found germinating cotton seeds exuded about seven and three times more amino acids and sugars at 18°C and 24°C, respectively, compared to 30°C. Accumulation of seed exudate at low temperatures was related to increased disease *in vitro* and in the soil.

Weinhold *et al.* (1972) demonstrated the importance of nutrients on disease severity. The capacity of inoculum of *R. solani* to cause disease when grown on a nutrient medium deficient in nitrogen was greatly increased when sand was moistened with a solution of nitrogen; asparagine, $(NH_4)_2HPO_4$ or KNO_3. When inoculum was prepared on a medium deficient in a carbon source, addition of an external source of glucose increased virulence. Armentrout *et al.* (1987) also found disease severity decreased when inoculum of *R. solani* was grown on a medium having low glucose levels.

VI. Interaction with other pests.

Seedling disease caused by *R. solani* has been shown to be influenced by other early season pests of cotton. Neal and Newsom (1951) in the Red River Valley of Louisiana observed more soreshin symptoms in plots not treated for thrips than plots treated to control tobacco thrips (*Frankliniella fusca*). Similarly, Colyer *et al.* (1991b) found a correlation between thrips infestation, stand reduction and disease index. *Rhizoctonia* spp. made up 45% of the fungi isolated from seedlings.

Field surveys indicated that plants having severe symptoms of seedling disease caused by *R. solani* had a higher root knot nematode galling index and were shorter than less severely diseased seedlings (Reynolds and Hanson, 1957). Controlled environmental studies confirmed these observations (Cauquil and Shepherd, 1970; Carter, 1975). Carter (1975) also demonstrated that as the coarseness of a soil increased, *M. incognita* damage and damage associated with *R. solani* increased as measured by seedling weight or a disease index. This was not simply an effect of root pruning since root pruning did not by itself increase disease caused by *R. solani* (Carter 1981). Brodie and Cooper (1964) also found cotton seedlings were more susceptible to *R. solani* in the presence of *M. hapla*, *M. arenaria*, *Rotylenchulus reniformis*, or *Hoplolaimus tylenchiformis*.

R. solani increased Verticillium wilt when the pathogens were inoculated simultaneously or when seedlings were already infected by *R. solani* for *G. hirsutum* and *G. barbadense* (Khoury and Alcorn, 1973).

VII. Influence of cultural practices on disease

Cotton is frequently produced as a monoculture. Johnson *et al.* (1978) found *R. solani* and *T. basicola* made up a greater proportion of the pathogens isolated from diseased hypocotyls from fields with histories of cotton production compared to *Pythium ultimum* and *Fusarium* spp.

Herbicide use in cotton production systems has been suggested as having an impact on seedling disease caused by *R. solani*. Pinckard and Standifer (1966) reported that trifluralin decreased cotton stands and increased disease severity and isolation frequency of *R. solani*. In controlled environmental studies, Neubauer and Avizohar-Hershenson (1973) reported trifluralin increased susceptibility of the host; plants transplanted from soil treated with the herbicide into soil infested with *R. solani* had a greater disease severity that plants transplanted from soil not treated with the herbicide. In addition, the saprophytic activity of the pathogen, recovery of the pathogen from bean stem segment baits, was found to increase in herbicide treated soil. El-Khadem *et al.* (1979) found trifluralin and dinitramine increased damping-off caused by *R. solani*. However, they reported that only the saprophytic activity of the pathogen was increased, not host susceptibility. El-Khadem *et al* (1984), in field studies, were unable to demonstrate any effect of the pre-emergence herbicides (trifluralin, dinitramine, fluometuron, diuron, dalapon, or prometrynon) on seedling disease caused by *R. solani*. Miller *et al.* (1979) also were unable to demonstrate an effect for trifluralin on Rhizoctonia seedling damage. El Khadem and Papavizas (1984) found that EPTC and linuron applied at two times the recommended rate reduced postemergence but not preemergence damping off caused by *R. solani*. Reduction in disease was attributed to a reduction in saprophytic growth of the pathogen.

The use of winter legume cover crops has been found to increase seedling disease caused by *R. solani*. Rothrock *et al.* (1995) reported that incorporation of a winter hairy vetch, *Vicia villosa* (Roth.), cover crop as a green manure generally increased populations of *R. solani* in soil at cotton planting and resulted in increased recovery of *R. solani* AG 4 from seedlings. Rickerl *et al.* (1992) also noted increased saprophytic activity of *R. solani* by a stem baiting technique following hairy vetch or crimson clover (*Trifolium incarnatum* L.) cover crops.

Soil solarization, which is not used widely for cotton production, was effective in reducing or eliminating populations of a number of soilborne plant pathogens, including *R. solani* (Pullman *et al.* 1981).

There has been a dramatic reduction in tillage practices for cotton production in recent years in many areas to reduce soil erosion, timeliness of field operations, or to conserve energy. These include no tillage operation, tillage in the fall (stale seedbed), and various minimum-tillage techniques. With many of these tillage systems researchers and growers have observed a reduction in stand and early season growth. In some cases this has been associated with an increase in disease symptoms (Colyer and Vernon, 1993). Research is needed to quantify the influence of tillage systems on seedling disease incidence and severity caused by *R. solani*. Factors that should be examined are the impact of tillage systems on the inoculum density and distribution of the pathogen and the importance of changes in soil physical and chemical factors, such as soil temperature and water, on disease incidence and severity.

VIII. Management of Seedling Disease

A number of management practices are common for the control of seedling diseases of cotton. Management practices include planting high quality seed and adjusting planting depth according to soil water content, but not planting deeper than 5 cm. As a result of the strong relationship between stand establishment and soil temperature, soil temperature at planting is one of the primary considerations in planting. It is recommended that cotton be planted only after the average daily soil temperatures reach 20°C at planting depth for several consecutive days and there is a favorable weather forecast.

There is little evidence that resistance to *R. solani* exists for species of cotton. Fulton *et al.* (1973) screened more than 600 lines of several cotton species for resistance to four seedling disease pathogens and found no appreciable resistance to *R. solani*. Stanton *et al.* (1994) screened two hundred A-genome cotton (*Gossypium arboreum* and *Gossypium herbaceum*) lines. Variation was found among accessions in disease severity, but no useful levels of resistance were found. As a group asiatic cottons were more susceptible than a *G. hirsutum* control. El-Samra *et al* (1981) found differences in infection among four Egyptian cultivars of *G. barbadense*. Rizk and Mohammed (1982) were able to quantify differences in these Egyptian lines in terms of rate and extent of lesion development, supporting the reported differences in susceptibility among these lines. Breeding for resistance to cotton seedling diseases in general is often based on stand or seedling vigor. The MAR (multi-adversity resistance) system is one example of a breeding program trying to incorporate seedling disease resistance into new cultivars (Bird, 1982). The MAR breeding program utilizes the rate of germination in cold soils, seedcoat resistance to mold growth, growth in soils infested with the seedling disease pathogens *R. solani* and *Pythium* spp., and resistance to bacterial blight to select for new lines. The assumption is that resistance to diseases will be increased with the selection for resistance to adverse conditions and selected diseases.

Fungicides are commonly applied for the control of seedling diseases. These compounds are most frequently applied as seed treatments. Several fungicides with different spectrums of activity are often applied to the seed to increase the pathogens suppressed. Additional fungicide treatments may be applied in the planter box or in the planting furrow at the time of seeding (Minton and Garber, 1983; Hillocks *et al.*,1988; Colyer *et al.* 1991a). Registered fungicides for cotton that have activity against *R. solani* include carboxin, demosan, pentachloronitrobenzene, myclobutanil, and triadimenol. Extensive research also has been conducted on the use of biological seed or in-furrow treatments for the control of seedling diseases. Dagger, a strain *Pseudomonas fluorescens*, is registered and was sold as an in-furrow product for the control of seedling diseases (Hagedorn *et al.* 1993). Kodiak, a strain of *Bacillus subtilis,* is registered and sold as a seed treatment (Backman *et al*, 1994). Currently Kodiak is applied in combination with a number of chemical fungicides on most of the cotton sold in the United States. Numerous other bacterial and fungal agents have been evaluated for the control of seedling diseases in cotton. Two of the most extensively evaluated genera of fungi are *Gliocladium* and *Trichoderma* (Lewis and Papvizas, 1991).

IX. REFERENCES

Armentrout VN & Downer AJ (1987) Infection cushion development by *Rhizoctonia solani* on cotton. Phytopathology 77:619-623.

Armentrout VN, Downer AJ, Grasmick DL & Weinhold AR (1987) Factors affecting infection cushion development by *Rhizoctonia solani* on cotton. Phytopathology 77:623-630.

Arndt CH (1957) Temperature as a factor in the infection of cotton seedlings by ten pathogens. Plant Dis. Rep. Suppl. 246:63-81.

Backman PA, Brannen PM, & Mahaffee WF (1994) Plant response and disease control following seed inoculation with *Bacillus subtilis*. In: Ryder MH, Stephens PM & Bowen GD (eds.) Improving Plant Productivity with Rhizosphere Bacteria (pp 3-8) CSIRO Division of Soils, Adelaide.

Bird LS (1982) The MAR (multi-adversity resistance) system for genetic improvement of cotton. Plant Dis. 66:172-176.

Brodie BB & Cooper WE (1964) Relation of parasitic nematodes to postemergence damping-off of cotton. Phytopathology 54:1023-1027.

Brookhouser LW, Hancock JG & Weinhold AR (1980) Characterization of endopolygalacturonase produced by *Rhizoctonia solani* in culture and during infection of cotton seedlings. Phytopathology 70:1039-1042.

Brookhouser LW & Weinhold AR (1979) Induction of polygalacturonase from *Rhizoctonia solani* by cotton seed and hypocotyl exudates. Phytopathology 69:599-602.

Brown EA & McCarter SM (1976) Effect of a seedling disease caused by *Rhizoctonia solani* on subsequent growth and yield of cotton. Phytopathology 66:111-115.

Carling DE, Rothrock CS, MacNish GC, Sweetingham MW, Brainard KA & Winters SA (1994) Characterization of anastomosis group 11 (AG 11) of *Rhizoctonia solani*. Phytopathology 84:1387-1393.

Carter WW (1975) Effects of soil texture on the interaction between *Rhizoctonia solani* and *Meloidogyne incognita* on cotton seedlings. J. Nematol. 7:234-236.

Carter WW (1981) The effect of *Meloidogyne incognita* and tissue wounding on severity of seedling disease of cotton caused by *Rhizoctonia solani*. J. Nematol. 13:374-376.

Cauquil J & Shepherd RL (1970) Effect of root-knot nematode-fungi combinations on cotton seedling disease. Phytopathology 60:448-451.

Colyer PD & Vernon PR (1993) Effect of tillage on cotton seedling diseases. J. Prod. Agric. 6:108-111.

Colyer PD, Micinski S & Nguyen KT (1991a) Effect of planting date on the efficacy of an in-furrow pesticide and the development of cotton seedling disease. Plant Dis. 75:739-742.

Colyer PD, Micinski S & Vernon PR (1991b) Effect of thrips infestation on the development of cotton seedling diseases. Plant Dis. 75:380-382.

Davis RG (1975) Microorganisms associated with diseased cotton seedlings in Mississippi. Plant Dis. Rep. 59:277-280.

El-Khadem M & Papavizas GC (1984) Effect of the herbicides EPTC and linuron on cotton diseases caused by *Rhizoctonia solani* and *Fusarium oxysporum* f.sp. *vasinfectum*. Plant Pathol. 33:411-416.

El-Khadem M, El-Kazzaz MK & Hassan MA (1984) Influence of different pre-emergence herbicides on cotton diseases caused by *Rhizoctonia solani* and *Fusarium oxysporum* f.sp. *vasinfectum*. Plant Soil 79:29-36.

El-Khadem M, Zahran M & El-Kazzaz MK (1979) Effect of the herbicides trifluralin, dinitramine and fluometuron on Rhizoctonia disease in cotton. Plant Soil 51:463-470.

El-Samra IA, El-Faham YM & Kamara AM (1981) Selective induction of infection cushions by *Rhizoctonia solani* in relation to host responses. Phytopath. Z. 102:122-126.

Ford RE, Bissonnette HL, Horsfall JG, Millar RL, Schlegel D, Tweedy BG & Weathers LG (1981) Plant Pathology in China, 1980. Plant Dis. 65:706-714.

Fulton ND & Bollenbacher K (1959) Pathogenicity of fungi isolated from disease cotton seedlings. Phytopathology 49:684-689.

Fulton ND, Waddle BA & McCarver T (1973) Searching for possible tolerance in cotton to seedling disease. Arkansas Farm Res. 22:10.

Hagedorn C, Gould WD & Bardinelli TR (1993) Field evaluations of bacterial inoculants to control seedling disease pathogens on cotton. Plant Dis. 77:278-282.

Hayman DS (1969) The influence of temperature on the exudation of nutrients from cotton seeds and on preemergence damping-off by *Rhizoctonia solani*. Can. J. Bot. 47:1663-1669.

Hillocks RJ (1992) Cotton Diseases. CAB International, Oxon.

Hillocks RJ, Chinodya R, & Gunner R (1988) Evaluation of seed dressing and in-furrow treatments with fungicides for control of seedling disease in cotton caused by *Rhizoctonia solani*. Crop Prot. 7:309-313.

Huisman OC (1988) Colonization of field-grown cotton roots by pathogenic and saprophytic soilborne fungi. Phytopathology 78:716-722.

Hunter RE (1974) Inactivation of pectic enzymes by polyphenols in cotton seedlings of different ages infected with *Rhizoctonia solani*. Physiol. Plant Pathol. 4:151-159.

Hunter RE (1978) Effects of catechin in culture and in cotton seedlings on the growth and polygalacturonase activity of *Rhizoctonia solani*. Phytopathology 68:1032-1036.

Hunter RE & Guinn G (1968) Effect of root temperature on hypocotyls of cotton seedlings as a source of nutrition for *Rhizoctonia solani*. Phytopathology 58:981-984.

Hunter RE, Halloin JM, Veech JA & Carter WW (1978a) Terpenoid accumulation in hypocotyls of cotton seedlings during aging and after infection by *Rhizoctonia solani*. Phytopathology 68:347-350.

Hunter RE, Halloin JM, Veech JA & Carter WW (1978b) Exudation of terpenoids by cotton roots. Plant Soil 50:237-240.

Hunter RE, Staffeldt EE & Maier CR (1960) Effects of soil temperature on the pathogenicity of *Rhizoctonia solani* isolates. Plant Dis. Rep. 44:793-795.

Johnson LF, Baird DD, Chambers AY & Shamiyeh NB (1978) Fungi associated with postemergence seedling disease of cotton in three soils. Phytopathology 68:917-920.

Johnson LF, Chambers AY & Measells JW (1969) Influence of soil moisture, temperature, and planting date on severity of cotton seedling blight. Univ. Tenn. Agric. Exp. Stn. Bull. 461. 28 pp.

Khadga BB, Sinclair JB & Exner BB (1963) Infection of seedling cotton hypocotyl by an isolate of *Rhizoctonia solani*. Phytopathology 53:1331-1336.

Khoury FY & Alcorn SM (1973) Influence of *Rhizoctonia solani* on the susceptibility of cotton plants to *Verticillium albo-atrum* and on root carbohydrates. Phytopathology 63:352-358.

Lewis JA & Papavizas GC (1991) Biocontrol of cotton damping-off caused by *Rhizoctonia solani* in the field with formulations of *Trichoderma* spp. and *Gliocladium virens*. Crop Prot. 10:396-402.

Lisker N & Meiri A (1992) Control of *Rhizoctonia solani* damping-off in cotton by seed treatment with fungicides. Crop Prot. 11:155-159.

Luke WJ, Pinckard JA & Wang SC (1974) Basidiospore infection of cotton bolls by *Thanatephorus cucumeris*. Phytopathology 64:107-111.

Mace ME, Bell AA & Stipanovic RD (1974) Histochemistry and isolation of Gossypol and related terpenoids in roots of cotton seedlings. Phytopathology 64:1297-1302.

Maier CR (1959) Cultural and pathogenic variability of *Rhizoctonia solani* isolates from cotton-growing areas of New Mexico. Plant Dis. Rep. 43:1063-1066.

McCarter SM & Roncadori RW (1971) Influence of low temperature during cottonseed germination on growth and disease susceptibility. Phytopathology 61:1426-1429.

Melero-Vara JM & Jimenez-Diaz RM (1990) Etiology, incidence, and distribution of cotton seedling damping-off in southern Spain. Plant Dis. 74:597-600.

Miller JH, Carter CH, Garber, RH & DeVay JE (1979) Weed and disease responses to herbicides in single- and double-row cotton (*Gossypium hirsutum*). Weed Sci. 27:444-449.

Minton EB & Garber RH (1983) Controlling the seedling disease complex of cotton. Plant Dis. 67:115-118.

Moustafa-Mahmoud SM, Sumner DR, Ragab MM & Ragab MM (1993) Interaction of fungicides, herbicides, and planting date with seedling disease of cotton caused by *Rhizoctonia solani* AG 4. Plant Dis. 77:79-86.

Neal DC (1942) *Rhizoctonia* infection of cotton and symptoms accompanying the disease in plants beyond the seedling stage. Phytopathology 32:641-642.

Neal DC (1944) Rhizoctonia leaf spot of cotton. Phytopathology 34:599-602.

Neal DC & Newsom LD (1951) Soreshin of cotton and its relationship to thrips damage. Phytopathology 41:854.

Neubauer R & Avizohar-Hershenson Z (1973) Effect of the herbicide, Trifluralin, on Rhizoctonia disease in cotton. Phytopathology 63:651-652.

Pinckard JA & Luke WJ (1967) *Pellicularia filamentosa*, a primary cause of cotton boll rot in Louisiana. Plant Dis. Rep. 51:67-70.

Pinckard JA & Standifer LC (1966) An apparent interaction between cotton herbicidal injury and seedling blight. Plant Dis. Rep. 50:172-177.

Pullman GS, DeVay JE, Garber RH & Weinhold AR (1981) Soil solarization:effects of Verticillium wilt of cotton and soilborne populations of *Verticillium dahliae*, *Pythium* spp., *Rhizoctonia solani*, and *Thielaviopsis basicola*. Phytopathology 71:954-959.

Ranney CD (1962) Fungi involved in the seedling disease complex of cotton in the Yazoo-Mississippi delta. Plant Dis. Rep. 46:122-123.

Ray WW & McLaughlin JH (1942) Isolation and infection tests with seed- and soil-borne cotton pathogens. Phytopathology 32:233-238.

Reynolds HW & Hanson RG (1957) Rhizoctonia disease of cotton in presence and absence of the cotton root-knot nematode in Arizona. Phytopathology 47:256-261.

Rickerl DH, Curl EA, Touchton JT & Gordon WB (1992) Crop mulch effects on *Rhizoctonia* soil infestation and disease severity in conservation-tilled cotton. Soil Biol. Biochem. 24:553-557.

Ridgway RL, Bell AA, Veech JA & Chandler JM (1984) Cotton protection practices in the USA and World. In: Kohel RJ and Lewis CF (eds.) Cotton. (pp 265-365) ASA Inc, Madison.

Rizk RH & Mohamed HA (1982) Histopathological studies of cotton seedlings infected with *Rhizoctonia solani*. Agric. Res. Rev. 2:83-91.

Rothrock CS, Kirkpatrick TL, Frans RE & Scott HD (1995) The influence of winter legume cover crops on soilborne plant pathogens and cotton seedling diseases. Plant Dis. 79:167-171.

Roy KW & Bourland FM (1982) Epidemiological and mycofloral relationships in cotton seedling disease in Missisippi. Phytopathology 72:868-872.

Shao FM & Christiansen MN (1982) Cotton seedling radicle exudates in relation to susceptibility to *Verticillium* wilt and *Rhizoctonia* root rot. Phytopath. Z. 105:351-359.

Smirnova AA (1980) Protection of cotton in the USSR. Outlook on Agriculture. 10:206-207.

Sneh B, Ichielevich-Auster M & Shomer I (1989) Comparative anatomy of colonization of cotton hypocotyls and roots by virulent and hypovirulent isolates of *Rhizoctonia solani*. Can. J. Bot. 67:2142-2149.

Stanton MA, Rothrock, CS, & Stewart J McD. (1994) Response of A-genome cotton germplasm to the seedling disease pathogens, *Rhizoctonia solani* and *Pythium ultimum*. Genetic Resources and Crop Evolution 41:9-12.

Veech JA (1976) Localization of peroxidase in *Rhizoctonia solani*-infected cotton seedlings. Phytopathology 66:1072-1076.

Watkins GM (ed.) (1981) Compendium of Cotton Diseases. APS Press, St. Paul, Minnesota.

Weinhold AR (1977) Population of *Rhizoctonia solani* in agricultural soil determined by a screening procedure. Phytopathology 67:566-569.

Weinhold AR & Bowman T (1974) Repression of virulence in *Rhizoctonia solani* by glucose and 3-O-methyl glucose. Phytopathology 64:985-990.

Weinhold AR, Dodman RL & Bowman T (1972) Influence of exogenous nutrition on virulence of *Rhizoctonia solani*. Phytopathology 62:278-281.

Weinhold AR & Motta J (1973) Initial host responses in cotton to infection by *Rhizoctonia solani*. Phytopathology 63:157-162.

V.4 FLOWERBULBS DISEASES INCITED BY *RHIZOCTONIA* SPECIES

GERDA DIJST AND JOHANNES H. M. SCHNEIDER

DLO Institute for Plant Protection Research (IPO-DLO), P.O.Box 9060, 6700 GW Wageningen, THE NETHERLANDS

I. THE DIFFERENT PATHOGENS

Flower bulbs may need several cropping seasons to grow sufficiently in size to propagate and produce bulb flowers. As a result, even little damage per season adds to considerable economical losses in the total production of flowers and new bulbs, which both are important export products. Several diseases can be incited by *R. crocorum* (see section II), *R. tuliparum* (see section III) and different anastomosis groups (AGs) of *R. solani* (see sections IV.A-F).

II. *RHIZOCTONIA CROCORUM*

R. crocorum is a soil borne plant pathogen with a wide host range. It causes 'copper web' on crocus and 'violet root rot' on various herbaceous plants e.g. Medicago and carrots. Old literature, reviewed by Moore (1939), describes the disease as most prevalent in saffron plantations in France and Spain at that time. In addition, the fungus incites disease of tulip, lily, iris, muscari and narcissus (Moore, 1939). In the recent Dutch reports on flower bulb diseases (Boerema and Hamers, 1988; 1989; Anonymous, 1983, 1995a) *R. crocorum* nor its synonyms are mentioned. This implies that the disease is not of economic importance under commercial cropping conditions in the Netherlands now.

Helicobasidium purpureum Pat. was demonstrated as the teleopmorph of *R. crocorum* Buddin and Wakefield (1927; 1929). For synonyms of both the anamorph and teleomorph see Boerema *et al.* (1987) and Andersen and Stalpers (1994). Historically, *R. crocorum* is the type species of *Rhizoctonia* but presently, however, it is not regarded as representative for the genus *Rhizoctonia* (Sneh *et al.*, 1991). The septal morphology (Moore, 1985) and the 5S RNA sequence (Walker and Doolittle, 1983) relate *R. crocorum* to the Ustomycetes, rather than to the Basidiomycetes.

III. *RHIZOCTONIA TULIPARUM*

'Gray bulb disease' is ascribed to *R. tuliparum* (Whetzel and Arthur, 1924; Boerema, 1963). Andersen and Stalpers (1991), however, considered the specimen (type not extant), which they examined synonymous with *R. solani*. Both the isolates of *R. tuliparum* examined by Burton and Coley-Smith (1985) as well as other isolates of *R. tuliparum* examined by Schneider (unpublished.) were binucleate and thus not *R. solani*. Andersen (pers. comm.) considered an isolate representing *R. tuliparum* in the Netherlands, not *R. solani*. Hence, it is suggested to critically revise this species. The teleomorph has not been perceived yet, which may contribute to confusion.

R. tuliparum has a wide host range, but seems most prevalent in flower bulbs (Gladders and Coley-Smith, 1978; Dickens, 1979, Boerema, 1963; Boerema and

B. Sneh et al. (eds.),
Rhizoctonia Species: Taxonomy, Molecular Biology, Ecology, Pathology and Disease Constrol, 279–288.

Hamers, 1988; 1989). Elsewhere, the fungus has been well described (Mordue, 1974), as well as the disease and disease symptoms (Moore, 1939; Dickens, 1979; Anonymous, 1983, 1995a). The host range is determined by the soil temperature rather than by host resistance (Gladders and Coley-Smith, 1978). Of the tested winter crops, *Allium porrum* and *Triticum aestivum* were not affected. Infection cushion formation and subsequent host infection have been described by Gladders and Coley Smith (1978; 1979). Shoots and storage organs are rotted by the pathogen, but not the root plate and the roots (Gladders and Coley-Smith, 1978). Since the fungus attacks the host during the winter, the disease occurs in distinct patches in spring. Sclerotia of the fungus are easily disseminated by agricultural practices and can survive in soil for at least 10 years (Coley-Smith *et al.*, 1979). Sclerotia are sensitive to soil fungistasis and rarely germinate in natural soil at temperatures above 10°C (Gladders and Coley-Smith, 1978).

For many years, the disease has effectively been controlled by fungicides and soil disinfestation means, e.g., formalin, brassicol, and PCNB. For environmental reasons, most of these chemicals are no longer allowed. Now, at planting, bulbs and soil can be treated with the fungicides Rizolex or Monarch (Anonymous, 1995b). Advises comprise: to carefully remove and destroy diseased plants and their neighbors in the field, to more intensively peel bulbs from infested fields and destroy the discarded plant material, and, if the soil type allows it, to eradicate soil infestation by a six week period of continuous and complete water inundation (Anonymous, 1995b). Development of alternative control measures warrants research on crop rotation with long intervals and of biocontrol. Gladders and Coley-Smith (1980) described the invasion and loss of viability of sclerotia by soil fungi amongst which *Trichoderma* spp.. *In vitro*, the antagonist *Coniothirium minitans* was unable to invade the sclerotia, but *Verticillium biguttatum* effectively parasitizes *R. tuliparum in vitro* (Morris *et al.*, 1995).

IV. *RHIZOCTONIA SOLANI*

A. *R. solani* AGs: diseases and symptoms

'Rhizoctonia-disease' can be incited by different AGs of *R. solani* and causes underground infection and damage in sprouts, stems, leaves and bulbs. Gladiolus are affected by AG 2-2 (Takano and Fuji, 1972). Tulip, iris, lily, hyacinth and anemone can be damaged by isolates of AG 2-t, AGs 2-1, 4 and 5 (Schneider *et al.*, 1995a). Disease severity, however, depends on the temperature (see chapter IV.2), the host type (see chapters IV.1, 2 and 3), the cultivar (Van Keulen and Van Aartrijk, 1993; Schneider *et al.*, 1995a), the pathogens' AG (subgroup) and the isolate (see chapter IV.2), crop management (see chapters IV.4 and IV.6) and the soil (see chapters IV.5 and IV.6).

Sprout infection at early plant growth, hampers plant emergence and causes bare patches. In tulip, symptoms appear as underground lesions and rot on young sprouts, leaves, stems and new daughter bulbs (MacLean, 1948; Nakatomi and Kaneko, 1971; Doornik, 1981a; Gullino and Garibaldi, 1987). In iris, lesions and rot develop on the neck and bulb (Sonderman and MacLean, 1949, Doornik, 1981a), appearing black-brown or gray-brown when incited by AGs 2-t or 4 respectively (Dijst, pers. comm.). In lilies, symptoms are reported as affected sprouts, reduced growth, yellow coloration and root rot (Bald *et al.*, 1955; McWorther, 1957; Doornik, 1981a). In hyacinth and anemone sprouts are affected, but no disease could be incited in narcissus (Doornik, 1981a; Schneider *et al.*, 1993).

In field grown tulip and iris in The Netherlands, the disease incidence is already known for long (Van Poeteren, 1928; Jaarsveld, 1952) and disease incidence increased over the last 30 years. In October, bulbs are planted at soil temperatures around 12°C. The soil can be covered with a layer of straw to prevent against winter frosts and wind erosion. It is unknown how soon infection starts after planting. In March, when soil temperatures rise again and sprouts appear above soil, empty patches and patches of reduced sprout growth become visible as a result from underground sprout infection. These patches are caused by 'cold-preferring isolates' (Doornik, 1980, 1981a) recently named AG 2-t (Schneider *et al.*, 1994). In tulip, patches are surrounded by grown plants with leaves, damaged as they develop from the affected outside of the bulb. In lightly affected iris plants, new leaves may escape damage as they develop from the inside of the bulb.

Infections established after the shoots are fully developed, do not show above ground. Until harvest in June (tulip) or August (iris), when the Dutch soil temperatures rise to 19°C (max 21°C). At higher temperatures, AG 2-t can still cause damage and spread (Doornik, 1980, 1981; Schneider *et al.*, 1993). Yet, these higher temperatures are most favorable for pathogenicity by 'warmth-preferring' AGs such as AG 4 (Schneider *et al.*, 1993). In field plots, artificially infested with AG 4, disease incidence increased from 0 to 96% after temperatures rose above 13°C (Doornik, 1980). *R. solani* affected sprouts and bulbs are quite prone to secondary invasions by *Botrytis cinerea* or *Pseudomonas* sp. Therefore, extended damage by *R. solani* beyond the borders of above ground visible patches, and the importance of the warmth-preferring pathogens in field crops, are hard to estimate.

Flowers can be produced in the field, or forced in the greenhouse from greenhouse or field raised plants. Transplants from the field can become diseased in the warm greenhouses although no disease appeared until then in the field.

B. *R. solani* AGs: temperature, pathogenicity and pathogen characterization

Depending on the season, isolates of AGs 2-t, 4 and 5 were retrieved more or less frequently from field grown tulip and iris bulbs (Schneider *et al.*, 1993). Pathogenicity tests were carried out at 9 and 18°C in the greenhouse, including sets of isolates of each of these AGs, of AG 2-1, previously associated with the disease in tulip (Nakatomi and Kaneko, 1971; Boerema and Hamers, 1988), of AG 2-2, previously associated with disease in gladiolus (Takano and Fuji, 1972), of AG 3, and one isolate of AG 6 and AG BI each (Schneider *et al.*, 1993). In general, our results with AGs 2-t and 4 were in line with previous reports on 'cold-' and 'warmth-preferring' isolates (Doornik, 1980, 1981a).

The 'warmth preferring' AGs 4 and 5 were pathogenic at 18 and not at 9°C. AG 4 damaged tulip severely, iris moderately, and hyacinth mildly, whereas lily was affected by some of the isolates. AG 5 slightly damaged tulip, iris and hyacinth. At 18°C AGs 3 and 6 did not incite any disease, all three tested isolates of AG 2-2 affected iris, two of them affected hyacinth and only the most aggressive isolate affected tulip. At 18°C the AG BI isolate incited disease in tulip, iris and hyacinth to a moderate extent, which is the first pathogenicity record for this group.

The so-called 'cold preferring' AG 2-t is pathogenic at least from 5 to 20°C, and causes severe disease near 6-12°C (Doornik, 1981a). This was confirmed in our pathogenicity tests, yet, quite many isolates still incited severe damage at 18°C. Two isolates of AG 2-1 (originating from pea in Japan and from cruciferous crops in The Netherlands) slightly affected tulip and iris at both 9 and 18°C but six other isolates of AG 2-1 (from cruciferous crops in the Netherlands) were unable to affect tulip and iris. Lily and hyacinth were not affected by AG 2-1. In lily AG 2-t caused growth reduction at 9°C and sometimes at 18°C. In hyacinth some of the AG 2-t isolates caused mild symptoms at 18 but not at 9°C.

In pot experiments in the greenhouse at 18°C, both AGs 2-t and 2-1 caused severe damping off in cruciferous seedlings (cauliflower, fodder radish, white mustard, oil-seed rape and swede). Disease rating in seedlings of lettuce, and sugarbeet ranged from zero to heavily affected, depending on the isolate tested. No disease established in seedlings of Tagetes, tomato, potato, maize and wheat (Schneider *et al.*, 1994, 1995a, unpublished). Yet, in older plants of sugarbeet AG 2-t incited disease (Van den Boogert, Personal communication).

To identify field isolates and distinguish AG 2-t from AG 2-1, pathogenicity tests in greenhouses seem inadequate. First, because of the variable results with the different AG 2-1 isolates. Secondly, since AG 2-t isolates vary largely in aggressiveness at higher temperatures and, in addition cultivars varied in susceptibility per isolate tested (Schneider *et al.*, 1995a). Also fluctuations in temperature can affect inoculum potential of isolates in the soil (Doornik, 1980). Colonies of AGs 2-t and 2-1 equally extended on agar plates at 5-35°C with an optimum near 23°C, confirming Doornik (1981a). A test on thiamin requirement is also useless, since both AGs 2-t and 2-1 appeared autotrophic (Ogoshi and Ui, 1979; Schneider, unpublished).

AG 2-t cannot be distinguished from AG 2-1 on the basis of anastomosis. Hyphal fusion frequency (FF) showed that AG 2-t forms a distinct group (FF $>50\%$) that AG 2-t differs from AG 2-2 (FF $<30\%$), that AG 2-t is unclearly related to AG 2-1 (2 $<$FF $>75\%$) and that AG 2-1 forms a highly variable group (1 $<$FF $<$82).

Clear distinction between AGs 2-t and 2-1 isolates from The Netherlands succeeded with pectin zymograms, which yielded a constant pattern for AG 2-t but a highly variable pattern for these AG 2-1 isolates (Schneider *et al.*, 1994, 1995a). However two Japanese AG 2-1 isolates (originating from pea and tulip) yielded the same pattern as the AG 2-t isolates. PCR on rITS regions yielded small differences between AG 2-t and AG 2-1 from the Netherlands but not between AG 2-t and the Japanese AG 2-1. In contrast to previous references (Cruckshank, 1990; Loerakker, unpublished), Schneider *et al.* (1994, 1995a) named the 'cold-preferring' tulip pathogen 'AG 2-t' because the variability within AG 2-1 warrants further research and because AG 2-t is a distinct group within AG 2.

C. *R. solani* AGs: stimulation of inoculum potential and spread

Especially at less favorable temperatures, the inoculum potential of *R. solani* AG 2-t in soil is affected by the nutritional state and the amount of inoculum, the extent of soil suppressiveness, and stimulatory neighboring plants (Doornik, 1980, 1981a; Dijst *et al.*, 1993, 1995). The same holds for AG 4, although the role of inoculum density appears much less important. The conclusions are based on experiments under standardized conditions (Oyarzun *et al.*, 1994a) with AG 4 at 18°C and with AG 2-t at 12°C.

These experiments showed that, in order to incite severe disease, a good nutritional basis appears more important than the amount of inoculum (Momberg *et al.*, 1995; Dijst *et al.*, 1995). In these experiments iris plants promoted the spread of AG 2-t and AG 4 through soil and helped the pathogens to overcome the natural soil suppressiveness. In unplanted natural soil, we detected no spread upon Juncus baiting, a technique advised by Doornik (1981b). Yet, in line with Doornik (1980, 1981a), we found that hyphae readily spread along smooth surfaces like pot walls. In contrast to Blair (1943) we did not detect a change in pathogenicity at increasing distance from the inoculum within five weeks.

Host plants differ in their stimulatory effect on the spread of the pathogen and its disease. Doornik (1980) recorded in pot experiments at 20°C less spread in tulip and lilies than in iris and less disease stimulation by amending the soil with parts of tulip and lily than with iris parts or oats. Affection of the bulbs is more superficial for tulip and lily but readily penetrates deep in the bulbs of iris. Yet, in all three crops, higher plant densities promoted disease spread (Doornik, 1980). Also in the field, disease spreads faster within than between rows (Schneider *et al.*, 1993).

These results support the hypothesis that healthy host plants may serve as an exogenous nutritional source, stimulating spread and disease severity (Papavizas, 1969; Weinholt *et al.*, 1972; Doornik, 1980). Some other results, however, are less easily understood from this point of view. Monoculture in iris and hyacinth incited disease decline for AG 2-t, faster at 17 than at 9°C, and above that, disease incidence revived in monocultures at 17°C by an 9°C interval (Doornik, 1980). In iris, cooling the bulbs before planting, enhanced shoot growth and increased disease severity, which might be explained as increased vulnerability based on stimulatory plant exudates (Doornik, 1981a). It may, however, as well result from a lack of time for soil organisms to settle and compete for nutrients with the pathogens at faster growing infested sprouts. Under normal growth conditions, young sprouting iris bulbs and 6 weeks old grown iris plants were equally susceptible (Dijst, unpublished).

Yet, in line with the hypothesis, we found that removal of the inoculum (infested oat kernels), six days after inoculation, reduced the spread in unplanted soil in the next four weeks. This effect was stronger in natural than in sterilized soil and also stronger in tests with AG 4 at 18°C than in bioassays with AG 2-t at 12°C, a difference for which a temperature effect on the general soil antagonism may well account.

D. *R. solani* AGs: survival, crop rotation and disease dynamics

Doornik (1982) reported that infection can result from inoculum in the soil or from colonized bulbs. Furthermore, that in tulip, latent infection is hard to detect and light symptoms do not always show at harvest. Detection by plating out on agar media often failed, although the examined bulbs remained a source of infection at a next planting. In storage, the infectivity of colonized iris bulbs remained higher at the usual storage temperature of 30°C than at lower temperatures. In soil, the infectivity of buried infested bulbs, survived for at least two years at depth of 40-60 cm and, shorter at more shallow depths (Muller, unpublished.). Doornik (1980) showed, that fallow periods reduce the inoculum potential of the soil, whereas alternative hosts like radish and pea have no effect. According to our pathogenicity tests also cruciferous fodder crops can hold a risk.

Reports from growers on patch dynamics in the field often seem contradictive. Also reports on disease dynamics in rotation and in monoculture systems differ. In The Netherlands, tulip and iris are usually cropped in at least 1:3 rotation with other crops. In these practices, the patches come and go irregularly and differ in size. Growers report that in some fields, patches always reappear in the same part, whereas in other fields patches appear and vanish randomly. To obtain a better view on patch dynamics, we studied disease incidence in experimental fields without pesticide use. Geostatistical analysis of the data on disease incidence showed that the applied sampling grid could have been reduced 90% (correlation coefficient 67%), up to one sample per m per row to study disease occurrence (Schneider and Kocks, 1995b).

Monoculture with iris or hyacinth caused decline of naturally and artificially (AG 2-t) incited disease, both in pot experiments at constant temperatures as well as in field plots (Doornik, 1980, 1981a). In pot experiments, the same author also observed that saprophytic activity in each crop first increased and later declined, dropped at soil disturbance during harvest and, at immediate replanting, recovered but remained at a much reduced level in the second crop. Recently, also Schneider observed a disease decline in a field trial with a rotation of tulip-iris-tulip-tulip (1991-1995). In this sandy field, both naturally and artificially established patches gradually disappeared. However, in the last year, all over the field, very light symptoms developed from which AG 2-t could be isolated. At the harvest of that last season, both AGs 2-t and (the lightly pathogenic) 5 could be isolated from bulbs from all over the field, although AG 5 much more frequently than AG 2-t. These observations in bulbs do not correspond with a report on the occurrence of AG 5 and AG 3 in potato (Jager and Velvis, 1995).

The decline of AG 2-t incited disease in monocultures of susceptible bulbous crops, may be explained from competition for infection stimulatory nutrients from the host plants, as suggested before (Doornik, 1980). The previous section of this chapter mentioned the stimulatory effects of host plant material on AG 2-t and the dependency of AG 2-t on nutrients to incite severe disease. Therefore, ongoing research studies our hypothesis that the susceptible bulb crops stimulate the population and spread of both AG 2-t and of other microorganisms, including the lightly pathogenic AG 5, which compete, not for space, but for the same infection stimulatory nutrients from the host plants, which are essential for AG 2-t to incite severe disease. The facts that disease decline establishes more quickly at higher temperatures (Doornik, 1980) and that AG 5 is pathogenic in bulbs at higher temperatures only (Schneider *et al.*, 1993) strengthen the hypothesis.

E. *R. solani* AGs: soil suppressiveness

We determined that 'the soil' has an important effect on spread and severity of *Rhizoctonia*-disease and should be taken into account in site-specific crop management. To compare and analyze the effect of natural soils on Rhizoctonia-disease, bioassays were carried out under controlled temperature and soil humidity with AG 2-t at 12°C and with AG 4 at 18°C (Oyarzun and Dijst, 1991; Oyarzun *et al.*, 1994a; Dijst *et al.*, 1993, 1995). A great diversity in natural disease suppression was found for soil samples from different locations. Test results were in line with the growers' information on disease occurrence at the sampled sites. For some samples we measured the same difference in suppressiveness in three successive years. This implies that stable differences in soil suppressiveness might occur at a broader scale.

Especially if the differences could be correlated indirectly to chemical or physical soil characteristics which, in general, largely determine the soil biology.

A general antagonism, based on the soil microbial activity seems to play a major role in mediating soil suppressiveness (Dijst *et al.*, 1995). Several findings strengthen this hypothesis: Sterilization through gamma-irradiation eradicated any suppressive effect of the soil samples. Also, larger and stronger differences in suppressiveness are measured at higher (AG 4 tested at 18°C) than at lower temperatures (AG 2-t tested at 12°C). Enrichment of the soil by amending it with straw some months prior to the testing, significantly increased the natural suppressiveness against disease spread from a point inoculation. In the field, specific antagonism is unlikely to play an important role, since in only one out of sixty soil samples tested, some specific antagonistic effect against *R. solani* was detected although the bioassay showed a diversity in disease suppression in this set of soil samples.

F. *R. solani* AGs: regular control measures and new perspectives

In this high value bulb industry, every disease occurrence demands action for control. The bulbs need several growing seasons to grow, propagate and develop a primordium for flower production. Therefore, every little damage per season adds to considerable losses in the complete production of bulbs and flowers. Advises for control (Anonymous, 1995b) comprise for field cropping, a fungicide treatment of the bulbs, soil and the straw cover at planting and, spraying fungicides on and around disease patches immediately upon symptom appearance. Greenhouse cropping are advised to treat the bulbs, disinfect soil, boxes or trays, to plant the bulbs partly above soil to avoid leaf damage and to discard contaminated soil and plant residues.

However, often these measures are insufficient. Chemical control is not always effective and its use is restricted by environmental issues of soil and groundwater quality. Resistant cultivars are not available, highly susceptible cultivars are often economically too attractive to be abandoned and systematic knowledge on effects of rotation and fodder crops are lacking. Water inundation of the fields, is ineffective against diseases incited by *R. solani* and *Pythium* in bulbs (Van Aartrijk, unpublished.; Van Os, 1995). This is in line with our finding that sterilization eradicates any natural suppression against Rhizoctonia-disease, as previously also reported for foot rot in pea (Oyarzun *et al.*, 1994b). It implies more generally, that cultural practices, like soil disinfestation or inundation, which negatively affect the soil microbial activity, hold a risk for Rhizoctonia-disease.

These facts warrant development of new and integrated control strategies. Chet *et al.* (1982) reported successful control in greenhouse cropped iris by combining bulb treatment with the antagonist *Trichoderma harzianum* in combination with a treatment of the soil with PCNB or solarization. For field croppings, our current efforts include identification of alternative hosts and application of *V. biguttatum*: on one hand to avoid and treat plant residues that hold a risk for survival or stimulation of spread and infection and, on the other hand to seek enhancement of the soil suppressiveness. This approach is based on the successful results against AG 3 in seed potatoes by the new cultural method of 'root cutting' or 'green crop harvesting' in combination with this antagonist (Dijst *et al.*, 1986, 1990; Mulder *et al.*, 1992; Van den Boogert and Deacon, 1994).

The fact that 'the soil' has an important effect on spread and severity of Rhizoctonia-disease, should be taken into account in the development of an integrated site-specific crop management and in the choice of sites for highly susceptible cultivars. The cropping history and the processes that are inherent to the development of the soil, give to each field its unique intrinsic properties and consequently its unique degree of receptivity (ranging from conducive to suppressive) for a soil borne disease (Alabouvette *et al.*, 1982). In general, as a result, the risk for damage can differ between equally infested fields (Oyarzun *et al.*, 1994b, 1995).

Development of preventive assessments on the risk factor 'soil' may serve the choice of land sites upon its degree of suppressiveness. However, it warrants additional research on simplification of the available bioassay on suppressiveness, on adequate soil-sampling protocols and on assessment of economic risk levels. If applicable, the use of soil characteristics as an indicator for soil suppressiveness, would be easier than a bioassay, but warrants additional scientific proof.

This aspect of risk-assessment appears increasingly important because of the fast growing number of growers that hire or collaborate with different (unknown) land each year. Some propagators specialize and thus, need different fields each year. Flower producers often delegate their last season for bulb growth to countries with a different climatic rhythm (Eastern Europe, Israel, Australia, etc.) which sets the formation of the flower primordium at an unusual time so the grower can meet the market competitively.

V. References

Alabouvette C, Couteaudier Y & Louvet J (1982) Comparaison de la réceptivité de différents sols et substrats de culture aux fusarioses vasculaires. Agronomie 2:1-6.

Anonymous (1983) Ziekten en afwijkingen in bolgewassen. Part 1: Liliaceae. Second edition. Bulb Research Centre (LBO), Lisse, the Netherlands.

Anonymous (1995a) Ziekten en afwijkingen in bolgewassen. Part 2: Amaryllidaceae, Araceae, Begoniaceae, Cannaceae, Compositae, Iridaceae, Oxalidaceae and Ranunculaceae. Second edition Bulb Research Centre (LBO), Lisse, the Netherlands.

Anonymous (1995b) Gewasbeschermings Adviezen 1995, Dienst Landbouw Voorlichting (DLV), The Netherlands, p. 40-42.

Andersen TF & Stalpers JA (1994) A check-list of *Rhizoctonia* epithets. Mycotaxon 51:437-457.

Bald JG, Kofranek AM & Lunt OR (1955) Leaf scorch and *Rhizoctonia* on croft lillies. Phytopathology 45:156-162.

Blair I.D (1943) Behaviour of the fungus *Rhizoctonia solani* Kühn in the soil. Ann. Appl. Biol. 30:118-127.

Boerema GH (1963) De veroorzaker van de "kwade-grondziekte" bij bloembolgewassen: *Rhizoctonia* tuliparum Whetzel and Arthur. Verslagen en Mededelingen van de Plantenziektenkundige Dienst No. 141 (Annual Report Plant protection Service The Netherlands):179-182.

Boerema GH, Loerakker WM & Hamers MEC (1987) Check-list for scientific names of common parasitic fungi. Supplement Series 2a (additions and corrections): Fungi on field crops: beet and potato; caraway, flax and oilseed poppy. Neth. J. Plant Pathol. 93:1.

Boerema GH & Hamers MEC (1988) Check-list for scientific names of common parasitic fungi. Series 3a: Fungi on bulbs: Liliaceae. Neth. J. Plant Pathol. 94:1-32.

Boerema GH & Hamers MEC (1989) Check-list for scientific names of common parasitic fungi. Series 3b: Fungi on bulbs: Amaryllidaceae and Iridaceae. Neth. J. Plant Pathol. 95:3.

Boerema GH & Hamers MEC (1990) Check-list for scientific names of common parasitic fungi. Series 3c: Fungi on bulbs: 'additional crops' belonging to the Araceae, Begoniaceae, Compositae, Oxalidaceae and Ranunculaceae. Neth. J. Plant Pathol. 96:1.

Buddin W & Wakefield EM (1927) Studies on *Rhizoctonia crocorum* (Pers.) DC. and *Helicobasidium purpureum* (Tul.) Pat.. Trans. Brit. Mycol. Soc. 12:116-140.

Buddin W & Wakefield EM (1929) Further notes on the connections between *Rhizoctonia crocorum* (Pers.) DC. and *Helicobasidium purpureum* (Tul.) Pat.. Trans. Brit. Mycol. Soc. 14:97-99.

Burton B & Coley Smith JR (1985) Antibiotics in sclerotia and mycelium of *Rhizoctonia* species. Trans. Brit. Mycol. Soc. 85:447-453.

Chet I, Elad Y, Kalfon A, Hadar Y & Katan J (1982) Integrated control of soilborne and bulbborne pathogens in iris. Phytoparasitica 10:229-236.

Coley-Smith JR, Humphreys-Jones DR & Gladders P (1979). Long-term survival of *Rhizcotonia tuliparum*. Plant Pathol. 28:128-130.

Cruickshank RH (1990) Pectic zymograms as criteria in taxonomy of *Rhizoctonia*. Mycol. Res. 94:938-946.

Dickens JSW (ed.) (1979) Diseases of Bulbs. Ministry of Agriculture, Fisheries and Food, Reference book HPD 1, London.

Dijst G, Bouman A, Mulder A & Roosjen J (1986) Effect of haulm destruction and additional root-cutting on the incidence of black scurf and skin damage, flexibility of harvest period and yield of seed potatoes in field experiments. Neth. J. Plant Pathol. 92:287-303.

Dijst G (1990) Effect of volatile and unstable exudates from underground potato plant parts on sclerotium formation by *Rhizoctonia solani* AG 3 before and after haulm destruction. Neth. J. Plant Pathol. 96:155-170.

Dijst G, Schneider JHM, Schilder MT & Fokkema NJ (1993) Comparison of soil receptivity to *Rhizoctonia solani* AG's in cauliflower and iris. Proceedings of the 6th Int. Congr. Plant Pathol. Montreal, Canada, 28 July - 6 Augustus (16.4.9) p. 282.

Dijst G, Schilder MT, Schneider JHM & Oyarzun PJ (1995) Population and disease dynamics and suppression of *Rhizoctonia solani* AG 2-t and AG 4 in Iris in natural soils under standardized conditions. 2nd Int Symp. on *Rhizoctonia*, 27-30 June Noordwijkerhout, the Netherlands. P-6-4.

Doornik AW (1980) Some factors affecting the parasitic and saprophytic activity of *Rhizoctonia solani*. Acta Hort. 109:387-394.

Doornik AW (1981a) Temperature dependence of the pathogenicity of several isolates of *Rhizoctonia solani* in some bulb crops as an intrinsic property of the isolate. Neth. J. Plant Pathol. 87:139-147.

Doornik AW (1981b) Comparison of methods for detection of *Rhizoctonia solani* in soil. . Neth. J. Plant Pathol. 87:173-176.

Doornik AW (1982) Effect of storage duration and temperature on the survival of *Rhizoctonia solani* in tulip and iris bulbs. . Neth. J. Plant Pathol. 88:185-190.

Gladders P & Coley-Smith JR (1978) *Rhizcotonia tuliparum*: a winter active pathogen. Trans. Brit. Mycol. Soc. 71:129-139.

Gladders P & Coley-Smith JR (1977) Infection cushion formation in *Rhizoctonia tuliparum*. Trans. Brit. Mycol. Soc. 68:115-118.

Gladders P & Coley-Smith JR (1979) Host infection by *Rhizoctonia tuliparum*. Trans. Brit. Mycol. Soc. 72:251-260.

Gladders P & Coley-Smith JR (1980) Interactions between *Rhizoctonia tuliparum* sclerotia and soil microorganisms. Trans. Brit. Mycol. Soc. 74(3):579-586.

Gullino ML & Garibaldi A (1987) Principal fungal diseases of bulbous flowering plants, gladiolous and tulip. Panorama Floricolo 12:5-7.

Jaarsveld A (1952) Annu. Rep. Plant Prot. Serv., Wageningen, The Netherlands 1951-1952 201 pp.

Jager G & Velvis H (1995) Dynamics of *Rhizoctonia solani* (black scurf) in successive potato crops. Eur. J. Plant Pathol. 101:467-478.

Loerakker WM & Van Dreeven F (1985) Anastomosis groups of *Rhizoctonia solani* Kühn occurring in the Netherlands. Plant Protection Service Wageningen, The Netherlands, Report 163 (Annual Report 1984):16-21.

MacLean NA (1948) Rhizoctonia rot of tulips in the Pacific Northwest. Phytopathology 38:156-157.

McWorther FP (1957) Association between *Rhizoctonia* and yellow coloration of easter lily bulbs. Phytopathology 47:447 448.

Mombarg HJM, Schilder MT & Dijst G (1995) Effect of soil sterilization and iris bulbs on the population and disease dynamics of *Rhizoctonia solani* AG 2-tulip and AG 4 in a sandy soil. Acta Botanica Neerlandica 44:78.

Moore WC (1939) Diseases of bulbs. Ministry of Agriculture, Fisheries and Food, Bull. 117, London.

Moore RT (1985) The Challenge of the dolipore/parenthesome septum. In: Moore D, Casselton LA, Wood DA & Frankland JA (eds.). Developmental Biology of Higher Fungi. (pp 175-212) Cambridge.

Mordue JEM (1974) *Rhizoctonia tuliparum*. CMI descriptions of Pathogenic Fungi and Bacteria No. 407. CMI, Kew.

Morris RAC, Coley-Smith JR & Whipps JM (1995) The ability of the mycoparisite *Verticillium biguttatum* to infect *Rhizoctonia solani* and other plant pathogenic fungi. Mycol. Res. 99:997-1003.

Mulder A, Turkensteen LJ & Bouman A (1992) Perspectives of Green-Crop-Harvesting to control soil-borne and storage diseases of seed potatoes. Neth. J. Plant Pathol. 98:103-114.

Muller PJ (1969) *Rhizoctonia solani* Kühn as a parasite of tulips. Mededel. Rijksfac. Landbouw. Gent 34:839-846.

Naito S & Kanematsu S (1995) Characterization and Pathogenicity of a New Anastomosis Subgroup AG 2-3 of *Rhizoctonia solani* Kühn Isolated from leaves of Soybean. Ann. Phytopathol. Soc. Jpn. 60:681-691.

Nakatomi Y & Kaneko H (1971) Ecology and control of leaf blight of tulip. Plant prot. Jpn. 25:191-194.

Ogoshi A & Ui T (1979) Specificity in vitamin requirement among anastomosis groups of *Rhizoctonia solani* Kühn. Ann. Phytopathol. Soc. Jpn. 45:47-53.

Oyarzun PJ & Dijst G (1991) Assessment of site-specific receptivity of soils to soil-borne diseases of pea and cauliflower. In: Beemster ABR, Bollen GJ, Gerlagh M, Ruissen MA, Schippers B & Tempe, A (eds.). Biotic Interactions and Soil-borne Diseases. Proc. 1st Conf. Eur. Founda. Plant Pathol. Wageningen, The Netherlands 1990. (pp 322-328) Elsevier Press, Amsterdam:

Oyarzun PJ, Zalm AJ van der, Roelofsen HW & Dijst G (1994a) A computerized system to control soil water potential and soil temperature in soil inoculum potential assays of soil-borne pathogens. Plant & Soil, 166:125-135.

Oyarzun PJ, Dijst G & Maas PWTh (1994b) Determination and analysis of soil receptivity to *Fusarium solani* f. sp. *pisi* causing dry root rot of peas. Phytopathology 84:834-842.

Oyarzun PJ, Gerlagh M, Dijst G & Maas PWTh (1995) Middelen voor het beheersen van bodemgebonden ziekten zijn al in de bodem aanwezig: Bodemreceptiviteit ten aanzien van voet-en wortelrot bij erwt. Gewasbescherming 26:119-128.

Papavizas GC (1969) Survival of Root infecting fungi in soil. XI. Survival of *Rhizoctonia solani* as affected by inoculum concentration and various amendments. Phytopathol. Z. 64:101-111.

Pitt D (1964) Studies on sharp eyespot disease of cereals. II. Viability of sclerotia: persistence of the causal fungus Rhizoctonia solani Kühn. Ann. Appl. Biol. 54:231-240.

Sanford GB (1952) Persistence of *Rhizoctonia solani* Kühn in soil. Can. J. Bot. 30:652-664.

Schneider JHM, Dijst G & Fokkema NJ (1993) Ecological and epidemiogical aspects of *Rhizoctonia solani* in flower bulbs. Proc. 6th ICPP. Montreal, Canada, 28 July 28 - 6 August: (Abst. 16.4.8) p 282.

Schneider JHM, Schilder MT, Dijst G & Keijer J (1994) Characterization of *Rhizoctonia solani* anastomosis group 2 (AG 2) in tulip. Fifth International Mycol. Congr. August, Vancouver Canada.

Schneider JHM, Schilder MT, Keijer J & Dijst G (1995a) Anastomosis behaviour, pathogenicity and pectic zymograms of *Rhizoctonia solani* AG 2 in tulip. 2nd Int. Symp. on *Rhizoctonia*, 27-30 June, Noordwijkerhout, The Netherlands. P-1-8 p 30.

Schneider JHM & Kocks CG (1995b) Mapping Rhizoctonia disease using geostatistics to reduce sampling. 2nd Int. Symp. on *Rhizoctonia*, 27-30 June, Noordwijkerhout, The Netherlands. P-5-5 p 85.

Sneh B, Burpee L & Ogoshi A (1991) Identification of Rhizoctonia Species. APS. Press St. Paul, Minnesota. 133pp.

Sonderman CH & MacLean NA (1949) Rhizoctonia neck and bulb rot of iris in the Pacific Northwest. Phytopathology 39:174-175.

Takano K & Fuji N (1972) On the sheath blight of gladiolus, a new disease. Ann. Phytopath. Soc. Jpn 38:192.

Van den Boogert PHJF & Deacon JW (1994) Biotrophic mycoparasitism by *Verticillium biguttatum* on *Rhizoctonia solani*. Eur. J. Plant Pathol. 100:137-156.

Van Keulen H & Van Aartrijk J (1993) Ziektegevoeligheid van cultivars van bloembolgewassen. Milieuplatform Bloembollensector, Hillegom, 43 pp.

Van Os (1995) *Pythium* aantasting: het gevolg van een verstoord biologisch evenwicht. Bulb Research Centre (LBO), Lisse, Netherlands. 37 pp.

Van Poeteren N (1928) Annu. Rep. Plant Prod. Serv. Wageningen, The Netherlands, 27 pp.

Walker WF & Doolittle WF (1983) 5S RNA sequences from eight basidiomycetes and fungi imperfecti. Nucleic Acids Res. 11:7625-7630.

Weinholt AR, Dodman R & Bowman T (1972) Influence of exogenous nutrition on virulence of *Rhizoctonia solani*. Phytopathology 62:278-281.

Whetzel HH & Arthur JM (1924) The grey bulb-rot of tulips. Phytopathology 14:30-31.

Whipps JM & Gerlach M (1992) Biology of *Coniothyrium minitans* and its potential for use in disease biocontrol. Mycol. Res. 96:897-907.

V.5 FORAGE AND OILSEED LEGUME DISEASES INCITED BY *RHIZOCTONIA* SPECIES.

SHEAU-FANG HWANG[1], RONALD J. HOWARD[2] AND KAN-FA CHANG[2]

[1]*Alberta Environmental Center, Bag 4000, Vegreville, Alberta, CANADA T9C 1T4.*, [2]*Crop Diversification Centre-South, SS #4, Brooks, Alberta, CANADA T1R 1E6.*

I. ALFALFA (*MEDICAGO SATIVA* L.)

Since alfalfa is a perennial crop, harvesting, winter injury and insect pests and parasitic nematodes can cause repeated wounds, through which several pathogens can gain access. Diseases caused by *Rhizoctonia solani* Kühn reduce its stand, yield, forage quality and persistence of alfalfa, and will also lessen its resistance to environmental stresses and its ability to utilize water and nutrients.

A. Symptoms and epidemiology

R. solani causes four main types of disease symptoms in alfalfa: seedling blight, stem and leaf blights, crown bud rot, and root canker.

1. Seedling blight Seed and seedling diseases can pose a significant threat to alfalfa production in some parts of the world. *Rhizoctonia* contributes significantly to stand losses and poor establishment in alfalfa fields (Hancock, 1983; Seif El-Nasr and Leath, 1983). Pre- or post-emergence damping-off is the most common result when young seedlings are attacked.

2. Stem and leaf blights These symptoms are prevalent in the southeastern United States (Graham *et al.*, 1978). During prolonged periods of hot and wet weather, *R. solani* attacks leaves and stems, which quickly collapse to form a water-soaked mass of tissue. Large areas of dead plants may look like they have been killed by scalding water. The fungus spreads among adjacent plant parts within a few days. Dead leaves stick to neighboring leaves, and stems and are covered by mycelial strands. Less severely damaged leaves and stems have large, irregularly shaped lesions that are usually gray with a dark, reddish-brown border.

3. Crown bud rot Crown bud rot caused by *R. solani* can be a major problem wherever alfalfa is grown in North America. The disease is irreversible and develops most rapidly during the first 2 months of annual plant growth when soil temperatures are 9-13°C (Hawn, 1958). There is a significant correlation between disease severity and plant age. This disease causes alfalfa stands to progressively deteriorate after the first growing season (Graham *et al.* 1979, Gray *et al.* 1980).

Crown bud rot causes dark brown or black lesions on tender bud tissues (Hawn *et al.*, 1981). Crown buds die, then the fungus grows into the crown and upper root areas, thus preventing further production of vegetative buds. Large portions of crown tissue may become necrotic, thereby limiting the amount of healthy tissue from which crown buds can form. Symptoms range from partial rotting of crowns to dead plants. Roots of these plants may appear healthy.

B. Sneh et al. (eds.),
Rhizoctonia Species: Taxonomy, Molecular Biology, Ecology, Pathology and Disease Constrol, 289–301.

4. Root canker *R. solani* causes dark, sunken, usually circular cankers along the taproots and larger lateral roots (Smith, 1943). Cankers usually occur where new rootlets emerge. The fungus is active when the soil temperatures exceed 20°C. In 1965 in western Queensland (Purss, 1965), a form of root canker resembling Phytophthora root rot was found to be caused by *R. solani*. Considerable damage occurred when soil temperatures were abnormally high.

B. Disease management

1. Overseeding Seedling diseases are only occasionally considered economically significant problems of alfalfa in the United States (Graham *et al.*, 1979). Overseeding is commonly recommended (Tesar and Jackobs, 1972) to compensate for seedling stand losses, and may be partially responsible for the lack of concern amongst growers for seedling diseases in their alfalfa crops.

2. Resistant cultivars Economic control of alfalfa diseases is usually obtained through plant resistance. Currently, Anchor, Beaver, Drylander, Roamer, Thor and Trek cultivars are resistant to *R. solani* crown bud rot (Horricks, 1981). Low to moderate resistance to damping off caused by AG 4 isolates has been obseved in some alfalfa cultivars, revealing the potential for the development of resistant cultivars (Kulik and Daery, 1995; Kulik *et al.*, 1995). Adapted varieties will decrease damage by Rhizoctonia crown and root rots, if used in conjunction with cultural practices that enhance vigorous plant growth.

3. Biological control Chet and Baker (1980) discussed biological control of *R. solani* in alfalfa using *Trichoderma harzianum* Rifa,, which parasitizes and lyses *R. solani* cells. Barnes *et al.* (1981) suggested that the nematode *Aphelenchus avenae* Bastian, has potential for biological control of *R. solani* in alfalfa fields. Sitrit *et al.* (1993) reported that *Rhizobium meliloti* Dangeard, during symbiosis on alfalfa roots, had antifungal activity against *R. solani*. To date, no *Rhizobium* bacteria have demonstrated the ability to control plant diseases.

4. Chemical control Hancock (1983) reported high yields of alfalfa in plots treated with benomyl. Benlate T (benomyl + Thiram), applied as a seed dressing at 1500ppm, promoted seedling emergence and survival when grown in *Rhizoctonia*-infested soil (Fathi, 1977).

5. Cultural control Alfalfa crops should be properly managed against *R. solani* crown bud rot by maintaining good plant vigor and adequate fertilization. *R. solani* does not invade healthy roots, but penetrates only through wounds. Reduced vigor predisposes affected plants to cold injury and to infection by fungi that cause crown rot (Hawn, 1958). *Rhizoctonia* becomes active when environmental conditions are favorable i.e. warm temperatures and moist soil.

Proper management includes allowing alfalfa to recover in the spring before cutting and leaving about 25 cm of re-growth before the end of the growing season to renew root reserves. Autumn harvests should be done after the plants become dormant (Horricks, 1981). If foliar blight develops, early cutting is recommended to reduce the effects of damage.

Perennial cropping enriches soil populations of root pathogens to which these crops are especially susceptible. *Rhizoctonia* overwinters in diseased tissue or in the soil as a saprophyte (Horricks, 1981). To reduce the population density of *R. solani* in the field, alfalfa should be rotated with crops which are not infected by the same AG isolates, such as cereals, for 2 or 3 years to reduce population build-up of the pathogen in the soil.

II. CLOVER (*TRIFOLIUM* SPP.)

A. Symptoms and epidemiology

R. solani isolates of AGs 2-1 and 2-2 are highly virulent on clover (Wong and Sivasithamparam, 1985), and cause seedling blight, damping-off and root rot. *Rhizoctonia* causes substantial damage to several types of clover. Subterranean clover, *Trifolium subterraneum* L., which is the most important pasture crop in Western Australia, (Gladstones, 1975) has suffered severe yield losses as a result of root rot. Red clover (*Trifolium pratense* L.) is susceptible to various soil-borne fungal pathogens that can kill new seedlings (Leath, 1989, Nelson and Campbell, 1993). White or ladino clover (*Trifolium repens* L.), is a useful pasture legume in the wetter coastal areas of southwestern Australia and southeastern United States, but its persistence falls far short of its potential because of Rhizoctonia root rot (Chow, 1960).

1. Seedling blight and damping-off *R. solani* can prevent seed germination or may attack the root collar of the emerging clover seedlings and cause damping-off (Maughan and Barbetti,1983). Dead seedlings disappear, creating gaps in the rows. When the plants reach the 4th or 5th leaf stage, they usually become resistant to damping-off and seedling blight.

2. Root rot Complexes of pathogenic, root-invading fungi have caused declining productivity and poor longevity in red clover (Leath, 1989). Poor persistence has plagued white clover, which usually declines or disappears within 4-5 years when grown with tall fescue (*Festuca arundinacea* Schreb) (Kochhar, 1974). In irrigated areas, white clover production has declined substantially because of root rot (Maughan and Barbetti, 1983). Ludbrook *et al.* (1953) found *R. solani* to be the main cause of 'bare-patch disease' and associated production loss problems in subterranean clover pastures in South Australia. Root rot of subterranean clover pastures in Western Australia is also induced by a complex of interacting fungal pathogens, such as *R. solani*, *Fusarium avenaceum* (Fr.) Sacc., *F. oxysporum* Schlecht., *Pythium irregulare* Buisman, and *Phoma medicaginis* Malbr. & Roum. (Wong *et al.*, 1984, 1985).

Infected subterranean clover may appear healthy, but does not recover after grazing. Barbetti (1984) and Barbetti *et al.* (1986) showed a high negative correlation between the severity of root rot and plant size and weight. Affected white clover plants usually died, but surviving plants showed rotting and stripping of all feeder roots, and severe rotting of the tap roots and stolons (Maughan and Barbetti, 1983).

Rhizoctonia root rot incidence was higher when the soil was relatively dry (Maugham and Barbetti, 1983; Crowder and Craigmiles, 1960). At lower soil moisture levels (20 to 50% water holding capacity), saprophytic activity of *Rhizoctonia* was significantly higher than at high soil moisture (70 to 90% WHC) (Papavizas and Davey, 1962. At 30% of soil saturation, *Rhizoctonia* hyphae were not affected by bacteria, but complete lysis of mycelium occurred at 80% (Kovoor, 1954). Root rot in subterranean clover was generally most severe at 20 to 25°C, but *Rhizoctonia* caused severe root rot over a wide temperature range (Wong *et al.*, 1984). Barbetti (1990) found less severe root rot symptoms at high soil pH levels.

B. Management

1. Cultural control Proper management techniques, such as use of sound seed, lack of weed competition, good seedbed preparation, and shallow seeding can reduce seedling blight of clover. It is important to inoculate seeds with beneficial nitrogen-fixing bacteria when this crop is grown for the first time. Pastures have often been re-seeded to maintain productivity and prevent a buildup of weeds and grasses. In some situations, however, seedling survival after re-seeding has been very poor due to severe root rot (Barbetti, 1984).

Disease caused by *R. solani* is more severe in clovers and alfalfa when they are seeded after clover or alfalfa, so this type of rotation should be avoided (Barbetti, 1991). Complete removal of subterranean clover from a pasture for one or two consecutive seasons significantly reduced tap and lateral root disease (Barbetti, 1991).

2. Chemical control Fungicidal seed treatments, such as benomyl or prochloraz, controlled seedling blight (Leath *et al.*, 1973; Greenhalgh and Clarke, 1985; Nan *et al.*, 1991). However, fungicide soil drenches have not effectively controlled root rot in established subterranean clover pastures in Western Australia (Barbetti, 1984). An effective seed treatment may only help in the initial re-seeding and re-establishment of subterranean clover.

3. Resistance Daliak and Dinninup cultivars are moderately resistant to root rot (Gillespie, 1980). No cultivars fully resistant to root rot have been identified to date.

III. Lupin (*Lupinus* spp.)

A. Symptoms and epidemiology

All lupines are susceptible to *R. solani*, but white lupin is more susceptible than the other types due to its low alkaloid content (Wink, 1984). In lupin, *R. solani* causes seed rot, root lesions, root decay, reduced nodulation, reduced root growth, damping-off, stem nipping, stem lesions, and apical bud mortality (Leach and Clapham, 1992). Leach and Clapham (1992) reported that AG 5 isolates are the major pathogen of white lupin. Strains of AG 1 and AG 4 also infect lupin, but cause only minor damage.

1. Seedling damping-off *R. solani* attacks seedlings before emergence and often kills large numbers of them by causing soil line stem cankers after they emerge. Older plants either sustain slight injury and show no top symptoms or they may be dwarfed or slowly killed. *R. solani* AG 4 causes sunken reddish-brown hypocotyl lesions on seedlings (Kalis-Kuznia *et al.*, 1991). In severe cases, these lesions can enlarge and completely girdle the stem. Mature plants are more resistant to infection by *R. solani*.

2. Root rot Weimer (1952) reported that Rhizoctonia root rot was the most widely distributed and destructive disease of lupin. Leach and Clapham (1992) showed that isolates of AG 5 can cause serious losses in lupin by reducing nodule formation, attacking roots and stems, and by killing the apical bud. This disease usually causes a reddish rot that may involve a part or all of the underground parts of the plant. Diseased plants usually have yellowish green foliage and may not set pods. These diseases seem to have little economic importance, but severity may increase if lupines are grown on the same site too often (Roemer, 1988).

Under humid spring conditions, root and stem lesions caused by *Fusarium* spp. and *Pythium* spp. were often found in conjunction with *Rhizoctonia* spp. (Roemer,

1988). Interactions between pathogens often increased disease incidence and severity and influenced plant growth.

3. Terminal bud mortality Lupin plants affected by terminal bud killing do not produce any pods, resulting in seed yield loss. Leach and Clapham (1992) observed up to 85% incidence of this symptom and concluded that it could become a major problem under high inoculum pressure. Terminal bud killing was observed in equal amounts in both rotated and monocropped plots. Terminal bud mortality is primarily caused by AG 5.

B. Disease management

1. Cultural control Good production practices, such as proper seedbed preparation and seeding depth can reduce damage caused by *R. solani* (Putnam *et al.*, 1991). The seedbed should be uniform, firm, level and free of depressions, to obtain good soil to seed contact. Planting seeds at depth of 3.75 cm, into moist soil, allows seedlings to emerge rapidly, and thereby reduces the duration they remain vulnerable to soil-borne pathogens. Deep planting of lupin seed should be avoided.

When lupines were monocropped root and stem lesions caused significant yield losses, whereas suitable crop rotations considerably reduced disease severity (Roemer, 1988). However, when potato was the previous crop, a high level of damping-off and other root disorders were observed (Leach and Clapham, 1992) as isolates of AG 5 infect both crops, and lupines are more susceptible to isolates of this AG. Rotation with small grains and corn can help reduce inoculum buildup. However, when soil inoculum density is high, crop rotation may not significantly reduce the populations of these organisms (Kalis-Kuznia *et al.*, 1991).

2. Chemical control In greenhouse experiments, Benlate (benomyl) was found to be the best seed dressing fungicide to control the root-rot caused by *R. solani* (El-Said *et al.*, 1975). However, it has not proven effective in the field (Kalis-Kuznia *et al.*, 1991). Some fungicides may negatively affect nodulation. Nodulation can be improved by applying the nitrogen-fixing bacteria inoculum to the soil separately. Where root rot and damping-off have been problems in the past, growers should increase the planting rate to compensate for the effects of seedling mortality (Kalis-Kuznia *et al.*, 1991).

IV. SOYBEAN [*GLYCINE MAX* (L.) MERRILL]

A. The causal organism

R. solani causes seed decay, pre- and post-emergence damping-off, hypocotyl rot, root rot, basal stem rot, and foliar blight of soybean (Boosalis, 1950; Hepperly *et al.*, 1982; Joye, 1986; O'Neill *et al.*, 1977; Tachibana *et al.* 1971). One or more of these diseases have been reported in all soybean-growing areas of the world (Sinclair and Backman, 1989).

Most isolates of *R. solani* obtained from soybean have been classified as either AG 1 or AG 4 (Anderson, 1982; Jones and Belmar, 1989; Yang *et al.*, 1990a,b) although AGs 2-2, 3 and 5 have also been from soybean (Nelson *et al.*, (1996). All foliar isolates appear to belong to AG 1-IA (Yang *et al.*, 1989). The pathogen produces on the host, microsclerotia (0.1-0.2mm) or macrosclerotia (2-7mm) (Verma and Thapliyal, 1976). Seed decay and seedling blight or post-emergence damping-off are caused by AG 2-2 (Liu and Sinclair, 1991). These isolates show sharply concentric zonation, which is the result of aggregation of small sclerotia.

AG 4 isolates cause root rot but not damping-off (Hepperly *et al.*, 1982). AG 4 isolates exhibit faint zonation, which is the result of periodic changes in mycelial density in culture, without apparent sclerotia. Different ecological conditions between the two locations may account for the geographical distribution of anastomosis groups (Galindo *et al.*, 1983; Ogoshi, 1987). All isolates obtained from soybean hypocotyl and root lesions in Ohio were of AG 2-2IIIB, while isolates from the same crops and plant parts in Zaire were of AG 4 (Muyolo *et al.*, 1993). AG 5 causes damping off, but is generally less virulent than AG 2-2 (Nelson *et al.*, 1996).

B. Symptoms and epidemiology

1. Seedling blights When environmental conditions are favorable, *R. solani* infects the germinating seed, which may result in seedling blight or death of the seedling. Seedling blight creates a major constraint on soybean production; it reduced stands by an average of 10% in Iowa (Tachibana, 1968) and up to 40% in fields in southern Brazil (Lehman *et al.*, 1974).

After emergence, the fungus attacks the root collar and water-soaked reddish lesions appear near the soil line. The initial infection may develop dark brown or reddish lesions just below the soil line (Sinclair and Backman, 1989). Stem lesions or cankers may cause plants to break off in mid-season. Infected stands may appear thin or spotty and may require replanting. *R. solani* may delay soybean emergence and reduce hypocotyl elongation rate (Grau and Martinson, 1979). This delay extends the duration of seedling vulnerability to soil-borne plant pathogens, thus increasing the probability of damping-off and seedling root rots. *R. solani* causes stunting in certain soybean cultivars, especially when it grows on the seed coat tissues of emerging seedlings. The hypocotyl-stunting syndrome has been reproduced by water extracts of *R. solani*-infected seed coats, which suggests the possibility of a phytotoxin or growth regulator activity (Aoki *et al.*, 1963). Young bean seedlings are susceptible to *R. solani*, while older plants become resistant (Bateman and Lumsden, 1965). The host also produces phytoalexins, such as glyceollin, which are toxic to the invading fungus (Pierre and Bateman, 1967; Wyss *et al.*, 1989).

2. Root rot *R. solani* causes brown to reddish-brown decay on the outer cortical layer of the stem base and lateral roots (Sinclair and Backman, 1989). Lesions on the hypocotyl and lower stem do not extend above the soil line. Root rot is generally first noticed when upper leaves are chlorotic and plant growth is generally stunted, wilted or dead early in the growing season. Root hair and lateral roots are more susceptible to fungal infection than the main root (Orellana *et al.*, 1976). The decay may later develop into a sunken canker and girdle the entire stem. In severely affected plants, the lower part of the taproot and the secondary root system are usually killed. As the soil dries out, these plants wilt and die. With adequate rainfall, plants may recover, but without rain they remain unproductive. The disease pattern may occur as a large spot die, but more often individual plants throughout the field are killed.

R. solani can also invade nodules and impair nitrogen fixation, caused by phytotoxic compounds (Orellana and Mandava, 1983). Naiki and Ui (1981) found that isolates of AG 4 reduced nodule formation and yield. Root hair decay and secondary root infection reduced main sites of nitrogen fixation on the root rhizoplane, which impaired plant growth and resulted in yield losses (Orellana *et al.*, 1976).

Although root rot is generally associated with young plants, older plants may also die if infected, especially in moisture stress conditions (Sinclair and Backman, 1989).

Yield loss is most closely related to mid-season kill because, at this stage of growth, plants do not have any appreciable seed development or yield. Moreover, adjacent plants lose their ability to branch out and cannot compensate for mid-season plant losses. Late-killed plants complete most of their seed development before death, so yield usually is not drastically reduced.

Both *Fusarium oxysporum* Schlechtend.:Fr. and *R. solani* were isolated from rotted soybean roots which suggests that *F. oxysporum* was either a secondary invader or co-invader with *R. solani* (Datnoff and Sinclair, 1988). Root rots may involve more than one pathogen, and, if two or more pathogens interact, disease symptoms are often more severe than those caused by a single pathogen (Datnoff and Sinclair, 1988; Zambolim *et al.*, 1983).

Root rot and hypocotyl rot epidemics of soybean are favored by wet weather and soil temperature range of 18°C (Abawi, 1989) to 25-29°C (Sinclair and Backman, 1989). Cool followed by warm weather is most favorable for disease development.

3. Aerial blight Aerial blight can affect stems, leaves and pods of soybean. Early symptoms include small gray to brown spots containing pinhead-sized sclerotia in the center (Joye *et al.*, 1990; Stroube, 1954). Infected leaves become water-soaked, greenish- to reddish-brown to black, and eventually defoliated. Diseased tissues fall out, giving leaves a ragged appearance. Severely affected plants may shed leaves on which sclerotia were produced. Symptoms ranging from small spots to complete blighting may develop in pods.

In the United States, aerial blight of soybean, caused by *R. solani* AG 1-IA, was first reported by Stroube (1954) in Louisiana. Since then, the incidence of the disease has steadily extended to more than half of the annual 750,000ha in Louisiana (Joye *et al.*, 1990). This is a major disease in soybeans grown in warm, humid areas, particularly where rice is included in the rotation (Anderson, 1982; O'Neill *et al.*, 1977; Verma and Thapliyal, 1976).

Rain-splashed soils containing sclerotia and mycelium fragments are sources of primary inoculum. The fungus survives as sclerotia or as a saprophyte and infects plant roots under favorable conditions. Seedling infection may play a role in the early stages of a foliar blight epidemic. Yang *et al.* (1990a,b,c, 1991) suggested that before canopy closure, rain hits the soil directly and may distribute the inoculum from soil to leaves. Canopy closure may prevent the direct impact of rain against the soil and thus reduce inoculum dispersal. The pathogen spreads in the canopy by means of conspicuous aerial mycelia, which grow from leaf to leaf and plant to plant, spreading the disease and forming distinct disease foci within the canopy during flowering. This is when the disease first becomes obvious. Rainfall during this stage may encourage a disease outbreak.

During warm, moist weather, aerial blight spreads very rapidly and can quickly defoliate plants, causing high yield losses (Atkins and Lewis, 1954; Joye, 1986). The disease requires free moisture to spread (Joye, 1986; Ogoshi, 1987; Yang *et al.*, 1990c). At mean air temperatures of 25-26°C and relative humidities greater than 80%, aerial blights are most severe, especially under rainy conditions (Galindo *et al.*, 1983; Sinclair and Beckman, 1989). Yang *et al.* (1990c) considered rainfall frequency and distribution during the growing season as determining factors for an aerial blight outbreak.

4. Web blight Foci of web blight have a large number of diseased leaves with small circular lesions initiated by airborne microsclerotia (Yang *et al.*, 1989; 1990a,b,c, 1991). The web blight pathogen infects soybean pods, and the incidence of diseased seeds has been as high as 5%. Lesions also form on cotyledons, with mycelial strands occasionally developing between cotyledons. Seedlings with infected shoot apices stop growing, become stunted, and cotyledons remain green. During the growing season, the fungus produces small sclerotia which easily become airborne or rain-splashed and allow it to quickly spread and increase in a field. Moreover, web blight has a seedborne phase, which facilitates its establishment in new fields, and a soilborne phase, which could help the pathogen survive in new fields after its introduction (Galindo *et al.*, 1983a, b).

C. Disease management

Rhizoctonia diseases have been managed mainly with cultural practices, biological control and partial protection by fungicides (Anderson, 1982; Leach and Garber, 1970).

1. Cultural control Cultural practices such as the use of disease-free seed, rhizobial inoculation, proper row spacing, stubble removal and crop rotation are effective in reducing damping-off.

a. Seed inoculation -Good quality seed with a high germination rate should be planted into a warm, well-drained, fertile seedbed. *R. solani* infection can reduce plant growth, N fixation and yield in soybeans. These effects can be alleviated by increased nodulation when efficient *Rhizobium* strains are used. Under field conditions, the yield potential in soybeans is affected by the susceptibility of both the host plant and the *Rhizobium* strain to *R. solani* (Ehteshamul-Haque and Ghaffar, 1993). Treated seed should be planted 4-5 cm deep. Excessive irrigation 2 weeks after planting should be avoided.

b. Row spacing - Since drought stresses soybean plants greatly and cannot be forecast at planting time, a 50 cm row spacing is the best generic recommendation to combat both aerial blight and drought stress. Joye (1986) reported a highly significant correlation between row spacing and soybean defoliation caused by aerial blight. In a field experiment in Louisiana, aerial blight severity decreased as row width increased from 25 to 100 cm, but increased row width restricted yield in two out of three years. Where aerial blight occurs, narrower row spacing compensates for the loss of leaf area, and higher yields can be expected.

c. Stubble removal - Pathogenic strains of *R. solani* can overwinter in parasitized, diseased soybean debris, or can survive in saprophytically-colonized residues of other crops. Disease severity has been closely correlated with the host residue density incorporated into the soil because the propagule density of *R. solani* is influenced by the presence of the susceptible host (Ploetz *et al.*, 1985). Stubble-planted plots had more diseased seedlings than disked plots, plow-and-disk plots, and burn-and-disk plots. During cultivation, soil can be ridged around the base of plants, thereby promoting root growth above diseased areas of the plant, such as the lower stem and roots. Mulching has been reported to reduce web blight (Galindo *et al.*, 1983b).

d. Crop rotation - Monocropping provides a continuous supply of substrate on which *R. solani* may persist and multiply (Whipps and Lynch, 1986); therefore, monoculturing of soybeans should be avoided. Several crops (e.g. cotton and rice) serve as alternative hosts to *Rhizoctonia*; therefore, these crops should be avoided in the rotation pattern.

2. **Chemical control** Both pentachloronitrobenzene (PCNB) and chloroneb (1,4-dichloro-2,5-dimethoxy benzene) are very selective against *R. solani* (Kataria and Grover, 1975). Kucharek (1982) reported that chloroneb seed treatment reduced disease severity and increased stand and yield of soybean, although it also reduced nodulation. Where soybeans have been previously grown, sufficient populations of *R. japonicum* Kichner) Buchanan have survived for at least 10 to 11 years (Kucharek, 1982). A reservoir of nodulating bacteria should exist, so nodulation should not be as severely affected by a chloroneb treatment. Chloroneb, which is a systemic fungicide, translocates into the hypocotyl and cotyledons and so protects plants against later infections.

Carboxin-PCNB seed treatments have increased plant stands, height, number of pods per plant, and yield and reduced disease severity in *R. solani*-infested soil when compared to an untreated control (Abou-Zeid *et al.*, 1987). Metalaxyl, Thiram, Captan, Bavistin, and Vitavax 200F have also been shown to reduce root rot in soybean (Abou-Zeid *et al.*, 1987; Kucharek and Cullen, 1989). Rovral was also effective in control of aerial blight. Benlate combined with Brestan or Rovral, can be applied effectively at much lower rates (Horn and Fontenot, 1981).

3. **Biological control** Seed treatment with *Trichoderma pseudokoningii* Rifai and *T. harzianum* Rifai, originally isolated from the rhizosphere of soybean plants, improved soybean seedling emergence where the pathogen was present and also was parasitic to *R. solani* in vitro (Wu, 1982). Seed treatment with *Bacillus megaterium* de Bary reduced symptoms cause by *R. solani* on soybean plants, enhanced plant and root growth, reduced propagule density in soil, and improved plant growth in soil infested with the pathogen (Liu and Sinclair, 1993). Ehteshamul-Haque and Ghaffar (1993) found that *Rhizobium* spp. and *Bradirhizobium* spp. reduced *R. solani* infection. *Chaetomium cupreum* Amas also inhibited *R. solani*, but has proven somewhat toxic to germinating soybean seedlings (Yeh, 1980).

4. **Resistance** Cultivars vary in susceptibility to pre-emergence damping-off caused by *R. solani* (Lewis and Papavizas, 1977). Sharma (1987) found that the soybean cultivars Amsoy, Clark and Williams were most susceptible, whereas Hood and Delmar were the least susceptible. The cultivars Asgrow 7986, Centennial, Pella, Hardee, RA 606 and Vickery were resistant to root rot and moderately resistant to hypocotyl rot (Abou-Zeid *et al.*, 1987; Muyolo *et al.*, 1993). Field resistance to aerial blight has been found in cultivar Leflore (Hartwig *et al.*, 1985). Commercial soybean cultivars with resistance to web blight have also been developed (Patel, 1989). Cultivars Gregg, Centennial, Hardee, and RA606 were resistant and Davis and FFR646 were susceptible to web blight.

V. REFERENCES

Abawi GS (1989) Root rots. In: Schwartz HF & Pastor-Corrales MA (eds.) Bean production problems in the Tropics (pp 105-157). 2nd ed. CIAT, Cali, Colombia.

Abou-Zeid NM, Abada KA & Ragab MM (1987) Effect of different fungicides *in vitro* and *in vivo* on the control of soybean damping-off disease. Ag. Res. Rev. 65:263-269.

Anderson NA (1982) The genetics and pathology of *Rhizoctonia solani*. Ann. Rev. Phytopathol. 20:329-347.

Aoki H, Sassa T & Tamura T (1963) Phytotoxic metabolites of *Rhizoctonia solani*. Nature 200:575.

Atkins JG & Lewis WD (1954) Rhizoctonia aerial blight of soybeans in Louisiana. Phytopathology 44:215-218.

Barbetti MJ (1984) Seed treatments for control of subterranean clover root rot. Aust. Plant Pathol. 13:43-45.

Barbetti MJ (1990) Effect of adding lime to potting soil on the pathogenicity of four root pathogens of subterranean clover. Phytophylactica 22:245-249.

Barbetti MJ (1991) Effect of clover-free rotations upon the severity of root rot and yield in regenerating subterranean clover pastures. Aust. J. Agric. Res. 42:1195-1204.

Barbetti MJ, Sivasithamparam K & Wong DH (1986) Root rot of subterranean clover. Rev. Pl. Pathol. 65:287-295.

Barnes GL, Russell CC, Foster WD & McNew RW (1981) *Aphelenchus avenae,* a potential biological control agent for root rot fungi. Plant Dis. 65:423-424.

Bateman DF & Lumsden RD (1965) Relation of calcium content and nature of the pectic substances in bean hypocotyls of different ages to susceptiblity to an isolate of *Rhizoctonia solani*. Phytopathology 55:734-738.

Boosalis MG (1950) Studies on the parasitism of *Rhizoctonia solani* Kühn on soybeans. Phytopathology 40:820-831.

Chet I & Baker R (1980) Induction of suppressiveness to *Rhizoctonia solani* in soil. Phytopathology 70:994-998.

Chow HS (1960) Relationship of soil fungi and *Sitonia* species to incidence of root rot of ladino white clover. M.Sc. Thesis. Univ. New Hampshire. Durham. 42 pp.

Crowder LV & Craigmiles JP (1960) The effect of soil temperature, soil moisture, and flowering on the persistency and forage production of white clover stands. Agron. J. 52:382-385.

Datnoff LE & Sinclair JB (1988) The interaction of *Fusarium oxysporum* and *Rhizoctonia solani* in causing root rot of soybeans. Phytopathology 78:771-777.

Ehteshamul-Haque S & Ghaffar A (1993) Use of rhizobia in the control of root rot diseases of sunflower, okra, soybean and mungbean. J. Phytopathol. 138:157-163.

El-Said HM, Ashoer WE, Zayed MA & Aly AZ (1975) Studies on the control of lupine root-rot disease incited by *Rhizoctonia solani* Kühn and *Fusarium oxysporum* Schlecht. Agri. Res. Rev. 53:105-121.

Fathi SM (1977) Studies on damping-off disease of alfalfa in Saudi Arabia. Proc. 1st Conference on the biological aspects of Saudi Arabia, January 15-17, 1977.

Galindo JJ, Abawi GS, Thurston HD & Galvez G (1983a) Source of inoculum and development of bean web blight in Costa Rica. Plant Dis. 67:1016-1021.

Galindo JJ, Abawi GS, Thurston HD & Galvez G (1983b) Effect of mulching on web blight of beans in Costa Rica. Phytopathology 73: 610-615.

Gillespie DJ (1980) Western Australia Department of Agriculture, Annual Report 1979/80, pp. 40-44.

Gladstones JS (1975) Legumes and Australian agriculture. J. Aus. Inst. Agric. Sci. 41:227-240.

Graham JH, Stuteville DL, Frosheier FI & Erwin DC (1979) A Compendium of Alfalfa Diseases. Amer. Phytopath. Soc., St. Paul, Minnesota 65 pp.

Grau CR & Martinson CA (1979) Inhibition of soybean hypocotyl elongation by *Rhizoctonia solani*. Phytopathology 69:706-709.

Gray FA, Haaland RL, Clark EM & Ball DM (1980) Diseases of alfalfa in Alabama. Plant Dis. 64:1015-1017.

Greenhalgh FC & Clarke RG (1985) Use of fungicides to study significance and etiology of root rot of subterranean clover in dryland pastures of Victoria. In: Parker CA, Rovira AD, Moore KJ, Wong PTW & Kollmorgen JF (eds.) Ecology and Management of Soilborne Pathogens (pp 234-236). APS Press, St. Paul, Minnesota.

Hancock JG (1983) Seedling diseases of alfalfa in California. Plant Dis. 67:1203-1208.

Hartwig EE, Young LD & Edwards CJ, Jr. (1985) Registration of "Leflore" soybean. Crop Sci. 25:1985.

Hawn EJ (1958) Studies on the epidemiology of crown bud rot of alfalfa in southern Alberta. Can J. Bot. 36:239-250.

Hawn EJ, Berkenkamp WB & Lebeau JB (1981) Evaluation of losses in alfalfa hay production caused by crown rot. Can. J. Plant Path. 3:103-105.

Hepperly PR, Mignucci JS, Sinclair JB, Smith RS & Judy WH (1982) Rhizoctonia web blight of soybean in Puerto Rico. Plant Dis. 66:256-257.

Horn NL & Fontenot MF (1981) Control of aerial blight with foliar sprays. Fung. Nemat. Tests. 36:95.

Horricks JS (1981) Forage Diseases. Alberta Agriculture. Agdex 120/20-4. pp. 81-86.

Jones RK & Belmar SB (1989) Characterization and pathogenicity of *Rhizoctonia* spp. isolated from rice, soybean, and other crops grown in rotation with rice in Texas. Plant Dis. 73:1004-1010.

Joye GF (1986) Management of Rhizoctonia aerial blight of soybean and biology of sclerotia of *Rhizoctonia solani* Kühn. Ph. D. thesis. Louisiana State University, Baton Rouge. 91 pp.

Joye GF, Berggren GT & Berner DK (1990) Effects of row spacing and within-row plant population on Rhizoctonia aerial blight of soybean and soybean yield. Plant Dis. 74:158-160.

Kalis-Kuznia R, Stewart E & Memonuck RA (1991) Lupin diseases and control measures. In: Meronuck RA, Meredith H & Putnam DH (eds.) Lupin Production and Utilization Guide (pp 16-17). CAPAP. Univ. Minn., St. Paul, Minnesota.

Kochhar M (1974) Phytotoxic and competitive effects of tall fescue on ladino clover as modified by ozone and/or *Rhizoctonia solani*. PhD. Thesis. N. Carolina St. Univ. Raleigh, North Carolina, USA. 71 pp.

Kataria HR & Grover RK (1975) Effect of chloroneb (1,4-dichloro-2,5-dimethoxy benzene) and pentachloronitrobenzene on metabolic activities of *Rhizoctonia solani* Kühn. Indian J. Exp. Biol. 13:281-285.

Kovoor ATA (1954) Some factors affecting the growth of *Rhizoctonia bataticola* in the soil. J. Madras Univ. 24:47.

Kousik CS, Snow JP & Berggren GT (1994) Factors affecting infection cushion development by *Rhizoctonia solani* AG1-IA and -IB on soybean leaves. Plant Pathol. 43:237-244.

Kucharek TA (1982) Effect of chloroneb seed treatment fungicide on soybean yield and nodulation. Soil and Crop Sci. Soc. Fla. Proc. 41:59-62.

Kucharek TA & Cullen RE (1989) Control of diseases caused by *Rhizoctonia solani* and *Sclerotium rolfsii* in soybeans, beans, peanuts, and cucumbers with flutolanil. Soil and Crop Sci. Soc. Fla Proc. 48:31-34.

Kulik MM & Dery PD (1995) Frequency of plants in populations of annual and perrenial medics (*Medicago* spp.) that were resistant to damping off incited by AG 4 isolates of Rhizoctonia aolani. Z. Pfl. Krankh. 102:392-398.

Kulik MM, Dery PD & Duglass LW (1995) Frequency of plants in populations of alfalfa (*Medicago sativa*) resistant to pre- and post emergence damping off incited by AG 4 isolates of *Rhizoctonia aolani*. Z. Pfl. Krankh. 102:51-57.

Leach LD & Garber RH (1970) Control of *Rhizoctonia solani*. In: Parmeter JR, Jr. (ed.) *Rhizoctonia solani*: Biology and Pathology (pp 189-198). Univ. California Press, Berkeley, California.

Leach SS & Clapham WM (1992) *Rhizoctonia solani* on white lupine. Plant Dis. 76:417-419.

Leath KT (1989) Diseases and forage stand persistance in the United States. In: Marten GC, Matches AG, Barnes RF, Brougham RW, Clements RJ & Sheath GW (eds.) Persistence of Forage Legumes (pp 465-478). American Society of Agromony, Madison, Wisconsin 572 pp.

Leath KT, Zeiders KE & Byers RA (1973) Increased yield and persistence of red clover after a soil drench application of benomyl. Agr. J. 65:1008-1009.

Lehman PS, Mochado CC, Tarrago MT & Correo CF (1976) Evaluation of recommended soybean, maize and sorghum cultivars and soybean lines for resistance to "dead patch". Summa Phytopathol. 2:47-49.

Lewis JA & Papavizas GC (1977) Factors affecting *Rhizoctonia solani* infection of soybeans in the greenhouse. Plant Dis. Rep. 61:196-200.

Liu ZL & Sinclair JB (1991) Isolates of *Rhizoctonia solani* anastomosis group 2-2 pathogenic to soybean. Plant Dis. 75:682-687.

Liu ZL & Sinclair JB (1993) Colonization of soybean roots by **Bacillus megaterium** B153-2-2. Soil Biol. Biochem. 25:849-855.

Ludbrook WV, Brockwell J & Riceman DS (1953) Bare-patch disease and associated problems in subterranean clover pastures in South Australia. Aust. J. Agric. Res. 4:403-414.

Maughan RD & Barbetti MJ (1983) Rhizoctonia root rot of white clover. Aust. Plant Pathol. 12:13-14.

Muyolo NG, Lipps PE & Schmitthenner AF (1993a) Anastomosis grouping and variation in virulence among isolates of *Rhizoctonia solani* associated with dry bean and soybean in Ohio and Zaire. Phytopathology 83:438-444.

Muyolo NG, Lipps PE & Schmitthenner AF (1993b) Reactions of dry bean, lima bean, and soybean cultivars to Rhizoctonia root and hypocotyl rot and web blight. Plant Dis. 77:234-238.

Naiki T & Ui T (1981) *Rhizoctonia solani* Kühn root rot of bean, soybean and adzuki bean seedlings. Rev. Plant Pathol. 61:168.

Nan ZB, Skipp RA & Long PG (1991) Use of fungicides to assess the effects of root disease: Effects of Prochloraz on red clover and microbial populations in soil and roots. Soil Biol. Biochem. 23:743-750.

Nelson B, Helms T, Christianson T & Kural I (1996) Characterization and pathogenicity of Rhizoctonia from soybean. Plant Dis. 80:74-80.

Nelson SC & Campbell CL (1993) Disease progress, defoliation, and spatial pattern in a multiple pathogen disease complex on white clover. Phytopathology 83:419-429.

O'Neill NR, Rush MC, Horn NL & Carver RB (1977) Aerial blight of soybeans caused by *Rhizoctonia solani*. Plant Dis. Rep. 61:713-717.

Orellana RG & Mandava NB (1983) m-Hydroxyphenylacetic and m-methoxyphenylacetic acids of *Rhizoctonia solani*: their effect on specific root-nodule activity and histopathology in soybean Phytopathol. Z. 107:159-167.

Orellana RG, Sloger C & Miller VL (1976) *Rhizoctonia-Rhizobium* interactions in relation to yield parameters of soybean. Phytopathology 66:464-467.

Papavizas GC, Adams PB, Lumsden RD, Lewis JA, Dow RL, Ayers WA & Kantzes JG (1975) Ecology and epidemiology of *Rhizoctonia solani* in field soil. Phytopathology 65:871-877.

Papavizas GC & Davis CB (1962) Activity of *Rhizoctonia* in soil affected by carbon dioxide. Phytopathology 52:759.

Patel MV (1989) Soybean disease reaction. Plant Disease Views and Reviews, Mississippi Cooperative Extension Service, Mississippi State University.

Pierre RE & Bateman DF (1967) Induction and distribution of phytoalexins in *Rhizoctonia*-infected bean hypocotyls. Phytopathology 57:1154-1160.

Ploetz RC, Mitchell DJ & Gallaher RN (1985) Population dynamics of soiborne fungi in a field milticropped to rye and soyubeans under reduced tillage in Florida. Phytopathology 75:1447-1451.

Purss GS (1965) Diseases of lucerne. Qd. Agr. J. 91:196-206.

Putnam DH, Field LA, Gunsolus J, Ayisi K & Hardman L (1991) Production factors for white lupin. In: Meronuck RA, Meredith H & Putnam, DH (eds.) Lupin Production and Utilization Guide (pp 6-15). CAPAP. Univ. Minn., St. Paul, Minnesota.

Roemer P (1988) Diseases and pests of *Lupinus mutabilis* in Germany. In: Sec. Intern. Lupin Ass. (ed.) Lupin Newsletter, (pp 28-32) AVDA, Menendez Pidal, Cordoba, Spain.

Seif El-Nasr HI & Leath KT (1983) Crown and root fungal diseases of alfalfa in Egypt. Plant Dis. 67:509-511.

Sharma KP (1987) Response of soybean cultivars to different levels of *Rhizoctonia solani* inoculum. J. Inst. Agric. Animal Sci. 8:89-97.

Sinclair JB & Backman PA, (eds.) (1989) Compendium of Soybean Diseases. 3rd ed. APS Press, St. Paul, Minnesota, 106 pp.

Sitrit Y, Barak Z, Kapulnik Y, Oppenheim AB & Chet I (1993) Expression of *Serratia marcescens* chitinase gene in *Rhizobium meliloti* during symbiosis on alfalfa roots. Molecular-Plant Microbe Interactions 6:293-298.

Smith OF (1943) Rhizoctonia root canker of alfalfa (*Medicago sativa*). Phytopathology 33:1081-1085.

Stuckley RE & McCarter-Zorner NJ (1982) Effect of soybean seed treatments on emergence and yield, 1981. Fung. Nemat. Tests 37:174-175.

Tachibana H, Jowett D & Fehr WR (1971) Determination of losses in soybeans caused by *Rhizoctonia solani*. Phytopathology 61:1444-1446.

Tachibana H (1968) *Rhizoctonia solani* root rot epidemic of soybeans in Central Iowa in 1967. Plant Dis. Rep. 52:613-614.

Tesar MB & Jackobs JA (1972) Establishing the stand. In: Hanson CH (ed.) Alfalfa Science and Technology. Agronomy 15 pp. 415-435. Amer. Soc. of Agr., Madison, Wisconsin. 812 pp.

Verma HS & Thapliyal PN (1976) Rhizoctonia aerial blight of soybean. Indian Phytopathol.29:389-391.

Whipps, JM & Lynch JM (1986) The influence of the rhizosphere on crop productivity. In: Marshall KC (ed.) Advances in Microbial Ecology vol. 9 (pp 187-243). Plenum Press, New York.

Weimer JL (1952) Disease of cultivated lupins in the south-east. Fmrs' Bull. 2053. USDA, Washington.

Wink M (1984) Chemical defense of Leguminosae. Are quinolizidine alkaloids part of the antimicrobial defense system of lupins? Z. Naturforsch. 39c:548-552.

Wong DH, Barbetti, MJ & Sivasithamparam K (1984) Effects of soil temperature and moisture on the pathogenicity of fungi associated with root rot of subterranean clover. Aust. J. Agric. Research 35:675-684.

Wong DH, Barbetti MJ & Sivasithamparam K (1985) Pathogenicity of *Rhizoctonia* spp. associated with root rots of subterranean clover. Trans. Br. Mycol Soc. 85:156-158.

Wong DH and Sivasithamparam K (1985) Rhizoctonia spp. associated with root rots of subterranean clover in Western Australia. Trans. Brit. Mycol. Soc. 85:21-27.

Wu WS (1982) Seed treatment by applying *Trichoderma* spp. to increase the emergence of soybeans. Seed Sci. Technol. 10:557-563.

Wyss P, Boller T & Wiemken A (1989) Glyceollin production in soybean during the process of infection by *Glomus mosseae* and *Rhizoctonia solani*. Ag. Ecosystems and Environment. 29:451-456.

Yang XB, Snow JP & Berggren GT (1988) Survey of reproductive forms of *Rhizoctonia solani* on soybean in Louisiana. Plant Dis. 72:644.

Yang XB, Snow JP & Berggren GT (1989) Morphogenesis of microsclerotia and "sasakii-type" sclerotia in *Rhizoctonia solani* Kühn, anastomosis group one intraspecific group IA and IB. Mycol. Res. 93:429-434.

Yang XB, Berggren GT & Snow JP (1990a) Seedling infection of soybean by isolates of *Rhizoctonia solani* AG 1, causal agent of aerial blight and web blight of soybean. Plant Dis. 74:485-488.

Yang XB, Berggren GT & Snow JP (1990b) Types of Rhizoctonia foliar blight on soybean in Louisiana. Plant Dis. 74:501-504.

Yang XB, Berggren GT & Snow JP (1990c) Effects of free moisture and soybean growth stage on focus expansion of Rhizoctonia aerial blight. Phytopathology 80:497-503.

Yang XB, Snow JP & Berggren GT (1991) Patterns of Rhizoctonia foliar blight on soybean and effect of aggregation on disease development. Phytopathology 81:287-293.

Yeh CC (1980) Effect of *Chaetomium cupreum* on seed germination and antagonism to other seedborne fungi of soybean. Plant Dis. 64:468-470.

Zambolim L, Schenck NC & Mitchell DJ (1983) Inoculum density, pathogenicity, and interactions of soybean root-infecting fungi. Phytopathology 73:1398-1402.

V.6 ORNAMENTAL DISEASES INCITED BY *RHIZOCTONIA* SPP.

D. MICHAEL BENSON AND D. KELLY CARTWRIGHT

Department of Plant Pathology, North Carolina State University
Raleigh, NC 27695-7629 USA

I. INTRODUCTION

Both multinucleate, *Rhizoctonia solani*, and to a much lesser extent binucleate *Rhizoctonia* spp. fungi cause diseases in ornamental crops grown in greenhouses, nurseries, and the landscape. This chapter includes information on some of the most important ornamental diseases, crops attacked, symptoms, pathogens, epidemiology, and a discussion of control strategies somewhat unique to ornamentals. Although *Rhizoctonia* spp. are primarily pathogens of ornamentals, some may be useful biological control agents for ornamental diseases (see below) and others are endophytes of orchids important for plant development (Sneh *et al.*, 1991).

II. DISEASES

Ornamentals diseases caused by *R. solani* include pre- and post-emergence damping-off, stem rot, foliar blight and web blight. Survival of *R. solani* as sclerotia and mycelia in plant debris coupled with a rapid growth rate and colonization of organic matter are key elements in disease development. In addition to diseases caused by *R. solani*, binucleate *Rhizoctonia* spp. (BNR) cause web blight on woody ornamentals such as azalea (*Rhododendron* spp.), cotoneaster, and Japanese holly (*Ilex crenata)*.

A. Pre-emergence damping-off

Since *R. solani* attacks juvenile tissues, ornamentals grown from seed are particularly vulnerable to infection. Pre-emergence damping-off alone no doubt accounts for a major portion of the total disease loss inflicted by *R. solani* on herbaceous and woody ornamentals propagated from seed. Although germinating seeds may require several days to emerge, the first 24-72 hours after sowing are critical for pre-emergence damping-off.

B. Post-emergence damping-off

In greenhouse ornamentals, the use of pathogen-free soilless mixes may reduce pre-emergence damping-off. However, inoculum of *R. solani* may still be introduced to seeded flats in the potting shed, during transport to the greenhouse bench, or after the flat has been benched. Under these circumstances, post-emergence damping-off may develop as seedlings emerge. Symptoms develop initially as water-soaked lesions at the seedling base that may eventually girdle and collapse the seedling in a few days.

C. Stem rot

Rhizoctonia stem rot is a common disease problem in propagaton and production of many potted, herbaceous ornamentals. Cuttings under propagation, for example, are especially susceptible to this disease. Infection at the soil line results in discrete lesions that expand rapidly and girdle the stem. Hyphae of *R. solani* may be visible on infected tissue under conditions of free moisture.

B. Sneh et al. (eds.),
Rhizoctonia Species: Taxonomy, Molecular Biology, Ecology, Pathology and Disease Constrol, 303–314.

D. Foliar blight

Splash dispersal of soilborne inoculum of *R. solani* may incite foliar blight in the canopy of bedding plants and other herbaceous plants in the greenhouse and landscape. Symptoms initially develop as water-soaked lesions on stems and leaves progressing rapidly in succulent plants to the complete collapse of the plant. Under high humidity and warm temperatures, ectotrophic mycelium can be observed on the collapsed foliage as well as ramifying into the canopy of adjacent healthy plants. On less succulent foliage, such as *Peperomia obtusifolia*, concentric rings may result from diurnal expansion of lesions (Chase, 1987).

E. Web blight

First proposed by Weber (1939) as a common name for a foliar blight of beans caused by *R. solani*, web blight has since been used as a standard name for a foliar disease of azalea (*Rhododendron* spp.) incited by binucleate *Rhizoctonia* spp. (Benson, 1987). Web blight develops during conditions of high relative humidity, warm temperatures and reduced sunlight (Weber and Roberts, 1951; Wehlburg and Cox, 1966; Frisina and Benson, 1987). On azalea, symptoms begin as small water-soaked lesions on leaves that are tan with a regular outline. When petioles become infected, leaves abscise but remain matted to the stem by the webbing of ectrotrophic mycelial growth (Frisina and Benson; 1987). Typically, leaves in the interior portion of the canopy are infected while the outer canopy leaves remain healthy. Inoculum for web blight development may be splashed to the foliage of plants, grow as hyphae up the stem from infested debris on the soil surface, or move plant to plant by hyphal growth during epidemics (Frisina and Benson, 1989).

III. Pathogens, isolate characteristics, and host range

Several *Rhizoctonia* spp. have been reported as pathogens on ornamentals. The most common species is *R. solani* (Farr *et al.,* 1989). In addition, binucleate *Rhizoctonia* spp. including *R. ramicola* (Weber and Roberts, 1951) and unidentified binucleate *Rhizoctonia* spp. (Chase, 1991; Frisina and Benson, 1987) have been reported on woody ornamentals.

A. *Rhizoctonia solani*

Although *R. solani* is a common pathogen of many ornamentals, only a limited amount of information is available on characterization of ornamental isolates. Stephens *et al.* (1982) collected 46 isolates from bedding plants and field-grown nonbedding plants in Michigan, USA. All bedding plant isolates were AG 4 and were more virulent on bedding plants than other hosts. AG 4 isolates of *R. solani* also were found on zinnia (Lumsden and Locke, 1989), poinsettia (Cartwright and Benson, 1994), New Guinea impatiens (Castillo and Peterson, 1990), carnation (Trujillo *et al.*, 1988) and hibiscus (Bolkan and Ribeiro, 1985). In Hawaii, AG 2-2 was more virulent on carnation than the more common AG 4 isolates (Trujillo *et al.,* 1988). Isolates of *R. solani* from *Impatiens sp., Ilex crenata,* and *Pittosporium* sp. with foliar blight symptoms belonged to AG 1 (Frisina and Benson, 1987).

Chase (1991) studied 309 isolates of *Rhizoctonia* spp. from ornamentals in Florida, USA, and found 129 multinucleate and 180 binucleate isolates. Thirty-three percent of the multinucleate isolates were from roots, 24% from stems, and 40% from leaves. Multinucleate isolates were pathogenic on the original host and many caused foot rot of pothos *(Epipremnum aureum)* in comparative inoculation trials (Chase,

1991). The major anastomosis group was AG 4 although some isolates from *Ficus, Nerium* and *Pilea* were AG 1.

The host range of *R. solani* on ornamentals is enormous with over 360 genera listed in the USA (Farr, *et al.*,, 1989) (Table 1). The host range will probably expand as more plant species are introduced to the ornamental industry.

TABLE 1. Host range for *Rhizoctonia solani* on ornamentals

Abelia	Bauhinia	Columnea	Erodium
Abelmoschus	Begonia	Convolvulus	Erysimum
Acacia	Berberis	Coreopsis	Eucharis
Acalypha	Bougainvillea	Cornus	Euonymus
Achillea	Brassica	Cortaderia	Eupatorium
Acoelorrhaphe	Brassolaeliocattleya	Cosmos	Euphorbia
Aconitum	Brunfelsia	Cotoneaster	Exacum
Acorus	Buddleja	Crassula	Fatshedera
Actinidia	Buxus	Crinum	Fatsia
Adiantum	Caesalpinia	Crossandra	Feijoa
Aechmea	Caladium	Crotalaria	Festuca
Aeschynanthus	Calathea	Ctenanthe	Ficus
Agapanthus	Calendula	Cuphea	Fittonia
Ageratum	Callistemon	Cycas	Frafaria
Aglaonema	Callistephus	Cyclamen	Gaillardia
Agrostis	Calluna	Cymbidium	Gardenia
Ajuga	Calpogonium	Cynoglossum	Gazania
Alcea	Camellia	Cyperus	Gelsemium
Allamanda	Campanula	Dactylis	Gerbera
Allium	Capsicum	Dahlia	Gladiolus
Alocasia	Caragana	Daphne	Gloriosa
Aloe	Carissa	Datura	Gynura
Alsophila	Carthamus	Davallia	Gypsophila
Alternanthera	Caryota	Delphinium	Hedera
Alyssum	Cassia	Dendrobium	Helianthus
Amaranthus	Catharanthus	Deschampsia	Heliconia
Ammophila	Cattleya	Dianthus	Hemerocallis
Anchusa	Celosia	Dichondra	Hemigraphis
Anthemis	Centaurea	Dieffenbachia	Hibiscus
Anthoxanthum	Cephalanthus	Dietes	Hippeastrum
Anthurium	Cerastium	Digitalis	Holcus
Antirrhinum	Cereus	Dimorphotheca	Hordeum
Aphelandra	Chamaecyparis	Dizygotheca	Hosta
Aquilegia	Chamaedorea	Doryopteris	Hoya
Arabis	Cheiranthus	Dracaena	Hydrangea
Arachis	Chenopodium	Echeveria	Hygrophila
Aralia	Chionanthus	Elaeagnus	Iberis
Araucaria	Chlorophytum	Elettaria	Ilex
Ardisia	Chrysalidocarpus	Elymus	Impatiens
Arecaceae	Chrysanthemum	Epidendrum	Indigofera
Asclepias	Cissus	Epipremnum	Ipomoea
Asparagus	Citrus	Episcia	Iresine
Asplenium	Clarkia	Equisetum	Iris
Aster	Coccoloba	Eragrostis	Ixora
Aucuba	Codiaeum	Eranthemum	Jasminum
Avena	Coffea	Eriophyllum	Juniperus

TABLE 1 cont'd

Kalanchoe	Nasturtium	Polypodium	Silphium
Koeleria	Nematanthus	Polyscias	Sinningia
Laeliocattleya	Nepeta	Poncirus	Solanum
Lagenaria	Nephrolepis	Portulaca	Soleirolia
Lagerstroemia	Nerium	Potamogeton	Spathiphyllum
Lamium	Nicotiana	Potentilla	Stellaria
Lantana	Ocimum	Primula	Stipa
Lathyrus	Oenothera	Prunus	Strelitzia
Lavatera	Oncidium	Pteris	Stromanthe
Leea	Ophiopogon	Pueraria	Synadenium
Lespedeza	Opuntia	Punica	Syngonium
Leucophyllum	Orchidaceae	Pyracantha	Syringa
Leucothoe	Oxalis	Ranunculus	Tagetes
Ligustrum	Paeonia	Reseda	Taraxacum
Lilium	Panicum	Rhaphiolepis	Taxus
Limonium	Papaver	Rhapis	Tecomaria
Linaria	Parthenocissus	Rheum	Tephrosia
Linum	Paspalum	Rubus	Thuja
Liriope	Passiflora	Ruellia	Thymus
Lobelia	Pelargonium	Rumex	Trachelospermum
Lobularia	Pellaea	Rumohra	Trachymene
Lonicera	Pellionia	Russelia	Tradescantia
Lotus	Pennisetum	Sabal	Trifolium
Lupinus	Pentas	Saccharum	Tropaeolum
Lychnis	Peperomia	Sagina	Tulipa
Lysimachia	Persea	Saintpaulia	Vaccinium
Lythrum	Petunia	Salix	Valeriana
Magnolia	Phalaenopsis	Salvia	Vanda
Malpighia	Phalaris	Sambucus	Veitchia
Malva	Phanerophlebia	Sanchezia	Verbena
Mandevilla	Philodendron	Sansevieria	Veronia
Maranta	Phlox	Sarracenia	Veronica
Matthiola	Phoenix	Saururus	Viburnum
Melaleuca	Photinia	Saxifraga	Vicia
Mentha	Phytolacca	Scaevola	Vinca
Mentzelia	Pilea	Schefflera	Viola
Michelia	Pinus	Schizanthus	Vitis
Mikania	Piqueria	Schlumbergera	Vriesea
Mirabilis	Pisum	Scutelleria	Vulpia
Monstera	Pittosporum	Sedum	Zantedeschia
Mucuna	Platycerium	Selaginella	Zebrina
Muhlenbergia	Platycodon	Senecio	Zingiber
Musa	Plectranthus	Shepherdia	Zinnia
Myrica	Podocarpus	Sicana	
Nandina	Polianthes	Siderasis	
Narcissus	Polygonum	Silene	

Compiled by Kala C. Parker, North Carolina State University. After Farr *et al.* (1989).

B. *Rhizoctonia ramicola*

Weber and Roberts (1951) described a web blight (silky thread blight) disease of *Elaeqnus pungens* in Florida, USA, caused by a new species of binucleate *Rhizoctonia* sp. they named *R. ramicola.* Leaf lesions consisted of small necrotic spots with a tan center and purple margins on leaves that became brittle. Tender twigs, petioles and leaf blades were infected but not roots. The fungus penetrated petioles through lenticles and basal cells of scales. Stomata on leaves also served as infection courts where intercellular hyphae colonized the spongy mesophyll cells. Disease development was greatest between 25 and 30°C when relative humidity was above 75%. In culture, no sclerotia developed nor were sclerotia found associated with the disease on naturally-infected plants (Weber and Roberts, 1951). In addition to *E. pungens, R. ramicola* was observed on *Ilex crenata, Eyrthrium herbacea, Pittosporium tobira, Lagerstroemia indica* and *Feijoa* near Gainesville, Florida, USA (Weber and Roberts, 1951). Inoculation trials with *R. ramicola* resulted in web blight development in *Gelsemium sempervirens, Rhododendron indicum,* and *Pyracantha formosana.*

The teleomorph of *R. ramicola,* found on *Pittosporium tobira,* was named *Ceratobasidium ramicola* Tu, Roberts, & Kimbrough, sp. nov (Tu *et al.*,, 1969). Burpee, *et al.,* (1980) placed *Ceratobasidium cornigerum* (syn. = *C. ramicola)* in CAG 2 which became AG-A of Ogoshi (1985).

C. Other binucleate *Rhizoctonia* spp.

In North Carolina nurseries, web blight caused by binucleate *Rhizoctonia* (BNR) is a common and important disease of azaleas particularly on cultivars in the Satsuki and Kurume hybrid group (Frisina and Benson, 1987). Binucleate *Rhizoctonia* spp. also were isolated from *Ilex crenata, Pittosporum* sp. and *Cotoneaster* sp. with symptoms of web blight in nurseries (Frisina and Benson, 1987) and from *Juniperus* spp. without symptoms (Fravel, *et al.,* 1982). Only BNR were isolated from azalea, *Ilex* spp. and *Philodendron* spp. in Florida, USA (Chase, 1991). Isolates of *Rhizoctonia* spp. from foliage of *Raphiolepsis indica* from Alabama, USA and from *Juniperus chinensis, Pittosporium tobira* and *Ligustrum* sp. from Florida, also proved to be BNR (Frisina and Benson, 1987). A BNR AG-E caused up to 19% mortality in long leaf nursery beds in Florida (English *et al.,* 1986). Among the BNR isolates from ornamentals in Florida, Chase (1991) found 57% from foliar samples and 43% from roots with the majority of isolates from woody ornamentals.

Burpee *et al.* (1980) developed an anastomosis system for BNR and placed foliar isolates from ornamentals in AGs -A, -E, or -S as revised by Ogoshi (1985). Only two of 59 isolates tested were AG-S, from the foliage of *Pittosporium,* and *Cinnamomun* in Florida. BNR isolates from roots of ornamentals were either AG-E or AG-R (Burpee, *et al.,* 1980). Several foliar BNR isolates from woody ornamentals in North Carolina, USA were AG-E or AG-S, based on results with tester isolates (Burpee, *et al.,* 1980; Frisina and Benson, 1987).

BNR isolates from azalea were about half as virulent as foliar isolates of *R. solani* from other woody hosts in web blight severity on azalea (Frisina and Benson, 1987). Even though multinucleate isolates of *R. solani* were more virulent on azalea, only BNRs were isolated from azaleas with natural web blight infections.

IV. Survival and Epidemiology

Rhizoctonia spp. survive as sclerotia in soil and as hyphae associated with organic debris. In the bedding plant industry where transplants are produced in soilless medium in plug trays under greenhouse conditions *R. solani* is a major damping-off pathogen (Cline *et al.,* 1988). In poinsettia production where cuttings are produced in greenhouses in polyfoam or rockwool rooting cubes, *R. solani* causes stem rot.

Survival of *R. solani* in landscape beds and field production areas for ornamentals is a common phenomenon. However, as the level of crop management intensifies from the nursery to the greenhouse with the concomitant use of soilless media, endemic and epidemic losses to *R. solani* and *Rhizoctonia* spp. should be greatly diminished, but this is not the case. Losses to *R. solani* continue in the most sophisticated operations.

Stephens *et al.* (1983) did not find *R. solani* or *Pythium* spp. in samples of potting components such as peat, vermiculite, perlite, or steam-treated soil in greenhouses producing bedding plants. However, both pathogens were recovered from dust and potting mix samples collected from floors and walkways in the greenhouse and from used flats. We have found *R. solani* surviving in wooden slats of lath used for benchtops (Cartwright, *unpublished*).

Although binucleate *Rhizoctonia* spp. were not recovered from stockpiles of pine bark potting medium or other potting components, the fungus was present in the potting medium of as many as 57% of the samples from flats in which azalea cuttings 'Gumpo' were propagated in a greenhouse (Frisina and Benson, 1989). Once rooted cuttings were transplanted to containers and placed in the nursery, spatial pattern analysis of diseased plants suggested that spatial pattern of diseased plants was best described as random (Frisina and Benson, 1989). The random pattern of web blight suggests that infected cuttings from propagation were the inoculum source. As the season progressed, plant-to-plant spread of the pathogen by direct mycelial growth increased as aggregation of diseased plants occurred. Interestingly, inoculum density in the container was not correlated with incidence of web blight. Environmental conditions in the nursery each year was the apparent limiting factor in disease development.

V. Control

Although *R. solani* and binucleate *Rhizoctonia* spp. attack numerous ornamentals, control strategies are similar for both (Chase, 1987; 1991; Frisina and Benson, 1988). The most important consideration is the judicious use of preventative measures, regardless of venue. An effective program can encompass primarily, or in combination, cultural practices, fungicides, biological controls, and eradicants.

A. Cultural practices

The principal areas involved here are sanitation, regulation of environmental conditions and to some degree, the use of suppressive soil mixes. Practiced diligently and used in combination, these measures can prevent most disease epidemics in ornamentals caused by *Rhizoctonia*.

1. Sanitation. This is the most critical method for control of *Rhizoctonia,* especially in greenhouse ornamentals. Because of its ubiquitous nature, care must be taken to limit initial infection by, and subsequent spread of *R. solani*. Hoses should be elevated off the floor, crops cycled expeditiously and never retained more than one season, infected plants promptly rogued and properly disposed (i.e. no 'plant dumps'), benches scrubbed and cleaned with a mild disinfectant or bleach (5.25% sodium hypochlorite) between crops, and weeds and debris removed from under benches and in close proximity to greenhouse entrances (Ecke *et al.*, 1990; Jones, 1994a; b; Jones and Averre, 1992; Jones and Benson, 1994). Containers should be disinfested before reuse and a pasteurized, prebagged, commercial potting mix or other well-aerated mix, free of pathogens, used for plant propagation and growth (Ecke *et al.*, 1990; Jones and Benson, 1994; Stephens *et al.*, 1983). Reused potting mixes, weather-exposed bulk media ingredients, and unsanitary mixing areas, equipment, and transport equipment are potential inoculum sources which lead to rapid spread of *R. solani* to multiple greenhouses and nursery sites. Where applicable, all procedures mentioned above should be practiced in nursery operations also.

2. Sanitation during propagation. Because *Rhizoctonia* is usually more severe on young, succulent ornamentals, especially strict sanitation should be practiced during propagation. *Rhizoctonia* can be spread on seed (Stephens *et al.*, 1983) and possibly vegetative cuttings, thus care should be taken to obtain both from reputable dealers. In addition, all plant material should be inspected upon arrival for signs of obvious infection, poor quality, or copious amounts of organic debris. eeds should be planted shallow and incubated under adequate light and temperature for germination and growth (Jones and Averre, 1992). Cuttings should be placed in a recommended propagation medium soon after arrival and on-site cuttings taken with a clean, sterile scalpel or knife with frequent dipping in a mild disinfectant or bleach. Containers for transporting cuttings should be disinfested and cuttings kept from contact with bench surfaces/floors. Watering during propagation should be minimized to initiate quicker 'hardening-off' by the plant and provide a less favorable environment for spread of *Rhizoctonia* spp.

3. Regulation of environmental conditions In general, diseases caused by *R. solani*, including those on ornamentals, are favored by higher soil moisture and air humidity although moisture above ~80% in the soil typically restricts *R. solani* (Baker and Martinson, 1970). Rhizoctonia stem rot of poinsettia, caused by *R. solani,* is favored by a moisture holding capacity (MHC) in soil between 50 and 80% (Jones and Benson, 1994). Overwatering, which leads to a saturated growing medium, increased humidity in and around the plant canopy, plant stress, and a very favorable infection court for *R. solani,* should be avoided. Weather conditions can also affect disease epidemics, even in the greenhouse. During periods of overcast and/or rain during the summer, we have observed epidemics of *R. solani* in poinsettia cuttings under mist in which plant to plant spread by *R. solani* was extremely rapid, occurring via aerial mycelium throughout the canopy (Cartwright and Benson, *unpubl.*). Rhizoctonia aerial blight of foliage plants and web blight of woody ornamentals are also favored by high moisture in the potting medium and high humidity in the plant canopy (Benson, *et al.*, 1986; Chase, 1987; Frisina and Benson, 1987). A strict watering regime targeted at the specific ornamentals should be maintained. Drip tubes, mist nozzles, or timed sprinklers (fixed irrigation) are probably the best strategy. Proper plant spacing and ventilation to increase air flow through the canopy are also important to prevent foliar infection by *Rhizoctonia* in greenhouses and nurseries (Benson *et al.*, 1986; Chase, 1987).

Temperature can have a marked effect on Rhizoctonia diseases in ornamentals. some of these effects are isolate dependent (Bateman, 1970; Chase, 1991) but, typically, disease is more severe as temperatures increase. For example, soil temperatures of 32°C reduced aerial blight of Boston fern by *R. solani* as opposed to soil temperatures of 27°C. In the same study, however, air temperatures ranging from 35 to 38°C were favorable for disease development (Chase and Conover, 1987). Foot rot of pothos, caused by *R. solani*, was more severe when both daily maximum air and soil temperatures were 30°C but decreased as maximum temperatures reached 35°C (Chase and Poole, 1990). The tendency of Rhizoctonia diseases to be worse during warmer temperatures is also evidenced by the lack of, or less severe disease during the winter on greenhouse ornamentals (Cartwright and Benson, *unpublished*; Chase, 1987). High temperatures combined with high moisture are very favorable for foliar disease development as well. These conditions, for example, favored epidemics of web blight on azalea in the nursery and greenhouse (Frisina and Benson, 1987). To lessen plant stress and reduce favorable conditions, ornamentals should be grown as near to their optimum growth temperature as possible. Especially during the summer months, adequate cooling systems should be used in greenhouses, maximum ventilation utilized, and plants properly spaced to open canopies, thus reducing both temperatures and humidity.

4. Suppressive soil mixes Many agricultural soils are naturally suppressive to plant pathogens (Chet and Baker, 1980; 1981) and much work has focused on suppressive soils to control soilborne pathogens, including *R. solani*. In ornamentals, use has focused on amendments such as composted bark added to growing media to induce suppressive characteristics (Hoitink, 1980). Composted hardwood bark (CHB) amended with sphagnum peat suppressed Rhizoctonia and Pythium crown and root rots of poinsettia as effectively as steam-treated medium drenched with fungicides (Daft *et al.*, 1979). Damping-off of *Celosia,* caused by *R. solani,* was reduced with a medium composed of at least 50% CHB. Composted pine bark was not as effective as CHB in *R. solani* suppression (Hoitink, 1980) and the addition of proportionately more organic material (i.e. peat) eliminated the suppressive effect of composted bark. Likewise, Stephens and Stebbins (1985) demonstrated that fungicides were more effective in controlling damping-off of *Impatiens* when applied to a CHB medium than to peat or composted pine bark. The CHB medium enhanced previously ineffective fungicides, was naturally suppressive alone, and maintained natural suppressiveness in the presence of fungicides. Though mechanisms may vary depending on bark components and method of utilization, greater aeration, antagonistic microorganisms, and fungicidal properties may all be important in suppression (Chet and Baker, 1981; Daft *et al.*, 1979; Hoitink, 1980). To date, bark amendments have found their greatest utility in the containerized nursery industry.

B. Resistance

Few disease-resistant cultivars have been introduced to the ornamental industry. Since most ornamentals are developed solely for appearance, and the rapid production cycles and annual influx of new cultivars allow very susceptible types to be quickly removed from production and replaced with other cultivars or ornamentals, this is unlikely to change in the near future.

Resistance based on plant age has been successful in ornamental production. Over time, seedlings and cuttings become more resistant to infection by *R. solani* due to the formation of calcium pectate in stems (Bateman, 1970; Bateman and Lumsden, 1965) and to lignification (Stockwell and Hanchey, 1987).

C. Chemical control

By and large, fungicides offer safe, effective, and economical methods for control of *R. solani* in ornamentals. Presently in the USA, the most effective and widely used fungicides for control of Rhizoctonia diseases in ornamental production are probably iprodione (Chipco 26019), thiophanate methyl (Cleary's 3336, Domain) and chlorothalonil (Daconil 2787) (Jones, 1993; Jones and Benson, 1993). Benson (1991) demonstrated that stem rot of poinsettia can be effectively controlled during propagation with drenches or sprays of iprodione, chlorothalonil, quintozene, thiophanate methyl, flutolanil, or benomyl. Powell (1988) also controlled stem rot with drenches of quintozene or benomyl, however, only benomyl provided acceptable control without phytotoxic effects. In other work, captan, benomyl, and etradiazole plus thiophanate methyl (Banrot) incorporated into a soilless mix controlled *R. solani* on *Zinnia*, *Celosia*, and *Impatiens* (Daughtrey and Wells, 1980a,b). On foliage ornamentals, etridiazol plus thiophanate methyl, iprodione, or PCNB controlled soilborne Rhizoctonia diseases while foliar sprays of chlorothalonil, iprodione, thiophanate methyl plus mancozeb, or vinclozolin effectively control of Rhizoctonia aerial blight or leaf spot (Chase, 1987). Frisina and Benson (1988) reported that drenches of iprodione and sprays of benomyl, benodanil, or iprodione limited web blight of azalea and binucleate *Rhizoctonia* spp. in the greenhouse. In the nursery, sprays of benomyl, chlorothalonil, iprodione, or thiophanate-methyl plus mancozeb controlled web blight (Benson *et al.*, 1986). Preventative fungicide applications used during propagation or at first symptom are the most effective way to control *Rhizoctonia* epidemics.

D. Biological control

The use of biological control agents for greenhouse ornamental diseases and particularly Rhizoctonia diseases will become more important in the near future. The principle reasons for this are the more controlled environments, uniform growing media, high monetary value of crops, grower concern over management of diseases with fungicides, and continued restrictions of chemical pesticides. An example of this is the new EPA worker protection standards in the USA which severely limit reentry into fungicide-treated greenhouses. Many biocontrol agents have been evaluated against *R. solani*, however, relatively few such as *Gliocladium virens*, *Trichoderma* spp., *Pseudomonas cepacia*, *Paecilomyces lilacinus,* and binucleate *Rhizoctonia* fungi, have been investigated for use in ornamentals (Cartwright and Benson, 1995; Elad *et al.*, 1981; Harris, *et al.*, 1994; Lumsden and Locke, 1989). Lumsden and Locke (1989) tested 20 isolates of *G. virens,* one of which gave effective and consistent control of damping-off of *Zinnia* caused by *Pythium ultimum* and *R. solani*. A 7 day incubation period in the potting medium prior to planting was required for *G. virens*, but control lasted as long as two months. A commercial product containing *G. virens* is now available in the USA (SoilGard®, WR Grace, Columbia, Maryland, USA) for control of damping-off diseases in greenhouse ornamentals (see also chapter VI.B7). Harris *et al.*, (1994) were able to reduce damping-off, caused by *R. solani*, on *Celosia*, *Viola*, and *Petunia* with BNR isolates (see also chapter VI.B4). At least two isolates increased shoot weights of bedding plants and were better than another biocontrol agent, *Bacillus subtilis*. Elad *et al.*, (1981) controlled stem rot of carnation caused by *R. solani* in carnation fields with an isolate of *Trichoderma harzianum*. Cartwright and Benson (1995a,b) controlled *R. solani* on poinsettia cuttings as much as 100% with a strain of *P. cepacia* or *P. lilacinus* applied to foam propagation cubes used for rooting cuttings. BNR fungi have recently been used to control *R. solani* on *Impatiens* and poinsettia (Cartwright and Benson, *unpublished*).

E. Eradicants

Eradication of *R. solani* inoculum, via several methods, is an option in ornamental production whereas it is not in other venues. Although not widely used today in the USA, steaming of potting media and mixes reduces damping-off in greenhouses and nurseries. Aerated steam at ~60 to 90°C for 30 minutes will eliminate most pathogens without affecting natural antagonists (Averre *et al.*, 1993; Jones and Averre, 1992; Leach and Garber, 1970). Solarization, the process in which temperatures reach at least 60°C during the summer in closed greenhouses, also reduces inoculum on greenhouse benches, in pots, on tools, or in shallow soil mixes (Averre *et al.*, 1993; see also chapter VI.A2). Methyl bromide alone or with chloropicrin can reduce inoculum, although alternatives are under investigation as the USA is a participant in the Montreal Accord that bans methyl bromide after the year 2000. Tools, equipment, and containers can also be disinfested with steam, solarization, methyl bromide, alcohol, or bleach (Averre *et al.*, 1993).

VI. Outlook

Rhizoctonia spp. will continue to cause disease problems in ornamentals as new crops are introduced and as changes in production practices occur that result in favorable environments for disease development. Although ornamental uses of benomyl (Benlate) have been withdrawn in the USA, several new fungicides will be introduced over the next few years that are effective at low doses and safer in the environment. The outlook for biological control agents is promising, but the biggest obstacle to adoption will be the economics of commercialization.

VII. Acknowledgements

The authors gratefully acknowledge the assistance of Kala Parker in compilation of the table. This work was supported by the North Carolina Agricultural Research Service, North Carolina State University. Use of tradenames does not imply endorsement by the North Carolina Agr. Res. Serv. of the products named nor criticism of similar ones not mentioned. This chapter contains information involving pesticides. It does not contain recommendations for their use, nor does it imply that the uses discussed here have been registered.

VIII. References

Averre CW, Jones RK, Shoemaker PB & Skroch WA (1993) Sanitizing greenhouses and plant beds. North Carolina Agricultural Chemicals Manual. College of Agricultural and Life Sciences, North Carolina State University, Raleigh.

Baker R & Martinson CA (1970) Epidemiology of diseases caused by *Rhizoctonia solani*. In: Parmeter JR Jr. (ed.) *Rhizoctonia solani*, Biology and Pathology. (pp 172-188) University of California Press, Berkeley, Los Angeles, and London.

Bateman DF (1970) Pathogenesis and disease. In: Parmeter JR Jr. (ed.) *Rhizoctonia solani*, Biology and Pathology. (pp 161-171) University of California Press, Berkeley, Los Angeles, and London.

Bateman DF & Lumsden RD (1965) Relation of calcium content and nature of the pectic substances in bean hypocotyls of different ages to susceptibility to an isolate of *Rhizoctonia solani*. Phytopathology 55:734-738.

Benson DM (1985) Common names for plant diseases: Azalea (*Rhododendron* spp.). Plant Dis. 69:651-652.

Benson DM (1991) Control of Rhizoctonia stem rot of poinsettia during propagation with fungicides that prevent colonization of rooting cubes by *Rhizoctonia solani*. Plant Dis. 75:394-398.

Benson DM, Jones RK & Frisina T (1986) *Rhizoctonia* web blight. In: Coyier DL and Roane MK (eds.) Compendium of rhododendron and azalea diseases. (pp 20-21) APS Press. St. Paul, Minnesota.

Bolkan HA & Ribeiro WR (1985) Anastomosis groups and pathogenicity of *Rhizoctonia solani* isolates from Brazil. Plant Dis. 69:599-601.

Burpee LL (1992) Assessment of resistance to *Rhizoctonia solani* in tall fescue based on disease progress and crop recovery. Plant Dis. 76:1065-1068.

Burpee LL, Sanders PL, Cole Jr, H & Sherwood RT (1980) Anastomosis groups among isolates of *Ceratobasidium cornigerum* and related fungi. Mycologia 72:689-701.

Cartwright DK & Benson DM (1995a) Biological control of Rhizoctonia stem rot of poinsettia in polyfoam rooting cubes with *Pseudomonas cepacia* and *Paecilomyces lilacinus*. Biol. Control 5:237-244.

Cartwright DK & Benson DM (1995b) Optimization of biological control of Rhizoctonia stem rot of poinsettia by *Paecilomyces lilacinus* and *Pseudomonas cepacia*. Plant Dis. 79:301-308.

Castillo S & Peterson JL (1990) Cause and control of crown rot of New Guinea impatiens. Plant Dis. 74:77-79.

Chase AR (1987) Compendium of Ornamental Foliage Plant Diseases. APS Press. St. Paul, Minnesota.

Chase AR (1991) Characterization of Rhizoctonia species isolated from ornamentals in Florida. Plant Dis. 75:234-238.

Chase AR & Conover CA (1987) Temperature and potting medium effects on growth of Boston fern infected with *Rhizoctonia solani*. HortSci. 22:65-67.

Chase AR & Poole RT (1990) Effect of air and growing medium temperatures on Rhizoctonia foot rot of *Epipremnum aureum*. J. Env. Hort. 8:139-141.

Chet I & Baker R (1980) Induction of suppressiveness to *Rhizoctonia solani* in soil. Phytopathology 70:994-998.

Chet I & Baker R (1981) Isolation and biocontrol potential of *Trichoderma hamatum* from soil naturally suppressive to *Rhizoctonia solani*. Phytopathology 71:286-290.

Cline MN, Chastagner GA, Aragaki M, Baker R, Daughtrey ML, Lawson RH, MacDonald JD, Tammen JF & Worf GL (1988) Current and future research direction of ornamental pathology. Plant Dis. 72:926-934.

Daft GC, Poole HA & Hoitink HAJ (1979) Composted hardwood bark: a substitute for steam sterilization and fungicide drenches for control of poinsettia crown and root rot. HortScience 14:185-187.

Daughtrey ML & Wells SL (1980a) Control of damping-off with fungicides incorporated into peat-lite mix. Fungic. and Nematic. tests. 36:123.

Daughtrey ML & Wells SL (1980b) Phytotoxicity of fungicide combinations incorporated into peat-lite mix. Fungic. and Nematic. tests. 36:124.

Ecke P, Jr., Matkin OA & Hartley DE (1990) The Poinsettia Manual. Paul Ecke Poinsettias. Encinitas, California.

Elad Y, Hadar Y, Hadar E, Chet I & Henis Y (1981) Biological control of *Rhizoctonia solani* by *Trichoderma harzianum* in carnation. Plant Dis. 65:675-677.

English JT, Ploetz RC & Barnard EL (1986) Seedling blight of longleaf pine caused by a binucleate *Rhizoctonia solani*-like fungus. Plant Dis. 70:148-150.

Farr DF, Bills GF, Chamuris GP & Rossman AY (1989) Fungi on plants and plant products in the United States. APS Press, St. Paul, Minnesota 1252 pp.

Fravel DL, Benson DM & Bruck RI (1983) Edaphic parameters associated with shore juniper decline. Phytopathology 73:204-207.

Frisina TA & Benson DM (1987) Characterization and pathogenicity of binucleate *Rhizoctonia* spp. from azaleas and other woody ornamentals with web blight. Plant Dis. 71:977-981.

Frisina TA & Benson DM (1988) Sensitivity of binucleate *Rhizoctonia* spp. and *R. solani* to selected fungicides in vitro and on azalea under greenhouse conditions. Plant Dis. 72:303-306.

Frisina TA & Benson DM (1989) Occurrence of binucleate *Rhizoctonia* spp. on azalea and spatial analysis of web blight in container-grown nursery stock. Plant Dis. 73:249-254.

Harris AR, Schisler DA, Neate SM & Ryder MH (1994). Suppression of damping-off caused by *Rhizoctonia solani*, and growth promotion, in bedding plants by binucleate *Rhizoctonia* spp. Soil Biol. Biochem. 26:263-268.

Hointink HAJ (1980) Composted bark, a lightweight growth medium with fungicidal properties. Plant Dis. 64:143-147.

Jones RK (1993) Commercial floral crop disease control. North Carolina Agricultural Chemicals Manual. College of Agricultural and Life Sciences, North Carolina State University, Raleigh.

Jones RK (1994a) Diseases of carnations and dianthus. Plant Pathology information note. North Carolina Cooperative Extension Service. No. 25.

Jones RK (1994b) Diseases of snapdragons. Plant Pathology information note. North Carolina Cooperative Extension Service. No. 26.

Jones RK & Averre CW (1992) Damping-off in seed beds, flower and vegetable seedlings. Plant Pathology information note. North Carolina Cooperative Extension Service. No. 14.

Jones RK & Benson DM (1993) Commercial landscape and nursery crops disease control. North Carolina Agricultural Chemicals Manual. College of Agricultural and Life Sciences, North Carolina State University, Raleigh.

Jones RK & Benson DM (1994) Disease management for poinsettias. Plant Pathology information note. North Carolina Cooperative Extension Service. No. 28.

Leach LD & Garber RH (1970) Control of *Rhizoctonia.* In: Parmeter JR (ed.) *Rhizoctonia solani*, Biology and Pathology. (pp 189-198) University of California Press, Berkeley, Los Angeles, and London.

Lumdsen RD & Locke JC (1989) Biological control of damping-off caused by *Pythium ultimum* and *Rhizoctonia solani* with *Gliocladium virens* in soilless mix. Phytpathology 79:361-366.

Powell CC Jr. (1988) The safety and efficacy of fungicides for use in Rhizoctonia crown rot control of directly potted unrooted poinsettia cuttings. Plant Dis. 72:693-695.

Sneh B, Burpee L & Ogoshi A (1991) Idenitification of *Rhizoctonia* Species. APS Press St. Paul, Minnesota 133 pp.

Stephens CT & Stebbins TC (1985) Control of damping-off pathogens in soilless container media. Plant Dis. 69:494-496.

Stephens CT, Herr LF, Schmitthenner AF & Powell CC (1982) Characterization of *Rhizoctonia* isolates associated with damping-off of bedding plants. Plant Dis. 66:700-703.

Stephens CT, Herr LF, Schmitthenner AF & Powell CC (1983) Sources of *Rhizoctonia solani* and *Pythium* spp. in a bedding plant greenhouse. Plant Dis. 67:272-275.

Stockwell V & Hanchey P (1987) Lignification of lesion borders in *Rhizoctonia*-infected bean hypocotyls. Phytopathology 77:589-593.

Trujillo EE, Shimabuku R, Cavin CA & Aragaki M (1988) *Rhizoctonia solani* anastomosis groupings in carnation fields and their pathogenicity to carnation. Plant Dis. 72:863-865.

Tu CC, Roberts DA & Kimbrough JW (1969) Hyphal fusion, nuclear condition, and perfect stages of three species of *Rhizoctonia.* Mycologia 61:775-783.

Weber GF (1939) Web-blight, a disease of beans caused by *Corticium microsclerotia.* Phytopathology 29:559-575.

Weber GF & Roberts DA (1951) Silky threadblight of *Elaeagnus pungens* caused by *Rhizoctonia ramicola* n. sp. Phytopathology 41:615-621.

Wehlburg C & Cox RS (1966) Rhizoctonia leaf blight of azalea. Plant Dis. Rep. 50:354-355.

V.7 PEANUT DISEASES INCITED BY *RHIZOCTONIA* SPECIES

TIMOTHY B. BRENNEMAN

Department of Plant Pathology, University of Georgia, Coastal Plain Experiment Station, Tifton, GA 31793-0748, USA

I. INTRODUCTION

Rhizoctonia solani Kühn can cause damage to all parts of the peanut (*Arachis hypogaea* L.) plant and has been reported from most countries where this crop is grown. Where conditions are favorable for disease development, yield and quality losses can be severe. In Georgia, USA, alone, losses were estimated at $44 million in 1988 (Thompson, 1988). Although losses are influenced by environment, genotypes, cultural practices, and other factors, *Rhizoctonia*-induced diseases are consistently responsible for significant damage to peanut crops. Early planting in cool, wet soils is conducive to seed decay and/or seedling damping-off due to *Rhizoctonia* alone or in concert with other pathogens such as *Pythium*, *Rhizopus*, *Fusarium*, and *Aspergillus*.

II. SYMPTOMS

Damage to seedlings is characterized by distinct sunken lesions on hypocotyls that can be fatal if they girdle the plant. Usually the plants survive, although their subsequent growth and development can be adversely affected. Factors such as plant age, hypocotyl exudates, plant anatomical differences, etc. all influence the pathogens attack on peanut seedlings (Reddy and Rao, 1981). Slight to severe necrosis of root systems also has been attributed to *Rhizoctonia*, and disease severity on tap roots shows a strong positive correlation to yield loss (Turner and Backman, 1988). However, the most severe damage results from colonization of stems, pegs, and the peanut pods. Rhizoctonia limb rot was reported in Georgia in 1982 (Thompson, 1982) as an emerging problem. Lesions on stems are characteristically light tan with a dark border, sometimes in a pattern of alternating rings. They can occur at any point on lower limbs, particularly those in contact with the soil or those physically damaged by machinery traffic. Lesions may remain localized or may girdle the stem, thus causing the loss of all pods formed on that stem beyond the point of girdling. The pods themselves are sometimes rotted with no effect on other parts of the plant. Symptoms vary from superficial brown or dark brown cankers to complete degradation of both the shell and the seed. Intact but discolored seed will reduce the quality and value of the crop. Other pathogens are frequently involved in pod rot, and calcium deficiency is strongly associated with predisposing pods to infection, somewhat analogous to a subterranean "blossom-end rot" on tomato (Csinos and Gaines, 1986).

Another significant source of loss is degradation of peanut gynophores (pegs) by *R. solani*. This can occur at any time from peg initiation until harvest. The tips of developing pegs can be rotted as they contact the soil and the meristems are aborted. Other pegs may be severed by *Rhizoctonia* late in the season resulting in the loss of pods at harvest that may or may not be otherwise damaged. Infected pegs are brown and shredded with only the vascular elements remaining. Variability in peg strength

B. Sneh et al. (eds.),
Rhizoctonia Species: Taxonomy, Molecular Biology, Ecology, Pathology and Disease Constrol, 315–320.

among peanut genotypes has been quantified and stronger pegs tend to develop fewer lesions from *Rhizoctonia* and other pathogens (Thomas, 1977).

Peanut leaves are also susceptible to *Rhizoctonia*. This is observed most frequently in irrigated peanuts with lush growth and may result in brownish mats of colonized foliage. A true foliar blight of peanut has been associated with *R. solani*

AG 1 (Littrell, 1974) but is not commonly observed. *R. solani* AG 1 sometimes causes a foliar blight in disease screening nurseries on certain genotypes indicating they must be very susceptible to the disease. In comparative studies with *R. solani* AG 4, AG 1 was a less virulent pathogen of peanut seed and seedlings, but was usually a more virulent pathogen of peanut foliage (Cedeno, 1976). The optimum temperature for growth of *R. solani* AG 1 and AG 4 was about 30°C (Cedeno, 1976), although the optimum for *R. solani* as a group has ranged from 18-32°C (Sherwood, 1970).

III. The Causal Organisms

Rhizoctonia solani Kühn [teleomorph, *Thanatephorus praticola* (Kotila) Flentje] AG 4 is the predominant *Rhizoctonia* parasitizing peanut. It is capable of infecting all organs of the peanut plant as described previously, although the true aerial blight of foliage is caused by *R. solani* AG 1. Several binucleate *Rhizoctonia* spp. (teleomorph, *Ceratobasidium* spp.) have also been recovered from peanut seeds. These include isolates of AGs -A, -E, -F, and -R (synonyms, CAGs 2, 3, 4 and 5 respectively), which were avirulent or only weakly virulent to peanut seed and seedlings (Bell and Sumner, 1984a). Isolates of AG-E (=CAG 3) have been recovered from pods (Baird *et al.*, 1991) and from lesions on stems damaged by tractor traffic, whereas lesions on nondamaged stems yielded only *R. solani* AG 4 (Brenneman and Sumner, 1990). A survey of diseased plants and soil from peanut fields in Texas and Oklahoma indicated that the great majority of isolates were *R. solani* AG 4 (Woodard and Jones, 1983). Populations of *R. solani* in the soils ranged from 0 to 14 propagules/100g soil, similar to that reported for cotton and potato fields in California (Weinhold, 1977). However, soil populations are not always correlated with actual disease severity (Sumner *et al.*, 1991).

Significant genetic variability has been found among isolates within AG 4. Two "homogeneous groups" have been described by Kuninaga and Yokosawa (1984) based on colony morphology and DNA/DNA reassociation kinetics, while Vilgalys has described an additional genetically distinct subgroup within *R. solani* AG 4 from North America. Characterization of isolates of *R. solani* AG 4 infecting different parts of the peanut plant and at different times of the year, i.e., early season hypocotyl versus late season stem infections, using rDNA restriction analysis by PCR fingerprinting has demonstrated that most isolates from peanut belong to AG 4 HG-I. The relative distribution of these groups was stable through the growing season. Work is underway to further clarify whether the different sub groups differ in their virulence, fungicide sensitivity, and saprophytic survival (M. Cubeta, N.C. State Univ., pers. comm.).

IV. Epidemiology and Management

Although isolates of AG 4 are found primarily on plants from the Chenopodiaceae, the Leguminosae, and the Solanaceae (Ogoshi, 1987), they have a wide host range. They also have excellent saprophytic ability which makes them difficult pathogens to manage. AG 4 isolates readily colonizes peanut pods which have been shown to be an effective source of inoculum for future epidemics (Bell and Sumner, 1984b). The pathogen will survive in colonized pods in soil for at least two years, although viability of the fungus drops sharply after 6 months (Baird *et al.*, 1993a). Detached pods lost during harvest contain higher populations of AG 4 than those pods successfully harvested mechanically (Baird *et al.*, 1993b). Other studies have shown that *R. solani* AG 4 can survive in fallow soil for up to 283 days and cause disease in subsequently planted crops (Bell and Sumner, 1987). Fungicides applied to peanut foliage, either by tractor-mounted sprayers or by injection into irrigation water, have had variable effects on populations of *Rhizoctonia* in the soil (Sumner and Littrell, 1989; Sumner *et al.*, 1991). Flutolanil inhibits *Rhizoctonia* and was registered for use on peanut in the United States in 1995. Foliar applications have been shown to decrease populations of AG 4, as well as *Trichoderma* spp., in peanut shells (Baird *et al.*, 1993b). Propiconazole did not alter AG 4 populations in shells, even when flooded through the peanut canopy via irrigation water (Baird *et al.*, 1991).

Many peanut growers utilize short rotations, which exacerbates control problems. Rotations with grass crops usually reduce peanut diseases incited by *Rhizoctonia*, but effects have been variable. One explanation may be the presence of weeds in the rotational crops that serve as hosts. Cotton (*Gossypium hirsutum* L.) is regaining popularity in the southeastern United States and production is increasing rapidly. Cotton is very susceptible to AG 4, and rotation studies have shown limb rot in peanuts rotated with cotton to be similar in severity to peanuts grown continuously (Brenneman *et al.*, 1996). Early studies by Hallock (1975) showed no difference in Rhizoctonia pod rot severity in peanuts rotated with several other crops or due to various plowing dates. Recent shifts toward no-till or minimum till production systems have also raised concerns since studies have shown that crop debris increases disease severity (Boyle, 1956). However, some studies have shown equivalent (Minton *et al.*, 1990) or less limb rot in minimum till compared to conventional tillage systems (Minton *et al.*, 1991).

The extensive use of irrigation also has been linked to the increased severity of Rhizoctonia limb rot in peanut (Melouk *et al.*, 1995). In Georgia for example, the irrigated acres rose approximately 700% from 1980 to 1988, the same interval that limb rot went from just being noticed to being one of the most damaging peanut diseases in the state. Irrigation has been shown to increase limb rot by up to 100% (Brenneman, 1992), and the effect is most severe with more frequent watering (Barnes *et al.*, 1990b). Fewer applications of water and reduced watering late in the season are recommended to aid in control.

Past efforts to breed crops with resistance to *R. solani* have met with limited success, but that is improving partially due to a better understanding of AGs and subgroups (Ogoshi, 1987). Various sources of peanut germplasm have been screened for their susceptibility to *R. solani*, and degrees of resistance have been reported. Woodard and Jones (1980) and Kahn and Wadud Mian (1974) identified plant introductions and varieties with resistance to seed and seedling diseases. Branch and Brenneman (1993) reported reduced Rhizoctonia limb rot in advanced Georgia breeding lines, and the cultivar Georgia Browne (GA T-2741) has since been released. Virginia 81-B also has shown partial resistance to limb rot (Barnes *et al.*, 1990a). Mechanisms of resistance are not fully elucidated, but, in the case of limb rot, may be related to reduced vegetative growth, upright vegetative growth, or actual physiological or metabolic resistance.

Currently the best control option is to use one of the recently registered fungicides effective against *R. solani*. Pentachloronitrobenzene (PCNB) was used previously but had limited activity, and some *R. solani* isolates from peanut are reported to have reduced sensitivity to PCNB (Filonow, 1984). Several biosynthesis inhibitors (EBIs) have shown potential, but the only one currently registered for use on peanut in the USA is tebuconazole. Although not as active on *R. solani* as they are on *Sclerotium rolfsii*, good disease control and significant yield increases have been observed with several EBIs (Brenneman *et al.*, 1991; Csinos *et al.*, 1987). Flutolanil was registered on peanut in the USA in 1995 and offers good control as well as being a tool to help manage potential resistance to EBIs. Peanut yield increases of 800-1500 kg/ha are not unusual with these fungicides, but it is difficult to separate the impacts of *R. solani* and *S. rolfsii* (Melouk and Backman, 1995). One concern with late season fungicide sprays utilizing tractor-mounted sprayers is an increase in limb rot due to mechanical damage to the plants (Brenneman and Sumner, 1990).

Biological control agents may play a more active role in managing Rhizoctonia diseases in the future, particularly for protection of peanut seed and seedlings. This has been demonstrated with the fluorescent pseudomonads (Savithiry and Gnanamanickam, 1987) and *Trichoderma* (Greer, 1978). Work has been done to evaluate the interaction of *Rhizoctonia* and other peanut rhizosphere inhabitants as influenced by various pesticides (Sunar and Chohan, 1969), but further studies are needed.

Fungicides have given only sporadic control of peanut pod rot. This disease is thought to be of complex etiology and probably associated more with nutrient imbalances in the geocarposphere than with the inoculum potential of *R. solani* (Csinos and Gaines, 1986). Various pathogens, including AG 4, have been associated with these symptoms, but the primary pathogen is frequently considered to be *Pythium myriotylum* Dresch. The most consistent control measure has been the application of gypsum to the pegging zone during fruiting (Garren, 1964; Csinos and Gaines, 1986). However, this may vary with soil type, pathogen complex, cultivar, and/or other factors because gypsum has not always provided control of Rhizoctonia pod rot caused by AG 4 (Filonow *et al.*, 1988).

V. References

Baird RE, Bell DK, Sumner DR, Mullinix BG & Culbreath AK (1993a) Survival of *Rhizoctonia solani* AG 4 in residual peanut shells in soil. Plant Dis. 77:973-975.

Baird RE, Brenneman TB, Bell DK, Culbreath AK & Mullinix BG (1993b) The peanut shell mycobiota of detached vs. mechanically harvested pods either treated or not treated with flutolanil. Plant Dis. 77:405-408.

Baird RE, Brenneman TB, Bell DK & Murphy AP (1991) The effects of the fungicide propiconazole (Tilt) on the groundnut shell mycobiota. Mycol. Res. 95:571-576.

Barnes JS, Csinos AS & Branch WD (1990a) Sensitivity of *Rhizoctonia solani* isolates to fungicides and evaluation of peanut cultivars to Rhizoctonia limb rot. Peanut Sci. 18:62-65.

Barnes JS, Csinos AS & Hook JE (1990b) Effects of fungicides, cultivars, irrigation, and environment on Rhizoctonia limb rot of peanut. Plant Dis. 74:671-676.

Bell DK & Sumner DR (1984a) Rhizoctonia diseases. In: Porter DM, Smith DH & Rodriguez-Kabana R (eds.) Compendium of Peanut Diseases (pp 23-25). APS Press, St. Paul. Minnesota.

Bell DK & Sumner DR (1984b) Unharvested peanut pods as a potential source of inoculum of soilborne peanut pathogens. Plant Dis. 68:1039-1042.

Bell DK & Sumner DR (1987) Survival of *Rhizoctonia solani* and other soilborne basidiomycetes in fallow soil. Plant Dis. 71:911-915.

Boyle LW (1956) Fundamental concepts in the development of control measures for southern blight and root rot on peanuts. Plant Dis. Rep. 40:661-665.

Branch WD & Brenneman TB (1993) White mold and Rhizoctonia limb rot resistance among advanced Georgia breeding lines. Peanut Sci. 20:124-126.

Brenneman TB (1992) Effects of irrigation on yield and Rhizoctonia limb rot in Southern Runner peanut at two harvest dates. Proc. Amer. Peanut Res. Educ. Soc. 24:4.

Brenneman TB, Minton NA, Baker SH, Herzog GA & Gascho GJ (1996) Effects of a cotton-peanut rotation with and without rye on diseases, nematodes and crop yields. Proc. Amer. Peanut Res. Educ. Soc. 27: (in press).

Brenneman TB, Murphy AP & Csinos AS (1991) Activity of tebuconazole on *Sclerotium rolfsii* and *Rhizoctonia solani*, two soilborne pathogens of peanut. Plant Dis. 75:744-747.

Brenneman TB & Sumner DR (1990) Effects of tractor traffic and chlorothalonil applied via ground sprays or center pivot irrigation systems on peanut diseases and pod yields. Plant Dis. 74:277-279.

Cedeno LF (1976) Growth and anastomosis grouping of *Rhizoctonia solani* isolates and pathogenicity to *Arachis hypogaea* and other hosts. M.S. Thesis, Univ. Georgia, Athens, GA.

Csinos AS & Gaines TP (1986) Peanut pod rot complex: A geocarposphere nutrient imbalance. Plant Dis. 70:525-529.

Csinos AS, Kvien CS & Littrell RH (1987) Activity of diniconazole on foliar and soil-borne diseases of peanut. Appl. Agric. Res. 2:113-116.

Filonow AB (1984) Tolerance of *Rhizoctonia solani* to pentachloronitrobenzene and sensitivity of *Pythium myriotylum* to metalaxyl and their pathogenicity to peanut. Phytopathology 74:810.

Filonow AB, Melouk HA, Martin M & Sherwood J (1988) Effect of calcium sulfate on pod rot of peanut. Plant Dis. 72:589-593.

Garren KH (1964) Landplaster and soil rot of peanut pods in Virginia. Plant Dis. Rep. 48:349-352.

Greer JE (1978) Antagonistic reactions of *Trichoderma harzianum* toward *Rhizoctonia solani* and *Sclerotium rolfsii*. M.S. Thesis, Univ. Georgia, Athens, GA.

Hallock DL (1975) Effect of plowing date and certain crop rotations on peanut productivity. Peanut Sci. 2:81-83.

Khan AR & Wadud Mian MA (1974) *In vitro* screening of *Arachis hypogaea* against *Rhizoctonia solani* and *Sclerotium rolfsii*. Indian Phytopathology 27:78-81.

Kuninaga S & Yokosawa R (1984) DNA base sequence homology in *Rhizoctonia solani* Kühn IV. Genetic relatedness within AG 4. Ann. Phytopathol. Soc. Jpn., 322-330.

Littrell RH (1974) Rhizoctonia foliar blight of peanut. Proc. Amer. Peanut Res. Educ. Soc. 6:62 (Abstr.).

Melouk HA & Backman PA (1995) Management of soilborne fungal pathogens. In: Melouk HA & Shokes FM (eds.) Peanut Health Management (pp 75-82) APS Press, St. Paul.

Minton NA, Csinos AS, Lynch RE & Brenneman TB (1991) Effects of two cropping and two tillage systems and pesticides on peanut pest management. Peanut Sci. 18:41-46.

Minton NA, Csinos AS & Morgan LW (1990) Relationship between tillage and nematicide, fungicide, and insecticide treatments on pests and yield of peanuts double-cropped with wheat. Plant Dis. 74:1025-1029.

Ogoshi A (1987) Ecology and pathogenicity of anastomosis and intraspecific groups of *Rhizoctonia solani* Kühn. Annu. Rev. Phytopathol. 25:125-143.

Reddy MN & Rao AS (1981) Disease cycle involved in damping-off of groundnut caused by *Rhizoctonia solani*. Plant Soil 62:465-467.

Savithiry S & Gnanamanickam SS (1987) Bacterization of peanut with *Pseudomonas flourescens* for biological control of *Rhizoctonia solani* and for enhanced yield. Plant Soil 102:11-15.

Sherwood RT (1970) Physiology of *Rhizoctonia solani*. In: Parmenter JR Jr (ed.) Biology and Pathology of *Rhizoctonia solani*. (pp 69-92) Univ. California Press, Berkeley.

Sumner DR, Brenneman TB & Harrison GW (1991) Populations of fungi in soil after chemigation with chlorothalonil and tebuconazole via center-pivot irrigation. Plant Dis. 75:999-1004.

Sumner DR & Littrell RH (1989) Effects of chemigation with chlorothalonil and diniconazole on soil fungi and pod, peg and stem diseases of peanut. Plant Dis. 73:642-646.

Sunar MS & Chohan JS (1971) Influence of chemical sprays on the rhizosphere mycoflora of groundnut (*Arachis hypogaea* L.). Indian J. Agric. Sci. 41:29-37.

Thomas RJ (1977) Peanut peg strength and anatomy as related to disease resistance. M.S. Thesis, Texas A&M University.

Thompson SS (1982) Rhizoctonia limb rot disease. Proc. Amer. Peanut Res. Educ. Soc. 14:88 (Abstr.).

Thompson SS (1988) Peanut disease losses in 1988. In: Rice A (ed.) 1988 Georgia Peanut Research-Extension Report (p 198) Univ. Georgia Coop. Res.-Ext. Publication No. 3.

Turner JT & Backman PA (1988) Severity, distribution and losses from taproot cankers caused by *Rhizoctonia solani* in peanuts. Peanut Sci. 15:73-75.

Weinhold AR (1977) Population of *Rhizoctonia solani* in agricultural soils determined by a screening procedure. Phytopathology 67:566-569.

Woodard KE & Jones BL (1980) Screening peanut plant introductions in controlled environment chambers for resistance to *R. solani*. Plant Dis. 64:949-950.

Woodard KE & Jones BL (1983) Soil populations and anastomosis groups of *Rhizoctonia solani* associated with peanut in Texas and New Mexico. Plant Dis. 67:385-387.

V.8 RHIZOCTONIA DISEASE ON POTATO

GILBERT J BANVILLE[1], DONALD E CARLING[2] AND BARBARA E OTRYSKO[1]

[1]*Ministère de l'Agriculture, des Pecheries et de l'Alimentation du Québec, Station de recherches Les Buissons C.P. 455, Les Buissons, Québec G0H 1H0, CANADA.*
[2]*University of Alaska, Fairbanks, Alaska 99645, USA.*

I. IMPORTANCE

The Rhizoctonia disease complex of potatoes comprises two distinct phases: infection of growing plants (Rhizoctonia canker) and infestation of daughter tubers by sclerotia (black scurf). Either or both phases may be present in potato crops. Both are economically important and each has been the object of much research. However, caution must be used when interpreting data reported in the literature because not all authors define which phase they are dealing with when talking about the "Rhizoctonia disease".

The Rhizoctonia disease complex is present everywhere in the world that potatoes are grown, although symptoms or signs associated with either phase may be more or less visible or important depending on the inoculum potential and on local climatic conditions.

Rhizoctonia disease is generally ignored until an extreme situation occurs, which in itself is exceptional. Nevertheless, in its many aspects and its ubiquitous presence, it is one of the most serious potato diseases. Marketable yield losses of up to 30% are often cited, although losses of 10 to 15% are more common (Carling *et al.*, 1989; Little *et al.*, 1988; Read *et al.*, 1989). A realistic evaluation of the incidence of Rhizoctonia infection in a potato field is difficult because the most distinctive symptoms occur beneath the soil surface and the impact on yield is not seen until harvest.

II. CAUSAL AGENT

The causal agent of the Rhizoctonia disease complex on potato is *Rhizoctonia solani* Kühn [teleomorph *Thanatephorus cucumeris* (Frank) Donk]; and Rhizoctonias other than *R. solani* are believed to play little or no role (Carling and Leiner, 1990b). *R. solani* is divided into anastomosis groups (AGs) (see chapter 1.A.2) and many of the twelve known AGs have been reported to be associated with potato plants in nature. These include AG 1 (Abe and Tsuboke, 1978; Chang and Tu, 1980); AG 2 (subgroups -1 and -2:) (Chand and Logan, 1983; Chang and Tu, 1980); AG 3 (Bandy *et al.*, 1988; Banville, 1989; Bains and Bisht, 1995; Carling *et al.*, 1989; Chand and Logan, 1983; Otrysko *et al.*, 1985;) AG 4 (Anguiz and Martin, 1989; Gudmestad *et al.*, 1989; International Potato Center Annual Report, 1983); AG 5 (Bandy *et al.*, 1984), and AG 9 (Carling *et al.*, 1987). In addition, the virulence of strains representing the above, plus all other AGs of *R. solani*, as well as other multinucleate and binucleate *Rhizoctonia* species, have been evaluated in growth chamber studies

B. Sneh et al. (eds.),
Rhizoctonia Species: Taxonomy, Molecular Biology, Ecology, Pathology and Disease Constrol, 321–330.

(Carling and Leiner, 1990a; Carling *et al.*, 1994; MacNish *et al.*, 1995; Otrysko *et al.*, 1985).

These and other observations illustrate that isolates of selected AGs, most notably AG*s* 3, 4, 5 and 8, are capable of doing moderate to extensive damage to potato plants, whereas other AGs are capable only of inflicting minor injury or no damage at all. Although strains of AG 4 are pathogenic on potato, reports regarding the extent of damage to potato plants are mixed (Anguiz and Martin, 1989; Carling and Leiner, 1990a; Gudmestad *et al.*, 1989). However, it does seem clear that AG 4 requires warmer temperatures than many other AGs and thus it will tend to play a more important role at lower elevations in warm temperate or even tropical environments (Anguiz and Martin, 1989).

Strains of AG*s* 3, 5 and 8 each attack subterranean organs of potato plants, but there are clear differences between the three types. AG 3 is far more aggressive than either AG 5 or AG 8 and indiscriminately attacks roots, stolons and the subterranean portion of the main stems (Carling and Leiner, 1990a). Although AG 3 is virulent across a broad range of temperatures (5°C to 25°C), it is most aggressive at 10°C to 15°C, where other AGs generally become less damaging (Carling and Leiner, 1990a). AG 5 is similar to AG 3 in that it is also capable of damaging roots, stems, and stolons but generally it is far less aggressive and inflicts significantly less damage than AG 3 (Bandy *et al.*, 1988; Carling and Leiner, 1990a;b). Unlike AG 3, AG 5 tends to cause more damage at higher temperatures (Carling and Leiner, 1990a). Attack by AG 8 on potato is limited almost exclusively to root tissues (Carling and Leiner, 1990a; Hide and Firmager, 1990). Infection by AG 8 has been observed in growth chamber studies, but not in the field. However, growth chamber evidence illustrates that AG 8 may pose a threat to potatoes growing in the field.

Traditionally, AG 3 has been known as the host-specific type of *R. solani* causing disease of potato. It is by far the most important cause of black scurf because most of the sclerotia isolated from potato tubers belong to AG 3 (Abe and Tsuboki, 1978; Carling and Leiner, 1986; Chand and Logan, 1983; Otrysko *et al.*, 1985). However, in recent years, strains of AG 3 that cause a leaf spot disease of tobacco have been reported (Meyer *et al.*, 1990). These two sub-groups of AG 3 are clearly different from each other. One is a pathogen on foliar tissues, whereas the other attacks subterranean organs. Furthermore, strains of AG 3 that attack tobacco are not pathogenic to potato and vice versa (Stevens Johnk *et al.*, 1993).

III. Symptoms, signs and disease cycle

A. Damage to the plant

Rhizoctonia disease of potato occurs primarily beneath the soil surface where most symptoms and signs can be observed. Meristematic and young succulent subterranean tissues are most susceptible to attack. The developing tips of sprouts are attacked by the fungus prior to emergence and, when sprout tips are killed, plants will often make multiple attempts to emerge (Baker, 1970). Light brown to black lesions of varying size form on sprouts, stolons and roots and each of these organs may be girdled or killed by such lesions.

Often, no symptoms or signs of Rhizoctonia disease can be found on the aerial portion of the plant and most of those that do appear are secondary responses to damage inflicted below the soil surface. Sometimes aerial tubers form in the axils of branches and petioles (Banville, 1989; Hide *et al.*, 1973). Leaf rolling, stunting and rosetting may occur in the upper portions of the plant, as well as purple pigmentation due to accumulation of anthocyanins in the uppermost leaves (Baker, 1970). Late in the growing season, hymenia frequently form on the lower portion of aerial stems near the soil surface. Hymenia are superficial and ephemeral and, like sclerotia, do not cause any physical damage to the surface on which they form. They are often, but not always, connected to the fungal strain attacking the subterranean portion of the plant (Carling and Leiner, 1986).

B. Damage to the progeny tubers

The general effect of Rhizoctonia disease on tubers is a change in their number, size distribution, and a degradation of quality because of the production of misshapen tubers and tubers with growth cracks (Banville, 1989; Carling *et al.*, 1989; Hide *et al.*, 1973; Hide *et al.*, 1985a;b; Hide *et al.*, 1989a;b; Hide and Horrocks, 1994; Scholte, 1989). This combination of factors can result in lower marketable yields. Specific gravity and chip color of tubers may also be negatively affected (Hide and Horrocks, 1994; Otrysko and Banville, 1992; Scholte, 1989). Since growth of an infected plant is delayed in all phases (Banville, 1989; Hide *et al.*, 1985b), it may be more difficult to kill the vines, thus leaving the plant more prone to infection by late blight or to attack by aphids, which are important vectors of potato viruses. However, none of these effects is specific to Rhizoctonia disease, making direct cause-effect estimates difficult.

Tubers can be affected in several additional ways, although lesions of the type that appear on sprouts and roots are not known to form on tubers. Russetted, scabby areas induced by *R. solani* sometimes appear on the tuber surface (Baker, 1970). Additionally, the fungus is known to invade injury sites created by insects or other means, producing characteristic pits, cracks and furrows (Ramsey, 1917).

C. Black scurf

The most obvious and specific signs of Rhizoctonia disease are sclerotia that commonly form on tubers produced by either healthy or infected plants. These sclerotia are brown to black, flat or raised, irregularly shaped masses of fungal mycelium that form on tuber surfaces penetrating not much more than 5 to 6 cells deep in the periderm, which itself is made up of a total of 8 to 14 cell layers (Chand and Logan, 1984). Sclerotia may range from less than one to several mm across. Often, individual sclerotia grow together to form continuous masses of tissue covering large areas of the tuber surface. Sclerotia adhere tightly to the skin and are often referred to as "the dirt that won't wash off". However, the formation of sclerotia does not result in physical damage to the tuber.

IV. ECONOMIC IMPACT

The economic impact of black scurf depends on the intended market for the potatoes. Processors of French fries and chips do not even take black scurf into account in their rating of tuber defects since sclerotia are generally superficial and are eliminated in the peeling operation. As regards the fresh table market, tubers with sclerotia are less attractive and do not command premium prices, even though consumers will generally tolerate a small amount of scurf. Most important is the economic impact on the seed market. The presence of sclerotia on seed lots has caused complete blockage of international shipments at the port of entry because the standards of tolerance of the importing countries were lower than those of the exporting countries. The costs for the exporting countries are enormous with repercussions on future trade relations. Until universally recognized tolerance standards are accepted in international trade, the old question remains: How much scab and scurf should we tolerate on seed potatoes? (Waid, 1928).

V. INFECTION

Infection of sprouts follows the typical pattern for infection by all *Rhizoctonia* species, i.e. with the formation of infection cushions or lobate appressoria. Penetration pegs or hyphae force their way between epidermal cells (Chand *et al.*, 1985). The infection process results in the various symptoms mentioned above. These morphological alterations of the plant and progeny tubers may result from the plant's inability to absorb the excess photosynthate that results when the flow to the tubers is interrupted by lesions on stolons (Hartill, 1989). Hormonal agents may also be involved (Scholte, 1989).

In culture filtrates, *R. solani* produces phytotoxic metabolites that have been identified as phenylacetic acid and its meta- and para-hydroxy derivatives (Aoki *et al.*, 1963; Nishimura and Sabaki, 1963). Phenylacetic acid, a well-known growth regulator, inhibits growth of roots and seedlings, but does not cause necrosis (Aoki *et al.*, 1963). Its meta-hydroxy derivative induces many of the symptoms associated with Rhizoctonia disease on potato, including necrosis and stolon pruning, and has been used to evaluate breeding lines for resistance to infection (Frank and Francis, 1976).

It has often been stated that Rhizoctonia disease occurs predominantly under cold and wet conditions. This is inexact as it refers primarily to the observation of plant infections and poor, uneven stands, which are generally more severe under these conditions. Infection, however, can and does occur whenever the inoculum and disease potentials are high. The disease potential of a potato plant varies over time as the various tissues and organs mature and manifest differential susceptibility. Thus, a plant may escape sprout nipping and stem canker and yet be severely affected by stolon infection. In fact, careful examination of apparently healthy stems may reveal stolon "scars", the sites where stolon initials were killed before they could develop.

VI. INFESTATION

When the mother plant begins to senesce, infestation of the progeny tubers by sclerotia begins. Since it is primarily this "inactive", but more obvious, state of *R. solani* on seed tubers that is responsible for reduction in crop quality and for perpetuating the disease in succeeding crops, understanding the trigger(s) for sclerotial deposition would be a major contribution towards controlling this disease in potatoes. The black scurf stage may occur on progeny tubers from healthy, uninfected plants as well as on progeny tubers from infected plants, but the absence of sclerotia does not mean that the causal agent is not present.

Few sclerotia form before the onset of plant senescence (Otrysko *et al.*, 1988; Spencer and Fox, 1979) and most form during the three-week period that follows vine-kill or natural plant senescence (Chand and Logan, 1984; 1986; Gudmestad *et al.*, 1979; Otrysko *et al.*, 1988; Spencer and Fox, 1979).

Dijst and co-workers in Holland (1986, 1990), as well as Otrysko *et al.* in Canada (1988), demonstrated that pulling haulms and separating progeny tubers from senescing plant tissues decreased infestation by sclerotia, thus permitting tubers to be left in the soil to mature without becoming infested to the same extent during the same time period. Both groups postulated that sclerotial formation was inhibited during active plant and tuber growth, and that inhibition disappears gradually with the death of the mother plant. Dijst (1990) further demonstrated that both inhibitory and stimulatory volatile exudates from growing tubers were involved. When the inhibitory substances were trapped by KOH, the non-trapped volatile tuber exudates from growing plants were as stimulatory as those from plants after vine-kill. The chemical nature of these substances has not been elucidated.

VII. Epidemiology

Both tuberborne and soilborne inoculum can initiate infection of potato plants, although tuberborne inoculum is considered to be more important (Adams *et al.*, 1980; Bolkan 1976; Frank and Leach, 1980; Hide *et al.*, 1973; Little and Cook, 1988; Weinhold *et al.*, 1982). This may be true because tuberborne sclerotia are ideally situated on the mother tuber in intimate proximity to the emerging sprouts (Adams *et al.*, 1980). The roots and stolons also grow through this area. Theoretically, soilborne inoculum may be as important as tuberborne inoculum with regard to the infestation of progeny tubers by sclerotia, but this is difficult to demonstrate because the increased inoculum potential with tuberborne inoculum necessarily increases the inoculum potential of the surrounding soil. Given that most potato soils contain a mixture of different AGs, it is perhaps surprising that almost all sclerotia obtained from potato tubers belong to AG 3 (Abe and Tsuboki, 1978; Carling and Leiner 1986; Chand and Logan, 1983; Otrysko *et al.*, 1985).

Disease develops during the entire season as new tissues, especially stolons, become available for infection (Hide *et al.*, 1985a). Depth of planting thus affects the severity of infection, particularly under cool, wet conditions, which prolong the period during which emergence takes place.

An integrated approach must be used in the management of the Rhizoctonia disease complex as no single practice is effective (Powelson *et al.*, 1993).

A. Cultural practices

1. Scurf-free seed The key lies in the production and use of scurf-free seed tubers because of their importance in disseminating the disease and adding to the pool of soilborne inoculum. Planting seed potatoes with as little as 5% of the tuber covered with sclerotia often leads to serious damage (Powelson *et al.*, 1993). As mentioned above, immature tubers are rarely infested with scurf but they are difficult to harvest without skinning and bruising which increases the likelihood of secondary rots and may seriously reduce their storability. Techniques such as the "green crop harvesting" method (Mulder *et al.*, 1992) need to be developed for use on all types of terrains and soils, and this technology needs to be transferred to commercial growers. It is a promising stop-gap measure until a better understanding of the substances that trigger sclerotial formation enable better control measures to be developed.

2. Crop rotations Control of all phases of the disease begins with planting scurf-free seed tubers, but this in itself will not guarantee freedom from disease. The soilborne inoculum needs to be reduced and the measures needed to achieve this are the same as those that contribute to maintaining sustainable production, namely, promoting soil microbial life by appropriate cultural practices, especially the use of crop rotations.

Crop rotation is addressed by several authors, but results are sometimes contradictory. All agree that potatoes should not precede potatoes. Thus, monoculture should be avoided (Blodgett, 1939; Frank and Murphy, 1977; Hide and Read, 1991; Scholte, 1989; Simons, 1990). However, no single rotation sequence exists. Ideally, three-year rotations or longer should be used. Carling *et al.* in Alaska (1986) recovered isolates of AG 3 from soil free of potato for two seasons but not from soil free of potato for five years or more. The sequence of the crops in the rotation, whether or not they are plowed under as green manures, as well as depth of plowing, may be more important than the actual frequency of potatoes in a given rotation (Leach *et al.*, 1993; Specht and Leach, 1987). Recommendations to growers most often mention grasses or cereals as good rotation crops and caution against sugar beets or solanaceous crops (Evans and Howard, 1994; Powelson *et al.*, 1993). Crops plowed under as green manures are generally beneficial as they increase microbial activity in the soil with its concomitant increase in competition, predation and parasitism. Also there is evidence that green manures augment the saprophytic isolates of *R. solani* to the detriment of pathogenic ones (Specht and Leach, 1987). Plowing under mature straw preceding a potato crop is almost always detrimental. Research is needed on the development of economical rotation systems that are suited to local production conditions.

3. **Practices which favor rapid emergence** Mature tissues express greater resistance to infection than tissues in a formative stage (Hide *et al.*, 1985a), thus all practices which favor rapid emergence will generally aid in diminishing sprout nipping and stem canker infections. These practices include: warming the seed, planting in warm soil (15°C), seeding shallow and covering the tuber with no more than 5 cm of soil. Care must be taken with respect to the physiological age of the seed tubers, because physiologically old seed planted in cold soil may result in premature tuber formation whereas in warm soils, too rapid emergence may occur to the detriment of root system development. The latter may cause severe problems if a dry growing season follows. In areas where potatoes are produced under irrigation, pre-irrigating dry soils is a recommended practice but again, care must be taken, as a period of prolonged cool weather following planting could result in an increase of disease.

B. Chemical control

From a global standpoint, chemical control of the Rhizoctonia disease complex in potato is not a sound approach and using fungicides at planting offers no more control than using disease-free seed tubers (Powelson *et al.*, 1993). For a discussion of chemical disease control see chapter VI.D.

C. Biological control

Perhaps the area of most active research is currently in the field of biological control (see chapter VI.B). Various bacteria (Meshram, 1984; Schmiedeknecht, 1993), fungi (Beagle-Ristaino and Papavizas, 1985; Jager and Velvis, 1985, 1986; Murdoch and Leach, 1993), hypovirulent or non pathogenic strains of *R. solani* (Bandy and Tavantzis, 1990), non pathogenic binucleate *Rhizoctonia*-like species (Escande and Echandi, 1991a;b), as well as microfaunal predators (Bollen *et al.*, 1991) have been studied with respect to controlling the Rhizoctonia disease of field-grown potatoes. Some of these show promise, but as yet (1995), no commercially available biological control agents are used by potato growers. However, much "natural" biological control is probably active under field conditions and cultural control measures such as crop rotations and organic amendments will augment any naturally occurring systems.

D. Resistance

1. Germplasm resistance to infection Growing resistant cultivars would obviously go a long way towards controlling Rhizoctonia disease problems in the field. Differences in susceptibility can be seen (Frank and Francis, 1976; Scholte, 1989), but no resistant cultivars have been identified as of 1995. One might hypothesize that a certain level of horizontal resistance is present, but no genetic studies have been published. Resistance to the infection stage may be assessed either by the inoculation of seedlings or plants in the field (Frank *et al.*, 1976; Frank and Wilson, 1972; Leach and Webb, 1993; Otrysko and Banville, 1992) or by exposing the seedlings to the toxin m-hydroxy phenylacetic acid (Frank and Francis, 1976). Although stem lesion assessments are easier and quicker to carry out under greenhouse conditions and correlations between stem lesions and yield have been made (Hide and Horrocks, 1994; Read *et al.*, 1989; Weinhold *et al.*, 1982), it is best to carry out yield determinations in the field, in the various size categories, during the final stages of evaluation (Little *et al.*, 1988).

2. Resistance to infestation Formation of sclerotia is controlled by a confluence of many factors; some known, e.g. the length of time tubers are left in the soil after plant senescence, and others unknown, e.g. what triggers sclerotial formation. Care must be taken to reproduce all of these factors when trying to assess potato cultivars and lines for resistance. This is not an easy task and, to date, no resistant cultivars have been identified. Furthermore, genetic studies are sorely lacking for either phase of the disease complex.

VIII. CONCLUSIONS

In summary, growing potatoes with minimal infection by *Rhizoctonia* is possible simply by using sclerotia-free seed tubers; however, the same cannot be said about the infestation of progeny tubers. This latter area is where a major research effort probably would have a significant impact on management of the disease.

In preparing to undertake research on disease management, it is of utmost importance to understand the cumulative effect of various factors, such as cultivar, physiological age of the seed, soil type, nitrogen fertilization, and dates of planting, vine killing and harvesting, on the receptivity of progeny tubers to sclerotial formation. "If we are to determine the actual cause of sclerotial development, we must conceive of an intricate set of conditions, which collectively, have an influence in one direction" (Allington, 1936). One may also need to take into account the probable effects of favorable or adverse climatic conditions on the phenomenon of receptivity.

The key to understanding the mechanism of tuber infestation may be found in laboratory, greenhouse or growth chamber studies where reproducible results are possible. However, the verification of hypotheses will have to be done in the field, the universal laboratory, where the most challenging compromise is to manage the environment and to preserve it at the same time.

IX. References

Abe H & Tsuboki K (1978) Anastomosis groups of isolates of *Rhizoctonia solani* Kühn from potatoes. Bull. Hokkaido Prefect. Agric. Exp. Stn. 40:61-70.

Adams MJ, Hide GA & Lapwood DH (1980) Relationships between disease levels on seed tubers, on crops during growth and in stored potatoes. I. Introduction and black scurf. Potato Res. 23:201-214.

Allington WB (1936) Sclerotial formation in *Rhizoctonia solani* as affected by nutritional and other factors. Phytopathology 26:831-844.

Anguiz R & Martin C (1989) Anastomosis groups, pathogenicity and other characteristics of *Rhizoctonia solani* isolated from potatoes in Peru. Plant Dis. 73:199-201.

Aoki H, Sassa T & Tamura T (1963) Phytotoxic metabolites of *Rhizoctonia solani*. Nature 4906:575.

Bains PS & Bisht VS (1995) Anastomosis group identity and virulence of *Rhizoctonia solani* isolates collected from potato plants in Alberta, Canada. Plant Dis. 79:241-242.

Baker KF (1970) Types of Rhizoctonia disease and their occurence. In: Parmeter Jr. JR (ed.) *Rhizoctonia solani*: Biology and Pathology, (pp 125-148) University of California Press, Berkeley.

Bandy BP & Tavantzis SM (1990) Effect of hypovirulent *Rhizoctonia solani* on Rhizoctonia disease, growth, and development of potato plants. Am. Potato J. 67:189-199.

Bandy BP, Leach SS & Tavantzis SM (1988) Anastomosis group 3 is the major cause of Rhizoctonia disease of potato in Maine. Plant Dis. 72:596-598.

Bandy BP, Zanzinger DH & Tavantzis SM (1984) Isolation of anastomosis group 5 of *Rhizoctonia solani* from potato field soils in Maine. Phytopathology 74:1220-1224.

Banville GJ (1989) Yield losses and damage to potato plants caused by *Rhizoctonia solani* Kühn. Am. Potato J. 66:821-834.

Beagle-Ristaino JE & Papavizas GC (1985) Biological control of Rhizoctonia stem canker and black scurf of potato. Phytopathology 75:560-564.

Blodgett FM (1939) The effects of some agronomic practices on the incidence of *Rhizoctonia*. Am. Potato J. 16:93-98.

Bolkan HA (1976) Seed tuber treatment for the control of black scurf disease of potatoes. N.Z.J. Exp. Agric. 4:357-361.

Bollen GJ, Middelkoop J & Hofman TW (1991) Effects of soil fauna on infection of potato sprouts by *Rhizoctonia solani*. In: Beemster ABR, Bollen GJ, Gerlagh M, Ruissen MA, Schippers B & Tempel A (eds.) Biotic Interactions and Soil-borne Diseases. Developments in Agric. and Managed-Forest Ecology 23 (pp 27-34) Elsevier, Amsterdam.

Carling DE & Leiner RH (1986) Isolation and characterization of *Rhizoctonia solani* and binucleate *R. solani*-like fungi from aerial stems and subterranean organs of potato plants. Phytopathology 76:725-729.

Carling DE & Leiner RH (1990a) Virulence of isolates of *Rhizoctonia solani* AG 3 collected from potato plant organs and soil. Plant Dis. 74:901-903.

Carling DE & Leiner RH (1990b) Effect of temperature on virulence of *Rhizoctonia solani* and other *Rhizoctonia* on potato. Phytopathology 80:930-934.

Carling DE, Leiner RH & Kebler KM (1986) Characterization of *Rhizoctonia solani* and binucleate *Rhizoctonia*-like fungi collected from Alaskan soils with varied crop histories. Can. J. Plant Pathol. 8:305-310.

Carling DE, Leiner RH & Kebler KM (1987) Characterization of a new anastomosis group (AG 9) of *Rhizoctonia solani*. Phytopathology 77:1609-1612.

Carling DE, Leiner RH & Westphale PC (1989) Symptoms, signs and yield reduction associated with Rhizoctonia disease of potato induced by tuberborne inoculum of *Rhizoctonia solani* AG 3. Am. Potato J. **66:693-701.**

Carling DE, Rothrock CS, MacNish GC, Sweetingham MW, Brainard KA & Winters SW (1994) Characterization of anastomosis group-11 (AG 11) of *Rhizoctonia solani*. Phytopathology 84:1387-1393.

Chand T & Logan C (1983) Cultural and pathogenic variation in potato isolates of *Rhizoctonia solani* in Northern Ireland. Trans. Br. Mycol. Soc. 81:585-589.

Chand T & Logan C (1984) Post-harvest development of *Rhizoctonia solani* and its penetration of potato tubers in Northern Ireland. Trans. Br. Mycol. Soc. 82:615-619.

Chand T & Logan C (1986) Development of *Rhizoctonia solani* on potato tubers. Ann. Biol. 2:32-37.

Chand T, Logan C & Fraser TW (1985) Modes of penetration of *Rhizoctonia solani* in potato sprouts. Ann. Biol. 1:1-6.

Chang YC & Tu CC (1980) Pathogenicity of different anastomosis groups of *Rhizoctonia solani* Kühn to potatoes. (Abstr.) J. Agric. Res. China 29:27-33.

Dijst G (1990) Effect of volatile and unstable exudates from underground potato plant parts on sclerotium formation by *Rhizoctonia solani* AG 3 before and after haulm destruction. Neth. J. Plant Pathol. 96:155-170.

Dijst G, Bouman A, Mulder A & Roosjen J (1986) Effect of haulm destruction supplemented by cutting off roots on the incidence of black scurf and skin damage, flexibility of harvest period and yield of seed potatoes in field experiments. Neth. J. Plant Pathol. 92:287-303.

Escande AR & Echandi E (1991a) Effect of growth media, storage environment, soil temperature and delivery to soil on binucleate *Rhizoctonia* AG G for protection of potato from Rhizoctonia canker. Plant Pathol. 40:190-196.

Escande AR & Echandi E (1991b) Protection of potato from Rhizoctonia canker with binucleate *Rhizoctonia* fungi. Plant Pathol. 40:197-202.

Evans IR & Howard RJ (1994) Rhizoctonia canker (black scurf). In: Howard RJ, Garland JA & Seaman WL (eds.) Diseases and Pests of Vegetable Crops in Canada (pp 236-237) Canadian Phytopathological Society and the Entomological Society of Canada.

Frank JA & Francis SK (1976) The effect of a *Rhizoctonia solani* phytotoxin on potatoes. Can. J. Bot. 54:2536-2540.

Frank JA & Leach SS (1980) Comparison of tuberborne and soilborne inoculum in the Rhizoctonia disease of potato. Phytopathology 70:51-53.

Frank JA & Murphy HJ (1977) The effect of crop rotations on Rhizoctonia disease of potatoes. Am. Potato J. 54:315-322.

Frank JA & Wilson DR (1972) Evaluation of three inoculation methods used to screen potatoes for resistance to *Rhizoctonia solani*. Plant Dis. Rep. 56:348-351.

Frank JA, Leach SS & Webb RE (1976) Evaluation of potato clone reaction to *Rhizoctonia solani*. Plant Dis. Rep. 60:910-912.

Gudmestad NC, Stack RW & Salas B (1989) Colonization of potato by *Rhizoctonia solani* as affected by crop rotation. In: Vos J & van Loon CD (eds.) The Effect of Crop Rotation on Potato Production in the Temperate Zone. (pp 247-252) Kluwer Academic Publishers, Boston.

Gudmestad NC, Zink RT & Huguelet JE (1979) The effect of harvest date and tuber-borne sclerotia on the severity of *Rhizoctonia* disease of potato. Am. Potato J. 56: 35-41.

Hartill WFT (1989) Some effects of *Rhizoctonia solani* on growth and yield of potatoes. Potato Res. 32:283-292.

Hide GA & Firmager JP (1990) Effects of an isolate of *Rhizoctonia solani* Kühn AG 8 from diseased barley on the growth and infection of potatoes (*Solanum tuberosum* L.). Potato Res. 33:229-234.

Hide GA & Horrocks JK (1994) Influence of stem canker (*Rhizoctonia solani* Kühn) on tuber yield, tuber size, reducing sugars and crisp colour in cv. Record. Potato Res. 37:43-49.

Hide GA & Read PJ (1991) Effects of rotation length, fungicide treatment of seed tubers and nematicide on diseases and the quality of potato tubers. Ann. Appl. Biol. 119:77-87.

Hide GA, Hirst JM & Stedman OJ (1973) Effects of black scurf (*Rhizoctonia solani*) on potatoes. Ann. Appl. Biol. 74:139-148.

Hide GA, Read PJ, Firmager JP & Hall SM (1989a) Stem canker (*Rhizoctonia solani*) on five early and seven maincrop potato cultivars. I. Infection of shoots, stolons and tubers. Ann. Appl. Biol. 114:255-265.

Hide GA, Read PJ, Firmager JP & Hall SM (1989b) Stem canker (*Rhizoctonia solani*) on five early and seven maincrop potato cultivars. II. Effects on growth and yield. Ann. Appl. Biol. 114:267-277.

Hide GA, Read PJ & Sandison JP (1985a) Stem canker (*Rhizoctonia solani*) of maincrop potatoes. I. Development of the disease. Ann. Appl. Biol. 106:413-422.

Hide GA, Read PJ & Sandison JP (1985b) Stem canker (*Rhizoctonia solani*) of maincrop potatoes. II. Effects on growth and yield. Ann. Appl. Biol. 106:423-437.

Jager G & Velvis H (1985) Biological control of *Rhizoctonia solani* on potatoes by antagonists. 4. Inoculation of seed tubers with *Verticillium biguttatum* and other antagonists in field experiments. Neth. J. Plant Pathol. 91:49-63.

Jager G & Velvis H (1986) Biological control of *Rhizoctonia solani* on potatoes by antagonists. 5. The effectiveness of three isolates of *Verticillium biguttatum* as inoculum for seed tubers and of a soil treatment with a low dosage of pencycuron. Neth. J. Plant Pathol. 92:231-238.

Leach SS & Webb RE (1993) Evaluation of potato cultivars, clones and a true seed population for resistance to *Rhizoctonia solani*. Am. Potato J. 70:317-328.

Leach SS, Porter GA, Rourke, RV & Clapham WM (1993) Effects of moldboard plowing, chisel plowing and rotation crops on the Rhizoctonia disease of white potato. Am. Potato J. 70:329-337.

Little G & Cooke LR (1988) The effect of fungicides applied to seed potato tubers on *Rhizoctonia solani* stem canker and black scurf. Tests of Agrochemicals & Cultivars 9 (Ann. Appl. Biol. 112 Suppl.) pp 34-35.

Little G, Marquinez R & Cooke LR (1988) The response of twelve potato cultivars to infection with *Rhizoctonia solani*. Tests of Agrochemicals & Cultivars 9 (Ann. Appl. Biol. 112 Suppl.) pp 88-89.

MacNish GC, Carling DE, Sweetingham MW, Ogoshi A & Brainard KA (1995) Characterization of anastomosis group 10 (AG 10) of *Rhizoctonia solani*. Australasian J. Plant Pathol. 25:252-260.

Meshram SU (1984) Suppressive effect of *Azotobacter chroococcum* on *Rhizoctonia solani* infestation of potatoes. Neth. J. Plant Pathol. 90:127-132.

Meyer JC, Van Wyk RJ & Phillip AJL (1990) Rhizoctonia leaf spot of tobacco in South Africa. Plant Pathol. 39:206-207.

Mulder A, Turkensteen LJ & Bouman A (1992) Perspectives of green crop-harvesting to control soilborne and storage diseases of seed potatoes. Neth. J. Plant Pathol. 98 (Suppl. 2):103-114.

Murdoch CW & Leach SS (1993) Evaluation of *Laetisaria arvalis* as a biological control agent of *Rhizoctonia solani* on white potato. Am. Potato J. 70:625-634.

Nishimura S & Sabaki M (1963) Isolation of the phytotoxic metabolites of *Pellicularia filamentosa*. Ann. Phytopathol. Soc. Jpn. 28:228-234.

Otrysko BE & Banville GJ (1992) Effect of infection by *Rhizoctonia solani* on the quality of tubers for processing. Am. Potato J. 69:645-652.

Otrysko BE, Banville GJ & Asselin A (1985) Appartenance au groupe anastomotique AG 3 et pouvoir pathogène d'isolats de *Rhizoctonia solani* obtenus de sclérotes provenant de la surface de tubercules de pomme de terre. Phytoprotection 66:17-21.

Otrysko BE, Banville GJ & Asselin A (1988) Influence du degré de dépendance des tubercules-fils de pommes de terre vis-à-vis de la plante-mère sur leur infestation par *Rhizoctonia solani*. Potato Res. 31:617-625.

Powelson ML, Johnson KB & Rowe RC (1993) Management of diseases caused by soilborne pathogens. In: Rowe RC (ed.) Potato Health Management, (pp 153-154) APS Press, St. Paul, MN, USA.

Ramsey GB (1917) A form of potato disease produced by *Rhizoctonia*. J. Agr. Res. 9:421-426.

Read PJ, Hide GA, Firmager JP & Hall SM (1989) Growth and yield of potatoes as affected by severity of stem canker (*Rhizoctonia solani*). Potato Res. 32:9-15.

Schmiedeknecht G (1993) Biologische Bekampfung von *Rhizoctonia solani* Kühn an Kartoffelpflanzen Durch Mikrobielle Antagonisten. Arch. Phytopath. Pflanz. 28:311-320.

Scholte K (1989) Effects of soil-borne *Rhizoctonia solani* Kühn on yield and quality of ten potato cultivars. Potato Res. 32:367-376.

Simons SA (1990) Effects of different cropping sequences on soil-borne inoculum of *Rhizoctonia solani* measured using mini tubers of *Solanum tuberosum*. Brighton Crop Protection Conf. Pests and Diseases pp 777-782.

Specht LP & Leach SS (1987) Effects of crop rotation on *Rhizoctonia* disease of white potato. Plant Dis. 71:433-437.

Spencer D & Fox RA (1979) The development of *Rhizoctonia solani* Kühn on the underground parts of the potato plant. Potato Res. 22:29-39.

Stevens Johnk J, Jones RK, Shew HD & Carling DE (1993) Characterization of populations of *Rhizoctonia solani* AG 3 from potato and tobacco. Phytopathology 83:854-858.

Waid CW (1928) How much scab and scurf (*Rhizoctonia*) may certified seed potatoes carry? Am. Potato J. 5:232-234.

Weinhold AR, Bowman T & Hall DH (1982) *Rhizoctonia* disease of potato: effect on yield and control by seed tuber treatment. Plant Dis. 66:815-818.

V.9 RICE DISEASES INCITED BY *RHIZOCTONIA* SPECIES

TERUYOSHI HASHIBA AND TAKASHI KOBAYASHI

Faculty of Agriculture, Tohoku University, 1-1 Tsutsumidori-Amamiyamachi Sendai 981, JAPAN

I. RICE SHEATH BLIGHT

A. Introduction

Rice sheath blight disease occurs throughout temperate and tropical rice production regions. The disease was first reported in Japan (Miyake, 1910), and subsequently found in many countries. The affected area ranges from some 32 to 50 % of the total fields cultivated with rice plants throughout the world. The higher the temperature and humidity, the more rapidly the disease spreads. The disease is most severe in systems where early-maturing, short-culmed and high-tillering rice cultivars are densely planted, and heavily fertilized. Highly resistant rice cultivars have not been found and cultural measures for the control of rice sheath blight are not practical. Accordingly, chemical control is the most effective counter-measure practiced against this disease.

B. The rice sheath blight pathogen

The causal fungus of the rice sheath blight disease was first named *Sclerotium irregulare* I. Miyake by Miyake (1910), but it was identified as *Hypochnus sasakii* Shirai by Sawada (1912). This fungus was first found by Sasaki on leaves of the camphor tree with web blight symptoms and described by Shirai (1906). Many researchers had long used the name of *Hypochnus sasakii* when *Pellicularia sasakii* (Shirai) S. Ito. was proposed by Ito (1955). However, it has long been recognized that the mycelial state of this fungus had many similarities to *Rhizoctonia solani* Kühn, and researchers now generally accept that its anamorph is *R. solani*, and that its teleomorph is *Thanatephorus cucumeris* (Frank) Donk. The fungus is synonymous with *R. solani* group *I* of Schultz (1936), group *A* of Richter and Schneider, (1953), AG 1 of Parmeter *et al.*, (1969), and the sasaki type Watanabe and Matsuda, (1966; Ogoshi, 1972; 1975).

The fungus forms sclerotia and basidiospores. The hymenia are white and powdery, and are formed on the host plants only around the booting stage of rice plants, usually on the healthy outer surface of sheaths or on blades of young leaves. Basidia and basidiospores are formed during the night hours. The basidiospores are hyaline, oval or elongated, and measure 6-11$\times$5-8μm. They can initiate infection, but do not play an important role in the epidemiology of this disease (Chin, 1976; Kozaka, 1970, 1975).

Sclerotia are about 2-4mm in diameter, semi-spherical, oval or spherical, and hyaline at first and become dark-brown during maturation. The structure and the formation of its mature sclerotia of have been studied by many investigators (Endo, 1931; Fukano, 1932; Hashiba and Yamaguchi, 1971; Hashiba *et al.*, 1972; Hashiba and Mogi, 1972; 1973; 1975; Hashiba and Yamada, 1981; Hashiba, 1982; Hemmi and Endo, 1929; Inoue and Uchino, 1963; Nakata and Kawamura, 1937; Nonaka and Kaku, 1973). Mature sclerotia consist of two layers, outer and inner layer (Fig. 1). at

B. Sneh et al. (eds.),
Rhizoctonia Species: Taxonomy, Molecular Biology, Ecology, Pathology and Disease Constrol, 331–340.

the early stage of sclerotial development they are composed of compact masses of short hyphal cells with thin cell walls (ca. 0.09μm; Fig. 2a). Sclerotial size increases during the early stage of development and reaches its maximum size after 30h, when pigmentation is initiated (Fig. 2b). At this stage, cells of the outer layer start to become empty. After initiation of sclerotia formation, the number of cells increases rapidly and reaches to about 1.3×10^6 cells after 40h. In the central mass cell width increase rapidly (to ca. 15μm) at about 40h, when browning is completed. Cell walls become thicker with increasing age,. Initially, sclerotial density is relatively high and they sink in water. After 15 days, the outer layer cells become empty, and sclerotia become buoyant in water. Cell wall thickness of sclerotia reach about 0.15μm after 15 days. At this stage, there is a well-defined mass of living cells in the center, surrounded by an outer layer of empty cells (Fig. 1). Most of the sclerotia which mature in the field float on the water surface after one month (Hashiba *et al.*, 1972). During development of sclerotia, metabolic changes occur and the number of nuclei per cell is reduced from 12 to 4.5 in 20h after induction of sclerotial development (Hashiba and Staples, 1976).

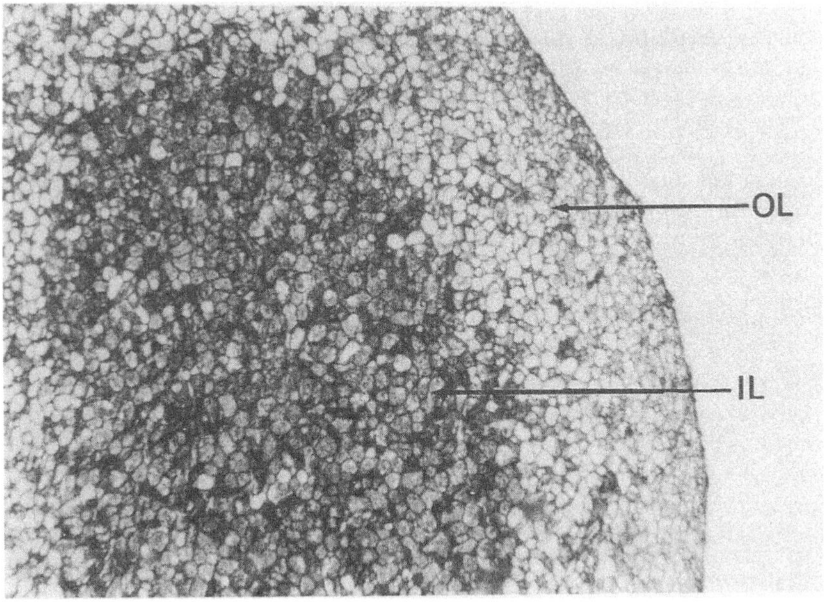

FIGURE 1. Structural feature of a *Rhizoctonia solani* AG 1-IA buoyant sclerotium formed on a rice plant. Arrows indicate (the inner layer) (IL) and (the outer layer)(OL).

C. Symptoms of rice sheath blight

The initial symptoms are usually lesions on sheaths of the lower leaves. The lesions are first greenish-gray and ellipsoid, 2-3cm or longer. The center of the lesion becomes grayish-white with a narrow blackish brown margin. Under favorable conditions of high humidity (ca. 95 % RH), and high temperature (28-32°C) runner hyphae spread rapidly to the upper parts of the plant (Fig. 3), and to adjacent plants. Severely diseased plants often snap at lesion sites on upper plant or lodge entirely on the ground.

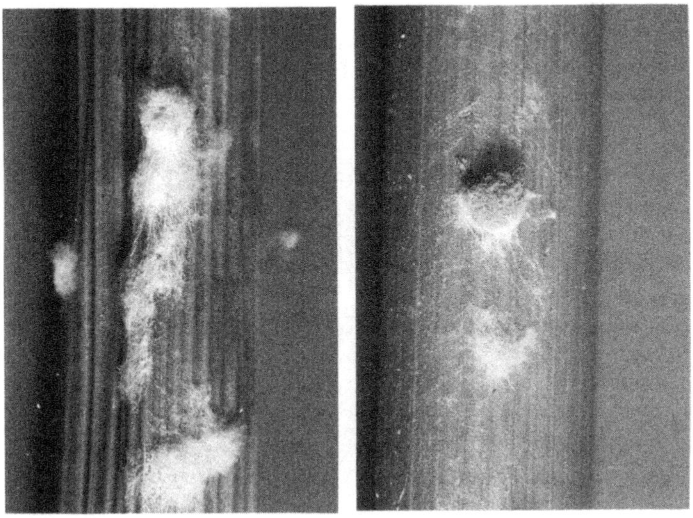

FIGURE 2. Development of sclerotia of *Rhizoctonia solani* AG-1 IA. Left - Sclerotia at early stage of development are white and composed of compact masses of hyphal cells. Right - Thirty hours after sclerotial initiation, the surface of sclerotia begins to turn brown.

FIGURE 3. Typical rice sheath blight lesions caused by *Rhizoctonia solani* AG 1-IA. Left - Lesions on rice leaf sheath during early internode elongation stage. Right - Natural infections caused by *R. solani* AG 1-IA in a rice cultivar at early heading stage.

D. Disease cycle.

Sclerotia form on the outer margin of lesions on leaf sheaths and can remain viable for more than 21 months under favorable conditions (Endo, 1931; Hashiba and Mogi, 1973; Kozaka, 1975). The fungus is capable to survive in soil as sclerotia or rarely as mycelia. The sclerotia fall down to soil (Fig. 4a), and can persist there after harvest (Fig. 4b). During puddling and leveling of the paddy fields in the next spring, most

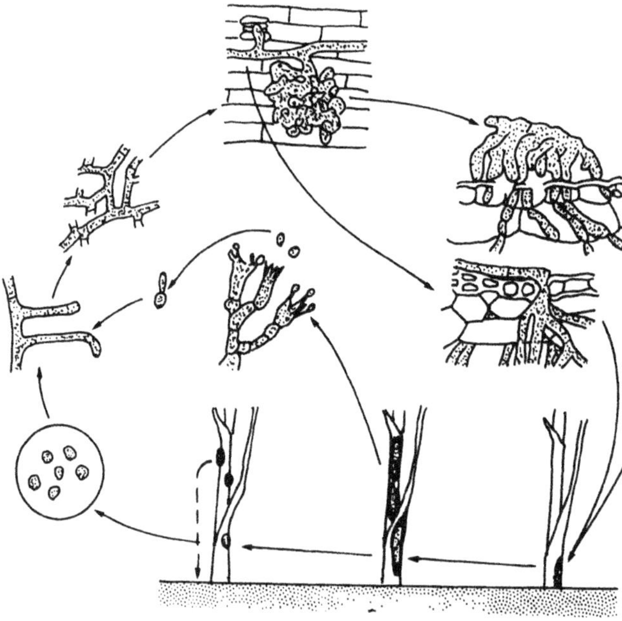

FIGURE 4. Disease cycle of *Rhizoctonia solani* AG 1-IA. (a) Lesions developed on the upper leaf sheaths and sclerotia are formed on the upper leaf sheaths. (b) Sclerotia overwinter in plant debris or in soil. (c) Young hyphae grow from a sclerotium. (d) Older hyphae. (e) Hyphal penetration into the host tissue. (f) Infection cushion and hyphal invasion through the cuticle and cells. (g) Penetration through a stomata. (h) Primary lesions are formed. (i) Lesions are developed upwards. (j) Basidiospores are formed on sexual fruiting structures. (k) A germinating basidiospore.

sclerotia rise to the water surface float (Fig. 5c) and play a major role as primary inoculum. The buoyant sclerotia adhere to the rice plants (Fig. 5d), germinate and initiate infection (Figs. 4h and 5e). Sclerotia germinate over a temperature range of 16-32°C, with an optimum at 28-30°C. High relative humidity >95 % is required for their germination (Hemmi and Endo, 1933; Kitani *et al.*, 1958; Kozaka *et al.*, 1957). Primary infection severity is closely correlated with the number of sclerotia adhering to the plants (Hori and Anraku, 1971; Kitani *et al.*, 1958; Kozaka, *et al.*, 1957; Yamaguchi *et al.*, 1971; Damicone *et al.*, 1993). Mycelia grown from sclerotia produce infection cushions, from which small penetration pegs arise for penetrating through epidermal cell walls (Fig. 3f) (Marshall and Rush, 1980; Matsumoto and Hirane, 1933; Kozaka, 1961; Dodman *et al.*, 1968). Stomatal infections are quite common on the inner surface of the sheath but seldom occur on the outer surface (Fig.

4g). Soon after the primary lesions are formed (Fig. 4h), mycelia rising from the lesions spread rapidly upwards (Figs. 4i and 5f) or to the lateral sides directly through air spaces or interstices between sheaths and along the outer surface of the plants (Kozaka, 1961; Yoshimura, 1954). Secondary hyphal spread occurs rapidly at the above mentioned favorable conditions (Hashiba, 1984; Kozaka, 1975). When vertical development of lesions is vigorous sclerotia and hymenia are formed on the leaf sheaths or leaf blades. Secondary infection is mostly through mycelia and rarely through basidiospores (Fig. 4j). Mature sclerotia drop to the ground and can over winter on, or in soil.

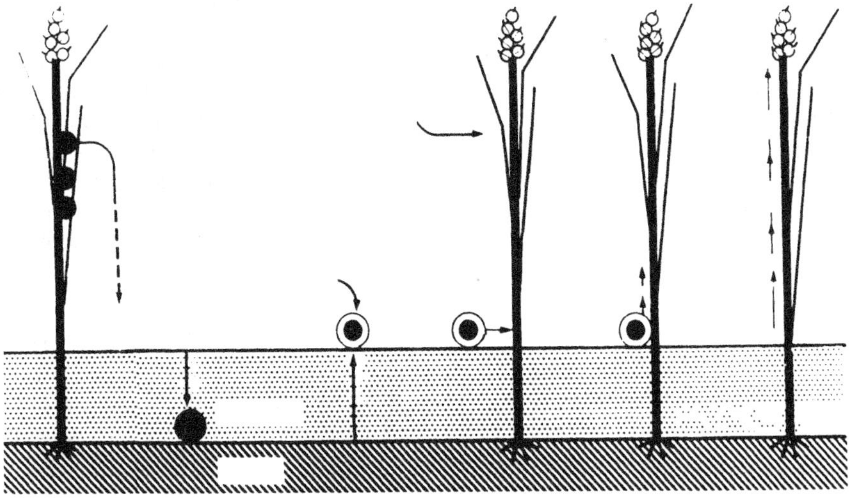

FIGURE 5. The infection process of *Rhizoctonia solani* causing rice sheath blight. (a) Lesions develop on the upper leaf sheaths. (b) Sclerotia initially sink in water. (c) After 15 days, sclerotia become buoyant in water. (d) The floating sclerotia contact with the rice plant. (e) Sclerotia germinate and initiate infection. (f) Primary lesions are formed, and develop upwards.

E. Disease severity assessment.

Various methods for the estimation of sheath blight severity have been evaluated (Ono, 1953; Yoshimura, 1954). Two phases of disease development were distinguished: vertical development i.e. infection of the upper leaf sheaths, and horizontal development, i.e. infection of neighboring stems (Kozaka *et al.*, 1960; Iwata, 1957). A simple method to estimate the disease severity is to use vertical and horizontal development of the lesions (Hashiba *et al.*, 1977; 1981; 1982; 1983; Hashiba, 1984).

Vertical development of the disease (X) is expressed by the relative length of the uppermost lesions to the plant length.

$$X = (\text{Length of the uppermost lesions / Plant length}) \times 100 \qquad (1)$$

Vertical development of the lesions (X) was examined under different temperature and relative humidity conditions. A model curve of vertical development was constructed from the relation between the daily temperature and relative humidity between rice hills and susceptibility of leaf sheath.

Horizontal development (A) is expressed by the percentage of diseased hills. A model curve of horizontal development (A) was calculated from the relation between daily temperature and relative humidity between rice hills and quantity of sclerotia per unit area in the paddy fields.

Whole disease severity (D) at the maturing stage can be represented as the product of the incidence of diseased hills estimated by (X) and (A):

$$D = (1.62X\text{-}32.4) \times A/100 \qquad (2)$$

The loss of fully ripened kernels has been estimated at 8.5 g per $3.3m^2$ for each additional 1 % of disease incidence, from which the following equation was derived:

$$L = (41.31X\text{-}826.2) \times A/100 \qquad (3)$$

where L represents yield loss in kg per ha.

F. Forecasting the disease.

A computerized forecasting system (BLIGHTAS) was developed to simulate the incidence of sheath blight and the resulting reductions in grain yield (Hashiba, 1984; Hashiba and Ijiri, 1989; Kobayashi *et al.*, 1995). This system consists of three major programs, an analytical program of diurnal variations of temperature and relative humidity, an analytical program of annual variations of temperature and relative humidity, and a forecasting program for the sheath blight disease. BLIGHTAS is designed to provide a 2-month forecast of the development of sheath blight in the fields. Data accumulated for seven and three years in Japan and the International Rice Research Institute (IRRI), respectively, indicated good agreement between the actual yield reductions or disease development and those predicted by BLIGHTAS (Fig. 6). BLIGHTAS will be applied for the development of new strategies to enhance chemical and biological control of this disease.

G. Management of the disease.

1. Host resistance. Studies on the mechanisms of host resistance are lacking due to the lack of resistant rice cultivars to *R. solani* (Marshall and Rush, 1980). Among over 6,000 rice cultivars collected in more than 40 countries of the world, no rice cultivar exhibited clear resistance to rice sheath blight (Hashiba, 1984). It has been shown that susceptibility of rice to this disease is greatly influenced by physiological, morphological and ecological factors. Short-culmed and high-tillering rice cultivars provide a favorable micro climate for disease development at an earlier stage of plant growth than tall varieties with few tillers (Lee and Rush, 1983; Marchetti, 1983). The meteorological conditions after the heading stage of early-maturing rice cultivars are more favorable for disease development than those of late-maturing cultivars. Heavy nitrogen fertilizer application results in luxuriant vegetative growth, which results in severe outbreaks of the disease similar to that of dense planting (Hashiba *et al.*, 1977; Lee and Rush, 1983; Kozaka, 1961; Loo *et al.*, 1963; Ou, 1985).

2. Chemical control At present, there are no rice cultivars that possess resistance against the sheath blight and the present methods of rice cultivation provide favorable conditions for the development of this disease. Chemical control is the most effective counter-measure against sheath blight. Organosynthetic chemicals such as validamycin, mepronil, flutolani, pencycuron and diclomezine were developed in succession. These fungicides possess high activities, both protective and curative, against sheath blight. In addition, these fungicides were not phytotoxic to rice, and toxicities to mammals and fish are extremely low. Thus, at present, these fungicides are relatively safe and widely applied to rice fields.

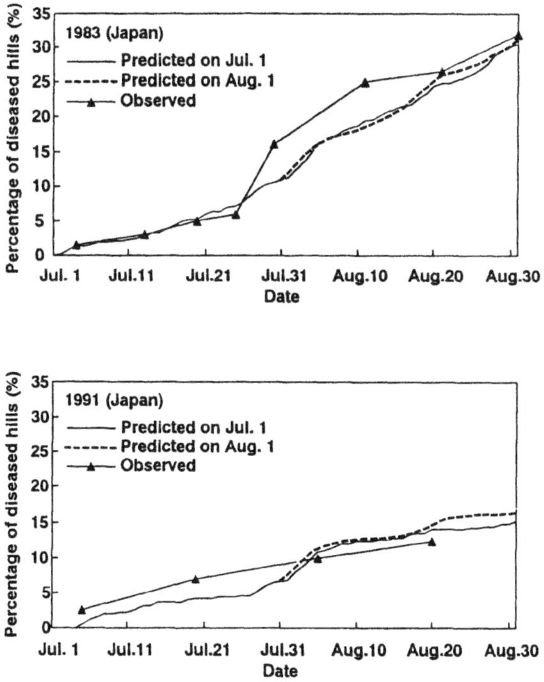

FIGURE 6. Comparison between horizontal lesions development observed in the field (in 1983 (top) and 1991 (bottom) and that predicted by BLIGHTAS. Predicted (1) indicates the predicted progress by BLIGHTAS on July 1. Predicted (2) shows the predicted progress on August 1.

Timing of fungicides applications is critical for obtaining effective sheath blight control. A computerized forecasting system has been designed to forecast a 2-month progress of disease. This information enables to determine the most suitable timing and amount of fungicide to be applied (Hashiba and Ijiri, 1989; Kobayashi *et al.*, 1995).

II. BROWN SHEATH BLIGHT

R. solani AG 2-2IIIB, is the causal agent of brown sheath blight (Oniki, 1976, 1977; Watanabe, 1977). The lesions develop at the heading stage. The central part of the lesion is similar to that of sheath blight, but areas surrounding the lesions are widely brown (Fig. 7). Sclerotia do not form on leaf sheaths, but are occasionally formed inside the leaf sheaths. The sclerotia are dark-brown and ellipsoidal around 1 mm long. Growth occurs over a temperature range of 5-38°C with the optimum temperature range for mycelial growth is 28-30°C. This fungus has high saprophytic ability in soil.

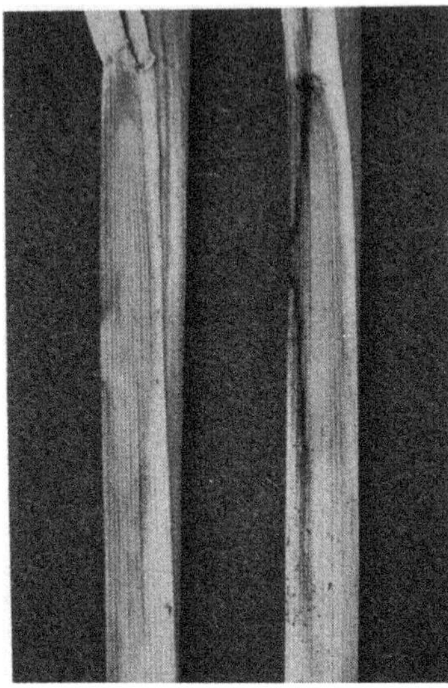

FIGURE 7. Typical brown sheath blight lesions caused by *Rhizoctonia solani* AG 2-2IIIB. The central part of brown sheath blight, but areas surrounding the lesions are widely brown. (Courtesy, Y. Kadowaki).

III. SHEATH SPOT

The rice sheath spot disease was first discovered in the USA by Tullis (1934). Nakada (1939) indicated that this fungus is *Rhizoctonia oryzae* Ryker et Gooch. The teleomorph confirmed in detail with *Waitea circinata* Warcuo et Talbot described on rice in Japan (Oniki *et al*; 1985). The disease has been reported in the southern USA, southeast Asia and Japan and is of minor economic importance (Atkins, 1974; Ou, 1985; Ryker and Gooch, 1938).

The disease appears after heading. The lesions are typically oval and 1-3 cm long and 1-2 cm wide, with a uniformly darkened surface devoid of obvious aerial mycelium. The lesions do not extend rapidly under field conditions. The lesions are limited to the outer rice sheath (Ryker and Gooch; 1938), and the fungus does not invade the culm (Marshall and Rush; 1980). The sclerotial masses of *R. oryzae* are not formed on the host, but inside the leaf sheath tissue and between leaf sheaths. They are short-column and flat or ellipsoid. The viability of over wintered sclerotia is lower than brown sclerotium disease.

IV. ACKNOWLEDGMENTS

The authors wish sincerely to thank Dr. T. Ishikawa, Teikyo University, Japan for helpful suggestions and critical reading of the manuscript, Dr. I. Havukkala, National Institute of Agrobiological Resources, Japan for critical reading of the manuscript.

V. REFERENCES

Atkins JG (1974) Rice disease of the Americas. A review of literature US Dept. Agric., Agric. Hdbk. 448:106 pp.

Belmar SB, Jones RK & Starr JL (1987) Influence of crop rotation on inoculum density of *Rhizoctonia solani* and sheath blight incidence in rice. Phytopathology 77:1138-1143.

Chin KM (1976) Occurrence of *Thanatephorus cucumeris*, the perfect stage of *Rhizoctonia solani*, on rice in West Malaysia. Mardi Res. Bull. 4:99-101.

Damicone JP, Parel, MV & Moore, WF (1993) Density of sclerotia of *Rhizoctonia solani* and incidence of sheath blight in rice fields in Mississippi. Plant Dis. 77:257-260.

Dodman RL, Barker KR & Walker JC (1968) A detailed study of the different modes of penetration by *Rhizoctonia solani*. Phytopathology 58:1271-1276.

Endo S (1931) Studies on Sclerotium disease of the rice plant. V. Ability of overwintering of certain important fungi causing Sclerotium disease of the rice plant and their resistance to dry conditions. Forsch. Gebiet Pfl-Krankh. Kyoto Imp. Univ. 1:149-167.

Fukano S (1932) Cytological studies in *Hypochnus sasakii* Shirai, causing a sclerotial disease of rice plant. Sci. Bull. Fac. Agric. Kyushu Imp. Univ. 5:117-136.

Hashiba T (1982) Sclerotial morphogenesis in the rice sheath blight fungus (*Rhizoctonia solani*). Bull. Hokuriku Nat. Agr. Exp. Sta. 24:29-83.

Hashiba T (1984) Estimating method of severity and yield loss by rice sheath blight disease. Bull. Hokuriku Nat. Agr. Exp. Sta. 26:115-164.

Hashiba T & Yamaguchi T (1971) Effect of temperature and humidity on the germination of sclerotium of *Pellicularia sasakii* Shirai) S. Ito. Proc. Assoc. Pl. Prot. Hokuriku 19:6-10.

Hashiba T & Mogi S (1972) Effect of nitrogen-carbon ratios on growth of mycelium and production of sclerotia by *Pellicularia sasakii* (Shirai) S. Ito. Proc. Assoc. Pl. Prot. Hokuriku 20:45-50.

Hashiba T & Mogi S (1973) The number and germination ability of sclerotia of *Pellicularia sasakii* (Shirai) S. Ito in no-cultivating paddy fields. Proc. Assoc. Pl. Prot. Hokuriku 21:6-8.

Hashiba T & Mogi S (1975) Development changes in sclerotia of the rice sheath blight fungus. Phytopathology 65:159-162.

Hashiba T & Staples RC (1976) Nucleic acid metabolism in developing sclerotia of the rice sheath blight fungus (*Rhizoctonia solani* Kühn). J. Gen. Microbiol. 96:239-245.

Hashiba T & Yamada M (1981) Viability of sclerotia formed on rice plants and on culture media by *Rhizoctonia solani*. Ann. Phytopathol. Soc. Japan 47:464-471.

Hashiba T & Ijiri T (1989) Estimation of yield loss and computerized forecasting system (BLIGHTAS) for rice sheath blight disease. Trop. Agr. Res. Ser. 22:163-171.

Hashiba T, Yamaguchi T & Mogi S (1972) Biological and ecological studies on the sclerotium of *Pellicularia sasakii* (Shirai) S. Ito. I. Floating on the water surface of sclerotium. Ann. Phytopathol. Soc. Japan 38:414-425.

Hashiba T, Yamaguchi T & Mogi S (1977) Quantitative changes in nitrogen and starch contents in rice sheaths during vertical disease development of sheath blight caused by *Rhizoctonia solani* Kühn. Ann. Phytopathol. Soc. Jpn. 43:1-8.

Hashiba T, Uchiyamada H & Kimura K. (1981) A method to estimate the disease incidence based on the height of the infected parts in rice sheath blight disease. Ann. Phytopathol. Soc. Jpn. 47:194-198.

Hashiba T, Koike K & Yamada M (1983) A method for estimation of the yield loss by rice sheath blight disease based on height of the lesions and percentage of affected hills in rice. Ann. Phytopathol. Soc. Jpn. 49:143-147.

Hashiba T, Koike K, Yuno I & Yamada M (1982) Model of vertical development of sheath blight lesions on rice plants caused by *Rhizoctonia solani* Kühn. Ann. Phytopathol. Soc. Jpn. 48:499-505.

Hemmi T & Endo S (1929) Studies on sclerotium disease of the rice plant, II. Agric. Hort. Japan 4:21-33.

Hemmi T & Endo S (1933) Studies on sclerotium disease of the rice plant. IV. On the relation of temperature and period of continuous wetting to infection of the rice plant by *Hypochnus sasakii* Shirai Forsch. Gebiet. Pfl-Krankh. Kyoto Imp. Univ. 1:111-125.

Hori M (1969) On forecasting the damage due to sheath blight of rice plants and the critical point for judgment the necessity of chemical control the disease. Rev. Plant Prot. Res. 2:70-73.

Hori M & Anraku M (1971) Studies on the forecasting techniques of sheath blight of rice plant. Plant Dis. Insect Pests 24:1-92.

Inoue Y & Uchino K (1963) Studies on sheath blight of rice plant caused by *Pellicularia sasakii* (Shirai) S. Ito. I. Ecology of damage and chemical control. Minist. Subsidized Exp. in Yamaguchi Pref. (Plant Dis. Insect Pests), Agric. For. Fish. Res. Council, Japan 4:1-131.

Ito S (1955) Mycological flora of Japan. II. Basidiomycetes. 4:107-108.

Iwata K (1957) Horizontal and Vertical development of rice sheath blight. Proc. Assoc. Pl. Prot. Hokuriku 5:15.

Kitani K, Inoue Y & Shigemura Y (1958) Relationships between density of planting and disease infection in *Pellicularia* sheath blight of rice plants. Pl. Dis. and Insect Pest Forecast Service Rep. 61:39-69.

Kobayashi T, Ijiri T, Mew TW, Maningas G & Hashiba T. (1995) Computerized forecasting system (BLIGHTASIRRI) for rice sheath blight disease in the Philippines. Ann. Phytopathol. Soc. Jpn. 61:562-568.

Kozaka T (1961) Ecological studies on sheath blight of rice caused by *Pellicularia sasakii* (Shirai) S. Ito and its chemical control. Chugoku Agr. Res. 20:1-133.

Kozaka T (1975) Sheath blight in rice plants and its control. Rev. Plant Prot. Res. 8:69-80.

Kozaka T , Sonku Y & Yunoki T (1957) Studies on sclerotial blight of rice plants caused by *Hypochnus sasakii*. II. Experimental observation on the preconditions of the primary infection. Bull. Chugoku Agr. Exp. Stn. 3:407-421.

Kozaka T, Yunoki T & Sonku Y (1960) Studies on the sclerotial blight of rice plants caused *Hypochnus sasakii* Shirai. IV. Relationship between susceptibility of leaf-blades and sheaths influenced by age and the upward development of the disease. Chugoku Agr. Res. 4:283-293.

Kuramoto T & Yamaguchi T (1969) Floating of sclerotium of *Pellicularia sasakii* (Shirai) S. Ito. Proc. Assoc. Pl. Prot. Hokuriku 17:3-5.

Lee FN (1980) Number, viability, and buoyancy of *Rhizoctonia solani* sclerotia in Arkansas rice fields. Plant Dis. 64:298-300.

Lee FN & Rush MC (1983) Rice sheath blight: A major rice disease. Plant Dis. 67:829-832.

Loo EP, Choun CQ & Lee DC (1963) Studies on *Rhizoctonia* blight of rice. Acta Phytophylactica Sinica 2: 431-440.

Marchetti MA (1983) Potential impact of sheath blight on yield and milling quality of short-statured rice lines in the southern United States. Plant Dis. 67:162-165.

Marshall DS & Rush MC (1980) Relation between infection by *Rhizoctonia solani* and *R. oryzae* and disease severity in rice. Phytopathology 70:941-946.

Marshall DS & Rush MC (1980) Infection cushion formation on rice sheaths by *Rhizoctonia solani*. Phytopathology 70:947-950.

Matsumoto T & Hirane S (1933) Physiology and parasitism of the fungi generally referred to as *Hypochnus sasakii* Shirai. III. Histological studies in the infection by the fungus. Jour. Soc. Trop. Agr. 5(4):367-373.

Miyake I (1910) Studien über die Pilze der Reispflanze in Japan. J. Coll. Agric., Imp. Univ. of Tokyo 2:237-276.

Nakata K & Kawamura E (1937) Studies on sclerotia disease in rice. Noji Kairyo Shiryo, Bureau Agric. Minist. Agric. For. Japan 139:1-176.

Nonaka F & Kaku H (1973) Anatomical studies on the sclerotia of rice plant fungi. On the sclerotia formed in field. Agric. Bull. Saga Univ. 34:35-40.

Ogoshi A (1972) Some characters of hyphal anastomosis group in *Rhizoctonia solani* Kühn. Ann. phytopathol. Soc. Jpn. 38:117-122.

Ogoshi A (1975) Grouping of *Rhizoctonia solani* Kühn and their perfect stage. Rev. Plant Prot. Res. 8:93-103.

Ogoshi A (1976) Studies on the grouping of *Rhizoctonia solani* Kühn with hyphal anastomosis and on the perfect stages of groups. Bull. Nat. Inst. Agr. Sci. C 30:1-63.

Oniki M (1977) On the perfect stage of cultural type IIIB of *Rhizoctonia solani* Kühn. Proc. Assoc. Pl. Prot. Kyushu 23:17-22.

Ono K (1953) Damage of sheath blight disease to rice. J. Plant Prot. 7:99-103.

Ou SH (1985) Rice diseases. 2nd ed. Commonwealth Mycological Institute, Kew, England.

Parmeter, JR Jr., Sherwood RT & Platt WD (1969) Anastomosis grouping among isolate of *Thanatephorus cucumeris*. Phytopathology 59:1270-1278.

Richter H & Schneider R (1953) Untersuchungen zur morphologischen und biologischen differenzierung von *Rhizoctonia solani* K. Phytopathol. Z. 20:165-226.

Ryker TC & Gooch FS (1938) Rhizoctonia sheath spot of rice. Phytopathology 28:233-246.

Sawada K (1912) On the --shirakinubyo-- of camphor. Special Bulletin of the Formosan Agr. Exp. Sta. Spec. Bull. 4. 805 pp.

Schultz H (1936) Vergleichende Untersuchungen zur Ökologie, Morphologie, und Cystematik des --Vermehrungspilzes--. Arb. Biol. Reichsanst. für Land-und Forstwirtsch. 22:1-41.

Shirai M (1906) On *Hypochnus sasakii* n. sp. Botanical Magazine, Tokyo 20:319-323.

Watanabe B & Matsuda A (1966) Studies on the grouping of *Rhizoctonia solani* Kühn pathogenic to upland crops. Bull. App. Exp. (Plant Dis. and Insect Pests) 7:1-131.

Watanabe B, Oniki M & Nonaka F (1977) Brown sheath blight of rice. Proc. Assoc. Pl. Prot. Kyushu 23:22-25.

Yamaguchi T, Iwata K & Kuramoto T (1971) Study on forecasting the sheath blight of rice plant caused by *Pellicularia sasakii* (Shirai) S. Ito. I. Relation between hibernated sclerotium and disease outbreak. Bull. Hokuriku Agric. Exp. Stn. 13:15-34.

Yoshimura S (1954) On the scale for estimating degree of severity of sheath blight by *Hypochnus sasakii* Shirai in rice plant. Ann. Phytopathol. Soc. Japan 19:58-60.

V.10 SUGAR BEET DISEASES INCITED BY *RHIZOCTONIA* SPP.

LEONARD J. HERR

Department of Plant Pathology, The Ohio State University, Ohio Agricultural Research and Development Center, Wooster, Ohio 44691, USA

I. INTRODUCTION

Rhizoctonia diseases of seedlings and mature plants of sugar beet have been recognized and described since the early days of plant pathology (Kühn, 1858; Duggar, 1899; Edson, 1915). Despite a century and a half of investigation, economic losses caused by Rhizoctonia diseases to the sugar beet crop are estimated to average 2% in the USA (Whitney and Duffus, 1986). Consequently, Rhizoctonia diseases of sugar beet continue to be the subject of research on etiology, epidemiology, pathogen/strain characterization, ecology, and control strategies (Kotila, 1947; Boosalis and Scharen, 1959; Naiki and Ui, 1969; Naito *et al.*, 1976; Roberts and Herr, 1979; Herr and Roberts, 1980; Naito and Sugimoto, 1981; Hyakumachi and Ui, 1982; Ogoshi, 1985; Herr, 1987; 1988; Kataria and Hoffmann, 1988; Windels and Nabben, 1989; Ruppel and Hecker, 1974; Rush *et al.*, 1994).

A. Violet root rot

1. Background Kühn (1858), in the first published account of root rot on sugar beet, reported that *Rhizoctonia medicaginis* DC. caused violet root rot of sugar beet in Germany. According to Duggar (1915) and current authority (Moore, 1987), *R. medicaginis* is synonymous with *R. crocorum* (Pers.) DC. ex Fr. The current accepted teleomorph is *Helicobasidium brebissonii* (Desm.) Donk (=*H. purpureum* (Tul.) Pat.). For further remarks regarding the taxonomy of the pathogen see chapter I.A1.

2. Occurrence and symptoms Violet root rot occurs commonly on sugar beet in Europe and sporadically in the western states of the USA (Whitney and Duffus, 1986). The disease occurs as patches of affected plants within fields. Distinctive symptoms include purplish spots on roots and felt like, reddish-purple mycelial growth over the beet surface. Rotting of beets takes place beneath the enveloping mycelium (Whitney and Duffus, 1986).

3. Survival and control The pathogen persists in soil on roots of perennial crop and weed hosts and in crop residues. Avoidance of susceptible crops and eradication of weed hosts are recommended control measures (Whitney and Duffus, 1986).

4. Remarks Kühn (1858) regarded the mycelial morphology of *R. crocorum* and *R. solani* Kühn as generally similar, but described *R. solani* as a new species. The distinction between these two species has long been recognized (Duggar, 1915; Parmeter, 1970). Recently, Moore (1978) reported that *Rhizoctonia (Rhizoctonia*-like) fungi can be classified into five anamorph genera. (see chapter I.A1). Under Moore's system the *Rhizoctonia* spp. reported to be pathogens on sugar beet are classified into the following anamorph genera. *R. crocorum* is placed in the genus *Rhizoctonia, R. solani and R.zeae* are placed in the genus *Moniliopsis* and binucleae *Rhizoctonia* spp.

B. Sneh et al. (eds.),
Rhizoctonia Species: Taxonomy, Molecular Biology, Ecology, Pathology and Disease Constrol, 341–350.

with *Ceratobasidium* teleomorphs are placed in the genus *Ceratorhiza*. Although the validity of Moore's classification system is recognized, the older *Rhizoctonia* classification is currently predominantly in use (Sneh *et al.*, 1991). In this chapter on sugar beet the anamorph *Rhizoctonia* will be used for those fungi conforming to Ogoshi's (1985) criteria for characterizing this anamorph genus.

B. Seedling diseases

1. Background According to Edson (1915), Eidam in 1888 first reported the occurrence of Rhizoctonia disease on sugar beet seedlings. Duggar (1899) inoculated beet seedlings with *R. betae* (Kühn) Eidam, which caused damping-off. Edson (1915) considered *R. betae* to be indistinguishable from *R. solani* and *R. betae* was listed as a synonym of *R. solani* by Duggar (1915). The teleomorph of *R. solani* was assigned to *Corticium vagum* B. & C. (now designated as = *T. cucumeris*).

2. Occurrence and symptoms Seedling diseases of sugar beet are ubiquitous. In addition to *R. solani*, other soilborne pathogens, especially *Aphanomyces cochlioides* Drechs., *Pythium aphanidermatum* (Edson) Fitzp. and *P. ultimum* Trow, commonly cause more or less similar seedling disease symptoms. Seedling disease phases can be grouped as those resulting in stand losses, i.e., seed decay, pre-emergence and post-emergence damping-off and those primarily injuring seedling roots or hypocotyls resulting in deformed, stunted, poorly developed older plants (Windels and Jones, 1989). *R. solani* typically causes post-emergence damping-off by attacking the hypocotyl below ground Severely diseased seedlings collapse and die (Whitney and Duffus, 1986).

3. Survival and control *R. solani* and other pathogenic *Rhizoctonia* species survive over winter and between sugar beet crops as sclerotia (Boosalis and Scharen, 1959; Naiki and Ui, 1969; Roberts and Herr, 1979; Hyakumachi and Ui, 1982), and melanized mycelium (Boosalis and Scharen, 1959; Roberts and Herr, 1979) free (Naiki and Ui, 1969; Roberts and Herr, 1979) or embedded in soil organic debris (Boosalis and Scharen, 1959; Naiki and Ui, 1969; Roberts and Herr, 1979), originating from parasitized crop and weed hosts and/or from saprophytically colonized crop residues (Peltier, 1916; Daniels, 1963; Herr, 1976; 1987; Hyakumachi and Ui, 1982). Large sclerotia were reported to survive better than small sclerotia obtained from Japanese sugar beet fields (Naiki and Ui, 1969; Hyakumachi and Ui, 1982).

Rhizoctonia seedling diseases are exacerbated by conditions which retard seed germination, emergence and post-emergence growth (Whitney and Duffus, 1986). Consequently, disease control measures include those practices such as planting cultivars selected for rapid germination and vigorous growth of seedlings, and shallow depth of planting seeds to speed emergence. More damping-off of beet seedlings occurs following crops of legumes or other susceptible crops than following maize (Whitney and Duffus, 1986). The primary control is seed treatment with effective fungicides (Whitney and Duffus, 1986).

4. Remarks Prior to the development of *Rhizoctonia* anastomosis grouping (AG) systems, Houston (1945) reported that the *R. solani* strains causing seedling diseases of sugar beet differed in pathogenicity/virulence and cultural characteristics from those causing root rot on older beets. These results have been substantiated and clarified in subsequent studies. A number of AGs of *R. solani*, reported as only AG or AG intraspecific groups (Ogoshi 1985), preferred nomenclature now AG+subgroups

(Sneh *et al.*, 1991), are reported to be pathogenic on sugar beet seedlings (Naito *et al.*, 1976; 1980; Herr and Roberts, 1980; Herr, 1982; Windels and Nabben, 1989; O'Sullivan and Kavanagh, 1991; Rush *et al.*, 1994), as well as certain AG of BNR (Uchino *et al.*, 1983; Ogoshi, 1985; Kataria and Hoffmann, 1988; O'Sullivan and Kavanagh, 1991).

In early studies, most *R. solani* isolates from sugar beet were placed in AGs 2 and 4. AG 4 isolates are generally more virulent on beet seedlings compared to the AG 2 isolates, which primarily attack crowns and roots of older beets. Additional AGs have been isolated (AGs 1, 2-1, 2-2, 3 and 5) from diverse sugar beet production areas. Of these, AG 2-2 represents a more precise designation of most of the virulent AG 2 isolates occurring on sugar beet, whereas AG 2-1 (cruciferous strain) was rarely isolated and was much less virulent on seedlings (similar to AG 3) than AG 2-2. The AG 1 isolates reported from Minnesota (probably AG 1-IC) and Japan (AG 1-IB) differ in both subgroups and virulence on beet seedlings, with AG 1-IC less virulent than either AGs 4 or 2-2 isolates and AG 1-IB equally as virulent as AG 4 and AG 2-2. Virulence of AG 5 isolates varied from nonpathogenic on sugar beet seedlings in Texas (Rush *et al.*, 1994) to virulent, but less virulent than AGs 4, 2-2 and 1-1C in Minnesota (Windels and Nabben, 1989). Because isolates within AGs can vary in virulence (Herr and Roberts, 1980), adequate numbers must be tested for meaningful comparisons of virulence. Among the factors affecting the diversity found in the composition of *Rhizoctonia* AG and subgroup populations isolated from sugar beet crops are the sources of isolates, growth stage of plants, plant organs assayed, season of sampling and geographical location of crop production areas.

Uchino *et al.* (1983) isolated BNR of five AGs from diseased sugar beet seedlings raised in paper pot transplant nurseries. Isolates of AG-E were highly virulent to sugar beet seedlings, followed by AG-A, AG-K, AG-C and AG-I. Ogoshi (1985) and Kataria and Hoffmann (1988) reported that AG-A isolates caused damping-off of beet seedlings. O'Sullivan and Kavanagh (1991) isolated *R. cerealis* van der Hoeven [=AG-D (CAG-1)], from sugar beet seedlings and reported that these isolates were pathogenic to beet seedlings at 10 and 15°C. Recently, Kuznia and Windels (1994) reported that *R. zeae* (=WAG-Z), isolated from diseased beet seedlings and spring wheat, significantly reduced emergence and stands of sugar beet seedlings.

C. Crown and root rot (dry rot canker)

1. Background Duggar (1899) reported sugar beet root rot to be caused by *R. betae*, but later he relegated *R. betae* to synonymy with *R. solani* (Duggar, 1915). Subsequently, Richards (1921) described a dry rot canker type of root rot caused by *R. solani* on sugar beet. The distinctive dry rot canker (crater rot) symptoms distinguish this disease phase from the more common Rhizoctonia crown and root rot (C-R) disease phase.

2. Occurrence and symptoms Rhizoctonia C-R of sugar beet occurs worldwide (Whitney and Duffus, 1986). C-R is endemic and economically important in relatively humid regions of the USA; in more arid areas a related Rhizoctonia dry rot canker of sugar beet prevails. Average disease loss from C-R and dry rot canker is estimated to be 2% of the crop (Whitney and Duffus, 1986), however disease damage varies greatly (0-50%) from field to field.

C-R occurs on older plants, generally associated with the developing canopy. C-R symptoms often begin with invasion and development of black lesions on older petioles in contact with the soil. Rotting then progresses to the crowns and roots accompanied by wilting, collapse and formation of a rosette of yellowing and dying leaves on the soil surface. Less commonly, crowns and roots are invaded directly, resulting in the occurrence of large fissures in affected beets. The disease generally occurs within fields as elongated patches of dead and diseased plants. Dry rot canker disease symptoms occur on the beet root surface and consist of numerous, defined, localized, circular, sunken lesions varying in size from ca. 2-25 mm in diam.; with an alternating dark and light color concentric ring pattern. Beneath the lesions are deep cankers filled with mycelium and dry remains of host tissue, which are sharply delimited from adjacent healthy tissue (Richards, 1921).

3. Survival and control Growth and survival of *R. solani* propagules in soil affects inoculum densities and potentials for causing disease. Roberts and Herr (1979) investigated the dynamics of soil populations of *R. solani* from areas of healthy beets (AH) and areas of diseased beets (DA) within Ohio sugar beet fields and over winter. They hypothesized that the soil populations consisted of groups of propagules differing in capacities for survival relative to AH populations and the seasonal progression of DA populations. Accordingly, the consistently low AH populations were said to consist of "residual survival propagules", i.e., survival forms probably sustained by saprophytic growth and symptomless parasitism (Daniels, 1963) between periods of active parasitism. The very high populations of *R. solani* within DA (Aug.-Oct.) consisted primarily of "parasitically evanescent propagules" dependent upon active parasitism for survival. The reports of Boosalis and Scharen (1959) and Hyakumachi and Ui (1982) confirm that parasitic activity can sustain very high soil populations of *R. solani*. Between the Oct. and Dec. assays of DA, a precipitous decline in *R. solani* soil populations occurred through elimination of the parasitically evanescent propagules by low temperature limitation of parasitism. The surviving DA soil populations, still several orders of magnitude greater than the AH population, were considered to consist, largely, of "parasitically augmented propagules". These would be mostly propagules surviving saprophytically in tissues of parasitized plants over winter until the advent of warmer temperatures. After overwintering, the June DA populations declined to levels similar to AH populations (i.e., the residual survival propagules). The report of Hyakumachi and Ui (1982) of a rapid decrease in germination of *R. solani* sclerotia from autumn to spring in sugar beet fields supports the results of Roberts and Herr (1979). Furthermore, Herr (1976) reported that little decrease was noted in overwinter survival of *R. solani* in field collected diseased beets buried in soil until the advent of warmer temperatures (May-June), when a major decrease in recovery of *R. solani* occurred. The reduction in survival was associated with increased microbial and faunal activity and advanced decomposition of beet residues.

The AH and DA, within fields initially cropped to sugar beet (Roberts and Herr, 1979), were assayed during rotations to other crops and recropping to sugar beet (Herr, 1987). Comparisons of *R. solani* populations over all of the sampling dates supports the hypothesis that low soil populations were the norm in these rotations and high soil populations were the exception. High soil populations of *R. solani* were associated with parasitic activity in patches of diseased beet plants or saprophytic activity on tomato residues following tomato harvest and did not necessarily coincide with the initial locations of either DA or AH (Herr, 1987). Monitoring May-June populations of AH and DA, which later coincided with patches of diseased beets failed to give predictive evidence (forecasts) of the subsequent increases in parasitically sustained populations of *R. solani* and disease. These results lead to the conclusion that "The initiation of disease patches in these sugar beet fields was governed by factors other than inoculum density *per se*" (Roberts and Herr, 1979; Herr, 1987). Furthermore, evidence reported by Herr and Roberts (1980) suggests that this lack of obvious correlation of inoculum density and disease initiation could not be attributed to the localized occurrence of strains of *R. solani* particularly virulent to sugar beet. Hyakumachi *et al.* (1983) reported that in an area of high disease incidence (DI), sclerotial formation was low in the fall, but germination was high the following spring, whereas in an area of low DI, fall sclerotial formation was high, but spring germination was low. They concluded that "...... the difference in the DI in different areas is due to the difference of activity of sclerotia as inoculum". Naiki and Ui (1969) demonstrated that many of the sclerotia isolated from soil, which failed to germinate on water agar, did germinate when placed in contact with host plants. Such sclerotia would not be detected in agar plate assays and therefore, the apparent inoculum density would not reflect the actual inoculum potential for causing disease. Microbially-induced fungistasis was suggested as a probable cause for the observed inhibition of sclerotial germination (Naiki and Ui, 1972). Based on the foregoing reports it appears that the nature and rates of changes in *R. solani* population densities within population survival cycles are affected by crops and cropping sequences; crop residues, including their disposition and decomposition; associated soil microbial and faunal populations; and environmental conditions.

C-R disease of sugar beet is notoriously difficult to control, because of its unpredictable occurrence and variable intensities of disease losses in different seasons. Crop rotations affect C-R and seedling diseases. Losses can either be reduced (e.g., preceding beets with maize and other nonhosts) or increased (e.g., preceding beets with tomato in Ohio (Herr, 1987) or wheat in Texas [Rush *et al.*, 1994]). Hilling of beets with soil during cultivation should be avoided, because covering beet crowns with soil exacerbates disease (Windels and Jones, 1989). Fungicides offer a measure of disease control (Ruppel and Hecker, 1987), but are not generally considered practical and no fungicides are currently registered for this use on sugar beet in the USA. Recent field experiments with applied biotic agents (Herr, 1988) demonstrated possibilities for effective biological control of C-R. Resistant varieties potentially afford the most promising and effective means of C-R disease control. Recently, Ruppel and Hecker (1994) reported the performance of a three-way cross hybrid beet which closely approached commercial acceptance in yields and percentage sucrose content (see also chapter VI.C1).

4. Remarks *R. solani* AG 2 (AG 2-2) is the major AG causing C-R of sugar beet worldwide, however other AGs (1, 4 and 5) are isolated at much lower frequencies (Naito *et al.* 1976; Herr and Roberts, 1980; Windels and Nabben, 1989; Rush *et al.*, 1994). In Japan, Naito *et al.*, (1976) investigated nearly 400 isolates obtained from petioles, crowns and roots. From petioles they obtained ca. 31% AG 1, 41% AG 2-2, 22% AG 4 and 5% AG 5 isolates; whereas, from roots and crowns 85% of isolates were AG 2-2, with the remainder consisting of AGs 1, 4 and 5. In virulence assays, AG 2-2 isolates were highly virulent on petioles, crowns and roots. In contrast, AGs 1, 4 and 5 isolates were moderately virulent on petioles and considerably less virulent on crowns and roots than AG 2-2 isolates. The histopathological observations of Naito and Sugimoto (1981) provide evidence of behavioral differences among these *R.*

solani AGs in invasion and tissue colonization capabilities, which relate to their relative virulence on sugar beet. Primary invasion sites for AG 2-2 include the lower surface of petioles in contact with soil, natural cracks in the crown, lenticels on the taproot, and lateral root development sites. Isolates of AGs 2-2 and 1 directly penetrate petioles through stomata. AG 4 isolates form infection cushions whereas isolates of AG 5 form lobate appressoria. Following penetration, AG 2-2 isolates progressively invade and colonize the vascular tissue whereas, invading hyphae of AGs-1, 4 and 5 are limited to the cortex.

A discussion of the relationships of the group designations, AG and subgroup is necessary to clarify my use of these terms. Ogoshi (1985) recognized that the use of only AG designations (AG 1 through AG 11 plus AG-BI) could not adequately describe the pathological, physiological and morphological characteristics of the *R. solani* (*T. cucumeris*) complex. He assigned AG designations first followed by sub or sub-sub designations, e.g., AGs 1-IA, 1-IB, 1-IC; 2-1, 2-2IIIB and 2-2IV. These combinations are designated as subgroups as distinguished from AG designations. For example, the use of simply AG 1 should be avoided because it fails to distinguish among three very different AG subgroups, i.e., AG 1-IA, sasaki type; AG 1-IB, web blight type; AG 1-IC, salt and pepper cultural type. Further, recent reports of AG 2-2 on sugar beet generally do not distinguish between AG 2-2IIIB, and AG 2-2IV. AG 2-2IIIB has major hosts in Graminae (Ogoshi, 1985), including maize and Fabaceae, including soybean (Liu and Sinclair, 1991) whereas, AG 2-2IV, has major hosts in Chenopodiaceae (Ogoshi, 1985), including sugar and garden beets. More complete host range information for AG 2-2IIIB and AG 2-2IV should be obtained from field AG subgroup population assays in various geographical areas and from pathogenicity tests of identified isolates on various crops. Note should be taken that the subgroups system is an evolving not a static system (Ogoshi, 1985), which is attested to by past and probable future additions of newly described AG and subgroups to this system. On a more intensive scale, determination of individual clones of AG subgroups within fields and regions can provide information on clone diversity and clues to their sources of origin and survival (Ogoshi and Ui, 1983).

Although it has been suggested that dry rot canker is apparently caused by strains of *R. solani* different from those causing C-R (Whitney and Duffus, 1986), the only relevant research appears to be LeClerg's (1939) comparative studies of factors affecting pathogenicity and physiology of C-R and dry rot canker isolates. Isolates causing dry rot canker should be re-examined and identified to AG+subgroups.

D. Foliar blight (see also chapter IV.3)

1. Background The first mention of the perfect stage of *R. solani*, [*C. vagum* B. & C. var. *solani* (now *T. cucumeris*)], on beet (garden beet) was apparently by Peltier (1916). The perfect stage was briefly described, but no disease symptoms were reported. *T. cucumeris* may form hymenia on soil and apparently healthy plants in the field, supposedly using the plant surfaces only as support (Naito, 1984). However, Windels *et al.* (1994; and Windels, personal comm.) reported that the presence of hymenia of *T. cucumeris* (= *R. solani* AGs 3 and 5) on petioles of healthy sugar beet was associated with reduced sucrose percentage and amounts of recoverable sucrose compared to healthy beets without hymenia present. This presumed host response to "superficial" fungal hymenia merits further investigation. Kotila (1947) first reported a foliage blight disease of sugar beet caused by the teleomorph of *R. solani*, then named *Pellicularia filamentosa* (Pat.) Rogers (now *T. cucumeris*).

2. Occurrence and symptoms Foliar blight of sugar beet occurs sporadically in the USA and thus is not considered important economically (Whitney and Duffus, 1986). In Japan, foliar blight occurs frequently and is of economic importance. Foliar blight occurrence is favored by hot, humid weather (Kotila, 1947; Naito 1984). Two phases of foliar blight have been described. An early phase, in which the heart leaves in the rosette are attacked reducing them to stubs consisting of deformed portions of leaf blades with blackened, irregular margins. Shallow black cankers may occur on petioles. Older leaves are not affected (Kotila, 1947). In Ohio, no hymenia of the perfect stage have been found associated with this early phase (Herr, 1982). The second disease phase begins with the production of hymenia on healthy tissues adjacent to disease lesions on the underside of the older petioles (Herr, 1982; Naito, 1984). Hymenia appear as appressed, white to beige colored, superficial membranous to felty fungus growths, loosely attached to healthy tissue. Basidiospores, produced and disseminated from petiole hymenia, initiate the leaf blade foliar blight phase. Initial leaf blade lesions are small, circular, necrotic spots (ca. 1 mm diam.) with a brown border (Naito, 1984). Subsequently, large, irregular blighted areas, up to half the leaf blade in size, initially appearing water-soaked, then becoming blackened, develop with hymenia on healthy tissue immediately adjacent to the lesions (Kotila, 1947). Hymenia are found mostly on the undersides of leaves but may also be found on young, healthy petioles of heart leaves (Herr, 1982). Older leaf lesions dry out and disintegrate, giving these leaves a ragged or shredded appearance (Kotila, 1947).

3. Survival and control Because foliage blight originates from C-R infected beets (Herr, 1982; Naito, 1984), survival between crops is similar to that previously described for C-R. Basidiospores are short-lived. They may be disseminated for distances up to 70m (Naito, 1984). The possible role of foliar blight on the prevalence and persistence of *R. solani* (*T. cucumeris*) in beet fields in subsequent seasons has not been fully determined (Naito, 1984). In Japan, serious outbreaks of foliar blight have been controlled with fungicidal sprays (Naito, 1984; 1990).

4. Remarks Kotila (1947) considered the strains of *R. solani* causing foliar blight of sugar beet to be distinct from C-R strains because, in his pathogenicity tests, foliar isolates did not cause root rot. In Japan, foliar isolates consisted of AGs 1 and 2-2, mainly AG 2-2 (Naito, 1984). Naito (1990) reported that AG 1-IB caused an early foliar disease by hyphal invasion. Basidiospores of AG 2-2IV from hymenia formed on petioles of beets with root rot or on soil, were the primary source of inoculum for foliage blight (Naito, 1990). Secondary spread within fields was mostly due to basidiospores (Naito, 1984). In Ohio, Herr (1982) reported hymenial isolates from petioles and leaf blades belonged predominately to AG 2. The early "heart leaf stub" phase of the disease was observed in several seasons in Ohio in the absence of the older leaf blight phase and, when assayed, yielded AG 4 isolates. Whitney and Duffus (1986) reported the foliar *R. solani* isolates belong in AG 4. A hypothesis is advanced that differences in reported AGs (in USA) are due to failure to distinguish between the early phase, primarily caused by AG 4, and the older leaf blight phase, primarily caused by AG 2-2. This hypothesis is based on the premise that AG 4 isolates primarily attack sugar beet seedlings and juvenile tissues, and generally are excluded from older tissues by "mature tissue resistance" (See Chapter on Potato). This hypothesis is consistent with the report by Naito *et al.* (1980) that AG 4 isolates were confined to the primary cortex of surviving diseased seedlings and subsequently, following sloughing-off of the cortex, failed to cause root rot of mature plants.

Further, the histopathological observations of Naito and Sugimoto (1981) indicate that AG 4 isolates differ from AG 2-2 isolates in their mode of tissue penetration and their general failure to invade and colonize vascular tissues.

Within the last 30 years considerable advancement has taken place in our understanding of the *Rhizoctonia* spp. associated with Rhizoctonia diseases of sugar beet, of their roles in the etiology of the various disease phases, of the nature and distribution of pathogen survival structures (propagules). Among the challenges for future research are: improvement in our understanding of the nature of pathogenesis and host/tissue resistance; elucidation of the role and ecology of associated soil microbial and faunal populations in relation to pathogen survival and disease, and comprehensive investigations of the distribution and population genetics of *Rhizoctonia* spp. Such additions to our knowledge will facilitate our abilities to manage Rhizoctonia diseases of the sugar beet crop by effective, economical, and non-polluting means.

II. References

Boosalis M & Scharen AL (1959) Methods for microscopic detection of *Aphanomyces euteiches* and *Rhizoctonia solani* and for isolation of *Rhizoctonia solani* associated with plant debris. Phytopathology 49:192-198.

Daniels J (1963) Saprophytic and parasitic activities of some isolates of *Corticium solani*. Trans. Br. Mycol. Soc. 46:485-502.

Duggar BM (1899) Three important fungous diseases of sugar beet. New York (Cornell) Agr. Exp. Sta. Bull. 163:335-363.

Duggar BM (1915) *Rhizoctonia crocorum* (Pers.) D. C. and *R. solani* Kühn (*Corticium vagum* B. & C.) with notes on other species. Ann. Missouri Botan. Garden 2:403-458.

Edson HA (1915) Seedling diseases of sugar beets and their relation to root-rot and crown rot. J. Agr. Res. 4:135-168.

Herr LJ (1976) In field survival of *Rhizoctonia solani* in soil and in diseased sugarbeets. Can. J. Microbiol. 22:983-988.

Herr LJ (1982) Characteristics of hymenial isolates of *Thanatephorus cucumeris* on sugar beets in Ohio. Plant Dis. 66:246-249.

Herr LJ (1987) Populations of *Rhizoctonia solani* in soil under crops in rotation with sugar beet. Ann. Appl. Biol. 110:17-24.

Herr LJ (1988) Biocontrol of Rhizoctonia crown and root rot of sugar beet by binucleate *Rhizoctonia* spp. and *Laetisaria arvalis*. Ann. Appl. Biol. 113:107-118.

Herr LJ & Roberts DL (1980) Characterization of *Rhizoctonia* populations obtained from sugarbeet fields with differing soil textures. Phytopathology 70:476-480.

Houston BR (1945) Culture types and pathogenicity of isolates of *Corticium solani*. Phytopathology 35:371-393.

Hyakumachi M & Ui T (1982) The role of overwintered plant debris and sclerotia as inoculum in the field occurred with sugarbeet root rot. Ann. Phytopathol. Soc. Jpn. 48:628-633.

Hyakumachi M, Yamamoto Y & Ui T (1983) Survival and pathogenicity of sclerotia produced by sugar beet root rot fungus (*Rhizoctonia solani* AG 2 Type 2) in the areas with different degree of disease incidence. Ann. Phytopathol. Soc. Jpn. 49:18-21.

Kataria HR & Hoffmann GM (1988) A critical review of plant pathogenic species of *Ceratobasidium* Rogers. J. Plant Dis. and Prot. 95:81-107.

Kotila JE (1947) Rhizoctonia foliage blight of sugar beets. J. Agr. Res. 74:289-313.

Kühn JG (1858) Die Krankheiten der Kulturegewächse, ihr ursachen and ihr Verhütung. Gustar Bosselmann, Berlin.

Kuznia RA & Windels CE (1994) *Rhizoctonia zeae* pathogenic to spring wheat and sugar beet seedlings. Phytopathology 84:1159. (Abstr.)

LeClerg EL (1939) Studies on dry-rot canker of sugar beets. Phytopathology 29:793-800.

Liu Z & Sinclair JB (1991) Isolates of *Rhizoctonia solani* anastomosis group 2-2 pathogenic to soybean. Plant Dis. 75:682-687.

Moore RT (1987) The genera of *Rhizoctonia*-like fungi: *Ascorhizoctonia*, *Ceratorhiza* gen. nov., *Epulorhiza* gen. nov., *Moniliopsis*, and *Rhizoctonia*. Mycotaxon 29:91-99.

Naiki T & Ui T (1969) On the survival of the sclerotia of *Rhizoctonia solani* Kühn in soil. Mem. Fac. Agr. Hokkaido Univ. 6:430-436.

Naiki T & Ui T (1972) The microorganisms associated with the sclerotia of *Rhizoctonia solani* Kühn in soil and their effects on the viability of the pathogens. Mem. Fac. Agr. Hokkaido Univ. 8:252-265.

Naito S (1984) Studies on foliage blight of sugar beets. Hokkaido Natl. Agr. Exp. Sta. Res. Bull. 139:145-188.

Naito S (1990) Ecological role of basidiospores of *Thanatephorus cucumeris* (Frank) Donk in the incidence of foliage blight of sugar beets in Japan. JARQ, Japan Agr. Res. Quarterly 23:268-275.

Naito S & Sugimoto T (1981) Histopathological observation on root rot of sugar beets by different anastomosis groups of *Rhizoctonia solani* Kühn. Hokkaido Natl. Agr. Exp. Sta. Res. Bull. 131:95-110.

Naito S, Yamaguchi T, Sugimoto T & Fujisawa I (1976) Studies on root rot of sugar beets. VI. Anastomosis groups of *Rhizoctonia solani* Kühn isolated from diseased petioles, crowns and roots. Proc. Sugar Beet Res. Assoc., Japan 17:37-44.

Naito S, Kanazawa K & Sugimoto T (1980) Relationship between paper-pot transplanting system and occurrence of Rhizoctonia root rot of matured sugar beets. Proc. Sugar Beet Res. Assoc., Japan 22:18-24.

Ogoshi A (1985) Anastomosis and intraspecific groups of *Rhizoctonia solani* and binucleate *Rhizoctonia*. Fitopatol. Bras. 10:371-390.

Ogoshi A & Ui T (1983) Diversity of clones within an anastomosis group of *Rhizoctonia solani* Kühn in a field. Ann. Phytopathol. Soc. Jpn. 49:239-245.

O'Sullivan E & Kavanagh JA (1991) Characteristics and pathogenicity of isolates of *Rhizoctonia* spp. associated with damping-off of sugar beet. Plant Pathol. 40:128-135.

Parmeter JR Jr (1970) *Rhizoctonia solani*: Biology and Pathology, University of Berkeley Press, Berkeley, Los Angeles, London.

Peltier GL (1916) Parasitic *Rhizoctonias* in America. Illinois Agr. Exp. Sta. Bull. 189:281-390.

Richards BL (1921) A dry rot canker of sugar beets. J. Agr. Res. 22:47-52.

Roberts DL & Herr LJ (1979) Soil populations of *Rhizoctonia solani* from areas of healthy and diseased beets within four sugarbeet fields differing in soil texture. Can. J. Microbiol. 25:902-910.

Ruppel EG & Hecker RJ (1987) Prelayby applications of experimental fungicides for suppressing Rhizoctonia root rot in sugar beet. Plant Dis. 71:694-698.

Ruppel EG & Hecker RJ (1994) Rhizoctonia root rot on sugarbeet cultivars having varied degrees of resistance. J. Sugar Beet Res. 31:135-142.

Rush CM, Carling DE, Harveson RM & Mathieson JT (1994) Prevalence and pathogenicity of anastomosis groups of *Rhizoctonia solani* from wheat and sugar beet in Texas. Plant Dis. 78:349-352.

Sneh B, Burpee LL & Ogoshi A 1991) Identification of *Rhizoctonia* Species. APS Press, St. Paul Minnesota. 133pp.

Uchino H, Ogoshi A & Kanzawa K (1983) Anastomosis groups of binucleate *Rhizoctonia* isolated from diseased sugar beet seedlings. Mem. Fac. Agr. Hokkaido Univ. 13:494-499.

Whitney ED & Duffus JE (1986) Compendium of beet diseases and insects. APS Press, St. Paul, MN, USA.

Windels CE & Jones RK (1989) Seedling and root diseases of sugar beets. Minn. Ext. Serv. AG-FO-3702:1-3.

Windels CE & Nabben DJ (1989) Characterization and pathogenicity of anastomosis groups of *Rhizoctonia solani* isolated from *Beta vulgaris*. Phytopathology 79:83-88.

Windels CE Kuznia RA & Call J (1994) First report of *Thanatephorus cucumeris* (=*Rhizoctonia solani* AG 3 and AG 5) on sugar beet. Phytopathology 84:1161 (Abstr.).

V.11 *RHIZOCTONIA* ASSOCIATED WITH FOREST TREES

ARI M. HIETALA[1] AND ROBIN SEN[2]

[1]*Finnish Forest Research Institute, P.O. Box 18, FIN-01301 Vantaa, Findland and* [2]*Department of Biosciences, Division of General Microbiology, P.O. Box 56, FIN-00014 University of Helsinki, FINLAND*

I. INTRODUCTION

Many *Rhizoctonia* species are economically important plant pathogens in agriculture and thus receive considerable attention. However, relatively little is known about these fungi and their effects on trees in forestry production. There are well documented cases of *Rhizoctonia*-linked diseases of tree seedlings cultivated in forest nurseries although information concerning their interactions with forest trees in natural ecosystems is limited and fragmentary.

The early descriptions of *Rhizoctonia* spp. associated with tree species were mainly based on a few anamorphic traits. More recently, a re-evaluation of these species using presently accepted criteria (Parmeter and Whitney, 1970; Ogoshi, 1975) has resulted in the exclusion of many taxa from the genus *Rhizoctonia* (Andersen and Stalpers, 1994).

In this review, the conventions of recently published checklists for *Rhizoctonia* are followed (Sneh *et al.*, 1991; Andersen and Stalpers, 1994). Where possible, the present taxonomic position of the earlier described species will be indicated.

II. *RHIZOCTONIA* IN FOREST TREE NURSERIES

In a comprehensive nursery study, Saksena and Vaartaja (1961) grouped *Rhizoctonia* species into three classes that caused (a) root and hypocotyl rot and damping-off, (b) varying root disorders with intracellular chlamydospores (= monilioid hyphae) infections and (c) no apparent disease symptoms following root penetration. Later reports also confirmed their role in shoot diseases of trees.

A. *Rhizoctonia* as a damping-off agent

Damping-off, either as pre- or post-emergence is a common disease of seeds, germlings and young seedlings of many plants including woody species. It has long been known that anamorphs attributed to *Rhizoctonia solani* Kühn (multinucleate, see Parmeter and Whitney, 1970) are a causal agent of damping-off in conifer seedlings (Wiant, 1929). World-wide, the seeds and young seedlings of a number of conifer and broad-leaved species (e.g. *Pinus*, *Picea*, *Larix*, *Ulmus*, *Eucalyptus* spp.) are known to be attacked (Hartley, 1921; Roth and Riker, 1943; Wright, 1944; Vaartaja and Cram, 1956; Saksena *and* Vaartaja 1960, 1961; Vaartaja *et al.*, 1961; Vaartaja, 1967; Gomez-Nava, 1967; Sharma *et al.*, 1984; Perrin and Sampangi, 1986). Another multinucleate species, *R. endophytica* var. *filicata* Saks. and Vaar (= *R. zeae* Voorhees) (see Sneh *et al.*, 1991), was also reported as a damping-off pathogen on conifer seedlings (Saksena and Vaartaja 1961; Vaartaja, 1967)

B. Sneh et al. (eds.),
Rhizoctonia Species: Taxonomy, Molecular Biology, Ecology, Pathology and Disease Constrol, 351–358.

The anastomosis groups (AGs) (see Chapter I.A2) of *R. solani* associated with trees have only been determined in a few studies. The most common, AG 4, has been isolated from *Pinus banksiana* Lamb. and *Picea glauca* Voss (Anderson, 1982), nursery soils (Camporota and Perrin, 1994) and the soil amendment, pine bark mulch (Huang and Kuhlman, 1990). Isolates from nursery soils, representing AGs 4 (the most common group), 5 and 1-2, but not AG 2-2, were very aggressive damping-off pathogens on *Pinus nigra* Arnold and the AG 4 isolate from pine bark mulch caused damping-off on *Pinus elliottii* Engelm. var. *elliottii*. Which AGs were involved in the early studies of Saksena and Vaartaja (1961)? The most common species, *Ceratobasidium praticola* (Kotila) Olive, causing damping-off, is presently included in the *Thanatephorus cucumeris* complex (and suggested to be *T. praticola*, Sneh *et al.*, 1991), despite sterigmatic characteristics previously regarded as specific, and now forms AG 4 of *R. solani* (Ogoshi, 1975). In examining the AG affinities of many earlier studied isolates, Parmeter *et al.* (1969) confirmed that three *C. praticola* isolates and one *R. solani* isolate (Saksena and Vaartaja, 1961), *R. solani* var. *cedri deodarae* (Castellani, 1934b) and five North American conifer isolates from different sources were all representatives of AG 4. This highlights the common occurrence of AG 4 on trees and seedlings.

Binucleate *Rhizoctonia* spp. causing damping-off in conifer seedlings include *R. callae* Cast. (Saksena and Vaartaja, 1961), *R. endophytica* var. *endophytica* Saks. and Vaar. (= AG-A, see Ogoshi *et al.*, 1983) (Saksena and Vaartaja, 1961; Gomez-Nava, 1967), and a *Rhizoctonia* sp. AG-E (=CAG 3, see Ogoshi *et al.*, 1983) (Huang and Kuhlman 1990).

The damping-off pathogens, *Rhizoctonia, Pythium* and *Fusarium* are ubiquitous inhabitants in forest nursery soils (e.g. Perrin and Sampangi 1986) and reports indicate that seedlings grown directly in seed beds are more affected by damping-off than containerized seedlings that are normally cultivated in soil-free growth media. The activities of these pathogens are strongly dependent on prevailing soil conditions, e.g. pH, temperature, moisture etc. (Roth and Riker 1943; Camporota and Perrin 1994). Soil fumigation and seed treatments with fungicides have been recommended in nursery manuals as measures for broad-spectrum control. Potential damping-off pathogens, including *R. solani* (AG 4), may also be introduced into soils via organic amendments (Huang and Kuhlman, 1990). Huang and Kuhlman (1991) subsequently formulated a mixture of organic and inorganic materials which stimulated antagonistic fungi and reduced the populations of *R. solani* and *Pythium aphanidermatum* (Edson) Fitzp. enabling effective control of damping-off both in fumigated and non-fumigated soils.

B. *Rhizoctonia* as a root pathogen of older seedlings

Descriptions of *Rhizoctonia* spp. causing root damage in older conifer seedlings are mainly those of Saksena and Vaartaja (1960, 1961). Fungi were isolated from various conifer species displaying damping-off symptoms, mycorrhizal rootlets or apparently non-diseased seedlings. Isolates of *R. solani*, *R. endophytica* var *endophytica* (= AG-A), *R. endophytica* var. *filicata* (= *R. zeae*) and *R. callae*, causing damping-off, and *R. repens* Bernard (binucleate, see Sneh *et al.* 1991) also induced stunted shoot and root growth in seedlings of *Pinus sylvestris* L., *P. resinosa* Ait., *Ulmus americana* L., *Caragana arborescens* Lam. and *Thuja occidentalis* L. A distinguishing characteristic of *R. endophytica* var. *endophytica*, *R. callae* and *R. repens* was the production of monilioid hyphae in root cortical cells of all test plants.

However, the results obtained from these tests conducted under sterile conditions were only considered indicative of the potential pathogenicity of these isolates (Saksena and Vaartaja, 1961).

More recently, root dieback of both containerized and bare-rooted Norway spruce [*Picea abies* L. (Karst.)] and Scots pine (*P. sylvestris*) seedlings has been a considerable problem in Nordic nurseries (Venn *et al.*, 1986; Lilja *et al.*, 1992). The visual symptoms: wilting of young shoots, hanging tops, retarded height growth, discoloration of the foliage and partial or total death of the root system often appear after midsummer, on first year seedlings or occasionally later during the second growing season. Surveys for fungi present in the diseased roots and pathogenicity testing suggest that root dieback is a complex phenomenon involving both *Rhizoctonia* spp. and *Pythium* spp. (Venn *et al.*, 1986; Lilja *et al.*, 1992). Pathogenicity experiments performed under nursery conditions implicated the supporting sand beds as an inoculum source for *Rhizoctonia* spp. in container seedling production (Venn *et al.*, 1986).

Based on conserved anamorphic and induced teleomorphic characteristics, the *Rhizoctonia* sp. causing damage on nursery tree stocks in Norway and Finland represents a single species possessing uninucleate cells (Hietala *et al.*, 1994; see preface). Basidial characteristics refer it to the teleomorph genus *Ceratobasidium*, possibly *C. bicorne* Erikss. and Ryv., which is exceptional since all the known anamorphs of *Ceratobasidium* species are binucleate (see Sneh *et al.*, 1991). However, there was no affinity with the other *Ceratobasidium* AG testers, and similar analysis with *C. bicorne*, only available as herbarium material, was not possible (Hietala *et al.*, 1994). In pathogenicity tests with Norway spruce (Hietala, 1995) and Scots pine (Sen and Hietala, unpublished), uninucleate *Rhizoctonia* isolates attacked all parts of the root system, although the root tips were most heavily infected resulting in a stunted root system morphology (see preface). Cortical cells infected with monilioid cells were occasionally observed.

In the Finnish surveys, non-pathogenic binucleate *Rhizoctonia* spp. were also isolated from the roots of similarly diseased and healthy-looking conifer seedlings (Lilja *et al.*, 1992; Lilja 1994). In a case study, binucleate *Rhizoctonia* spp. were found to co-exist with the uninucleate *Rhizoctonia* sp. in the same root system and were divided into four anastomosis groups including AG-I of the genus *Ceratobasidium* and three anastomosis groups not related to known anastomosis groups of *Ceratobasidium* (Hietala, 1995). In pathogenicity tests on Norway Spruce (Hietala, 1995) and Scots pine (Sen and Hietala, unpublished) seedlings, isolates representing these groups infected only basal root regions and cortical cells filled with monilioid hyphae were commonly observed. Although the infection had no effect on root growth of Norway spruce seedlings, isolates representing AG-I and two other AGs significantly increased shoot and root growth of Scots pine seedlings. A non-orchid mycorrhizal relationship could be hypothesized with the latter conifer species (see chapters V.14 and VI. B4) although fungal infection levels were low (<20 % of root length infected).

C. Other *Rhizoctonia* associated with tree roots

Castellani (1934b) described three species, *R. pini-insignis* Cast., *R. quercus* Cast. and *R. fraxini* Cast., from tree roots. The former two were respectively isolated from nursery grown *Pinus insignis* Dougl. (= *Pinus radiata* D. Don) and *Quercus pedunculata* Ehrh. and the latter from apparently naturally growing *Fraxinus excelsior* L. Burpee *et al.* (1980) found *R. fraxini* and *R. pini-insignis* to be binucleate and *R. quercus* uninucleate although Ogoshi *et al.* (1983) could not confirm a uninucleate condition in the latter species. Unfortunately, the pathogenicity of these species was not tested on the respective hosts (Castellani 1934a).

D. *Rhizoctonia* associated with shoot diseases

In India, *R. solani* has been identified as the causal agent of web blight in a number of nursery-grown broad-leaved trees including *Eucalyptus*, *Acacia* and *Albizia* species (Sharma *et al.*, 1984; Sharma and Sankaran 1984; Sankaran *et al.*, 1986; Mehrotra, 1990). The symptoms vary depending on the host, but a common feature is the premature defoliation of mycelium covered diseased leaves of young plants (Mehrotra, 1990). Isolates of *R. solani* obtained from the different hosts could be separated into three morphological groups, but cross inoculations of different hosts indicated no host specificity (Mehrotra, 1990). Another shoot disease caused by *R. solani* was reported in the same study; top flagging of khasi pine (*Pinus kesiya* Royle ex Gordon). Diseased nursery seedlings, 15 - 25 cm in height, displayed withered tops which later drooped and turn ash-colored. Infected needles were loosely bound with hyphae and the stems were also attacked. In both these shoot diseases, the respective host trees were most susceptible to *R. solani* over the wet season when the prevailing humid conditions enabled foliage colonization from soil inoculum sources via the stems (Mehrotra, 1990). Certain weed species were also highly susceptible to the disease and were suspected of being an additional source of infection. Disease control measures involve immediate removal of affected seedlings, weeding and a reduction in sowing densities.

Needles of four- to five-year-old Norway spruce seedlings have been reported to be killed by a binucleate *Rhizoctonia* sp. in Norwegian forest nurseries (Roll-Hansen and Roll-Hansen, 1968). A binucleate *Rhizoctonia* sp. AG-E (=CAG 3), was shown to be associated with foliar blight of nursery grown longleaf pine (*P. palustris* Mill., English *et al.*, 1986) and loblolly pine (*P. taeda* L., Runion and Kelly, 1993) in USA. On longleaf pine, the symptoms include necrosis of needles and terminal buds which are especially severe in sandy soils where sand accumulates around the needle bases and terminal buds. The sand possibly acts as an inoculum source and provides warm and humid conditions that apparently favour disease development. Affected needles of loblolly pine are characteristically bound in a mycelial webbing, turn grey and eventually drop-off. The severity of disease generally increases in seedbeds following crown closure which allow vegetative spread of the fungus via shoot contact (Runion and Kelly, 1993). Fungicide application does to some extent control the disease (Runion *et al.*, 1994). Prior to sowing, seedbed soils are routinely fumigated with methyl bromide (English *et al.*, 1986). As shown by Huang and Kuhlman (1990), organic amendments represent a potential source for introducing AG-E into fumigated soils.

III. *RHIZOCTONIA* IN FOREST HABITATS

Limited observations on the occurrence of the perfect states of *Rhizoctonia* confirm that the genus does thrive in forest habitats. Eriksson and Ryvarden (1973) reported the occurrence of *Ceratobasidium cornigerum* (Bourdot) Rogers on a variety of substrates and particularly the fresh bark of detached branches of both conifer and deciduous trees. Other *Ceratobasidium* species observed on bark of trees include *C.*

pseudocornigerum M.P. Christ. (Eriksson and Ryvarden, 1973; Breitenbach and Kränzlin, 1986; Kotiranta and Saarenoksa, 1990) and *C. stridii* Erikss. and Ryv. (Eriksson and Ryvarden, 1973). The green needles of standing pines and the upper soil litter layer were implicated as a source of the orchid endophyte, *R. goodyera-repentis* Costantin and Dufour (teleomorph, *C. cornigerum*), in baiting experiments using the compatible orchid host (Downie, 1943). Understory plants have been found to host *T. cucumeris* (Poldmaa *et al.* 1982; Kotiranta and Saarenoksa, 1993) and *C. bicorne* (Eriksson and Ryvarden (1973).

A. *Rhizoctonia* spp. as disease agents in forests

The nuts of beech (*Fagus sylvatica* L.) have been reported to be attacked by *R. solani* in France (Perrin and Muller, 1979) and Germany (Dubbel, 1989). In Denmark, beechnuts are often similarly affected by *Rhizoctonia* (J. Koch, personal communication) and the isolates appear to be binucleate (Koch and Hietala, unpublished). The soilborne nature of *R. solani* was demonstrated in comparisons of seed from the forest floor with those collected from nets suspended above the ground (Dubbel, 1989). The disease is characterized by a decay of the cotyledon bearing tissue within the seed resulting in reduced germination or death after cotyledon emergence (Perrin and Muller, 1979). Higher disease incidence is related to increased pH and the level of soil organic material but soil cultivation before the mast is a good disease control strategy. It is recommended that beechnuts are rapidly harvested and where possible incubated before storage at 34 - 37°C at 10 % relative humidity for 24 hours. This procedure significantly reduced disease incidence (Perrin and Muller, 1979).

Surface sterilized seeds of Norfolk Island pine *Araucaria heterophylla* (Salisbury) Franco, growing in Egypt, were found to harbor *Rhizoctonia solani* while still within cones on the tree (El-Lakany *et al.*, 1981). The fungus was isolated from the endosperm and embryo tissues and was also implicated as the principle seedborne pathogen causing pre- and post emergence damping-off in this and two other *Araucaria* species (Kamara *et al.*, 1981).

In two tropical Indian forests, *R. solani* has been identified as the causative agent of a root rot disease on seedlings of *Phyllanthus emblica* L., *Quercus serrata* Thunb. and *Citrus* sp. (Chakravarty and Mishra, 1982). Disease symptoms, on mainly one to six-month-old seedlings, included either wilting or, in severe cases, stem collapse due to rotting of tissues at the base of the seedlings. The high temperatures and rainfall prevalent between May and August coincided with the highest disease incidence.

IV. FUTURE PROSPECTS

A range of *Rhizoctonia* species are reported to be associated with trees both in nurseries and forest habitats. However, there is still a clear need for further identification of the putative *R. solani* anamorph. This was highlighted by Parmeter *et al.* (1967) who observed that mycelial and cultural characteristics of some *Rhizoctonia* species with a *Ceratobasidium* perfect state resemble those of *R. solani* so closely that only numbers of nuclei per cell or the perfect state distinguish them. Burpee *et al.* (1980) also stated that "many of the binucleate isolates in AGs-E -F and -R (=CAGs 3, 4 and 5) were morphologically indistinguishable from isolates of *R. solani*". Yet, in the majority of papers reviewed the observed anamorph was assigned to *R. solani* although neither the teleomorph nor the nuclear condition of cells were described. As induction of the teleomorph is problematic in many species, the nuclear condition is a minimum diagnostic characteristic that can easily be determined using a number of nuclear stains (see Sneh *et al.*, 1991).

Few studies have been reported on a number of *R. solani* anastomosis groups (see Chapter I.A2), particularly AG 4, that are associated with nursery tree seedlings. The AGs represent independent genetic units that often show different host preferences and levels of pathogenicity (see Chapter II.1). As the AG testers of *R. solani* and also *Ceratobasidium* are standardized and available from culture collections (e.g. American Type Culture Collection), the affinities of unknown multi- and bi-nucleate *Rhizoctonia* isolates associated with all plants, including trees, can be readily determined. This is of particular importance in enabling valid comparisons of the increasing data on the effects of *Rhizoctonia* spp. in nursery tree production and in natural forest ecosystems.

It is clear that *Rhizoctonia* spp. are associated with forest trees but are mainly pathogens of young seedlings in forest nurseries. Increasing global pressure for re-forestation will require expanded nursery seedling production where conventional chemical disease control practices have often proved ineffective. Due to their broad-spectrum toxicity, the use of many of the pesticides and fumigants e.g. benomyl and methyl bromide is being phased-out leading the way to the development of biological control solutions. For the successful application of this alternative control strategy further detailed information is needed on the *Rhizoctonia* spp. involved in seedling diseases, their alternative host preferences (e.g. weed species) and interactions with other pathogens (e.g. *Pythium* and *Fusarium* spp.) and beneficial micro-organisms (e.g. *Trichoderma* spp. and mycorrhizal fungi, see chapters VI.B2, VI.B4 and VI.B7). Re-forestation of former agricultural land also may present additional problems relating to the potential broad host-range of pathogenic *Rhizoctonia* species.

These challenges will require increased effort and co-operation between researchers in fields such as *Rhizoctonia* taxonomy and genetics, plant/tree physiology and pathology, agronomy, silviculture, molecular biology and biotechnology.

V. Acknowledgements

We thank Dr. Kari Korhonen for valuable comments on the manuscript and the financial support of Suomen Luonnonvarain Tutkimussäätio and the Academy of Finland.

VI. References

Andersen TF & Stalpers JA (1994) A check-list of *Rhizoctonia* epithets. Mycotaxon 51:437-457.

Anderson NA (1982) The genetics and pathology of *Rhizoctonia solani*. Annu. Rev. Phytopathol. 20:329-347.

Burpee LL, Sanders PL & Cole H Jr (1980) Anastomosis grouping among isolates of *Ceratobasidium cornigerum* and related fungi. Mycologia 72:689-701.

Breitenbach J & Kränzlin F (1986) Pilze der Schweiz. Band 2. Nichtblätterpilze. Verlag Mykologia, Luzern.

Camporota P & Perrin R (1994) Survey for damping-off in forest nurseries in France. Preliminary results. In: Perrin R & Sutherland JR (eds.) Diseases and insects in Forest nurseries. Les Colloques no 68 (pp. 51-58) INRA, Paris.

Castellani E (1934a) Ricerche preliminari sulla biologia di alcune Rizottonie. Ann. R. Ist. Sup. Agrario e Forestale. Vol. 5. 33 pp.

Castellani E (1934b) Ricerche morfologico-sistematiche su alcune Rizottonie. Ann. R. Ist. Sup. Agrario e Forestale. Vol. 5. 16 pp.

Chakravarty P & Mishra RR (1982) Studies in Forest pathology III. Root rot due to *Rhizoctonia solani* Kühn. Eur. J. For. Path. 12:308-316.

Downie DG (1943) Source of the symbiont of *Goodyera repens*. Trans. Bot. Soc. Edinb. 33:383-390.

Dubbel V (1989) Die Bedeutung des Bodenkontaktes für die Qualität des Buchensaatgutes. Forst und Holz 44:512-516.

El-Lakany MH, Kamara AM, Badran OA & Attia YG (1981) Seed pathology of *Araucaria* spp. II. Fungal species associated with *Araucaria heterophylla* seed. Aust. For. Res. 11:275-281.

English JT, Ploetz RC & Barnard EL (1986) Seedling blight of longleaf pine caused by a binucleate *Rhizoctonia solani* -like fungus. Plant Dis. 70:148-150.

Eriksson J & Ryvarden L (1973) The Corticiaceae of North Europe. Vol. 2, Aleurodiscus-Confertobasidium. Fungiflora, Oslo.

Gomez-Nava M (1967) Comparative morphological study of specimens of *Rhizoctonia* isolated from forest nursery beds. Bol. Tec. Inst. Nac. Invest, For. Mexico No. 21. 34 pp.

Hartley C (1921) Damping-off in forest nurseries. U.S. Dept. Agr. Bul. 934. 99 pp.

Hietala AM (1995) Uni- and binucleate *Rhizoctonia* spp. coexisting on the roots of Norway-spruce seedlings suffering from root dieback. Eur. J. For. Path. 25:136-144.

Hietala AM, Sen R & Lilja A (1994) Anamorphic and teleomorphic characteristics of a uninucleate *Rhizoctonia* sp. isolated from the roots of nursery grown conifer seedlings. Mycol. Res. 98:1044-1050.

Huang JW & Kuhlman EG (1990) Fungi associated with damping-off of slash pine seedlings in Georgia. Plant Dis. 74:27-30.

Huang JW & Kuhlman EG (1991) Mechanisms inhibiting damping-off pathogens of slash pine seedlings with a formulated soil amendment. Phytopathology 81:171-177.

Kamara AM, El-Lakany MH, Badran OA & Attia YG (1981) Seed pathology of *Araucaria* spp. A survey of seed-borne fungi associated with four *Araucaria* spp. Aust. For. Res 11: 269-274.

Kotiranta H & Saarenoksa R (1990) Reports of Finnish corticolous Aphyllophorales (Basidiomycetes). Karstenia 30:43-69.

Kotiranta H & Saarenoksa R (1993) Rare Finnish Aphyllophorales (Basidiomycetes) plus two new combinations in *Efibula*. Ann. Bot. Fennici 30:211-249.

Lilja A (1994) The occurrence and pathogenicity of uni- and binucleate *Rhizoctonia* and Pythiaceae fungi among conifer seedlings in Finnish forest nurseries. Eur. J. For. Path. 24:181-192.

Lilja A, Lilja S, Poteri M & Ziren L (1992) Conifer seedling root fungi and root dieback in Finnish nurseries. Scand. J. For. Res. 7:547-556.

Mehrotra MD (1990) *Rhizoctonia solani*, a potentially dangerous pathogen of khasi pine and hardwoods in forest nurseries in India. Eur. J. For. Path. 20:329-338.

Ogoshi A (1975) Grouping of *Rhizoctonia solani* Kühn and their perfect stages. Rev. Pl. Prot. Res. 8:93-103.

Parmeter JR Jr, Sherwood RT & Platt WD (1969) Anastomosis grouping among isolates of *Thanatephorus cucumeris*. Phytopathology 59:1270-1278.

Parmeter JR Jr & Whitney HS (1970) Taxonomy and nomenclature of the imperfect state. In: Parmeter JR Jr (ed.) *Rhizoctonia solani*, Biology and Pathology. (pp 7-19) University of California Press, Berkeley.

Parmeter JR Jr, Whitney HS & Platt WD (1967) Affinities of some *Rhizoctonia* species that resemble mycelium of *Thanatephorus cucumeris*. Phytopathology 57:218-223.

Perrin R & Muller C (1979) La pourriture des faines causée par *Rhizoctonia solani* Kühn: Incidence de cette maladie après les fainées de 1974 et 1976. Traitement curatif des faines en vue de la conservation. Eur. J. For. Path. 9:89-103.

Perrin R & Sampangi R (1986) La fonte des semis en pépinière forestière. Eur. J. For. Path. 16:309-321.

Poldmaa P, Parmasto I & Parmasto E (1982) Distribution of *Thanatephorus cucumeris* in Estonia and its variability (in Russian with an English summary). Eesti NSV Teaduste Akadeemia Toimetised. Köide Bioloogia 31:119-125.

Roll-Hansen F & Roll-Hansen H (1968) A species of *Rhizoctonia* DC. ex Fr. damaging spruce plants in nurseries in southern Norway. Meddel. fra Det Norske Skogf. 82:421-440.

Roth LF & Riker AJ (1943) Seasonal development in the nursery of damping-off of red pine seedlings caused by *Pythium* and *Rhizoctonia*. J. Agric. Res. 67:417-431.

Runion GB & Kelley WD (1993) Characterization of a binucleate *Rhizoctonia* species causing foliar blight of loblolly pine. Plant Dis. 77:754-755.

Runion GB, Kelley WD, Land DH, Gilly SP & Sharp DJ (1994) Chemical control of *Rhizoctonia* seedling blight of longleaf pine. South. J. Appl. For. 18:5-9.

Saksena HK & Vaartaja O (1960) Descriptions of new species of *Rhizoctonia*. Can. J. Bot. 38:931-943

Saksena HK & Vaartaja O (1961) Taxonomy, morphology and pathogenicity of *Rhizoctonia* species from forest nurseries. Can. J. Bot. 39:627-647.

Sankaran KV, Balasundaran M & Sharma JK (1986) Seedling diseases of *Azadirachta* in Kerala, India. Eur. J. For. Path. 16:324-328.

Sharma JK & Sankaran KV (1984) *Rhizoctonia* web blight of *Albizia falcataria* in India. Eur. J. For. Path. 14:261-264.

Sharma JK, Mohanan C & Florence JM (1984) Nursery diseases of *Eucalyptus* in Kerala. Eur. J. For. Path. 14:77-89.

Sneh B, Burpee L & Ogoshi A (1991) Identification of *Rhizoctonia* species. The APS Press, St. Paul Minnesota 133 pp..

Vaartaja O (1967) Damping-off pathogens in South Australian nurseries. Phytopathology 57:765-768.

Vaartaja O & Cram WH (1956) Damping-off pathogens of conifers and of *Caragana* in Saskatchewan. Phytopathology 46:391-397.

Vaartaja O, Cram WH & Morgan GA (1961) Damping-off etiology in forest nurseries. Phytopathology 51:35-42.

Venn K, Sandvik M & Langerud BR (1986) Nursery routines, growth media and pathogens affect growth and root dieback in Norway spruce seedlings. Meddel. Norsk Inst. Skogforsk. 39:313-328.

Wiant JS (1929) The *Rhizoctonia solani* damping-off of conifers, and its control by chemical treatments of the soil. N.Y. (Cornell) Agr. Exp. Sta. Mem. 124. 64 pp.

Wright E (1944) Damping-off in broadleaf nurseries of the Great Plains region. J. Agric. Res. 69:77-94.

V.12 BIOLOGY OF TURFGRASS DISEASES INCITED BY *RHIZOCTONIA* SPECIES

LEON L. BURPEE[1] AND S. BRUCE MARTIN[2]

[1]*Department of Plant Pathology, University of Georgia, Georgia Station, Griffin, GA 30223 USA,* [2]*Clemson University, Pee Dee Research and Education Center, Route 1, Box 531, Florence, SC 29501-9603 USA*

I. INTRODUCTION

Turfgrasses are those members of the family Poaceae (Gramineae) characterized by a dense foliar canopy and tolerance to mowing. Approximately 40 species in 18 genera represent two physiological groups commonly referred to as cool-season and warm-season grasses. Cool-season grasses are adapted to subarctic and temperate regions, with optimal growth occurring at 15-25°C. These species, known as C_3 grasses, undergo photosynthetic carbon fixation through the Calvin cycle. In contrast, warm-season grasses are known as C_4 grasses because they fix carbon via the Hatch-Slack pathway. These grasses are adapted to tropical and subtropical regions, with optimal growth occurring at 25-35°C (Smiley *et al.*, 1992).

More than 100 infectious agents are known to incite diseases of turfgrasses (Couch, 1985). Most of these pathogens are fungi, with *Rhizoctonia* spp. included among the most virulent. As turfgrass pathogens, *Rhizoctonia* spp. have a broad host range, inciting diseases in both cool- and warm-season grasses (Burpee and Martin, 1992). This chapter is presented as a review of information on the biology of these diseases.

II. RHIZOCTONIA-INCITED TURFGRASS DISEASES

A. Rhizoctonia blight

In 1984, the term "Rhizoctonia blight" was adopted by plant pathologists to describe a turfgrass disease incited by *Rhizoctonia solani* Kühn (Couch, 1985). Previously, the disease was known as "brown patch", which was considered to be less descriptive, as several diseases and conditions of turf may result in patches of brown grass.

Rhizoctonia solani incites diseases of both cool- and warm-season turfgrasses. The primary cool-season species include bentgrass (*Agrostis* spp.), ryegrass (*Lolium* spp.), fescue (*Festuca* spp.), and bluegrass (*Poa* spp.). The major warm season species are zoysiagrass (*Zoysia* spp.), centipedegrass [*Eremochloa ophioroides* (Munro) Hack], St. Augustinegrass [*Stenotaphrum secundatum* (Walter) Kuntze], and bermudagrass (*Cynodon* spp.). All are attacked by *R. solani*, although symptoms, disease severity, and turf recuperative potential may vary among species and genotypes within species.

Cool-season turfgrasses are most susceptible when they are under stress from heat, and consequently they are commonly attacked during summer months. The warm-season grasses, however, are most susceptible when their growth rates are reduced by cool temperatures; thus, they are commonly attacked as they approach winter dormancy and when they resume growth in the spring.

B. Sneh et al. (eds.),
Rhizoctonia Species: Taxonomy, Molecular Biology, Ecology, Pathology and Disease Constrol, 359–368.

1. **Rhizoctonia blight of cool-season grasses.** Symptoms of Rhizoctonia blight vary among cool-season grasses, weather conditions, and with turfgrass maintenance practices. However, much of the variation in symptoms is due to differences in the height of cut from mowing. Irregular tan or brown patches in stands of grasses may range in size from a few centimeters up to several meters in diameter. On grasses that are maintained at low heights of cut (<13 mm), such as bentgrass, ryegrass, and bluegrass on golf course tees and putting greens, patches frequently possess a dark gray or purplish ring along the patch margin. This symptom is diagnostic, and is referred to as a "smoke ring". The smoke ring border (<5 cm wide) is the result of visible mycelium giving a grayish cast to newly colonized leaves at the border of the patch. As dew evaporates, the margins of the patches usually assume a dark brown or sometimes a wilted purple appearance, due to fresh, water-soaked "active" lesions on the leaves. Within a day or so, the infected leaves assume a light brown color.

When heights of cut are greater than 13 mm, the smoke ring symptom may not be apparent, as leaves are wider and the turf canopy is less dense than on grasses maintained at low heights of cut. Under these conditions, disease is initially apparent as lesions on leaves. These lesions are irregular in shape, and are silver-gray to light brown in color. A thin, nondiscrete, dark brown or purple border surrounds the lesions. As disease progresses, overall symptoms may change from leaf spotting to foliar blight, in which whole leaves become necrotic. When disease is severe, leaves within the patches may be more or less uniformly blighted. However, aggregations of symptomatic leaves may assume more or less circular shapes, with relatively green, asymptomatic leaves in the center of the patches. These leaves develop from meristems that have escaped infection.

Severe outbreaks of disease are associated with extended periods of high air temperatures and relative humidity or extended periods of leaf wetness. These conditions favor growth of the pathogen as well as limit the ability of the cool-season grasses to recover by growth and replacement of affected leaf tissue. Optimally, periods of 48 hours or more of continuous leaf wetness and air temperatures of 26-28°C lead to severe disease. Leaves are colonized by *R. solani* most rapidly at temperatures of 29-32°C (Dickinson, 1930; Couch, 1995), but colonization is slowed significantly at temperatures greater than 32°C. Disease severity is also increased by high nitrogen fertility, particularly if it is not balanced with potassium and phosphorous (Bloom and Couch, 1960). Under balanced fertility levels, disease is more severe when soil pH is low. When nitrogen is high and unbalanced with potassium and phosphorous, disease is more severe when soil pH is high (Bloom and Couch, 1960). Effects of soil moisture as predisposing factors have not been adequately studied.

Several anastomosis groups (AG) of *R. solani* have been reported from isolations from cool-season grasses, including AGs 1, 2, 4 and 5 (Martin *et al.*, 1984). Recently, an effort has been made to further characterize isolates to the subgroup level or beyond, and patterns are beginning to emerge. For instance, tall fescue (*Festuca arundinacea*) is known to be attacked most commonly by AG 1, but isolates belonging to AG 2-2 are also collected from this host (Martin *et al.*, 1984). Creeping bentgrass (*Agrostis palustris*) is commonly attacked by AG 2-2, but isolates belonging to AG 1 are sometimes collected (Burpee, Martin and Dernoeden, unpublished). Isolates belonging to AG 2-2 from cool-season grasses probably belong to subgroup AG 2-2

IIIB ("mat rush" types) which infect mainly shoots and leaves of the Poaceae. Isolates grouped in AG 4 or 5 have been infrequently collected from cool-season grasses.

There is little information on fungal survival in turfgrass ecosystems, as well as plant colonization and pathogenesis by *R. solani* anastomosis groups that attack cool-season grasses. Even less information is available for warm-season grass-*Rhizoctonia* pathosystems. Penetration and colonization of turfgrass leaf tissue is probably similar to the processes that occur in rice (Marshall and Rush, 1980) at least for AG 1-IA, the "sasakii" type strain, which attacks many cool-season grasses as well as causing sheath blight of rice. In rice, the fungus grows longitudinally along the surface of leaf sheaths as "runner hyphae", before differentiation into lobate appressoria and/or infection cushions. Lobate appressoria are composed of short branches of hyphae with about 2 to 5 swollen apical lobes. Penetration pegs differentiate from the apical lobes. Infection cushions are formed as aggregates of hyphae into structures about 100-400 mm in diameter and 80-170 mm in height. Penetration pegs are formed from specialized cells in infection cushions that are aligned perpendicular to the plant surface (Marshall and Rush, 1980).

In cool-season grasses, *R. solani* survives in soil and thatch as sclerotia, or bulbils, and as colonized plant matter (Shurtleff, 1953; Martin *et al.*, 1983). As air temperatures reach 18-20°C, the fungus begins growth from the dormant state and begins colonization of leaves as air temperatures approach 24°C. Colonization and infection of leaves occurs at temperatures up to 32°C (Dickinson, 1930). The fungus can penetrate leaves through stomatal openings or through the leaf cuticle and cell walls. This is accomplished by direct growth of the fungus hyphae through stomata or by differentiation of hyphae into lobate appressoria and/or infection cushions. Penetration pegs are formed from appressoria and infection cushions, and epidermal cells are probably penetrated by mechanical and enzymatic mechanisms (Bateman, 1970).

From studies on other hosts, it is probable that after leaves of turfgrasses are penetrated, *R. solani* hyphae grow intra- and intercellularly, and destroy tissue by enzymatic degradation of middle lamellae and cell walls.

There is limited information available on the survival and saprophytic growth of *R. solani* in turfgrass pathosystems. Sclerotia or bulbils of the fungus have been observed attached to leaf sheaths and blades of blighted *Agrostis* spp., and in greenhouse inoculations, sclerotia were formed on both adaxial and abaxial sides of leaf sheaths, in the crown, and near the point of root initiation in the crown (Shurtleff, 1953). In tall fescue turf, *R. solani* was isolated from naturally colonized organic debris from the thatch layer (Martin *et al.*, 1983). Microscopic observations indicated that the mycelium usually originated from colonized organic debris, but occasionally small sclerotia were discerned.

2. Rhizoctonia blight of warm-season grasses. The warm-season grasses are most susceptible to *R. solani* when their growth rates are reduced by cool temperatures; thus, they are commonly attacked as they approach winter dormancy and when they resume growth in the spring. Rhizoctonia blight has been reported on St. Augustinegrass, centipedegrass, zoysia, and bermudagrass. In Japan, Rhizoctonia blight in zoysia is known as "large patch". Isolates of *R. solani* from warm-season grass hosts have been characterized consistently as AG 2-2; however, recent evidence indicates that they are distinct from other known subgroups within AG 2-2. That is, they differ from subgroups AG 2-2IIIB ("mat rush types") and AG 2-2IV (commonly causing root rot of Chenopodiaceae).

Liu and Sinclair (1992) studied genetic variation among AG 2-2 and found that this subgroup was composed of genetically distinct groups based on isozyme polymorphism and DNA patterns generated by polymerase chain reaction (PCR). Turfgrass isolates from St. Augustinegrass were characterized as subgroup AG 2-2d. In a separate study, Stevens-Johnk and Jones (1993) found that AG 2-2 isolates from St. Augustinegrass were distinct from subgroups AG 2-2IIIB and AG 2-2IV, using fatty acid profiles. They designated their turf subgroups as AG 2-2 subgroup "turf". Isolates of AG 2-2 from other warm-season grasses appear similar to those from St. Augustinegrass by cultural appearance (dark brown, abundant aerial mycelium, little or no bulbil formation) and will anastomose frequently with them (Haygood and Martin, 1990). However, detailed genetic studies comparing AG 2-2 among warm-season grass host sources have not been conducted.

Symptoms of Rhizoctonia blight in warm-season grasses differ from symptoms in cool-season grasses. Patches of affected turf may enlarge to over 5 meters in diameter. These patches frequently recur in the same location over years, and also may enlarge in size within and between years. These large patches are more frequently observed in cool weather conditions in the spring as the warm-season grasses break dormancy and resume growth. In the autumn, symptoms may become apparent if weather conditions are cool, humid or wet, but patches are usually not as distinct as in spring. There frequently is a sharp demarcation between the edge of affected grass shoots and healthy appearing turf, and leaf lesions on the upper lamina of leaves are absent. Smoke ring symptoms are not induced; however, affected shoots at the margins of patches frequently are a distinct yellow, and become brown as the disease progresses to affect new shoots. Basal areas on leaf sheaths of affected grass shoots at the margins of the patches are rotted, with a slimy, water-soaked lesion that is brown, tan or purplish in color. The darker tinted lesions are most apparent during prolonged periods of cool, wet weather which continuously favors disease development. Brown or tan lesions are more common when conditions intermittently favor disease development.

Rhizoctonia blight in all of the warm-season grasses is favored by wet soil conditions, and poorly drained areas are more prone to patch development than well-drained locations. Microscopic examination of necrotic tissues in the lower leaf sheaths reveals typical *Rhizoctonia* mycelium, with differentiation of hyphae into "runner" hyphae and hyphae aggregated into infection cushion-like structures. Roots and rhizomes are not infected, but stolons may occasionally develop lesions.

Persistence of patch symptoms depends on the severity of the infections, management practices and the inherent recuperative potential of the warm-season grass species. Generally, zoysiagrass recovers very slowly, and patch symptoms may persist throughout the summer. Centipedegrass also may develop persistent symptomatic patches, due to death of stolons. Bermudagrass and St. Augustinegrass generally recover rapidly as weather and soil conditions favor growth of these grasses.

B. Rhizoctonia leaf and sheath spot

Rhizoctonia leaf and sheath spot refers to diseases of cool- and warm-season turfgrasses that are caused by *Rhizoctonia zeae* Voorhees and *R. oryzae* Ryker &

Gooch. Symptoms vary somewhat from those of Rhizoctonia blight in grasses, but enough overlap occurs that diagnosis strictly by symptoms is not advised.

In cool-season grasses, such as tall fescue, individual leaves develop nondiscrete leaf lesions that initially are water-soaked and light tan, or reddish-brown in color. A reddish brown border surrounds the lesions and separates healthy appearing tissue from diseased tissue. Leaf lesions are not discrete, and may expand to eventually blight whole leaves. On grasses maintained at low heights of cut, such as *Agrostis* spp. on putting greens, water-soaked patches may develop that are up to 50 cm in diameter. These patches are virtually indistinguishable from those induced by *R. solani,* except that distinct smoke ring symptoms are rare. Occasionally, under conditions of prolonged high relative humidities, cream-colored aerial mycelium may develop and be mistaken for *Pythium* spp. Other symptoms that have been noted for this disease include yellow, light brown or dark brown rings or arcs in turf maintained at low heights of cut.

Symptoms in warm-season grasses include basal rotting of leaf sheaths under conditions of warm, wet weather. Patches of affected turf may develop, but usually are smaller and less distinct than those induced by *R. solani* AG 2-2. Lesions on basal leaf sheaths are at first water-soaked, and become tan or light brown to reddish-brown in color. An extreme form of this disease has been observed in South Carolina (USA), with extensive blighting of centipedegrass that was stressed by prolonged periods of high heat (>32°C) and low soil moisture, followed by abundant rainfall. Infections rapidly rot leaves, which eventually turn almost black in color, killing entire lawns when most severe (Martin, unpublished).

In cool- and warm-season grasses, Rhizoctonia leaf and sheath spot is favored by weather conditions similar to those favoring Rhizoctonia blight in cool-season grasses: extended periods of leaf wetness and daytime temperatures of 28-33°C with night temperatures greater than 28°C. Rhizoctonia leaf and sheath spot may be favored under somewhat warmer conditions than Rhizoctonia blight, as the optimum temperature for growth of *R. zeae* and *R. oryzae* in culture is approximately 32°C. In South Carolina, *R. zeae* and *R. oryzae* were most frequently isolated from centipedegrass and St. Augustinegrass during the summer months, in contrast to isolations of *R. solani* AG 2-2 from these hosts primarily in the spring (Haygood and Martin, 1990).

R. oryzae and *R. zeae* were collected in tall fescue thatch as free sclerotia and isolated from colonized organic debris (Martin *et al.*, 1983). Additional information on the survival, saprophytic growth and pathogenesis of *R. oryzae* and *R. zeae* is needed and awaits investigations.

C. Yellow patch (cool-weather brown patch)

Yellow patch is a leaf and sheath blight of cool- and warm-season grasses caused by *R. cerealis* Van der Hoeven (Burpee, 1980). Susceptible grasses include species of *Agrostis, Poa, Festuca, Zoysia* and *Cynodon* (Couch, 1985). The infection process has not been studied in detail, but it is assumed to be similar to that described for *R. solani* on rice (Marshall and Rush, 1980). Infected grasses exhibit foliar chlorosis and necrosis in patches ranging from a few centimeters to over a meter in diameter (Burpee, 1980). Distinct foliar lesions are not common. Patches often develop into chlorotic or necrotic rings due to growth of symptomless foliage in the patch center. These rings may resemble fairy rings, but they lack the dense root zone hyphae often associated with the latter. *R. cerealis* has not been isolated from roots of turfgrasses, and is believed to be primarily a foliar pathogen. The fungus also causes sharp eyespot disease of cereal crops (Boerema and Verhoeven, 1977).

Isolates of *R. cerealis* are white to buff on potato-dextrose agar, with hyphae 2.8-6.2 mm in diameter (Boerema and Verhoeven, 1977). Vegetative hyphal cells are predominantly binucleate. Sclerotia, formed through the aggregation of monilioid cells, are 0.3-1.2 mm in diameter, globose to irregular, and white to yellow, turning brown with age (Boerma and Verhoeven, 1977). Some isolates fail to form sclerotia in culture (Burpee, 1980). Optimum temperature for growth in culture is approximately 23°C (Burpee, 1980).

Isolates of *R. cerealis* have been assigned to the anastomosis group AG-D (Ogoshi *et al.*, 1979) which is equivalent to the group designated CAG-1 by Burpee *et al.* (1980). This anastomosis group is homogeneous, including isolates from grasses and cereals in North America, Europe and Asia.

The teleomorph of *R. cerealis* is *Ceratobasidium gramineum* (Ikata & T. Matsura) Oniki, Ogoshi, & Araki (Oniki *et al.*, 1986). This stage has been observed rarely in nature, and the potential of basidiospores as pathogenic propagules is unknown.

D. Other Rhizoctonia-incited diseases

Binucleate *Rhizoctonia* spp. that do not fit the species concept of *R. cerealis* or anastomose with tester isolates from AG-D have been isolated from turfgrasses (Burpee and Goulty, 1984; Martin and Lucas, 1984). The few isolates tested were nonpathogenic or weakly virulent; however, more research is required on the pathology of these fungi given their ubiquitous nature. For example, a binucleate species designated *Rhizoctonia* AG-Q has been characterized as causing a disease of zoysiagrass in Japan with symptoms similar to yellow patch (Oniki *et al.*, 1986). The taxonomic relationship between this species and *R. cerealis* is unclear.

III. Disease management

Management strategies for Rhizoctonia diseases of turfgrasses can be classified under the categories of chemical, cultural, or biological control. Although an integrated approach is advised (Burpee, 1993), the value of a turfgrass sward often dictates the adoption of a particular type of control. For example, fungicide applications at 10-21 day intervals may be economically sound for control of Rhizoctonia blight on golf or bowling greens, but not on most home lawns. In the latter case, emphasis may be placed on overseeding with a disease resistant cultivar and implementing cultural controls such as increasing the mowing height and decreasing fertilizer applications.

A. Chemical control

Fungicides registered in the USA for control of Rhizoctonia blight of turfgrasses include dithiocarbamate, benzimidazole, triazole, pyrimidine, carboximide, dicarboximide, and various aromatic compounds (Couch, 1995). In general, the systemic fungicides (benzimidazoles, triazoles, pyrimidines and carboximides) provide longer term control than contact materials. Nine or more fungicide applications may be required to provide acceptable control of Rhizoctonia blight on susceptible grasses grown in regions where disease conducive conditions persist for several months (Burpee, unpublished). Disease forecasting models have been used successfully under experimental conditions to predict Rhizoctonia infection periods in creeping bentgrass and reduce the frequency of required fungicide applications (Schumann *et al.*, 1994). These models, in conjunction with improved pathogen detection methods (Miller and Martin, 1988) should increase precision in determining optimum timing for fungicide sprays.

Fungicide resistance has not become a problem with *Rhizoctonia* spp. in turfgrass, despite repeated use of at-risk materials such as benzimidazoles and triazoles. Sensitivity differences exist among species, e.g. *R. zeae* and *R. cerealis* are significantly less sensitive to benzimidazole fungicides than *R. solani* (Van der Hoeven and Bollen, 1980; Martin *et al.*, 1984) but differences among strains and anastomosis groups have not been great enough to result in control failures.

B. Cultural control

Strategies for cultural control of Rhizoctonia blight include manipulations of common turfgrass management practices such as fertilizer applications and irrigation schedules. Nitrogen (N) is the most important element affecting disease severity, with higher levels of disease observed as N is increased (Bloom and Couch, 1960; Burpee, 1995). Applications of slowly available, organic forms of N may predispose turf to less disease than faster available forms like urea or ammonium nitrate (Nelson and Craft, 1990; Soika and Sanders, 1991) but this has not been substantiated consistently by experimental results (Peacock and Daniel, 1992). Applications of phosphorus (P) and potassium (K) have decrease severity of Rhizoctonia blight in cool season grasses (Bloom and Couch, 1960) but only when nitrogen levels were relatively high. Limited information is available on the disease suppressive effects of other elements.

Timing of irrigation events plays a significant role in limiting the severity of Rhizoctonia blight, especially in cool season grasses. Long periods of foliar wetness produced by irrigation in late afternoon or early evening increase disease severity over that produced by irrigation in the morning (Fidanza, 1995). Saturated soils predisposed warm season grasses more than cool season grasses to severe Rhizoctonia blight (Green *et al.*, 1994) probably because the fungus infects the former hosts near the soil surface (crowns and leaf sheaths) rather than higher in the foliar canopy.

C. Biological control

Research on suppression of Rhizoctonia blight by living organisms includes the development of resistant cultivars of turfgrasses and studies of interactions between *R. solani* and antagonistic organisms. Resistance to *Rhizoctonia* spp. in turfgrasses has not been studied extensively. However, field and greenhouse observations indicate that susceptibility is not equivalent across species or subspecies selections. Among bentgrasses for example, *Agrostis palustris* has exhibited less severe Rhizoctonia blight than other *Agrostis* species (Doney *et al.*, 1992). In the southeast USA, differences in susceptibility of zoysiagrass and bermudagrass are significant enough that the latter is recommended as a replacement for zoysia in lawns and golf fairways where Rhizoctonia blight has been a problem (Burpee, unpublished).

Differences in susceptibility to *R. solani* have been observed among cultivars and lines of bentgrasses (Colbaugh and Engelke, 1993; Doney *et al.*, 1993), tall fescue (Burpee, 1992; Clarke *et al.*, 1985), perennial ryegrass (Vincelli and Powell, 1992) and zoysiagrass (Metz *et al.*, 1994). Although not studied in detail, these differences do not appear to be the result of a hypersensitive response to the pathogen in resistant plants. The observed resistance is probably controlled by several genes affecting various functions. For example, the density of foliar canopies among cultivars has been correlated with susceptibility to *R. solani* in tall fescue (Yuen *et al.*, 1994a). Cultivars with more dense canopies exhibited higher disease severity compared to cultivars with less dense canopies. This does not rule out the possibility however that quantitative forms of resistance to *Rhizoctonia* spp. exist in grasses. Detailed studies on differences in infection period, lesion expansion and infection efficiency of *Rhizoctonia* spp. on cultivars are warranted.

Biological suppression of Rhizoctonia blight has been demonstrated experimentally with various bacterial (Zimmerman and Hagedorn, 1994) and fungal (Burpee and Goulty, 1984; Yuen *et al.*, 1994b) inoculants. In addition, applications of some composted organic materials have suppressed the disease on cool- season grasses (Nelson and Craft, 1990; Soika and Sanders, 1991). The biological nature of this suppression remains unclear however, because of the nutritional effects of composts on plant growth and development. Commercial formulations of bacterial- and fungal-based biocontrols are currently being tested on turfgrasses (Burpee and Martin, unpublished).

Fungal endophytes in species of *Festuca* and *Lolium* confer resistance in host grasses to various herbivorous insects (Siegel *et al.*, 1985). Evidence of endophyte-mediated resistance to pathogens is less clear. Gwinn and Gavin (1992) found that seedling emergence of tall fescue cv. Kentucky 31 in soil infested with *R. zeae* was correlated with the percentage of endophyte infested seed. The greatest emergence occurred at high levels of endophyte infestation. In contrast, Burpee and Bouton (1993) noted no significant difference between endophyte-infected (EI) and endophyte-free (EF) clones of tall fescue with regard to progress of Rhizoctonia blight or recovery from blight in controlled environment studies. Similar effects were observed when EI and EF plants of the cv. Arid were compared. Further research is required to more clearly elucidate the effect of endophytes on Rhizoctonia-incited diseases of seedling and mature turfgrass plants.

IV. References

Bateman DF (1970) Pathogenesis and disease. In: Parmeter JR (ed.) Biology and Pathology of *Rhizoctonia solani* (pp 161-171) University of California Press, Berkeley.

Bloom JR & Couch HB (1960) Influence of environment on diseases of turfgrasses. I. Effect of nutrition, pH, and soil moisture on Rhizoctonia brown patch. Phytopathology 50:532-534.

Boerema GH & Verhoeven AA (1977) Checklist for scientific names of common parasitic fungi. Ser. 2b. Fungi on field crops: Cereals and grasses. Neth. J. Plant Pathol. 83:165-204.

Burpee LL (1980) *Rhizoctonia cerealis* causes yellow patch of turfgrass. Plant Dis. 64:1114-1116.

Burpee LL (1992) Assessment of resistance to *Rhizoctonia solani* in tall fescue based on disease progress and crop recovery. Plant Dis. 76:1065-1068.

Burpee LL (1993) A Guide to Integrated Control of Turfgrass Diseases. GCSAA Press, Lawrence, KS.

Burpee LL (1995) Interactions among mowing height, nitrogen fertility and cultivar affect the severity of Rhizoctonia blight of tall fescue. Plant Dis. 79:721-726.

Burpee LL & Bouton JH (1993) Effect of eradication of the endophyte *Acremonium coenophialum* on epidemics of Rhizoctonia blight in tall fescue. Plant Dis. 77:157-159.

Burpee LL & Goulty LG (1984) Suppression of brown patch disease of creeping bentgrass by isolates of nonpathogenic *Rhizoctonia* spp. Phytopathology 74:692-694.

Burpee LL & Martin SB (1992) Biology of *Rhizoctonia* species associated with turfgrasses. Plant Dis. 76:112-117.

Clarke BB, Funk CR, & Halisky PM (1985) Development of *Festuca arundinacea* Schreb. cultivars with improved resistance to *Rhizoctonia solani* Kühn. In: Lemaire L (ed.) Proc. Int. Turfgrass Res. Conf., 5th. (pp 641-646) INRA Publications, Paris.

Colbaugh PF & Engelke MC (1993) Rhizoctonia blight on N.T.E.P. bentgrasses 1992. Biol. Cult. Tests Control Plant Dis. 8:115.

Couch HB (1985) Turgrass (several cultivated spp.). Plant Dis. 69:672-675.

Couch HB (1995) Diseases of Turfgrasses. 3rd ed. Krieger Publishing Co., Malabar, FL.

Dickinson LS (1930) The effect of air temperatures on the pathogenicity of *Rhizoctonia solani* parasitizing grasses on putting-green turf. Phytopathology 20:597-608.

Doney JC, Vincelli PC & Powell AJ (1992) Reaction of bentgrasses to dollar spot and brown patch, 1991. Biol. Cult. Tests Control Plant Dis. 7:101.

Doney JC, Vincelli PC & Powell AJ (1993) Reactions of bentgrasses to dollar spot and brown patch, 1992. Biol. Cult. Tests Control Plant Dis. 8:118.

Fidanza MA (1995) Integrated strategies for brown patch *(Rhizoctonia solani* Kühn) management through environmental monitoring and cultural methods. Ph.D. Diss. University of Maryland, College Park.

Green DE, Fry JD, Pair JC & Tisserat NC (1994) Influence of management practices on Rhizoctonia large patch disease in zoysiagrass. HortSci. 29:186-188.

Gwinn KD & Gavin AM (1992) Relationship between endophyte infestation level of tall fescue seed lots and *Rhizoctonia zeae* seedling disease. Plant Dis. 76:911-914.

Haygood RA & Martin SB (1990) Characterization and pathogenicity of species of *Rhizoctonia* associated with centipedegrass and St. Augustinegrass in South Carolina. Plant Dis. 74:510-514.

Liu Z & Sinclair JB (1992) Genetic diversity of *Rhizoctonia solani* anastomosis group 2. Phytopathology 82:778-787.

Marshall DS & Rush MC (1980) Infection cushion formation on rice sheaths by *Rhizoctonia solani*. Phytopathology 70:947-950.

Martin SB, Campbell CL & Lucas LT (1983) Horizontal distribution and characterization of *Rhizoctonia* spp. in tall fescue turf. Phytopathology 73:1064-1068.

Martin SB & Lucas LT (1984) Characterization and pathogencity of *Rhizoctonia* spp. and binucleate *Rhizoctonia*-like fungi from turfgrasses in North Carolina. Phytopathology 74:170-175.

Martin SB, Lucas LT & Campbell CL (1984) Comparative sensitivity of *Rhizoctonia solani* and *Rhizoctonia*-like fungi to selected fungicides in vitro. Phytopathology 74:778-781.

Metz SP, Colbaugh PF & Engelke MC (1994) Rhizoctonia blight on inoculated zoysiagrasses. Biol. Cult. Tests Control Plant Dis. 9:158.

Miller SA & Martin RR (1988) Molecular diagnosis of plant disease. Annu. Rev. Phytopathol. 26:409-432.

Nelson EB & Craft CM (1990) Application of top-dressings amended with composts and organic fertilizers for the suppression of brown patch (*Rhizoctonia solani*) on a creeping bentgrass putting green. Phytopathology 80:121.

Ogoshi A, Oniki M, Sakai R & Ui T (1979) Anastomosis grouping among isolates of binucleate *Rhizoctonia*. Trans. Mycol. Soc. Jpn. 20:33-39.

Oniki M, Ogoshi A & Araki T (1986) *Ceratobasidium setariae, C. cornigerum,* and *C. gramineum*, the teleomorphs of the pathogenic binucleate *Rhizoctonia* fungi from gramineous plants. Trans. Mycol. Soc. Jpn. 27:147-158.

Peacock CH & Daniel PF (1992) A comparison of turfgrass response to biologically amended fertilizers. HortScience 27:883-884.

Schumann GL, Clarke BB, Rowley LV & Burpee LL (1994) Use of environmental parameters and immunoassays to predict Rhizoctonia blight and schedule fungicide applications on creeping bentgrass. Crop Prot. 13:211-218.

Shurtleff MC (1953) Brown patch of turf caused by *Rhizoctonia solani* Kühn. Ph.D. Diss. University of Minnesota, St. Paul.

Siegel MR, Latch GCM & Johnson MC (1985) *Acremonium* fungal endophytes of tall fescue and perennial ryegrass: Significance and control. Plant Dis. 69:179-183.

Soika MD & Sanders PL (1991) Effect of various nitrogen sources, organic amendments, and biological control agents on turfgrass quality and disease development, 1990. Biol. and Cult. Tests Control Plant Dis. 6:91.

Smiley RW, Dernoeden PH & Clarke BB (1992) Compendium of Turfgrass Diseases. 2nd ed. American Phytopathology Society, St. Paul, MN.

Stevens-Johnk J & Jones RK (1993) Differentiation of populations of AG 2-2 of *Rhizoctonia solani* by analysis of cellular fatty acids. Phytopathology 83:278-283.

Van der Hoeven EP & Bollen GJ (1980) Effect of benomyl on soil fungi associated with rye. 1. Effect on the incidence of sharp eyespot caused by *Rhizoctonia cerealis*. Neth. J. Plant Pathol. 86:163-280.

Vincelli PC & Powell AJ (1992) Reaction of perennial ryegrass varieties to brown patch. Biol. and Cult. Tests Control Plant Dis. 7:111.

Yuen GY, Giesler LJ, & Horst GL (1994a) Influence of canopy structure on tall fescue cultivar susceptibility to brown patch disease. Crop Prot. 13:439-442.

Yuen GY, Craig ML & Giesler LJ (1994b) Biological control of *Rhizoctonia solani* on tall fescue using fungal antagonists. Plant Dis. 78:118-123.

Zimmerman KK & Hagedorn (1994) Biological control of brown patch, 1992. Biol. and Cult. Tests Control Plant Dis. 9:155.

V.13. VEGETABLE DISEASES INCITED BY *RHIZOCTONIA* SPP.

CHIN-CHYU TU, TING-FANG HSIEH AND YIH-CHANG CHANG

Taiwan Agricultural Research Institute, 189 Chung-cheng Road, Wu-feng 413, Taichung, TAIWAN, REPUBLIC OF CHINA

I. INTRODUCTION

Among *Rhizoctonia* species, the following have been reported to cause diseases on vegetable crops. *R. solani* AGs 1 to AG 5 and binucleate *Rhizoctonia* AGs-E, -F and -R. Different symptoms are caused by different specified *Rhizoctonia* isolates. For examples, aerial blight or web blight are caused by *R. solani* AG 1 (Yang *et al.*, 1990), AG 2-3 (Naito and Kanematsu, 1994) and AG 3 (Date *et al.*, 1984). Damping-off, stem or root rot are mainly caused by AG 2 (Naiki and Kanoh, 1978; Liu and Sinclair, 1991), AG 4 (Win and Sumner, 1988) and AG 5 (Naiki and Ui, 1981). The hypocotyl rot is induced by AGs 1, 2, 4, and E (Warren *et al.*, 1972; Galindo *et al.*, 1982; Ploetz *et al.*, 1985). In order to control diseases caused by different AGs of *Rhizoctonia* spp., the ecological manner of different *Rhizoctonia* isolates will be seriously considered and realized. In this paper, the *Rhizoctonia* diseases of major vegetable crops are demonstrated.

II. RHIZOCTONIA DISEASES OF VEGETABLE CROPS

The host plants, disease symptoms, anastomosis groups, and representative references are listed in table 1.

A. Papilionaceae

The Papilionaceae family is of great importance in providing some of the major vegetable crops in the world. Peas and beans are included in this family. Rhizoctonia disease was first described on common beans and lima beans in the southern USA (Weber, 1939). It has also been found in the north of USA (Dillard, 1987). Other records include Puerto Rico, Nicaragua, El Salvador, Costa Rica (Echandi, 1965; Galindo *et al.*, 1983a; b), Panama, Venezuela, Brazil, Argentina, the Philippines, India, Sri Lanka and Japan (Snowdon, 1991).

R. solani persists in soil in the form of sclerotia and may also produce the sexual stage. Cool and moist conditions tend to favor seedling blight, basal stem rot and root rot , which are characteristic symptoms in the temperate growing areas. Most of the *R. solani* isolates associated with Rhizoctonia diseases of bean belong to AG 4, but the isolates of AGs 1 and 2 have also been reported (Galindo *et al.*, 1982). Pods in contact with the ground may also be invaded (Dillard, 1987). In the humid tropics, however, the fungus is also capable to attack the aerial parts of mature bean plants, causing 'web blight' or 'aerial blight' (Weber, 1939; O'Neill *et al.*, 1977; Galindo *et al.*, 1983a; Dillard, 1987; Yang *et al.*, 1990). Initial infection may be incited by airborne basidiospores (Echandi, 1965), the sclerotia and hyphal fragments, which are dispersed by wind, rain, irrigation water, animals and farm machinery (Weber, 1939), and soil splash (Galindo *et al.*, 1983a).

B. Sneh et al. (eds.),
Rhizoctonia Species: Taxonomy, Molecular Biology, Ecology, Pathology and Disease Constrol, 369–377.

TABLE 1. List of vegetable crop diseases caused by *Rhizoctonia* isolates.

Host	Symptoms	Anastomosis group	Reference
Papilionaceae			
Soybean (*Glycine max*. L.)	aerial blight web blight	AG 1-IA AG 1-IB	O'Neill *et al*., 1977; Yang *et al*., 1990
	hypocotyl rot	AG 4, AG-E (=CAG 3)	Ploetz *et al*., 1985
	stem rot, root rot	AG 2-2IIIB	Liu & Sinclair, 1991
	foliar blight	AG 2-3	Naito & Kanematsu, 1994
	damping-off, root, stem decay, bud, leaf blight		Sinclair & Backman, 1989
Red bean or azuki bean [*Phaseolus angularis* (Willd) W.F. Wight]	root rot	AG 4	Liu, 1991
Snap bean (*P. vulgaris* L.)	hypocotyl rot	AGs 1, 2, 4	Warren *et al*., 1972; Galindo *et al*., 1982
	root rot	AGs 2-2, 4	Sumner, 1985
	web blight	AG 1-IB	Echandi, 1965; Galindo *et al*., 1983ab
Lima bean (*P. lunatus* L.)	pod rot, bean rot	AG 1	Dillard & Wallis, 1986
	web blight	AG 1-IB	Weber, 1939; Dillard, 1987
	root rot	AGs 2-2, 4	Sumner, 1985
		AGs-E, -R (CAGs 3, 5)	Sumner, 1985
Southern pea or cowpea (*Vigna unguiculata* (L.) Walp.)	root rot	AGs 2-1, 2-2, 4, AGs-E, -R (CAGs 3, 5)	Sumner *et al*., 1978; Sumner, 1985
Pea, French bean (*Pisum sativum* L.)	epicotyl rot, tip blight		McCoy & Kraft, 1984
		AGs-E, -F, -R (CAGs 3, 4, 5)	Burpee *et al*., 1980
Solanaceae			
Tomato (*Lycopersicon esculentum* Mill.)	soil rot		Murphy *et al*., 1984
	fruit rot		Schroeder & Provvidenti, 1961; Strashnov *et al*., 1985
	leaf blight	AG 3	Date *et al*., 1984
Pepper (*Capsicum frutescens* L.)	damping-off		Katan & Eshel, 1974
Eggplant (*Solanum melongena* L.)	brown spot	AG 3	Kodama *et al*., 1982

TABLE 1. Cont'd

Host	Symptoms	Anastomosis group	Reference
Radish (*Raphanus sativus* L.)	root canker	AGs -1, 2-2	Grisham & Anderson, 1983
		AGs-E, -F, -R (CAGs 3, 4, 5)	Burpee *et al.*, 1980
	brown spot, brown lesion	AGs 2-2, 4 ,2-1	Homma & Ishii, 1984
Rape (*Brassica campestris* L.)	root rot		Barkenkamp & Vaartnou, 1972
	damping-off	AGs 2-1, 4	Kataria *et al.*, 1991
Cauliflower (*B. oleracea* var. *botrytis* L.)	damping-off		Kundu & Nandi, 1985
Cabbage (*B. oleracea* var. *capitota* L.)	foliar blight	AG 1	Abawi & Martin, 1985
	damping-off, wire stem, root rot, bottom rot		Baker, 1970
	rot	AG 1-IB	Mukou *et al.*, 1975
Kale (*B. alboglabra* Bailey)	damping-off	AG 4	Huang & Yang, 1992
Umbelliferae			
Celery (*Apium graveolens* L.)	crater rot		Houston & Kendrick, 1949; Pieczarka, 1981
Carrot (*Daucus carota* var. *sativa)*	crown rot canker	AG 2	Mildenhall & Williams, 1970
	crater rot	AG 2-2	Igarashi & Baba, 1976
	damping-off damping-off & canker	AGs 1, 4, 2-2	Grisham & Anderson, 1983
Compositae			
Lettuce (*Lactuca sativa* L.)	bottom rot	AG 1-IB	Pieczarka & Lorbeer, 1975; Mahr *et al.*, 1986; Wareing *et al.*, 1986
	bottom rot	AGs 1-IB, 1-IC, 2-1	Herr, 1992
Chenopodiaceae			
Spinach (*Spinacia oleracea* L.)	damping-off, foot rot	AGs 1, 2-1, 2-2, 4, 5	Naiki & Kanoh, 1978
Cucurbitaceae			
Cucumber (*Cucumis sativus* L.)	damping-off, fruit rot, soil rot, belly rot	AG 4	Fukutomi & Takada, 1979
			Sumner & Smittle, 1976; Lewis & Papavizas, 1980
Pumpkin (*Cucurbita pepo* L.)	spongy rot		Gangopadhyay & Sharma, 1976
Muskmelon (*Cucumis melo* var. *reticulatus)*	seed rot		Maholay & Sohi, 1983

TABLE 1. Cont'd

Host	Symptoms	Anastomosis group	Reference
Bittergourd (*Momordica charantia* L.)	seed rot		Maholay & Sohi, 1983
Liliaceae			
Onion (*Allium cepa* L.)		AGs-E,-F,-R (CAG 3,-4, 5)	Burpee *et al.*, 1980
	damping-off	AG 4	Yamamoto & Uehara, 1972

The symptoms of soybean diseases caused by *R. solani* include damping-off, stem and root decay (Sinclair and Backman, 1989), stem and root rot (Liu and Sinclair, 1991), hypocotyl rot (Ploetz *et al.*, 1985), and aerial or web blight (O'Neill *et al.*, 1977; Yang *et al.*, 1990).

Root rot, hypocotyl rot, and web blight symptoms of snap beans were observed in the field. Web blight symptoms were caused by *R. solani* AG 1-IB (Echandi, 1965; Galindo *et al.*, 1983a; b), while root and hypocotyl rot were caused by AGs 1, 2, and 4 (Warren *et al.*, 1972; Galindo *et al.*, 1982; Sumner, 1985).

R. solani AG 1 also caused pod or bean rot (Dillard and Wallis, 1986) or web blight (Weber, 1939; Dillard, 1987) of lima bean , and AG 2-2 and AG 4 caused root rot (Sumner, 1985). The root rot of adzuki bean was caused by AG 4 (Liu, 1991). *R. solani* AGs 2-1, 2-2, and 4, as well as binucleate *Rhizoctonia* AG-E (=CAG 3) and AG-R (=CAG 5) were highly virulent on cowpea cultivars (Sumner, 1985). McCoy and Kraft (1984) found that *R. solani* is the cause of cotyledon rot, tip blight, and epicotyl rot of peas. Moreover, binucleate *Rhizoctonia* AG-E, AG-F (=CAG 4) and AG-R were also virulent and caused seed, root and/or hypocotyl decay of pea (Burpee *et al.*, 1980).

B. Solanaceae

The Solanaceous fruit vegetables include tomatoes, peppers and eggplants. Soil rot or fruit rot is of particular importance on tomatoes, and heavy losses have been reported on the fresh and market fruit (Baker, 1970) and also in the consignments for processing (Schroeder and Provvidenti, 1961). Most reports are from the southern USA, but the disease also occurs in many other countries, including Israel (Strashnov *et al.*, 1985), India, Iran and Brazil (Snowdon, 1991). Peppers and eggplants are occasionally affected. Under wet conditions, *R. solani* AG 4 can attack fruits in contact with the ground by direct penetration or via injuries (Murphy *et al.*, 1984; Strashnov *et al.*, 1985). Less frequently, the infection may occur as a result of soil being splashed on to low-hanging fruits (Baker, 1970), or alternatively from airborne basidiospores (Schroeder and Provvidenti, 1961). The pathogen can penetrate into healthy fruit, which are in contact with the diseased ones (Strashnov *et al.*, 1985). The pathogen was also the major causal organism of damping-off in directly seeded pepper (Katan and Eshel, 1974).

C. Cruciferae

Cabbage, kale, cauliflower, rape, and Chinese cabbage belong to the family Cruciferae. Rhizoctonia disease has been reported on cabbage in Germany, various parts of the USA (Abawi and Martin, 1985), and Western Canada (Berkenkamp and Vaartnou, 1972). It has also been observed on cauliflower in Italy (D'ercole and

Nipoti, 1978), and on Chinese cabbage in the USA, South Korea and West Indies (Rajnauth, 1987).

R. solani is known to cause several diseases such as damping-off, wire stem, root rot, bottom rot , and basal rot in cabbages and other crucifers (Baker, 1970). Isolates of *R. solani* AG 1 caused the aerial blight and rot of cabbages (Mukou *et al.*, 1975; Abawi and Martin, 1985). Isolates of AG 4 caused damping-off of kale and cauliflower (Kundu and Nandi, 1985; Huang and Yang, 1992) and root rot of rape (Berkenkamp and Vaartnou, 1972).

D. Cucurbitaceae

The Cucurbitaceae family includes a number of important crop plants. Cucumber, melon, watermelon, squash, and pumpkin are popular vegetables in many countries. Cucumber thrives well in the tropics and subtropics but has also become important in cooler latitudes where it is grown under glass or plastic structures (Snowdon, 1991). Melon and watermelon are climacteric fruits.

Diseases caused by *R. solani* occur in warm and moist growing regions. They are occasionally found on melon and watermelon, but are more important on cucumbers (Sumner and Smittle, 1976; Fukutomi and Takada, 1979), for example, in the southern USA (Lewis and Papavizas, 1980) and parts of the USSR. Pumpkin (Gangopadhyay and Sharma, 1976), bittergourd and muskmelons (Maholay and Sohi, 1983) are attacked in India. Symptoms have been reported as damping-off, fruit rot, soil rot and belly rot on cucumbers (Sumner and Smittle, 1976; Fukutomi and Takada, 1979; Lewis and Papavizas, 1980), seed rot on bittergourd and muskmelons (Maholay and Sohi, 1983), damping-off on watermelon (Hopkins and Elmstrom, 1974), and spongy rot on pumpkin (Gangopadhyay and Sharma, 1976).

E. Other vegetable crops

Celery and carrot belong to the Umbelliferae family. Lettuce belongs to the Compositae family. Spinach is a member of the Chenopodiaceae family. Onion belongs to the Liliaceae family. These vegetable crops also have been reported to be attacked by *R. solani* in many regions. Crater rot of celery has been reported in the UK and the USA (Houston and Kendrick, 1949; Pieczarka, 1981). Plant tissues below ground level may be directly invaded by this fungus, which cause oval brown sunken lesions on the lower parts of the stalks (Houston and Kendrick, 1949). However, damping-off, canker, crown rot, crater rot of carrot are caused by *R. solani* AGs 1, 2-2, and 4 (Mildenhall and Williams, 1970; Igarashi and Baba, 1976; Grisham and Anderson, 1983).

Bottom rot of lettuce appears to be of particular importance in lettuce grown under glass or plastic structures in Europe, whereas in North America it has also caused losses in outdoor crops. The disease has been studied in Austria, France, the Netherlands (Snowdon, 1991), the UK (Wareing *et al.*, 1986), and the USA (Pieczarka and Lorbeer, 1975; Mahr *et al.*, 1986). Bottom rot is a disease complex, often involving several fungi and bacteria (Pieczarka and Lorbeer, 1975; Wareing *et al.*, 1986).

Damping-off and foot rot diseases of spinach caused by *R. solani* AGs 1, 2-1, 2-2, 4, and 5 were reported in Japan (Naiki and Kanoh, 1978). Burpee *et al.* (1980) reported that binucleate *Rhizoctonia* AGs -E, -F, and -R (CAGs 3, 4, and 5) were virulent on onion.

III. Concluding remarks

R. solani causes severe diseases on economically important vegetable crop plants, especially in the subtropic area. Different isolates of *R. solani* of the different anastomosis groups cause various symptoms on a variety of host plants. Understanding of the anastomosis groups as the particular causal agents and their ecology in soil will help us to develop effective control measures to suppress the diseases. Some methods have been used to control Rhizoctonia diseases on vegetable crops. They are: a) cultural control: shallow planting (Leach and Garber, 1970), crop rotation (Burke and Draft, 1974), deep plowing (Lewis and Papavizas, 1980), organic amendments (Kundu and Nandi, 1985) are recommended to control Rhizoctonia diseases of vegetable crops; b) chemical control: fungicides that are effective against *R. solani* and frequently used include PCNB (McCarter and Barksdale, 1977; Sciumbato and Hegwood, 1979), benomyl (Hopkins and Elmstrom, 1974), chloroneb (Hopkins and Elmstrom, 1974; Sumner and Smittle, 1976; McCoy and Kraft, 1984), chlorothalonil (Sumner and Smittle, 1976), iprodione (Mahr *et al.*, 1986), and Tolchofos-methyl (Win and Sumner, 1988; Kataria *et al.*, 1991); c) plant resistance: physical (Bassi *et al.*, 1979; McCoy and Kraft, 1984; Yang *et al.*, 1992) and chemical (Zaumeyer and Meiners, 1975; Prasad and Weigle, 1976) properties of vegetable crops correlated with resistance to Rhizoctonia diseases have been studied; d) biological control: biological agents, such as *Tricoderma* spp. (Harman *et al.*, 1980; Liu, 1991), *Gliocladium virens* (Tu and Vaartaja, 1981; Lumsden and Locke, 1989), *Laetisaria arvalis* (Conway, 1986; Lewis and Papavizas, 1992), *Bacillus cereus* (Wu, 1992), a nonpathogenic isolate of *R. solani* AG 4 (Ichielevich-Auster *et al.*, 1985) and one binucleate *Rhizoctonia* spp. (Cardoso and Echandi, 1987) could reduce the Rhizoctonia diseases in the greenhouse and/or field tests; and e) integrated control: Lewis and Papavizas (1980) demonstrated an integrated pest management system to reduce cucumber fruit rot severity caused by *R. solani*.

Two strategies of disease control are required for reducing the primary inoculum source in the soil and decreasing the disease outbreak. The first strategy includes: a) to obtain the pathogen-free seeds or propagative stocks, b) to treat the seed and soil with fungicides, c) to apply routinely various sanitation processes, d) to use some of cultural control practices, and e) to employ biocontrol agents for disease control. The second strategy includes: a) planting disease resistant cultivars, b) application of fungicides in the cultural system, and c) combination of different methods to achieve integrated control. However, more extensive studies on the ecology of *Rhizoctonia* isolates are still needed to improve the methods for control of Rhizoctonia diseases.

IV References

Abawi GS & Martin SB (1985) Rhizoctonia foliar blight of cabbage in New York State. Plant Dis. 69:158-161.

Baker KF (1970) Types of Rhizoctonia diseases and their occurrence. In: Parmeter JR Jr.(ed.) *Rhizoctonia solani*: Biology and Pathology. (pp 125-148) University of California Press, Berkeley. CA, USA.

Bassi A Jr., Moore EL & Batson WE Jr. (1979) Histopathology of resistant and susceptible tomato fruit infected with *Rhizoctonia solani*. Phytopathology 69:556-559.

Berkenkamp B & Vaartnou H (1972) Fungi associated with rape root rot in Alberta. Can. J. Plant Sci. 52:973-976.

Burke DW & Kraft JM (1974) Responses of beans and peas to root pathogens accumulated during monoculture of each crop species. Phytopathology 64:546-549.

Burpee LL, Sanders PL, Cole H Jr. & Sherwood RT (1980) Pathogenicity of *Ceratobasidium cornigerum* and related fungi representing four anastomosis groups. Phytopathology 70:843-846.

Cardoso JE & Echandi E (1987) Biological control of Rhizoctonia root rot of snap bean with binucleate *Rhizoctonia*-like fungi. Plant Dis. 71:167-170.

Conway KE (1986) Use of fluid-drilling gels to deliver biological control agents to soil. Plant Dis. 70:835-839.

Date H, Yagi S, Okamoto Y & Oniki M (1984) On the leaf blight of tomatoes by *Thanatephorus cucumeris* (Frank) Donk (*Rhizoctonia solani*). Ann. Phytopath. Soc. Japan 50:339.

D'ercole N & Nipoti P (1978) La rizottoniosi del cvolfiore. 3. Modalita di penetrazione dell'agente patogeno (*Rhizoctonia solani* Kühn). Informatore Fitopatologico 28:21-23.

Dillard HR (1987) Characterization of isolates of *Rhizoctonia solani* from lima beans grown in New York State. Phytopathology 77:748-751.

Dillard HR & Wallis G (1986) Rhizoctonia pod and bean rot of baby lima bean in New York State. Phytopathology 76:651 (Abstr.).

Echandi E (1965) Basidiospore infection by *Pellicularia filamentosa* (= *Corticium microsclerotia*), the incitant of web blight of common bean. Phytopathology 55:698-699.

Fukutomi M & Takada H (1979) Ultrastructure of infection process of *Rhizoctonia solani* Kühn in cucumber hypocotyls. Ann. Phytopathol. Soc. Japan 45:453-462.

Galindo JJ, Abawi GS & Thurston HD (1982) Variability among isolates of *Rhizoctonia solani* associated with snap bean hypocotyls and soil in New York. Plant Dis. 66:390-394.

Galindo JJ, Abawi GS, Thurston HD & Galvez G (1983a) Source of inoculum and development of bean web blight in Costa Rica. Plant Dis. 67:1016-1021.

Galindo JJ, Abawi GS, Thurston HD & Galvez G (1983b) Effect of mulching on web blight of beans in Costa Rica. Phytopathology 73:610-615.

Gangopadhyay S & Sharma RK (1976) Spongy rot of pumpkin. Indian Phytopathol. 29:423-424.

Grisham MP & Anderson NA (1983) Pathogenicity and host specificity of *Rhizoctonia solani* isolated from carrots. Phytopathology 73:1564-1569.

Harman GE, Chet I & Baker R (1980) *Trichoderma hamatum* effects on seed and seedling disease induced in radish and pea by *Pythium* spp. or *Rhizoctonia solani*. Phytopathology 70:1167-1172.

Herr LJ (1992) Characteristics of *Rhizoctonia* isolates associated with bottom rot of lettuce in organic soils in Ohio. Phytopathology 82:1046-1050.

Homma Y & Ishii M (1984) Anastomosis groups of *Rhizoctonia solani* Kühn responsible for various symptoms of the root rot of Japanese radish. Bull. Shikoku. Agric. Exp. Stn. 42:1-11.

Hopkins DL & Elmstrom GW (1974) Chemical control of watermelon damping-off and seedling wilt. Plant Dis. Reptr. 58:114-117.

Houston BR & Kendrick JB (1949) A crater spot of celery petioles caused by *Rhizoctonia solani*. Phytopathology 39:470-474.

Huang JW & Yang SH (1992) A baiting technique for assay of *Rhizoctonia solani* in kale nurseries. Plant Pathol. Bull. (Taiwan) 1:26-30.

Ichielevich-Auster M, Sneh B, Barash I & Koltin Y (1985) Suppression of damping-off caused by *Rhizoctonia* spp. by a nonpathogenic isolate of *R. solani*. Phytopathology 75:1080-1084.

Igarashi F & Baba T (1976) On the crater rot of carrot (preliminary report). Ann. Phytopathol. Soc. Japan 42:97.

Katan J & Eshel Y (1974) Effect of the herbicide diphenamid on damping-off disease of pepper and tomato. Phytopathology 64:1186-1192.

Kataria HR, Verma PR & Gisi U (1991) Variability in the sensitivity of *Rhizoctonia solani* anastomosis groups to fungicides. J. Phytopathol. 133:121-133.

Kodama T, Horimoto K & Ogoshi A (1982) On the brown spot of eggplants caused by *Thanatephorus cucumeris* (Frank) Donk (*Rhizoctonia solani*) AG 3. Ann. Phytopathol. Soc. Japan 48:356.

Kundu PK & Nandi B (1985) Control of Rhizoctonia disease of cauliflower by competitive inhibition of the pathogen using organic amendment in soil. Plant and Soil 83:357-362.

Leach LD & Garber RH (1970) Control of *Rhizoctonia*. In: Parmeter JR Jr.(ed.) *Rhizoctonia solani*: Biology and Pathology. (pp. 125-148) University of California Press, Berkeley. CA, USA.

Lewis JA & Papavizas GC (1980) Integrated control of Rhizoctonia fruit rot of cucumber. Phytopathology 70:85-89.

Lewis JA & Papavizas GC (1992) Potential of *Laetisaria arvalis* for the biocontrol of *Rhizoctonia solani*. Soil Biol. Biochem. 24:1075-1079.

Liu SD (1991) Biological control of adzuki-bean root rot disease caused by *Rhizoctonia solani*. Plant Prot. Bull. (Taiwan) 33:63-71.

Liu Z & Sinclair JB (1991) Isolates of *Rhizoctonia solani* anastomosis group 2-2 pathogenic to soybean. Plant Dis. 75:682-687.

Lumsden RD & Locke JC (1989) Biological control of damping-off caused by *Pythium ultimum* and *Rhizoctonia solani* with *Gliocladium virens* in soilless mix. Phytopathology 79:361-366.

Maholay MN & Sohi HS (1983) Rhizoctonia seed rot of bittergourd and muskmelon. Indian J. Mycol. & Plant Pathol. 13:1-3.

Mahr SER, Stevenson WR & Sequeira L (1986) Control of bottom rot of head lettuce with iprodione. Plant Dis. 70:506-509.

McCarter SM & Barksdale TH (1977) Effectiveness of selected chemicals in controlling Rhizoctonia fruit rot of tomato in greenhouse and field tests. Plant Dis. Reptr. 61:811-815.

McCoy RJ & Kraft JM (1984) Resistance in *Pisum sativum* to epicotyl rot caused by *Rhizoctonia solani*. Plant Dis. 68:491-493.

Mildenhall JP & Williams PH (1970) Rhizoctonia crown rot and cavity spot of muck-grown carrots. Phytopathology 60:887-890.

Mukou H, Kagiwata T & Sueyama K (1975) Rot of cabbage by *Rhizoctonia solani*. Ann. Phytopath. Soc. Japan 41:82. (Abstr.)

Murphy JB, McFerran J & Goode MJ (1984) The effects of genotype and ethephon on Rhizoctonia soil rot of processing tomatoes. HortScience 19:676-677.

Naiki T & Kanoh M (1978) Grouping of *Rhizoctonia solani* Kühn causing root diseases of spinach in plastic house cropping. Ann. Phytopath. Soc. Japan 44:554-560.

Naiki T & Ui T (1981) Rhizoctonia root rot of bean, soybean and adzuki bean seedlings. Mem. Fac. Agric. Hokkaido Univ. 12:262-269.

Naito S & Kanematsu S (1994) Characterization and pathogenicity of a new anastomosis subgroup AG 2-3 of *Rhizoztonia solani* Kühn isolated from leaves of soybean. Ann. Phytopath. Soc. Japan 60:681-690.

O'Neill NR, Rush MC, Horn NL & Carver RB (1977) Aerial blight of soybeans caused by *Rhizoctonia solani*. Plant Dis. Reptr. 61:713-717.

Pieczarka DJ (1981) Shallow planting and fungicide application to control Rhizoctonia stalk rot of celery. Plant Dis. 65:879-880.

Pieczarka DJ & Lorbeer JW (1975) Microorganisms associated with bottom rot of lettuce grown on organic soil in New York State. Phytopathology 65:16-21.

Ploetz RC, Mitchell DJ & Gallahan RN (1985) Characterization and pathogenicity of *Rhizoctonia* species from a reduced-tillage experiment multicropped to rye and soybean in Florida. Phytopathology 75:833-839.

Prasad K & Weigle JL (1976) Association of seed coat factors with resistance to *Rhizoctonia solani* in *Phaseolus vulgaris*. Phytopathology 66:342-345.

Rajnauth G (1987) Web blight, an important disease of been and pak-choi in Trinidad. Tropical Agriculture 64:356-358.

Schroeder WT & Provvidenti R (1961) Rhizoctonia fruit rot of processing tomatoes. Plant Dis. Reptr. 45:160-163.

Sciumbato GL & Hegwood CP Jr. (1979) Use of elevated fungicide rates to control cucumer fruit rot under multiple harvesting conditions,. Plant Dis. Reptr. 63:482-485.

Sinclair JB & Backman PA (eds.) (1989) Compendium of soybean diseases 3rd ed. APS, St. Paul, Minnesota. 106pp.

Snowdon AL (1991) A colour atlas of post-harvest diseases & disorders of fruits & vegetables. Vol. 2: Vegetables. Wolfe Scientific, Aylesbury, England.

Strashnov Y, Elad Y, Sivan A, Rudich Y & Chet I (1985) Control of *Rhizoctonia solani* fruit rot of tomatoes by *Trichoderma harzianum* Rifai. Crop Protection 4:359-364.

Sumner DR (1985) Virulence of anastomosis groups of *Rhizoctonia solani* and *Rhizoctonia*-like fungi on selected germ plasm of snap bean, lima bean, and cowpea. Plant Dis. 69:25-27.

Sumner DR, Johnson AW, Glaze NC & Dowler CC (1978) Root diseases of snapbean and southern pea in intensive cropping systems. Phytopathology 68:955-961.

Sumner DR & Smittle DA (1976) Etiology and control of fruit rot of cucumber in single harvesting for pickles. Plant Dis. Reptr. 60:304-307.

Tu JC & Vaartaja O (1981) The effect of the hyperparasite (*Gliocladium virens*) on *Rhizoctonia solani* and on Rhizoctonia root rot of white beans. Can. J. Bot. 59:22-27.

Wareing PW, Wang ZN, Coley-Smith JR & Jeves TM (1986) Fungal pathogens in rotted basal leaves of lettuce in Humberside and Lancashire with particular reference to *Rhizoctonia solani*. Plant Pathol. 35:390-395.

Warren HL, Helfrich RM & Blount VL (1972) Evaluation of lima bean varieties and selections for resistance to *Rhizoctonia solani*. Plant Dis. Reptr. 56:268-270.

Weber GF (1939) Web-blight, a disease of beans caused by *Corticium microsclerotia*. Phytopathology 29:559-575.

Win HH & Sumner DR (1988) Root rot induced in snap bean by *Rhizoctonia solani* AG 4 and AG 2 type 2 in conservative tillage following corn. Plant Dis. 72:1049-1053.

Wu WS (1992) The application of bioagents to control *Rhizoctonia solani* and other soil-borne plant pathogens. Plant Pathol. Bull. (Taiwan) 1:1-12.

Yamamoto DT & Uehara H (1972) Seedling damping-off of welsh onion and the control. Plant Prot., Japan 26:153-159.

Yang J, Verma PR & Lees GL (1992) The role of cuticle and epidermal cell wall in resistance of rapeseed and mustard to *Rhizoctonia solani*. Plant and Soil 142:315-321.

Yang XB, Berggren GT & Snow JP (1990) Seedling infection of soybean by isolates of *Rhizoctonia solani* AG 1, causal agent of aerial blight and web blight of soybean. Plant Dis. 74:485-488.

Zaumeyer WJ & Meiners JP (1975) Disease resistance in beans. Annu. Rev. Phytopathol. 13:313-334.

V.14 THE MYCORRHIZAL SPECIES OF *RHIZOCTONIA*

TORBEN F. ANDERSEN[1] AND HANNE N. RASMUSSEN[2]

[1]*Botanical Laboratory, University of Copenhagen, Gothersgade 140, DK-1123 Copenhagen K, DENMARK.* [2]*Research Centre Aarslev, Danish Institute of Plant and Soil Science, Kirstinebjergvej 10, DK-5792 Aarslev, DENMARK.*

I. INTRODUCTION

The form genus *Rhizoctonia* DC. not only contains a number of species known to be troublesome pathogens, but also some that are beneficial to plants through the formation of mycorrhiza. Most plants living in symbiosis with *Rhizoctonia* are found in the family Orchidaceae; other examples are members of *Gentiana* (Gentianaceae) and bryophytes belonging to the genus *Aneura*. These types of mycorrhiza appear to be physiologically similar to orchid mycorrhiza (Stahl, 1949), which is by far the best described.

All orchids are mycorrhizal plants and usually associated with species of *Rhizoctonia* although a few, mainly chlorophyll deficient species host other fungi, such as *Armillaria mellea* (Vahl:Fr.) Kumm. (Terashita, 1985) [see also Currah and Zelmer (1992)]. Numerous strains of *Rhizoctonia* species have been isolated from orchids, and a certain level of specificity between orchid and fungus is suggested.

Data mainly relating to the orchid *Rhizoctonia* interaction will be presented in this chapter. Beginning with a brief historic outline, there will be a description of the structures and functions of the mycorrhiza, followed by a section on the taxonomy of the endophytes. Patterns of specificity in the association with orchids are discussed. A short review of other mycorrhizal interactions involving *Rhizoctonia* will be given and, finally, the most important issues for future research will be evaluated.

II. HISTORY OF ORCHID MYCORRHIZA

According to Ramsbottom (1923), anatomical drawings by Link (1840) were the first recorded observations of hyphal structures in the cells of an orchid seedling. However, Link did not describe the structures as hyphae and in our opinion his illustration could represent starch grains.

Irmisch (1853) later explicitly described fungal hyphae in the rhizome system of the chlorophyll deficient orchid *Corallorhiza innata* R.Br. (= *C. trifida* Chatelain). Although the nutritional status of such orchids was extensively discussed, no linkage with the presence of intracellular hyphae was made.

A functional relationship between orchid nutrition and the presence of endophytic hyphae was first suggested by Frank (1891). Based on experience with other types of mycorrhiza, Frank assumed that the relationship was mutualistic, and suggested that the endophytes would be difficult or impossible to cultivate outside of the orchid because of specific nutritional requirements. However, isolation of orchid endophytes and the *in vitro* synthesis of mycorrhizal seedlings was achieved by Bernard (1909)

B. Sneh et al. (eds.),
Rhizoctonia Species: Taxonomy, Molecular Biology, Ecology, Pathology and Disease Constrol, 379–390.

and Burgeff (1909). Bernard (1909) referred the orchid endophytes to *Rhizoctonia* in describing the three first species, i.e., *R. repens*, *R. mucoroides*, and *R. lanuginosa*. Other anamorphic species occurring as orchid endophytes were subsequently described by Costantin and Dufour (1920), Burgeff (1932; 1936) and Curtis (1939). The teleomorphs of a number of rhizoctonias associated with orchids were induced and described by Warcup and Talbot (1966; 1967; 1971; 1980), thus providing a better basis for discussion of specificity patterns between orchids and endophytes (Warcup, 1973; 1981). Recently, several new taxa of *Rhizoctonia* from a number of North American orchids have been described or validated (Currah, 1987; Currah *et al.*, 1987; 1988; 1990; Currah and Zelmer, 1992; Zelmer and Currah, 1996).

III. PHYSIOLOGY OF ORCHID MYCORRHIZA

A. *Rhizoctonia* and germination of orchid seeds.

All orchids require fungal infection to complete their life cycle (Rasmussen, 1994). Orchid seedlings are heterotrophic to varying extents and obtain carbohydrates and other organic nutrients from the endophyte. In some orchid species there are mechanisms to delay germination until the presence of mycelia of a compatible endophyte induces or significantly increases germination. The primary signals have not been identified, but an array of compounds, such as ethylene and various vitamins, stimulate seed germination in orchids (Rasmussen, 1995). They are produced by strains of *Rhizoctonia* in culture (Harvais and Pekkala, 1975; Hanke and Dollwet, 1976; Barroso *et al.*, 1986b), and could be involved in the interaction.

After Bernard's and Burgeff's initial success with symbiotic seed germination *in vitro*, techniques were developed to produce many orchid species on nutrient substrates without fungi (asymbiotic), primarily by Knudson (e.g. 1922). However, in recent years the symbiotic propagation technique has regained popularity because many orchids are difficult to propagate asymbiotically, and the mycorrhizal plants develop more rapidly and are more readily established after replanting (e.g., Zettler and McInnis, 1992). This has resulted in a renewed interest in the physiology of the orchid - *Rhizoctonia* association.

B. Invasion of orchid tissues

Compatible infection by *Rhizoctonia* in orchid tissue is restricted to the roots and rhizomes that are located below ground or, in epiphytic orchids, prostrate on the substrate. Cortical tissues are the main sites of infection. Precocious infection of the orchid seed is apparently checked by production of phenolics in the suspensor region (Rasmussen, 1990). There are few other signs of defense reactions in the infected tissue and no cell mortality arises.

Infection follows simple cell wall penetration by solitary hyphae which occurs mostly through absorptive structures such as the suspensor end of the seed, rhizoids of seedlings or root hairs. In a few cases, compatible infection takes place through epidermal cells other than the root hair cells (Rasmussen, 1995). No infection cushions are formed, although the same strains may produce appressoria during pathogenic infection of other plant hosts (Williamson and Hadley, 1970).

C. Histology of infected tissue

The intracellular hyphae form dense coils (pelotons) that may occupy major parts of the plant cell lumen (Fig. 1). After a period of build-up the peloton is broken down to an aggregated clump. Infection and lysis may take place repeatedly in the same cell and tissues, suggesting that there is a rhythmic or seasonal alternation between periods conducive to intracellular hyphal growth and stages of breakdown (digestion).

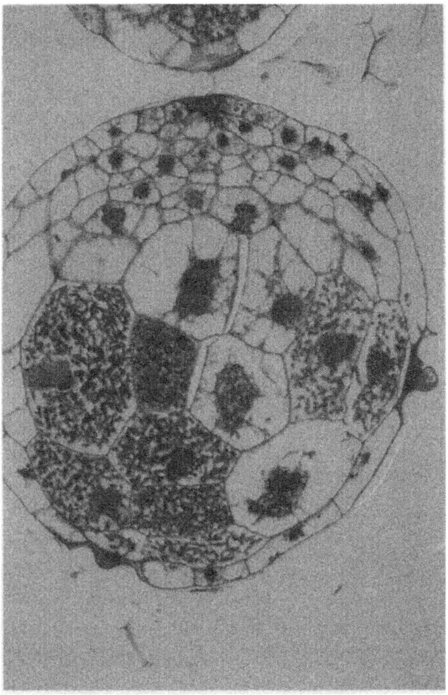

FIGURE 1. Section of a seedling of *Dactylorhiza majalis* (Rchb.) Hunt & Summer. symbiotically grown with an *Epulorhiza* sp. The large, basal cells of the seedling are heavily infected with hyphae. Some cells are newly infected (middle right) others are filled with living hyphae (bottom left), and in some cells the pelotons are in the process of collapsing (bottom right). Nuclei sre enlarged in the infection zone. The apical part has small cells developing into a meristematic zone (top) that remains uninfected. Stained with PAS-ABB. x187 (ondias x56) Photo: HN Rasmussen.

Shortly before infection, the size of the host cell nucleus increases considerably due to endoreduplication and amplification of the DNA (Williamson and Hadley, 1969; Nagl, 1972) and nuclear bodies develop that may suggest an accumulation of messenger RNA (Barroso and Pais, 1990). The invading hyphae appear to establish a close contact with the nucleus of each infected cell. Plasmalemma of the infected cell invaginates in front of the advancing hyphal tip to form a vast sequestration membrane sheathing the hyphae as they traverse cytoplasm, central vacuole and nucleus of the cell. This exceedingly complicated system of membranes provides extensive, short range contact between the symbionts.

During initial infection there are indications of high hyphal secretory activity (Hadley *et al.*, 1971; Rasmussen, 1995). High polyphenol oxidase activity is noted initially in the hyphae and later in the interfacial space (Pais and Barroso, 1983; Barroso *et al.*, 1986a). The fungal cytoplasm becomes loaded with glycogen (Fig. 1), particularly in dilated portions of the peloton hyphae (Rasmussen, 1990).

The infected plant cytoplasm contains numerous mitochondria and profusely developed ER, probably involved in the production of hydrolytic enzymes (Dörr and Kollmann, 1969; Barroso *et al*, 1986a). In the early stages of lysis the hyphal cells vacuolate and the dolipore septa begin to disintegrate (Richardson *et al.*, 1992). During subsequent hyphal degeneration, acid phosphatase activity is detected inside the hyphal wall. As the hyphal cytoplasm disorganize and the cells collapse, the periplasmic canals in which they lie fuse so that the hyphal residue aggregates in larger clumps which remain in the plant cells.

D. Regulation

Phytoalexins such as orchinol and hircinol exhibiting strong fungicidal activity have been found in orchid tissues, and are induced following infection by *Rhizoctonia* and other fungal endophytes found in association with orchids (Nobecourt, 1923). However, 24 unspecific saprophytic and semi-parasitic soil fungi did not induce phytoalexin production in challenged orchid tissue (Gäumann *et al.*, 1960). Some *Rhizoctonia* spp., including strains of *Ceratobasidium cornigerum* (Bourdot) D.P. Rogers and *Thanatephorus cucumeris* (Frank) Donk, tolerate fairly high concentrations of phytoalexins (up to 150mg/l) (Pegg and Beardmore, according to Hadley and Pegg, 1989).

E. Metabolic exchange

The existence of orchids with highly deficient photosynthetic capacity (e.g. Leake, 1994) strongly supports a role of orchid mycorrhiza in supplementing or wholly replacing photosynthesis. Transfer of carbohydrates and phosphate through a mycelium of *Rhizoctonia* into orchid seedlings has been demonstrated *in vitro* by means of isotope labeling (Smith, 1967; Alexander *et al.*, 1984; Alexander and Hadley, 1985). The fact that many orchid seedlings require amino acids and organic compounds other than soluble carbohydrates for their growth, suggests that the fungal partner provides a variety of substances. Stimulation of orchid seedling growth may occur as early as 48 hours after infection but it is unclear whether this occurs before lysis of hyphae begins (Purves and Hadley, 1976). While necrotrophic transfer obviously takes place, there is presently no indication that it is preceded by a phase of biotrophic interchanges.

Experiments designed to demonstrate a transfer of carbohydrates from orchid to fungus were unsuccessful (Hadley and Purves, 1974; Purves and Hadley, 1976). Outgrowth of hyphae from an infected seedling onto a substrate devoid of nutrients has been advanced in favour of this kind of transfer (Smith, 1967). However, this could be explained by re-allocation within the living part of the mycelium. The present evidence does not suggest any nutritional benefits for the fungus.

IV. Taxonomy of Rhizoctonias Mycorrhizal with Orchids

The majority of fungi associated with orchids can be referred to *Rhizoctonia* sensu lato, i.e. sterile mycelia capable of entering living plant tissue and producing sclerotia (DeCandolle, 1815). In some strains sclerotia have not been observed in culture, but monilioid cells that are considered to be the precursors of sclerotia are always present. Many of the strains associated with orchids have been given anamorph names (Bernard, 1909; Costantin and Dufour, 1920; Burgeff, 1932; 1936; Curtis, 1939; Currah *et al.*, 1988; 1990; Currah and Zelmer, 1992; Zelmer and Currah, 1996). Until recently all were assigned to *Rhizoctonia* s.l., but the five anamorph genera with certain teleomorph affinity established by Moore (1987), has made it possible to refer orchid endophytes to three of these, *Ceratorhiza*, *Epulorhiza*, and *Moniliopsis*. The anamorph genus *Opadorhiza* T.F. Andersen and R.T. Moore is proposed elsewhere in this book to accommodate strains with *Sebacina* teleomorphs (see chapter I.A1). A number of strains from Australian orchids will probably be referable to this genus. No orchid endophytes have so far been referred to the genera *Ascorhizoctonia*, *Rhizoctonia* s.str. or *Chrysorhiza* T.F. Andersen and J.A. Stalpers, proposed for the anamorphs of *Waitea* (see chapter I.A3).

Although it is fairly easy to refer a strain to form genus, it is extremely difficult to distinguish between species. The few mycelial characters available have been found to vary significantly between hyphae of the same mycelium, sclerotia and monilioid cells are often very variable, and substrate composition influences mycelial structures (Andersen, 1990). Furthermore, a recent survey has demonstrated that most names within *Rhizoctonia*, including most of the names of orchid endophytes, are either illegitimate or of doubtful application (Andersen and Stalpers, 1994). Type specimen or illustration is rarely preserved, and the description is inadequate for a positive identification. Only six taxa with orchid association are presently acceptable (Table 1), whereas 18 names of orchid endophytes referred to *Rhizoctonia* are disregarded for various reasons and can only be re-established on the basis of validation (Table 2).

TABLE 1. Accepted *Rhizoctonia* s.l. associated with orchids. Condensed from the data of Downie (1959), Currah & Zelmer (1992) and Zelmer & Currah (in press). Host species are mentioned. When more species in a genus host the fungus, only generic name is given.

Name & Author	Host
Ceratorhiza pernacatena Zelmer & Currah	*Platanthera praeclara* Sheviak & Bowles
Moniliopsis anomala Burgeff ex Currah	*Coeloglossum viride* (L.) Hartm.
Moniliopsis solani (Kühn) Moore	*Dactylorhiza purpurella* (T. & T.A. Steph.) Soó
	Spiranthes and *Vanilla*
Epulorhiza albertaensis Currah & Zelmer	*Platanthera orbiculata* (Pursh) Lindl.
E. anaticula (Currah) Currah	*Platanthera obtusata* (Banks ex Pursh) Lindl.
E. calendulina Zelmer & Currah	*Platanthera*, *Calypso* and *Coeloglossum*
	Amerorchis rotundifolia (Banks) Hulten

In some investigations, reproductive structures have been induced in culture making the strains referable to species within the teleomorphic genera *Ceratobasidium*, *Thanatephorus*, *Tulasnella*, and *Sebacina* (Warcup and Talbot, 1966; 1967; 1971; 1980; Currah, 1987; Warcup, 1991) (Table 3). The best results obtained in fruiting experiments is only 62% (Warcup, 1981), which demonstrates the necessity to maintain a parallel identification of anamorphs.

TABLE 2. ejected *Rhizoctonia* s.l. names associated with orchids. Condensed from data in Andersen & Stalpers (1994).

Names & Authors	Reason for rejection
Ceratorhiza goodyerae-repentis (Costantin & Dufour) Moore	Doubtful taxon
Epulorhiza repens (Bernard) Moore	Doubtful taxon
Rhizoctonia aerea Burgeff	Doubtful taxon
R. arachnion Burgeff	Doubtful taxon
R. asclerotica Burgeff	Invalid name
R. borealis J.T. Curtis	Invalid name
R. floccosa Burgeff	Doubtful taxon
R. gracilis Burgeff	Invalid name
R. lanuginosa Bernard	Doubtful taxon
R. monilioides J.T. Curtis	Invalid name
R. neottiae Wolff ex Burgeff	Invalid name
R. psycodis Thomas	Provisional name
R. robusta Burgeff	Invalid name
R. sclerotica Burgeff	Invalid name
R. stahlii Burgeff	Invalid name
R. suavis Thomas	Provisional name
R. subtilis Burgeff	Invalid name
R. subtilis v. *nigra* J.T. Curtis	Invalid name

At species level, very few connections have been established between anamorph and teleomorph taxa. Only three anamorph-teleomorph species relations have reached some degree of recognition, in spite of the fact that the anamorphs lack a proper nomenclatural basis (Andersen and Stalpers, 1994). *Moniliopsis* (*Rhizoctonia*) *solani* (Kühn) Moore, the anamorph of *Thanatephorus cucumeris*, (is treated in chapter I.A1). *Ceratorhiza goodyerae-repentis* (Costantin and Dufour) Moore is tentatively considered as the anamorph of *Ceratobasidium cornigerum* (Bourdot) Rogers. All discussions of this relationship are based on application of *C. goodyerae-repentis* in the sense of Downie (1941), which mainly relies on an assumed host specificity to the orchid *Goodyera repens* (L.) R.Br. Furthermore, *C. cornigerum* may be a collective species containing more than one anamorph taxon (e.g. Burpee *et al.*, 1980). *Epulorhiza repens* (Bernard) Moore now considered to be the anamorph of *Tulasnella deliquescens* (Juel) Juel was previously thought to be *Tulasnella calospora* (Boudier) Juel due to a misidentification by Warcup and Talbot (1967) (Roberts, 1994). *T. deliquescens* is considered to be one of the most widespread orchid symbionts, as it has been found in a number of Australian orchid genera, and has been tentatively identified from several holarctic species (Warcup, 1981; Rasmussen, 1995). However, other species of *Tulasnella* have been described based on isolates from Australian orchids with limited anamorphic data. Many of the orchid endophytes from the holarctic region, supposedly referable to *Tulasnella deliquescens*, may prove to belong to some of these taxa.

TABLE 3. Teleomorphs of *Rhizoctonia* strains associated with orchids. Condensed from Warcup & Talbot (1967; 1971; 1980), Warcup (1981), and Currah (1987). For further details, see Warcup (1981). Host species are mentioned. When more species in a genus host the fungus, only generic name is given.

Names & Authors	Host
***Ceratobasidium* D.P. Rogers**	
C. angustisporum Warcup & Talbot	*Pterostylis mutica* R.Br.
C. cornigerum (Bourdot) Rogers	*Goodyera repens* (L.) R.Br., *Pterostylis* and *Prasophyllum*
C. globisporum Warcup & Talbot	*Trichoglottis australiensis* Dockr. *Calanthe triplicata* (Willem.) Ames
C. obscurum D.P. Rogers	*Acianthus reniformis* (R.Br.) Schlechter
C. papillatum Warcup & Talbot	*Thrixspermum congestum* (F.M. Bail.) Dockr., *Sarcochilus*
C. sphaerosporum Warcup & Talbot	*Pomatocalpa macphersonii* (F.Muell.) T.E. Hunt and various members of Sarcanthinae
***Thanatephorus* Donk**	
T. anomala (Talbot) J.A. Stalpers & T. F. Andersen	*Prasophyllum machrostachyum* R.Br. var. *ringens* (Reichb.f.) A.S. George
T. cucumeris (Frank) Donk	*Prasophyllum odoratum* Rogers *Pterostylis foliata* Hook.f.
T. gardneri Warcup	*Rhizantella gardneri* Rogers
T. orchidicola Warcup & Talbot	*Orchis mascula* (L.) L. *Coeloglossum viride* (L.) Hartm.
T. pennatus Currah	*Calypso bulbosa* (L.) Oakes
T. sterigmaticus (Bourdot) Talbot	*Thelymitra antennifera* Hook.f.
***Tulasnella* J. Shroeter**	
T. allantospora Wakefield & Pearson	*Corybas dilatatus* (Rupp & Nicholls) Rupp
T. asymmetrica Warcup & Talbot	*Thelymitra* and *Dendrobium*
T. cruciata Warcup & Talbot	*Acianthus caudatus* R.Br. and *Thelymitra*
T. deliquescens (Juel) Juel	*Dactylorhiza*, *Diuris*, *Acianthus*, *Caladenia*, *Thelymitra*, *Lyperanthus* and *Dendrobium*
T. irregularis Warcup & Talbot	*Dendrobium dicuphum* F. Muell.
T. violea (Qu'let) Bourdot & Galzin	*Thelymitra aristata* Lindl.
***Sebacina* Tulasne**	
S. vermifera Oberwinkler	*Glossodia major* R.Br. *Microtis*, *Acianthus*, *Caladenia*, *Elythranthera*, *Eriochilus* and *Prasophyllum*.

V. Specificity between Rhizoctonias and orchids

It was originally assumed that a highly specific relationship existed between the orchids and their endophytes, and fungi were often named after the host of the initial isolate, e.g. M.R. (Mycelium Radicis) *(Habenaria) psychodis* (Burgeff, 1909). Later it was recognized that plant-fungal relation was not highly specific in general, although fungal type preference seems to occur in many orchids (Burgeff, 1936; Curtis, 1939; Warcup, 1971; 1981). However, beside the difficulties in identifying imperfect strains there are three major issues that confound the evaluation of specificity.

a. Isolation procedures have varied considerably, some placing fairly large pieces of orchid root tissue on an agar plate (Curtis, 1939) while others used meticulous preparation of single pelotons from root tissue, enabling a critical distinction between epiphytes and inter-, and intracellular endophytes (Warcup and Talbot, 1967, 1971, 1980; Rasmussen *et al.*, 1990a). Only the latter can be considered mycorrhizal.

b. Isolates have mainly been derived from roots of adult orchids. However, at the photoautotrophic stage the plants are no longer totally dependent on the endophyte, and they have been shown to accommodate a variety of fungal strains (Warcup, 1981; Alexander and Hadley, 1983). Adult plant material is thus less useful for an evaluation of specificity patterns. Seeds may be responsive to a range of fungi during and immediately after germination, but this gives no clue to the subsequent success in seedling development (Warcup, 1973). It is known that fungi stimulating the highest germination rate *in vitro* are often different from those that support the best protocorm growth [Figs. 5.2 and 5.3 in Rasmussen (1995)]. The most critical phase concerning symbiont compatibility is during seedling development and the transition from seedling to a leaf-bearing adult (Muir, 1989), also characterized by a shift from holomycotrophy to partial mycotrophy and photoautotrophy.

c. Laboratory experiments involving *in vitro* formation of mycorrhizal associations do not correspond closely to natural conditions and, as a result, atypical mycorrhizal associations that would not occur in nature may be established (Rasmussen, 1995). In cases when fungal strains have been isolated from orchid seedlings found underground the identification of the host is more or less uncertain (Rasmussen *et al.*, 1990a). A new 'seed packet' method has been developed (Rasmussen and Whigham, 1993) which allows both a controlled study of germination under natural conditions, and of a fungal association with seeds of known origin under natural conditions that are relevant for the species in question (Rasmussen and Whigham, 1993; Masuhara and Katsuya, 1994).

In studies of Australian orchids (Warcup 1971; 1973; 1981), careful isolation; teleomorph identification of strains (Warcup and Talbot, 1967; 1971; 1980), and germination tests *in vitro*, revealed some degree of specificity between certain fungi and genera or species of orchids, particularly in species belonging to *Pterostylis*, *Caladenia* and *Thelymitra*. The species grew adjacent to each other in the same habitat, but hosted three different endophytes, belonging to *Ceratobasidium*, *Sebacina* and *Tulasnella*, respectively (Warcup, 1973). Both species - species, species - genus, and genus - genus specificity levels were demonstrated (Warcup, 1981).

Corresponding studies on orchids from the northern hemisphere have proven to be largely inconclusive (Downie, 1941; Hadley 1970; Currah *et al.*, 1988; 1990; Muir, 1989). There are indications that some of the orchids are not highly selective with respect to the endophyte, e.g. *Dactylorhiza purpurella* (T. and T.A. Steph.) Soó (Downie, 1959), and furthermore, a number of orchid species and genera host members of *Tulasnella* (*Epulorhiza*) (Currah *et al.*, 1990; Rasmussen *et al.*, 1990a; 1990b; Currah and Zelmer, 1992; Rasmussen, 1995; Zelmer and Currah, 1996). Another notable observation is the apparent lack in the northern hemisphere of orchid endophytes belonging to *Sebacina*. The Canadian strain tentatively identified as *Sebacina* sp. from *Platanthera orbiculata* (Pursh) Lindl. (Currah *et al.*, 1990) can now be referred to *Tulasnella* (Andersen, 1996).

VI. *RHIZOCTONIA* MYCORRHIZAL WITH OTHER PLANTS

There are few records of *Rhizoctonia* in mycorrhizal relationships with non-orchid plant groups. The fungi such as *R. pini-insignis* (nom. inval.), isolated from roots of *Pinus radiata* Don (=*Pinus insignis* Dougl.), may be involved in a mycorrhizal relationship as it improved growth of the test plants, lupin and pea (Castellani, 1935). However, the best documented case is the bryophyte genus *Aneura* (Rayner, 1927; Stahl, 1949), where both achlorophyllous and green species form mycorrhiza with *Rhizoctonia*. The mycorrhiza was found to be strikingly similar to the orchid mycorrhiza with respect to penetration, formation of hyphal coils in cortical cells, and subsequent hyphal digestion (Rayner, 1927). Stahl (1949) isolated a number of *Rhizoctonia* strains from the moss, and referred them tentatively to taxa described by Burgeff (1932; 1936). Unidentified septate fungi with a similar infection pattern have been found in gametophytes of *Psilotum nudum* (L.) Beauv. (Bierhorst, 1953). It is possible that they belong to *Rhizoctonia* (Harley, 1969).

In the Gentianaceae most species host mycorrhizal fungi, and many species show similarities to the orchids with respect to seed size (Harley, 1969). Mycorrhiza of some *Gentiana* species investigated by Neumann (1934) were of the orchid type, and one of the mycelia did belong to *Rhizoctonia*. *Gentiana* seeds are difficult to germinate *in vitro* and no successful synthesis of mycorrhiza was reported.

Species belonging to the Burmanniaceae and Triuridaceae have been claimed to form mycorrhiza with *Rhizoctonia* (Harley 1969), but no certain identification has been possible, and this association remains unclear.

VII. CONCLUSION

There is a great need for more taxonomic work on the mycorrhizal species of *Rhizoctonia*. The six accepted species associated with orchids are hardly enough to accommodate the diversity of endophytes derived from this large plant family. An important step would be to carefully describe new taxa or continue to validate previously introduced taxa (Currah *et al.*, 1990; Currah and Zelmer, 1992; Zelmer and Currah, 1996). Anamorph-teleomorph species relations are still poorly known and, whenever possible, descriptions of new anamorph taxa should be based on material with known teleomorph origin. When a larger number of clearly defined, recognizable taxa are available, a better understanding of patterns of specificity can be achieved.

Among Northern temperate species, the identity of the endophytes, particularly in the rhizomatous genera such as *Cypripedium*, *Epipactis* and *Cephalanthera*, is still largely unknown. Studies of the endophytes of some of the basal groups within the orchidaceae would be particularly interesting from an evolutionary point of view.

The chemical background for recognition and response, i.e. the communication between the two organisms, is an issue which has so far received very little attention with respect to the orchid-*Rhizoctonia* interaction. There are also interesting questions to be raised with respect to the associations of *Rhizoctonia* to non-orchid plant groups.

In conclusion, it seems that much remains to be learned and that we are only beginning to explore the diversity of mycorrhizal relationships between *Rhizoctonia* and plants.

VIII. References

Alexander C & Hadley G (1983) Variation in symbiotic activity of *Rhizoctonia* isolates from *Goodyera repens* mycorrhizas. Trans. Br. mycol. Soc. 80:99-106.

Alexander C & Hadley G (1985) Carbon movement between host and mycorrhizal endophyte during the development of the orchid *Goodyera repens* Br. New Phytol. 101:657-665.

Alexander C, Alexander IJ & Hadley G (1984) Phosphate uptake in relation to mycorrhizal infection. New Phytol. 97:401-411.

Andersen TF (1990) A study of hyphal morphology in the from genus *Rhizoctonia*. Mycotaxon 37:25-46.

Andersen TF (1996) A comparative taxonomical study of *Rhizoctonia* s.l. employing morphological, ultrastructural and molecular methods. Mycol. Res. (in press).

Andersen TF & Stalpers JA (1994) A check-list of *Rhizoctonia* epithets. Mycotaxon 51:437-457.

Barroso J & Pais MSS (1990) Nuclear features in infected roots of *Ophrys lutea* Cav. (Orchidaceae). New Phytol. 105:67-70.

Barroso J, Chaves Neves H & Pais MSS (1986a) Ultrastructural, cytochemical and biochemical aspects related to the formation of *O. lutea* endomycorrhizae. In: Gianinzzi-Pearson V & Gianinazzi S (eds.) Physiological and Genetical Aspects of Mycorrhizae (pp 265-268) INRA, Paris.

Barroso J, Chaves Neves H & Pais MSS (1986b) Production of Indole-3-ethanol and indole-3-acetic acid by the mycorrhizal fungus of *Ophrys lutea* (Orchidaceae). New Phytol. 103:745-749.

Bernard N (1909) L'evolution dans la symbiose des orchidées et leur champignons commensaux. Ann. Sci. Nat. Paris 9. sér. 9:1-196.

Bierhorst DW (1953) Structure and development of the gametophyte of *Psilotum nudum*. Amer. J. Bot. 40:649-658.

Burgeff H (1909) Die Wurzelpilze der Orchideen, ihre Kultur und ihre Leben in der Pflanze. Gustav Fischer, Jena. .

Burgeff H (1932) Saprophytismus und Symbiose. Studien an tropischen Orchideen. Gustav Fischer, Jena.

Burgeff H (1936) Samenkeimung der Orchideen. Gustav Fischer, Jena.

Burpee LL, Sanders PC, Cole HJr & Sherwood RT (1980) Anastomosis groups among isolates of *Ceratobasidium cornigerum* and related fungi. Mycologia 72:689-701.

Castellani E (1935) Ricerche preliminari sulla biologia di alcune Rizottonie. Annali R. Ist. sup. agr. for. naz., Ser. 2a, 5:29-61.

Costantin J & Dufour L (1920) Sur la biologie du *Goodyera repens*. Rev. Gen. Bot. 32:529-533.

Currah RS (1987) *Thanatephorus pennatus* sp. nov. isolated from mycorrhizal roots of *Calypso bulbosa* (Orchidaceae) from Alberta. Can. J. Bot. 65:1957-1960.

Currah RS & Zelmer C (1992) A key and notes for the genera of fungi mycorrhizal with orchids and a new species in the genus *Epulorhiza*. Rep. Tottori Mycol. Inst. 30:43-59.

Currah RS, Sigler L & Hambleton S (1987) New records and new taxa of fungi from the mycorrhizae of terrestrial orchids of Alberta. Can. J. Bot. 65:2473-2482.

Currah RS, Hambleton S & Smreciu A (1988) Mycorrhizae and mycorrhizal fungi of *Calypso bulbosa*. Amer. J. Bot. 75:739-752.

Currah RS, Smreciu EA & Hambleton S (1990) Mycorrhizae and mycorrhizal fungi of boreal species of *Platanthera* and *Coeloglossum* (Orchidaceae). Can. J. Bot. 68:1171-1181.

Curtis JT (1939) The relation of specificity of orchid mycorrhizal fungi to the problem of symbiosis. Amer. J. Bot. 26:390-399.

DeCandolle AP (1815) Flore Française 3. ed. Vol. V. Desray, France.

Downie DG (1941) Notes on the germination of some British orchids. Trans. Bot. Soc. Edinburgh 33:94-103.

Downie DG (1959) The mycorrhiza of *Orchis purpurella*. Trans. Bot. Soc. Edinburgh 38:16-29.

Dörr I & Kollmann R (1969) Fine structure of mycorrhiza in *Neottia nidus-avis* (L.) L.C.Rich. (Orchidaceae). Planta 89:372-375.

Ellis JB (1883) North American Fungi, Ser. I, no. 977.

Frank AB (1891) Über die auf Verdauung von Pilzen abziehende Symbiose der mit endotrophen Mykorrhizen begabten Pflanzen, sowie der Leguminosen und Erlen. Ber. Deutsch. Bot. Ges. 9:244-253.

Gäumann E, Nüesch J & Rimpau RH (1960) Weitere Untersuchungen über die chemischen Abwehrreaktionen der Orchideen. Phytopath. Z. 38:274-308.

Hadley G (1970) Non-specificity of symbiotic infection in orchid mycorrhiza. New Phytol. 69:1015-1023.

Hadley G & Pegg GF (1989) Host-fungus relationships in orchid mycorrhizal systems. In:Pritchard HW (ed.) Modern methods in orchid conservation (pp 57-71) Cambridge University Press, Cambridge.

Hadley G & Purves S (1974) Movement of carbon 14 from host to fungus in orchid mycorrhiza. New Phytol. 73:475-482.

Hadley G, Johnson RPC & John DA (1971) Fine structure of the host-fungus interface in orchid mycorrhiza. Planta 100:191-199.

Hanke M & Dollwett HHA (1976) The production of ethylene by certain soil fungi. Sci. Biol. J. 2:227-230.

Harley JL (1969) Biology of mycorrhiza 2. ed. Leonard Hill, London.

Harvais G & Pekkala D (1975) Vitamin production by a fungus symbiotic with orchids. Can. J. Bot. 53:156-163.

Irmisch T (1853) Beiträge zur Biologie und Mophologie der Orchideen. Ambrosius Abel, Leipzig.

Knudson L (1922) Non-symbiotic germination of orchid seeds. Bot. Gaz. 73:1-25.

Leake JR (1994) The biology of myco-heterotrophic ('saprophytic') plants. New Phytol. 127:171-216.

Link HF (1840) Icones selectae anatomico-botanicae II:10 + Tab. VII. C. G. Lüderitz Verlag, Berlin.

Masuhara G & Katsuya K (1994) *In situ* and *in vitro* specificity between *Rhizoctonia* spp. and *Spiranthes sinensis* (Persoon) Ames. var. *amoena* (M. Bieberstein) Hara (Orchidaceae). New Phytol. 127:711-718.

Moore RT (1987) The genera of *Rhizoctonia*-like fungi:*Ascorhizoctonia*, *Ceratorhiza* gen. nov., *Epulorhiza* gen. nov., *Moniliopsis*, and *Rhizoctonia*. Mycotaxon 29:91-99.

Muir HJ (1989) Germination and mycorrhizal fungus compatibility in European orchids. In:Pritchard HW (ed.) Modern methods in orchid conservation (pp 39-56) Cambridge University Press, Cambridge.

Nagl W (1972) Evidence of DNA amplification in the orchid *Cymbidium in vitro*. Cytobios 5:145-154.

Neumann G (1934) Über die Mykorrhiza in der gattung *Gentiana*. Zbl. Bakt. II 89:433-458.

Nobecourt P (1923) Sur la production d'anticorps par les tubercules des Ophrydees. Comp. R. Seanc. Acad. Sci. (Paris) 17:1055-1057.

Pais MSS & Barroso J (1983) Localization of polyphenoloxidases during the establishment of *Ophrys lutea* endomycorrhizas. New Phytol. 95:219-222.

Purves S & Hadley G (1976) The physiology of symbiosis in *Goodyera repens*. New Phytol. 77:689-696.

Ramsbottom J (1923) Orchid mycorrhiza. Trans. Brit. mycol. Soc. 8:28-61.

Rasmussen HN (1990) Cell differentiation and mycorrhizal infection in *Dactylorhiza majalis* (Rchb.f.) Hunt & Summerh. (Orchidaceae) during germination *in vitro*. New Phytol. 116:137-147.

Rasmussen HN (1994) The roles of fungi in orchid life history. In:Pridgeon A (ed.) Proc. 14th WOC, Glasgow (pp 130-137) Her Majesty's Stationary Office, London.

Rasmussen HN (1995) Terrestrial orchids from seed to mycotrophic plant. Cambridge University Press, Cambridge.

Rasmussen HN & Whigham D (1993) Seed ecology of dust seeds *in situ*: a new study technique and its application in terrestrial orchids. Amer. J. Bot. 80:1374-1378.

Rasmussen HN, Andersen TF & Johansen B (1990a) Temperature sensitivity of *in vitro* germination and seedling development of *Dactylorhiza majalis* (Orchidaceae) with and without a mycorrhizal fungus. Plant Cell and Env. 13:171-177.

Rasmussen HN, Andersen TF & Johansen B (1990b) Light stimulation and darkness requirement for the symbiotic germination of *Dactylorhiza majalis in vitro*. Physiol. Plant. 79:226-230.

Rayner MC (1927) Mycorrhiza, chapter IX. New Phytol. 26:22-45.

Richardson KA, Peterson RL & Currah RS (1992) Seed reserves and early symbiotic protocorm development of *Platanthera hyperborea* (Orchidaceae). Can. J. Bot. 70. 291-300.

Roberts P (1994) Long-spored *Tulasnella* species from Devon, with additional notes on allantoid-spored species. Mycol. Res. 98:1235-1244.

Smith SE (1967) Carbohydrate translocation in orchid mycorrhizas. New Phytol. 66:371-378.

Stahl M (1949) Die Mykorrhiza der Lebermoose mit besonderer Berucksichtigung der thallosen Formen. Planta 37:103-148.

Terashita T (1985) Fungi inhabiting wild orchids in Japan (III). A symbiotic experiment with *Armillaria mellea* and *Galeola septentrionalis*. Trans. mycol. Soc. Japan 26:47-53.

Warcup JH (1971) Specificity of mycorrhizal association in some Australian terrestrial orchids. New Phytol. 70:41-46.

Warcup JH (1973) Symbiotic germination of some Australian terrestrial orchids. New Phytol. 72:387-392.

Warcup JH (1981) The mycorrhizal relationships of Australian orchids. New Phytol. 87:371-381.

Warcup JH (1991) The *Rhizoctonia* endophytes of *Rhizanthella* (Orchidaceae). Mycol. Res. 95:656-659.

Warcup JH & Talbot PHB (1966) Perfect states of some rhizoctonias. Trans. Brit. mycol. Soc. 49:427-435.

Warcup JH & Talbot PHB (1967) Perfect states of rhizoctonias associated with orchids. New Phytol. 66:631-641.

Warcup JH & Talbot PHB (1971) Perfect states of rhizoctonias associated with orchids II. New Phytol. 70:35-40.

Warcup JH & Talbot PHB (1980) Perfect states of rhizoctonias associated with orchids III. New Phytol. 86:267-272.

Williamson B & Hadley G (1969) DNA content of nuclei in orchid protocorms symbiotically infected with *Rhizoctonia*. Nature 222:582-583.

Williamson B & Hadley G (1970) Penetration and infection of orchid protocorms by *Thanatephorus cucumeris* and other *Rhizoctonia* isolates. Phytopathology 60:1092-1096.

Zelmer CD & Currah RS (1995) *Ceratorhiza pernacatena* and *Epulorhiza calendulina* spp. nov.:mycorrhizal fungi of terrestrial orchids. Can. J. Bot. 73:1981-1985.

Zettler LW & McInnis TMJr (1992) Propagation of *Platanthera integrilabia* (Correll) Luer, an endangered terrestrial orchid, through symbiotic seed germination. Lindleyana 7:154-161.

V.15 PHOTOGRAPHS OF SYMPTOMS INCITED BY *RHIZOCTONIA* SPP. ON VARIOUS CROPS

AUTHORS OF SECTION V

B. Sneh et al. (eds.),
Rhizoctonia Species: Taxonomy, Molecular Biology, Ecology, Pathology and Disease Constrol, 391–403.

FIGURE 1. Uninucleate *Rhizoctonia* sp. hyphae (Dapi staining), Curtesy, Robin Sen and Ari Hietala.

FIGURE 2. Binucleate *Rhizoctonia* sp. hyphae (Dapi staining), Curtesy, S. Bruce Martin.

FIGURE 3. Multinucleate *Rhizoctonia solani*. hyphae (Dapi staining), Curtesy, S. Bruce Martin.

FIGURE 4. Typical damping off symptoms of cabbage (*Brassica oleracea*) incited by *Rhizoctonia solani* (AG 1-IA). Curtesy, Baruch Sneh

FIGURE 5. Hypocotyl rot of canola (*Brassica napus* L.) Caused by *Rhizoctonia solani* (AG 2-1) (increasing disease severity, from left to right). Curtesy, Prithwi Verma.

FIGURE 6. Stem and root rot of garden pea (*Pisum sativum* L.) caused by *Rhizoctonia solani* (AG 4). Curtesy, Chin-Chyu Tu.

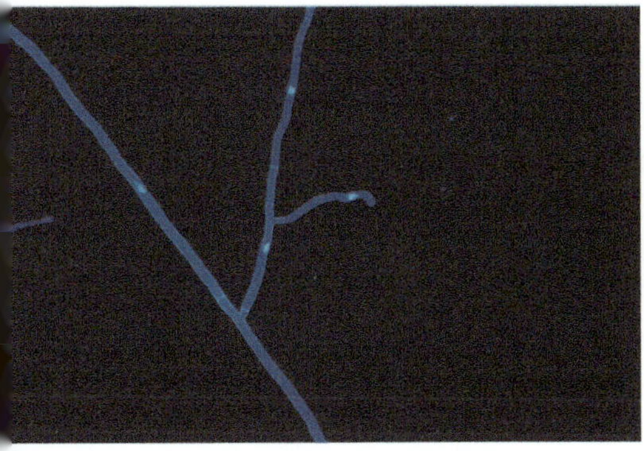

1

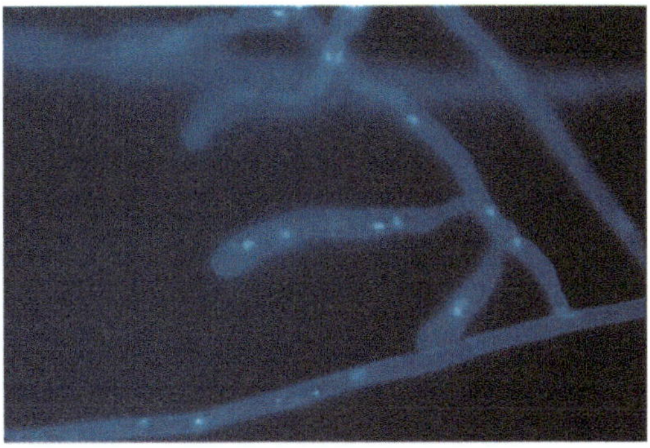

2

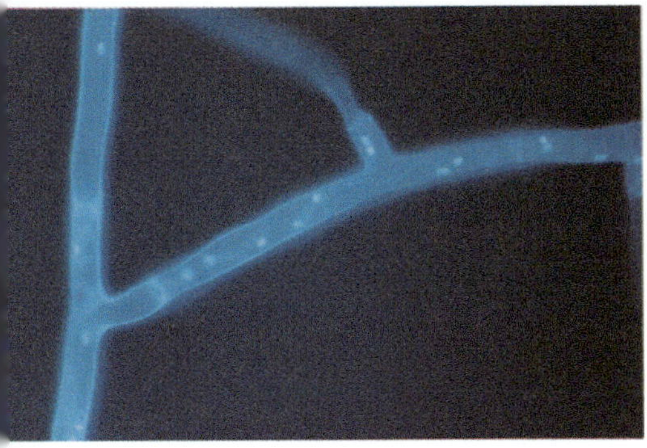

3

4

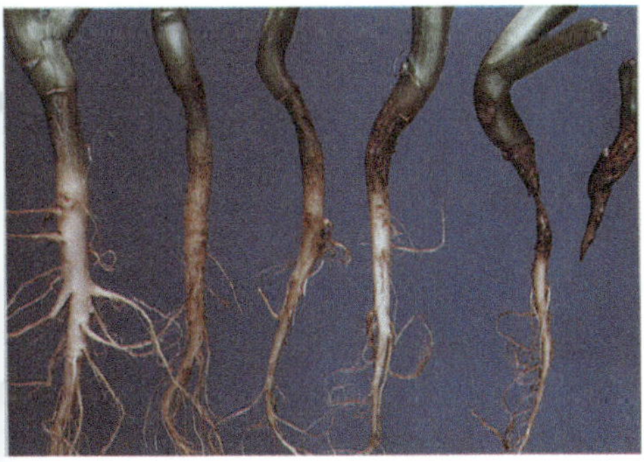

5

6

FIGURE 7. Wire stem of chinese kale (*Brassica oleracea* L. var. *oblongata*). incited by *Rhizoctonia solani* (AG 4) Curtesy, Chin-Chyu Tu.

FIGURE 8. Stem rot of peanut (*Arachis hypogaea* L.) incited by *Rhizoctonia solani* (AG 4). Curtesy, Timothy Brenneman.

FIGURE 9. Limb rot of peanut (*Arachis hypogaea* L.) caused by *Rhizoctonia solani* (AG 4). Curtesy, Timuthy Brenneman

FIGURE 10. Limb rot of peanut (*Arachis hypogaea* L.) caused by *Rhizoctonia solani* (AG 4). Curtesy, Timuthy Brenneman

FIGURE 11. Black scurf of a potato (*Solanum tuberosum* L.) tuber incited by *Rhizoctonia solani* (AG 3). Curtesy, Gilbert J. Banville and Barbara E. Otrysko.

FIGURE 12. Scots pine (*Pinus sylvestris* L.) seedlings Inoculated (7 weeks after sowing) with a uninucleate *Rhizoctonia* isolate. Roots and shoots Stunting symptoms are clearly seen (after 5 months) on the right seedling as compared to the noninoculated control one on the left. Curtesy, Ari M. Hietala and Robin Sen.

7

8

9

10

11

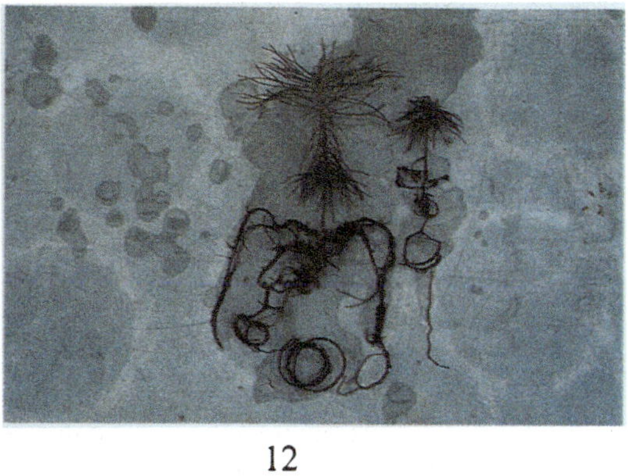

12

FIGURE 13. Stem rot of impatiens (*Impatiens balsamia* L.) incited by *Rhizoctonia solani* (AG 4) Curtesy, D. Michael Benson.

FIGURE 14. *Rhizoctonia* patch in tulip beds caused by *Rhizoctonia solani* (AG 2-T). Curtesy, Hans Schneider.

FIGURE 15. Iris bulbs infected with *Rhizoctonia solani* (AG 2-T). Curtesy, Gerda Dijst.

FIGURE 16. Bare patch in a young wheat (*Triticum aestivum* L.) crop incited by *Rhizoctonia solani* (AG 8) Curtesy, Gordon MacNish.

FIGURE 17. Wheat (*Triticum aestivum* L.) seedlings infected with *Rhizoctonia solani* (AG 8) grown in undisturbed soil cores removed from: center of the patch (left), patch edge (center), and outside of the patch (right). Curtesy, Gordon MacNish.

FIGURE 18. Brown "spear tip" symptoms on wheat (*Triticum aestivum* L.) crown roots and cortical rot leading to "spear tip" incited by *Rhizoctonia solani* (AG 8) Curtesy, R. James Cook and Mark Mazzola.

13

14

15

16

17

18

FIGURE 19. Yellow patch of creeping bentgrass (*Agrostis pulustris* Hunds.) incited by *Rhizoctonia cerealis* (binucleate *Rhizoctonia* sp., AG-D). Curtesy, Lee L. Burpee

FIGURE 20. A lesion on a leaf of tall fescue (*Festuca arundinacea* L.) incited by *Rhizoctonia solani* (AG 1). Curtesy, Lee L. Burpee.

FIGURE 21. Typical rice (*Oryza sativa* L.) sheath blight lesions incited by *Rhizoctonia solani* (AG 1-IA). Note the sclerotium on the central sheath. Curtesy, Teruyoshi Hashiba.

FIGURE 22. Typical rice (*Oryza sativa* L.) brown sheath blight lesions incited by *Rhizoctonia solani* (AG 2-2 IIIB) Curtesy, Teruyoshi Hashiba.

FIGURE 23. Patch (area) of sugar beet (*Beta vulgaris* var. *saccharifera* Alef.) crown and root rot caused by *Rhizoctonia solani* (AG 2-2IV) Curtesy, Leonard J. Herr.

FIGURE 24. Hymenium of *Thanatephorus cucumeris* (*Rhizoctonia solani* AG 2-2IV) on healthy sugar beet (*Beta vulgaris* var. *saccharifera* Alef.) leaf petiole tissue (whitish membranous area) surrounded by diseased (blackened petiole tissue). Curtesy, Leonard J. Herr.

19

20

21

22

23

24

FIGURE 25. Secondary infection of a sugar beet (*Beta vulgaris* var. *saccharifera* Alef.) leaf with basidiospores of *Thanatephorus cucumeris* (*Rhizoctonia solani*, AG 2-2IV). Curtesy, Shigeo Naito.

FIGURE 26. Foliage blight of sugar beet (*Beta vulgaris* var. *saccharifera* Alef.) heart leaves. Young heart leaves have blackened leaf margins and miss happened leaves reduced to stubs. Irregular ragged margin expanded leaves occur in the outer leaves of the rosette. Curtesy, Leonard J. Herr.

FIGURE 27. Primary and secondary infection of bean (*Phaseouls vulgaris* L.) leaves with basidiospores of *Thanatephorus cucumeris* (*Rhizoctonia solani*, (AG 2-2IV). Curtesy, G. Godoy-Lutz and J.R. Steadman.

FIGURE 28. Typical web blight symptoms on bean (*Phaseouls vulgaris* L.) leaves incited by *Rhizoctonia solani* (AG 1-IB, microsclerotia). Curtesy, G. Godoy-Lutz and J.R. Steadman.

FIGURE 29. Lesions on a bean (*Phaseouls vulgaris* L.) pod caused by the web pathogen *Rhizoctonia solani* (AG 1-IB microsclerotia). Curtesy, G. Godoy-Lutz and J.R. Steadman.

FIGURE 30. Blemished and unblemished bean (*Phaseouls vulgaris* L.) seeds infected by the web pathogen *Rhizoctonia solani* (AG 1-IB microsclerotia). Curtesy, G. Godoy-Lutz and J.R. Steadman.

25

26

27

28

29

30

FIGURE 31. Disease severity index of lesions incited by *Rhizoctonia solani* (AG 4) on bean (*Phaseouls vulgaris* L.) hypocotyls. From right to left: 0 (no symptoms) to 5 (stem collapsed), Curtesy, Baruch Sneh.

FIGURE 32. Protection of potato (*Solanum tuberosum* L.) stems in a naturally infested field against *Rhizoctonia solani* (AG 3) by the introduction of a non pathogenic isolate of *R. solani* (AG 4). The three stems on the left are from treated plots, while the three stems with lesions on the right are from untreated control plots, Curtesy, Baruch Sneh.

FIGURE 33. Growth promotion of radish (*Raphanus sativus* L.) incited by a non pathogenic isolate of *Rhizoctonia solani* (AG 4) Curtesy, Baruch Sneh.

FIGURE 34. Growth promotion of carrot (*Daucus carota* L.) incited by a non pathogenic isolate of *Rhizoctonia solani* (AG 4) Curtesy, Baruch Sneh.

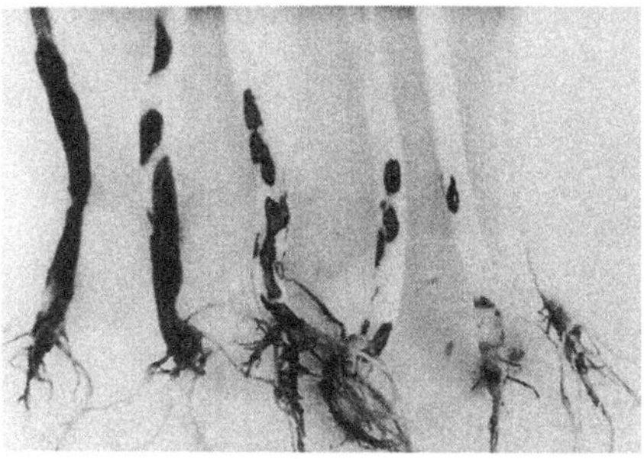

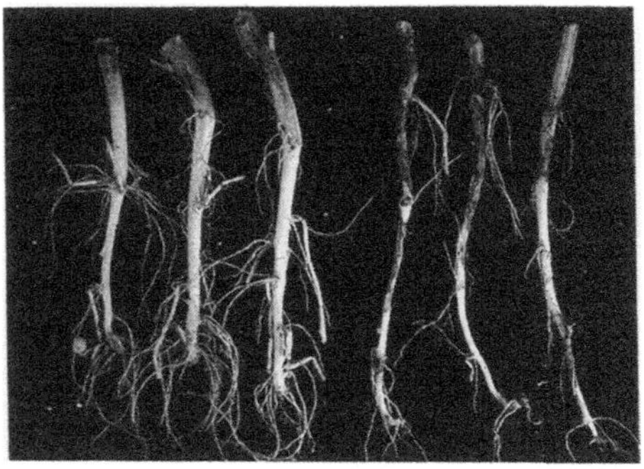

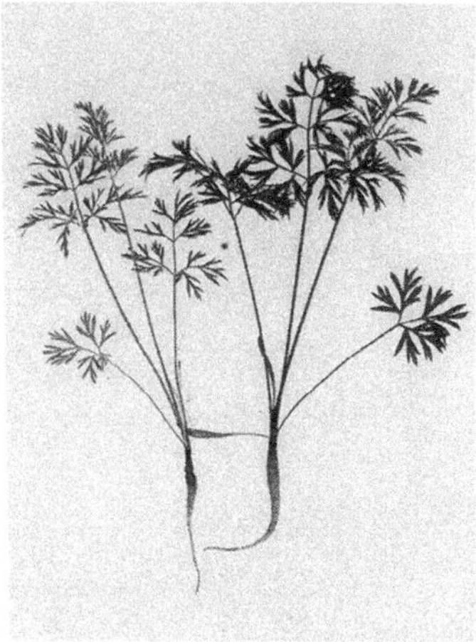

Section **VI.**

CONTROL OF DISEASES CAUSED BY RHIZOCTONIA SPECIES

VI.A1 CULTURAL METHODS AND SOIL NUTRIENT STATUS IN LOW AND HIGH INPUT AGRICULTURAL SYSTEMS, AS THEY AFFECT *RHIZOCTONIA* SPECIES

ARIENA H.C. VAN BRUGGEN, NIKLAUS J. GRÜNWALD AND MARK BOLDA

Department of Plant Pathology, University of California, Davis CA 95616, USA

I. INTRODUCTION

For the purpose of this chapter, low-input crops are defined as crops receiving limited or no synthetic external inputs in the form of pesticides and fertilizers. However, they do receive regular inputs of organic amendments in the form of cover crops, compost or manure to maintain soil fertility, high levels of activity by the soil microflora and fauna, and a good soil structure. Low-input crops are typically grown in organic or biological and integrated farming systems (National Research Council, 1989). Organic or biological farming systems are defined as systems where synthetic fertilizers and pesticides are not used. Instead, biological and cultural methods of pest and disease control are emphasized (National Research Council, 1989). Integrated farming systems may apply pesticides and fertilizers but at a reduced rate compared to conventional farming systems. High-input crops are considered to be crops that receive pesticides and fertilizers as currently used in conventional farming systems. Organic amendments are usually not applied (except in areas with excess manure). In addition to these differences in organic and synthetic inputs, organic and integrated farming systems often differ from conventional farming systems in their tillage operations. On many organic and integrated farms, minimum- tillage practices are utilized, i.e., soil cultivation to a depth of about 15 cm instead of 30 cm or more on conventional farms (El Titi and Landes, 1990). On the other hand, no-till practices are more commonly used on currnet conventional farms because of the need for some tillage for weed control on organic farms. Finally, crop rotations are often longer and spatial diversity greater in organic and integrated farming systems than in conventional farming systems (van Bruggen, 1995).

There are numerous reports in the literature showing that these differences in cultural methods when applied individually could influence saprophytic growth and infection of *Rhizoctonia* species. However, there are only few reports on differences in disease severity or populations of *Rhizoctonia* species between different farming systems. Therefore, differences in *Rhizoctonia* diseases between agronomic treatments applied in conventional farms and experiment stations will be included in this review. We will focus on agronomic practices that are different in conventional and alternative farming systems, i.e., crop rotation, organic amendments, soil tillage, pesticides, and fertilizer use. This review will cover effects of these practices on various *Rhizoctonia* species.

Some aspects of this review pertaining to the ecology of *Rhizoctonia* have been discussed in several excellent papers or book chapters since the publication of the

B. Sneh et al. (eds.),
Rhizoctonia Species: Taxonomy, Molecular Biology, Ecology, Pathology and Disease Constrol, 407–421.

book on the biology and pathology of *R. solani* (Parmeter, 1970). In particular, the ecology, survival, and population genetics and biology of *R. solani* have been reviewed by Ogoshi (1987), Butler (1993), Anderson (1982), and Vilgalys and Cubeta (1994).

II. COMPARATIVE STUDIES OF LOW AND HIGH INPUT CROPPING SYSTEMS

Comparisons of organic, integrated and conventional farming systems indicate that different management practices can result in qualitatively different soil environments. Organic and reduced-input systems have been found to have a thicker topsoil layer, lower soil bulk density, greater water holding and cation exchange capacities, and higher organic matter contents and associated microbial biomass and enzymatic activities, relative to conventional systems (Doran *et al.* 1987). Soil nitrate contents tend to be lower (Doran *et al.* 1987) and nitrogen mineralization rates higher (Drinkwater *et al.* 1995) in organic or reduced-input systems compared to conventional systems (van Bruggen, 1995).

The rhizosphere microflora of organic or integrated farms has been compared to that of conventional farms in a few studies only. Populations of total fungi and bacteria were significantly higher in organically than in conventionally farmed areas in Australia (Sivapalan *et al.*, 1993). Fluorescent pseudomonads and actinomycetes were more numerous in the organically managed areas on three of the five sampling dates. Populations of fluorescent pseudomonads and actinomycetes were also higher in tomato rhizosphere soils from organic than in those from conventional farms in California (Workneh and van Bruggen, 1994). Differences observed between microbial communities from organic and conventional soils were ascribed to higher percentages of cellulolytic and hemicellulolytic actinomycetes in organically managed soils, possibly due to recent additions of cellulosic and hemicellulosic materials in the form of green manure or compost on the organic farms (Workneh and van Bruggen, 1994). The diversity of actinomycetes was also higher in soils from organic than in those from conventional farms (Workneh and van Bruggen, 1994). Similarly, the diversity of fungal species was higher in organic than in conventional farms in Australia (Sivapalan *et al.*, 1993), and the diversity of fungal genera was higher on roots of crops grown in rotation than on roots of crops grown as monocrops in Japan (Nitta, 1991). Fungal diversity was also increased when farm yard manure or crop residues were added to monocropped or rotation fields (Nitta, 1991).

Increased diversity of soil fauna in organic or low-input farming systems has also been demonstrated in some studies (El Titi and Ipach 1989; El Titi and Landes, 1990; El Titi and Richter, 1987), and greater abundance and biomass of the soil fauna in many other studies (El Titi and Ipach 1989; Foissner, 1992). In particular, the numbers and biomass of earthworms, protozoa, collembola, predatory nematodes and mites, fungus gnats, and carabid beetles were increased in several organic or integrated farming systems relative to conventional systems (El Titi and Ipach 1989; Foissner, 1992). The most important factors affecting the soil fauna seemed to be the quantity and quality of organic matter input (El Titi and Ipach, 1989, Foissner, 1992) and soil tillage, with a richer soil fauna in no-till or reduced tillage, low-input fields than in ploughed, high-input fields (El Titi and Ipach, 1989).

All these changes in soil physical and biological properties in organic and integrated relative to conventional farms may affect incidence and severity of root diseases including those caused by *Rhizoctonia* species (Table 1). Sharp eyespot and footrot on wheat and rye (caused by *R. cerealis* = AG-D) were diminished in two organic farming systems compared to two neighboring conventional farming systems which were initiated five years prior to the time when disease observations were made (Daamen *et al.*, 1989, Piorr and Hindorf, 1986). When fungi were isolated by dilution plating from soil from an organic farm and a nearby conventional farm, *R. solani* was only detected in soil from the conventional farm (Sivapalan *et al.*, 1993). No explanation was presented for the higher disease or pathogen levels in the conventional farms. However, the microflora (fungi, actinomycetes, and bacteria) and micro- and mesofauna (nematodes, fungus gnats, midges, springtails, and mites) were more numerous and diverse in organic and low-input farms than in their conventional counterparts (El Titi and Ipach, 1989; Sivapalan *et al.*, 1993).

TABLE 1: Relative disease severity[a] in organic, integrated, and conventional farms.

Experiment	Crop	Disease	Pathogen	Org.	Int.	Conv.	Reference
DFS[b]	Wheat	Sharp eyespot	*R. cerealis*	1	2	2	Daamen *et al.*,1989
Rheinland[c]	Wheat	Sharp eyespot	*R. cerealis*	2-3	-	3	Piorr & Hindorf, 1986
	Rye	Foot rot	*R. cerealis*	1-2	-	2	Piorr & Hindorf, 1986
SAFS[d]	Tomato	Root lesions	*R. solani* AG 4	1	2	2	M. Bolda, unpubl.
Growers' fields[e]	Rice	Aggregate sheath spot	*R. oryzae sativae*	2	-	1	R.K. Webster, unpubl.

[a] 1 = low, 2 = moderate, 3 = severe, summarized from several years of observations.

[b] Development Farming Systems at Nagele, the Netherlands; not replicated, two years of observations; fungicides applied in both conventional and integrated farming systems; no pesticides applied in the organic system.

[c] Conversion study from a conventional to a biodynamic farm, compared with a conventional farm near Moers, Germany; not replicated, five years of observations; no pesticides applied in biodynamic farm.

[d] Sustainable Agriculture Farming Systems at the University of California at Davis; four blocks; two years of observations; no fungicides applied to any of the systems; no pesticides applied in the organic system.

[e] Surveys of organic and conventional rice farms in the Central Valley of California

At the University of California, Davis, we are studying the effects of farming practices on soilborne pathogens and root diseases in the Sustainable Agriculture Farming Systems (SAFS) experiment. This experiment was initiated in 1989, and entails a comparison of organic, reduced-input, and conventional farming practices in four-year rotations of tomato-safflower-corn-wheat/bean. The organic and reduced-input rotations also include annual winter covercrops of oat/vetch (Temple *et al.*, 1994). In the first four years of the project, significant differences developed in soil microbiological and microfaunal characteristics (Scow *et al.*, 1994). Microbial biomass levels were consistently higher and plant parasitic nematode numbers lower in organic and low-input plots than in conventional plots. In 1994 and 1995, *Rhizoctonia* populations were monitored once a month in all fields that were cropped with tomato and safflower, respectively, in the summer season. *Rhizoctonia* spp. were recovered from twenty bulked soil samples per plot using a wetsieving method and placing organic debris on water agar. Most (about 95%) of the *Rhizoctonia* isolates were pathogenic to tomato seedlings. At the end of the tomato growing season, populations of *R. solani* (primarily AG 4) were significantly greater in the conventional than in the organic and reduced-input plots (Fig. 1).

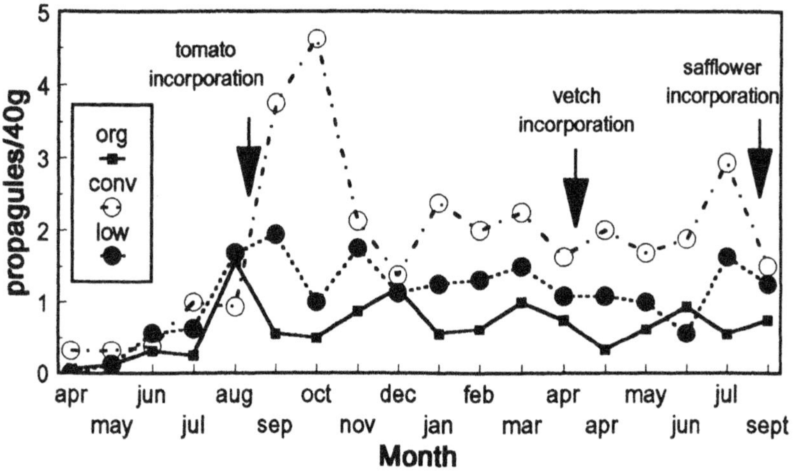

FIGURE 1. Mean number of colony forming units of *Rhizoctonia* spp. per 40 g of dry soil from organic (org.), conventional (conv.), and low-input (low) plots (4 replications) of the Sustainable Agriculture Farming Systems project at the University of California at Davis in 1994 and 1995. Tomatoes were grown in the summer of 1994, vatch (mixed with volunteer oats) in the fall, winter, and spring of 1994/95, and safflower in summer 1995. Soil was sampled once a month but twice in April of 1995.

On the other hand, an ongoing study into rice diseases on organic and conventional farms indicates that aggregate sheath spot [caused by *R. oryzae-sativae* (AG-Bb)] is more severe in organic farms compared to conventional farms (R.K. Webster, personal communication). On conventional farms, rice stubble has commonly been burned at the end of the season, while the stubble is plowed under in organic farms, possibly resulting in a build-up of inoculum. Another difference is that rice is grown in monoculture on conventional farms and is rotated with vetch in organic farms (National Research Council, 1989). The common practice of straw burning in conventional farms will be abandoned in the near future due to environmental concerns. It is not known if aggregate sheath spot might then become more severe in conventional farms. Monoculture may either result in increased disease compared to rotated rice or in reduced disease due to specific disease suppression as observed for *R. solani* in radishes and potatoes.

Because only a few studies have been conducted on the relative intensity of *Rhizoctonia* diseases on low- versus high-input crops, we will review the effects of individual cropping practices on *Rhizoctonia* diseases in the following sections.

III. Crop Rotation

In recent decades, crop rotations have become shorter and field and farm sizes have increased to such an extent that large-scale monocropping is becoming a common practice. Intensive cropping systems have favored epidemic development of many plant diseases including those caused by *R. solani* and *R. cerealis* (Daamen *et al.*, 1989; Hide and Read, 1991; Huiskamp and Lamers, 1992). On the other hand, a decline in *Rhizoctonia* diseases has sometimes been observed in monocultures, for example of radishes (Wijetunga and Baker, 1979) and potatoes (Jager and Velvis, 1980). Disease decline or suppression by monoculture will not be discussed in this chapter since that topic will be covered in chapter IV.6.

Although *Rhizoctonia* species have a wide host range, individual anastomosis groups (AGs of *R. solani*) or their subgroups often have much more limited host ranges (Butler, 1993). AG 1 causes seed and hypocotyl rot and aerial blights on many plant species (Anderson, 1982), but subdivisions of AG 1 are more specific: AG 1-IA causes sheath blight on rice while AG 1-IB causes web blight on legumes or bottom rot on lettuce (Ogoshi, 1985; Herr, 1993). AG 2 isolates cause cankers on root crops and root diseases on crucifers and beans, but there are two subgroups (AGs 2-1 and 2-2) which have a narrower host range (Anderson, 1982). AG 3 is primarily pathogenic on potato (Anderson, 1982; Parmeter, 1970), while AG 4 has a broad host range, causing seed and hypocotyl rot. AG 8 and AG 10 primarily cause root rot and on cereal crops (Ogoshi *et al.*, 1987). Since a mix of AGs varying in virulence to a particular host commonly occurs in soil, crop rotation may shift this mix of AGs towards isolates that are less virulent to that host (Herr, 1993).

Population densities of *R. zeae*, with a host range restricted to corn and other gramineous hosts, increased more after corn than after peanut, whereas *R. solani* AG 2 or AG 4, both with a wider host range, were not affected by crop rotation (Sumner and Bell, 1986). Population densities of *R. solani* AG 1-1 and sheath blight severity were higher when rice was rotated with soybean (both susceptible to AG 1-1) compared to a rotation with bermudagrass pasture, and highest when rice was grown in alternate years compared to once every three years in the rotation with soybean (Belmar *et al.*, 1987). Several different *Rhizoctonia* species and anastomosis groups were isolated from infected foliage of the various crops in the rotation, some isolates being more virulent on rice (AG 1-1A and unknown AGs) others more virulent on soybean (AG 1-1B and other unknown AGs; Jones and Belmar, 1989).

In addition to the susceptibility of a preceding crop to Rhizoctonia strains potentially damaging to the subsequent crop, other factors need to be considered when cropping sequences are selected (Ruppel, 1985; Rush and Winter, 1990). One of these factors is the residual nitrate in soil after the preceding crop, since this may be negatively correlated with *Rhizoctonia* disease severity in the following crop (Rush and Winter, 1990). Another factor could be the amount of residual organic debris remaining from the previous crop, which may be positively correlated to *Rhizoctonia* infection (Rush and Winter, 1990). In that respect, the time of tillage after the preceding crop, and thus the time allowed for decomposition, may also affect disease severity in the following crop (Specht and Leach, 1987). Potato stem lesion severity caused by *R. solani* AG 3 was lower when the previous crop, in particular Japanese millet, was tilled under in August rather than in November (Specht and Leach, 1987).

IV. Organic Amendments

Suppressive soil amendments are covered in chapter VI.A3. However, since use of organic amendments is essential in low-input cropping systems, this topic will also be discussed to a certain extent in this chapter.

Organic amendments such as manure, compost, or cover crops, could either have a positive or negative effect on Rhizoctonia diseases depending on the material used and its state of decomposition. Most experiments have been conducted in the greenhouse and relatively few in the field. Cellulose-based amendments generally were most successful in disease suppression (Papavizas, 1970). However, pure cellulose amendments increased root rot by *R. solani* on white mustard in pot experiments, while oat straw meal and alfalfa meal reduced the disease (Bochow *et al.*, 1970). Addition of neem oil cake and *Gliricidia* leaves reduced *R. solani* on rice both in pot and field experiments (Baby and Manibhushanrao, 1993). Amendments with cellulose, rice stubble or water hyacinth reduced saprophytic growth of *R. solani* and damping-off of cauliflower in potted soil (Kundu and Nandi, 1985). Addition of ammonium nitrate to each of these amendments increased saprophytic growth and infection as the C:N ratio decreased from about 60-100 to 20% (Kundu and Nandi, 1985). The reduction in disease severity by high C:N residues was attributed to increases in antagonistic actinomycete and bacterial populations and competition for nitrogen. Although the C:N ratio had a large effect on suppression of *R. solani* when this ratio was varied by the addition of inorganic nitrogen, the natural C:N ratio of various organic amendments did not affect the saprophytic or parasitic activity of *R. solani* (De Swardt and Pauer, 1978). Straw of oats, wheat, alfalfa, and cotton (C:N ratio ranging from 12 to 88) suppressed the saprophytic and parasitic growth of *R. solani* on cotton segments and cotton seedlings, respectively, whereas soybean debris (C:N ratio = 19) had no effect (De Swardt and Pauer, 1978).

The stage of decomposition is perhaps the most important determinant of the effect organic amendments will have on saprophytic growth and infection by *Rhizoctonia* species. Fresh amendments generally increased inoculum density and disease severity (Wall, 1984), whereas partially decomposed or composted organic matter often suppressed the pathogen (Chung *et al.*, 1988a). *R. solani* diseases on bean, cotton and radish were suppressed by composted sewage sludge added 2-4 weeks before planting, but the disease on cotton was not affected by the amendment when it was applied one week before planting (Lumsden *et al.*, 1983). Addition of a relatively large amount of cellulose to composted hardwood bark negated the suppressive effect of this amendment, indicating that the reason for conduciveness in fresh organic matter may be the presence of a large proportion of available cellulose (Chung *et al.*, 1988b). On the other hand, when available nutrient levels in organic amendments become too low to support high microbial activity (as in dark peat), disease suppression may be reduced (Inbar *et al.*, 1991). However, storage of composted hardwood bark or organic household waste (up to 23 weeks) did not adversely affect suppression of *Pythium* and *Rhizoctonia* damping-off in pot experiments (Schueller *et al.*, 1989). Little work has been done on disease suppression by compost in the field. In one experiment, composted cattle manure increased *Rhizoctonia* root rot on bean seedlings over noncomposted manure (Voland and Epstein, 1994). Extensive literature is available about the effects of cover crop amendments on *Rhizoctonia* diseases in the field. For example, maximum disease suppression was observed when cotton was planted 30-40 days after incubation of soil with debris of various crops (De Swardt and Pauer, 1978). Emergence of cotton seedlings was reduced due to infection by *R. solani* when leguminous cover crops (vetch or clover) were killed with paraquat 2-3 weeks before planting (Rickerl *et al.*, 1992). A longer period of decomposition of debris may be required (see also under the Cultivation section in this review).

Organic amendments can also be combined with application of biocontrol agents such as *Trichoderma* or *Gliocladium* to control *Rhizoctonia* diseases. This strategy was successful in the case of sheath blight control on rice (Baby and Manibhushanrao, 1992).

Regular additions of organic amendments enhance the numbers and activity of the soil fauna, in particular bacteria-feeding protozoa and nematodes, and fungivorous collembola (springtails), mites and nematodes (El Titi and Ipach, 1989, Foissner, 1992). *R. solani* can be partially controlled by predatory activity of mycophagous nematodes or collembola (Barnes *et al.*, 1981; Lartey *et al.*, 1994). Furthermore, earthworm populations can be enhanced by organic amendments. A reduction in Rhizoctonia bare patch disease on wheat was associated with the presence of earthworms (Stephens *et al.*, 1993). The mechanisms underlying this suppression of Rhizoctonia bare patch are not known. Both increased microbial activity and improvement in soil structure after repeated organic amendments may have contributed to the reduction in this disease.

Since the use of organic amendments is characteristic for organic and often for low-input farming systems, lower incidence and severity of root diseases in those systems compared to conventional farming systems may be partially due to long-term use of organic amendments.

V. Cultivation

Different cultivation practices affect the soil ecosystem in different ways. To reduce soil erosion and conserve water and fuel, minimum- soil cultivation is often used in reduced-input farming systems instead of plowing or disking (Cook, 1994). Implementation of conservation tillage practices depend on the use of herbicides to prepare strips of land for planting (Cook and Veseth, 1991; Cook, 1994). Herbicides themselves can affect incidence and severity of Rhizoctonia diseases (see next section).

Probably, the largest differences in saprophytic and parasitic activity of *Rhizoctonia* species can be seen in the effects of till versus no-till systems (Weller *et al.*, 1986; Cook, 1994). Minimum- or no-till cultivation may lead to increased disease severity by pathogens that survive better when infested crop debris remains on or near the soil surface. Root rot and bare patch disease of wheat caused by *R. solani* AG 8 are favored by reduced- or no-till treatments in the Pacific Northwest of the U.S. and in Australia (Weller *et al.*, 1986; Rovira, 1986). It was suggested that fragmentation of residues by tillage reduced the inoculum potential of *R. solani*. However, repeated use of no-till practices over a period of eleven years led to suppression of bare patch disease to levels comparable in tilled plots, presumably due to changes in microbial communities in soil (Roget, 1995).

Soilborne root diseases can also be exacerbated if a cash crop is planted too soon after strip-application of herbicides in a green manure crop or strip-tillage of stubble from the previous crop (Cook and Haglund, 1991; Smiley *et al.*, 1992). Similarly, Rhizoctonia root rot was more severe in Australia after grasses were sprayed with herbicides immediately before planting of cereals, than if spraying was done two to three weeks before planting (Roget *et al.*, 1987).

Colonization of plant debris killed by herbicides leads to a temporary increase in the inoculum potential of *R. solani*. These results highlight the need for integration of cultural methods and for careful timing of any method with respect to all other cultural practices. Chisel plowing (30cm) significantly reduced incidence and severity of stem lesions on potato caused by *R. solani* AG 3 compared to mold board plowing (22cm) (Leach *et al.*, 1993). This reduction in disease by moldboard plowing was attributed to low inoculum densities in the upper soil layer by burial of sclerotia to depths that are inhibitory (Leach *et al.*, 1993) and lower survival of inoculum at greater depth (Ruppel, 1991). *Rhizoctonia* root rot on wheat seedlings in undisturbed soil cores in which the top layers were removed, mixed and returned was reduced by deep mixing but not by shallow mixing (MacNish, 1984). Deep plowing also reduced Rhizoctonia fruit rot of cucumber (Lewis and Papavizas, 1980) and root and hypocotyl rot on snap bean (Lewis *et al.*, 1983).

Cultivation practices have a pronounced effect on soil physical properties. Intensive tillage generally leads to soil compaction. Soil compaction led to higher root rot severity of white beans caused by *R. solani* (Tu and Tan, 1991). Thus, cultural practices such as cover cropping and mulching that reduce soil compaction might help in controlling *Rhizoctonia* disease.

A less recognized effect of cultural practices is associated with the aerial phase of some Rhizoctonia diseases. Sclerotia, mycelium, and basidiospores of *R. solani* can all be sources of inoculum for bean web blight in Costa Rica (Galindo *et al.*, 1983b). Mulching with rice husks or sowing into a living mulch of natural vegetation, later cutting the vegetation and leaving the residue as mulch, were highly effective in controlling web blight because it greatly reduced splash dispersal of inoculum (Galindo *et al.*, 1983a). These cultural treatments were considerably better than fungicides.

As mentioned in previous sections of this review, abundance of Collembola and other members of the soil fauna can help reduce disease. Clearly, farming systems can have an impact on abundance of Collembola and subsequently on Rhizoctonia disease, since no-tillage systems generally have more Collembola than soils conventionally tilled (Rickerl *et al.*, 1989). Tillage system also affects abundance of earthworms and *Aporrectodea trapezoides* significantly reduced *R. solani* disease rating on wheat seedlings (Stephens *et al.*, 1993).

VI. Pesticides

Pesticides such as fumigants, nematicides, soil applied insecticides, herbicides, and fungicides all can interact with disease incidence due to a soil-borne pathogen (Bollen, 1993; Rodriguez-Kabana and Curl, 1980). Pesticides used at the recommended rate can have both positive and negative effects on plant-pathogen interactions (Lévesque and Rahe, 1992).

One of the most striking effects of pesticides is the predisposition of crop plants to disease. This effect is most pronounced for herbicides, presumably because they act on plants directly, often damaging roots, increasing exudation of nutrients from roots, or influencing resistance mechanisms (Bollen, 1993). For instance, glyphosate blocks the production of phenolics involved in disease resistance (Amrhein *et al.*, 1981), thus making plants more susceptible to attack by various root pathogens, presumably due to suppression of defense mechanisms in the crop and/or increased leakage of nutrients from treated plants (Johal and Rahe, 1988; Levesque and Rahe, 1992). Similarly, trifluralin has been shown to enhance susceptibility of various crops to *R. solani* (Altman and Rovira, 1989). Since herbicides are not used in organically grown crops, this could be one of the reasons for a reduction in root diseases in organic compared to conventional farms.

Timing of herbicide applications can also affect plant pathogens. *Rhizoctonia* root rot on barley was more severe and grain yield lower when intervals were short between treating weeds and volunteer wheat or barley with the herbicide glyphosate and planting spring barley by direct drilling into the residue (Smiley *et al.*, 1992). Disease was most severe when glyphosate was applied 2-3 days before planting and least severe when glyphosate was applied the previous autumn.

The use of herbicides or other chemicals to kill potato shoots just before harvest time can also lead to increased incidence and severity of black scurf (AG 3) on potato tubers (Mulder *et al.*, 1992). During tuber growth, volatile substances are produced by the tubers that are either stimulatory or inhibitory to sclerotium formation. When the shoots are killed, the inhibitory substances are no longer produced and sclerotia formation is initiated (Dijst, 1990). In addition, populations of *R. solani* can increase due to saprophytic growth in the dying and dead haulms. Recently, a new technique was developed in Europe to reduce the tuber phase of this disease, namely 'green-crop-harvesting' (Mulder *et al.*, 1992). With this harvesting technique, the roots are undercut, the living haulms are pulled and removed from the plant row, and the potato tubers lifted onto the soil surface (reducing the concentrations of stimulatory volatiles). A fungicide or biological control agent (*Verticillium biguttatum*) can be applied, and then the tubers are covered again with soil to promote suberization. After about ten days, the tubers are harvested again. This proved to be an effective, integrated method to control black scurf on potatoes (Mulder *et al.*, 1992).

Besides an indirect effect on disease through increased susceptibility of the host, pesticides can also have a direct effect on pathogens such as *Rhizoctonia*. Pathogens can be stimulated at a low dose and inhibited at a high dose of the pesticide. For example, *R. solani* was found to grow normally at low dosages of the herbicide pymarin, and was increasingly inhibited in its growth with increasing concentrations (El-Abyad and Abu-Taleb, 1991). Aldrin, dieldrin, and lindane aggravated damping-off of pea seedlings caused by *Pythium ultimum* and *R. solani* (although all insecticides were fungistatic *in vitro*), while aldicarb, phorate and carbofuran increased damping-off of sugar beet seedlings caused by *R. solani* (Kataria *et al.*, 1989)

Different pesticides can also interact and lead to unforeseen results. One example is the synergistic action of several herbicides with the fungicide cyproconazole. Infection by *R. cerealis* (AG-D) on wheat was reduced more, when leaves were sprayed with the fungicide cyproconazole in combination with one of the herbicides DNOC, dicamba, ioxynil, or bromoxynil, than when either of the pesticides were used alone (Kataria and Gisi, 1990). Thus, some herbicides might predispose fungi to effects of a fungicide.

The use of a selective fungicide can also lead to replacement of sensitive pathogens by tolerant pathogens that can exploit the opened niche. For instance, benomyl used in cereals to control *F. culmorum* on rye can lead to the replacement of the sensitive pathogen *F. culmorum* by a tolerant one, namely *R. cerealis* (Van der Hoeven and Bollen, 1980). Conversely, the fungicide PCNB is effective in controlling *R. solani*, and can lead to increases in incidence and severity of disease caused by *Pythium* and *Fusarium* species (Rodriguez-Kabana and Curl, 1980).

Finally, pesticides can reduce populations of natural predators or parasites of plant pathogenic fungi, including *Rhizoctonia*, leading to increased disease incidence. A dramatic increase in infections by *R. solani* AG 3 on potato stolons and tubers was observed when insecticide-nematicides were used to control golden nematodes (*Globodera rostochiensis*). This surge in *Rhizoctonia* was attributed to a reduction in collembola populations and to a lesser extent populations of fungal feeding nematodes (Hofman, 1988). Collembola can also control *Rhizoctonia* disease on cotton (Curl *et al.*, 1988; Lartey *et al.*, 1994). Furthermore, the most effective biocontrol of *R. solani* on cotton occurred when fungal biocontrol agents (including *Laetisaria arvalis*) were combined with collembola (Lartey *et al.*, 1994). Similarly, the fungivorous nematode *Aphelenchus avenae* was able to reduce virulence and incidence of *R. solani* (Barnes *et al.*, 1981). From this discussion emerges the possible danger of using insecticides and nematicides that subsequently kill non-target organisms such as collembola, which in turn can lead to higher disease incidence by fungi such as *Rhizoctonia*.

VII. FERTILIZERS

A major difference in the management of fertility in organic and conventional farms is the use of synthetic fertilizers - usually NPK formulations - in conventional farms and the use of organic fertilizers (i.e., green manures, compost, etc.) in organically managed farms. Both synthetic and organic amendments can have an impact on disease incidence and severity caused by *Rhizoctonia* (see also chapter VI.A3).

It has long been known that fertilizer applications can have a profound effect on disease severity of *Rhizoctonia* (Engelhard, 1989; Parmeter, 1970). All essential elements can influence disease incidence and/or severity (Engelhard, 1989; Huber, 1978, 1980). Response to a particular nutrient can be different as one goes from deficiency, to adequate supply, to excess, toxicity, and/or salinity. Generally, deficiencies of potassium, nitrogen, or calcium increase *Rhizoctonia* incidence or severity (Baker and Martinson, 1970), but disease potential can also increase with excessive nitrogen fertilization (Baker and Martinson, 1970). Applications of NP, PK or NPK (applied as NH_4NO_3, NaH_2PO_4, and K_2SO_4 respectively) were effective in reducing disease severity of root rot caused by *R. solani* on field peas (Srihuttagum and Sivasithamparam, 1991). Huber (1980) reported that NH_4-N and K increase, while NO_3-N, and P decrease disease caused by *R. solani* on legumes and cereals, indicating that the kind of N fertilizer can make a difference. Papavizas *et al.* (1975) reported a positive correlation between saprophytic activity of *R. solani* (sugarbeet seed colonization and greenhouse disease severity indices) and total inorganic N and NH_4-N. Addition of ammonium sulfate, urea, ammonium nitrate or sodium nitrate all reduced *Rhizoctonia* root rot or bare patch of cereals (MacNish 1985; Brennan, 1991), while Pumphrey *et al.* (1987) reported that N in the form of ammonium nitrate and ammonium sulfate had no effect on *Rhizoctonia* root rot. Similarly, Peacock and Daniel (1992) reported that infection of turfgrass plants with *Rhizoctonia* species was unaffected by fertilizer treatments (urea, triple superphosphate and K_2SO_4). However, Taya (1988) reported that urea increased while superphosphate decreased dry root rot incidence caused by *R. bataticola* on chickpea. Finally, Walia and colleagues (1992) reported that all sources of N (urea, ammonium sulphate, sodium nitrate) and P (calcium monophosphate, sodiumphosphate) increased seedling mortality of cowpea due to *R. solani*. With such differing results in mind, Brennan (1991) stated that whether N is having a direct effect on the pathogen (*R. solani* causing bare patch) or an indirect effect through plant vigor is unresolved.

VIII. Summary and Conclusions

Root diseases caused by *Rhizoctonia* species were usually less severe in organic or low-input cropping systems than in conventional cropping systems. Above-ground diseases such as aggregate sheath spot on rice may be more severe in organic farms, especially if stubble is burned on conventional farms. The reduction in root disease severity in organic compared to conventional farms can be ascribed to longer rotations, regular applications of organic amendments, and abstinence from or reductions in pesticide use. Reduced tillage which can result in increases in root diseases caused by *Rhizoctonia* species in conventional agriculture (Cook and Haglund, 1991) did not pose a problem in integrated farming systems as long as other cultural practices such as appropriate rotations and use of organic amendments were implemented (El Titi and Richter, 1987). Moreover, when no-till practices were carried out over a period of ten years, the initial increase in bare patch disease of wheat was reversed into a suppression of this disease (Roget, 1995). The exact mechanisms of root disease control in organic and integrated farming systems are not known. However, it is generally assumed that organic amendments reduce root diseases by increasing the general level of microbial activity resulting in increased competition and/or antagonism in the rhizosphere. Relatively little attention has been paid to the role of the soil fauna in disease suppression. Since there are such striking differences in soil fauna between organic and conventional farming systems (El Titi and Ipach, 1989; Foissner, 1992), plant pathologists need to collaborate with nematologists and soil ecologists to study the role of the soil fauna in disease suppression (van Bruggen, 1995).

IX. References

Altman J & Rovira AD (1989) Herbicide-pathogen interactions in soil-borne root diseases. Can. J. Plant Pathol. 11:166-172.

Amrhein N, Deus B, Gehrke P, Hollander H, Schab J, Schultz A & Steinrücken HC (1981) Interference of glyphosate with the shikimate pathway. Proc. Plant Growth Regul. Soc. Am. 8: 99-106.

Anderson NA (1982) The genetics and pathology of *Rhizoctonia solani*. Ann. Rev. Phytopathol. 20:329-347.

Baby UI & Manibhushanrao K (1993) Control of rice sheath blight through the integration of fungal antagonists and organic amendments. Trop. Agric. 70:240-244.

Baker R & Martinson CA (1970) Epidemiology of diseases caused by *Rhizoctonia solani*. In: Parmeter JR (ed.). *Rhizoctonia solani*, Biology and Pathology (pp172-188) University of California Press, Berkeley.

Barnes GL, Russell CC & Foster WD (1981) *Aphelenchus avenae*, a potential biological control agent for root rot fungi. Plant Dis. 65: 423-424.

Belmar SB, Jones RK & Starr JL (1987) Influence of crop rotation on inoculum density of *Rhizoctonia solani* and sheath blight incidence in rice. Phytopathology 77:1138-1143.

Bochow H, Hentschel KD & Schmidt HH (1970) The influence of organic substances on the parasitic acitivity of *Rhizoctonia solani* in the soil. Arch. Pflanzenschutz 6:125-133.

Bollen GJ (1993) Mechanisms involved in non-target effects of pesticides on the incidence of soil-borne pathogens. In: Altman J (ed.). Pesticide interactions in crop production: beneficial and deleterious effects. CRC Press, Florida.

Brennan RF (1991) Effect of nitrogen and residual copper on the occurrence of rhizoctonia bare patch in wheat grown near Esperance, Western Australia. Austr. J. Exp. Agric. 31:259-62.

Butler EE (1993) *Rhizoctonia solani*. In: Lyda SD & Kenerley CM (eds.). Biology of Sclerotial-forming Fungi. (pp 87-112) Texas A&M University, College Station, Texas.

Chung YR, Hoitink HAJ, Dick WA & Herr LJ (1988a) Effects of organic matter decomposition level and cellulose amendment on the inoculum potential of *Rhizoctonia solani* in hardwood bark media. Phytopathology 78:836-840.

Chung YR, Hoitink HAJ & Lipps PE (1988b) Interactions between organic-matter decomposition level and soilborne disease severity. Agric. Ecosyst. Environm. 24:183-193.

Cook RJ (1994) Problems and progress in the biological control of wheat take-all. Plant Pathol. 43:429-437.

Cook RJ & Haglund WA (1991) Wheat yield depression associated with conservation tillage caused by root pathogens in the soil not phytotoxins from the straw. Soil Biol. Biochem. 23:1125-1132.

Cook RJ & Veseth RJ (1991) Wheat Health Management. APS Press: St. Paul, Minnesota.

Curl EA, Lartey R & Petersen CM (1988) Interactions between root pathogens and soil microarthropods. Agric. Ecosyst. and Environ. 24:249-261.

Daamen RA, Wijnands FG & van der Vliet G (1989) Epidemics of diseases and pests of winter wheat at different levels of agrochemical input. A study on the possibilities for designing an integrated cropping system. J. Phytopathology 125:305-319.

De Swardt GJ & Pauer GDC (1978) The Effect of plant material on the saprophytic and parasitic activity of *Rhizoctonia solani*. Phytophylactica 10:103-106.

Dijst G (1990) Effect of volatile and unstable exudates from underground potato plant parts on sclerotium formation by *Rhizoctonia solani* AG 3 before and after haulm destruction. Netherl. J. Plant Pathol. 96:155-170.

Doran JW, Fraser DG, Culik MN & Liebhardt WC (1987) Influence of alternative and conventional agricultural management on soil microbial processes and nitrogen availability. American Journal of Altern. Agric. 2:99-106.

Drinkwater LE, Letourneau DK, Workneh F, van Bruggen AHC & Shennan C (1995) Fundamental differences between conventional and organic tomato agroecosystems in California. Ecol. Applic. 5:1098-1112.

El-Abyad MS & Abu-Taleb AM (1991) Growth activities of the sugarbeet pathogens *Fusarium solani* (Mart.) Sacc., *Rhizoctonia solani* Kühn and *Sclerotium rolfsii* Sacc. under pymarin stress. Zentralbl. Mikrobiol. 146:419-424.

El Titi A & Ipach U (1989) Soil fauna in sustainable agriculture: Results of an integrated farming system at Lautenbach, F.R.G. Agric. Ecosyst. Environ. 27:561-572.

El Titi A & Landes H (1990) Integrated Farming System of Lautenbach: A practical contribution toward sustainable agriculture in Europe. In: Edwards CA Lal R Madden P Miller RH & House G (eds.). Sustainable Agricultural Systems. (pp 265-286) Soil and Water Conserv. Soc., Ankeny, Iowa.

El Titi A & Richter J (1987) Integrierter Pflanzenschutz im Ackerbau: Das Lautenbach Projekt. III. Schaedlinge und Krankheiten 1979-1983. Z. Planzenkrank. und Planzenschutz 94:1-13.

Engelhard AW (1989) Soilborne Plant Pathogens: management of diseases with macro- and microelements. APS Press, St. Paul, Minnesota.

Foissner W (1992) Comparative studies on the soil life in ecofarmed and conventionally farmed fields and grasslands of Austria. Agric. Ecosyst. Environ. 40:207-218.

Galindo JJ, Abawi GS, Thurston HD & Gálvez G (1983a) Effect of mulching on web blight of beans in Costa Rica. Phytopathology 73:610-615.

Galindo JJ, Abawi GS, Thurston HD & Gálvez G (1983b) Source of inoculum and development of bean web blight in Costa Rica. Plant Dis. 67: 1016-1021.

Herr LJ (1993) Host sources, virulence and overwinter survival of *Rhizoctonia solani* anastomosis groups isolated from field lettuce with bottom rot symptoms. Crop Prot. 12:521-526.

Hide GA & Read J (1991) Effects of rotation length, fungicide treatment of seed tubers and nematicide on diseases and the quality of potato tubers. Ann. Appl. Biol. 119:77-87.

Hofman TW (1988) Effects of granular nematicides on the infection of potatoes by *Rhizoctonia solani*. PhD Thesis, Agric. Univ. Wageningen, the Netherlands.

Huber DM (1978) Disturbed mineral nutrition. In: Horsfall JG & Cowling EB (eds.). Plant Disease - An Advanced Treatise. How Plants Suffer from Disease Vol. III (pp 163-181) Academic Press, New York.

Huber DM (1980) The role of mineral nutrition in defense. In: Horsfall JG Cowling EB (eds.). Plant Dis. - An Advanced Treatise. How Plants Defend Themselves Vol. V. (pp 381-406) Academic Press, New York.

Huiskamp T & Lamers JG (1992) Effects of cropping frequency on peas, *Vicia faba, Phaseolus vulgaris*, forage maize, fibre flax and onions. Verslag Proefstation Akkerb. Groentet. Vollegrond 143 (82 pp).

Inbar Y, Boehm MJ & Hoitink HA (1991) Hydrolysis of fluorescein diacetate in sphagnum peat container media for predicting suppressiveness to damping-off caused by *Pythium ultimum*. Soil Biol. Biochem. 23:479-483.

Jager G & Velvis H (1980) Onderzoek naar het voorkomen van *Rhizoctonia*-werende aardappel-percelen in noord-Nederland. Inst. Bodem vruchtbaarheid Rep. 1-80:66 pp.

Johal GS & Rahe JE (1988) Glyphosate, hypersensitivity and phytoalexin accumulation in the incompatible bean anthracnose host-parasite interaction. Physiol. Mol. Plant Pathol. 32:267-281.

Jones RK & Belmar SB (1989) Characterization and pathogenicity of *Rhizoctonia* spp. isolated from rice, soybean, and other crops grown in rotation with rice in Texas. Plant Dis. 73:1004-1010.

Kataria HR & Gisi U (1990) Interactions of fungicide-herbicide combinations against plant pathogens and weeds. Crop Prot. 9:403-409.

Kataria HR, Singh H & Gisi U (1989) Interactions of fungicide-insecticide combinations against *Rhizoctonia solani in vitro* and in soil. Crop Prot. 8:399-404.

Kundu PK & Nandi B (1985) Control of Rhizoctonia disease of cauliflower by competitive inhibition of the pathogen using organic amendments in soil. Plant and Soil 83:357-362.

Lartey RT, Curl EA, Peterson CM & Williams JC (1994) Interactions of mycophagous collembola and biological control fungi in the suppression of *Rhizoctonia solani*. Soil Biol. Biochem. 26:81-88.

Leach SS, Porter GA, Rourke RV & Clapham WM (1993) Effects of moldboard plowing, chisel plowing and rotation crops on the rhizoctonia disease of white potato. Amer. Potato J. 70:329-377.

Lévesque CA & Rahe JE (1992) Herbicide interactions with fungal root pathogens, with special reference to glyphosate. Annu. Rev. Phytopathol. 30:579-602.

Lewis JA, Lumsden RD, Papavizas GC & Kantzes JG (1983) Integrated control of snap bean diseases caused by *Pythium* spp. and *Rhizoctonia solani*. Plant Dis. 67:1241-1244.

Lewis JA & Papavizas GC (1980) Integrated control of Rhizoctonia fruit rot of cucumber. Phytopathology 70:85-89.

Lumsden RD, Lewis JA & Millner PD (1983) Effect of composted sewage sludge on several soilborne pathogens and diseases. Phytopathology 73:1543-1548.

MacNish GC (1984) The use of undisturbed soil cores to study methods of controlling rhizoctonia patch of cereals. Plant Pathol. 33:355-359.

MacNish GC (1985) Methods of reducing rhizoctonia patch of cereals in Western Australia. Plant Pathol. 34:175-181.

Mulder A, Turkesteen LJ & Bouman A (1992) Perspectives of green-crop-harvesting to control soil-borne and storage diseases of seed potatoes. Netherl. J. Plant Pathol. 98:103-114.

National Research Council (1989) Alternative Agriculture. National Academy Press, Washington, D.C.

Nitta T (1991) Diversity of root fungal floras: Its implications for soil-borne diseases and crop growth. Japan Agricultural Research Quarterly 25:6-11.

Ogoshi A (1987) Ecology and pathogenicity of anastomosis and intraspecific groups of *Rhizoctonia solani* Kühn. Annu. Rev. Phytopathol. 25:125-143.

Papavizas GC (1970) Colonization and growth of *Rhizoctonia solani* in soil. In: Parmeter JR Jr (ed.). *Rhizoctonia solani*, Biology and Pathology. (pp 108-122) University of California Press, Berkeley, CA.

Papavizas GC, Adams PB, Lumsden RD, Lewis JA, Dow RL, Ayers WA & Kantzes JA (1975) Ecology and epidemiology of *Rhizoctonia solani* in field soil. Phytopathology 65:871-877.

Parmeter JR Jr (ed.) (1970) *Rhizoctonia solani*, Biology and Pathology. University of California Press. Berkeley, CA 255 pp.

Peacock CH & Daniel PF (1992) A comparison of turfgrass response to biologically amended fertilizers. HortScience 27:883-884.

Piorr HP & Hindorf H (1986) The implication of Plant Dis. and pests during the conversion from conventional to biological agriculture. In: Vogtmann H Boehncke E & Fricke I (eds.) The Importance of Biological Agriculture in a World of Diminishing Resources (pp 421-435) Verlagsgruppe Witzenhausen, Germany, 1986.

Pumphrey FV, Wilkins DE, Hane DC & Smiley RW (1987) Influence of tillage and nitrogen fertilizer on rhizoctonia root rot (bare patch) of winter wheat. Plant Dis. 71:125-127.

Rickerl DH, Curl EA & Touchton JT (1989) Tillage and rotation effects on Collembola populations and *Rhizoctonia* infestation. Soil and tillage Res. 15:41-49.

Rickerl DH, Curl EA, Touchton JT & Gordon WB (1992) Crop mulch effects on *Rhizoctonia* soil infestation and disease severity in conservation-tilled cotton. Soil Biol. Biochem. 24:553-557.

Rodriguez-Kabana R & Curl EA (1980) Nontarget effects of pesticides on soilborne pathogens and disease. Annu. Rev. Phytopathol. 18:311-332.

Roget DK (1995) Decline in root rot in wheat caused by *Rhizoctonia solani* AG 8 in a tillage and rotation trial at Avon, South Australia. Abstract No. P-7-6, International Symposium on *Rhizoctonia*. Noordwijkerhout, The Netherlands.

Roget DK, Venn NR & Rovira AD (1987) Reduction of rhizoctonia root rot of direct-drilled wheat by short-term chemical fallow. Austral. J. Experim. Agric. 27:425-430.

Rovira AD (1986) Influence of crop rotation and tillage on Rhizoctonia bare patch of wheat. Phytopathology 76:669-673.

Rovira AD & McDonald HJ (1986) Effects of the herbicide chlorsulfuron on Rhizoctonia bare patch and take-all of barley and wheat. Plant Dis. 70:879-882.

Ruppel EG (1985) Susceptibility of rotation crops to a root rot isolate of *Rhizoctonia solani* from sugar beet and survival of the pathogen in crop residues. Plant Dis. 69:871-873.

Ruppel EG (1991) Survival of *Rhizoctonia solani* in fallow field soil and buried sugarbeet roots at three depths. J. Sugar Beet Res. 28: 141-153.

Rush CM & Winter SR (1990) Influence of previous crops on *Rhizoctonia* root and crown rot of sugar beet. Plant Dis. 74:421-425.

Schueller C, Biala J, Bruns C, Gottschall R, Ahlers S & Vogtmann H (1989) Suppression of root rot on peas, beans and beetroots caused by *Pythium ultimum* and *Rhizoctonia solani* through the amendment of growing media with composted organic household waste. J. Phytopathol. 127:227-238.

Scow KM, Somasco O, Gunapala N, Lau S, Venette R, Ferris H, Miller R & Shennan C (1994) Transition from conventional to low-input agriculture changes soil fertility and biology. Calif. Agric. 48:20-26.

Sivapalan A, Morgan WC & Franz PR (1993) Monitoring populations of soil microorganisms during a conversion from a conventional to an organic system of vegetable growing. Biol. Agric. Hortic. 10:9-27.

Smiley RW, Ogg AG Jr. & Cook RJ (1992) Influence of glyphosate on rhizocotnia root rot, growth, and yield of barley. Plant Dis. 76:937-942.

Specht LP & Leach SS (1987) Effects of crop rotation on *Rhizoctonia* disease of white potato. Plant Dis. 71:433-437.

Srihuttagum M & Sivasithamparam K (1991) The influence of fertilizers on root rot of field peas caused by *Fusarium oxysporum*, *Pythium vexans* and *Rhizoctonia solani* inoculated singly or in combination. Plant and Soil 132:21-27.

Stephens PM, Davoren CW, Doube BM, Ryder MH, Benger AM & Neate SM (1993) Reduced severity of *Rhizoctonia solani* disease on wheat seedlings associated with the presence of the earthworm *Aporrectodea trapezoides* (Lumbricidae). Soil Biol. Biochem. 25:1477-1484.

Sumner DR & Bell DK (1986) Influence of crop rotation on severity of crown and brace root rot caused in corn by *Rhizoctonia solani*. Phytopathology 76:248-252.

Taya RS, Tripathi NN & Panwar MS (1988) Influence of soil type, soil moisture and fertilizers on the severity of chickpea dry root-rot caused by *Rhizoctonia bataticola* (Taub.) Butler. Indian J. Mycol. Pl. Pathol. 18:133-136.

Temple SR, Somasco OA, Kirk M & Friedman D (1994) Conventional, low-input and organic farming systems compared. Calif. Agric. 48:14-19.

Tu JC & Tan CS (1991) Effect of soil compaction on growth, yield and root rots of white beans in clay loam and sandy loam soil. Soil Biol. Biochem. 23:233-238.

Van Bruggen AHC (1995) Plant Disease severity in high-input compared to reduced-input and organic farming systems. Plant Dis. 79:976-984.

Van der Hoeven EP & Bollen GJ (1980) Effect of benomyl on soil fungi associated with rye, 1. Effect on the incidence of sharp eyespot caused by *Rhizoctonia cerealis*. Neth. J. Pl. Path. 86: 163-180.

Vilgalys R & Cubeta MA (1994) Molecular systematics and population biology of *Rhizoctonia*. Annu. Rev. Phytopathol. 32:135-155.

Voland RP & Epstein AH (1994) Development of suppressiveness to diseases caused by Rhizoctonia solani in soil amended with composted and noncomposted manure. Plant Dis. 78:461-466.

Walia GS, Sunder S & Grover RK (1992) Influence of nutrients on pathogenic behaviour of *Rhizoctonia solani* on cowpea. Indian J. Mycol. Pl. Pathol. 22:170-177.

Wall RE (1984) Effects of recently incorporated organic amendments on damping-off of conifer seedlings. Plant Dis. 68:59-60.

Weller DM, Cook RJ, MacNish G, Bassett EN, Powelson RL, & Petersen RR (1986) Rhizoctonia root rot of small grains favored by reduced tillage in the Pacific Northwest. Plant Dis. 70:70-73.

Wijetunga C & Baker R (1979) Modeling of phenomena associated with soil suppressive to *Rhizoctonia solani*. Phytopathology 69:1287-1293.

Workneh F & van Bruggen AHC (1994) Microbial density, composition, and diversity in organically and conventionally managed rhizosphere soil in relation to suppression of corky root of tomatoes. Appl. Soil Ecol. 1:219-230.

VI.A2 SOIL SOLARIZATION FOR THE CONTROL OF DISEASES CAUSED BY *RHIZOCTONIA SPP.*

JAACOV KATAN

Department of Plant Pathology and Microbiology, The Hebrew University of Jerusalem, Faculty of Agriculture, Rehovot 76100, ISRAEL.

I. Introduction

Diseases caused by *Rhizoctonia* spp. are usually difficult to control, as is common with soilborne pathogens. This stems from the fact that the pathogen's life cycle is connected with the soil. It is difficult to reach the pathogen at all sites therein, especially when it is embedded in the debris of infected tissues. Moreover, drastic methods of control, although effective, may also adversely affect nontarget beneficial organisms. In severe cases they may even increase disease incidence (Baker and Cook, 1974).

Rhizoctonia spp. pathogens can cause different types of diseases, e.g. damping-off at the seedling stage, root rots, stem cankers, and foliar and fruit diseases. Controlling the soilborne inoculum is not sufficient to control all of these diseases, raising the need for integrated approaches. Seedling diseases, which are limited to several weeks of plant growth, are usually easier to control than soilborne diseases of mature plants. This, because in the latter case, the soilborne inoculum has to be controlled to considerable soil depths and the plant has to be protected for the entire growth season. When choosing control methods, many factors need to be considered, such as effectiveness of control, yield increase and improvement of quality, long-term effect, possible negative side effects, technological feasibility, environmental impact and cost.

Many methods have been developed for the control of diseases caused by *Rhizoctonia* spp. These include fungicide application, breeding for resistant cultivars, sanitation, crop rotation, biological control and soil disinfestation. The need for many methods of control, stems from the fact that usually none of them are perfect, nor can any one be used in all circumstances. Moreover, the life cycle of the pathogen may vary in different crop systems, thus requiring different control strategies. Therefore, any new control method, even if restricted in its use, is of value since it adds to our rather limited arsenal of control methods. This is especially true for novel nonchemical methods of control which are needed to replace toxic and hazardous pesticides. In recent years, considerable efforts have been devoted to the development of novel, environmentally acceptable methods.

The concept of soilborne diseases control has changed over the last decades. In the past, control of these pathogens concentrated on eradication of pathogen populations. Later, it has been realized that effective control could be achieved by interrupting the disease cycle via manipulation of the pathogen population, plant resistance or the microbial balance, leading to economic reduction of disease level, rather than absolute control. The integrated pest management concept encompasses many elements; it is in essence a holistic approach, the ultimate goal of which is achieved one step at a time

B. Sneh et al. (eds.),
Rhizoctonia Species: Taxonomy, Molecular Biology, Ecology, Pathology and Disease Constrol, 423–432.

(Katan, 1996). In this context, soil solarization can play a significant role. *Rhizoctonia* spp. may produce different types of inocula in different crop systems, such as sclerotia, mycelia, fungal structures embedded in the plant tissues, etc. These structures may survive in different niches, e.g. soil, propagation material, plant debris attached to the greenhouse structure, etc. This chapter will concentrate on the control of soilborne inoculum of *Rhizoctonia* spp. by soil solarization. However, solarization can potentially play a role in controlling *Rhizoctonia* spp. in other ways: sanitation (Besri, 1991), sanitation of the greenhouse structure by space solarization (Shlevin *et al.*, 1995), production of pathogen-free propagation material, such as bulbs and nurseries (Chet *et al.*, 1982, Dupp and Barnaart, 1992; Nataraj, 1993; Annesi and Motta, 1994) and the control of fruit rots (Keinath, 1995).

Soil disinfestation is one of the approaches to control root diseases. It is especially common with intensive and high-value crops, e.g. crops grown in greenhouses. It is a sophisticated, expensive but effective control method, which has great advantages, but also limitations. The basic principle is to eradicate the harmful agents in soil before planting, using drastic chemical or physical means. The method has the advantage of controlling a wide spectrum of pests: pathogens, weeds and arthropod pests, in addition to *Rhizoctonia* spp. There are three approaches to soil disinfestation. The first two: steaming and fumigation, were developed over 100 years ago, and until recently were the only approaches (Chen *et al.*, 1991; Newhall, 1955). The third approach is soil solarization (also called solar heating), which is relatively new (Katan *et al.*, 1976; Katan and DeVay, 1991). In this chapter, some current developments in soil solarization, with special reference to *Rhizoctonia* spp. will be described and discused.

II. Principles of soil solarization

The principles of soil solarization have been described in detail in many publications, including reviews (e.g. Katan, 1981; Stapleton and DeVay, 1986; Borges, 1990; Davis, 1991; Gaur and Pery, 1991; Stapleton and DeVay, 1995; Souza, 1994; Katan, 1996) and two books (DeVay *et al.*, 1991; Katan and DeVay, 1991). Mulching (covering) the soil with transparent (not black) polyethylene or polyvinyl sheets is, at present, the most common means of solarization of soils, although there are exceptions. Effective control by solarization, provided that climatic conditions are adequate, can be achieved under the following conditions: (a) Soil mulching should be carried out during periods of high temperature and intense solar irradiation. (b) The soil should be kept moist to increase the thermal sensitivity of resting structures and to improve heat conduction. (c) A thin polyethylene tarp (25-50 mm) is recommended, because it is both cheaper and somewhat more effective than thicker ones. Since the lower soil layer is less intensively heated than the upper ones, the mulching period should be sufficiently long--usually 4 weeks or more--to achieve pest control at all desired depths. Usually, the longer the mulching period, the deeper its effectiveness and the higher the pathogen- killing rates.

Typical maximal temperatures in solarized plots in which effective disease and weed control have been obtained are within the range of 45-50 and 38-45°C at depths of 10 and 20 cm, respectively, although temperatures of 5-10°C higher have also been recorded. Temperatures in solarized soils are 5-15°C higher than those in the comparable non-solarized ones.

III. Control of Rhizoctonia diseases by solarization

Rhizoctonia solani is relatively sensitive to elevated temperatures (Pullman *et al.*, 1981 b; Fig. 1) and is likely to be effectively controlled by solarization. This has been indeed demonstrated in many studies in both artificially and naturally infested soils. Populations of *R. solani* in soil, as assessed by various methods, have been reduced to various levels (even down to 0%) by solarization (e.g. Al-Raddad, 1979; Elad *et al.*, 1980; Pullman *et al.*, 1981a; Ki and Kim, 1985; Lee, 1985; Hernandez *et al.*, 1987; Vannacci and Triolo, 1989; Kaewruang *et al.*, 1989a,b; Triolo *et al.*, 1989; Fahim *et al.*, 1990; Dwivedi, 1993;, Keinath, 1995). For example, solarization reduced the number of propagules of *R. solani* from 4 in 100g soil to under detectable levels (Pullman et al., 1981a) in percentage of organic matter colonized by the fungus by 83-96% (Keinath, 1995). Usually, satisfactory disease reduction is also obtained by solarization, reaching 90% reduction or more in certain cases. This has been

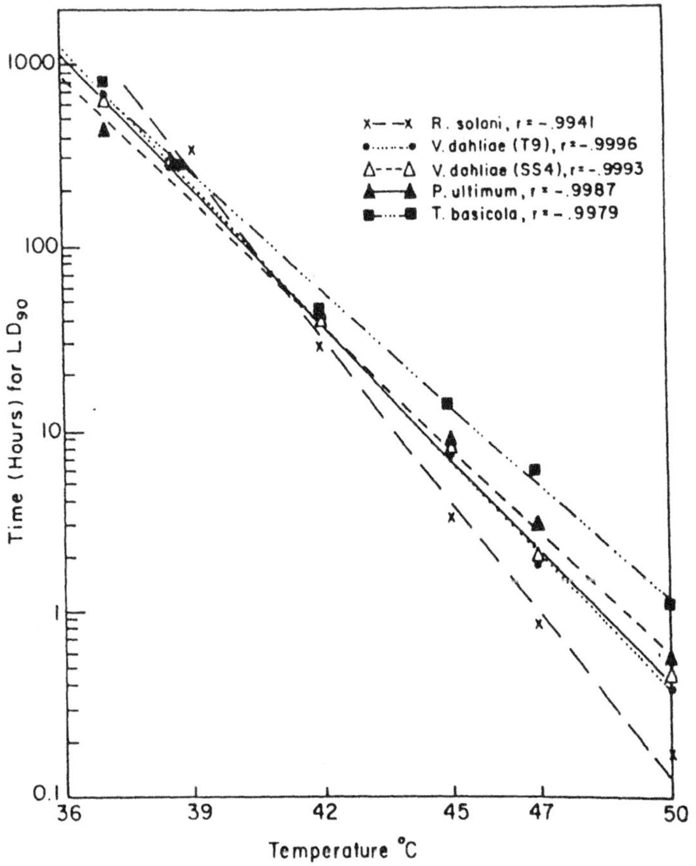

FIGURE 1. Time and temperature exposures required to kill 90% of the propagules of *Rhizoctonia solani*, *Verticillium dahliae* (two pathotypes), *Pythium ultimum* and *Thielaviopsis basiola* on potato dextrose agar medium (After Pullman *et al.*, 1981b).

demonstrated with a variety of crops, such as iris bulbs (Chet *et al.*, 1980), beans (Tamietti *et al.*, 1987; Garibaldi and Tamietti, 1989), onions (Katan *et al.*, 1980), *Pinus nigra* (Annesi and Motta, 1994), strawberries (Abdul Razik *et al.*, 1989; Fahim *et al.*, 1990), spinach (Horiuchi, 1991), lettuce (Triolo *et al.*, 1989), radish (Vannacci *et al.*, 1987; Triolo *et al.*, 1989), cotton (Melero-Vara *et al.*, 1995), cucumber (Hassan and Younis, 1984), gerbera (Kaewruang *et al.*, 1989b) *Gypsophila* (Gamliel *et al.*, 1993) and lupin (Osman and Sahab, 1983). The incidence of Rhizoctonia on potato shoots was reduced by solarization (Grinstein *et al.*, 1979; Elad *et al.*, 1980; Abdul-Rahman, 1987), as was that on potato tubers in one case (Elad *et al.*, 1980), but not in another (Abdul-Rahman, 1987).

In arecent comprehensive study carried out for three years (Keinath, 1995), solarization reduced inoculum density of *R. solani* as well as belly rot severity on pickling cucumber in two, out of three years. Garibaldi and Tamietti (1989) improved the solarization process by closing the glass or plastic-house. This approach has potentially wide applicability and should be considered in cooler regions (Garibaldi and Gullino, 1991). Frequently, disease control is accompanied by an increase in plant biomass and yield, as well as in weed control. It is interesting to note that solarization for the control of *R. solani* was also successfully attempted by exposing soil to natural solar irradiation (Nataraj, 1993), a simple method which has been known for decades (Groshevoy, 1939) but unfortunately neglected.

In certain cases, solarization improves product quality in addition to yield, as seen, for example, with *Gypsophila* (Gamliel *et al.*, 1993). Solarization of soil infested with *R. solani* and planted with radish resulted in a shift towards higher grades of marketable roots (Vannacci *et al.*, 1987). Similarly, in gerberas solarization reduced root rots, and increased the yield (by 250%) and flower diameter (Kaewruang *et al.*, 1989a). Combining solarization with an application of *Trichoderma harzianum* increased the weight of iris bulbs in soil infested with *R. solani* (Chet *et al.*, 1982).

A long-term effect of solarization on disease control and on yield increase, extending for a second or even a third crop, has been observed with the control of Rhizoctonia diseases in several crops (Osman and Sahab, 1983; Garibaldi and Tamietti, 1989), as well as in cooler regions, as shown with *Verticillium dahliae* on potato in Idaho (Davis and Sorensen, 1986).

Several publications list studies which have been performed to control Rhizoctonia diseases (and other pests) by solarization (e.g. Stapleton and DeVay, 1986; Davis, 1991; Souza, 1994). In the latter publication, a list of 187 references, including nine dealing with the control of *R. solani* by solarization, was given.

IV. Control of Other Pests by Solarization

Soil solarization under the proper conditions can control a wide spectrum of pests. Even if solarization is initially intended to control Rhizoctonia diseases in certain crops, additional pathogens and many weeds are also controlled, to various levels, hence increasing the economic value of this method. The controlled pathogens include fungi such as species or pathotypes of *Verticillium*, *Fusarium*, *Sclerotinia*, *Sclerotium*, *Pythium* and *Rosellinia*, as well as many nematodes. This additional benefit of soil solarization should be taken into consideration. Weed control is one of the visible results of solarization. Annual weeds are usually more sensitive to solarization than perennial ones (Rubin and Benjamin, 1983; Elmore, 1991). As expected, some pathogens are usually not controlled by solarization, e.g. *Macrophomina phaseolina* (Mihail and Alcorn, 1984).

V. Pathogen Control: Physical and Microbial Processes During and After Solarization

The reduction in disease incidence in plants growing in solarized soils, as with any soil treatment, results from the effects exerted on each of the three living components involved in disease (host, pathogen and surrounding microorganisms), as well as on the physical and chemical environment, which in turn, affects the activity and interrelationships of the organisms (Stapleton and DeVay, 1986; Katan *et al.*, 1989; DeVay and Katan, 1991). Although these microbial processes occur primarily during solarization, they may continue, to various extents and in different ways, after the removal of the polyethylene sheets and planting. The most pronounced effect of soil mulching with polyethylene is a physical one, i.e. an increase in soil temperatures, for several hours daily during the solarization period. However, other accompanying processes, such as shifts in the microbial populations, changes in the chemical composition and physical structure of the soil (Chen *et al.*, 1991), high moisture levels maintained by the polyethylene mulch, and changes in gas composition of the soil should also be considered when analyzing mechanisms of disease control. Hence, in the long run, solarization (or any type of soil disinfestation) may lead to soil suppressiveness, reflected by slow reinfestation, if beneficial biological processes are enhanced, or by rapid reinfestation, if antagonists are adversely affected, the latter resulting in the creation of a "biological vacuum". Studies have frequently shown that solarized soils become hostile and less receptive to pathogen reinfestation. This can be regarded as induced suppressiveness (Hardy and Sivasithamparam, 1985; Greenberger *et al.*, 1987; Gamliel and Katan, 1993). However, the possible creation of a biological vaccuum by soil solarization, under certain conditions, should not be excluded (Greenberger *et al.*, 1987; DeVay and Katan, 1991).

Thermal death or inactivation of a population of an organism depends on both temperature and exposure time, which are inversely related. In many cases, heat-mortality curves are exponential. Some studies investigating the time-temperature relationship of thermal killing have obtaind straight lines by plotting the logarithms of the number of survivors against the exposure time units at a given temperature. Pullman *et al.* (1981b) obtained a linear relationship by plotting the logarithm of the time required to kill 90% of the propagules of various soilborne fungal pathogens against temperature (Fig. 1). They found that temperatures of 39°C and 50°C killed cultures of *R. solani* in 14 days and 10 minutes, respectively.

Microbial processes induced by solarization may contribute to disease control (in addition to the physical effect of heat), since the impact of any lethal agent in the soil extends beyond target organisms. Such processes may be especially useful where the cumulative effect of heat may be insufficient for disease control, e.g. in deeper soil layers or during marginal seasons. Biological control may operate at any stage of pathogen survival or disease development, during or after solarization, via antibiosis, lysis, parasitism or competition. Solarization may affect inoculum density, inoculum potential or both. The mechanisms of biological control, which may be created or stimulated by solarization (or any disinfestation method), have been described alsewhere (Katan, 1981; Katan *et al.*, 1989; DeVay and Katan, 1991 and others). Garibaldi and Tamietti (1989) found that solarization significantly reduced incidence of Rhizoctonia disease on bean. The effect lasted for two to four successive bean crops grown in the same plot. This suggests that recolonization of *R. solani* in solarized soil is slowed, apparently due to a beneficial microbial shift in the solarized soil. Similarly, the buildup of *R. solani* populations in solarized soil was slower than in nontreated soil, suggesting a beneficial microbial shift (Elad *et al.*, 1980). Solarization increased the proportion of bacteria and fungi antagonistic to *R. solani* (Kaewruang *et al.*, 1989a).

Antogonism may be improved further if the pathogen's propagules are weakened by sublethal heating and become vulnerable to microbial activity, as shown with *F. oxysporum* f. sp. *niveum* (Freeman and Katan, 1988). Disease incidence in watermelon seedlings inoculated with heat-treated conidia of *F. oxysporum* f. sp. *niveum* was reduced by 35-82%.

VI. Increased growth response

The phenomenon of increased growth response (IGR) denotes the improvement in plant growth when disinfestation is carried out in soils free of known pathogens. This phenomenon was discovered in both artificially heated and in fumigated soils several decades ago. It has also been observed in solarized soils (Chen and Katan, 1980; Stapleton and DeVay, 1984; Stapleton *et al.*, 1985; Chen *et al.*, 1991; Gamliel and Katan, 1991; Grunzweig *et al.*, 1993). The mechanisms which can explain IGR are either chemical (release of mineral nutrients or growth factors; nullification of toxins) or biological (elimination of minor or unknown pathogens, stimulation of beneficial microorganisms). Indeed, higher concentrations of mineral and organic substances were found in solutions of solarized soils. Such a phenomenon represents a potentially beneficial side effect of *Rhizoctonia* spp. control by solarization.

VII. Application

The major topics related to the application and implementation are: mulching technology, plastic chemistry, economic analysis and extension for introducing the method into new regions. Soil solarization is being investigated in more than 40 countries (Katan, 1996), and in some of them, e.g. Israel, Italy, Japan, Greece, Egypt, Iraq, Mexico and the USA, it is applied by growers for specific crops. The technological aspects of soil solarization are described elsewhere (Grinstein and Hetzroni, 1991). With strip application, each solarized bed is surrounded by infested, nonsolarized soil. It is therefore less effective as compared to mulching as the whole plot.

VIII. Future studies and improvements

1. Solarization could be extended to cooler regions or marginal seasons via a variety of approaches, e.g. soil solarization in a closed greenhouse (Garibaldi and Gullino, 1991) and the use of novel plastic materials, or of double layers of plastic (Ben-Yephet *et al.*, 1987; Tamietti *et al.*, 1987) and combining with other methods.

2. Solarization can be combined with other methods of control, such as biocontrol agents or pesticides at *reduced* dosages to obtain improved control with a wider range of target pests and climatic regions and to extend its effectiveness over longer periods. When applicable, soil solarization should be incorporated into integrated control systems. Combining solarization with the application of the

biocontrol agent *Trichoderma harzianum* was more effective in delaying the buildup of *R. solani* populations in a potato field (Elad *et al.*, 1980) and in improving iris bulb quality (Chet *et al.*, 1982). Combining solarization with compost or manure is an especially attractive option (Gamliel and Stapleton, 1993).

3. Further studies should concentrate on elucidating the biological mechanisms involved in solarization mode of action. One should be alert to possible negative side-effects.

4. Solarization has a great potential for producing healthy propagation material, e.g. bulbs (Chet *et al.*, 1982), nurseries (Duff and Barnaat, 1992; Nataraj, 1993; Annesi and Motta, 1994), seeds, etc. The possibility of producing seedlings or transplants in nurseries, using solarized suppressive soils should be considered. Emphasis should also be placed on the solarization of container media and potting mixes (Gamliel *et al.*, 1989; Duff and Barnaart, 1992; Kaewruang *et al.*, 1989b).

5. Studies should be performed on the potential of employing used polyethylene (Al-Raddad, 1979; Avissar *et al.*, 1986). This may provide an extremely inexpensive material for solarization.

IX. General remarks

Soil solarization, like all control methods, has both advantages and limitations. As compared with fumigation, solarization is a nonchemical method, simple and less expensive, which appears to cause no major undesirable disturbances in the soil's biological balance. It is, however, restricted to certain climatic areas and seasons and ties up the field for several weeks. Solarization should not be regarded as a magic tool to cure all ills, but rather as an additional option for control to be used only under the suitable climatic, biological and economic conditions and to be integrated into disease management. Soil solarization is a relatively new method. Many of its features, potential benefits and drawbacks have yet to be explored. The potential benefits have to be adopted and adapted while avoiding the harmful side effects. The search for new, nonchemical methods of control which are effective, economical and have minimal undesirable side effects is a continuing task. Additional nonchemical options of control, to be used alone or in combination with other methods, are necessary in this era in which many effective pesticides have been banned or restricted in use. The development of new methods requires interdisciplinary research and international cooperation. Last but not least, the successful introduction of soil solarization (or any new method of pest management) depends on its proper implementation and introduction to the growers, as well as on economic considerations.

X References

Abdul Rahman M & Katan J (1987) Effect of solarization and ethylene dibromide on root knot and Rhizoctonia canker in potato. Biol. Cultural Control 2:19.

Abdul Razik A, Grinstein A, Zeydan O, Rubin B, Tal A & Katan J (1989) Soil solarization and fumigation of strawberry plots. Acta Hort. 265:586-590.

Al-Raddad AMM (1979) Soil Disinfestation by Plastic Tarping. MSc thesis. Univ. Jordan.

Al-Samarria H, El-Bahaldi AH & Al-Rawi F (1988) Comparison between effects of soil disinfestation methods on some pathogens of cucumber. Arab J. Pl. Prot. 6:106-112 (Abstr. from Rev. Pl. Pathol. 69:546).

Annesi T & Motta E (1994) Soil solarization in an Italian forest nursery. Eu. J. For. Pathol. 24:203-209.

Avissar R, Naot O, Mahrer Y & Katan J (1986) Field aging of transparent polyethylene mulches: II Influence on the effectiveness of soil heating. Soil Sci. Am. J. 50, 205-209.

Baker KF & Cook RJ (1974) Biological Control of Plant Pathogens. Freeman, San Francisco.

Ben-Yephet Y, Stapleton JJ, Wakeman RJ & DeVay JE (1987) Comparative effect of soil solarization with single and double layers of polyethylene film on survival of *Fusarium oxysporum* f. sp. *vasinfectum*. Phytoparasitica 15:181-187.

Besri M & Diop M (1985) Control of *Didymella lycopersici* in tomato by storing the supports in plastic tunnels: new application of the solar heating or solarization. Rev. Hort. 58:99-102.

Borges MLV (1990) Solarizacao do soloe protecao das culturas. Rev. de Cienc. Agrar. 13:91-93.

Chen Y & Katan J (1980) Effect of solar heating of soils by transparent polyethylene mulching on their chemical properties. Soil Sci. 130:271-277.

Chen Y, Gamliel A, Stapleton JJ & Aviad T (1991) Chemical, physical and microbial changes related to plant growth in disinfested soils. In: Katan J & DeVay JE (eds) Soil Solarization (pp 103-129) CRC Press, Boca Raton, FL.

Chet I, Elad Y, Kalfon A, Hadar Y & Katan J (1982) Integrated control of soilborne and bulbborne pathogens in iris. Phytoparasitica 10:229-236.

Davis JR (1991) Soil solarization: pathogen and disease control and increases in crop yield and quality: short and long-term effects and integrated control. In: Katan J & DeVay JE (eds) Soil Solarization (pp 39-50) CRC Press, Boca Raton, FL.

Davis JR & Sorensen LH (1986) Influence of soil solarization at moderate temperatures on potato genotypes with differing resistance to *Verticillium dahliae*. Phytopathology 78:1956-1661.

DeVay JE & Katan J (1991) Mechanisms of pathogen control in solarized soils. In: Katan J & DeVay JE (eds) Soil Solarization (pp 87-101) CRC Press, Boca Raton, FL.

DeVay JE, Stapleton JJ & Elmore CL (1991) (eds) Soil Solarization. Food & Agricultural Organization, Rome.

Duff JA & Barnaart A (1992) Solarization controls soilborne fungal pathogens in nursurey potting mixes. Australasian Pl. Pathol. 21:20-23.

Dwivedi SK (1993) Soil solarization adversely affects some fungal pathogens causing wilt disease of Guava (*Psidium guazava* L.). Soil Biol. Biochem. 25: 1635-1636.

Elad E, Katan J & Chet I (1980) Physical, biological, and chemical control integrated for soilborne diseases in potatoes. Phytopathology 70:428-422.

Elmore, CL (1991) Weed control by soil solarization. In: Katan J & DeVay JE (eds). Soil Solarization (pp 61-72) CRC Press, Boca Raton, FL.

Fahim MM, Attia MF & Abada KA (1990) Effect of solarization on fungi causing root and crown-rot disease of strawbery. Proc. 8th Congr. Mediter. Phytopath. Union. pp 393-395.

Freeman S & Katan J (1988) Weakening effect of propagules of *Fusarium* by sublethal heating. Phytopathology 87:1656-1661.

Gamliel A & Katan J (1991) Involvement of fluorescent pseudomonads and other microorganisms in increased growth response of plants in solarized soils. Phytopathology 81:494-502.

Gamliel A & Katan J (1993) Supression of major and minor pathogens by fluorescent pseudomonads in solarizaed and nonsolarized soils. Phytopathology 83:68-75.

Gamliel A & Stapleton JJ (1993) Effect of chicken compost or ammonium phosphate and solarization on pothogen control, rhizosphere microorganisms, and lettuce growth. Pl. Dis. 77:886-891.

Gamliel A, Katan J, Chen Y & Grinstein A (1989) Solarization for the recycling of container media. Acta Hort. 255:181-188.

Gamliel A, Hadar E & Katan J (1993) Improvement of growth and yield of *Gypsophila paniculata* by solarization or fumigation of soil or container medium in continuous cropping systems. Pl. Dis. 77:933-938.

Garibaldi A & Gullino ML (1991) Soil solarization in southern European countries, with emphasis on soilborne disease control of protected crops. In: Katan J & DeVay JE (eds) Soil Solarization (pp 227-235) CRC Press, Boca Raton, FL.

Garibaldi A & Tamietti G (1989) Solar heating: Recent results obtained in northern Italy. Acta Hort. 255:125-130.

Gaur HS & Perry RN (1991) The use of soil solarization for control of plant parasitic nematodes. Nematol. Abstr. 9:153-167.

Greenberger A, Yogev A & Katan J (1987) Induced suppressiveness in solarized soils. Phytopathology 77:1663-1667.

Grinstein A & Hetzroni A (1991) The technology of soil solarization. In: Katan J & DeVay JE (eds) Soil Solarization (pp 159-170) CRC Press, Boca Raton, FL.

Grinstein A, Orion D, Greenberger A & Katan J (1979) Solar heating of the soil for the control of *Verticillium dahliae* and *Pratylenchus thornei* in potatoes. In: Schippers B & Gams W (eds) Soilborne Plant Pathogens (pp 431-438) Academic, London.

Groshevoy SE (1939) Disinfection of seed-bed soil in cold frames by solar energy. Abstr. Rev. Appl. Mycol. 18:635-636.

Grunzweig JM, Rabinowitch HD & Katan J (1993) Physiological and developmental aspects of increased plant growth in solarized soils. Ann. Appl. Biol. 122:579-591.

Hardy GE & Sivasithamparam K (1985) Soil solarization effects on Fusarium wilt of carnation and Verticillium wilt of eggplants. In: Parker CA (ed) Biology and Management of Soilborne Plant Pathogens (pp 279-281) APS, Press, St Paul.

Hassan MS & Younis MA (1984) Cucumber cultivation with soil solarization and plastic mulching. Arab J. Plant Prot. 2:63-69.

Hernandez JH, LLovet G & Perez MJ (1987) Preliminary study of soil disinfection using solar heat for tomato plantings. 7th Congr. Phytopathol. Union p. 59.

Horiuchi (1991) Solarization for greenhouse crops in Japan. In: DeVay JE, Stapleton JJ & Elmore CL (eds) Soil Solarization (pp 16-27) Food & Agricultural Organization, Rome.

Kaewruang W, Sivasithamparam K & Hardy GE (1989a) Use of soil solarization to control root rots in gerberas (*Gerbara jamesomii*). Biol. Fert. Soils 8:38-47.

Kaewruang W, Sivasithamparam K & Hardy GE (1989b) Effect of solarization of soil within plastic bags on root rot of gerbera (*Gerbera jamesomii* L.). Plant & Soil 120:303-306.

Katan J (1981) Solar heating (solarization) of soil for control of soilborne pests. Annu. Rev. Phytopathol. 19:211-236.

Katan J (1996) Soil solarization: integrated control aspects. In: Hall R (ed) Stategies for Managing Soilborne Plant Pathogens. APS Press, ST Paul, MN (in press).

Katan J & DeVay JE (1991) (eds) Soil Solarization. CRC Press, Boca Raton, FL.

Katan J, Greenberger A, Alon H & Grinstein A (1976) Solar heating by polyethylene mulching for the control of diseases caused by soilborne pathogens. Phytopathology 76:683-688.

Katan J, Rotem I, Finkel Y & Daniel J (1980) Solar heating of the soil for the control of pink root and other soilbone diseases in onions. Phytoparasitica 8:39-50.

Katan J, DeVay JE & Greenberger A (1989) The biological control induced by soil solarization. In: Tjamos EC & Beckman CH (eds) Vascular Wilt Diseases of Plants (pp 293-299) Springer-Verlag, Berlin.

Keinath AP (1995) Reduction in inoculum density of *Rhizoctonia solani* and control of belly rot on pickling cucumber with solarization. Plant Dis. 79:1213-1219.

Ki KU & Kim KC (1985) Possibility of soil solarization in Korea. Korean J. Plant Prot. 24:107-114.

Lee FN (1985) Effect of soil solarization with clear plastic and shallow flood on the survival of *Rhizoctonia solani* sclerotia. Phytopathology 75:1291 (Abstr.).

Melero-Vara JM, Blanco-Lopez MA, Bejarano-Alcazon J & Jimenez Diaz RM (1995) Control of Verticillium wilt of cotton by means of soil solarization and tolerant cultivars in southern Spain. Plant Pathol. 44:250-260.

Mihail JD & Alcorn SM (1984) Effects of soil solarization on *Macrophomina phaseolina* and *Sclerotium rolfsii*. Plant Dis. 68:156-159.

Nataraj T (1993) Effect of exposure of nursurey soil to sunlight on the control of collar rot of coffee. Ind. Coffee 57:9-11 (Abstr. Rev. Pl. Pathol 73:287)

Newhall AG (1955) Disinfestation of soil by heat, flooding and fumigation. Bot. Rev. 21:189-250.

Osman AR & Sahab AF (1983) Control of *Rhizoctonia solani* by soil solarization. Act. Hort. 152:245-251.

Pullman GS, DeVay JE, Garber RH & Wemhold AR (1981a) Soil solarization: effects on Verticillium wilt of cotton and soilborne populations of *Verticillium dahliae*, *Phthium* spp. *Rhizoctonia solani* and *Thielaviopsis basicola*. Phytopathology 71:954-959.

Pullman GS, DeVay JE & Garber RH (1981b) Soil solarization and thermal death: a logarithmic relationship between time and temperature for four soilborne plant pathogens. Phytopathology 71:959-964.

Rubin B & Benjamin AD (1983) Solar heating of the soil: effect on weed control and on soil-incorporated herbicides. Weed Sci. 31:819-825.

Shlevin E, Katan J, Mahrer Y & Kritzman G (1995) Space solarization for sanitation of inocula of plant pathogens in the greenhouse structure. Phytopathology 85:1209 (Abstr.).

Souza NL de (1994) Solarizacoa do solo. Summa Phytopathologica 20:3-15.

Stapleton JJ & DeVay JE (1984) Thermal components of soil solarization as related to changes in soil and root microflora and increased plant growth response. Phytopathology 74:255-259.

Stapleton JJ & DeVay JE (1986) Soil solarization: a nonchemical approach for management of plant pathogens and pests. Crop Prot. 5:190-198.

Stapleton JJ & DeVay JE (1995) Soil solarization: a natural mechanism of integrated pest management. In: Reuveni R (ed) Novel Approaches to Integrated Pest Management (pp 309-322) Lewis Publishers, Boca Raton, FL.

Stapleton JJ Quick J & DeVay JE (1985) Soil solarization: effect on soil properties, crop fertilization and plant growth. Soil Biol. Biochem. 17:369-373.

Tamietti G, Valentino D & Garibaldi (1987) Solar heating in northern Italy to control soilborne pathogens on vegetable crops. Proc. 7th Congr. Mediter. Phytopath. Union. pp 73-75.

Triolo E, Vannacci G & Materazzi A (1989) Soil solarization in vegetable crop production. 3. Control of *Rhizoctonia solani* Kuhn. Difesia delle Piante 12:1-2 (Abstr. Rev. Pl. Pathol. 70:862).

Vannacci G & Triolo E (1989) Soil solarization against *Sclerotinia* and *Rhizoctonia solani* in Italy (1989). Phytoparasitica 17:150 (Abstr.).

Vannacci G, Materazzi A & Triolo E (1987) Effects of solar heating on soil-borne *Rhizoctonia solani* in greenhouses. Proc. 7th Congr. Phytopathol. Union. pp 55-57.

VI.A3 SUPPRESSIVE SOIL AMENDMENTS FOR THE CONTROL OF *RHIZOCTONIA SPECIES*

DON M. HUBER[1] AND DONALD R. SUMNER[2]

[1]*Department of Botany & Plant Pathology, Purdue University, West Lafayette, IN 47907-1155,* [2]*Department of Plant Pathology, University of Georgia Coastal Plain Experiment Station, Tifton, GA 31793-0748. USA*

I. INTRODUCTION

Disease suppression in cultivated soil is generally influenced more by cropping and management practices than by soil type (Huber and Watson, 1970). Crop rotation and, in some cases, a specific crop sequence have been standard recommendations for the control of diseases caused by *Rhizoctonia*. The immediately preceding crop in the rotation, or organic amendment, has a much greater effect on the severity of diseases caused by *Rhizoctonia solani* than crops several years preceding (Afanasiev and Morris, 1950; Huber, 1963; Kommedahl, 1952). Where crop rotation is not an effective or economic option, various soil amendments have been proposed to reduce disease severity.

Soil incorporation of green manure crops, rotation crop residues, farmyard manure, or various other organic materials has been frequently recommended to prevent the increase of soilborne pathogens in newly cultivated soil and to produce conditions less favorable for pathogens in established crop land (Butler, 1961; Huber and Watson, 1970; Papavizas *et al.*, 1962; Patrick and Toussoun, 1965). Manure, sewage sludge, chitin, food processing wastes and composts all have been recognized for their potential disease suppressive effects (Baker and Martinson, 1970; Garrett, 1956; Huber, 1991; Huber and Watson, 1970; Sneh *et al.*, 1972; Ylimaki *et al.*, 1983). Increased emphasis on resource utilization and environmentally sensitive disposal of various agricultural and industrial wastes have encouraged the use of these materials as nutrient resources for crop production as well as for disease suppression (Huber *et al.*, 1993a).

II. SUPPRESSIVE ORGANIC AMENDMENTS:

A. Crop residues:

Organic and inorganic amendments reported to influence diseases caused by *R. solani* are listed in Table 1. Root rot of bean (*Phaseolus vulgaris*) was reduced markedly when residues with a high carbon to nitrogen ratio such as sawdust; pine shavings; cellulose; wheat bran; and mature sorghum, soybean, barley, oat, corn or wheat residues were applied to soil (Davey and Papavizas, 1960; Snyder *et al.*, 1959). Amendment of soils with sugar increased disease severity, and high rates of nitrogen fertilizer nullified the beneficial effects of most carbonaceous materials with a high C:N ratio (Burke, 1955; Huber and Watson, 1970, 1974). Crop residues high in nitrogen such as bean straw, bean seed meal, tomato, lettuce, alfalfa and green barley hay generally increased early infection of beans (*P. vulgaris* L.), but reduced the total infection by the end of the season (Burke, 1955; Papavizas and Davey, 1960; Snyder

B. Sneh et al. (eds.),
Rhizoctonia Species: Taxonomy, Molecular Biology, Ecology, Pathology and Disease Constrol, 433–443.

et al., 1959). Green corn and oat residues reduced the severity of root rot as well as the population of *Rhizoctonia* in soil (Papavizas and Davey, 1960).

TABLE 1. Amendments reported to affect diseases caused by *Rhizoctonia solani*.

Suppressive organic materials:		
Alfalfa hay, ground	Corn husks and immature residue	Peanut cake
Bagasse	Ellupa cake	Pine bark and shavings
Barley straw, mature	Food processing wastes	Sawdust
Cellulose	Formulated soil mixes	Sewage sludge
Chitin	Gingelly cake	Pine shavings
Composted:	Gliricidia, green leaves	Sorghum
Cattle manure	Manure	Soybean, mature
Grape marc	Neem green leaves and cake	Uramite
Potting mix	Oat hay and mature straw	Wheat straw
Sewage sludge	Organic household wastes	
Wood residue		

Predisposing organic materials:
Alfalfa, green
Barley, immature
Bean seed meal
Lettuce
Peat
Soybean green manure
Sugar
Tomato residue

Suppressive inorganic materials:	
Calcium	Manganese
Gypsum	Nitrate nitrogen
Lime	Sulfur
Magnesium	Zinc

Predisposing Inorganic Materials:
Ammonium nitrogen
Potassium

B. Manure

Manure has long been known to reduce the severity of root and cortical rots caused by *Rhizoctonia* (Baker and Martinson, 1970). Farmyard manure was more effective in reducing the severity of *R. solani* on potato tubers when potatoes were grown on manured fields previously cropped to rye or wheat compared with a previous crop of peas, rape or potato (Weber, 1977). Preplant applications of cattle manure reduced "winter-kill" of wheat caused by *R. cerealis* by as much as 85% and provided effective control of this cool temperature disease (Huber *et al.*, 1993b; Riegel and Huber, 1992).

C. Composts

Many composts have been proposed to reduce disease severity, and commercial scale formulation of composts suppressive to diseases caused by *Rhizoctonia* and other soilborne pathogens has been achieved (Grebus *et al.*, 1993; Hoitink *et al.*, 1993, 1997). The source of the compost, decomposition level, and secondary amendments are frequently critical to successful biological control so that disease management with composts is as much an art as a science at this time (Hoitink *et al.*, 1997). "Maturity" of the compost may be less important for disease control in field crops where additional stabilization may proceed in the field after application than in container-grown crops where delayed time for planting is not an economically viable option, and environmental conditions are not as conducive to stabilization.

Composted hardwood bark suppressed Rhizoctonia damping-off of cucumber and poinsettia (Stephens *et al.*, 1981; Daft *et al.*, 1979; Nelson and Hoitink, 1983). Milled pine bark reduced damping-off of slash pine seedlings by *R. solani* by as much as 50%, but it failed to protect seedlings from infection by binucleate *Rhizoctonia* (Huang and Kuhlman, 1991a, b). Composted sewage sludge reduced the incidence of damping-off of peas and cotton; root rot of beans, cotton and radish (Lewis *et al.*, 1992; Lumsden *et al.*, 1983; Mathout, 1987; Van Assche and Uytterbroeck, 1981) and Rhizoctonia brown patch of tall fescue (O'Neil, 1982).). Composted organic household wastes were suppressive to *R. solani* and reduced disease incidence on beans even after storage for prolonged periods (Schueler *et al.*, 1989).

In contrast to peat which increased damping-off in container media, composted grape marc (pulp) and composted separated cattle manure both suppressed bean and radish damping-off caused by *R. solani,* while providing the physical, chemical, and biological properties in the media for plant growth (Chen and Hadar, 1986; Chen *et al.*, 1992; Gorodecki and Hadar, 1990). Suppression was not a function of altered pH since both composts were highly suppressive even at pH7.

Disease suppressive effects also may be enhanced as specific microorganisms recolonize the substrate during the compost process (Hoitink *et al.*, 1997). Kuter *et al.* (1983) reported that compost produced out side where there was a large diversity of organisms was more consistently suppressive to *R. solani* than compost prepared in a partially enclosed facility. Composts naturally suppressive to *Pythium* have to be incubated for 4 to 8 weeks before they become suppressive to *R. solani* (Kuter, *et al.*, 1988. In contrast to fresh wastes, mature compost (low in cellulose) is not readily colonized by *R. solani* (Chung *et al.*, 1988).

Inoculation of composts with specific fungi or bacteria has been proposed to reduce variability in suppressiveness to *Rhizoctonia* (Hoitink *et al.*, 1991, 1997; Phae *et al.*, 1990). The addition of recycled potting mix to fresh mix provided a source of high-temperature inoculum for antagonistic microorganisms. However, although widely used in Australia, it rarely reduced the severity of *Rhizoctonia* on *Celosia argentae* in bedding soils (Schisler and Ryder, 1991). Several bacteria isolated from potting media were suppressive to *C. argentae* damping-off caused by *R. solani* (AG 8) and were proposed as protective amendments to nursery potting media (Harris *et al.*, 1994).

D. Formulated soil amendments

The general chemical characteristics of organic materials added to infested soils generally are more important than their C:N ratio (Huber and Watson, 1970; Huber, 1963, 1991; Huber *et al.*, 1965; Tyner, 1948); and the addition of inorganic nutrients can drastically change the effect of an organic amendment on disease (Huang and Huang, 1993; Huber and Watson, 1974; Sun and Huang, 1985). The formulation of agricultural and industrial organic wastes with inorganic fertilizer materials has produced suppressive soil amendments which can control disease while also promoting plant growth (Huang and Huang, 1993; Sun and Huang, 1985). A mixture of bagasse, rice husks, oyster shell powder, urea, potassium nitrate, calcium superphosphate, and mineral ash (silicon dioxide, calcium oxide, magnesium oxide, aluminum oxide, and ferrous oxide) reduced the incidence of Rhizoctonia bean blight and rice sheath blight with subsequent yield increases of 22 to 54% (Sun and Huang, 1985). A formulated soil amendment consisting of milled pine bark, glycerine, and mineral nutrients (SF-21) controlled post-emergence damping-off by both *R. solani* and binucleate *Rhizoctonia* on slash pine seedlings (Huang and Huang, 1993).

III. Suppressive inorganic amendments

Specific inorganic nutrients appear to be increasingly useful for the control of diseases caused by *Rhizoctonia* (Baker and Martinson, 1970). Nitrogen has a greater effect than any other element on individual soil fungi (Kaufman and Williams, 1964); however, the form of nitrogen is more important than the rate of nitrogen applied. Nitrate sources of nitrogen generally reduce the severity of *Rhizoctonia* while ammonium sources can increase disease severity. Ammonium and inhibition of nitrification increased the Rhizoctonia potato stolon canker, but decreased Verticillium wilt (Huber, 1989; Huber and Watson, 1974). Ammonium nitrogen increased Rhizoctonia cotton diseases at low temperatures, but reduced them at temperatures closer to optimal for plant growth and nitrification (Ramasami and Shanmugam, 1976).

The positive correlation between nitrification rate and disease suppression may indicate the importance of a specific available nitrogen form to the plant or pathogen after residue incorporation (Huber and Watson, 1970, 1974); however, the use of specific nitrogen forms for disease control must take into account residual soil nitrogen, preferential utilization of nitrogen forms by the crop, and effects of different ammonium-to-nitrate ratios in soil. Most manure, compost and organic wastes stimulate nitrification to provide a predominantly nitrate source of nitrogen in soil for plant uptake (Huber, 1991; Huber *et al.*, 1965). The effect of organic amendments on nitrification may account for inconsistencies regarding the effects of specific residues on disease (Huber and Watson, 1970). Ammonium and Uramite (a urea-formaldehyde product) reduced the severity of *R. solani* when added with sawdust, a highly carbonaceous amendment (Davey and Papavizas, 1960); however, without the sawdust, ammonium nitrate resulted in the greatest disease reduction.

Phosphorus, sulfur, or calcium deficiencies predispose cotton to *Rhizoctonia*, and adequate fertilization with those nutrients will reduce disease severity (Bell, 1989). The application of high rates of $CaSO_4$ reduced peanut pod breakdown and increased the calcium content of pods while fertilization with magnesium sulfate and potassium sulfate, both of which increased the concentration of potassium in the pods, increased pod rot (Hallock and Garren, 1968). A concomitant application of calcium with nitrogen, magnesium or potassium reduced their effect on disease (Csinos and Bell, 1989; Csinos and Gaines, 1986). Amendments and environmental conditions which increase zinc availability have been correlated with reduced winter kill of wheat by *R. cerealis* and other diseases (Graham, 1983; Graham and Webb, 1993; Huber *et al.*, 1993b; Reigel and Huber, 1992).

IV. Mechanisms of control with suppressive amendments:

Earlier studies with crop rotations assumed that reduced disease resulted from decreased population densities of *Rhizoctonia*, presumably through "starvation" during the culture of "non-susceptible" crops. Although disease-suppression in cultivated soil is generally considered of microbial origin and is influenced more by cropping and management practices than by soil type (Huber *et al.*, 1966; Huber and Watson,

1970), the effect may or may not be correlated with the detectable pathogen populations in soil (Andersen and Huber, 1965).

The involvement of soil microbial populations in disease suppression has been established through enumeration of the general, and/or specific microflora before and after amendment incorporation, environmental conditions more or less conducive to increased microbial activity, and loss of suppressiveness after soil sterilization. Disease suppression has been enhanced by increasing the time between soil amendment and planting which resulted in increased microbial activity even though the specific amendment had no effect on survival of the pathogen (Lumsden *et al.*, 1983). The organic matter content of *Rhizoctonia*-suppressive soils and microbial activity were two times higher than those of conducive soils (Shim and Lee, 1990). Gamma irradiation significantly reduced the suppressiveness of composted wood residues (Nelson and Hoitink, 1983), grape marc, or cattle manure (Gorodecki and Hadar, 1990), indicating that suppressiveness is related to microbial activity; however, other mechanisms are also involved with cattle manure (Gorodecki and Hadar, 1990; Huber *et al.*, 1993b; Reigel and Huber, 1992). Reinoculation of sterilized compost with a small amount of non-sterile compost followed by a short incubation period restored suppressiveness of composted grape marc and cattle manure (Chen *et al.*, 1992).

It is now recognized that several different mechanisms of control are involved with crop rotation and suppressive amendments (Huber and Watson, 1970). The population of *Rhizoctonia* may, or may not, be correlated with disease severity since available techniques may not differentiate between inhibited virulence and absence of propagules. Thus, disease suppression may be as much from modifying the abiotic or biological soil environment to affect virulence of *Rhizoctonia* or to induce plant resistance as from changing the population of the pathogen in soil (Andersen and Huber, 1962, 1965; Hoitink *et al.*, 1997; Sadasivan and Subramanian, 1960; Weinhold *et al.*, 1969, 1972). Organic amendments may also influence disease because of their effect on the availability of nutrients to the pathogen or the plant (Huber, 1991; Huber and Watson, 1970; Rogers, 1969).

A. Reduced pathogen population density

Reduced disease severity by amending soil with manure and various crop residues has been correlated with the increased antagonistic soil microbial activity, which reduce *Rhizoctonia* population density through competition, antibiosis, hyperparasitism, or the production of fungi-toxic products (Boosalis, 1956; Hoitink *et al.*, 1997; Sneh and Henis, 1972; Weindling, 1934, 1946). Neem and peanut cake amendments under dry conditions, and ellupa, gingelly, and neem cake in conjunction with flooding were very effective in reducing the viability of *Rhizoctonia* causing blight of bean and sheath blight of rice (Lakshmanan and Nair, 1984). Soil amendment with green leaves of neem and *Gliricidie* also reduced sclerotial viability. The populations of *Rhizoctonia* were reduced 50-90 % four weeks after the application of a formulated soil amendment (Huang and Kuhlman, 1991a, b). Suppressiveness was associated with colonization by *Trichoderma harzianum* or *Penicillium oxalicum*; and inoculation of autoclaved suppressive soil with these microorganisms restored suppressiveness within seven days. Even though *Trichoderma* isolates that function in biocontrol of *Rhizoctonia* are capable of colonizing fresh as well as mature compost, both fungi grow compatibly until readily available substrates are exhausted at which time *Rhizoctonia* sclerotia can be hyperparasitized and destroyed (Chen *et al.*, 1988; Cruz *et al.*, 1993; Nelson and Hoitink, 1983).

There was a direct relationship between hyphal growth rate and soil pH in Hawaiian soils suppressive to *R. solani* (Ko and Ho, 1983; Kobayashi and Ko, 1985). Suppression was induced in conducive soils proportionate to the amount of suppressive soil added, and lowering soil pH irreversibly counteracted suppression. Although autoclaving also nullified soil suppressiveness, suppression of *R. solani* was the result of a combination of nonspecific microorganisms and abiotic factors. Reduced disease with composted sewage sludge was correlated with increased total nitrogen, phosphorus, magnesium, calcium and organic matter provided by the compost; and reduced inoculum density of *Rhizoctonia* (based on beet seed colonization) (Lewis *et al.*, 1992).

High temperatures generated during the composting process generally kill most pathogens and weed seeds (Hoitink *et al.*, 1997; Yuen and Raabe, 1984). This high temperature during the decomposition period significantly reduces the inoculum potential of *R. solani* and provides a suppressive environment for container-grown crops (Ylimaki *et al.*, 1983). Sclerotia of *Rhizoctonia* in potato and other food processing wastes were eradicated one week after the composted material reached 70°C. A quality, homogeneous product can be produced if compost piles are turned frequently to provide optimum exposure to higher temperatures which promote the degradation process. Turning of the compost periodically was required to expose mycelium and sclerotia in surface layers to the high temperatures in the center of the compost windrows (Ylimaki *et al.*, 1983).

Suppression of *Rhizoctonia* and other soilborne pathogens with composts has also been attributed to a general increase of microbial activity in the rhizosphere and colonization by *Trichoderma* spp. (Chen *et al.*, 1988; Kwok *et al.*, 1987). Reduced damping-off after soil amendment with a compost of cow, chicken and sheep manure and peat moss was correlated with an increased population of rhizosphere bacteria antagonistic to *Rhizoctonia* and a marked increase in the percentage of siderophore producers (Alvarez *et al.*, 1995). The mechanism of control of *R. solani* by composts is different from the control of diseases caused by *Pythium*, *Phytophthora*, or *Fusarium* (Hoitink *et al.*, 1997). Specific suppression of *Rhizoctonia* induced by a very narrow group of microorganisms was described by Hoitink *et al.* (1991). *Trichoderma* species interact with various bacteria to reduce *Rhizoctonia* damping-off (Kwok *et al.*, 1987).

The saprophytic growth of *Rhizoctonia* decreased with increasing incubation time of chitin in soil (Henis *et al.*, 1967; Sneh *et al.*, 1971). Inhibition was greatest with chitin amendment as compared with various plant residues and appeared to result from accumulation of antifungal substances (polar lipisd) produced during chitin degradation (Sneh *et al.*, 1972). Viability of the pathogen in buckwheat-stem segments subjected to decomposing cabbage tissue volatiles was decreased by 56%. In contrast, plant tissues with a low C:N ratio (<15) such as vegetable crop and immature grain crop residues (corn, oat, rye, barley, sorghum, buckwheat) formed volatile amine compounds which stimulated growth and pigmentation of *R. solani* (Lewis and Papavizas, 1974). Decomposition of several other residues with a high C:N ratio produced volatiles which did not influence the fungus in any way.

B. Disease suppression without reduction in population density of *Rhizoctonia*

1. Altered virulence of *Rhizoctonia*. Disease suppression may occur by modifying the microbial environment to reduce virulence of *R. solani*, or by increase of plant resistance by reducing the pathogen's population levels (Andersen and Huber, 1962, 1965; Easton, 1964; Elmer, 1942; Huber, 1963; Lockwood, 1977). Although most soil fungi were influenced by the present and previous crop, the population density of (actively growing) *Rhizoctonia* was remarkably stable in studies of 18 crop sequence-rotations with beans, even though Rhizoctonia root and hypocotyl rot severity was considerably influenced by cropping sequence. The consistent isolation of *R. solani* in high frequency, and its compatibility with most other soil fungi, indicate that *Rhizoctonia* is an indigenous soil inhabitant (Andersen and Huber, 1962; Huber, 1963).

In a study of inoculum density-disease incidence relationships of *R. solani* in soil under controlled temperature and moisture conditions, Kinsbursky and Weinhold (1988) found no relationship between inoculum density and disease incidence. They identified a disease suppressive soil which had a high native inoculum density and a consistently low infection incidence. Heat treatment at 55°C for one hour increased disease incidence, indicating that a microbial factor was involved in disease suppression, even though *Rhizoctonia* population levels remained stable. Similarly, no relationship was found between inoculum density and Early Alaska peas damping off severity (Lewis *et al.*, 1992).

Specific bacterial-fungal associations affected by crop sequence were correlated with disease control without a corresponding reduction in pathogen numbers in soil (Huber, 1963). The bacteria grew in the hyphosphere and reduced virulence to beans without greatly affecting the over-all growth or population of *Rhizoctonia* in soil, although older hyphae became necrotic (Andersen and Huber, 1965; Huber and Andersen, 1966). Fungistasis is another type of microbial interaction in which sclerotia are maintained in a dormant state (and thereby not virulent), yet the pathogen is not eradicated from the soil (Huber and Watson, 1970; Lockwood, 1977).

Polygalacturonase, an extracellular enzyme associated with disease caused by *Rhizoctonia* was greatly reduced in calcium solution, but enhanced by ammonium, potassium, and sodium solutions (Hallock and Garren, 1968). Virulence of *R. solani* was increased by organic sources of nitrogen such as asparagine (Weinhold *et al.*, 1969), and ammonium increased virulence more than nitrate nitrogen (Huber, 1991; Huber and Watson, 1974); but either form of nitrogen can increase disease severity (Weinhold *et al.*, 1972). The disease suppressive effect of 27 crop residue amendments on diseases caused by *Rhizoctonia* and other soilborne pathogens was correlated with their effect on nitrification (Huber and Watson, 1974; Huber *et al.*, 1965). Residues which enhanced nitrification generally reduced disease, while those inhibiting nitrification increased diseases caused by *Rhizoctonia* (Huber, 1991; Huber and Watson, 1974). Such a selective biological effect was consistent with the effects of the specific forms of nitrogen. Thus, the effect of the form of nitrogen on disease severity is either by modifying virulence of *Rhizoctonia* or increasing plant resistance, or both, since the populations of the pathogen are not affected significantly (Huber, 1963). Tyner (1948) also concluded that the general chemical properties of organic materials were more important in reducing disease severity than the carbon to nitrogen ratio.

2. Increased plant resistance. Altering plant nutritional status by organic or inorganic amendment may alter a disease response without affecting numbers of pathogen propagules in the soil (Huber and Watson, 1970). Disease control as a result of modified plant physiology may result in decreased infection or delayed pathogenesis after penetration (Graham and Webb, 1991; Huber and Watson, 1970; Huber *et al.*, 1993b). A correlation between suppression of disease severity by various cultural practices and the availability of various minor elements has been reported (Graham and Webb, 1991; Huber, 1989). Barnyard manure supplied enough zinc to wheat plants for increasing plant resistance to *R. cerealis* (Huber *et al.*, 1993b). Other cultural conditions in which zinc was readily available also reduced disease (Table 2). In leaf drop areas of tree wind breaks where wheat was only mildly damaged by *R. cerealis* "winter-kill", the wheat was heavily colonized by mycorrhizal fungi which increased zinc uptake while wheat growing beyond the leaf drop areas was non-mycorrhizal, low in tissue levels of zinc and severely diseased.

TABLE 2. Correlation of tissue zinc with Rhizoctonia "winter-kill" of wheat (after Riegel and Huber, 1992).

	Tissue zinc		% of wheat killed	
Condition	**Without**	**With(in)**	**Without**	**With(in)**
Barnyard manure	20	41	80	15
Sediment area	17	27	100	45
Tree leaf-drop area	19	34	65	20

Physiological Sufficiency Level = 20-150 mg/kg

Rhizoctonia winter-kill of wheat was reduced by preplant incorporation of zinc fertilizers, but not by foliage sprays, indicating an indirect effect through enhanced resistance to this foliage disease rather than direct toxicity of zinc to the pathogen (Reigel and Huber, 1992). Amendment with zinc fertilizers also reduce the severity of wheat root rot caused by *R. solani* by increasing plant resistance (Graham and Webb, 1991). Although manganese is toxic to *R. solani*, and the incidence of disease of cotton from this pathogen is reduced in soils with a high availability of Mn compared to the same soils treated with lime to immobilize Mn (Bell, 1989), the effect of manganese is more likely through increased resistance of the plant than through fungitoxicity (Graham and Webb, 1991).

V. Conclusions

Amendment through crop sequence, organic manuring or direct addition to soil plays a significant role in controlling the various diseases caused by *Rhizoctonia*. Recognition of the nutrient value and disease suppressive effects of many waste products has encouraged their application for crop production. Composting, enrichment with inorganic nutrients or inoculation with specific microorganisms have enhanced the disease suppressive effects of many materials. Compost and formulated soil amendments which reduce disease severity provide an optimum plant growth medium and have been commercialized for the containerized industry. Several distinct mechanisms of disease suppression have been observed following soil amendment in which disease suppression may occur with or without a reduction of pathogen population in the soil. Organic amendment and crop rotation influence disease severity by stimulating microbial activity responsible for soil "biological buffering", altering the nutritional environment of the plant and pathogen, and releasing fungitoxic products during decomposition in soil. The implementation of any particular amendment system has been primarily through empirical information since generalized concepts have been difficult to develop for this ubiquitous soilborne pathogen. Integration of specific amendments with other pest control and cultural programs can greatly enhance the efficiency and reliability of crop production.

VI. References

Afanasiev MM & Morris HE (1950) The effect of preceding crops and nutrients on the growth and seedling diseases of sugar beets in Montana. Phytopathology 40:963.

Alvarez MAB, Gagne S & Antoun H (1995) Effect of compost on rhizosphere microflora of the tomato and on the incidence of plant growth-promoting rhizobacteria. Appl. Env. Microbiol. 61:194-199.

Andersen AL & Huber DM (1965) The plate-profile technique for isolating soil fungi and studying their activity in the vicinity of roots. Phytopathology 55:592-594.

Baker R & Martinson CA (1970) Epidemiology of diseases caused by *Rhizoctonia solani*. In: Parmeter JR Jr (ed.) *Rhizoctonia solani*: Biology and Pathology (pp 172-188) Univ. California Press, Berkeley.

Bell A (1989) Role of nutrition in diseases of cotton. In: Engelhard A (ed.) Soilborne Plant Pathogens: Management of Diseases with Macro- and Microelements. (pp 167-204) APS Press, St. Paul, MN.

Boosalis MG (1956) Effect of soil temperature and green manure amendment of unsterilized soil on parasitism of *Rhizoctonia solani* by *Penicillium vermiculatum* and *Trichoderma* sp. Phytopathology 46:473-478.

Butler FC (1961) Root and foot rot diseases of wheat. Sci. Bul. No. 77, Agr. Res. Inst. Wagga Wagga, N.S.W. Australia.

Burke DW (1955) Soil microflora relationships in the development of bean root rot in Columbia basin soils. Diss. Abstr. 15:2390-2391.

Chen Y & Hadar Y (1986) Composting and use of agricultural wastes in container media. In: Compost: production, quality and use. Proc. (pp 71-77) Symp. Comm. European Communities, Directorate-Gen. Sci., Res. and Dev. Udine, Italy.

Chen W, Hoitink HAJ, Schmitthenner AF & Tuovinen OH (1988) The role of microbial activity on the suppression of damping-off caused by *Pythium ultimum*. Phytopathology 78:314-322.

Chen Y, Inbar Y & Hadar Y (1992) Composted residues reduce peat and pesticide use. Biocycle 33:48-51.

Chung YR, Hoitink HAJ & Lipps PE (1988) Interaction between organic-matter decomposition level and soilborne disease severity. Agric. Ecosyst. Environ. 24:183-193.

Cruz, de la J, Rey M, Lora JM, Hidalgo-Gallego A, Dominguez F, Pintor-Toro JA, Llobell A & Benitez T (1993) Carbon source control on ß-glucanases, chitobiase and chitinase from *Trichoderma harzianum*. Microbiologica 159:316-322.

Csinos AS & Bell DK (1989) Pathology and nutrition in the peanut pod rot complex. In: Engelhard A (ed.) Soilborne Plant Pathogens: Management of Diseases with Macro- and Microelements. (pp 124-136) APS Press, St. Paul, MN.

Csinos AS & Gaines TP (1986) Peanut pod rot complex: A geocarposphere nutrient imbalance. Plant Dis. 70:525-529.

Daft GC, Poole HA & Hoitink HAJ (1979) Composted hardwood bark: A substitute for steam sterilization and fungicide drenches for control of poinsettia crown and root rot. HortScience 14:185-187.

Davey CB & Papavizas GC (1960) Effect of dry mature plant materials and nitrogen on *Rhizoctonia solani* in soil. Phytopathology 50:522-525.

Easton GD (1964) The results of fumigating *Verticillium* and *Rhizoctonia* infested potato soils in Washington. Amer. Potato J. 41:296.

Elmer OH (1942) Effect of environment on the prevalence of soil-borne *Rhizoctonia*. Phytopathology 32:972-977.

Garrett SD (1956) Biology of Root Infecting Fungi. Cambridge Publishing Co.

Gorodecki B & Hadar Y (1990) Suppression of *Rhizoctonia solani* and *Sclerotium rolfsii* diseases in container media containing composted separated cattle manure and composted grape marc. Crop Prot. 9:271-274.

Graham RD & Webb MJ (1991) Micronutrients and disease resistance and tolerance in plants. In: Mortvedt JJ & Welch RM (eds.) Micronutrients in Agriculture, 2nd ed (pp 329-370) Soil Sci. Soc. Amer., Madison, WI.

Hallock DL & Garren KH (1968) Pod breakdown, yield, and grade of Virginia type peanuts as affected by Ca, Mg, and K sulfates. Agron. J. 60:253-257.

Harris AR, Schisler DA, Correll RL & Ryder MH (1994) Soil bacteria selected for suppression of *Rhizoctonia solani*, and growth promotion, in bedding plants. Soil Biol. Biochem. 26:1249-1255.

Henis Y, Sneh B & Katan J (1967) Effect of organic amendments on *Rhizoctonia* and accompanying microflora in soil. Can. J. Microbiol. 13:643-650.

Hoitink HAJ, Inhar Y & Boehm MJ (1991) Status of composted-amended potting mixes naturally suppressive to soilborne diseases of floricultural crops. Plant Dis. 75:869-873.

Hoitink HAJ, Boem MJ & Hadar Y (1993) Mechanisms of suppression of soilborne plant pathogens in compost-amended substrates. In: Hoitink HAJ & Keener HM (eds.) Design, Environmental, Microbiological and Utilization Aspects. (pp 601-621) Renaissance Publications, Worthington, OH.

Hoitink HAJ, Madden LV & Boehm MJ (1997) Relationships between organic matter decomposition level, microbial species diversity and soilborne disease severity. In: Hall R (ed.) Effect of Organic Matter on Soilborne Diseases. APS Press, St. Paul, MN.

Huang HC & Huang JW (1993) Prospects for control of soilborne plant pathogens by soil amendment. Curr. Topics Bot. Res 1:223-235.

Huang JW & Kuhlman EG (1991a) Formulation of a soil amendment to control damping-off of slash pine seedlings. Phytopathology 81:163-170.

Huang JW & Kuhlman EG (1991b) Mechanisms inhibiting damping-off pathogens of slash pine seedlings with a formulated soil amendment. Phytopathology 81:171-177.

Huber DM (1963) Investigations on root rot of beans caused by *Fusarium solani* f. sp. *phaseoli*. Ph-D thesis, Michigan State Univ., East Lansing, MI.

Huber DM (1989) Introduction. In: Engelhard A (ed.) Soilborne Plant Pathogens: Management of Diseases with Macro- and Microelements. (pp 1-8) APS Press, St. Paul, MN.

Huber DM (1991) The role of fertilizer and organic amendments in the control of plant disease. In: Pimentel D (ed.) Handbook of Pest Management in Agriculture Vol. 1(pp 357-394) CRC Press, Boca Raton, FL.

Huber DM & Andersen AL (1966) Necrosis of hyphae of *Fusarium solani* f. *phaseoli* and *Rhizoctonia solani* induced by a soil-borne bacterium. Phytopathology 56:1416-1417.

Huber DM, Andersen AL & Finley AM (1966) Mechanisms of biological control in a bean root rot soil. Phytopathology 56:953-956.

Huber DM & Watson RD (1970) Effect of organic amendment on soil-borne plant pathogens. Phytopathology 60:22-26.

Huber DM & Watson RD (1974) Nitrogen form and plant disease. Annu. Rev. Phytopathol. 12:139-165.

Huber DM Watson RD & Steiner GW (1965) Crop residues, nitrogen, and plant disease. Soil Sci. 100:302-308.

Huber DM, Sutton AL & Jones DD (1993) Nutrient management of manure to enhance crop production and protect the environment. In: Integrated Resource Management (pp 39-45) Amer. Soc. Agric. Engineers, St. Joseph, MI.

Huber DM, McCay-Buis TS, Riegel C, Graham RD & Robinson N (1993) Correlation of zinc sufficiency with resistance of wheat to Rhizoctonia winter-kill. Abstr. 115, 6th Int. Cong. Plant Pathol., Montreal, Canada.

Kaufman GG & Williams LE (1964) Effect of mineral fertilization and soil reaction on soil fungi. Phytopathology 54:134-139.

Kinsbursky RS & Weinhold AR (1988) Influence of soil on inoculum density-disease incidence relationships of *Rhizoctonia solani*. Phytopathology 78:127-130.

Ko WH & Ho WC (1983) Screening soils for suppressiveness to *Rhizoctonia solani* and *Pythium splendens*. Ann. Phytopathol. Soc. Japan 49:1-9.

Kobayashi N & Ko WH (1985) Nature of suppression of *Rhizoctonia solani* in Hawaiian soils Trans. Brit. Mycol. Soc. 84:691-694.

Kommedahl T (1952) The incidence of *Fusarium*- and *Rhizoctonia*-infected plants in wheat, corn, and oats. Phytopathology 42:468-469.

Kuter GA, Nelson EB, Hoitink HAJ & Madden LV (1983) Fungal populations in container media amended with composted hardwood bark suppressive and conducive to Rhizoctonia damping-off. Phytopathology 73:1450-1456.

Kwok OCH, Fahy PC Hoitink HAJ & Kuter GA (1987) Interactions between bacteria and *Trichoderma hamatum* in suppression of Rhizoctonia damping-off in bark compost media. Phytopathology 77:1206-1212.

Lakshmanan P & Nair MC (1984) Effect of soil amendments on the viability of sclerotia of *Rhizoctonia solani* in soil. Madras Agric. J. 71:526-529.

Lewis JA & Papavizas GC (1974) Effect of volatiles from decomposing plant tissues on pigmentation, growth, and survival of *Rhizoctonia solani*. Soil Sci. 118:156-163.

Lewis JA, Lumsden RD, Millner PD & Keinath AP (1992) Suppression of damping-off of peas and cotton in the field with composted sewage sludge. Crop Pro. 11:260-266.

Lockwood JL (1977) Fungistasis in soils. Biol. Rev. 52:1-43.

Lumsden RD, Lewis JA & Millner PD (1983) Effect of composted sewage sludge on several soilborne pathogens and diseases. Phytopathology 73:1543-1548.

Mathout P (1987) Emploi de composts dans la lutte contre *Rhizoctonia* sp. Meded. Fac. Landb. Wetensch. Rijks. Gent 52:1127-1132.

Nelson EB & Hoitink HAJ (1983) The role of microorganisms in the suppression of *Rhizoctonia solani* in container media amended with composted hardwood bark. Phytopathology 73:274-278.

Patrick ZA & Toussoun TA (1965) Plant residues and organic amendments in relation to biological control. In: Baker KF & Snyder WC (eds.) Ecology of Soilborne Plant Pathogens: Prelude to Biological Control. (pp 440-459) Univ. Calif. Press, Berkeley.

Papavizas GC & Davey CB (1960) *Rhizoctonia* disease of bean as affected by decomposing green plant materials and associated microfloras. Phytopathology 50:516-522.

Papavizas GC, Davey CB & Woodard RS (1962) Comparative effectiveness of some organic amendments and fungicides in reducing activity and survival of *Rhizoctonia solani* in soil. Can. J. Microbiol. 8:915-922.

Phae CG, Saski M, Shoda M & Kubota H (1990) Characteristics of *Bacillus subtilis* isolated from composts suppressing phytopathogenic microorganisms. Soil Sci. Plan Nutr. 36:575-586.

Ramasami R & Shanmugam N (1976) Effect of nutrients on the incidence of *Rhizoctonia* seedling disease of cotton. Indian Phytopathol. 29:465-466.

Riegel C & Huber DM (1992) Variation in virulence and effects of associated organisms on Rhizoctonia winter-kill of wheat. Phytopathology 82:994.

Rogers PF (1969) Organic manuring for potato scab control and its relation to soil manganese. Ann. Appl. Biol. 63:371-378.

Sadasivan TS & Subramanian CV (1960) Interaction of pathogen, soil, other micro-organisms in the soil, and host. In: Horsfall JG & Dimond AE (eds.) Plant Pathology - An Advanced Treatise (pp 273-313). Academic Press, New York.

Schisler DA & Ryder MH (1991) Microbial recolonization and suppression of *Rhizoctonia solani* in a bedding-plant potting mix amended with recycled mix before aerated-steam treatment. Biol. Fertil. Soils 11:174-180.

Schueler C, Biala J & Vogtmann H (1989) Antiphytopathogenic properties of biogenic waste compost. Agric. Ecosyst. Environ. 27:477-482.

Shim JO & Lee MW (1990) Nature of suppressiveness and conduciveness of some plant pathogens in soils. Korean J. Mycol. 18:164-177.

Sneh B & Henis Y (1972) Production of antifungal substances active against *Rhizoctonia solani* in chitin-amended soil. Phytopathology 62:595-600.

Sneh B, Katan J & Y Henis (1971) Mode of inhibition of *Rhizoctonia solani* in chitin-amended soil. Phytopathology 61:1113-1117.

Snyder WC, Schroth MN & Christou T (1959) Effect of plant residues on root rot of bean. Phytopathology 49:755-756.

Stephens CT, Herr LJ, Hoitink HAJ & Schmitthenner AF (1981) Suppression of Rhizoctonia damping-off by composted hardwood bark medium. Plant Dis. 65:796-797.

Sun SK & Huang JW (1985) Formulated soil amendment for controlling Fusarium wilt and other soilborne diseases. Plant Dis. 69:917-920.

Tyner LE (1948) Effect of crop debris, plant roots, and crop sequence on the microbial flora of the soil in relation to root rot in cereal crops. Can. J. Res. 26:86-93.

Van Assche C & Uyllerbroeck P (1981) The influence of domestic waste compost on plant diseases. Acta Hort. 126:169-178.

Weber Z (1977) The occurrence of soil fungi in potato fields. Acta Mycol. 13:125-132.

Weindling R (1934) Studies on a lethal principle effective in the parasitic action of *Trichoderma lignorum* on *Rhizoctonia solani* and other soil fungi. Phytopathology 24:1153-1179.

Weindling R (1946) Microbial antagonism and disease control. Soil Sci. 61:23-30.

Weinhold AR, Dodman RL & Bowman T (1969) Virulence of *Rhizoctonia solani* as affected by nutrition of the pathogen. Phytopathology 59:1601-1605.

Weinhold AR, Dodman RL & Bowman T (1972) Influence of exogenous nutrition on virulence of *Rhizoctonia solani*. Phytopathology 62:278-281.

Ylimaki AA, Toivianan A, Kallio H & Tikanmaki E (1983) Survival of some plant pathogens during industrial-scale composting of wastes from a food processing plant. Ann. Agric. Fenniae 22:77-85.

Yuen GY & Raabe RD (1984) Effects of small-scale aerobic composting on survival of some fungal plant pathogens. Plant Dis. 68:134-136.

VI.B1 ANTIBIOTICS AND SIDEROPHORE PRODUCING BACTERIA

YOSHIHISA HOMMA

Hokkaido National Agricultural Experiment Station, Hitsujigaoka-1, Toyohira, Sapporo, 062 JAPAN

I. RHIZOBACTERIA: ANTIBIOTIC AND SIDEROPHORE PRODUCING BACTERIA

Rhizobacteria are active in plant rhizospheres and a certain proportion of them are capable to defend the roots against attack by deleterious and pathogenic microorganisms naturally, or as introduced biocontrol agents. Strains of certain rhizobacteria isolated from roots exhibit beneficial activity by controlling soilborne diseases and by promoting plant growth (Schroth and Hancock, 1981; Rovira *et al.*, 1992). Many strains of rhizobacteria which are effective in biocontrol have been reported to produce antibiotics, siderophores, bacteriocins and cyanide. The role of these secondary metabolites in biocontrol has been demonstrated (Weller, 1988; Fravel, 1988).

Pseudomonads, which a major component of the rhizobacterial populations, are good candidates for biocontrol for the following reasons: a) many of the secondary metabolites produced by pseudomonads inhibit other microorganisms (Leisinger and Margraff, 1979), b) a higher proportion of pseudomonads are inhibitory to other microorganisms than that of other soil bacteria (Rovira and Ridge, 1973), c) Pseudomonads are selectively stimulated in the rhizosphere (Weller, 1984), d) Pseudomonads can extensively colonize the endorhizosphere (Van Peer *et al.*, 1990).

II. BIOCONTROL OF *RHIZOCTONIA SOLANI* WITH RHIZOBACTERIA.

A. Antagonism of bacteria in the life cycle of *R. solani*

The basic considerations on *R. solani* have been discussed by Parmeter (1970). *R. solani* is pathogenic to a wide range of crop plants. The strains are extremely variable in specialized responses to host plants and in symptom production. It is important to determine the successful measures by considering the characteristics of the pathogens during its life cycle. The pathogen grows competitively and saprophytically in soil and forms sclerotia as a resting structures in soil. The pathogen expands rapidly from a food base, colonizes roots, grows externally on root surfaces, and invades into epidermal cells from infection cushion or lobate appressoria.

Soil bacteria may affect the formation and longevity of sclerotia, their germination and mycelial saprophytic growth in soil (Leach and Garber, 1970). On the other hand, rhizobacteria may affect the parasitic activity of the pathogen (Schroth and Hancock, 1981); the growth in the rhizosphere, the external growth of hyphae on root surface, the formation of infection structures for invasion into the epidermal cells and multiplication. The bacteria have a high potential to reduce the inoculum level of the pathogen, to suppress its saprophytic and pathogenic activity, and result in the biocontrol of *R. solani*.

B. Sneh et al. (eds.),
Rhizoctonia Species: Taxonomy, Molecular Biology, Ecology, Pathology and Disease Constrol, 445–453.

B. Biocontrol of *R. solani* during the infection process

Rhizobacteria have been reported to have a high potential to suppress the infection process of *R. solani* in the infection courts. For example, in the case of damping-off, the infection court includes the seed coat, endosperm, embryo, the emerging radicle, cotyledons, hypocotyl etc. (Paulitz, 1992). The activity of the pathogen in the infection court includes saprophytic growth in rhizosphere, external growth on root surface, formation of infection structures, invasion into epidermal cells, parasitic growth in the host tissues and translocation into rhizosphere. It is important to consider the strategies of biocontrol of *R. solani* to suppress the activity in the infection court by the antibiotic and siderophore producing bacteria.

III. Mechanisms of biocontrol with rhizobacteria

Rhizobacteria colonize roots and defend the infection courts against attack by pathogens. Biocontrol efficacy is essentially attributed to the competitive ability of the agent and its population density in the rhizosphere. Various mechanisms have been reported to be involved in biocontrol of soilborne plant pathogens and in plant growth promotion (Weller, 1988). These include the role of antibiotics, siderophores, HCN, induced resistance, and plant growth promoting substances. Defining these mechanisms allows for more efficient selection of new strains and improved effectiveness of the biocontrol agents.

The advance in molecular biology in recent years provided more specific and accurate tools to detect microbial inhibitory secondary metabolites and study the genetics of biocontrol, in addition to traditional methods, as reviewed by Fravel (1988) and Tomashow (1991).

A. Examples of biocontrol mediated by antibiotic and siderophore production

Pseudomonas spp. have been shown to produce the following antibiotics: pyrrolnitrin and pyoluteorin, implicated in control of damping-off of cotton caused by *R. solani* and *Pythium ultimum* (Howell and Stipanovic, 1979, 1980; Kraus and Loper, 1989), of cucumber caused by *P. ultimum* (Kraus and Loper, 1991, 1992) of radish caused by *R. solani* (Homma and Suzui, 1989; Homma *et al.*, 1989); phenazine-1-carboxylic acid and its derivatives, in suppression of take-all caused by *Gaeumannomyces graminis* var. *tritici* (Brisbane and Rovira, 1988; Gurusiddaiah *et al.*, 1986; Thomashow and Weller, 1988); oomycin A, suppressive to *P. ultimum* on cotton (Howie and Suslow, 1991); 2,4-diacetylphloroglucinol, against *Thielaviopsis basicola* in tobacco and against *G. graminis* var. *tritici* in wheat (Keel *et al.*, 1990, 1992; Vincent *et al.*, 1991); and pyocyanin, on *R. solani* AG 2-1 in canola (Dahiya, 1992) and against *R. solani* AG 1 causing the rice sheath blight (Rosalis *et al.*, 1995).

Siderophores produced by fluorescent pseudomonads have received much attention, largely because of their proposed role in the biocontrol of soilborne plant pathogens and in disease suppressive soils (Leong, 1986; Neilands and Leong, 1986; Schippers *et al.*, 1987; Loper and Buyer, 1991). The fluorescent pseudomonads are characterized by the production of yellow-green pigments, termed pyoverdins or pseudobactins, that function as siderophores (Cox, 1980; Teinze *et al.*, 1981). Siderophores are low molecular weight compounds that are produced under conditions limiting iron availability, chelate the ferric ion (Fe^{3+})with a high specific activity, and serve as the transport of Fe^{3+} into a microbial cell (Neilands, 1981).

The role of siderophore in biocontrol has been evaluated in studies that: a) varied the iron availability of the soil, b) added purified siderophores or synthetic iron chelates to soil, or c) compared the biocontrol activities of parental and derivative strains that are deficient in siderophore biosynthesis (Loper and Buyer, 1991). The following evidence support the role of siderophores produced by *Pseudomonas* spp. in biocontrol of wilt diseases caused by *Fusarium oxysporum*, of damping-off diseases caused by *P. ultimum*, and in plant growth promotion, which is attributed in part to the biocontrol of minor soilborne pathogens (Kloepper and Schroth, 1981; Schippers *et al.*, 1987).

Biocontrol of Rhizoctonia diseases by antibiotic and siderophore producing bacteria is summarized in Table 1; The antibiotics produced by the strains of Pseudomonas such as pyrrolnitrin, pseudane, phenazine and phloroglucinol have been considered to be important in suppression of biocontrol of *R. solani*. There are few reports on the role of siderophores in the biocontrol of *R. solani*; Additional research is needed to clarify the role of siderophore in biocontrol of *R. solani*.

B. Correlation between antibiotic production and disease suppression

Evidence for pyrrolnitrin production in culture by effective biocontrol isolates of Pseudomonas spp. was obtained from studies on biological control of seedling damping-off of cotton (Howell and Stipanovic, 1979; Kraus and Loper, 1989), radish (Homma *et al.*, 1989), cucumber (Kraus and Loper, 1991, 1992) and sugar beet (Nagahama *et al.*, 1988; Ohto *et al.*, 1990; Homma *et al.*, 1990), and Fusarium-wilt of bottle gourd (Kijima and Arie, 1987) and adzuki bean (Hasegawa *et al.* 1990). The role of the antibiotic pyrrolnitrin in disease suppression was demonstrated by Homma and Suzui (1989) and Howell and Stipanovic (1979): a) production of the antibiotic in culture and high activity of the antibiotic against target pathogens, b) efficacy of seed coating with purified antibiotics in disease suppression and reduction of the activity of the pathogen at the infection court, c) correlation between disease reduction *in vivo* and pyrrolnitrin production *in vitro* among naturally occurring strains and chemically-induced mutants and iv) detection of pyrrolnitrin from rhizosphere soil. Within limits, this evidence was useful to demonstrate the role of antibiotic production in biological control using traditional methods.

The role of bacterial antibiotics in biocontrol of soilborne pathogens was supported by a correlation between antibiotic production and pathogen inhibition *in vitro* and disease suppression *in vivo* among naturally occurring strains and chemically generated mutants. Homma and Suzui (1989) reported that strains RB425 and RB3292 of *P. cepacia* (=*Burkholderia cepacia* Burkh. Yabuuchi *et al.*), known producers of the antibiotics pyrrolnitrin and two pseudane compounds, were both suppressive to damping-off of radish caused by *R. solani* AG 4 when the bacteria were applied on radish seeds. Two other naturally-occurring strains of *P. cepacia* A2 and A4 , known producers of pyrrolnitrin but not pseudanes, also were suppressive to damping-off of radish. A fifth strain, *P. cepacia* ATCC No. 25416, produced no detectable level of any of the three antibiotics and not suppressive to damping-off of radish. Three mutants of strain RB425, induced by nitrosoguanidine produced no detectable levels of either pyrrolnitrin or the two pseudanes and provided no suppression of damping-off. Tests with derivative strains resistant to rifampicin and nalidixic acid shown that the mutants were not significantly different from the parent in the ability to colonize radish roots.

TABLE 1. Biocontrol of Rhizoctonia diseases by antibiotic producing bacteria

Host plants	Diseases	Pathogen	Bacteria	Metabolites	References
Cotton	damping off	*R. solani*	*P. fluorescens*	Pyrrolnitrin	1, 2
Cucumber	damping off	*R. solani*	*P. fluorescens*	Pyrrolnitrin	3
Radish	damping off	*R. solani*	*P. cepacia*	Pyrrolnitrin	4
Sugar beet	damping off	AG 2-2, 4	RB425	Pseudanes Phenols	5
Sugar beet	damping off	*R. solani* AG 2-2	*P. cepacia* 1218	Pyrrolnitrin	6, 7
Canola	damping off brown girdl- ing root rot	*R. solani* AG 2-1	*P. fluorescens* JAG-1	Pyrrolnitrin Pyocyanine Phenazine	8
Cotton	damping off	*R. solani*	*P. fluorescens* BL 915	Pyrrolnitrin	9
Rice	sheath blight	*R. solani* AG 1	*P. cepacia* In-b-6854,6858 *P. putida* In-b-1521	Pyrrolnitrin Pyocyanine Phloroglucinol	10

1. Howell & Stipanovic (1979); 2. Kraus & Loper (1989);
3. Kraus & Loper (1991; 1992); 4. Homma *et al.* (1989); 5. Homma *et al.* 1990)
6. Nagahama *et al.* (1988); 7. Ohto *et al.* (1990); 8. Dahiya (1992);
9 .Hill *et al.* (1994); 10. Rosales *et al.* (1995)

C. Suppression of *R. solani* in the rhizosphere by application of purified antibiotics to seeds

Damping-off of cotton caused by *R. solani* and *P. ultimum* was suppressed by seed bacterization with *P. fluorescens* strain Pf-5 (Howell and Stipanovic, 1979, 1980). The strain produced two antibiotics, pyrrolnitrin and pyoluteorin. Both inhibited hyphal growth of *R. solani* and *P. ultimum* respectively. Seed treatment with 200μg purified pyrrolnitrin per 30 seeds or 10μg pyoluteorin per seeds resulted in an increase in number of surviving seedlings from 13 to 70% and from 33 to 65%, respectively, while seed treatment with bacterial culture resulted in an increase from 30 to 79%. This method was applied to study the role of antibiotics produced by *P. cepacia* RB425 in suppression of damping-off of radish caused by *R. solani* (Homma and Suzui, 1989). The strain produced the antibiotics pyrrolnitrin and pseudane in culture (Homma *et al.*, 1989). Seed coating with 1.0μg pyrrolnitrin per seed resulted in reduction of 50% of damping-off and in suppression of root colonization by *R. solani*. The effect was equivalent to that achieved by seed coating with 10^7 cfu/seed. Coating with 40μg pseudane had no effect on the suppression. One of the reasons for the success of this method could be the stability of pyrrolnitrin in soil; pyrrolnitrin persisted for up to 30 days in moist natural soil with no measurable loss in inhibitory activity (Howell and Stipanovic, 1979).

D. Detection and measurement of antibiotics in rhizosphere

Direct evidence for the role of antibiotic production has been obtained from detection and determination of the relevant antibiotics in rhizosphere soil inoculated with strains of *Pseudomonas* sp.. The development of analytical techniques improved the ability to detect and estimate the antibiotics at effective concentration *in situ*. Recently, Tn*5* generated mutants which are deficient in production of antibiotics and colonize root in comparison with the parent strain was used; the antibiotic phenazine-1-carboxylic acid and acetylphloroglucinol were detected and determined in wheat rhizosphere inoculated with *P. fluorescens* 2-79 (Thomashow *et al.* 1990) and with *P. fluorescens* CHA0 (Keel *et al.*, 1992), respectively. Rhizosphere colonization by the mutants was similar to that of the parent strain, and the production of the antibiotic in rhizosphere soil was restored by complementation of the mutants with DNA from the parent genomic library.

Pyrrolnitrin was extracted with 80% acetone from rhizosphere soil of radish and sugar beet seedlings bacterized with *P. cepacia* RB425 (Homma *et al. unpublished*). Partially purified materials were tentatively identified to pyrrolnitrin based on: a) activity against *R. solani*, b) Rf-values on TLC, and c) coloration on TLC when sprayed with a 1% methanol solution of p-dimethyl-amino-benzaldehyde. Pyrrolnitrin could not be detected from sugar beet rhizosphere soil inoculated with a NTG-mutant of RB425 which is deficient in pyrrolnitrin production. Nakayama *et al.* (personal communication) tried to detect pyrrolnitrin from rhizosphere soil of sugar beet by using amberlite XAD-4 macroreticular resins as an absorbent and determined the amount of the antibiotic in the rhizosphere by HPLC in relation to the population size. Kempf *et al.* (1993) also detected pyrrolnitrin in cotton rhizosphere and barley spermosphere inoculated with *Pseudomonas* sp. Further study is required to demonstrate pyrrolnitrin production *in situ* by these strains at a concentration adequate to suppress pathogens in the spermosphere or rhizosphere.

Concentrations produced in the rhizosphere by *Pseudomonas* spp. are adequate to influence microbial interactions (Bakker *et al.*, 1988; Leong, 1986; Schippers *et al.*, 1987) This evidence points out an opportunity to antagonize certain soilborne pathogen population during saprophytic phases of their life cycles through manipulation of the available iron concentration in the rhizosphere (Loper and Buyer, 1991).

E. Evidence for the involvement of antibiotics in biocontrol using generated mutants

Several recent investigations confirmed that Tn*5* insertion mutants of *Pseudomonas* strains defective in antibiotic production have lost most their capability to suppress disease development. Antibiotic production and disease suppression were restored in the mutants by complementation with genes encoding for antibiotic production obtained from the wild-type genomic library.

The role of pyrrolnitrin in biological control has not yet been demonstrated in detail by using Tn5 mutants. Tn5 insertion mutants of pyrrolnitrin producers yielded mutants deficient in antibiotic production and they had lost their antagonistic ability *in vitro*. An additional unidentified antibiotic produced by *Pseudomonas* spp. effective against *R. solani* and *P. ultimum* was reported by Kraus and Loper (1989). Tn5 insertion mutants of *P. fluorescens* Pf-5 were generated and screened for antagonism against *R. solani* and *P. ultimum*, or for pyrrolnitrin and pyoluteorin production.

Screening of 6286 mutants for pyrrolnitrin and pyoluteorin production yielded 13 PN^+PL^- mutants, 6 PN^-PL^- mutants, several overproducing mutants, and 1 PN^-PL^+ mutant. They suggested that these mutants will be useful in future studies determining the roles of specific antibiotics in the biocontrol activity of Pf-5.

Jayaswal *et al.* (1992) isolated a strain of *P. cepacia* producing pyrrolnitrin and unidentified volatile antifungal compounds which inhibited many important corn fungal pathogens *in vitro*. Only one mutant among 1500 of Tn*5-259* transconjugants was found to be defective in the production of pyrrolnitrin. This mutant had lost its inhibitory activity to the pathogens. Using flanking DNA of the mutated gene as a probe, they have isolated four overlapping cosmid clones from a genomic library of *P. cepacia*. However, they were unable to complement the mutant because of difficulty in mobilizing the cosmids from *Escherichia coli*.

F. The role of antibiotic production in biocontrol

In the study of mechanism of biocontrol, it has become apparent that biocontrol activity is not a simple process mediated by a single microbial metabolite. In some cases, successful biocontrol agents have been reported to produce several kinds of metabolites inhibitory, including antibiotics, siderophores and cyanide. Certain antibiotics were demonstrated to play an important role in the suppression of soilborne diseases using Tn5 mutants lacking antibiotic production, but the activity did not explain the whole phenomenon (Kraus and Loper, 1991; Keel *et al.*, 1990). Keel *et al.* (1990) suggested that disease suppression by a strain is a multifactorial mechanism which is mainly due to the main antibiotic production.

Biocontrol agents which produced pyrrolnitrin have also been reported to produce other metabolites which are probably concerned with biological control of soilborne plant pathogens; pseudanes and antifungal phenols by RB425 (Homma *et al.*, 1989), IAA (Nagahama *et al.*, 1988; Ohto *et al.*, 1990), pyoluteorin and phenols (Howell and Stipanovic, 1980); pyoluteorin, HCN, pyoverdine and an uncharacterized antibiotic (Kraus and Loper, 1992) and volatile antifungal compounds (Jayaswal *et al.*, 1992). Homma and Suzui (1989) noticed that pyrrolnitrin deficient mutant of *P. cepacia* exhibited some suppression of radish damping-off by bacterization. Although pyrrolnitrin may play an important role in the suppression, further research is required to evaluate the role of other antibiotics in addition to pyrrolnitrin.

On the other hand, Kraus and Loper (1992) reported lack of evidence for role of antifungal metabolite production by *P. fluorescens* Pf-5 in biological control of Pythium damping-off of cucumber. Tn5 insertion mutants of Pf-5 were screened for loss of pyoluteorin production of pyoverdine production, or for overproduction of pyoluteorin. These mutants were similar to Pf-5 in suppression of Pythium damping-off of cucumber. They concluded that these metabolites do not play an important role in biological control of Pythium damping-off of cucumber. The mechanism(s) by Pf-5 may be due to competition for nutrients or for infection sites rather than antifungal metabolite production.

IV. CONCLUSIONS

Biocontrol agents have been known to produce several kinds of inhibitory secondary metabolites against pathogens; antibiotics, siderophores, cyanide, and bacteriocins. The role of antibiotic production by biocontrol agents in disease suppression or in plant growth promotion has been confirmed by the use of modern tools of microbial genetics, recombinant-DNA techniques, and new detection techniques in combination with traditional research strategies. The development of analytical technology has improved ability to detect and estimate the effective concentration of antibiotics *in situ*, in relation to the population of biocontrol agents in the rhizosphere. After clarifying the role of these metabolites in biocontrol using antagonistic microorganisms, this research should be followed by a) using the metabolites as a marker in searching for effective strains and reducing labor costs in screening for biocontrol agents, and b) improving biocontrol agents by genetic modification of the strain. The possibility to improve biological control should be possible through improved analytical technology, DNA modifying technology, and analysis of the rhizosphere environments. Reference systems to enable a search for suitable biocontrol agents for control of certain disease might also be developed by searching for strains producing specific metabolites inhibitory against specific pathogens. For commercial biocontrol preparations, see Chapter VI.B7.

V. References

Arima K, Imanaka H, Kousaka M, Fukuda A & Tamura G (1964) Pyrrolnitrin, a new antibiotic substance, produced by *Pseudomonas*. Agr. Biol. Chem. 28:575-576.

Bakker PAHM, Weisbeek PJ & Schippers B (1988) Siderophore production by plant growth-promoting *Pseudomonas* spp. J. Plant Nutr. 11:925-933.

Brisbane PG & Rovira AD (1988) Mechanism of inhibition of *Gaeumannomyces graminis* var. *tritici* by fluorescent pseudomonads. Pl. Pathol. 37:104-111.

Cox CD (1980) Iron uptake with ferripyochelin and ferric citrate by *Pseudomonas aeruginosa*. J.Bacteriol. 142:581-587.

Dahiya JS (1992) Antibiotics production during control of brown girdling root-rot of *Brassica napus* (canola/rapeseed) by seed bacterization with *Pseudomonas fluorescens*. Ind. J. Exp. Biol. 30:715-718.

Fravel DR (1988) Role of antibiosis in the biocontrol of plant diseases. Annu. Rev. Phytopathol. 26:75-91.

Gurrusiddaiah S, Weller DM, Sarkar A & Cook RJ (1986) Characterization of an antibiotic produced by a strain of *Pseudomonas fluorescens* inhibitory to *Gaeumannomyces graminis* var. *tritici* and *Pythium* spp. Antimicrob. Ag. Chemoth. 29:488-495.

Hamdan H, Weller DM & Thomashow L (1991) Relative importance of fluorescent siderophores and other factors in biological control of *Gaeumennomyces graminis* var. *tritici* by *Pseudomonas fluorescens* 2-79 and M4-80R. Appl. Env. Microbiol. 57:3270-3277.

Harrison LA, Letendre L, Kovacevich P, Pierson E & Weller D (1993) Purification of an antibiotic effective against *Gaeumannomyces graminis* var. *tritici* produced by a biocontrol agent, *Pseudomonas aureofaciens*. Soil Biol. Biochem. 25:215-221

Hasegawa S, Kodama F, Nakajima M & Murooka H (1990) Isolation of *Pseudomonas* spp. and antifungal agents against phytopathogenic fungi. Bull. College Agr. Vert. Medec., Nippon Univ. 47:224-232.

Homma Y & Suzui T (1989) Role of antibiotic production in suppression of radish damping-off by seed bacterization with *Pseudomonas cepacia*. Ann. Phytopathol. Soc. Jpn. 55:643-652.

Homma Y, Sato Z, Hirayama F, Konno K, Shirahama H & Suzui T (1989) Production of antibiotics by *Pseudomonas cepacia* as an agent for biological control of soilborne plant pathogens. Soil Biol. Biochem. 21:723-728.

Homma Y, ChikuoY & Sayama M (1990) Effect of seed bacterization with strains of *Pseudomonas cepacia* differing antibiotic productivity on seedling damping-off and rhizomania of sugar beets. Ann. Phytopathol. Soc. Jpn. 56:405.

Howell CR & Stipanovic RD (1979) Control of *Rhizoctonia solani* on cotton seedlings with *Pseudomonas fluorescens* and with an antibiotic produced by the bacterium. Phytopathology 69:480-482.

Howell CR & Stipanovic RD (1980) Suppression of *Pythium ultimum*-induced damping-off of cotton seedlings by *Pseudomonas fluorescens* and its antibiotic, pyoluteorin. Phytopathology 70:712-715.

Howie WJ & Suslow TV (1991) Role of antibiotic biosynthesis in the inhibition of *Pythium ultimum* in the cotton spermosphere and rhizosphere by *Pseudomonas fluorescens*. MPMI 4:393-399.

Jayaswal RK, Fernandez MA, Visintin L & Upadhyay RS (1992) Transposon Tn-5-259 mutagenesis of *Pseudomonas cepacia* to isolate mutants deficient in antifungal activity. Can. J. Microbiol. 38:309-312.

Keel C, Maurhafer M, Oberhänsli T, Voisard C, Haas D & Défago G (1991) Role of 2,4-diacetylphloroglucinol in the suppression of take-all of wheat by a strain of *Pseudomonas fluorescens*. In: Beemster ABR , Bollen, GT, Gerlach M, Ruissen MA, Schippers B & Tempel A (eds.), Biotic Interactions and Soil-Borne Disease. (pp 335-339). Elsevier, Amsterdam.

Keel C, Schnider U, Maurhofers M, Voisard C, Laville J, Burger U, Wirthner P, Haas D & Défago G (1992) Suppression of root disease by *Pseudomonas fluorescens* CHA0: Importance of the bacterial secondary metabolite 2,4-diacetylphloroglucinol. MPMI 5:4-13.

Keel C, Wirthner PH, Oberhänsli TH, Voisard C, Burgaer V, Haas D & Defago G (1990) Pseudomonads as antogonists of plant pathogens in the rhizosphere: role of the antibiotic 2,4-diacetylphloroglucinol in the suppression of black root rot of tobacco. Symbiosis 9:327-342.

Kempf HJ, Sinterhauf S, Muller M, Becker JO & Pachlatko P (1993) Production of pyrrolnitrin by a biocontrol bacterium in the rhizosphere of cotton and in the spermosphere of barley. 6th Int. Congr. of Plant Pathology, P 266. (Abstr.).

Kijima T, & Arie T (1987) Biological control of Fusarium diseases by using companion plants inoculated with antibiotic bacteria (*Pseudomonas gladioli* M-2196). Plant Prot. 41:129-133.

Kloepper JW & Schroth MN (1981) Relationship of invitro antibiosis of plant growth-promoting rhizobacteria to plant growth and the displacement of root microflora. Phytopathology 71:1020-1024.

Kraus J & Loper JE (1989) Tn5 insertion mutants of *Pseudomonas fluorescens* Pf5 altered in production of the antibiotics pyrrolnitrin and pyoluteorin. Phytopathology 79:910.

Lambert B, Leyns F, Van Rooyen L, Gossele F, Papon Y & Swings J (1987) Rhizobacteria of maize and their antifungal activities. Appl. Envir. Microbiol. 53:1866-1871.

Leach LD & Garber RH (1970) Control of *Rhizoctonia*. In: Parmeter JR Jr (ed.) *Rhizoctonia solani*: Biology and Pathology. (pp 189-198). Univ. California Press, Berkeley, Los Angeles and London

Leisinger T & Margraff R (1978) Secondary metabolites of fluorescent pseudomonads. Microbiol. Rev. 43:422-445.

Leong J (1986) Siderophores: Their biochemistry and possible role in the biocontrol of plant pathogens. Annu. Rev. Phytopathol. 24:187-209.

Loper JE & Buyer JS (1991) Siderophores in microbial interactions on plant surfaces. MPMI 4:5-13.

Nagahama M, Nagai Y, Lee WH, Hasegawa S, Kobayashi K & Ogoshi A (1988) Studies on seed bacterization of sugar beets. (V) Identification of the strain of antagonistic bacteria and production of antifungal substances and the plant growth promoting substances. Ann. Phytopathol. Soc. Jpn. 54:116.

Neilands JB (1981) Microbial iron compounds. Annu. Rev. Biochem. 50:715-731.

Neilands JB & Leong SA (1986) Siderophores in relation to plant growth and disease. Annu. Rev. Plant Physiol. 37:187-208.

Ohto Y, Kobayashi K & Ogoshi A (1990) Studies on seed bacterization of sugar beets. (VII) Mechanism of suppression of preemergence damping-off of sugar beets caused by *Rhizoctonia solani* AG 4. Ann. Phytopathol. Soc. Jpn. 56:143.

Pameter JR Jr (ed.) (1970) *Rhizoctonia solani*: Biology and Pathology. Univ. California Press, Berkeley, Los Angeles and London, 255pp.

Paulitz T (1992) Biological control of damping-off diseases with seed treatments. In: Tjamos ES, Papavizas, GC & Cook RJ (eds.), Biological Control of Plant Diseases, (pp 145-155), Plenum Press, New York.

Pierson III LS & Thomashow L (1992) Cloning and heterologous expression of the phenazine biosynthetic locus from *Pseudomonas aureofaciens* 30-84. MPMI 4:330-339.

Rosales AM, Thomashow L, Cook RJ & Mew TW (1995) Isolation and identification of antifungal metabolites produced by rice-associated antagonistic *Pseudomonas* spp. Phytopathology 85:1028-1032.

Rovira AD & Ridge EH (1973) The use of a selective medium to study the ecology of *Pseudomonas* spp. in soil. Bull. Ecol. Res. Comm. (Stockholm) 17:329-335.

Rovira A, Ryder M & Harris A (1992) Biological control of root diseases with pseudomonads. *In:* Tjamos ES Papavizas, GC & Cook RJ (eds.), Biological Control of Plant Diseases, (pp 175-184) Plenum Press, New York.

Schippers B, Bakker AW & Bakker PAHM (1987) Interactions of deleterious and beneficial rhizosphere microorganisms and the effect of cropping practices. Annu. Rev. Phytopathol. 25:339-358.

Schroth MN & Hancock JG (1981) Selected topics in biological control. Annu. Rev. Microbiol. 35:453-476.

Shanahan P, O'Sullivan DJ, Simpson P, Glennon JD & O'Gara F (1992) Isolation of 2,4-diacetylphloroglucinol from a fluorescent pseudomonad and investigation of physiological parameters influencing its production. Appl. Env. Microbiol. 58:353-358.

Teinze M, Hoosain MB, Barnes CL, Leong J & van der Helm D (1981) Structure of ferric pseudobactin, a siderophore from a plant growth promoting *Pseudomonas*. Biochem. 20:6446-6457.

Thomashow L & Weller DM (1988) Role of a phenazine antibiotic from *Pseudomonas fluorescens* in biological control of *Gaeumannomyces graminis* var. *tritici*. J. Bacteriol. 170:3499-3508.

Thomashow L, Weller DM, Bonsall RF & Pierson III LS (1990) Production of the antibiotic phenazine-1-carboxylic acid by fluorescent *Pseudomonas* species in the rhizosphere of wheat. Appl. Env. Microbiol. 56:908-912.

Thomashow LS (1991) Molecular basis of antibiosis mediated by rhizosphere pseudomonads. In: Keel C, Koller B & Defago G (eds.), Plant Growth-Promoting Rhizobacteria, (pp 109-114) WPRS Bulletin.

Van Peer R, Punte HLM, de Weger LA & Schippers B (1990) Characterization of root surface and endorhizosphere pseudomonads in relation to their colonization of roots. Appl. Env. Microbiol. 56:2462-2471.

Vincent MN, Harrison LA, Brackin JM, Kovacevich PA, Mukerji P, Weller DM & Pierson EA (1991) Genetic analysis of the antifungal activity of a soilborne *Pseudomonas aureofaciens* strain. Appl. Env. Microbiol. 57:2928-2934.

Voisard C, Keel C, Haas D & Defago G (1989) Cyanide production by *Pseudomonas fluorescens* helps suppress black root rot of tobacco under gnotobiotic conditions. EMBO 8:351-358.

Weller DM (1984) Distribution of take-all suppressive strain of *Pseudomonas fluorescens* on seminal roots of winter wheat. Appl. Env. Microbiol. 48:897-905.

Weller DM (1988) Biological control of soil-borne plant pathogens in the rhizosphere with bacteria. Annu. Rev. Phytopathol. 26:378-389.

Weller DM, Howie WJ & Cook RJ (1988) Relationship between *in vitro* inhibition of *Gaeumannomyces graminis* var. *tritici* and suppression of take-all of wheat by fluorescent pseudomonads. Phytopathology 78:1094-1100.

VI.B2 BACTERIAL AND FUNGAL CELL-WALL HYDROLYTIC ENZYMES IN RELATION TO BIOLOGICAL CONTROL OF *RHIZOCTONIA SOLANI*

YIGAL ELAD

Department of Plant Pathology, Agricultural Research Organization, The Volcani Center, P.O. Box 6, Bet Dagan 50250, ISRAEL

I. INTRODUCTION

Cell-wall-degrading enzymes are produced by fungal parasites. Parasitism of *Rhizoctonia solani* was first reported by Weindling (1932), who observed the parasitic nature of *Trichoderma lignorum* on several plant pathogens. Weindling's (1932) report on the possible use of mycoparasites for the control of plant pathogens raised a great deal of interest in this area of research, including the role of hydrolytic enzymes in microbial interactions. It prompted reviews and discussions of the subject in numerous articles (Barnett, 1963; Boosalis, 1964; Barnett & Binder, 1973; Ayers & Adams, 1981; Wells, 1981; Baker, 1988; Ghaffar, 1988; Elad & Chet, 1995; see also chapter VI.B5). One of the features required for successful degradation of host fungi is the ability to produce the enzymes which degrade the polymers underlying the cell wall and membranes of the host. Boosalis (1956) estimated the frequency of *R. solani* parasitism by *Penicillium vermiculatum* in soil by observing the parasitism of *R. solani* hyphae *in situ*. Indeed, living fungal structures are often parasitized in natural soils (Boosalis, 1956; Hunter *et al.*, 1977; Sneh *et al.*, 1977). For this purpose, it may be assumed that the parasites produce the hydrolytic enzymes in natural niches where the pathogen is present. However, even with increasing evidence of natural mycoparasitism, the ecological significance of this phenomenon, at least in soil, is still ambiguous. Griffin (1972) maintained that mycoparasitism is of no ecological significance in soil, while others have suggested that mycoparasitism has considerable impact on soil fungal populations, by reducing the longevity of survival structures (Sneh *et al.*, 1977). In any case, one of the features required for successful degradation of host fungi is the ability to produce enzymes which degrade the polymers underlying the cell wall and membranes of the host. The studies related to these enzymes, as related to degradation of *R. solani*, will be described in this chapter.

As will be demonstrated, the biochemical basis has been thoroughly investigated for few systems only. Most information on hyphal interactions has been obtained by observing the approach and contact of paired colonies, of necrotrophic mycoparasites and *R. solani*, respectively, on nutrient-rich agar. However, behavior under such conditions *in vitro* may not reflect the actual behavior *in vivo*. The nomeclature and definitions of enzymes which degrade chitin, glucans, proteins and lipids have changed during the generations of work on the subject of this chapter. In some cases the enzymatic nomeclature is still under debate (for instance, regarding chitinases). The identifications of enzymes and their names throughout this chapter follow their presentation by the investigators in their relevant publications cited below.

B. Sneh et al. (eds.),
Rhizoctonia Species: Taxonomy, Molecular Biology, Ecology, Pathology and Disease Constrol, 455–462.

II. Hydrolytic enzymes produced by fungi which degrade hyphae of *R. solani*

Trichoderma spp. and *Gliocladium* spp. are common soil inhabitants, especially in soils and in organic matter. These fungi can live either as saprophytes or parasites on other fungi. Fungi of these two genera are well known for their antagonistic behavior, and have been used extensively in biocontrol trials (Cook & Baker, 1983; Papavizas, 1985; Tronsmo, 1986). The actual mechanisms of antagonism have not always been elucidated, but competition for nutrients, production of inhibitory compounds and of hydrolytic enzymes have frequently been described.

Isolates of *T. harzianum* - an effective biocontrol agent against *R. solani* and *S. rolfsii* produce extracellular β-1,3-glucanase and chitinase when grown on cell walls of *R. solani* (Chet *et al.*, 1979; Hadar *et al.*, 1979; Chet and Baker, 1981). Later, Elad *et al.* (1982) studied the enzymatic basis for this interaction in culture and in soil. *T. harzianum* excreted β-1,3-glucanase or chitinase into the medium when it was grown on laminarin and chitin, respectively. *T. harzianum* also showed high activity of both enzymes when it was grown on homogenized sclerotia. High lytic activity was detected in a wheat-bran culture of the fungus and in soil inoculated with this culture. In the presence of cycloheximide which inhibits enzymatic activity, antagonism was prevented (Elad *et al.*, 1983b).

Isolates of *T. harzianum* differ in the levels of hydrolytic enzymes produced when mycelia of *R. solani, S. rolfsii* and *P. aphanidermatum* are attacked in soil by this fungus. The variation in enzymatic activity corresponds with the ability of each of the *Trichoderma* isolates to control the respective soil-borne pathogen. The specific effect of lytic enzymes was also observed by means of different microscopic techniques. Interaction sites were stained by fluorescein isothiocyanate - conjugated lectins or Calcofluor White M2R New. The latter compound binds to the edges of polysaccharide oligomers (Kritzman *et al.*, 1978). The observation of fluorescence was indicative of the presence of localized cell wall lysis at the point of interaction between the antagonist and its host (Elad *et al.*, 1983b). The specific binding germ agglutinin to coiling zones indicated the presence of *N*-acetyl-D-glucosamine (GlcNac) oligomers at these sites (Elad *et al.*, 1983b). Thus, chitin fibrils were evidently exposed in the cell walls of the host as a result of extracellular β-1,3-glucanase excreted by the mycoparasite.

Ridout *et al.* (1988) also described a *T. harzianum* isolate which was able to coil round and penetrate hyphae of *R. solani* in dual-culture. This isolate produced chitinase, β-1,3-glucanase and protease when it was grown on *R. solani* cell-walls. In a similar system, with other isolates of the same species, Benhamou and Chet (1993) demonstrated in detail that attachment of hyphae of the parasite and the *R. solani* was followed by a series of degradation events in the host. The cytochemical localization of GlcNac residues with wheat germ agglutinin/ovomucoid-gold complex revealed that chitin breakdown occurred gradually, suggesting continuous production of the degrading enzymes by the antagonist. Disorganization of the cell-wall structure of *R. solani* in this system appear to be an early event that promoted internal changes such as retraction of the plasma membrane and cytoplasmic aggregation.

The chitinolytic system of *T. harzianum* was found to be a complex one, consisting of six distinct enzymes (Haran *et al.*, 1995). The system was found to be composed of two β-1,4-*N*-acetylglucoseaminidases and four endochitinases. All the chitinolytic enzymes were induced and excreted during growth of the antagonistic fungus on chitin as the sole carbon source in culture. The researchers suggested that the chitinolytic system of *T. harzianum* involves complementary modes of action of the six enzymes, all of which are required for effective parasitism against a broad spectrum of chitin-containing plant pathogens, including *R. solani*.

Detailed descriptions of mycoparasitism have been obtained by transmission electron microscopy during the interaction of *Trichoderma* spp. with either *R. solani* or *S. rolfsii* (Elad *et al.*, 1983a). Hyphae of the parasite contacted their host and then enzymatically digested its cell walls. Parasite organells (e.g., mitochondria, vesicles and dark osmophillic inclusions) accumulated in the parasitized cells. In response to the invasion, the host produced a sheath matrix that encapsulated the penetrating hyphae; later, the host cells became empty of cytoplasm and filled with the cells of the parasite (Elad *et al.*, 1983a). Therefore, it can be assumed that the *T. harzianum* isolates described above attack the host mycelium by first dissolving its cell walls in certain locations. They then continue this attack and follow it, sometimes, with hyphal penetration. The hyperparasite subsequently uses additional extracellular enzymes, such as lipases and proteases (Elad *et al.*, 1982). When attacked, the susceptible host hyphae reacted with rapid vacuolation, collapse, and disintegration (Chet *et al.*, 1981).

When *Gliocladium virens* was grown in liquid culture on *R. solani* cell walls it produced β-1,3-glucanase, *N*-acetylglucoseaminidase, lipases and proteinases (Vantilburg & Thomas, 1993). The enzyme β-1,3-glucanase had the greatest specific activity of any enzymes tested, at 8 h after inoculation. The authors concluded that this enzyme may be the first important enzyme in the *G. virens-R. solani* interaction. The various enzymes had either early or late peaks of accumulation in the liquid cultures. Therefore, the researchers concluded that in culture the extracellular proteins are regulated by a combination of gene regulation and protein degradation.

Geremia *et al.* (1992) identified a basic proteinase of *T. harzianum* that is induced in the presence of either autoclaved mycelia or cell walls of phytopathogenic fungi. This proteinase has been characterized biochemically and identified as serine proteinase. By using synthetic oligonucleotides deduced from peptide microsequencing data, the researchers were able to clone the corresponding cDNA from a library generated with the aid of poly(A)+ RNA from induced mycelia. The gene encoding for the proteinase was mapped to chromosome 6 by Southern analysis of chromosomes separated by pulse-field electrophoresis. Northern analysis showed that induction by cell walls occurs at the RNA level. Immunofluorescence experiments indicated that the proteinase is probably produced only in the area of direct physical contact with the pathogen during *in vitro* confrontation (Geremia *et al.*, 1992).

Studies on *Trichoderma* spp. revealed that production of hydrolytic enzymes is part of a complex process that includes several successive steps: (a) recognition, (b) attachment, (c) coiling. The earliest detectable interaction shows the growth of the mycoparasite hyphae directly towards its host (Barnett, 1964; Boosalis, 1964; Hadar *et al.*, 1979; Harman *et al.*, 1980; Chet & Baker, 1981; Chet *et al.*, 1981; Alexander, 1982; Elad *et al.*, 1983). Following these steps, an infrequent third stage may involve host mycelium penetration by the mycoparasite, apparently by partially degrading its cell walls (Elad *et al.*, 1983b).

Oomycetes cell walls are composed of β-glucans (Bullock *et al.*, 1986), cellulose, and <1.5% chitin. Basidiomycetes and ascomycetes contain relatively more chitin, as well as β-glucans, but no cellulose. As mentioned above, mycoparasites, such as *Trichoderma* spp. and *Pythium nunn*, produce hydrolytic enzymes: β-1,3-glucanase, cellulase and chitinase. As expected, different combinations of enzymes are produced, which are appropriate for substrate degradation. For example, large amounts of β-1,3-glucanase and chitinase were produced by *P. nunn* in liquid cultures containing cell walls of *R. solani* (Elad *et al.*, 1985). Cellulase, but not chitinase, was also produced in the presence of cell walls of two oomycetous pathogens. Apparently, oomycetous hosts of *P. nunn* do not produce cell wall-degrading enzymes. Similarly, *Pythium oligandrum* is known to show biological control activity against a number of important soil-borne plant pathogens (Lewis *et al.*, 1989). It has been described as an aggressive mycoparasite of a wide host range (Deacon & Henry, 1978; Lewis *et al.*, 1989). *P. oligandrum* produced β-1,3-D-glucanase, chitinase, lipases and proteases, as evidenced in other systems. However, metabolites which are inhibitory to certain host fungi *in vitro* were also detected (Lewis *et al.*, 1989).

Schizophyllum commune is capable of attacking 16 different fungal hosts, including *R. solani* (Chiu & Tzean, 1995). In dual-culture agar plates of this parasite with *R. solani*, the production of extracellular endo-b-1,3(4)-glucanase was markedly enhanced, whereas chitinase, *N*-acetylglucoseaminidase and glucosidase were either barely or not detected. Treatment of *R. solani* young hyphae with a crude preparation of endo-b-1,3(4)-glucanase led to the release of protoplasts. The purified enzyme was capable of hydrolyzing purified *R. solani* cell walls. Chiu & Tzean (1995) suggested that in the case of *S. commune* parasitizing *R. solani*, endo-b-1,3(4)-glucanase, in presence of trace amounts of certain wall-lytic enzymes such as cellulase, produced by the hyperparasite, may play a crucial role in host-parasite interactions.

In vitro production of b-1,3-glucanases and chitinases by *Stachybotrys elegans*, a mycoparasite of *R. solani*, was examined under various cultural conditions by Tweddell *et al.* (1994). Production of both enzymes was influenced by the carbon source and was stimulated by acidic pH and $NaNO_3$. Production of the enzymes was higher in the presence of host cell walls in the medium than with glucose. Protein electrophoresis detected, different isoforms of both enzymes, depending on the carbon source used. Tweddell *et al.* (1994) also reported that *S. elegans* culture filtrates, possessing b-1,3-glucanase and chitinase activities, were capable of degrading *R. solani* mycelium.

Degradation of sclerotia, which are resistant resting structures, pose a special challenge in the process of fungal attack by hyperparasites. Once the contact between the mycoparasite hyphae and the host sclerotium has been made, penetration of the melanized rind takes place. It is followed by colonization of the cortex and medulla. In *C. minitans*, *Verticillium biguttatum*, *Gliocladium*, and *Trichoderma* spp., both inter- and intracellular growth occur, aided by the production of induced b-1,3-glucanases and chitinases (Jones *et al.*, 1974; Elad *et al.*, 1982, 1983b; Artigues *et al.*, 1984). Activity of hydrolytic enzymes in sclerotia is inhibited by presence of phenols in the cell walls. However, the degradation of sclerotial cell walls was still effective.

III. Bacterial enzymes hydrolyzing *R. solani* cell walls

Streptomyces isolates were selected for their chitinolytic activity (Zeinet *et al.*, 1993). Culture filtrates of the most active chitinolytic isolate were tested for antifungal activity on several fungi including *R. solani*. The greatest inhibition of hyphal growth was found with *R. solani*. When crude enzyme produced by *Streptomyces* sp. was incubated with living mycelium of the pathogen, glucose and GlcNac were released from the mycelium.

Shapira *et al.* (1989) were able to control *R. solani* by means of chitinase expressed by an *Escherichia coli* clone harboring a DNA fragment carrying the *chi*A gene from *Serratia marcescence* in the plasmid pBR322. To obtain high levels of chitinase expression, the leftward operatory promoter of bacteriophage lambda was introduced upstream of the chiA gene. The resulting plasmid was introduced into *E. coli,* which produced high levels of the enzyme following induction. In chitinase preparation obtained from the cloned *E. coli*, viable cells, but not from those of *E. coli*-carrying the plasmid lacking the promoter, significantly reduced disease severity in cotton seedlings caused by *R. solani*.

Inbar & Chet (1991), suggested that chitinase production is involved in the control of *R. solani* and other plant pathogens. They worked with *Aeromonas caviae,* which controlled *R. solani* in cotton. This bacterium lysed living mycelia of different fungi, including *R. solani,* when grown on the mycelium as the sole carbon source. No b-1,3-glucanase was produced by the bacterium. The purified chitinase of *A. caviae* was found to act in an exo-splitting manner.

Chernin *et al.* (1995) reported on three strains of *Enterobacter agglomerans* which when. were grown in the presence of colloidal chitin as the sole carbon source in culture, where induced to produce proteins with chitinolytic activity. The proteins were identified following their separation by electrophoresis. Four enzymes were detected: two *N*-acetyl-b-D-glucoseaminidases, one endochitinase and one chitobiosidase. Chernin *et al.* (1995) showed that chitinolytic *E. agglomerans* strains suppressed *R. solani* in cotton, whereas mutants of one of them, which were deficient in chitinolytic activity, were incapable to protect the plants against the pathogen.

IV. Concluding remarks

It is quite obvious that true mycoparasitism involves the use of cell-wall-degrading enzymes by the mycoparasite for penetration. At the same time, and also later, lipases and proteinases are used for digestion of other components embedded in the cell walls and in the host cytoplasm. It may be assumed that enhanced production of hydrolytic enzymes by biocontrol agents can enhance their ability to effectively degrade *R. solani* cells, i.e., enhance their activity.

Further research on these aspects, as well as on microclimatic and nutritional influence on enzyme production and activity, is needed Understanding the relative importance of this phenomenon and of its role in mycoparasite ecology is essential for the more effective and extended use of mycoparasites in biological control. Similarly, research in pursuit to improve biocontrol agents may contribute to our understanding of the importance of cell-wall-degrading enzymes. Production of superior strains is achievable by methods such as: production of homokarions (in mycoparasite cells containing several nuclei that are usually of variable biotypes), mutation by chemical and physical, sexual genetic recombination (in fungi where it is possible), protoplast fusion, and transformation of DNA from and into mycoparasites. Such studies have already generated initial results with fungi such as *Trichoderma*, and are likely to contribute to our knowledge in the future.

Furthermore, activation of hydrolytic enzymes in plants previously susceptible to *R. solani* as well as to other pathogens is potentially significant for the prevention of disease development, as demonstrated by Roby *et al.* (1990). They analyzed transgenic tobacco plants containing a chimeric gene carrying chitinase fused to *gus*A gene. Chitinase promoter was activated by *R. solani* infection. Better protection of transgenic plants against diseases was obtained when they expressed both chitinase and b-1,3-glucanase than those expressing each of these enzymes separately (Zhu *et al.*, 1994; Jach *et al.*, 1996; see also chapter VI.C2).

Only few studies have demonstrated the importance of hydrolytic enzymes production in the actual process of biocontrol of *R. solani* in whole systems that involve the interacting microorganisms in soil and plants. However it may be assumed that production of these enzymes by bacteria or fungi is one of the most important factors which enables them to suppress the pathogen and eventually to digest its cells contents. Among other features are their ability to utilize complex or simple organic sources of nutrition, to compete for such nutrients and for space and other important resources, to produce compounds inhibitory to the pathogen or to other members of the surrounding microflora, and to establish themselves in the pathogen's niche. The integration of hydrolytic activity with that of inhibitory compounds may render the host fungus more susceptible to the biocontrol agent and thus, enhance its lytic activity (Di Pietro *et al.*, 1993).

V. ACKNOWLEDGEMENTS

Contribution from the Agricultural Research Organization, The Volcani Center, Bet Dagan, Israel, No. 1741-E, 1995 series. The author thanks Prof. Ilan Chet for his guidance and help throughout the long time cooperation in biocontrol of plant diseases. The manuscript is dedicated to the late Prof. Ralph Baker who inspired us all.

VI. REFERENCES

Alexander M (1982) Introduction to Soil Microbiology, Wiley, New York, 467 pp.

Artigues M, Daveet P & Roure C (1984) Comparison des aptitudes parasitaires de clones de *Trichoderma* vis-a-vis de quelques champignons a sclerotes. Soil Biol. Biochem. 16:413-417.

Ayers WA & Adams PB (1981) Mycoparasitism and its application to biological control of plant diseases. In: Papavizas GC (ed.) Biological Control in Crop Production (pp 91-103) Allanheld, Osmun and Co., Totowa, New Jersey.

Baker R (1988) Mycoparasitism: Ecology and physiology. Can. J. Plant Pathol. 9:370-379.

Barnett HL (1963) The nature of mycoparasitism by fungi. Annual Review of Microbiology 17:1-14.

Barnett HL & Binder FL (1973) The fungal host-parasite relationship. Annu. Rev. Phytopathol. 11:273-292.

Benhamou N & Chet I (1993) Hyphal interactions between *Trichoderma harzianum* and *Rhizoctonia solani* - ultrastructure and gold cytochemistry of the mycoparasitic process. Phytopathology 83:1062-1071.

Boosalis MG (1956) Effect of soil temperature and green-mature amendment of unsterilized soil on parasitism of *Rhizoctonia solani* by *Penicillium vermiculatum* and *Trichoderma* sp. Phytopathology 46:473-478.

Boosalis MG (1964) Hyperparasitism. Annu. Rev. Phytopathol. 2:363-376.

Bullock S, Adams PB, Willets HJ & Ayers WA (1986) Production of haustoria by *Sporidesmium sclerotivorum* in sclerotia of *Sclerotinia minor*. Phytopathology 76:101-103.

Carling DE, Brown MF & Millikan DE (1976) Ultrastructural examination of the *Puccinia graminis* - *Darluca filum* host-parasite relationship. Phytopathology 66:419-422.

Chernin L, Ismailov Z, Haran S & Chet I (1995) Chitinolytic *Enterobacter agglomerans* antagonistic to fungal plant pathogens. Appl. Environ. Microbiol. 61:1720-1726.

Chet I & Baker R (1981) Isolation and biocontrol potential of *Trichoderma hamatum* from soil naturally suppressive of *Rhizoctonia solani*. Phytopathology 71:286-290.

Chet I, Hadar Y, Elad Y, Katan J & Henis Y (1979) Biological control of soil-borne plant pathogens by *Trichoderma harzianum*. In: Schippers B & Gams W (eds.) Soil-Borne Plant Pathogens. (pp 585-592) Academic Press, London.

Chet I, Harman GE & Baker R (1981) *Trichoderma hamatum*: its hyphal interactions with *Rhizoctonia solani* and *Pythium* spp. Microbial Ecol. 7:29-38.

Chiu SC & Tzean SS (1995) Glucanolytic enzyme production by *Schizophyllum commune* Fr. during mycoparasitism. Physiol. Mol. Plant Pathol. 46:83-94.

Cook RJ & Baker KF (1983) The Nature and Practice of Biological Control of Plant Pathogens (539 pp), APS Press, St. Paul, Minnesota

Deacon JW & Henry CM (1978) Mycoparasitism by *Pythium oligandrum* and *P. acanthicum*. Soil Biol. Biochem. 10:409-415.

Di Pietro A, Lorito M, Hayers CK, Broadway RM & Harman GE (1993) Endochitinase from *Gliocladium virens*: Isolation, characterization, and synergistic antifungal activity in combination with gliotoxin. Phytopathology 73:308-313.

Elad Y, Barak R, Chet I & Henis Y (1983a) Ultrastructural studies of the interaction between *Trichoderma* spp. and plant pathogenic fungi. J. Phytopathol. 107:168-175.

Elad Y & Chet I (1995) Mycoparasitism. In: Kohmoto K, Singh US & Singh RP (eds.) Pathogenesis and Host Specificity in Plant Diseases: Histopathological, Biochemical, Genetic and Molecular Basis. Volume II: Eukaryotes (pp 289-307), Pergamon, Elsevier Science Ltd. Oxford, UK.

Elad Y, Chet I & Henis Y (1982) Degradation of plant pathogenic fungi by *Trichoderma harzianum*. Can. J. Microbiol. 28:719-725.

Elad Y, Chet I, Boyle P & Henis Y (1983b) Parasitism of *Trichoderma* spp. on *Rhizoctonia solani* and *Sclerotium rolfsii*, Scanning electron microscopy and fluorescence microscopy. Phytopathology 73:85-88.

Elad Y, Lifshitz R & Baker R (1985) Enzymatic activity of the mycoparasite *Pythium nunn* during interaction with host and non-host fungi. Physiol. Plant Pathol. 27:131-148.

Geremia RA, Goldman GH, Jacobs D, Montagu MV & Herrera-Estrella A (1992) Hydrolytic extracellular enzymes of *Trichoderma harzianum*: Specific induction of a basic proteinase. In: Jensen DF, Hockenhull J & Fokkema NJ (eds.) New Approaches in Biological Control of Soil-Borne Diseases. Bulletin WPRS 15:193-195.

Ghaffar A (1988) Biological control of Sclerotinia diseases. In: Mukerji KG & Garg KL (eds.) Biocontrol of Plant Diseases, Vol. I (pp 153-175) CRC Press, Inc., Boca Raton, FL.

Griffin DM (1972) Ecology of Soil Fungi, Chapman & Hall, London, 193pp.

Hadar Y, Chet I & Henis Y (1979) Biological control of *Rhizoctonia solani* damping-off with wheat bean culture of *Trichoderma harzianum*. Phytopathology 69:64-68.

Haran S, Schickler H, Oppenheim A & Chet I (1995) New components of the chitinolytic system of *Trichoderma harzianum*. Mycol. Res. 99:441-446.

Harman GE, Chet I & Baker R (1980) *Trichoderma hamatum* effects on seed and seedling disease induced in radish and pea by *Pythium* spp. or *Rhizoctonia solani*. Phytopathology 70:1167-1172.

Hunter WE, Duniway JM & Butler EE (1977) Influence of nutrition, temperature, moisture, and gas composition on parasitism of *Rhizopus oryzae* by *Syncephalis californica*. Phytopathology 67:664-669.

Inbar J & Chet I (1991) Evidence that chitinase produced by *Aeromonas caviae* is involved in the biological control of soil-borne plant pathogens by this bacterium. Soil Biol. Biochem. 23:973-978.

Jach G, Gorunhardt B, Longemann J, Pinsdorf E, Mondi J, Schell J & Maas, C (1996) Enahnced quantitative resistance agaist fungal diseases by combinatorial expression of different barley antifungal proteins in transgenic plants. Plant J. (In Press).

Jones D, Gordon AH & Bacon JSD (1974) Co-operative action by endo- and exo-(1->3)-glucanases from parasitic fungi in the degradation of cell-wall glucans of *Sclerotinia sclerotiorum* (Lib.) de Bary. Biochem. J. 140:47-55.

Kritzman G, Chet I, Henis Y & Huttermann A (1978) The use of the brightener "Calcofluor White M2R New" in the study of fungal growth. Isr. J. Bot. 27:138-146.

Lewis K, Whipps JM & Cooke RC (1989) Mechanisms of biological disease control with special reference to the case study of *Pythium oligandrum* as an antagonist. In: Whipps JM & Lumsden RD (eds.) Biotechnology of Fungi for Improving Plant Growth (pp 191-217) Cambridge University Press, London.

Papavizas GC (1985) *Trichoderma* and *Gliocladium*: biology, ecology, and potential for biocontrol. Annu. Rev. Phytopathol. 23:23-54.

Ridout CJ, Coley-Smith JR & Lynch JM (1988) Fractionation of extracellular enzymes from a mycoparasitic strain of *Trichoderma harzianum*. Enz. Microbiol. Technol. 10:180-187.

Roby D, Broglie K, Cressman R, Biddle P, Chet I & Broglie R (1990) Activation of a bean chitinase promoter in transgenic tobacco plants by phytopathogenic fungi. Plant Cell 2:999-1008.

Shapira R, Ordentlich A, Chet I & Oppenheim AB (1989) Control of plant diseases by chitinase expressed from cloned DNA in *Escherichia coli*. Phytopathology 79:1246-1249.

Sneh B, Humble SJ & Lockwood LJ (1977) Parasitism of oospores of *Phytopthora megasperma* var. *sojae, P. cactorum, Pythium* sp., and *Aphanomyces euteiches* in soil by oomycetes, chytridiomycetes, hyphomycetes, actinomycetes, and bacteria. Phytopathology 67:622-628.

Tronsmo A (1986) Use of *Trichoderma* spp. in biological control of necrotrophic pathogens. In: Fokkema N & van den Heuvel J (eds.) Microbiology of the Phyllosphere (pp 348-361) Cambridge University Press, London.

Tweddell RJ, Jabajihare SH & Charest PM (1994) Production of chitinases and β-1,3-glucanases by *Stachybotrys elegans*, a mycoparasite of *Rhizoctonia solani*. Appl. Environ. Microbiol. 60:489-495.

Vantilburg AUB & Thomas MD (1993) Production of extracellular proteins by the biocontrol fungus *Gliocladium virens*. Appl. Environ. Microbiol. 59:236-242.

Weindling R (1932) *Trichoderma lignorum* as a parasite of other soil fungi. Phytopathology 22:837-845.

Wells HD (1988) Trichoderma as a biocontrol agent. In: Mukerji KG & Garg KL (eds.) Biocontrol of Plant Diseases, Vol. I (pp 71-82) CRC Press, Inc., Boca Raton, FL.

Zeinet K, Nahid H & Fadida F (1993) Extracellular chitinase from *Streptomyces* sp. and its antifungal activities. Acta Pharm. Technol 35:135-143.

Zhu Q, Maher EA, Masoud S, Dixon RA & Lamb CJ (1994) Enahnced protection against fungal attack by constitutive co-expression of chitinase and glucanase genes in transgenic tobacco. Bio/Technology 12:807-811.

VI.B3 ACTINOMYCETES, CYANOBACTERIA AND ALGAE

MARTIN M. KULIK

Soybean and Alfalfa Research Laboratory Plant Sciences Institute, Agricultural Research Service, United States Department of Agriculture, Beltsville, Maryland 20705, U.S.A.

I. ACTINOMYCETES

A. Introduction

The actinomycetes or "ray fungi" are wide-spread, Gram-positive eubacteria belonging to the Actinomycetales, a large order that consists of approximately 60 genera. Genera of special interest include *Streptomyces*, many species of which produce antibiotics and other useful metabolites, as well as *Actinomyces*, *Corynebacterium*, *Mycobacterium*, and *Nocardia* (Logan, 1994).

B. Suppressive activity of actinomycetes towards *Rhizoctonia solani* Kühn.

The work by Tahvonen (1982) on Streptomyces spp. isolated from light-colored Finnish peat led to the development of a commercial product, Mycostop. This material, which contains living mycelium and spores of *S. griseoviridis* Anderson *et al.* is applied as a seed dressing or by soil treatment (Lahdenperä *et al.* 1991). It is recommended for the control of a number of seed- and soilborne plant pathogenic fungi, including *Rhizoctonia solani*. A group of antibiotics named validamycins and validoxyamines, is produced by *Streptomyces hygroscopicus* subsp. *limoneus* (Iwasa, *et al.*, 1971). Some of these compounds have been reported to exhibit suppressive activity towards *R. solani in vitro* (Asano *et al.*, 1987). In addition, a commercial preparation of validamycin A has been used to control certain diseases incited by *R. solani*, including sheath blight of rice (*Oryza sativa* L.). This antibiotic suppresses *R. solani* by inhibiting hyphal growth and causing abnormal development. Maximum control by validamycin A of rice sheath blight was achieved by spraying from two weeks before until about the time of heading. This material is also used in Japan and The Netherlands to control black scurf of potato (*Solanum tuberosum* L.) incited by *R. solani*. Validamycin can also control *R. solani*-incited damping-off of seedlings of rice and other plants (Yamamoto, 1985). Additional reports of the successful suppression of *R. solani* by actinomycetes are presented in Table 1.

C. Factors affecting the suppression of *R. solani* by actinomycetes

1. Addition of organic amendments (see also chapter VI.A3). It has long been recognized that the addition of various organic amendments to soil can cause significant changes in its microflora (Baker, 1991). Incubation of chitin in soil is known to considerably increase actinomycetes populations and consequently reduce the saprophytic and pathogenic activities of *R. solani* (Henis *et al.*, 1967). Naumann and Lange-de la Camp (1976) studied the influence of residues of clover (*Trifolium* sp.), grass, and 1:1 mixtures of the two, on the parasitic activity of *R. solani* and the populations of actinomycetes and other bacteria were determined.

B. Sneh et al. (eds.),
Rhizoctonia Species: Taxonomy, Molecular Biology, Ecology, Pathology and Disease Constrol, 463–471.

TABLE 1. Summary of reports in the literature of the suppressive activity of actinomycetes towards *Rhizoctonia solani* (Rs).

Actinomycete	Source of Isolate	Assay Method	Reference
Streptomyces spp.	Cropped and non-cropped soils in Australia	1) PDA[1]; 2) Rs-infested, steamed soil in pots	Broadbent *et al*. 1971
Non-identified (NI) isolates	Pasture and wheat[2] soils in Australia	PDA	Merriman *et al*. 1974a
S. griseus[3]	Pasture and wheat soils in Australia	Pepper[4] and wheat seed + antagonist in Rs-infested soil in pots	Merriman *et al*. 1974b
Streptomyces sp.	Rhizosphere of a *Eucalyptus* sp. in Brazil	*Eucalyptus* sp. + antagonist + Rs on PDA	dos Santos & de Melo 1993
Streptomyces spp.	Soybean[5] and wheat rhizosphere and non-rhizosphere soils in Egypt	Soil	Mohamed 1982
NI	Broad bean[6] rhizosphere soil in Egypt	Antagonist + Rs-infested soil + broad bean seed	Abdel-Moneim 1993
Streptomyces sp.	Egyptian soil	Agar	Kamel *et al*. 1993
Streptomyces spp.	Finnish peat	1) PDA; 2) Antagonist + cauliflower[7] seed + Rs + peat	Tahvonen 1982
S. griseus	Soil from a non-specified location	Antibiotic (bafilomycin) from antagonist + Rs on agar	Werner *et al*. 1984
S. ambofaciens[8]	Indian soils	Czapek-Dox agar	Reddi & Rao 1971
Streptomyces spp.	Rhizosphere of gram[9] in India	Czapek's agar	Mathur & Chauhan 1972
S. griseus & *S. ruber*[10]	Salt-affected soils in India	Agar	Gupta & Tandon 1977
S. arenae[11] & *S. chibaensis*[12]	Soil from a field of cauliflower in India	1) Agar; 2) Cauliflower seed + antagonist planted in Rs-infested soil	Kundu & Nandi 1984
NI	Rhizosphere of rice[13] in India	Agar	Kausar & John 1987
Steptomyces sp.	Cultivated fields in India	Agar	De & Gupta 1991
Micromonospora sp.	Soil in Japan	Antibiotic (dapiramicin) from the antagonist + Rs on an agar disk	Shomura *et al*. 1983

TABLE 1 cont.

Actinomycete	Source of Isolate	Assay Method	Reference
Micromonospora sp.	Soil in Japan	Antibiotic (dapiramicin B) sprayed on rice plants inoculated with Rs (sheath blight)	Nishizawa *et al.* 1983
Kitosatosporia cystarginea[14]	Not stated	Antibiotic (cystargin) inhibited Rs *in vitro*	Uramoto *et al.* 1988
Streptomyces sp.	Surface of a pear[15] in Japan	Potato sucrose agar and Bennett's medium	Matsuyama 1991
Streptomyces sp.	Not stated	Antibiotic (respirantin) inhibited Rs on agar	Urushibata 1993
Streptomyces sp.	Korean soil	Not stated	Park & Ryu 1992
S. violaceoniger[16]	Pepper field soil in Korea	Antibiotics (tubercidin & SF2A) on PDA	Hwang *et al.* 1994
Streptomyces sp.	Rs sclerotia on a potato[17] tuber in Northern Ireland	PDA	Ehteshamul-Haque *et al.* 1990
Streptomyces sp.	Not stated	1) Soybean seed + antagonist planted in a Rs-infested field; 2) Antagonist placed in furrow with okra[18] seed	Dela Peña *et al.* 1994
Streptomyces sp.	Not stated	1) PDA; 2) Agoho[19] seed coated with antagonist planted in pots of Rs-infested soil	Turhan & Grossman 1986
NI	Turkish soils	PDA	Turhan & Turhan 1989
Acrophialophora levis[20]; *Neocomospora vasinfecta* var. *africana*[21]; *S. nobilis*[22]; *S. ochraceiscleroticus*[23]	Different localities in Turkey	Antagonist + pepper seed planted in pots of Rs-infested soil	
S. hygroscopicus var. *geldanus*[24]	ational Center for Agricultural Utilization esearch, USDA[25], Peoria, Illinois, U.S.A.	Antibiotic (geldanamycin) in glucose-yeast extract agar	Rothrock & Gottlieb 1981
Streptomyces sp.	Soil in Mississippi, U.S.A	1) Cotton[26] seed treated with antagonist and planted in cups of Rs-infested soil; 2) Antagonist added directly to soil	Brown & Brown 1985

TABLE 1. cont'd. (Footnotes)

1. PDA = Potato dextrose agar	14. "*Kitosatosporia cystarginea*" Uramoto et al. = *Streptomyces species*
2. Wheat = *Triticum aestivum* L.	15. Pear = *Pyrus communis* L.
3. *Streptomyces griseus* Krainsky	16. *S. violaceoniger* Waksman & Curtis
4. Pepper = *Capsicum frutescens* L. var. *grossum* Bailey	17. Potato = *Solanum tuberosum* L.
5. Soybean = *Glycine max* (L.) Merr.	18. Okra = *Abelmoschus esculentus* (L.) Moench
6. Broad bean = *Vicia faba* L.	19. Agoho = *Casurina equisetifolia* L.
7. Cauliflower = *Brassica oleracea* var. botrytis L.	20. *Acrophialophora levis* Samson & T. Mahmood
8. *S. ambofaciens* Pinnert-Sindico	21. *Neocomospora vasinfecta* var. *africana* (von Arx) Cannon & Hawksworth
9. Gram = *Cicer arietinum* L.	22. *S. nobilis* Baldacci *et al.*
10. *S. ruber* Krainsky	23. *S. ochraceiscleroticus* Pridham
11. *S. arenae* Pridham *et al.*	24. "*S. hygroscopicus* var. *geldanus*" = *S. hygroscopicus* subsp. *geldanus* Dietz, which is an invalid taxon (ref. J. Antibiotics 23:442; 1970)
12. *S. chibaensis* Susuki *et al.*	25. USDA = United States Department of Agriculture
13. Rice = *Oryza sativa* L.	26. Cotton = *Gossypium hirsutum* L.

The observed reduction in the incidence of *R. solani*-incited disease could have been due to the promotion of the non-parasitic soil microflora, the production of volatile, inhibitory, decomposition products, and the possible stimulation of competitive, saprophytic strains of *R. solani*. Kundu and Nandi (1985) were able to effectively control damping-off of cauliflower incited by *R. solani* by adding powdered commercial cellulose, rice stubble, or water hyacinth (*Eichornia crassipes* Solms) biomass in combination with ammonium nitrate to the soil. Powdered cellulose was the most effective soil amendment for reducing damping-off.

Schisler and Ryder (1991) studied microbial recolonization and suppression of *R. solani* in a bedding plant potting mix amended with recycled mix prior to treatment with aerated steam. They stated that their results "may indicate that actinomycetes stably recolonized the aerated steam-treated and amended mix and that this colonization was at the expense of aerobic bacteria, including pseudomonads." de Brito Alvarez *et al.* (1995) studied the effect on several parameters of adding four commercial composts to an air-dried, non-sterile field soil. Numbers of actinomycetes four weeks after planting, in the soil amended with the four composts, ranged from 1.4 to 3.8/g, compared to 1.4/g for the controls.

2. Herbicide treatment of soil. Gausterov *et al.* (1972) studied the effect of five herbicides, Lasso, Treflan, Ramrod, Semeron, and Dymid, on the antibiotic activity of 43 strains of suppressive actinomycetes towards *R. solani* and five other plant pathogenic fungi. Using an agar assay, they found that the antibiotic activity of some of the actinomycetes was stimulated, while others were either inhibited or not affected at all. Stimulation and inhibition depended on the concentration of the herbicides. Lasso and Ramrod had almost no harmful effect on the antibiotic activity of the actinomycetes that were assayed in this study. Hamed *et al.* (1980) applied the herbicide Atrazine to pots of soil that were artificially infested with *R. solani*, 0, 10, 20, and 30 days before planting broad beans (*Vicia faba* L.) in them. Although the actinomycete population and other components of the soil microflora were not affected by the herbicide, they were adversely affected by its by-products during the early stages of plant growth. Other components of the soil microflora affected by the Atrazine by-products regained their former population levels one month after its application and complete disappearance from the soil, but the population of actinomycetes generally was more sensitive. Interestingly, Atrazine reduced the soil population of *R. solani* and the incidence of disease incited by this pathogen.

3. Other pesticide applications. Mahmoud *et al.* (1980) studied the effect of the fungicide Dithane A-40 and the insecticide Cyolane on the rhizosphere microflora and disease incidence of broad bean plants growing in *R. solani*-infested soil in pots. They found that Dithane A-40 stimulated the population of actinomycetes whereas Cyolane depressed it. This was reflected in the incidence of *R. solani*-incited damping-off in pots that had received the fungicide, i.e. 6.7%, compared to 49.9% for the Cyolane and 30.8% in the controls.

4. Other factors. Wong and Griffin (1974) were interested in the effect of soil moisture levels on the growth, antibiotic production, and antagonism of *Streptomyces* spp. Eighteen isolates of streptomycetes were obtained from garden and cultivated soils in semi-arid areas of Australia. They were cultured on various agars that varied in their osmotic potential. Nine isolates were able to grow at -100 bars and two of them could grow at -150 bars. As would be expected, growth and sporulation decreased as the osmotic potential increased. However, a potential between -5 and -10 bars was found to be optimal for one isolate. Furthermore, maximum production of a water-diffusible antibiotic by one isolate occurred between -20 and -35 bars. This increased antibiotic production at moderate osmotic potentials was associated with inhibition of *R. solani* on agar.

Suppressiveness was induced by Chern and Ko (1989) in soils from five districts in Hawaii, by infesting them with *R. solani* and planting radish (*Raphanus sativus* L.) seeds in them each week for eight to 13 weeks. Subjecting soil with induced suppressiveness to 96°C for 30 minutes reduced the microbial population and significantly increased *R. solani*-incited disease by 93%, compared to the nonheat-treated suppressive soil. Although there were no significant differences between the populations of actinomycetes, fungi in general, and *Trichoderma* spp. in suppressive and conducive soils, the population of other bacteria in suppressive soil was about four times greater than in conducive soil.

II. Cyanobacteria

A. Introduction

The cyanobacteria or blue-green algae (Cyanophyta) are photosynthetic, prokaryotic organisms exhibiting morphology that ranges from unicellular to filamentous. Approximately 150 genera comprising over 1,000 species have been described (Rippka *et al.*, 1979). They occur in freshwater, marine, and terrestrial (soil) habitats and, along with the true algae, comprise most of the world's biomass.

B. Production by cyanobacteria of antibiotics active against fungi

The literature contains numerous reports of the production of bactericidal and fungicidal compounds by cyanobacteria, based mainly on *in vitro* screening studies (Kulik, 1995). The objective of many of these studies was the discovery of new compounds of possible pharmaceutical interest. It appears however that only two reports have been published that concern cyanobacteria and *R. solani*. de Caire *et al.* (1990), using a culture of *Nostoc muscorum* Ag. isolated from a rice paddy in Argentina, evaluated extracellular products and cell extracts of this taxon for their effect on the growth of *R. solani* on PDA. The average growth of *R. solani* on the agar which contained extracellular products was inhibited 77% compared to the controls. The cell extracts however, only inhibited growth by 14 percent. Moore *et al.* (1991) were granted a patent for the use of scytophycins, produced by a strain of *Scytonema pseudohofmanni* Bharadwaja, that suppressed the growth of *R. solani* on agar.

III. Algae

A. Introduction

The total number of identified species of algae has been estimated at approximately 40,000 (Raven *et al.*, 1992). Although most algae live in freshwater or marine habitats, species belonging to 147 genera have been reported from soil (Metting, 1981). The algae are a rich and largely untapped source of a vast assortment of biologically active products (Radmer and Parker, 1994).

B. Production by algae of antibiotics active against fungi

As is the case with the cyanobacteria, there are many reports in the literature of compounds elaborated by algae that were active against bacteria and fungi (Kulik, 1995). Most of these studies were centered on the macroalgae or seaweeds. The main objective was to discover anti-bacterial compounds of pharmaceutical interest. However, there are reports that a number of algae produced compounds *in vitro* that inhibited the growth of four fungi, not including *R. solani*, two of which were human pathogens.

IV. Conclusions

Streptomyces spp. are a rich source of biologically active compounds that suppress the growth of *R. solani*. With few exceptions however, the vast amount of literature that has accumulated on these microorganisms consists of reported studies done either on agar or with pasteurized or sterilized soil. Furthermore, suppression of *R. solani* was often obtained only when the antagonist was added to the growing medium sometime prior to the introduction of the pathogen (Kulik, 1995).

Unlike the actinomycetes, the cyanobacteria (with two exceptions, i.e. de Caire *et al.*, 1990, and Moore *et al*, 1991) and the true algae have not been reported to produce compounds that can suppress *R. solani*. Given the fact however, that these microorganisms have been reported to produce antibiotics that effectively suppressed fungi other than *R. solani*, at least *in vitro* (Kulik, 1995), it is quite likely that many materials that would suppress *R. solani* (and other plant pathogens) are waiting to be discovered.

Because of the ever-increasing pressure from the general public, environmentalists, and government regulatory agencies to limit the use of chemical pesticides, plus their high cost, there is a timely need to find other means of protecting plants from pathogenic fungi. It is the opinion of the author that bioactive compounds produced by the actinomycetes, cyanobacteria, and true algae could provide effective and environmentally-safe means of controlling diseases incited by fungi such as *R. solani*, as well as other plant pathogenic fungi. The author would like to offer the following thoughts concerning the use of products elaborated by these microorganisms for controlling *R. solani* and other plant pathogenic fungi:

a) With three exceptions, all of the actinomycetes that have been reported to suppress *R. solani* were species of *Streptomyces*. This is in spite of the fact that there are approximately 60 other actinomycete genera that are recognized. The possible explanation for this situation is that *Streptomyces* spp. are generally the most common actinomycetes found in soils. It is the author's belief that attempts should be made to isolate other genera of actinomycetes from soils, since some of them would no doubt be found to be capable of suppressing *R. solani*, as well as other plant pathogenic fungi. The possibility of using rare actinomycetes as biocontrol agents has been mentioned by Tanaka and Omura (1993).

b) Since actinomycetes are relatively slow-growing organisms, their use as seed dressings usually has not been effective except under special conditions. This is evident from many reports stating that the antagonist has to be established in the soil sometime before the pathogen is introduced or before seeds are planted. This can be practical when seeds or plants are grown in containers, but it is impractical for row crops. In addition, since it appears that with the majority of actinomycetes, suppression of *R. solani* is due to antibiosis and not to competition or hyperparasitism, relying on a microorganism to elaborate a suppressive compound in natural soil with its competitive microflora may be risky. Also, it is still not known for certain whether actinomycetes produce antibiotics *in vivo*.

c) Culture filtrates of suppressive microorganisms should be concentrated before use by extracting the active principle with an appropriate solvent. The concentrate could be used as a seed treatment applied directly to the seeds as a slurry. Alternatively, the concentrate could be used to treat seeds by incorporating it in a seed pellet. Concentrated culture filtrates could also be absorbed on materials such as clay and used for in-furrow treatment at the time of planting seeds, or they could also be applied directly in the furrow.

V. Acknowledgments

I thank Dr. David P. Labeda, National Center for Agricultural Utilization Research, U. S. Department of Agriculture (USDA), Peoria, Illinois, and Dr. Jack A. Lewis, Biocontrol of Plant Diseases Laboratory, USDA, Beltsville, Maryland, for their careful review of the manuscript and their helpful suggestions for its improvement. Mention of a trademark or a proprietary product does not constitute a guarantee or warranty of the product by the USDA-ARS and does not imply its approval to the exclusion of other products that may also be suitable. I thank Dr. Gerda Dijst, DLO Res. Inst. for Plant Protection, Wageningen, The Netherlands, for translating a portion of the paper by Naumann and Lange-de la Camp (1976).

VI. References

Agrios GN (1978) Plant pathology. Academic Press, New York, 2nd ed., 703 pp.

Abdel-Moneim AA, Ali FS & Hammad AMM (1993) Interaction between rhizosphere microflora of *Vicia faba* and certain pathogenic root infecting fungi. Zentralbl. Mikrobiol. 148:39-47.

Asano N, Yamaguchi T, Kameda Y & Matsui K (1987) Effect of validamycins on glycohydrolases of *Rhizoctonia solani*. J. Antibiot. 40:526-532.

Asano N, Kameda Y, Matsui K, Horii S & Fukase H (1990) Validamycin H, a new pseudo-tetrasaccharide antibiotic. J. Antibiot. 43:1039-1041.

Baker R (1991) Diversity in biological control. Crop Prot. 10:85-94.

Broadbent P, Baker KF & Waterworth Y (1971) Bacteria and actinomycetes antagonistic to fungal root pathogens in Australian soils. Austral. J. Biol. Sci. 24:925-944.

Brown LR & Brown-Skrobot S (1985) Controlling plant fungi using streptomycetes grown on chitin. United States Patent No. 4,534,965, Patent and Trademark Office, U. S. Dept. of Commerce, Arlington, Virginia, 7 pp.

Chand T & Logan C (1984) Antagonists and parasites of *Rhizoctonia solani* and their efficacy in reducing stem canker of potato under controlled conditions. Trans. Brit. Mycol. Soc. 83:107-112.

Chern LL & Ko WH (1989) Characteristics of inhibition of suppressive soil created by monoculture with radish in the presence of *Rhizoctonia solani*. J. Phytopathol. 126:237-245.

De Brito Alvarez MA, Gagné S & Antoun H (1995) Effect of compost on rhizosphere microflora of the tomato and on the incidence of plant growth-promoting rhizobacteria. Appl. Env. Microbiol. 61:194-199.

De Caire GZ, de Cano MS, de Mulé MCZ & de Halperin DR (1990) Antimycotic products from the cyanobacterium *Nostoc muscorum* against *Rhizoctonia solani*. Phyton 51:1-4.

De K & Gupta MK (1991) Antifungal activity of some soil actinomycetes. Indian J. Microbiol. 31:53-54.

Dela Peña FA, Raymundo AK, Militante EP, Garcia MU & Dela Cruz LU (1994) Biological control of damping-off fungi of agoho (*Casuarina equisetifolia* L.) using antagonistic bacteria. Biotropia 7:1-11.

Ehteshamul-Haque S, Ghaffar A & Zaki MJ (1990) Biological control of root rot diseases of okra, sunflower, soybean and mungbean. Pak. J. Bot. 22:121-124.

Gausterov G, Brankova R & Vlahov S (1972) The influence of some herbicides on the development and the antibiotical activity of actinomycete antagonists with anti-fungal activity spectra. Symp. Biol. Hungar. 11:359-363.

Gupta RC & Tandon RN (1977) Growth inhibition of fungi by volatiles from *Streptomyces*. Trans. Brit. Mycol. Soc. 68:438-439.

Hamed AS, Mahmoud SAZ, Zaki MM & Sahab AF (1980) The effect of the herbicide Atrazine on rhizosphere microflora of broad bean plants, infected with *Fusarium oxysporum* f. *fabae* and *Rhizoctonia solani*. Zbl. Bakt II, Abt. 135:60-69.

Henis Y, Sneh B and Katan J (1967) Effect of organic amendments on *Rhizoctonia* and accompanying microflora in soil. Can. J. Microbiol 13:643-650

Hwang BK, Ahn SJ & Moon SS (1994) Production, purification, and antifungal activity of the antibiotic nucleoside, tubercidin, produced by *Streptomyces violaceoniger*. Can. J. Bot. 72:480-485.

Iwasa T, Higashide E, Yamamoto H & Shibata M (1971) Studies on validamycins, new antibiotics II. Production and biological properties of validamycins A and B. J. Antibiot. 24:107-113.

Kamel Z, Heikel N & Fahmy F (1993) Extracellular chitinase from *Streptomyces* species and its antifungal activity. Acta Pharmac. Turcica 35:135-143.

Kausar R & John VT (1987) Antagonism of rhizosphere and phyllosphere actinomycetes to rice pathogenic fungi - *in vitro* studies. Indian J. Bot. 10:55-58.

Kulik MM (1995) The potential for using cyanobacteria (blue-green algae) and algae in the biological control of plant pathogenic bacteria and fungi. Eur. J. Plant Pathol. 101:585-599.

Kundu PK & Nandi B (1984) Control of cauliflower damping-off by using antagonist coated seeds. Pedobiologia 27:43-48.

Kundu, PK & Nandi B (1985) Control of Rhizoctonia disease of cauliflower by competitive inhibition of the pathogen using organic amendments in soil. Plant Soil 83:357-362.

Lahdenperä M-L, Simon E & Uoti J (1991) Mycostop - A novel biofungicide based on *Streptomyces* bacteria. In: Beemster ABR, Bollen GJ, Gerlagh M, Ruissen MA, Schipper B, & Tempel A (eds.) Biotic Interactions and Soil-borne diseases (pp 258-263) Elsevier, Amsterdam.

Logan NA (1994) Bacterial systematics. Blackwell Scient. Publ., London, 263 pp.

Logan C, Little WG & Cooke LR (1984) Preliminary experiments with biological control of diseases of potato caused by *Rhizoctonia solani* in Northern Ireland. Proc. Crop Prot. in Northern Britain pp 120-125.

Mahmoud SAZ, Zaki MM, Hamed AS & Sahab AF (1980) Rhizosphere and soil microflora of *Vicia faba* var. *equina* as influenced by *Rhizoctonia solani*, inoculation and pesticide treatments. Egypt. J. Microbiol. 15:53-62.

Mathur S & Chauhan SK (1972) A comparative study of rhizosphere of gram (*Cicer arietinum* L.) in relation to its wilt diseases caused by different fungi. J. Indian Bot. Soc. 51:267-274.

Matsuyama N (1991) Purification and characterization of antifungal substance Ac-1 produced by a *Streptomyces* sp. AB-88M. Ann. Phytopathol. Soc. Japan 57:591-594.

Merriman PR, Price RD & Baker KF (1974a) The effect of inoculation of seed with antagonists of *Rhizoctonia solani* on the growth of wheat. Austral. J. Agric. Res. 25:213-218.

Merriman PR, Price RD, Kollmorgen JF, Piggott T & Ridge EH (1974b) Effect of seed inoculation with *Bacillus subtilis* and *Streptomyces griseus* on the growth of cereals and carrots. Austral. J. Agric. Res. 25:219-226.

Metting B (1981) The systematics and ecology of soil algae. Bot. Rev. 47:195-311.

Mohamed ZK (1982) Physiological and antagonistic activities of streptomycetes in rhizosphere of some plants. Egypt. J. Phytopathol. 14:121-128.

Moore RE, Patterson GML & Mynderse JS (1991) Scytophycins. United States Patent No. 4,996,229, Patent and Trademark Office, U. S. Dept. of Commerce, Arlington, Virginia, 16 pp.

Naumann K & Lange-de la Camp M (1976) Effect of plant residues on the parasitic activity of soilborne pathogens and the saprophytic microflora of the soil. Zbl. Bakt. Abt. II, Bd. 131:378-391 (In German with Eng. summ.).

Nishizawa N, Kondo Y, Koyama M, Omoto S, Iwata M, Tsuruoka T & Inouye S (1984) Studies on a nucleoside antibiotic, diapiramycin. II. Isolation, physico-chemical and biological characterization. J. Antibiot. 37:1-5.

Park K-S & Ryu J-C (1992) Studies on characterization of active substances from antagonistic *Streptomyces* sp. A-2 strain against soil-borne phytopathogen. J. Korean Soc. Soil Sci. Fert. 25:401-406 (In Korean with Eng. summ.).

Radmer RJ & Parker BC (1994) Commercial applications of algae: opportunities and constraints. J. Appl. Phycol. 6:93-98.

Raven PH, Evert RF & Eichorn SE (1992) Biology of Plants, 5th ed., Worth Publ., New York.

Reddi GS & Rao AS (1971) Antagonism of soil actinomycetes to some soil-borne plant pathogenic fungi. Indian J. Phytopathol. 24:649-657.

Rippka R, Deruelles J, Waterbury JB, Herdman M & Stanier RY (1979) Generic assignments, strain histories and properties of pure cultures of cyanobacteria. J. Gen. Microbiol. 111:1-61.

Rothrock CS & Gottlieb D (1981) Importance of antibiotic production in antagonism of selected *Streptomyces* species to two soil-borne plant pathogens. J. Antibiot. 34:830-835.

dos Santos TMC & de Melo IS (1993) Influence of antagonistic microorganisms on pre and post emergence damping-off in *Eucalyptus* caused by *Rhizoctonia solani* and *Cylindrocladium scoparium*. Summa Phytopathol. 19:127-129 (Portuguese with Eng. abstr.).

Schisler DA & Ryder MH (1991) Microbial recolonization and suppression of *Rhizoctonia solani* in a bedding-plant potting mix amended with recycled mix before aerated-steam treatment. Biol. Fert. Soils 11:174-180.

Shomura T, Nishizawa N, Iwata M, Yoshida J, Ito M, Amano S, Koyama M, Kojima M & Inouye S (1983) Studies on a new nucleoside antibiotic, dapiramicin. I. Producing organism, assay method and fermentation. J. Antibiot. 36:1300-1304.

Tahvonen R (1982) Preliminary experiments in the use of *Streptomyces* spp. isolated from peat in the biological control of soil and seed-borne diseases in peat culture. J. Scient. Agric. Soc. Finland 54:357-369.

Tanaka Y & Omura S (1993) Agroactive compounds of microbial origin. Annu. Rev. Microbiol. 47:57-87.

Turhan G & Grossmann F (1986) Investigation of a great number of actinomycete isolates on their antagonistic effects against soil-borne fungal plant pathogens by an improved method. J. Phytopathol. 116:238-243.

Turhan G & Turhan K (1989) Suppression of damping-off on pepper caused by *Pythium ultimum* Trow and *Rhizoctonia solani* Kühn by some new antagonists in comparison with *Trichoderma harzianum* Rifai. J. Phytopathol. 126:175-182.

Uramoto M, Itoh Y, Sekiguchi R, Shin-ya K, Kusakabe H & Isono K. (1988) A new fungal antibiotic, cystargin: Fermentation, isolation, and characterization. J. Antibiot. 61:1763-1768.

Urushibata I, Isogai A, Matsumoto S & Susuki A (1993) Respirantin, a novel insecticidal cyclodepsipeptide from *Streptomyces*. J. Antibiot. 46:701-703.

Werner G, Hagenmaier H, Drautz H, Baumgartner A & Zähner H (1984) Metabolic products of microorganisms. 224. Bafilomycins, a new group of macrolide antibiotics. Production, isolation, chemical structure and biological activity. J. Antibiot. 37:110-117.

Wong PTW & Griffin DM (1974) Effect of osmotic potential on streptomycete growth, antibiotic production and antagonism to fungi. Soil Biol. Biochem. 6:319-325.

Yamamoto H (1985) Development of validamycin, its controlling effect on rice sheath blight. Japan Pest. Inform. 47:17-22.

VI.B4 NON PATHOGENIC ISOLATES OF *RHIZOCTONIA* SPP. (np-R) AND THEIR ROLE IN BIOLOGICAL CONTROL

BARUCH SNEH

Department of Botany and Institute for Nature conservation Research,
The George S. Wise Faculty of Life Sciences, Tel Aviv University,
Ramat Aviv, ISRAEL 69978

I. HYPOVIRULENCE

Expression of virulence by populations of any pathogen is generally a continuum between two extremes; high virulence and avirulence. Low virulent to avirulent isolates are considered hypovirulent. At some point along this continuum however, hypovirulence ends and virulence begins (Van Alfen, 1982). Hypovirulence may be transmissible, involving dsRNA mycoviruses as in the case of *Cryphonectria parasitica* (Van Alfen, 1982), or non transmissible as in *Rhizoctonia* spp. (Ichielevich-Auster *et al.*, 1985a; Finkler *et al.*, 1985).

II. BIOCONTROL OF PLANT PATHOGENS WITH HYPOVIRULENT ISOLATES

Except for their incapability to cause appreciable disease symptoms, non pathogenic (or hypovirulent) isolates of the same plant pathogenic species (or genera) are expected to have similar characteristics to those of the respective virulent ones, regarding their competitive competence to colonize and occupy the same ecological niches on the plant surfaces. Therefore, a certain proportion of them may be considered to have the potential to compete successfully against their respective pathogens on host infection site niches (Sneh *et al.*, 1989a; Sneh, 1990), and/or protect the colonized plants against diseases caused by virulent isolates of the same species via other mechanisms such as induction of plant resistance (Ogawa and Komada, 1986), transmission of dsRNA mycoviruses (Van Alfen, 1982) and others. Fungal pathogens may also be genetically converted to non pathogenic epiphytic mutualists (Freeman and Rodriguez, 1993). Examples for disease suppression by nonpathogenic or hypovirulent isolates, have been reported for a considerable number of fungal (as well as bacterial and viral) plant pathogens as follows:

A. Transmissible hypovirulent isolates:

Gaeumannomyces graminis (Lapierre *et al.*, 1970), *Ustilago maidis* (Koltin and Day, 1976), *Cryphonectria parasitica* (Van Alfen, 1982), *Ceratocystis ulmi* (Pusey and Wilson, 1982), *Leucostoma persoonii* (Hammer *et al.*, 1989), *R. solani* (Castanho *et al.*, 1978).

B. Non transmissible hypovirulent isolates:

Rhizoctonia spp. (Burpee and Goulty, 1984; Ichielevich-Auster *et al.*, 1985b; Cardoso and Echandi, 1987; Sneh *et al.*, 1989a; Herr, 1988; Harris *et al.*, 1993b; 1993c), *Fusarium oxysporum* (Ogawa and Komada, 1986; Schneider, 1984; Tamietti and Allabouvette, 1986, Sneh *et al.*, 1987; Louter and Edgington, 1990), *Pythium* spp. (Deacon and Henry, 1978; Harris *et al.*, 1993c; Lifshitz *et al.*, 1984), *Alternaria* sp. (Spurr, 1977), *Penicillium funiculosum* (Lim and Robrbach, 1980), *Erysiphe*

B. Sneh et al. (eds.),
Rhizoctonia Species: Taxonomy, Molecular Biology, Ecology, Pathology and Disease Constrol, 473–483.

graminis (Martinelli, 1990) *Colletotrichum magna* (Freeman and Rodriguez, 1993), *Phytophthora parasitica* (Holmes and Benson, 1994), *Aspergillus flavus* (Cotty, 1994), *Septoria tritici* (Stein *et al.*, 1994) and *Gaeumannomyces graminis* (Duffy and weller, 1995). Thus, certain non pathogenic isolates are seriously considered and developed for use as biocontrol agents against their respective virulent plant pathogens.

III. Presence of hypovirulent *Rhizoctonia* spp. populations in soil

It is quite possible that non pathogenic *Rhizoctonia* spp. (np-R) isolates exist among populations of all the AG's. Np-R isolates have been reported for the following groups: *R. solani* AGs 1, 2, 3, 4, 5 (Ichielevich-Auster *et al.*, 1985a; Bandy and Tavantzis, 1990, Villajuan-Abgona *et al.*, 1993; 1996a). *R. zeae*, and Binucleate *Rhizoctonia*: AGs-A (CAG2), F (CAG-4) (Ichielevich-Auster *et al.*, 1985a), A, Ba, Bb (Villajuan-Abgona *et al.*, 1996a), Ba (Cubeta *et al.*, 1995), B(o), S (CAG 7) (Herr, 1988; Yuen *et al.*, 1994), O (Cubeta *et al.*, 1995), B(o), P (Sneh and Ichielevich-Auster *et al.*, 1995), G (Escande and Echandy, 1991), R (CAG 5) (Cardoso and Echandi, 1987), K, S (CAG-7) (Sneh *et al.*, 1991). In addition, the reported information for most or all of the isolates studied, belonging to the following AGs were not pathogenic: *R. solani* - AGs 10, BI, Binucleate *Rhizoctonia* spp. - AGs-H, J, L, M, N. Certain isolates of the following groups were reported to form mycorrhizal associations with orchids: Binucleate *Rhizoctonia*: AG A; R. repens: AGs Rr1 and Rr2, R. *anaticula*, and a variety of *Rhizoctonia teleomorph* species in addition to *Thanatephorus* spp. *Ceratobasidium* spp. *Waitea* spp. and *Tulasnella* spp. (cited in: Sneh *et al.*, 1991).

Hypovirulent or np-R isolates comprise a worldwide integral part of *Rhizoctonia* spp. populations in soil and may reach considerable proportions (10 to 30%) of the total *Rhizoctonia* spp. populations (Ichielevich-Auster *et al.*, 1985a; Cubeta *et al.*, 1995; Villajuan-Abgona *et al.*, 1996a). Factors affecting the presence and distribution of hypovirulent isolates of the various AGs in soils have not been studied. It is quite likely that their distribution in soil may be determined by the existing environmental conditions, including: crop history, nutrition and climate (temperature and humidity). Frequency of their isolation from soils depends also on the isolation method used. The plant debris particles procedure, may yield the highest proportion of hypovirulent isolates (Ichielevich *et al.*, 1985a; Villajuan-Abgona *et al.*, 1996a), because this method is less selective than the baiting methods using plants, plant-segments or seeds. The actual natural role and significance of np-R in soils have not been investigated. However, their prevalence in field soils suggest that they may play an important role in suppression of disease development. The fact that a certain proportion of them are capable to protect plants against pathogen(s) and that certain isolates may also promote plant growth, suggests that they may play an important beneficial role in agricultural field soils. Pathogenic *R. solani* populations dominated the center of the "bare patches" in wheat fields, while binucleate np-R dominated the clear-cut edges of the patches (Roberts and Sivasithamparam 1986), suggesting that binucleate np-R may play an important role in suppression of patch extension by antagonizing the pathogenic *R. solani*. Binucleate np-R populations suppressed damping off of rapeseed (Gugel *et al.*, 1987).

Genetic, as well as physiological factors are involved in determining the virulence (and hypovirulence) of *Rhizoctonia* spp. isolates. Pectolytic enzymes play an important role in pathogenicity of *Rhizoctonia* spp. (Bateman and Lumsden, 1965). Both hypovirulent and virulent *R. solani* (AG 4) isolates produced two endo-polygalacturonases and one pectin methyl esterase in culture. Pectin lyase was produced only by the virulent isolate (Marcus *et al.*, 1986), and cellulase production was lower by the hypovirulent one (Sneh *et al.*, 1989a). A PL difficient mutant of *Colletitrichum magna* had lost its pathogenicity (Wattad *et al.*, 1995).

Np-R isolates are frequently hyaline (Ichielevich-Auster *et al.*, 1989b; Villajuan-Abgona *et al.*, 1996c), while virulent ones are generally colored (brownish or grayish) (Sneh *et al.*, 1991) due to melanin depositions in their hyphal cell walls. Presence of melanin may be essential for pathogenicity. Melanin synthesis in appressoria was shown to be essential for the physical rigidity of the infection pegs in *Pyricularia oryzae*, *Colletotrichum lagenarium* and *C. lindemuthianum*, enabling them to penetrate into the host epidermis and advance through the plant tissues as a pathogen (Bell and Wheeler 1986). Pathogenicity of two different Tn-5 albino mutants, which have lost their pathogenicity, could be restored by transformation with complementary genes encoding for two different enzymes of the melanin synthetic pathway (Kubo *et al.*, 1991). It may be interesting to find out whether the lack of melanin in hyaline np-R isolates is involved in their reduced virulence (or avirulence). Hyphal tips of binucleate as well as multinucleate np-R isolates penetrated the epidermal host layers only intercellularly, but could not penetrate the cells intracellularly (Eayre and Echandi, 1988; Sneh *et al.*, 1989b).

IV. Np-R ISOLATES AND THEIR ROLE IN BIOLOGICAL CONTROL

Since 1984 reports have been accumulating regarding the different potentials of binucleate or multinucleate np-R isolates of several different AG's to protect seedlings of a wide host range against infection by different virulent *Rhizoctonia* spp. (Table 1). While isolate 521 (AG 4) did not protect against pathogens other then *Rhizoctonia* (Sneh *et al.*, 1989a), certain isolates protected also against *Pythium ultimum* var *sporangiiferum*. (Harris *et al.*, 1993c; Harris, 1994); *P. aphanidermatum* and *Pseudomonas syringae* pv. *lachrymans* (Sneh and Ichielevich-Auster, 1995). Some np-R isolates may provide a different degree of protection against different sub groups of pathogenic *Rhizoctonia*. Four binucleate np-R isolates (AG-G) protected cucumber and bean seedlings against pathogenic isolates of AG 4 HG-II and HG-III at a rate of 82-96%, while a considerably weaker protection, 17-53% was provided against AG 4 HG-I (Cubeta, personal communication). The role of np-R in biological control of *Rhizoctonia* has been recently reviewed by Herr (1995).

Some effective np-R isolates grow relatively slower than virulent ones (Ichielevich Auster *et al.*, 1985a). This may be a disadvantage for them as biocontrol agents. However, np-R isolates can be applied in colonized cereal grain close to the sown seeds. This rich food base increases tremendously their inoculum potential and provides them with a significant competitive advantage over the naturally existing virulent *Rhizoctonia* spp. propagules frequently associated with relatively exhausted plant debris particles in soil (Sneh *et al.*, 1986; Harris *et al.*, 1993a). Thus, the richer physiologically and stronger np-R hyphae grow out vigorously to become the pioneer colonizer of the germinating seeds, roots and hypocotyls and form dense mycelial mats earlier and at a higher inoculum potential than the virulent ones. The earlier the colonization with the np-R, the higher protection is obtained (Harris *et al.*, 1993c). However, in certain instances such as protection of sugar beet crowns, a better protection was obtained when the np-R inoculation was delayed by 5-10 weeks (Herr 1988). Np-R isolates survived on the crop roots up to the harvest time and thereafter in plant debris (Sneh *et al.*, 1986). Their survival in soil depended upon the type of substrate colonized (cereal grain or bean stem segments), soil depth and watering regime (Cubeta *et al.*, 1991). When fresh organic substrates and or germinating seeds are available, their propagule density is increased, while in their absence it is decreased. After 11-12 months none of the binucleate np-R incorporated in soil could be recovered (Cubeta *et al.*, 1991).

TABLE 1. Reported hypovirulent *Rhizoctonia* spp. isolates of various AG's and seedling protection against *Rhizoctonia* spp. and *Pythium* spp.

Hypovirulent AG's	Protection Against AG's	Crop plants
Binucleate:		
A (CAG 2)	2-2[1]; 4[1]	Cucumber[1]
Ba	2-2[1]; 3[2]; 4[1,2]	Cucumber[1,2], Bean[2], Potato[2]
Bb	-	-
B(o)	1-IA[3]; 1-IB[2,5]; 2-2[4] 4[1,5]; *Pythium*[5]	Cabbage[5], Carrot[5], Cucumber[1,5], Lettuce[5], Parsley[5], Radish[5], Sugar beet[4], Table beet[5], Tall fescue[3], Tomato[5]
F (CAG 4)	-	-
G	3[2,6]	Cucumber[2], Bean[2], Potato[2,6]
H,J,K,L,M,N	-	-
O	4[2]	Cucumber[2], Bean[2], Potato[2]
P	1-IA[5]; 4[5]; *Pythium*[5]	Cabbage[5], Carrot[5], Cucumber[5], Lettuce[5], Parsley[5], Radish[5], Table beet[5], Tomato[5]
R (CAG 5)	4[7]	Bean[7]
S (CAG 7)	-	-
Not identified[11]	4[6,8,9]; 8[8]; *Pythium*[8]	Bentgrass[9], Brussel sprouts[8], Celosia[8], Cucumber[6], Pepper[8], Petunia[8], Viola[8]
R. repens,	-	-
R. anaticula	-	-
Multinucleate:		
1,2	-	-
3[12]	3[10]	Potato[10]
4[12]	1-IA[6]; 4[11]; *R. zeae*[11]	Cabbage[11], Carrot[11], Cotton[11], Cucumber[6], Lettuce[11]
5,7,10,BI	-	-
R. zeae	-	-

[1]-Villajuan-Abgona *et al.*, (1996b); [2]-Cubeta *et al.* (1995); [3]-Yuen *et al.* (1994);
[4]-Herr, (1988; 1991); [5]-Sneh & Ichielevich-Auster, (1995);
[6]-Escande and Echandi (1991), Cubeta & Echandi (1991);
[7]-Cardoso & Echandi (1978); [8]-Harris *et al.* (1993c);
[9]-Burpee & Goulty; (1984); [10]- Bandy and Tavantzis (1990);
[11]-Ichielevich-Auster *et al.* (1985b), Sneh *et al.* (1986);
[12]-Isolates which have also induced plant growth.

V. MODES OF ACTION OF SEEDLING PROTECTION BY NP-R

As different np-R isolates have a different capability to protect seedlings against virulent isolates of *Rhizoctonia* spp., they may differ in their mode(s) of action. Seedling protection by an effective np-R isolate may be obtained by one or more mechanisms. By evaluating the following possibile mechanisms, the mode(s) of action involved in seedling protection by a certain np-R isolate may be determined:

A. **Direct interactions** (np-R vs. virulent):
  1. Transmission of a dsRNA mycovirus
  2. Direct antagonism:
    a. Competition for:
      i. Nutrients (plant exudates)
      ii. Infection sites
    b. Antibiosis
    c. Hyperparasitism

B. **Indirect interactions** (Induced resistance, via PR-proteins):
  1. Formation of physical barriers:
    a. Cuticle
    b. Lignification, Suberization
    c. Increased Ca^{2+} uptake (Ca-pectate)
  2. Production of lytic enzymes and inhibitory substances
    a. Fungal cell-wall lytic enzymes
    b. Polyphenols and Phytoalexins etc.
    c. Inhibitors of pathogen's enzymes (Pectinases, Cellulases etc.).

Castanho *et al.*, (1978) claimed that transmissible dsRNA mycoviruses like those described for *C. parasitica* (Van Alfen, 1982) are involved in hypovirulence and biocontrol in *Rhizoctonia* spp. Subsequent studies using the same isolates could not support this approach in *Rhizoctonia* (Finkler *et al.*, 1985; Kusik *et al.*, 1994). The isolates used by Castanho and Butler could not be considered for practical use. They were degenerate, grew slow and their survival was for very short.

Evidence for hyphal mycoparasitic interactions, hyphal interference or antibiosis between np-R isolates and pathogenic isolates of *Rhizoctonia* spp. could not be observed by different np-R isolates in several studies (Ichielevich Auster *et al.*, 1985b; Cardoso and Echandi 1987; Sneh *et al.*, 1989b; Harris *et al.*, 1993b; 1993c; Villajuan-Abgona, 1996c). However, some binucleate np R isolates directly parasitized hyphae of *Pythium ultimum* and *P. oligandrum in vitro* (Walther & Gindrat 1987). Such mycoparasitism could not be observed at first with two binucleate np-R isolates which protected pepper seedlings against *P. ultimum* (Harris *et al.*, 1993c). However, in a subsequent study (Siwek *et al.*, 1995), clear mycoparasitism of *P. ultimum* var *sporangiiferum* could be observed by a binucleate np-R isolate. Yet, as *Pythium* spp. propagules germinate and penetrate the embrio of germinating seeds very fast, more studies are required to clarify the role and significance of mycoparasitism in seedling protection by the binucleate np-R against damping off.

Np-R isolates densely colonize plant surfaces, forming dense mycelial mats on crowns, other underground plant organs (Sneh *et al.*, 1989a; Harris *et al.*, 1993c; Villajuan-Abgona, 1996c), as well as stems and leaves (Burpee and Goulty, 1984; Yuen *et al.*, 1994). This indicates that np-R's may compete physically with virulent isolates to recognize and occupy infection sites, as suggested by Sneh *et al.* (1989a;

1989b), or to compete for nutrients. Root exudates are essential for pathogenicity of *Rhizoctonia* spp. (Dodman *et al.*, 1968; Brookhouser and Weinhold 1979). Therefore, competition for nutrients has been suggested as a possible mechanism for seedling protection by np-R (Burpee and Goulty 1984; Cardoso and Echandi 1987; Harris *et al.*, 1993c). However, it has not been supported by appropriate experimental data. The study of Sneh *et al.*, (1989a) could not support the involvement of competition for nutrients in seedling protection by an np-*R. solani* isolate. Cotton seedlings colonized by this isolate released more ^{14}C labelled root exudates than the non colonized control ones. Also, addition of either root extracts or a combination of nutrients to the colonized roots could not nullify the protection effect.

Root exudates from bean seedlings colonized with a binucleate np-R inhibited virulent *R. solani in vitro*, and fewer infection cushions were formed by the pathogen on the np-R-colonized seedlings as compared to that on control seedlings (Cardoso and Echandi, 1987), The authors suggested that the np-R induced seedling resistance. In another system, inhibition of *R. solani* with extracts from seedlings colonized with the np- *R. solani* (isolate 521) was very weak as compared to inhibition by extracts from seedlings colonized with the virulent isolate (Sneh *et al.*, 1989a). Cuticle and lignification may be involved in bean hypocotyl resistance to infection by *R. solani* (Stockwell and Hanchey, 1983). On theother hand, there were no evidences that np-R could induce formation of thicker cuticlar layers, lignification and nor production of more polyphenols (Sneh *et al.*, 1989a). Cucumber seedlings colonized with a non pathogenic *Fusarium* isolate induced seedling resistance and increased root exudation levels as compared with non colonized ones. The increased levels of root exudates enhanced siderophore production by the root-colonizing *Pseudomonas putida* and resulted in improved protection *against Fusarium oxysporum* f. sp. *cucumerinum* (Park *et al.*, 1988).

Killing the hyphae of np-R colonizing seedlings, with sodium hypochlorite did not nullify the protective effect (Cardoso and Echandi, 1987; Sneh *et al.*, 1989a). It suggests that the mechanism involves induced resistance (Cardoso and Echandi, 1987) and/or physical competition for infection sites by the mycelium (although killed (Sneh *et al.*, 1989a; 1989b), due to the fact that nullification of the protection could be obtained only by removing the colonizing np-R hyphae. This indicates that the protection by this np-R isolate on this niche might probably not be caused by induced resistance, but by competition for infection sites. Different np-R isolates which protected seedlings against virulent *Rhizoctonia* spp. when applied in soil, may differ in their capability to induce seedling resistance to subsequent challenge inoculation of the upper hypocotyls or leaves. Certain isolates induced resistance and could protect the upper hypocotyls of cucumber seedlings when challenge inoculated with *R. solani*, *Pythium aphanidermatum*, or the foliar pathogen, *Pseudomonas syringae* pv. lachrymans (Sneh and Ichielevich-Auster, 1995). A hypovirulent isolate of *Rhizoctonia* sp. induced resistance in tomato against *Fusarium oxysporum* f. sp. *radicis lycopersici* (Louter and Edgington, 1990).

Inhibitory substances were involved in protection of orchid bulbs with a weak isolate of *R. repens* (Stoessl, 1983). It has been shown that mycorrhizal plants acquired resistance to pathogens and that PR proteins and phytoalexins are involved in the induced resistance (Reinecke and Kindl, 1994). However, the question of the establishment of mycorrhizal fungi in the roots has not yet been completely clarified despite the plant responses. Harley and Smith (1983) suggested that *Rhizoctonia* spp.

can destroy orcinol and hircinol (phytoalexins) in orchids and that in the symbiotic relationship, a balance between phytoalexin production by the plant and its inactivation by the fungus is probably established. Reinecke and Kindl (1994) have shown that there was an extremely low level of bibenzyl synthase (an important enzyme in hircinol synthetic pathway) in mycorrhitic orchid roots colonized with *Rhizoctonia*. Although the functional role of phytoalexins during developmental stages of mycorrhizal associations is yet unclear, they hypothesized that phytoalexin production by the plant is reduced, the establishment of mycorrhizal associations is progressing and the exchange of chemical signals between the orchid and *Rhizoctonia* silences the responses of both partners.

Np-R-seedling interactions differ from mycorrhizal associations of *Rhizoctonia* spp. with orchids. In the latter, the hyphae penetrate orchid seed and root cells (Uetake *et al.*, 1988), while the np-R hypahe grow only intercellularly in the epidermal layer of the host and they do not penetrate the host plant cells (Cardoso and Echandi, 1987; Sneh *et al.*, 1989b; Harris *et al.*, 1993c; Villajuan-Abgona *et al.*, 1996c).

Ca^{2+} content in plant tissues may influence pectins soluble/insoluble ratio and its increase may increase Ca-cross links in cell wall pectins and thus reduce pectin degradation by pectolytic enzymes of plant pathogens (McGuire and Kelman, 1986). As the Ca^{2+} content increases with plant age, resistance to *R. solani* increases (Bateman and Lumsden 1965). The 3-4 fold increase in Ca^{2+} content in cortex of seedlings colonized with np-R (Sneh *et al.*, 1989a) may thus contribute to the protection of the cortical tissues against penetration by virulent isolates. Production of pectolytic and cellulolytic enzyme-inhibitors may interfere with disease development caused by *Rhizoctonia* spp. A dialyzable heat-stable inhibitor of endo-PG was found in lesions caused by *R. solani* on bean hypocotyls (Albersheim and Anderson 1971). An increase in such proteins could not be detected in seedlings colonized by the np-R isolate (Sneh *et al.*, 1989a).

Recently, Villajuan-Abgona *et al.*, (1996c) observed that a protective binucleate np-R isolate (W-7, AG-A) degraded the cuticular layer of the epidermal surface in cucumber. The epidermal surface became covered with a thin layer of a viscous material positively stained with phenyl thionin which stains pectic substances. Such a layer was not formed on the control seedlings, nor on those infected by a virulent *R. solani*. The virulent isolate penetrated into the hypocotyl and roots and macerated the tissues, while the np-R isolate did not penetrate the epidermal cells. The authors suggested that the viscous layer could be formed by degradation products of the plant cell wall pectins by the np-R isolate. Pectin oligomers released by the activity of pectolytic enzymes of the np-R isolate may elicit plant defense reactions. Hyphal lysis of both virulent and hypovirulent isolates was also observed in association with the viscous layer in seedlings inoculated with both isolates. Hyphal lysis may be caused by the plant's cell wall lytic enzymes induced by the np-R isolate, and/or by fungal autolytic enzymes. The authors suggest that this barrier and the lytic enzymes prevent penetration of both the pathogen and the np-R into the plant epidermis. Recent findings of Tschen and Hsue (1995) support the involvement of chitinase in seedling protection with np-R isolates. In addition, the np-R isolate-hypocotyl interaction also reduced considerably the capability of the pathogenic *R. solani* isolate to form infection structures such as lobate appressoria and infection cusions (Villajuan-Abgona *et al.*, 1996c). Neither a viscous layer nor hyphal lysis could be observed by Sneh *et al.*, (1989b) on cotton seedlings colonized with a protective np-R. *solani* isolate. Transmission EM indicated that the np-R isolate degraded the cuticle in close contact with the hyphae, but the rest of the plant cell walls remained intact. This isolate did not induce seedlings to produce any more chitinases nor 1-3β glucanase than the non - colonized control ones (Sneh *et al.*, 1989a).

VI. PLANT GROWTH PROMOTION INDUCED BY NP-R

In addition to seedling protection, some np-R isolates were shown to promote plant growth even in the absence of the virulent isolate. This is sometimes expressed in yield increases of 15 to 60% or more, in various crops such as: radish, carrot, lettuce, cotton and wheat (Sneh *et al.*, 1986), pepper and ornamentals (Harris *et al.*, 1993c; Harris, 1994). Potato plants biomass was increased during the plant growth stages, flowering and tubers formation occerred 2-4 weeks earlier, but these were not reflected in a significant increase in mature tuber yields (Sneh *et al.*, 1986; Bandy and Tavantzis, 1990). However, when the crops were severely damaged by *Phytophthora infestans* or stormy weather, a couple of weeks before the due harvest time, tuber yields of the np-R-colonized potatoes were significantly higher than those of the control ones (Sneh, unpublished). Early (or green) harvesting is also practiced for escaping from black scurf of potato caused by *R. solani* AG 3 (Dijst, 1989; Malder *et al.*, 1992). The capabilities to increase yields and protection of seedlings against *Rhizoctonia* diseases may be of great value for development of microbial preparations based on highly effective np-R isolates for the agro-biotechnological industry and use in safer sustainable agricultural practices. For such purposes, it is highly important to preserve these cultures in proper conditions. Multiple transfers on agar may result in loss of the growth promotion property of such np-R isolates (Sneh, unpublished). Increase in $^{14}CO_2$ uptake, and photosynthetic CO_2 incorporation by the plant growth promoting np-*R. solani*-colonized seedlings (Sneh *et al.*, 1989a) may be involved in the increase in plant growth rate, expressed in higher crop yields (Sneh *et al* 1986).

VII. DROUGHT TOLERANCE

Certain np-R isolates increased seedling tolerance to low soil moisture levels (Sneh and Ichielevich-Auster, 1995). Further studies are required to evaluate the detailed processes involved in the induction of plant resistance to pathogens and for drought tolerance on the molecular level.

VIII. REFERENCES

Albersheim P & Anderson AJ (1971) Proteins from plant cell walls inhibit polygalacturonase secreted by plant pathogens. Proc. Natl. Acad. Sci. 68:1815-1819.

Bandy BP & Tavantzis SM (1990) Effect of hypovirulent *Rhizoctonia solani* on Rhizoctonia disease, growth and development of potato plants. Am. Potato J. 67:189-199.

Bateman DF & Lumsden R D (1965) Relation of calcium content and nature of the pectic substances in bean hypocotyls of different ages to susceptibility to an isolate of *Rhizoctonia solani*. Phytopathology 55:734-738.

Bell AA & Wheeler MH (1986) Biosynthesis and functions of melanins. Annu. Rev. Phytopathol. 24: 411-451.

Brookhouser LW & Weinhold AR (1979) Induction of polygalacturonase from *Rhizoctonia solani* by cotton seed hypocotyl exudates. Phytopathology 69:599-602.

Burpee LL & Goulty LG (1984) Suppression of brown patch disease of creeping bentgrass by isolates of nonpathogenic *Rhizoctonia* spp. Phytopathology 74:692-694.

Castanho B, Butler EE & Shepperd RS (1978) The association of ds-RNA with Rhizoctonia decline. Phytopathology 68:1511-1519.

Cardoso JE & Echandi E (1987) Nature of protection of bean seedlings from Rhizoctonia root rot by a binucleate *Rhizoctonia*-like fungus. Phytopathology 77: 1548-1551.

Cotty PJ (1994) Influence of field application of an atoxigenic strain of Aspergillus flavus on the populations of *A. flavus* infecting cotton balls and on the aflatoxin content of cottonseed. Phytopathology 84:1270-1277.

Cubeta MA & Echandi E (1991) Biological control of *Rhizoctonia* and *Pythium* damping off of cucumber: an integrated approach. Biol. Control 1:227-236.

Cubeta MA, Echandi E & Gumpertz ML (1991) Survival of binucleate *Rhizoctonia* species, biological control agents in soil and plant debris under field conditions. Biol. Control 1:218-226.

Cubeta MA, Echandi E & Sun ML (1995) Biological control of Rhizoctonia diseases with binucleates *Rhizoctonia* spp. International Symposium on *Rhizoctonia*, Facts and challenges in Pathology, Taxonomy, Ecology and Disease Control. June 27-30, Abstr. No. P-8- 16, June 27-30, Noordwijkerhout, The Netherlands.

Deacon JW & Henry CM (1978) Mycoparasitism by *Pythium oligandrum* and *P. achanticum*. Soil Biol. Biochem. 10:409-415.

Dodman RL, Barker KA & Walker JC (1968) A. detailed study of the different modes of penetration by *Rhizoctonia solani*. Phytopathology 58:1271-1275.

Dijst G (1989) The effect of chemical halum destruction and halum pulling on potato black scurf caused by *Rhizoctonia solani* AG 3 Ph.D. Thesis, Agricultural University, Wageningen, The Netherlands 70 pp.

Duffy BK & Weller DM (1995) Use of *Geaumannomyces graminis* var. *tritici* alone and in combination with Fluorescent *Pseudomonas* spp. to suppress take-all of wheat. Dlant Dis. 79:907-901

Eayre CG & Echandi E (1988) Light and scanning microscopy of hypersensitive lesions associated with protection of *Phaseolus vulgaris* from *Rhizoctonia solani* induced by an avirulent binucleate *Rhizoctonia*-like fungus. Phytopathology 78:1524 (Abstr.).

Escande AR & Echandi E (1991) Protection of potato from *Rhizoctonia* canker with binucleate *Rhizoctonia* fungi. Plant Pathol. 40:197-202.

Finkler A, Koltin Y, Barash I, Sneh B & Pozniak D (1985) Isolation of a virus from virulent strains of *Rhizoctonia solani*. J. Gen. Virol. 66:1221-1232.

Freeman S & Rodriguez RJ (1993) Genetic conversion of fungal plant pathogen to nonpathogenic, epiphytic mutualist. Science 260:75-78.

Gehlert R & Kindl H (1991) Induced formation of dihydrophenanthrenes and bibenzyl synthase upon destruction of orchid mycorrhiza. Phytochemistry 30:457-460.

Gugel RK, Yitbarek SM, Verma PR, Morrall RAA and Sadasivaia RS (1987) Ethiology of the Rhizoctonia root rot complex of canola in the Peace River region of Alberta. Can. J. Plant Pathol. 9:119-128.

Hammer S, Fulbright DW & Adams G (1989) Association of double stranded RNA with low pathogenicity of *Leucostoma persoonii*. Phytopathology 79:568-572.

Harley JL & Smith SE (1983) Mycorrhizal Symbiosis. Academic Press, London.

Harris AR (1994) Plant growth promotion and biological control of damping off in container-grown seedlings using soil bacteria and fungi. In: Ryder MH, Stephens PM & Bowen GD (eds.) Improving Plant Productivity With Rhizosphere Bacteria. (pp 18-23) Proc. 3rd Int. Workshop on Plant Growth-Promoting Rhizobacteria, Adelaide, S. Australia.

Harris AR, Schisler DA & Neate S (1993a) Culture of *Rhizoctonia solani* and binucleate *Rhizoctonia* spp. on organic substrates for inoculation of seedlings in containers. Soil Biol. Biochem. 25:337-341.

Harris AR, Schisler DA, Neate S & Ryder MH (1993b) Suppression of damping off caused by *Rhizoctonia solani* and growth promotion in bedding plant by binucleate *Rhizoctonia* spp. Soil Biol. Biochem. 25:263-268.

Harris AR, Schisler DA & Ryder MH (1993c) Binucleate *Rhizoctonia* isolates control *Pythium ultimum* var. *sporangiiferum* and promote growth in *Capsicum* and *Celosia* seedlings in pasteurized potting medium. Soil Biol. Biochem. 25:909-914.

Herr LJ (1988) Biocontrol of Rhizoctonia crown and root rot of sugar beet by binucleate *Rhizoctonia* spp. and *Latiseria arvalis*. Ann. Appl. Biol. 113:107-108.

Herr LJ (1991) Relationship of binucleate *Rhizoctonia* isolates used for biocontrol of Rhizoctonia crown rot of sugar beets to anastomosis systems. Can. J. Microbiol. 37:33-344.

Herr LJ (1995) Biological control of *Rhizoctonia solani* by binucleate *Rhizoctonia* spp. and hypovirulent *R. solani*. Crop Prot. 14:179-187.

Holmes KA & Benson DM (1994) Evaluation of *Phytophthora parasitica* var. *nicotianae* for biocontrol of *Phytophthora parasitica* on *Catharanthus roseus*. Plant Dis. 78:193-199.

Ichielevich-Auster M, Sneh B, Koltin Y & Barash I (1985a) Pathogenicity, host specificity and anastomosis groups of *Rhizoctonia* spp. isolated from soils in Israel. Phytoparasitica 13:103-112.

Ichielevich-Auster M, Sneh B, Koltin Y & Barash I (1985b) Suppression of damping off caused by *Rhizoctonia* spp., by nonpathogenic *R. solani*. Phytopathology 75:1080-1084.

Koltin Y & Day P (1976) Inheritance of killer phenotype and double stranded RNA in *Ustilago maidis*. Proc. Natl. Acad. Sci. USA 73:594-598.

Komada H (1975) Development of a selective medium for quantitative isolation of *Fusarium oxysporum* from natural soils. Rev. Plant Prot. Res. 8:114-125.

Kubo K, Nakamura H, Kobayashi K, Okuno T & Furusawa I (1991) Cloning of a melanin biosynthetic gene essential for appressorial penetration of *Colletotrichum lagenarianum*. Mol. Plant-Microbe Interact. 4:440-45.

Kusik CS, Snow JP & Valverde RA (1994) Comparison of double stranded RNA components and virulence among isolates of *Rhizoctonia solani* AG1-IA and AG1-IB. Phytopathology 84:44-49

Lapierre H, Lemaine J M, Juan B & Molin G (1970) Mise en evidence de particules virales associees a une perte de pathogenicite chez le pietin-echaudage de cereales *Ophiobolus graminis* Sacc. C.R. Seances Acad. Sci Ser. D, 271:1833-1836.

Lim TK & Rohrbach KG (1980) Role of *Penicillium funiculosum* strains in the development of pineapple fruit diseases. Phytopathology 70:663-665.

Lifshitz R, Sneh B & Baker R (1984) Soil suppressiveness to a pathogenic *Pythium* species. Phytopathology 74:1054-1061.

Louter JH & Edgington LV (1990) Interaction of cross-protection against Fusarium crown and root rot of tomato. Can. J. Plant Pathol. 12:283-288.

Malder A, Turkensteen LJ & Bauman A (1992) Perspectives of green crop-harvesting to control soil borne and storage diseases of seed potatoes. Neth. J. Plant Path. 98:103-114.

Marcus L, Barash I, Koltin Y, Sneh B & Finkler A (1986) Characterization of pectic enzymes produced by virulent and hypovirulent isolates of *Rhizoctonia solani*. Physiol. & Molecular Plant Pathol. 29:325-336.

Martinelli JA (1990) Induced resistance of barley (*Hordeum vulgare* L.) to powdery mildew (*Erysiphe graminis* DC: FR. f. sp. *hordei*) and its potential for crop protection. PhD Thesis Univ. of Cambridge.

McGuire RG & Kelman A (1986) Calcium in Potato tuber cell walls in relation to tissue maceration by *Erwinia carotovora* pv. atroseptica Phytopathology 76:401-406

Ogawa K & Komada H (1986) Induction of systemic resistance against Fusarium wilt of sweet potato with non-pathogenic *Fusarium oxysporum*. Ann. Phytopathol Soc. Jpn. 50:1-9.

Park S-C, Paulitz TC & Baker R (1988) Biocontrol of Fusarium wilt of cucumber resulting from interaction between *Pseudomonas putida* and nonpathogenic isolates of *F. oxysporum*. Phytopathology 78:190-94.

Pusey PL & Wilson CL (1982) Detection of double stranded RNA in *Ceratocystis ulmi*. Phytopathology, 72:423-428.

Reinecke T & Kindl H (1994) Inducible enzymes of the 9, 10-dihydro-phenanthrene pathway. Sterile orchid plants responding to fungal infection. Mol. Plant-Microbe Interact. 7:449-454.

Roberts DL & Sivasithamparam K (1986) Identity and pathogenicity of *Rhizoctonia* spp. associated with bare patch disease of cereals at a field site in Western Australia. Neth. J. Pl. Path. 92:185-195.

Schneider RW (1984) Effect of non-pathogenic strains of *Fusarium oxysporum* on celery root infection by *F. oxysporum* f. sp. *apii* and a novel use of Lineweaver Buck double reciprocal plot technique. Phytopathology 74:646-653.

Siwek K, Harris AR & Scott ES (1995) Hyphal interactions between binucleate *Rhizoctonia* and the plant pathogen *Pythium ultimum*. International Symposium on *Rhizoctonia*, Facts and challenges in Pathology, Taxonomy, Ecology and Disease Control. June 27-30, Abstr. No. P-4-2, June 27-30, Noordwijkerhout, The Netherlands.

Sneh B (1990) Mechanisms Involved in Protection of Infection Sites. In: Baker R & Dunn P (eds.) New Directions in Biological Control: Alternatives for Suppressing Agricultural Pests and Diseases. (pp 653-662) Alan Liss Inc. New York.

Sneh B, Burpee LL & Ogoshi A (1991) Identification of *Rhizoctonia* Species. APS Press, St. Paul Minnesota. 133pp.

Sneh B & Ichielevich-Auster M (1995) Induced resistance of cucumber seedlings caused by non pathogenic *Rhizoctonia* (np-R) isolates. International Symposium on *Rhizoctonia*, Facts and challenges in Pathology, Taxonomy, Ecology and Disease Control. Abstract No. P-8-3, June 27-30, Noordwijkerhout, The Netherlands.

Sneh B, Ichielevich-Auster M & Plaut Z (1989a) Mechanism of seedling protection induced by a hypovirulent isolate of *Rhizoctonia solani*. Can. J. Bot. 67:2135-2141.

Sneh B, Ichielevich-Auster M & Shomer I (1989b) Comparative anatomy of colonization of cotton hypocotyls and roots by virulent and hypovirulent isolates of *Rhizoctonia solani*. Can. J. Bot. 67:2142-2149.

Sneh B, Pozniak D & Salomon D (1987) Soil Suppressiveness to *Fusarium* wilt of melon, induced by repeated cropping of resistant varieties of melon. J. Phytopathology 120:347-354.

Sneh B, Zeidan M, Ichielevich-Auster M, Barash I & Koltin Y (1986) Increased growth responses induced by a nonpathogenic *Rhizoctonia solani* Can. J. Bot. 64:2372-2378.

Spurr WH (1977) Protective application of non-pathogenic *Alternaria* sp. isolates for control of tobacco brown spot disease. Phytopathology 67:128-132.

Stein T, Schuster S, Zilberstein A, Pnini-Cohen S & Eyal Z (1994) The suppression of pycnidial production on wheat following challenge inoculation by *Septoria tritici* isolates. (Abstr.) Phytopathology 84:1100.

Stoessl A (1983) Secondary plant petabolites in preinfectional and postinfectional resistance. In: Bailey JA and Deverall BJ (eds.), The Dynamics of Host Defence. (pp 71-122) Academic Press, London.

Stockwell V & Hanchey P (1983) The role of the cuticle in resistance of bean to *Rhizoctonia solani*. Phytopathology 73:1640-1642.

Tamietti G & Allabouvette C (1986) Resistance des sols aux maladies: XIII- Role des *Fusarium oxysporum* non pathogens dans les mechanisms de resistance d'une sol de Noirmoutier aux fusariose vasculaires. Agronomie 6:541-548.

Tschen, JSM. & Hsue CC (1995) Control of *Rhizoctonia solani* by means of cross protection. International Symposium on *Rhizoctonia*, Facts and challenges in Pathology, Taxonomy, Ecology and Disease Control. Abstract No. P-8-9, June 27-30, Noordwijkerhout, The Netherlands.

Uetake Y, Kobayashi K, Ogoshi A & Tsutsui K (1988) Identification of *Rhizoctonia* species isolated from orchids. Ann. Phytopathol. Soc. Japan 54:114.

Van Alfen NK (1982) Biology and potential for disease control with hypovirulence of *Endothia parasitica*. Annu. Rev. Phytopathol. 20:349- 362.

Villajuan-Abgona R, Katsuno N, Kageyama K & Hyakumachi M (1993) Isolation of hypovirulent strains of *Rhizoctonia* from soil and their cross protection effect. (Abstr.) Ann. Phytopath. Soc. Japan 59:61.

Villajuan-Abgona R, Katsuno N, Kageyama K & Hyakumachi M (1996a) Isolation and identification of hypovirulent *Rhizoctonia* isolates from soil. Plant Pathol. (in press).

Villajuan-Abgona R, Kageyama K & Hyakumachi M (1996b) Biocontrol of Rhizoctonia damping off of cucumber by non pathogenic binucleate *Rhizoctonia*. Plant Pathol. (In Press).

Villajuan-Abgona R, Kageyama K & Hyakumachi M (1996c) Biological control of Rhizoctonia damping off of cucumber by non-pathogenic binucleate *Rhizoctonia*. (Submitted).

Walther D & Gindrat D (1987) Antagonism of *Rhizoctonia* spp. to *Pythium oligandrum* and damping off fungi J. Phytopathology 119:248- 254.

Wattad C, Freeman S, Dinur A & Prusky D (1995) A nonpathogenic mutant of *Colletotrichum magna* is difficient in extracellular secretion of pectate lyase. Mol. Plant-Microbe Interact. 8:621-626.

Yuen GY, Craig ML & Giesler LJ (1994) Biological control of *Rhizoctonia solani* on tall fescue using fungal antagonists. Plant Dis. 78:118-123.

VI.B5 MYCOPARASITISM AND BIOCONTROL

PAUL H.J.F VAN DEN BOOGERT

DLO Research Institute for Plant Protection, P.O. Box 9060, 6700GW Wageningen, The Netherlands

I. SUMMARY

Exploitation of mycoparasites as biocontrol agents in agriculture is well-documented. A number of them have shown potential to control *Rhizoctonia solani* in a diversity of crops or sites, alone or integrated with chemical or mechanical techniques. In respect to the pathogen's behaviour on the plant surface, the biocontrol effect ranges from niche exclusion to niche replacement depending on the nature of the mycoparasite involved. These mutual non-exclusive phenomena interfere with colonization and survival of the pathogen, respectively. The practice of biocontrol of the soil-borne pathogenic fungus *R. solani* exploiting mycoparasites is based on this rationale. At a cellular level modes of action are diverse, ranging from direct cell-cell interaction to effects far beyond the site of confrontation. This indicates that mycoparasites may control *R. solani* not only directly but also beyond the areas of infection. Mycoparasites or their gene products may also enhance sensitivity of the target pathogen to fungitoxic compounds by causing increased uptake of fungicides via cell wall digestion. In addition, gene products from mycoparasites are attractive candidates for genetic engineering in planta or plant associated microbiota. Recent advances on cell, biochemical and molecular biology of mycoparasitism confirm the preposition that mycoparasites and their gene products have great potential for biorational control of soil-borne pathogens, including *R. solani*.

II. INTRODUCTION

Rhizoctonia diseases occur in a wide range of hosts in agricultural, horticultural and floricultural crops. Economic losses due to the disease vary in severity and predictability ranging from the acceptable levels of disease as a biological phenomenon to levels demanding attempts to control the pathogen. At present Rhizoctonia diseases on some hosts are treatable with fungicides (Kataria and Verma, 1990; Olaya *et al.*, 1994), while a very few diseases can be controlled by the use of resistant cultivars (Hecker and Ruppel, 1991). Many of the Rhizoctonia diseases are attenuated by cultural and cropping practices only (Mulder *et al.*, 1992; Scholte, 1992). For some high cash crops such as greenhouse crops or nursery crops expensive eradication of the pathogen is applied by heat or chemical treatment of the soil (Summer and Phatak, 1988). Also, alternative control strategies have been advanced using antagonistic organisms onto seeds, transplants, crops or on crop residues as protectants against *R. solani* (e.g. Van den Boogert and Luttikholt, 1995). The potential of mycoparasites for controlling soil-borne diseases has extensively been studied since the first description of mycoparasitism by *Trichoderma lignorum* (synonym of *T. viride*) on plant pathogenic fungi (Weindling, 1932). Reviews of mycoparasites as biological control agents of soil-borne plant pathogens have been published by Jeffries and Young (1994), Baker and Dickman (1993), Whipps (1991), Adams (1990), Chet (1990) and Papavizas (1985).

B. Sneh et al. (eds.),
Rhizoctonia Species: Taxonomy, Molecular Biology, Ecology, Pathology and Disease Constrol, 485–493.

In spite of an impressive research endeavour, the use of introduced mycoparasites has not yet achieved an economic disease control perspective (Adams, 1990). There are several reasons for this, of which the first and foremost is the lack of ecological knowledge of the pathosystem organisms. Not only efficacy or biological properties are important but also product-related demands determine their feasibility such as mass-producibility, and cost-effectiveness, all in perspective to their chemical alternatives.

In this chapter, ecological aspects of mycoparasite-pathogen interactions in a tripartite association with the crop plant will be discussed against a background of developing successful, reproducible and cost-effective disease control. Emphasis will be laid on the specific ecological niches mycoparasites possess in relation to *R. solani* and their significance to disease control. This chapter provides selected examples to illustrate concepts rather than an all-inclusive review of the literature.

III. DEFINITION OF MYCOPARASITES

Fungi living on other fungi are commonly termed as mycoparasites or fungicolous fungi (Jeffries and Young, 1994). The latter term embraces a very wide range of fungus-fungus relationship, including those of which no nutritional relationships has been demonstrated. According to the definition of the Federation of British Plant Pathologists, the term mycoparasite strictly addresses to fungi existing in close association with another fungus from which they derive some or all of their nutrients while yielding no benefit in return. Mycoparasites are usually subdivided into fungi with a biotrophic or a necrotrophic way of parasitism. Biotrophic mycoparasites obtain nutrients from living host cells, whilst necrotrophic mycoparasites kill host cells, sometimes in advance of direct contact and utilise their nutrients. In nature, most biotrophic and some of the necrotrophic fungi are totally dependent upon mycoparasitism for their growth and development: these have been termed ecologically obligate (Whipps, 1991; Jeffries and Young, 1994) or non-competitive (Deacon and Berry, 1992) mycoparasites (EOMs). Contrary, ecologically facultative mycoparasites (EFMs) have a primarily saprophytic way of life, using antibiosis or mycoparasitism to facilitate competition.

The efficacy of biocontrol by EFMs is determined, at least partially, by the intrinsic ability to utilize nutrients released from the host plant in competition with other micro-organisms, which potentially leads to exclusion of the pathogen (niche exclusion). Characteristically, EFMs, exemplified by the extensively studied species *Trichoderma* spp. and *Gliocladium* spp. (Papavizas, 1985), interfere with a wide host range (*R. solani*, *Fusarium* spp. and *Pythium* spp.) and utilize nutrients from a wide variety of crop plants, including their residues. Contrary, EOMs are by nature not competitive and their efficacy in biocontrol is determined by their intrinsic ability to retrieve nutrients from fungal host cells, which potentially leads to the replacement of the pathogen. EOMs, exemplified by *Verticillium biguttatum* W. Gams, possesses host specialization on *Rhizoctonia*, demands specific nutrition (e.g. biotin) and is unable to colonize stems and roots in the absence of living host mycelium (Van den Boogert, 1989a; 1989b; Morris *et al*, 1995a).

Based on their ecological behaviour towards host fungus and host plant and their significance for biocontrol, EOMs and EFMs can be represented in the following hypothetical models of tripartite pathosystems (Fig. 1). The triangular model A would apply to a tritrophic interaction pattern with EFMs, and model B to EOMs. Obligate mycoparasites need host fungi for their growth, whilst facultative mycoparasites are less dependent on host fungi and may rely on alternative nutrient sources available in the plant rhizosphere.

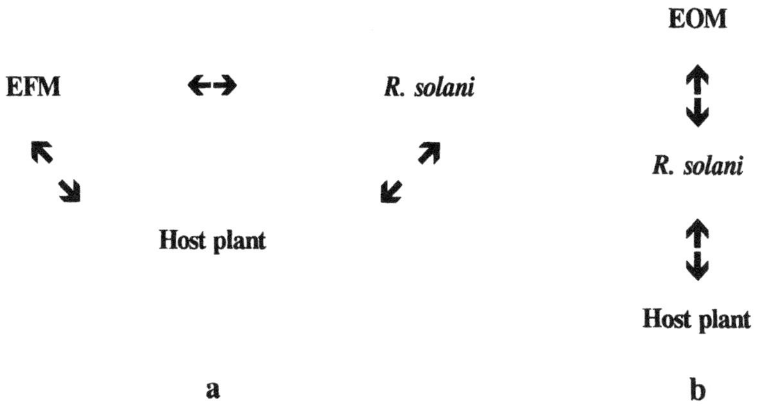

FIGURE. 1. Hypothetical models of tripartite pathosystems with: a. ecologically facultative (EFM) and b. obligate (EOM) mycoparasites.

IV. MODES OF INTERACTION

The significance of mycoparasites for disease control is determined by their specific properties to exclude or replace pathogenic fungi but also by the eco-physiological behaviour of the pathogen. Unlike most soil-borne fungi, *R. solani* is an ectotrophic organism that completes its entire life cycle at the outside of the host plant. It produces an extensive hyphal network on the below ground plant surface. Even at infection sites, the pathogen superficially penetrates susceptible host tissue. Such an ectotrophic growth habit, has an obvious advantage since hyphal networks are formed, through which food reserves are pooled for infection or survival structures (Deacon and Scott, 1985). On the other hand, through its extensiveness on below ground plant surfaces, hyphal networks of *R. solani* are easily accessible to soil-borne mycoparasites.

Mycoparasitism is a complex process which may affect contacting host cells or the hyphal network as a whole. At a cellular level, the generalized interaction between host and mycoparasite is as follows (Chet, 1987): 1. directed growth towards the host hyphae, presumably supported by a nutrient gradient; 2. Recognition and binding to the host surface by specific cell wall-bound lectins and carbohydrates; 3. formation of appressoria-like structures around *R. solani* hyphae varying from distinct coiling patterns to loose side branch contact; 4. disintegration of host cytoplasm followed by cell death. Steps 1-4 can easily be observed in dual culture on cellulosic substrates using light microscopy. There are two contrasting behavioural differences between EFMs and EOMs in dual culture with *R. solani*. *Gliocladium* spp. and *Trichoderma* spp. are able to inhibit radial growth extension following opposite inoculations (inhibition zone) or exhibit intense coiling patterns upon encounter, but penetration is a relatively rare event. At similar inoculations, however, the EOM *V. biguttatum* is unable to inhibit growth extension of *R. solani* but penetration of host cells and subsequent formation of internal haustoria-like structures is common. In addition, infection of the hyphal network by *V. biguttatum* affects *R. solani* beyond the site of parasitism, resulting in an almost complete suppression of sclerotium development (Fig. 2). This so-called 'distance effect' was not observed with EFMs (Van den Boogert and Deacon, 1994). The mechanism underlying the 'distance' effect is unknown, but can be ascribed to the creation of a nutrient sink imposed by the haustorial feeding of *V.*

biguttatum, which is consistent with biotrophy. To what degree indirect effects by mycoparasitism are operable *in situ* on the plant surface needs to be verified in future experiments. If *V. biguttatum* does so at a distance, as in culture, then effective biocontrol would not be dependent on parasitism of the whole hyphal network nor on the attendant problem of achieving uniform distribution of the mycoparasite in soil or on below-ground plant surfaces.

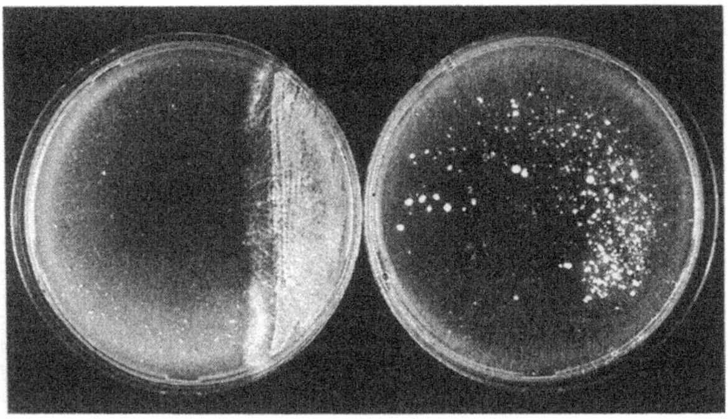

FIGURE 2. Hyphal network of *R. solani* on cellulose agar forming numerous sclerotia (left) and locally infected network showing almost complete suppression of sclerotia formation beyond the site of infection by *V. biguttatum* (right).

At a molecular level, several modes of action have been identified as having an essential role in the biocontrol activity of mycoparasites against *R. solani*, none of which are mutually exclusive. These include antibiotics (Papavizas, 1985), release of mycotoxicants, e.g. gliotoxin or gliovirin by *Gliocladium virens* (Howell *et al.*, 1993) or cell wall degrading enzymes (Tronsmo and Harman, 1993; Jabaji-Hare and Charest, 1994; see also chapter VI.B2) and other lytic enzymes (Fravel and Roberts, 1991). Only a few authors present conclusive evidence on the actual involvement of these mechanisms (Deacon and Berry, 1992). It is most probable that a combination of mechanisms operate during resource capture by mycoparasites. For instance, *in vitro* studies have shown that parasitism of *R. solani* hyphae by *V. biguttatum* involves the production of cell wall lytic enzymes (Van den Boogert *et al.*, 1989) possibly accompanied by the release of low-molecular antibiotics (Morris *et al.*, 1995b). On the other hand, the observed biocontrol of gliotoxin-deficient mutants of *G. virens* put into question the importance of mycotoxicants in biocontrol of *R. solani*-incited disease in cotton (Howell and Stipanovic, 1995). Conclusive proof must await the results of *in situ* experiments using mutants with disrupted genes coding for the relevant compounds in mycoparasitism (molecular postulate of Koch).

V. ISOLATION OF MYCOPARASITES AND THEIR TESTING FOR BIOCONTROL

The best way to increase mycoparasites to facilitate their isolation is on their natural substrates, e.g. *R. solani* mycelium or sclerotia. This ability of mycoparasites was exploited by using *R. solani*-precolonized agar plates (RP) for the selective isolation of mycoparasites from plant roots or from soil. RP appeared to be highly selective both for EOMs and EFMs and almost pure cultures may arise on RP after plating out infected plant parts or soil particles (Jager *et al.*, 1979; Van den Boogert, 1988; Mulligan and Deacon 1992). In an inventory study on the occurrence of mycoparasites associated with tuber-borne sclerotia from potato a number of mycoparasites were selected of which species frequency and composition depended on crop rotation and soil type. Among EOMs *V. biguttatum* appeared most frequently on tuber-borne sclerotia from potato cropping systems and among the EFMs, *Trichoderma* spp. and *Gliocladium* spp. were regularly observed, while *Hormiactis fimicola*, *Volutella ciliata* were relatively uncommon (Jager *et al.*, 1979; Van den Boogert and Saat, 1991). Novel mycoparasites or common fungi with so far unknown mycoparasitic properties are also expected to be isolated using *Rhizoctonia*-precolonized agar plates.

Considerations about where and when an isolated mycoparasite is most likely to function are of paramount importance when designing bio-assay procedures. Bio-assay systems for biocontrol efficacy by mycoparasites have been developed with different degrees of microbial diversity: monoxenic in Petri dishes to polyxenic systems with selected or naturally occurring biota. A mycoparasite with general competitive properties, such as EFMs, would be a good choice for biocontrol in environments of low biological diversity, e.g. in steamed, solarized or inundated soil or onto seeds. Such a mycoparasite must be able to compete with invasive micro-organisms using antibiotics, mycotoxicants or mycoparasitic abilities to exert competition. For example, if *T. harzianum* is used to treat cucumber seeds, biocontrol effect is marginal. However, if the organism is applied to the seed in a food base and if the seeds are then incubated for a few days at 100% relative humidity, prior to sowing, the bioprotectant grows and colonizes the seed surface giving an almost perfect biocontrol (Harman and Nelson, 1994). In a system of high biological diversity however, EOMs would be more likely to provide biocontrol rather than general competitors. For such a mycoparasite timing of application and accessibility of the plant to be protected are vital factors. For example, if *V. biguttatum* is used to treat seed tubers, the mycoparasite fails to colonize the root and stolon system in the absence of its host *R. solani* and the biocontrol effect is marginal (Van den Boogert, 1989b). However, if applied onto *Rhizoctonia*-infected progeny tubers and crop residues, *V. biguttatum* gives an appreciable biocontrol of black scurf disease. Accordingly, a minituber system have been developed with mini potato tubers and crop residues in field soil for testing *V. biguttatum* treatments alone or in combination with other natural or synthetic fungicides on tuber diseases and pathogen survival (Van den Boogert *et al.*, 1994; Van den Boogert and Luttikholt, 1995).

VI. PROGRESS AND DRAWBACKS IN BIOCONTROL WITH MYCOPARASITES

From an ecological point of view an ideal mycoparasite for biocontrol would possess similar environmental or nutritional requirements as the target fungus leading to a reproducible and cost-effective control. Also the candidate control agent must be implementable in the existing methods of crop production, including compatibility with current chemical or other biological control agents.

Biological control of *R. solani* has been the subject of extensive research, in bare soil and on various crops with a wide variety of mycoparasites: *Trichoderma harzianum* on iris bulbs (Chet *et al.*, 1982), bean, cotton and tomato (Elad *et al.*, 1980), on strawberry (Elad *et al.*, 1981), on radish (Henis *et al.*, 1978), *T. hamatum* in soil (Lewis and Papavizas, 1987) or on processed manure as soil amendment (Kok *et al.*, 1996) and on radish and pea (Harman *et al.*, 1980), *Trichoderma sp.* on cucumber (Lewis and Papavizas, 1980), *T. viride* on potato (Beagle-Ristaino and Papavizas, 1985), *Laetiaria arvalis* on sugar beet (Allen *et al.*, 1985; Larsen *et al.*, 1985), a soil-inhabiting basidiomycete on sugar beet

(Odvody *et al.*, 1980), *Pythium nunn* in soil (Lifshitz *et al.*, 1984), *Gliocladium roseum* on potato (Jager *et al.*, 1979), *G. virens* on cotton seedlings (Howell, 1993) and on white beans (Tu and Vaartaja, 1981), *V. biguttatum* (Jager *et al.*, 1991) and *Stachybotrys elegans* (Benyagoub *et al.*, 1994) on potato.

Some mycoparasites, e.g. *T. harzianum*, have potential for broad-spectrum control of fungal pathogens, but cannot be applied economically because of the excessive amounts of formulated product needed to obtain disease control (Adams, 1990). Others, e.g. *Trichoderma* spp., are only effective as biological control agents when applied to sterile soil or in soilless systems (Locke *et al.*, 1984; Marois and Locke, 1985). Although mostly successful in potato crop against black scurf disease, *V. biguttatum* requires specific conditions to render effect, e.g. temperature of at least 13°C and minimum air humidity of 97% RH. Green crop harvesting, provides an excellent momentum to gain access to progeny tubers just before tuber-borne sclerotia are prone to develop under conditions conducive for mycoparasitism by *V. biguttatum*. Successful field trials with the application of *V. biguttatum*, indicate the potential of this biocontrol system in seed potato production (Van den Boogert and Luttikholt, 1995). In addition, *V. biguttatum* is effective in disinfection of *R. solani* on progeny tubers (Velvis and Jager, 1988) and on potato crop residues (Table 1). Since the beginning of rational agriculture, diseased plants or infected crop residues have been routinely removed from the field as a general phytosanitary measure. As an alternative phytosanitary measure, crop residue treatment with *V. biguttatum* may affect renewal of viable *R. solani*-inoculum with consequences for inter-crop survival of the pathogen

TABLE 1. Effect of potato crop treatment with *Verticillium biguttatum* on survival of *Rhizoctonia solani*-inoculum on soil-borne crop residues.

	Potato crop residues with viable *R. solani* (%)	
Location (year)	**Untreated**	***Verticillium*-inoculated**
Rolde (1993)	19	3 **
Creil (1993)	24	1 **
Munnekezijl (1994)	69	19 **

** Significant reduction at $P < 0.01$ (t test).

VII. Future of bio-(rational) control

In the past, improvement of mycoparasites as biocontrol agents was achieved by genetic manipulation using physical (e.g. UV irradiation) or chemical mutagenesis to achieve random improvement for fungicide-resistant strains for integrated control (Ahmad and Baker, 1987; 1988). Also, protoplast fusion between different strains of *Gliocladium* and *Trichoderma* spp. can bring together desired biocontrol-related properties into the genome of viable progeny, e.g. rhizosphere competence. Recently, transformation became available as a third method to improve the genome in *Trichoderma* by integrating DNA-sequences of interest for biocontrol using PEG or biolistic protocols (Lorito *et al.*, 1993).

In search of understanding mycoparasitism at a molecular level, viz. the mycotoxicants gliotoxin and gliovirin by *G. virens* (Howell *et al.*, 1993), cell wall degradation by *T. harzianum* (Tronsmo and Harman. 1993), key gene products have been identified which are supposed to be responsible for the biocontrol effect. The genomes of *Trichoderma* spp. and *Gliocladium* spp. apparently contain a number of genes coding for potential natural fungicides. The chitinolytic and glucanolytic enzymes from *T. harzianum* are candidates par excellence for biorational control. These enzymes act synergistically with each other, with gliotoxin and with synthetic fungicides (Di Pietro *et al.*, 1993), and with the biocontrol bacterium *Enterobacter cloacae* (Lorito *et al.*, 1993). Mixtures of cell wall degrading enzymes alone or in combination with gliotoxin or low concentrations of synthetic fungicides have potential to be commercialized as novel bioprotectants against plant pathogenic fungi, including *R. solani*.

Another approach has been the successful introduction of the chitinase gene from the soil-borne bacterium *Serratia marcescens* into tobacco plants. The transgenic tobacco plant grew nearly as fast as unaffected plants and showed significantly better protection against *R. solani* infection than the control plants (Broglie *et al.*, 1991). Introduction of biocontrol genes into higher plants or specialized micro-organisms with enhanced rhizosphere competence to increase protection against fungal infection is an interesting possibility in the years to come.

Recently, developing techniques have great potential to identify the molecular basis for mycoparasitism. Identification of gene products involved with mycoparasitism may render new compounds with anti-fungal properties. The possibilities then exist to determine whether inactivation of genes or inclusion of extra copies behind a constitutive promoter may either reduce or improve biocontrol by the mycoparasite. Also, the design of transformant mycoparasites with (deficient or over-expressing) biocontrol genes enables us to do fundamental research on regulation of *R. solani* populations in tripartite associations with crop plants.

VIII. REFERENCES:

Ahmad JS & Baker R (1987) Competitive saprophytic activity of rhizosphere-competent mutants of *Trichoderma harzianum*. Phytopathology 77:358-362.

Ahmad JS & R Baker R (1988) Implications of rhizosphere competence of *Trichoderma harzianum*. Canadian Journal of Microbiology 34:229-234.

Adams BP (1990) The potential of mycoparasites for biological control of plant diseases. Annual Review of Phytopathology 28:59-72.

Allen MF, Boosalis MG, Kerr ED, Muldoon AE & Larsen HJ (1985) Population dynamics of sugar beets, *Rhizoctonia solani & Laetisaria arvalis*: Response of a host, plant pathogen and hyperparasite to perturbation in the field. Appl. and Env. Microbiol. 50:1123-1127.

Baker R & Dickman MB (1993) Biological control with fungi. In: Blaine Metting F. (ed.) Soil Microbial Ecology (pp 275-305) Marcel Dekker Inc., New York, Basel and Hong Kong.

Beagle-Ristaino JE & Papavizas GC (1985) Biological control of *Rhizoctonia* stem canker and black scurf in potato. Phytopathology 75:560-564.

Benyagoub M, Jabasi-Hare SH, Banville G & Charest PM (1994) *Stachybotrys elegans* A destructive mycoparasite of *Rhizoctonia solani*. Mycol. Res. 98:493-505.

Broglie K, Chet I, Holliday M, Cressman R, Biddle P, Knowlton S, Mauvais CJ & Broglie R (1991) Transgenic plants with enhanced resistance to the fungal pathogen *Rhizoctonia solani*. Science 254:1194-1196.

Chet I (1990) Biological control of soil-borne plant pathogens with fungal antagonists in combination with soil treatments. In: Hornby D (ed.) Biological Control of Soil-borne Plant Pathogens (pp 15-26) CAB International, Wallingford, UK.

Chet I (1989) Mycoparasitism, recognition, physiology, and ecology. In: Baker R. and Dunn PE (eds.) New Directions in Biological Control (pp 725-733) Alain R. Liss, New York.

Chet I (1987) *Trichoderma*: Application, mode of action, and potential as a biocontrol agent of soil-borne plant pathogenic fungi. In: Chet I (ed.) Innovative Approaches to Plant Disease Control (pp 137-160) J. Wiley and Sons, New York, Chichester, Brisbane, Toronto and Singapore.

Chet I, Elad Y, Kalfon A, Hadar Y & Katan J (1982) Integrated control of soil-borne and bulb-borne pathogens in iris. Phytoparasitica 10:229-236.

Deacon JW & Berry LA (1992) Modes of action of mycoparasites in relation to soil-borne plant pathogens. In: Tjamos EC, Papavizas GC and Cook RJ (eds.) Biological Control of Plant Diseases (pp 157-167) Plenum Press, New York and London.

Deacon JW, (1985) *Rhizoctonia solani* associated with crater disease (stunting) of wheat in South Africa. Trans. Brit. Mycol. Soc. 85:319-327.

Di Pietro A, Lorito M, Hayes CK, Broadway RM & Harman GE (1993) Endochitinase from *Gliocladium virens*: isolation, characterization, and synergistic antifungal activity in combination with gliotoxin. Phytopathology 83:308-313.

Elad Y, Chet I & Henis Y (1981) Biological control of *Rhizoctonia solani* in strawberry fields by *Trichoderma harzianum*. Plant and Soil 60:245-254.

Elad Y, Chet I & Katan J (1980) *Trichoderma harzianum*: A biocontrol agent effective against *Sclerotium rolfsii* and *Rhizoctonia solani*. Phytopathology 70:119-121.

Fravel DR & Roberts DP (1991) *In situ* Evidence for the role of glucose oxidase in the biocontrol of Verticillium wilt by *Talaromyces flavus*. Biocontrol Sci. and Technol. 1:91-99.

Harman GE & Nelson EB (1994) Mechanisms of protection of seed and seedlings by biological seed treatments: Implications for practical disease control. In: Trevor M (ed.) Seed treatment: Progress and prospects. BCPC Monograph 57:283-292.

Harman GE, Chet I & Baker R (1980) *Trichoderma hamatum* effects on seed and seedling disease induced in radish and pea by *Pythium spp.* or *Rhizoctonia solani*. Phytopathology 70:1167-1172.

Hecker RJ & Ruppel EG (1991) Registration of Rhizoctonia root rot resistant sugar beet germplasm FC 710. Crop Sci. 31:494.

Henis Y, Ghaffer A & Baker R (1978) Integrated control of *Rhizoctonia solani* damping-off of radish: Effect of successive plantings, PCNB and *Trichoderma harzianum* on pathogen and disease. Phytopathology 68:900-909.

Howell CR, Stipanovic RD & Lumsden RD (1993) Antibiotic production by strains of *Gliocladium virens* and its relation to the biocontrol of cotton seedling diseases. Biocontrol Science and Technology 3:435-441.

Howell CR & Stipanovic RD (1995) Mechanisms in the biocontrol of *Rhizoctonia solani*-induced cotton seedling disease by *Gliocladium virens* - Antibiosis. Phytopathology 85:469-472.

Jager G, Velvis H, Lamers JG, JG Mulder JG & Roosjen J (1991) Control of *Rhizoctonia solani* in potato by biological, chemical and integrated measures. Potato Res. 34:269-284.

Jager G & Velvis H (1988) Inactivation of sclerotia of *Rhizoctonia solani* on potato tubers by *Verticillium biguttatum*, a soil borne mycoparasite. Neth. J. Plant Pathol. 94:225-230.

Jager G, Hoopen A ten & Velvis H (1979) Hyperparasites of *Rhizoctonia solani* in Dutch potato fields. Neth. J. Plant Pathol. 85:253-268.

Jeffries P & Young TWK (1994) Interfungal parasitic relationships. CAB International, Wallingford, UK.

Kararia HR & Verma PR (1990) Efficacy of fungicidal seed treatments against pre-emergence damping off and post-emergence seedling root rot of growth chamber grown canola caused by *Rhizoctonia solani* AG 2-1 and AG 4. Can. J. of Plant Pathol. 12:409-416.

Kok CK, Hageman PEJ, Maas PWT, Postma J, Roozen NJM & van Vuurde JWL (1996) Processed manure as carrier to introduce *Trichoderma harzianum*: Population dynamics and biocontrol effect on *Rhizoctonia solani*. Biocontrol Sci. and Technol. (in Press).

Larsen HJ, Boosalis MG & Kerr ED (1985) Temporary depression of *Rhizoctonia solani* field populations by soil amendment with *Laetisaria arvalis*. Plant Dis. 69:347-350.

Lewis JA, Roberts DP & Hollenbeck MD (1991) Induction of cytoplasmic leakage from *Rhizoctonia solani* hyphae by *Gliocladium virens* and partial characterization of a leakage factor. Biocontrol Sci. and Technol. 1:21-29.

Lewis JA & Papavizas GC (1980) Integrated control of Rhizoctonia fruit rot of cucumber. Phytopathology 70:85-89.

Lewis JA & Papavizas GC (1987) Reduction of inoculum of *Rhizoctonia solani* in soil by germlings of *Trichoderma hamatum*. Soil Biol. Biochem. 19:195-201.

Lifshitz R, Dapler M, Elad Y & Baker R (1984) Hyphal interactions between a mycoparasite, *Pythium nunn* and several other soil fungi. Can. J. of Microbiol. 30:1482-1487.

Locke JC, Marois JJ & Papavizas GC (1984) Biological control of Fusarium wilt of greenhouse-grown chrysanthemums. Plant Dis. 69:167-169.

Lorito M, Hayes CK, Di Pietro A & Harman GE (1993) Biolistics transformation of *Trichoderma* spp. and *Gliocladium* spp. using plasmids and genomic DNA. Curr. Genet. 14:349-356

Lorito M, Hayes CK, Woo SL, Di Pietro A & Harman GE (1993) Antifungal synergistic interaction between chitinolytic enzymes from *Trichoderma harzianum* and *Enterobacter cloacae*. Phytopathology 83:721-728.

Marois JJ & Locke JC (1985) Population dynamics of *Trichoderma viride* in steamed plant growth medium. Phytopathology 75:115-118.

Morris RAC, Coley-Smith JR & Whipps JM (1995a) The ability of the mycoparasite *Verticillium biguttatum* to infect *Rhizoctonia solani* and other plant pathogenic fungi. Mycol. Res. 99:997-1003.

Morris RAC, Ewing D, Whipps JM & Coley-Smith JR (1995b) Antifungal hydroxymethyl-phenols from the mycoparasite *Verticillium biguttatum*. Phytochem. (in Press).

Mulder A, Turkensteen LJ & Bouman A (1992) Perspectives of green-crop-harvesting to control soil-borne and storage diseases of seed potatoes. Neth. J. Plant Pathol. 98 (Supplement 2):103-114.

Mulligan DFC & Deacon JW (1992) Detection of presumptive mycoparasites in soil placed on host-precolonized agar plates. Mycol. Res. 96:605-608.

Odvody GN, Boosalis MG & Kerr ED (1980) Biological control of *Rhizoctonia solani* with a soil-inhabiting basidiomycete. Phytopathology 70:655-658.

Olaya G, Abawi GS & Barnard J, (1994) Response of *Rhizoctonia solani* and binucleate *Rhizoctonia* to five fungicides and control of pocket rot and table beets with foliar sprays. Plant Dis. 78:1033-1037.

Papavizas GC (1985) *Trichoderma* and *Gliocladium*: Biology, ecology and potential for biocontrol. Annu. Rev. Phytopathol. 23:23-54.

Scholte K (1992) Effect of crop rotation on the incidence of soil-borne fungal diseases of potato. Neth. J. Plant Pathol. 98 (Supplement 2): 93-101.

Summer DR & Phatak SC (1988) Efficacy of metham-sodium applied through overhead sprinkling irrigation for control of soil borne fungi and root diseases of vegetables. Plant dis. 72: 160-166.

Tronsmo A & Harman GE (1993) Detection and quantification of N-Acetyl-J-D-glucosaminidase, chitobiosidase and endochitinase in solutions and on gels. Analyt. biochem. 208:74-79.

Tu JC & Vaartaja O (1981) The effect of the hyperparasite *Gliocladium virens* on *Rhizoctonia solani* and on Rhizoctonia root rot of white beans. Can. J. of Bot. 59:22-27.

Tweddell R, Jabaji-Hare SH & Charest PM (1994) Production of chitinases and b-1,3-glucanases by *Stachybotrys elegans*, a mycoparasite of *Rhizoctonia solani*. Appl. Env. Microbiol. 60:489-495.

Van den Boogert PHJF (1989a) Nutritional requirements of the mycoparasitic fungus *Verticillium biguttatum*. Neth. J. Plant Pathol. 95:149-156.

Van den Boogert PHJF (1989b) Colonization of roots and stolons of potato by the mycoparasitic fungus *Verticillium biguttatum*. Soil Biol. Biochem. 21:255-262.

Van den Boogert PHJF & Gams W (1988) Assessment of *Verticillium biguttatum* in agricultural soils. Soil Biol. Biochem. 20:899-905.

Van den Boogert PHJF & Saat TAWM (1991) Growth of the mycoparasitic fungus *Verticillium biguttatum* from different geographical origin at near minimum temperatures. Neth. J. Phytopathol. 97:115-124.

Van den Boogert PHJF & Deacon JW (1994) Biotrophic mycoparasitism by *Verticillium biguttatum* on *Rhizoctonia solani*. Eur. J. Plant Pathol. 100:137-156.

Van den Boogert PHJF & Luttikholt AJG (1995) Control of sclerotial renewal of *Rhizoctonia solani* by biological and cultural methods. In: Whipps JM and Gerlagh M (eds.) Biological control of sclerotial-forming pathogens. IOBC wprs Bull. 18(3): 67-73

Van den Boogert PHJF, Reinartz H, Sjollema KA & Veenhuis M (1989) Microscopic observations of the mycoparasite *Verticillium biguttatum* with *Rhizoctonia solani* and other soil-borne fungi. Antonie van Leeuwenhoek 56:161-174.

Van den Boogert PHJF, Kastelein P & Luttikholt AJG (1994) Green-crop-harvesting, a mechanical haulm destruction method with potential for disease control of tuber pathogens in potato. In: Trevor M (Ed) Seed treatment: Progress and Prospects. BCPC Monograph 57: 237-246.

Weindling R (1932) *Trichoderma lignorum* as a parasite of other soil fungi. Phytopathology 22: 837- 845.

Whipps JM (1991) Effects of mycoparasites on sclerotia-forming fungi. In: Beemster ABR, Bollen GJ, Gerlagh M, Ruissen MA, Schippers B and Tempel A (eds.) Biotic interactions and soil-borne diseases (pp 129-140) Elsevier, Amsterdam.

VI.B6 ROLE OF SOIL FAUNA IN BIOLOGICAL CONTROL OF *RHIZOCTONIA*

EARL A. CURL[1] AND ROBERT T. LARTEY[2]

[1]*Department of Plant Pathology, Auburn University, AL 36849 U.S.A. and* [2]*Department of Biochemistry and Cell Biology, State University of New York, Stony Brook, NY 11794 U.S.A.*

I. INTRODUCTION

Living biotic components most dramatically involved in rhizosphere phenomena that affect plant growth, health and productivity are found among the many species of microflora, microfauna and mesofauna. Successful biological control of root pathogens such as *Rizoctonia solani* will require a complete understanding of the interactive functions of microflora and fauna at the root surface. One of the criteria for effective biological control is the reduction of inoculum density. Micro- and mesoscopic species of soil fauna depend to a large extent on microflora food sources in the rhizosphere and during organic matter decomposition, thus creating the opportunity for the consumption of pathogen propagules and hyphae and reduction of inoculum potential as defined by Garrett (1970).

By definition, the term antagonism features competition, antibiosis, mycoparasitism and predation. In this chapter, which constitutes updating of previous background information (Curl, 1988; Curl *et al.*, 1988; Curl and Harper, 1990), we attempt to elevate the predation phenomenon to a more prominent position in the biological control process by focusing upon invertebrate activities that may contribute directly or indirectly to disease suppression.

II. THE CRITICAL SOIL FAUNA

Arbitrary grouping of the microfauna primarily refers to animals with body widths of $<100\mu m$, thus encompassing the large populations of free-living protozoa and nematodes; the mesofauna (largely microarthropods) generally measure 1-2mm. These invertebrates, because of their small body size and feeding habits, are most likely to respond to root exudates and biotic phenomena in the rhizosphere. The macrofauna, exemplified by large arthropods and the annelids, are important in organic matter fragmentation and decomposition but are largely inhabitants of surface litter and do not fit prominently into the interactive scheme within microniches of the rhizosphere.

Most protozoa are represented by flagellates and small amoebae, their distribution being determined by soil-pore size and water films (Bamforth, 1988). Both virgin and cultivated soils harbor relatively abundant and heterogeneous populations, particularly of the amoebae. Peak populations of amoebae may be expected to follow peaks in numbers and biomass of bacteria, especially after rainfall (Clarholm, 1981). High populations in plant rhizospheres in terms of R/S ratios may range from 2:1 for total protozoa to 20:1 for some species. Generally, bacteria and yeasts are preferred food sources for protozoa; however, certain species of mycophagous amoebae in agroecosystems can destroy a sufficient quantity of pathogenic fungal propagules to

B. Sneh et al. (eds.),
Rhizoctonia Species: Taxonomy, Molecular Biology, Ecology, Pathology and Disease Constrol, 495–506.

induce significant biological control of specific root diseases (Old and Chakraborty, 1986).

Population densities of free-living nematodes vary widely with season and kinds of plant cover. Summary estimates in terms of biomass (Hendrix *et al.*, 1986) showed nonparasitic nematodes more abundant in conventional tillage systems than in no-tillage agroecosystems; highest values were for bacterivores rather than fungivores or herbivores. Along with bacteria-feeding species, populations of mycophagous species of *Aphelenchoides, Aphelenchus, Ditylenchus*, and *Neotylenchus* are relatively abundant in various cropping systems (Nicholas, 1984).

Mesoscopic members of the Arthropoda commonly referred to as microarthropods, like other small soil animals, vary in population with season, soil chemical and physical factors, vegetative cover, tillage practices, and food availability. In conventional agroecosystems, the biomass of mites was estimated to be 118 mg and Collembola 17 mg dry wt m^{-2} (Hendrix *et al.*, 1986); these figures were higher in no-tillage systems. Andrén and Lagerlof (1983) have compared the relative abundance and biomass of various small-animal groups in Swedish experimental fields.

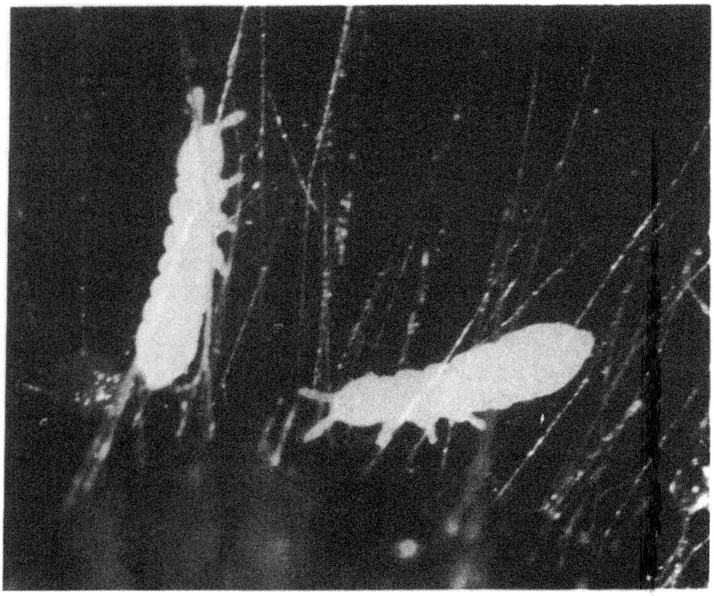

FIGURE 1. *Onychiurus encarpatus* (Collembola) feeding on a culture of *Rhizoctonia solani*.

The Collembola (Fig. 1) are the most abundant and most important of the orders of primitive insects in the Apterygota. They are well represented in agricultural soils to 15 cm depth, and may migrate to much greater depths depending upon soil type, moisture and cropping practices. Collembola were consistently more abundant in the rhizosphere of cotton plants in Alabama than 20 cm distance from roots, and a direct relationship was shown between increasing cotton-root density and collembolan populations (Wiggins *et al.*, 1979). High populations and activities of microarthropods can be expected to coincide with soil conditions and plant root systems that harbor high densities of the microflora.

III. Soil fauna and the inoculum potential

"The absolute inoculum potential is a measure of the maximum capacity of a pathogen population to infect fully susceptible plant tissue under optimum conditions for infection" (Mitchell, 1979). Thus, the inoculum potential will vary with the inherent nature of different soilborne pathogens to produce propagules, to survive in soil, and to infect host tissue. Logically, therefore, the feeding activities of protozoa, nematodes, and microarthropods in the rhizosphere could affect inoculum density, the potential for disease to occur, and the disease severity.

A. Amoebae: mycophagy and disease suppression

Old and Chakraborty (1986) thoroughly reviewed the biology and significance of mycophagous soil amoebae in relation to suppression of certain soilborne plant pathogens and disease. Prominent genera of fungus-feeding amoebae are *Vampyrella, Arachnula, Cashia, Thecamoeba, Saccamoeba, Gephyramoeba, Theratromyxa, and Mayorella*. Selectivity or preference for food sources (fungi, bacteria, and algae) is common among amoebae species. Feeding activity on fungal hyphae or spores is evidenced by perforation and lysis of cell walls (Homma *et al.*, 1979). Anderson and Patrick (1978) found that vampyrellid amoebae extracted from field soil in Ontario perforated chlamydospores and mycelium of *Thielaviopsis basicola* and conidia of *Cochliobolus sativus* resulting in reduced populations of both fungi. Chakraborty and Warcup (1985) then demonstrated suppression of take-all disease (*Gaeumannomyces graminis* var. *tritici*) of wheat in pot bioassays with introduced populations of *Gephyramoeba, Saccamoeba, and Thecamoeba*. The potential for amoebae species to suppress Rhizoctonia diseases has not yet received adequate attention.

B. Nematodes

Investigations of mycophagous nematodes for suppression of fungal pathogens and root disease have focused primarily upon species of *Aphelenchus and Aphelenchoides*. *Aphelenchus avenae*, commonly found associated with roots and rhizomes, can be reared *in vitro* on a wide range of soil fungi; over 92 species have been recorded as food sources. In laboratory tests (Mankau and Mankau, 1963) the common root-infecting fungi *R. solani, Verticillium albo atrum, Fusarium solani, and Thielaviopsis basicola* were highly favorable diets for reproduction and development of *A. avenae*. Abundantly sporulating fungi (normally saprophytes), including *Aspergillus terreus, Penicillium sp., Rhizopus nigricans, and Trichoderma viride*, were not favorable food sources; some species of *Pythium* and *Phytophthora* also were poor food sources and apparently were toxic. Other studies (Cayrol and Combettes, 1977) showed that both *A. avenae* and *Aphelenchoides composticola* were aggressive and inhibitory on cultures of *R. solani* but neither prevented growth of *P. ultimum*.

Several root-infecting fungi have been effectively controlled in greenhouse environments using relatively large populations of *A. avenae*. These include *R. solani* on bean (Barker, 1964) and *R. solani* and *Fusarium solani* on alfalfa seedlings (Barnes et. al., 1981). Damage to wheat by *Fusarium culmorum* was significantly reduced in greenhouse pots infested with *Aphelenchoides hamatus* (Rossner and Nagel, 1984). Thus, these limited tests under artificial conditions suggest a potential for disease reduction by the activities of mycophagous nematodes.

C. Arthropods as plant-pathogen suppressants

Many species of microarthropods (Acarina and Collembola) are mycophagous, feeding upon pathogens as well as saprophytes, but often showing distinct preferences for food sources. Bollen *et al.* (1991) tested combinations of a nematode (*Aphelenchus avenae*), a collembolan (*Tullbergia krausbaueri*), and mites (species of *Histiosoma, Tyrophagus, and Rhizoglyphus*) for control of *R. solani* on potato sprouts and found the mites to be less effective than the other agents. However, the known microphagous feeding habits and high populations of the Cryptostigmata in agroecosystems suggests a strong probability of influence on inoculum density and disease.

Collembola-plant pathogen interactions attracted no research attention until 1976 when field and laboratory studies were conducted on Collembola and rhizosphere microflora interactions in relation to cotton seedling disease potential in Alabama (Wiggins and Curl, 1979). In the ensuing years to the present, significant research progress has been made at several institutions, world-wide, as plant pathologists and soil scientists recognized the probability of relationships between these very active animals and certain biological control processes. Selected highlights are detailed in the following section.

IV. RESEARCH WITH COLLEMBOLA

A. The rhizosphere factor

Collembola, and other invertebrates that feed on the microflora, are not uniformly distributed through soil but tend to congregate around living roots, dead or dying roots, and organic matter particles where food sources are most abundant. The rhizosphere factor is important in relation to microarthropod potential to challenge plant pathogens at the infection site (Curl, 1982; Curl and Truelove, 1986). Laboratory tests showed that Collembola are attracted to cotton roots, particularly in slowly drying soil (Wiggins and Curl, 1979). Extensive field assessments in Alabama (Wiggins *et al.*, 1979) provided evidence that two collembolan species, *Proisotoma minuta and Onychiurus encarpatus*, were predominant in cotton rhizosphere; further, they were significantly more numerous in the rhizosphere than in root-free soil regardless of long-term fertilization regimes (Fig. 2). Direct observations with minirhizotrons in Michigan agroecosystems (Snider *et al.*, 1990) showed that *Folsomia candida* was much more numerous in the rhizospheres of soybean and sugar beet than in the rhizosphere of corn.

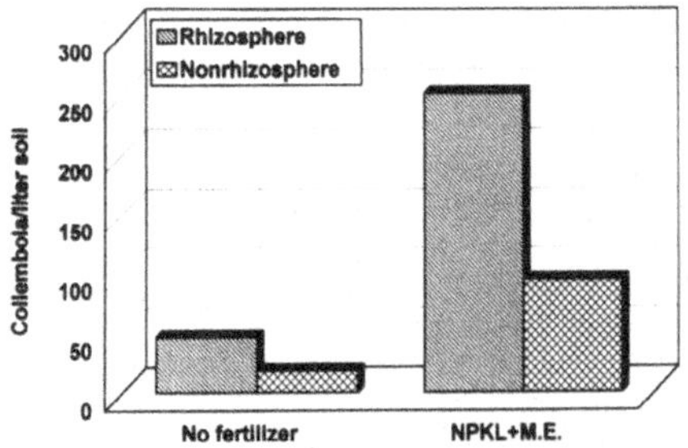

FIGURE 2. Relative populations of Collembola in cotton rhizosphere and nonrhizosphere soil in field plots subjected to long-term fertilization regimes. NPKL + M.E. = nitrogen, phosphorus, potassium, lime + minor elements.

B. Grazing activities and food preference

Though gut-content analyses of field-collected Collembola often reflect non-specific feeding habits, preferential feeding upon specific fungi has been demonstrated in laboratory experiments (McMillan, 1976; Visser and Whittaker, 1977; Shaw, 1988; Lartey *et al.*, 1989) and to a limited extent in field tests (Shaw, 1985). In laboratory feeding trials (Curl *et al.*, 1985), mixed populations of *Proisotoma minuta* and *Onychiurus encarpatus* (200 animals per culture) totally consumed surface colonies of pathogens, *Macrophomina phaseolina, Verticillium dahliae, and Fusarium oxysporum* f. sp. *vasinfectum*, and prevented further growth; the animals had little effect on *Sclerotium rolfsii* and the biocontrol agent *Trichoderma harzianum*, and actually were repelled by these two fungi. In further tests (Lartey *et al.*, 1989) we demonstrated a strong feeding preference by *P. minuta* and *O. encarpatus* for *R. solani* in the presence of *T. harzianum, Gliocladium virens*, or *Laetisaria arvalis*. When the pathogen was omitted in other tests with *P. minuta*, there was a clear choice of *L. arvalis* as preferred food over *T. harzianum* or *G. virens* and greater egg production occurred on the basidiomycete (Fig. 3). The animals showed no preference between *T. harzianum* and *G. virens* and were repelled by these two fungi when sporulation began (Lartey *et al.*, 1994). Compatibility of fungal food sources for collembolans also was reflected in the reproductive capacity of *P. minuta* in monocultures of *R. solani*, *T. harzianum*, *G. virens*, and *L. arvalis* (Fig. 4). Nakamura *et al.* (1991) named five collembolan species, including *P. minuta*, which grazed preferentially and multiplied on two-five fungal species out of six species tested. Chlamydospores of *Fusarium oxysporum* f. sp. *vasinfectum*, buried in collembolan-infested soil, were readily consumed but many remained viable in fecal pellets (Curl *et al.*, 1985). The animals (*P. minuta* and *O. encarpatus*) were not able to ingest the macrosclerotia of *Sclerotium rolfsii* or microsclerotia of *Verticillium dahliae* and *Macrophomina phaseolina*. Emerging germ tubes, however, were grazed thus reducing effectiveness of sclerotia as viable propagules. *Rhizoctonia* sclerotia were not included in these tests.

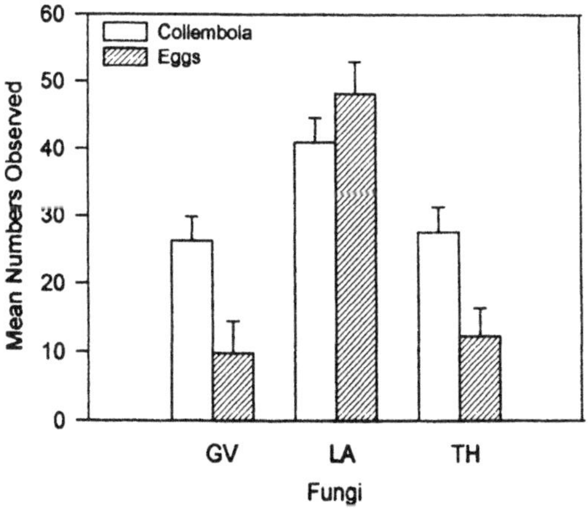

FIGURE 3. Response of Collembola (*Proisotoma minuta*) to three known biocontrol fungi as available food sources in laboratory culture: numbers of animals attracted to mycelial disks followed by egg production. GV = *Gliocladium virens*; LA = *Laetisaria arvalis*; TH = *Trichoderma harzianum*.

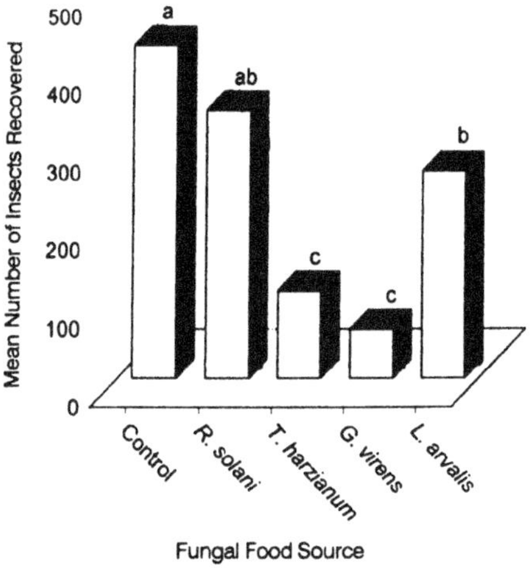

FIGURE 4. Reproductive capacity of Collembola (*Proisotoma minuta*) in cornmeal-soil monocultures of *Rhizoctonia solani* and three biocontrol fungi.

That grazing preferences extend to mycorrhizal symbionts was confirmed in laboratory experiments by Shaw (1985); *Lactarius rufus* was most highly preferred by *O. armatus* in tests with eight mycorrhizal and saprophytic fungi. Three of the least preferred fungi (*Hebeloma crustiliniforme, Paxillus involutus*, and *Rhizopogan luteolus*) are known to contain toxic compounds. Ectomycorrhizal fungi *Laccaria laccata, Pisolithus tinctorius, Suillius luteus, and Thelephora terrestris*, along with *R. solani*, were all grazed by *Proisotoma minuta* in vitro but *R. solani* was grazed more extensively than any of the other fungi (Hiol Hiol *et al.*, 1994).

Microarthropods, especially the Collembola, also are believed to affect the distribution and density of vesicular-arbuscular mycorrhizal (VAM) fungi (Rabatin and Stinner, 1988). Studies by Warnock *et al.* (1982) suggest that the collembolan *Folsomia candida* selectively consumes *Glomus fasciculatum*. In other feeding experiments (Moore *et al.*, 1985) with collembolans *F. candida, Onychiurus encarpatus, O. folsomi*, and *P. minuta*, only *F. candida* grazed the germ tubes of *G. fasciculatum* on aqueous agar, whereas all collembolans fed on *Gigaspora rosea*; none of the animals fed on hyphae of *Glomus mosseae* or *Gigaspora margarita*. Ponge (1991) cautioned that, while observations have revealed food preferences between species or groups of species occupying the same micro-habitat, each collembolan species within the habitat may not be specialized in feeding habits.

C. Biological control potential

Information derived from studies on feeding habits of Collembola led to a continuing investigation at the Alabama Agricultural Experiment Station to determine the biological control potential of laboratory reared populations of *Proisotoma minuta* and *Onychiurus encarpatus*.

Control of *R. solani* on cotton was significant with an animal population (mixed collembolan species) of 2400/kg in sterilized soil infested with pathogen mycelium at

0.01 g/kg (Curl, 1979); a higher population was required for suppression of higher inoculum densities. Subsequent tests in greenhouse pots with non-sterilized field soil supplemented with *Rhizoctonia* (chopped, infested oat grains) at 0.02 g/kg showed that collembolan populations of 1000 or 2000/kg were effective. Bollen *et al.* (1991) observed that mesofauna suppression of Rhizoctonia disease of potato sprouts was less effective under conditions of high inoculum pressure. Results of studies in Japan (Nakamura *et al.*, 1991) showed that *Folsomia hidakana* grazed and multiplied on *R. solani* and, when introduced into pots of soil, suppressed seedling diseases of radish, cucumber, cabbage and burdock.

Investigations in the Federal Republic of Germany (El Titi and Ulber, 1991) have extended collembolan (*Onychiurus fimatus*) suppression of root disease to include *Pythium ultimum* on sugar beet seedlings. Application of an insecticide to experimental plots resulted in reduced seedling emergence and increased infestation with *Pythium*, attributed to the exclusion of mycophagous Collembola. In Japan (Nakamura *et al.*, 1992), the collembolan *Sinella curviseta* from rhizospheres of vegetable crops grazed and multiplied on *Fusarium oxysporum* f. sp. *cucumerinum* and suppressed infection of cucumber.

D. Integration of biocontrol agents

Combinations of two or more biological control agents, which are compatible at the root-soil interface, should provide maximum protection from pathogen activity. Bollen *et al.* (1991) used combinations of mycophagous nematodes, springtails and mites from potato fields to demonstrate control of *R. solani* on potato sprouts in pots of sandy soil. A combination of the nematode *Aphelenchus avenae* and the collembolan *Folsomia fimetaria* was more effective than either alone in disease suppression. Similarly, *A. avenae* and *Tullbergia krausbaveri* together, or these two combined with mite species, provided superior disease control.

Collembolan *Proisotoma minuta* and the known biocontrol fungus *Laetisaria arvalis* were evaluated for suppression of *R. solani* on cotton seedlings in greenhouse tests (Lartey *et al.*, 1991). The combination of agents increased the pathogen-suppressive benefit in either sterilized or non-sterilized soil compared with either agent used alone (Table 1); the benefit increased further as numbers of animals applied were increased. Continuing studies of this type (Lartey *et al.*, 1994) also have demonstrated added benefit of combining *P. minuta* populations with *Trichoderma harzianum* or *Gliocladium virens* for control of *R. solani*; however, the *L. arvalis* + *P. minuta* treatment was most effective. The specific mechanisms that promote increased biocontrol capacity of mycophagous fauna + fungus combinations are not yet clearly defined.

V. Disease control by other fauna

While the foregoing focus has been upon the activities of amoebae, nematodes and microarthropods, other small animals which collectively contribute to soil mineralization processes may occur in large numbers in various cropping systems and indirectly affect plant pathogens (El Titi and Ulber, 1991). In this regard, relatively little is known about the Enchytraeidae (pot worms) and Lumbricidae (earthworms). Stephens *et al.* (1993) demonstrated the potential of the earthworm *Aporrectodea trapezoides*, applied to soil at the equivalent of 471m^{-2}, to reduce the root-disease rating of *R. solani* on wheat in Australia. Suppression of *Gaeumannomyces graminis* var. *tritici* on wheat (Stephens *et al.*, 1994) and *Plasmodiophora brassicae* on cabbage seedlings (Nakamura *et al.*, 1995) using earthworm species also has been demonstrated.

TABLE 1. Evaluation of *Laetisaria arvalis* and *Proisotoma minuta* (Collembola) used alone or combined to control *Rhizoctonia solani* on cotton seedlings in greenhouse tests (selected data from Lartey *et al.*, 1994).

Treatments	Disease index	
	Sterilized soil (RS = 10 mg/kg)	Nonsterilized soil (RS = 150 mg/kg)
O (Untreated soil)	0.48	0.91
RS	4.81	5.00
RS + LA	1.73	3.67
RS + PM	1.44	3.31
RS + LA + PM	0.23	1.42

RS = *Rhizoctonia solani* applied at two inoculum densities.
LA = *Laetisaria arvalis* applied as seed treatment.
PM = *Proisotoma minuta* applied at 1000/kg soil.
All values were significantly different from RS control (P=0.05, Dunnett's Test).

The larval forms of some macrofauna also may qualify, by feeding habits, as potential pathogen suppressants. Larvae of the sciarid fly (*Pnyxia scabiei*) has been found colonizing buried cotton-stem segments along with Collembola and mites (Benham and Hancock, 1980); none of the groups seemed implicated in reducing numbers of microsclerotia of *Verticillium dahliae*. Naito and Sugimoto (1994) observed that, when larvae-infested residues from decomposing sugar beet roots were added to soil with *R. solani*, root rot disease declined. When *P. scabiei* was partially eradicated by insecticide treatment, sclerotial numbers and disease severity of sugar beet increased. Naito and Makino (1995) counted up to 5,000 larvae of the sciarid fly per diseased sugar beet root infected with *R. solani*; the larvae were observed feeding on sclerotia and hyphae and were demonstrated to reduce damping-off of sugar beet.

VI. ADVERSITY FACTORS

The value of mycophagous fauna as potential biological control agents must be assessed with caution in view of certain adverse effects that may occur. Like fungal and bacterial agents, some species of mycophagous invertebrates may cause harmful effects to specific crops under conducive environmental conditions. A commonly cited adversity is the feeding by fauna species on mycorrhizal fungi resulting in reduced root colonization and function of the symbionts. Mycophagous amoebae (Chakraborty *et al.*, 1985) and nematodes, *Aphlenchoides* spp. (Riffle, 1975), have been implicated in the reduced colonization of pine roots by ectomycorrhizal fungi, *Rhizopogon luteolus* and *Suillus granulatus*. The nematode *Aphelenchus avenae*, used in relatively high populations (Hussey and Roncadori, 1981), suppressed cotton-plant response to endomycorrhizal fungi. Collembolans *Folsomia candida, Onychiurus encarpatus, O. folsomi* and *Proisotoma minuta* have been implicated in altering the establishment and performance of the VAM fungus *Gigaspora rosea* on leek seedlings (Warnock *et al.*, 1982). Findlay (1985) has cited a number of other examples of mycorrhiza-fauna interactions.

Root injury by some species of Collembola is believed, for the most part, to occur through indiscriminate feeding where preferred alternative food sources are not available (El Titi and Ulber, 1991; Curl *et al.*, 1988). However, certain species of *Onychiurus* in particular have been recognized as pests on sugar beets (Dunning and Baker, 1977; Ulber, 1980; Brown 1985) and seedlings of several other crops (Getzin, 1985), requiring serious chemical control measures.

VII. CONCLUSIONS AND CHALLENGES

Fungus-feeding amoebae, nematodes, and microarthropods collectively provide an enormous potential for natural suppression of inoculum density and disease incidence/severity. While laboratory and greenhouse experiments have demonstrated that specific populations of fungus-grazing animals can reduce inoculum density and disease in controlled environments, a more meaningful biological control might be induced through the management of indigenous populations, as visualized by Cook (1991) for the microflora. Soil invertebrate populations, like the microflora, respond to pesticides, soil fertility, organic matter, cultivation practices and crop species, thus affording opportunities for agroecosystem manipulation to promote fauna and crop benefit. Advanced technology for *in situ* observation of mesofauna in the rhizosphere (Lussenhop *et al.*, 1991; Lussenhop, 1992) has improved the reliability of population estimates in natural environments, essential for monitoring changes in indigenous fauna activities.

New challenges lie ahead for the exploitation of the "friendly fauna" for crop benefit. *R. solani* in particular seems to represent a logical prime target for biological control efforts in view of its suitability as a preferred food source for collembolan species. Not only is mycophagy a common trait, but collembolan predation upon plant parasitic nematodes also has been cited (Boosalis and Mankau, 1965). The advances made with mycophagous amoebae and nematodes that feed upon other pathogens should serve to stimulate further research on these groups as well. Fauna activity as a contributing mechanism in the "suppressive vs conducive soils" phenomenon (Schneider, 1982) deserves serious attention, as indicated by the implication of mycophagous amoebae by Old and Chakraborty (1986) in Australia.

Wilson and Backman (1996) have provided an update of biological control technology and cited critical references to the subject. As the commercial formulation and delivery of biocontrol agents make significant advances, it will become increasingly essential to assess the role of fungal and bacterial feeding fauna in the successes or failures of introduced biocontrol agents.

VIII. ACKNOWLEDGMENTS,

We thank those individuals from various institutions who kindly provided reprints of published investigations. Grateful acknowledgment also is extended to Karen Wood for consultation and invaluable aid in overall manuscript preparation.

IX. References

Anderson TR & Patrick ZA (1978) Mycophagous amoeboid organisms from soil that perforate spores of *Thielaviopsis basicola* and *Cochliobolus sativus*. Phytopathology 68:1618-1626.

Andrén O & Lagerlof J (1983) Soil fauna (microarthropods, enchytraeids, nematodes) in Swedish agricultural cropping systems. Acta Agric. Scand. 33:33-52.

Bamforth SS (1988) Interactions between protozoa and other organisms. In: Edwards CA, Stinner BR, Stinner D & Rabatin S (eds.) Biological Interactions in Soil (pp 229-234) Elsevier, Amsterdam.

Barker KR (1964) On the disease reduction and reproduction of the nematode *Aphelenchus avenae* on isolates of *Rhizoctonia solani*. Plant Dis. Rep. 48:428-432.

Barnes GL, Russell CC, Foster WD & McNew RW (1981) *Aphelenchus avenae*, a potential biological control agent for root rot fungi. Plant Dis. 65:423-424.

Benham GS & Hancock JG (1980) Colonization of buried cotton stems by microarthropods. Soil Biol. Biochem. 12:43-47.

Bollen GJ, Middelkoop J & Hofman TW (1991) Effects of soil fauna on infection of potato sprouts by *Rhizoctonia solani*. In: Beemster ABR *et al.* (eds.) Biotic Interactions and Soilborne Diseases (pp 27-34) Elsevier, Amsterdam.

Boosalis MG & Mankau R (1965) Parasitism and predation of soil microorganisms. In: Barker KF & Snyder W (eds.) Ecology of Soilborne Plant Pathogens (pp 374-391) University of California Press, Berkeley.

Brown RA (1985) Effects of some root-grazing arthropods on the growth of sugar-beet. In: Atkinson D, Read DJ & Usher MB (eds.) Ecological Interactions in Soil: Plants, Microbes and Animals (pp 285-295) Blackwell Scientific Publishers, Oxford.

Cayrol JC & Combettes S (1977) Etudes preliminaires sur les possibilities d'utilisation des nematodes mycophages comme agents de lutte biologique contre les "fontes de semis. Rev. Hortic., No. 180, p. 33.

Chakraborty S & Warcup JH (1985) Reduction of take-all by mycophagous amoebas in pot bioassays. In: Parker CA, Rovira AD, Moore KJ, Wong PTW & Kollmorgan JF (eds.) Ecology and Management of Soilborne Plant Pathogens (pp 107-109) APS Press, St. Paul, MN.

Chakraborty S, Theodorou C & Bowen GD (1985) The reduction of root colonization by mycorrhizal fungi by mycophagous amoebae. Can. J. Microbiol. 31:295-297.

Clarholm M (1981) Protozoan grazing of bacteria in soil: impact and importance. Microb. Ecol. 7:343-350.

Cook RJ (1991) Biological control of plant diseases: Broad concepts and applications. In: Komada H, Kiritani K & Bay-Petersen J (eds.) The Biological Control of Plant Diseases (pp 1-29) Food & Fertilizer Technology Center, Taiwan.

Curl EA (1979) Effects of mycophagous Collembola on *Rhizoctonia solani* and cotton seedling disease. In: Schippers B & Gams W (eds.) Soilborne Plant Pathogens (pp 253-269) Academic Press, London.

Curl EA (1982) The rhizosphere: Relation to pathogen behavior and root disease. Plant Dis. 66:624-630.

Curl EA (1988) The role of soil microfauna in plant-disease suppression. CRC Critical Reviews in Plant Scienes 7:175-196.

Curl EA & Harper JD (1990) Fauna-microflora interactions. In: Lynch JM (ed.) The Rhizosphere (pp 369-388) Wiley, Chichester.

Curl EA & Truelove B (1986) The Rhizosphere. Springer-Verlag, Berlin (p 228).

Curl EA, Gudauskas RT, Harper JD & Peterson CM (1985) Effects of soil insects on populations and germination of fungal propagules. In: Parker CA, Rovira AD, Moore KJ, Wong PTW & Kollmorgen JF (eds.) Ecology and Management of Soilborne Plant Pathogens (pp 20-23) Am. Phytopathol. Soc., St. Paul, MN.

Curl EA, Lartey R & Peterson CM (1988) Interactions between root pathogens and soil microarthropods. Agric. Ecosyst. & Env. 24:249-261.

Dunning RA & Baker AN (1977) Some sugar beet cultural practices in relation to incidence and damage by soil-inhabiting pests. Ann. Appl. Biol. 87:528-532.

El Titi A & Ulber B (1991) Significances of biotic interactions between soil fauna and microflora in integrated arable farming. In: Beemster ABR (ed.) Biotic Interactions and Soil-Bourne Diseases (pp 1-19) Elsevier, Amsterdam.

Findlay RD (1985) Interactions between soil micro-arthropods and endomycorrhizal associations of higher plants. In: Fitter AH, Atkinson D, Read DJ & Usher MB (eds.) Ecological Interactions in Soil: Plants, Microbes and Animals. (pp 319-331) Blackwell Scientific, Palo Alto, CA.

Garrett SD (1970) Pathogenic Root-Infecting Fungi. The University Press, Cambridge.

Getzin LW (1985) Chemical control of the springtail *Onychiurus pseudarmatus* (Collembola: Onychiuridae). J. Econ. Entomol. 78:1337-1340.

Hendrix PF, Parmalee RW, Crossley Jr. DA, Coleman DC, Odum EP & Groffman PM (1986) Detritus food webs in conventional and no-tillage agroecosystems. BioScience 36:374-380.

Hiol Hiol F, Dixon RK & Curl EA (1994) The feeding preference of mycophagous Collembola varies with the ectomycorrhizal symbiont. Mycorrhiza 5:99-103.

Homma Y, Sitton JW, Cook RJ & Old KM (1979) Perforation and destruction of pigmented hyphaae of *Gaeumannomyces graminis tritici* by vampyrellid amoebae from Pacific Northwest wheat field soils. Phytopathology 69:1118-1122.

Hussey RS & Roncadori RW (1981) Influence of *Aphelenchus avenae* on vesicular-arbuscular endomycorrhizal growth response in cotton. J. Nematology 13:48-52.

Lartey RT, Curl EA & Peterson CM (1994) Interactions of mycophagous Collembola and biological control fungi in the suppression of *Rhizoctonia solani*. Soil Biol. Biochem. 26:81-88.

Lartey RT, Curl EA, Peterson CM & Harper JD (1989) Mycophagous grazing and food preference of *Proisotoma minuta* (Collembola: Isotomidae) and *Onychiurus encarpatus* (Collembola: Onychiuridae). Environ. Entomol. 18:334-337.

Lartey RT, Curl EA, Peterson CM & Williams JC (1991) Control of *Rhizoctonia solani* and cotton seeding disease by *Laetisaria arvalis* and a mycophagous insect *Proisotoma minuta* (Collembola). Journal of Phytopathology 133:89-98.

Lussenhop J (1992) Mechanisms of microarthropods-microbial interactions in soil. Adv. Ecol. Res. 23:1-33.

Lussenhop J, Fogel R & Pregitzer K (1991) A new dawn for soil biology: video analysis of root-soil-microbial-faunal interactions. Agriculture, Ecosystems & Environment 34:235-249.

Mankau R & Mankau SK (1963) The role of mycophagous nematodes in the soil. I. The relationships of *Aphelenchus avenae* to phytopathogenic fungi. In: Doeskin J & Van der Drift J (eds.) Soil Fauna, Soil Microflora and their Relationships. (pp 271-280) North Holland, Amsterdam.

McMillan JH (1976) Laboratory observations on the food preference of *Onychiurus armatus* (Tullb.) Gisin (Collembola, Family Onychiuridae). Revue d'Ecologie at de Biologie du Sol 13:353-364.

Mitchell JE (1979) The dynamics of the inoculum potential of populations of soil-borne plant pathogens in the soil ecosystem. In: Schippers B & Gams W (eds.) Soil-borne Plant Pathogens (pp 3-20) Academic Press, London.

Moore Jc, St. John TV & Coleman DC (1985) Ingestion of vesicular-arbuscular mycorrhizal hyphae and spores by microarthropods. Ecology 66:1979-1981.

Naito S & Sugimoto T (1994) Some properties of disease suppresiveness induced by root-rotten sugarbeet. Ann. Rept. Plant Prot. North Japan 45:38-42.

Naito S & Makino S (1995) Control of Sclerotia of *Rhizoctonia solani* by a Sciarid fly, Pnyxia scabiei, in soil. JARQ 29:31-37.

Nakamura Y, Itakura J & Matsuzaki I (1991) Mycophagous meso soil animals from cropfields in Fukishima Pref. (In Japanese with English abstract). Edaphologia 45:49-54.

Nakamura Y, Matsuzaki I & Itakura J (1992) Effects of grazing by *Sinella curviseta* (Collembola) on *Fusarium* oxysporum f. sp. *cucumerinum* causing cucumber disease. Pedobiologia 36:168-171.

Nakamura Y, Itakura J & Matsuzaki I (1995) Influence of the earthworm *Pheretima hilgendorfi* (Megascolecidae) on *Plasmodiophora brassicae* clubroot galls of cabbage seedlings in pots. Edaphologia 54:39-41.

Nicholas WL (1984) The Biology of Free-Living Nematodes, 2nd ed. Clarendon Press, Oxford, 235 pp.

Old KM & Chakraborty S (1986) Mycophagous soil amoebae: Their biology and significance in the ecology of soil-borne plant pathogens. Progress in Protistology. Vol. 1 (163-194).

Ponge JF (1991) Food sources and diets of soil animals in a small area of Scots Pine litter. Geoderma 49:33-62.

Rabatin SC & Stinner BR (1988) Indirect effects of interactions between VAM fungi and soil-inhabitating invertebrates on plant processes. Agriculture, Ecosystems & Environment 24:135-146.

Riffle JW (1975) Two *Aphelenchoides* species suppress formation of *Suillus granulatus* ectomycorrhizae with *Pinus ponderosa* seedlings. Plant Dis. Reporter 59:951-955.

Rossner J & Nagel S (1984) Untersuchungen zur Okologie und vermehrung des mycohagen nematoden *Aphelenchoides hamatus Nematologica*, 30:90-98.

Schneider RW (ed.) (1982) Suppressive Soils and Plant Disease. The American Phytopathological Society, St. Paul, Minnesota.

Shaw PJA (1985) Grazing preferences of *Onychiurus armatus* (Insecta: Collembola) for mycorrhizal and saprophytic fungi of pine plantations. In: Fitter AH, Atkinson D, Read DH & Usher MB (eds.) Ecological Interactions in Soil: Plants, Microbes and Animals (pp 333-337) Blackwell Scientific, Oxford, England.

Shaw PJA (1988) A consistent hierarchy in the fungal preferences of the Collembola *Onychiurus armatus*. Pedobiologia 31:179-187.

Snider RJ, Snider RM & Smucker AJM (1990) Collembola populations and root dynamics in Michigan agroecosystems. In: Box JE & Hammond LC (eds.) Rhizosphere Dynamics. AAAS Selected Symposium 113:168-191.

Stephens PM, Davoren CW, Doube BM, Ryder MH, Benger AM & Neate SM (1993) Reduced severity of *Rhizoctonia* disease on wheat seedlings associated with the presence of the earthworms *Aporrectodea trapezoides* (Lumbricidae). Soil Biol. Biochem. **25**:1477-1484.

Stephens PM, Davoren CW, Doube BM & Ryder MH (1994) Ability of the lumbricid earthworms *Aporrectodea rosea* and *Aporrectodea trapezoides* to reduce the severity of take-all under greenhouse and field conditions. Soil Biol. Biochem. 26:1291-1297.

Ulber von B (1980) Untersuchungen zur Nahrungswahl von *Onychiurus fimatus* Gisin (Onychiuridae, Collembola), einer Aufgangsschädling der Zuckerrübre. Z. Angew. Entomol. 90:333-346.

Visser S & Whittaker JB (1977) Feeding preference for certain litter fungi by *Onychiurus subtenuis* (Collembola) Oikos 29:320-325.

Warnock AJ, Fitter AH & Usher MB (1982) The influence of a springtail *Folsomia candida* (Insecta, Collembola) on the mycorrhizal association of Leek *Allium porrum* and the vesicular-arbuscular mycorrhizal endophyte *Glomus fasciculatus*. New Phytol. 90:285-292

Wiggins EA & Curl EA (1979) Interactions of Collembola and microflora of cotton rhizosphere. Phytopathology 69:244-249.

Wiggins EA, Curl EA & Harper JD (1979) Effects of soil fertility and cotton rhizosphere on populations of Collembola. Pedobiologia 19:75-82.

Wilson M & Backman PA (1996) Biological control of plant pathogens. In: Ruberson JR (ed.) Handbook of Pest Management (In press) Marcel Dekker, New York.

VI.B7 INTRODUCED BIOCONTROL AGENTS TO SUPPRESS DISEASES CAUSED BY *RHIZOCTONIA*

JACK A. LEWIS[1] AND MARTIN M. KULIK[2]

[1]*Biocontrol of Plant Diseases Lab.*, [2]*Soybean and Alfalfa Research Laboratories Agricultural Research Service U. S. Department of Agriculture Beltsville, Maryland, USA 20705*

I. Introduction

The application or augmentation of preparations of biocontrol agents to suppress diseases caused by soilborne plant pathogens generally, and species of *Rhizoctonia* specifically, has received considerable attention during the past 18 years with regard to field or horticultural greenhouse applications. However, due to the space constraints imposed on this manuscript, certain limitations were placed upon papers cited in this brief review. Consequently, it was decided only to include information which has been published within the past 10 years. Apologies are given to investigators who published earlier work or to those whose work was inadvertently omitted. Much of this earlier information, specifically with reference to introduced biocontrol agents to control Rhizoctonia diseases, can be obtained from articles in recent books edited by Mukerji and Garg (1988), Baker and Dunn (1990), Hornby (1990), Beemster *et al.* (1991), Tjamos *et al.* (1992), and Lumsden and Vaughn (1993). This chapter is divided into the following sections: introduction, application of formulations prepared with biocontrol agent biomass, use of bulk formulations and infested grains, treatment of seeds and soils with fungal spores and bacterial cells, use of novel biocontrol agents, and conclusions. A chapter dealing with integrated control is presented in section VI.E.

II. Formulations with biomass of biocontrol fungi

The most recent advances in technology to control fungi that cause soilborne diseases involve potential commercial formulations which are effective, stable, easy to apply, safe, and acceptable to growers. Invariably, these formulations consist of powders or granules which contain biomass, abundant in propagules, of the biocontrol agents which are effective against plant pathogens. These formulations have to be applied to the appropriate niche at the proper time. In the industrialized countries, biomass is preferably prepared by deep-tank liquid fermentation (Papavizas et al, 1984; Mintz and Walter, 1993). This method is adapted to industrial conditions and uses readily-available, inexpensive substrates. After the incubation period, the biomass is then separated from the spent medium, dried, and milled.

The biomass powder can be mixed with a clay such as Pyrax® or formulated into other products (granules or prills of alginate, gelatinized starch and flour, gluten, or chitosan) for delivery. A Pyrax®/biomass formulation of isolates of *Trichoderma viride* Pers. ex Fries and *Gliocladium virens* Miller, Giddens & Foster applied as dusts to potato (*Solanum tuberosum* L.) seed pieces infested with sclerotia of *R. solani* Kühn before planting reduced disease incidence in the field by 50 and 55%,

B. Sneh et al. (eds.),
Rhizoctonia Species: Taxonomy, Molecular Biology, Ecology, Pathology and Disease Constrol, 507–514.

respectively (Beagle-Ristaino and Papavizas, 1985). In addition, the viability of sclerotia from tubers retrieved from the field was reduced by 54 to 89%.

Encapsulated or immobilized formulations are receiving increased attention (Scott, 1987). Research in this area is in the preliminary stages and is mostly concerned with laboratory and greenhouse research. However, commercial and field applications are being reported. The first commercially produced alginate prill to receive Environmental Protection Agency (EPA) registration for use in the United States was the result of a cooperative effort between the USDA Biocontrol of Plant Diseases Laboratory (BPDL) and W. R. Grace & Co. This product (GlioGard™) is an air-dried alginate prill consisting of biomass with abundant chlamydospores of an isolate (G1-21) of the biocontrol fungus *G. virens*, a food base (wheat [*Triticum aestivum* L.] bran) and a carrier/binder (sodium alginate). The formulation has been effective in commercial, horticultural greenhouses as a supplement to soilless mix to reduce damping-off diseases (Knauss, 1992).

In horticultural potting mixes, delivery of various isolates of *G. virens* as alginate prill reduced Rhizoctonia damping-off of zinnia (*Zinnia elegans* Jacq.), snapdragons (*Antirrhinum majus* L.), and other ornamentals in commercial greenhouses at various locations (Lumsden and Locke, 1989: Lumsden *et al.*, 1990). In addition, it was recently shown that alginate prills containing various food bases and biomass of isolate G1-21 of *G. virens* also prevented damping-off of zinnia caused by the pathogen (Lewis *et al.*, 1996). The application of biomass of isolate TRI-4 of *T. hamatum* (Bonord.) Bain and isolates G1-3 and G1-21 of *G. virens* in pregelatinized starch/corn flour granules (Dunkle and Shasha, 1988) was also shown, for the first time, to prevent Rhizoctonia damping-off of zinnia, eggplants (*Solanum melongena* L.), and peppers (*Capsicum anuum* L.) as transplants in soilless mix (Lewis *et al.*, 1995).

III. Grain and Bulk Formulations

Over the years, for laboratory and greenhouse research on biocontrol, mass production of inoculum for the application of biocontrol agents continues to emphasize solid (semi-solid) fermentations of various grains and agricultural by-products such as bran and straw which are inoculated with a biocontrol agent (especially a fungus) and allowed to incubate axenically for various periods of time. This system is particularly useful for tests which require minimum facilities for implementation. Solid fermentations are also useful for fungi which do not sporulate in liquid culture. It is also important to note that most biocontrol agents used in these preparations were isolates of *Trichoderma* spp. and *G. virens*. Although this approach may be appropriate in developing countries where agricultural wastes are available, elaborate facilities are limited, and labor is abundant (Carrizales and Jaffe, 1986), some commercial companies such as Sylvan of Kittanning, Pennsylvania, USA, are becoming more interested in this approach for application of biocontrol agents.

Wheat bran supplemented with peat and sawdust is still used to the greatest extent. Bran/peat substrates each containing 2-week old inoculum of various isolates of *Trichoderma* or *Gliocladium* were added to a potting mix infested with *R. solani* (Maplestone *et al.*, 1991). Six of the eight antagonists decreased damping-off of lettuce (*Lactuca sativa* L.) transplants caused by the pathogen and preparations of three of the isolates increased yield consistently compared to that of the pathogen-infested control. In a commercial test in England with lettuce, the crop was grown in field subplots infested with *R. solani* under polyethylene tunnels (Coley-Smith *et al.*, 1991). Three- and 14-day-old cornmeal/Perlite® preparations of a *T. viride* isolate were applied, but biocontrol of bottom rot caused by *R. solani* was poorer than that obtained with the fungicide tolclofos-methyl. However, the enhanced growth of the lettuce induced by treatment with *T. viride* produced marketable yields similar to those obtained with the fungicide. In Austria, using bran preparations of an isolate of *T. viride* added to fields, root rot of lettuce caused by *R. solani* was reduced, but the effect was not as great as that achieved with various seed-treatment fungicides (Bedlan, 1988).

In a study to evaluate the effect of an isolate of *T. harzianum* on tomato (*Lycopersicon esculentum* Mill.) fruit rot caused by *R. solani*, 8-day old bran/peat preparations of the antagonist were applied to fields before planting or 20 to 70 days after planting (Strashnov *et al.*, 1985). The most effective treatment was application of the preparation 70 days after planting. The disease incidence was only 12.2%, but was significantly less than the 18.3% Rhizoctonia disease incidence in the control plots. Also, a reduction in pathogen inoculum was observed. In a 2-year field experiment at Beltsville to reduce fruit rot of tomatoes, 3-day old bran preparations of actively-growing cultures of isolates of *T. hamatum* and *G. virens* were used. Disease incidence was not reduced from the 50% level of the control plots (Lewis *et al.*, 1990). Nevertheless, bran preparations of all the isolates significantly reduced the pathogen inoculum density. Bran preparations with 3-day-old actively-growing mycelium of 18 isolates of *Trichoderma* spp. and *G. virens* were applied to *R. solani*-infested field plots before planting to cotton (*Gossypium hirsutum* L.) seed over a 4-year period (Lewis and Papavizas, 1991). In all 4 years, the preparations containing an isolate of *T. hamatum* and an isolate of *G. virens* significantly prevented damping-off as well as saprophytic activity of the pathogen. In addition, in 3 of the 4 years plant stands (>80%) comparable to that of the noninfested plots were obtained.

As indicated earlier, solid formulations may have more impact when used in Asian and African countries to combat various diseases caused by *Rhizoctonia*. For example, in Egypt, isolates of *T. harzianum* Rifai and *G. penicilloides* Corda were grown on barley (*Hordeum vulgare* L.) straw for 2 weeks, and added to field soils naturally-infested with *R. solani* and planted with beans (*Phaseolus vulgaris* L.) and lentils (*Lens culinaris* Medik.) (Abd-El-Moity, 1992). Two isolates of *T. harzianum* reduced disease incidence to 60 and 42%, respectively, compared to 71% for the control plots. However, when the isolates were added together, disease incidence was reduced to 2%. In cotton fields, *T. harzianum* isolates and *G. penicilloides* were the most effective antagonists against *R. solani* with 66 to 85% reduction in disease, respectively.

In China, a variety of fungal and bacterial isolates were grown on rice (*Oryza sativa* L.) husks and composted sawdust for application to *R. solani*-infested seeding beds to evaluate the biocontrol of basal stem rot of chrysanthemum (*Chrysanthemum morifolium* [Ramat.] Hemsl.) caused by this pathogen (Tshen, 1992). Of the 16 antagonists investigated, isolates of *T. harzianum, Penicillium* sp., and *Bacillus subtilis* ([Ehren.] Cohn) were the most effective in reducing the disease more than 80%. Wheat bran/sawdust preparations of an isolate of *T. longibrachiatum* Rifai and *G. virens* significantly protected peanut (*Arachis hypogaea* L.) from Rhizoctonia disease in fields in India by 65 to 74% compared to the disease in the control plots (Sreenivaprasad and Manibushanrao, 1993). Coating seeds with spore suspensions also reduced root rot incidence, but to a lesser degree. Also, bran/sawdust preparations of these isolates reduced the inoculum density of the pathogen.

IV. Application of Bacterial Cells and Fungal Spores to Seed and Soil

For many years, the application of biocontrol fungi, especially of the genera *Trichoderma, Gliocladium,* and *Penicillium*, to control Rhizoctonia damping-off and root rot of crops has been attempted by using conidial suspensions of these fungi as seed treatments or soil drenches. Although the production of these propagules occurs most readily on agar surfaces, manually scraping conidia from agar is clearly not economically feasible for commercial purposes. Recently however, deep-tank liquid fermentation has been successfully used for the production of large amounts of *Trichoderma* conidia for seed application (Harman, 1990). In contrast, deep-tank fermentation for the commercial production of bacterial vegetative cells and endospores has been done for many years. The use of biological seed treatments to control fungal pathogens of seedlings has recently undergone small-scale commercial development (Campbell, 1989). However, irregular performance and competition with chemical seed dressings is delaying further development of this technology.

Examples of the application of bacterial suspensions in horticultural situations include the use of *Burkholderia cepacia* (Burkh, Yab. *et al.* [=*Pseudomonas cepacia* Burkh.]) suspensions. These are used to soak polyfoam rooting cubes prior to planting with poinsettia (*Euphorbia pulcherrima* Willd.) stems to control Rhizoctonia rot (Cartwright and Benson, 1995a). In some areas of China, *R. solani* (AG 4) appears to be more prevalent than *Pythium* spp. as a pathogen in seedling production systems. Bacterial suspensions of seven strains of various *B. subtilis* and *P. fluorescens* Migula isolates were used to coat seeds of cucumber (*Cucumis sativus* L.) , tomato, pepper, eggplant and Chinese cabbage (*Brassica chinensis* L.) seeds before the seeds were sown in starter beds used to produce seedlings for transplanting to the field. Metalaxyl and zineb were used as fungicide controls. Six of the seven bacterial isolates reduced Rhizoctonia caused disease of the seedlings comparable to that obtained with the fungicides (Zhang *et al.*, 1990).

In Germany, potato seed pieces were coated with a mixed suspension of cells of *Bacillus* sp., *Streptomyces* sp. and the fungus *G. roseum* Bain (Schmeiderknecht, 1993). Both stem canker and black scurf of tubers caused by *R. solani* were reduced up to 78%. Bacterial cells used to coat potato seed pieces have also been shown to reduce *R. solani* infection of young stems and stolons in Japan (Tanii *et. al.*, 1990). Isolates of unidentified, gram-negative rods as well as isolates of *B. cereus* Frank. & Frank. were particularity effective in reducing black scurf of tubers as well as damage to other vegetative parts of the plants.

In studies on *R. solani* of cotton in Virginia, suspensions of strains of *P. fluorescens* were applied to seed along with peat and planted at various locations (Hagedorn *et al.*, 1993). In addition, suspensions of the bacteria were applied to the seed furrows. Although damping-off was controlled, there was lack of consistency between locations and years. Also, seedling stand was not correlated with cotton yield. In-furrow application of the biocontrol agents was superior to seed treatment for disease control. In another field study, a strain of *B. subtilis* applied to peanut seed colonized the roots of the developing plants, reduced cankers caused by *R. solani*, improved seed germination and plant stand, increased nodulation by *Rhizobium* spp. and increased root growth (Turner and Backman, 1991).

In India, rice seeds were treated with suspensions of strains of *P. putida* (Trev.) Migula , *P. fluorescens, Bacillus* sp. or the fungicide validamycin as a control (Thara and Gnanamanickam, 1994). The seeds were sown in seed flats before transplanting to a field infested with *R. solani*. Two strains of *P. putida* and *P. fluorescens* were the most efficient in controlling rice sheath blight and resulted in 68 and 52% disease control, respectively, whereas the fungicide gave 27% disease control. Bacterial seed treatments controlled sheath blight in direct-seeded as well as transplanted rice (Gnanamanickam *et al.*, 1992). However, control was more pronounced in direct-seeded rather than transplanted seedlings. In addition, in three field tests, treatment with a *P. fluorescens* isolate reduced the severity of sheath blight by 20 to 42% with five rice cultivars. Also, seed bacterization increased plant height, number of tillers, and grain yield by 3 to 160%.

The application of fungal spores in selected situations has also continued to be evaluated in recent years. Biocontrol of poinsettia stem rot caused by *R. solani* in rooting cubes using a spore suspension of *Paecilomyces lilacinus* (Thom) Sams. has been demonstrated (Cartwright and Benson, 1995b). In fields in Nigeria, the foliage of field-grown cowpea (*Vigna unguiculata* [L.] Walp.) was inoculated with an isolate of *R. solani* that caused web blight. Twice thereafter, the plants were sprayed with a spore suspension of a *T. koningii* Ouden. isolate (Latunde-Dada, 1991). The spread of web blight was restricted and the yield levels of cowpea were significantly higher in the antagonist-treated plants than in the control plants. Moreover, the degree of control was similar to that achieved with a fungicide mixture containing benomyl. Spores of three isolates of *T. viride* were used to coat sesame (*Sesamum indicum* L.) seed in South Korea to evaluate the effectiveness of the biocontrol agents to control damping-off caused by *R. solani* in the field (Chung and Choi, 1990). When seeds were treated with spores or benomyl, seedling emergence was significantly increased compared with the nontreated control. In addition, the incidence of post-emergence damping-off as well as the inoculum density of the pathogen were decreased.

V. NOVEL MICROORGANISMS FOR THE CONTROL OF *R. SOLANI*

Although most application and augmentation technology for the control of Rhizoctonia diseases in the field and in potting mixes involves common soil fungi (e.g. *Trichoderma, Gliocladium*) and bacteria (e.g. *Bacillus, Pseudomonas*), recent research is elucidating the impact of little-known or difficult to isolate microorganisms for their potential in biocontrol of plant diseases. To date, most studies with these microorganisms have been limited to laboratory and greenhouse research and therefore will not be covered in this chapter.

Foremost among these microorganisms is the use of nonpathogenic, hypovirulent strains of *R. solani* and binucleate *Rhizoctonia* spp. to control diseases caused by pathogenic strains of *R. solani*. Research on the interactions between these groups of microorganisms and pathogenic isolates of *Rhizoctonia* is discussed in another chapter of this book. Over many years in The Netherlands, researchers have extensively investigated the application of the mycoparasite *Verticillium biguttatum* Gams, alone, or in combination with other biocontrol agents, for the reduction of the impact of *R. solani* on inciting disease of potatoes. The recent developments in these investigations are discussed at length in another chapter of this book. At this point, it is sufficient to indicate that formulations of these biocontrol agents are being developed for field applications.

Laetisaria arvalis Burdsall, a basidiomycete similar to *R. solani*, has also received some attention as a potential biocontrol fungus for the reduction of diseases caused by the pathogen in the field. This antagonist competes well for the ecological niches occupied by the pathogen. *Laetisaria arvalis* was applied as biomass containing sclerotia in a fluid gel to reduce rot of peppers caused by the pathogen in the field (Conway, 1986). When the antagonist was grown on rye (*Secale cereale* L.) seed and added to potato fields infested with *R. solani*, or applied as culture-grown mycelium to seed pieces, 5 years of field studies indicated that these treatments reduced infection by this pathogen of potato stems and stolons equal to that of a chemical treatment (Murdoch and Leach, 1993). In addition, sclerotia formation on the tubers was reduced. An isolate of *L. arvalis*, added to soil as a sand cornmeal inoculum reduced root-rot of corn (*Zea mays* L.) caused by *R. solani* (Sumner and Bell, 1994). However, *L. arvalis* isolates grown on various grains did not significantly prevent the occurrence of Rhizoctonia diseases of field-grown sugar beets (*Beta vulgaris* L.) caused by *R. solani* (Herr, 1988).

TABLE 1. Developed formulations to suppress Rhizoctonia diseases

Product	Antoganist	Host	Application	Source
Kodiak	*Bacillus subtilis*	beans, cotton, peanuts, soybeans	seed treatment	Gustafson, Inc. Plains, Texas, USA 75093
EPIC	*Bacillus subtilis*	cotton, legumes	seed treatment	Gustafson, Inc.
Blue Circle	*Burkholderia cepacia*	corn, melons, cotton, beans,	seed treatment	Stine Microbials Madison, Wisconsin, USA 53704
Intercept	*Burkholderia cepacia*	corn, melons,	seed treatment	Soil Technologies Fairfield, Iowa USA 52556
Anti-Fungus	*Trichoderma* spp.	lettuce	drench	Grondontsmettinger De Ceuster B-2860 St. Katalijn, Belgium
GlioGard	*Gliocladium virens*	bedding plants	prill mix	W. R. Grace-Conn. Columbia, Maryland USA 21044
SoilGard	*Gliocladium virens*	bedding plants	granules	W.R.Grace-Conn.
T-22G T-22HB	*Trichoderma harzianum*	turf	granules/spray	TGT, Inc. Geneva, New York USA 14456
Trichodex	*Trichoderma harzianum*	tomatoes, strawberries	drench/spray	Machteshim, Ltd. Beer Sheva, Israel

VI. CONCLUSIONS

Biological control of soilborne plant pathogens, including *Rhizoctonia*, is becoming a reality. However, it is doubtful that this approach alone, rather than its incorporation within an integrated system, will satisfactorily eliminate all disease. Moreover, the erratic and inconsistent results in fields have limited the introduction and acceptance of commerical biocontrol agents. These problems are due mainly to climatic and edaphic variables. Other obstacles to the implementation of biocontrol is the cultivation of an appropriate, effective biocontrol agent and the development of commercial products. Products currently available that are specific for the control of Rhizoctonia-caused diseases are listed in Table 1. Moreover, the number of these commercial products increases periodically. There is no doubt that the list will grow as the technological advances for product development increase.

This chapter has been an attempt to at least indicate the various biocontrol approaches followed in an attempt to solve a particular plant disease problem. It is hoped that it will be useful and will provide a stimulus for biocontrol research into the next century

VII. References

Abd-El-Moity TH (1992) The use of *Trichoderma* spp. to control soilborne plant pathogens in Egypt. In: Tjamos EC, Papavizas GC & Cook RJ (eds.) Biological Control of Plant Diseases: Progress and Challenges for the Future. (pp 255-265) Plenum Press, New York, NY.

Baker RR & Dunn PE (1990) New Directions in Biological Control. Alan R. Liss, Inc. New York, NY.

Beagle-Ristanio J & Papavizas GC (1985) Biological control of Rhizoctonia stem canker and black scurf of potato. Phytopathology 75:560-564.

Bedlan G (1988) Der Einsatz von *Trichoderma viride* Pers. gegen *Rhizoctonia solani* Kühn an Salat im Freiland. Pflanzenschutzberichte 49:27-33.

Beemster ABR, Bollen GJ, Gerlagh M, Russin MA, Schippers B & Tempel A (1991) Biotic Interactions and Soilborne Diseases. Elsevier Publishers, New York, NY.

Campbell R (1989) Biological Control of Microbial Plant Pathogens. Cambridge University Press, Cambridge, UK.

Carrizales V & Jaffe W (1986) Solid state fermentation: An appropriate technology for developing countries. Interscience 11:9-15.

Cartwright DK & Benson DM (1995a) Comparison of *Pseudomonas* species and application techniques for biocontrol of Rhizoctonia stem rot of poinsettia. Plant Dis. 79:301-308.

Cartwright DK & Benson DM (1995b) Optimization of biological control Rhizoctonia stem rot of poinsettia by *Paecilomyces lilacinus* and *Pseudomonas cepacia*. Plant Dis. 79:309-313.

Chung HS & Choi WB (1990) Biological control of sesame damping-off in the field by coating seed with antagonistic *Trichoderma viride*. Seed Sci. Technol. 18:451-459.

Coley-Smith JR, Ridout CJ, Mitchell CM & Lynch JM (1991) Control of bottom rot disease of lettuce (*Rhizoctonia solani*) using preparations of *Trichoderma viride, T. harzianum* or tolclofos-methyl. Plant Pathol. 40:359-366.

Conway KE (1986) Use of fluid-drilling gels to deliver biological control agents to soil. Plant Dis. 70:835-839.

Dunkle RL & Shasha BS (1988) Starch-encapsulated *Bacillus thuringiensis:* A potential new method for increasing environmental stability of entomopathogens. Environ. Entomol. 17:120-126.

Gnanamanickman SS, Candole BL & Mew TW (1992) Influence of soil factors and cultural practice on biological control of sheath blight of rice with antagonistic bacteria. Plant and Soil 144:67-75.

Hagedorn C, Gould WD & Bardinelli TR (1993) Field evaluations of bacterial inoculants to control seedling disease pathogens on cotton. Plant Dis. 77:278-282.

Harman GE (1990) Deployment tactics for biocontrol agents in plant pathology. In: Baker RR & Dunn PE (eds.) New Directions in Biological Control. (pp 779-792) Alan R. Liss, Inc., New York, NY.

Herr L (1988) Biocontrol of Rhizoctonia stem and root rot of sugar beet by binucleate *Rhizoctonia* spp. and *Laetisaria arvalis*. Ann. Appl. Biol. 113:107-118.

Hornby D (1990) Biological Control of Soil-borne Plant Pathogens. CAB International, Wallingford, UK.

Knauss JF (1992) *Gliocladium virens*, a new microbial for control of *Pythium* and *Rhizoctonia*. Florida Foliage 18:6-7.

Latunde-Dada AO (1991) The use of *Trichoderma koningii* in the control of web blight disease caused by *Rhizoctonia solani* in the foliage of cowpea (*Vigna unguiculata*). J. Phytopathol. 133:247-254.

Lewis JA, Barksdale TH & Papavizas GC (1990) Greenhouse and field studies on the biological control of tomato fruit rot caused by *Rhizoctonia solani*. Crop Prot. 9:8-14.

Lewis JA, Lumsden RD, Fravel DR & Shasha BS (1995) Application of pregelatinized starch flour granular formulations of biocontrol fungi to control damping-off diseases caused by *Rhizoctonia solani*. Biological Control 5:397-404.

Lewis JA, Lumsden RD & Locke JC (1996) Damping-off diseases caused by *Rhizoctonia solani* and *Pythium ultimum* as affected by alginate prills with biomass of biocontrol fungi and various organic food bases. Biocontrol Science and Technology (in press).

Lewis JA & Papavizas GC (1991) Biocontrol of cotton damping-off caused by *Rhizoctonia solani* in the field with formulations of *Trichoderma* spp. and *Gliocladium virens*. Crop Prot. 10:396-402.

Lumsden RD & Locke JC (1989) Biological control of damping-off caused by *Pythium ultimum* and *Rhizoctonia solani* with *Gliocladium virens* in soilless mix. Phytopathology 79:361-366.

Lumsden, RD, Locke JC, Lewis JA, Johnston SA, Peterson JL & Ristaino JB (1990) Evaluation of *Gliocladium virens* for biocontrol of Pythium and Rhizoctonia damping-off of bedding plants at four greenhouse locations. Biological and Cultural Tests 5:90.

Lumsden RD & Vaughn JL (1993) Pest Management: Biologically Bassed Technologies. Amer. Chem. Soc., Washington, DC.

Maplestone PA, Whipps, JM & Lynch JM (1991) Effect of peat-bran inoculum of *Trichoderma* species on biological control of *Rhizoctonia solani* in lettuce. Plant and Soil 136:257-263.

Mintz AS & Walter JF (1993) A private industry approach: Development of GlioGard for disease control in horticulture. In: Lumsden RD & Vaughn, JL (eds.) Pest Management: Biologically Based Technologies. (pp 398-403) Amer. Chem. Soc., Washington, D.C.

Mukerji KG & Garg KL (1988) Biocontrol of Plant Diseases, Vol. I, II. CRC Press, Boca Raton, FL.

Murdoch CW & Leach SS (1993) Evaluation of *Laetisaria arvalis* as a biological control agent of *Rhizoctonia solani* on white potato. Amer. Potato J. 70:625-634.

Papavizas GC, Dunn MT, Lewis, JA & Beagle-Ristaino, L (1984) Liquid fermentation technology for experimental production of biocontrol fungi. Phytopathology 74:1171-1175.

Schmiederknecht G (1993) Biologische Bekämpfung von *Rhizoctonia solani* Kühn an Kartoffelpflanzen durch mikrobielle Antagonisten. Archiv für Phytopathologie und Pflanzenschutz 28:311-320.

Scott CD (1987) Immobilized cells: a review of recent literature. Enzyme Microb. Technol. 9:66-73.

Sreenivasaprasad S & Manibushanao K (1993) Efficacy of *Gliocladium virens* and *Trichoderma longibrachiatum* as biological control agents of groundnut root and stem rot diseases. Int. J. of Pest Management 39:167-171.

Strashnov Y, Elad Y, Sivan A, Rerdick Y & Chet I (1985) Control of *Rhizoctonia solani* fruit rot of tomatoes by *Trichoderma harzianum* Rifai. Crop Prot. 4:359-364.

Sumner DR & Bell DK (1994) Survival of *Rhizoctonia* spp. and root diseases in a rotation of corn, snap bean, and peanut in microplots. Phytopathology 84:113-118.

Tanii D, Takeuchi T & Horita H (1990) Biological control of scab, black scurf and soft rot on potato by seed tuber bacterization. In: Hornby D (ed.) Biological Control of Soil-borne Plant Pathogens. (pp 143-164) C.A.B. International, Wallingford, UK.

Thara KV & Gnanamanickman SS (1994) Biological control of rice sheath blight in India: Lack of correlation between chitinase production by bacterial antagonists and sheath blight suppression. Plant and Soil 160:277-280.

Tjamos EC, Papavizas GC & Cook RJ (1992) Biological Control of Plant Diseases: Progress and Challenges for the Future. Plenum Press, New York, NY.

Tschen JSM (1992) Biological control of Rhizoctonia diseases by application of antagonists. J. Prot. in the Tropics 9:57-62.

Turner JT & Backman PA (1991) Factors relating to peanut yield increases after seed treatment with *Bacillus subtilis*. Plant Dis. 75:347-353.

Zhang BX, Ge QX, Chen DH, Wang ZY & He SS (1990) Biological and chemical control of root diseases on vegetable seedlings in Zhejiang Province, China. In: Hornby D (ed.) Biological Control of Soil-borne Plant Pathogens. (pp 181-196) C.A.B. International, Wallingford, UK.

VI.C1 AVAILABILITY OF GERMPLASM FOR RESISTANCE AGAINST *RHIZOCTONIA* SPP.

LEE PANELLA AND EARL G. RUPPEL

USDA-ARS Sugarbeet Research Crops Research Laboratory
1701Center Ave. Fort Collins, Colorado 80526 USA

I. INTRODUCTION

Farr and co-workers (1989) list over 500 genera of plants in *Fungi on Plants and Plant Products in the United States* that are hosts to *Rhizoctonia solani* Kühn or any of 21 other species of *Rhizoctonia* that are phytopathogenic. Agricultural, horticultural, ornamental, and some tree species are affected. Twenty-five years ago, Leach and Garber (1970) reviewed resistance to *Rhizoctonia* infection and concluded, "In general, while it has been possible to identify differences among cultivars or selections in susceptibility to *Rhizoctonia* infection, it is extremely rare that a high degree of resistance has been found or produced by selection or breeding within a susceptible host species." Today, the use of resistant cultivars provides effective and adequate protection against the various diseases caused by *Rhizoctonia* spp. In some cases, the use of a cultivar with a moderate level of resistance can be coupled with reduced fungicide applications to provide a sufficient level of disease management. A survey of *Crop Science* revealed a list of 59 cultivar, parental line, and germplasm registrations in nine crop species during last 10 years that have some resistance to diseases caused by *Rhizoctonia* spp. These are included with cultivar releases from other sources in Table 1. All of this information attests to the importance of germplasm with resistance to *Rhizoctonia spp.*

Resistant germplasm is the most effective and environmentally safe way to manage plant disease (Sherf and MacNab, 1986). The development of disease-resistant germplasm depends on many factors, including: the genetic control of host plant resistance; availability, efficacy, and cost (monetary and environmental) of chemical controls; efficacy, cost, and ease of cultural control practices; annual occurrence and geographic distribution of the disease; and economic importance (loss of yield potential) of the disease. Soilborne pathogens like *Rhizoctonia* usually are difficult to control chemically. Fumigation is expensive, providing only a temporary solution, especially with perennial plant species. In many crop species, the use of resistant germplasm, coupled with crop rotation and other cultural practices, provides excellent management of diseases caused by *Rhizoctonia* spp.

II. GENETIC CONTROL OF RESISTANCE

To accurately determine the degree of host plant resistance to a pathogen, it is important that the maximum plant response to the pathogen occurs over a sufficient period of time under uniform selection pressure (Nelson and MacKenzie, 1973). Once this is achieved, genetic control of host plant resistance can be determined and used to develop resistant germplasm. Monogenic host plant resistance, conditioned by a single dominant gene, is very easily backcrossed into existing cultivars; however, this type

B. Sneh et al. (eds.),
Rhizoctonia Species: Taxonomy, Molecular Biology, Ecology, Pathology and Disease Constrol, 515–527.

of resistance may not be as durable as a resistance conditioned by more than one gene (polygenic) (Nelson, 1973).

TABLE 1. List of registrations since 1985 of new cultivars that have resistance to diseases caused by *Rhizoctonia* spp.

Crop / Taxon	Number[1]	References
Birdsfoot trefoil *Lotus corniculatus*	2	(Beuselinck, 1994; Beuselinck and Steiner, 1994)
Creeping bentgrass *Agrostis stolonifera*	3	(Brauen *et al.*, 1993; Hurley *et al.*, 1994a; Engelke *et al.*, 1995)
Peanut *Arachis hypogaea*	1	(Branch, 1994)
Perennial ryegrass *Lolium perenne*	22	(Hurley *et al.*, 1987b, 1988, 1994b; Meyer *et al.*, 1987, 1988a, 1989, 1990; Russel *et al.*, 1987; Funk *et al.*, 1989, 1990; Pepin *et al.*, 1989; Bailey *et al.*, 1990; Meyer and Rose-Fricker, 1990a; Peterson *et al.*, 1991; Rose-Fricker and Meyer, 1991a, b; Rose-Fricker *et al.*, 1991, 1993a, 1995a, c, d; Ford *et al.*, 1995)
Potato *Solanum tuberosum*	1	(Reeves *et al.*, 1994)
Rice *Oryza sativa*	8	(McKenzie *et al.*, 1986; Nowick and Groth, 1989; Moldenhauer *et al.*, 1990; Bollich *et al.*, 1991; Xie *et al.*, 1992; Kalita *et al.*, 1993; Gravois *et al.*, 1994; Tseng and Oster, 1994)
Serica lespedeza *Lespediza cuneata*	1	(Mosijidis and Donnelly, 1989)
Sugarbeet *Beta vulgaris*	6	(Hecker and Ruppel, 1985, 1986, 1987, 1988, 1991; Panella *et al.*, 1995; Ruppel *et al.*, 1995)
Tall fescue *Festuca arundinacea*	17	(Hurley *et al.*, 1987a, 1990, 1992, 1995; Meyer *et al.*, 1988b, 1991; Bailey *et al.*, 1989; Meyer and Rose-Fricker, 1990b; Robinson *et al.*, 1990; Brilman *et al.*, 1991; Peterson *et al.*, 1993; Rose-Fricker and Meyer, 1993, 1994; Rose-Fricker *et al.*, 1993b, 1995b; Fraser *et al.*, 1994; Watson *et al.*, 1995)
Wheat *Triticum aestivum*	1	(Shaner *et al.*, 1986)

[1] This is the number of registrations. Often more than one germplasm is contained in a registration.

Genetic control of resistance to *Rhizoctonia* spp. has been studied in a number of crop species. Sugar beet (*Beta vulgaris* L.) is one of the crops for which we have a good deal of information on the genetic control of Rhizoctonia root- or crown-rot, caused by *R. solani* (AG 2-2). Seedling damping-off, which also affects sugar beet, primarily is caused by *R. solani* AG 4. Root-rot is endemic in sugar beet growing areas, and John Gaskill began a resistance breeding program for it in the late 1950s. He released the first two resistant cultivars in 1966 (Gaskill, 1968). Current germplasm has a level of resistance to root- and crown-rot in which there is no yield loss under heavy disease pressure and no loss of root or sucrose yield when disease pressure is mild or absent (Ruppel and Hecker, 1994). John Gaskill realized early that a natural field epiphytotic did not produce consistent, uniform disease pressure necessary for recurrent mass selection for Rhizoctonia resistance (Pierson and Gaskill, 1961). He and others performed mass selection or recurrent field selection (Hecker and Ruppel, 1977) under uniform, heavy disease pressure resulting from an artificially induced epiphytotic (Ruppel *et al.*, 1979; Schneider *et al.*, 1982).

Resistance to *R. solani* in sugar beet is polygenic, involving at least two loci, two or three alleles, and modifying genes in some populations (Hecker and Ruppel, 1975). Broad-sense heritability is about 0.65 with nonadditive components of the variance (Hecker and Ruppel, 1975). This agrees with a study by Hecker and Ruppel (1976) in which dominance effects were present in diploid, triploid, and tetraploid resistant hybrids. The relatively high heritability has allowed for development of increased host plant resistance to Rhizoctonia root- and crown-rot, and over 15 germplasm lines have been released in the last 10 years (Table 1). Resistance to *Rhizoctonia* has been incorporated into O-type maintainer, CMS female, and multigerm-pollinator germplasm and remains a very important means of reducing crop damage by this disease (see also chapter V.10).

Deakin and Dukes (1975) analyzed inheritance of resistance to *R. solani* in snap bean (*Phaseolus vulgaris* L.) and reported that resistance was highly heritable and could be transferred easily. They noted that resistance was linked closely to a gene(s) that determined pigmented seeds, which caused these cultivars to be unacceptable to the seed industry. They were unsure of the mechanism of resistance and discussed techniques to break the linkage to colored seeds. Subsequent investigators (Dickson and Boettger, 1977; Silva and Hartmann, 1982) indicated a more complex inheritance of Rhizoctonia resistance in snap beans. Dickson and Boettger (1977) reported that resistance was quantitatively inherited, and Silva and Hartmann (1982) concluded that there were two or three genes with additive gene action responsible. They agreed with Deakin and Dukes (1975) that resistance was highly heritable, but indicated polygenic control and additive gene action.

Clark and Block (1984) and Sloane and co-workers (1985) determined the inheritance of resistance to Rhizoctonia fruit (or belly) rot (caused by *R. solani*) in cucumber (*Cucumis sativus* L.). Sloane *et al.* (1985) used a wide-base population of pickling cucumber germplasm to estimate the narrow-sense heritability of resistance to belly rot at 0.24. Even though this is low to medium heritability, incorporation of resistance into commercial cultivars is possible, especially when using a breeding scheme that relies upon progeny testing (Dudley and Moll, 1969). Interestingly, there is one report of single, dominant-gene resistance to belly rot in cucumber (Clark and Block, 1984). This could be due to the *H* gene that controls heavy netting on the surface of the fruit, a highly undesirable horticultural characteristic (Sloane *et al.*, 1985).

There also are reports of single-gene resistance to Rhizoctonia fruit rot (caused by *R. solani*) of tomato (*Lycopersicon esculentum* Mill.) with either additive (Werner *et al.*, 1980) or dominant gene action (Johnson *et al.*, 1981). These results are somewhat ambiguous, and were obtained from different segregating populations (Werner *et al.*, 1980). Polygenic control of resistance could explain the results satisfactorily in some of the populations. In the populations showing monogenic control, host plant resistance might be due to mechanical resistance to Rhizoctonia fruit rot (monogenically inherited) conditioned by the skin toughness of the tomato fruit; skin toughness is also dependent on the environment (Werner *et al.*, 1980).

In most plant species, genetic control of host plant resistance to *Rhizoctonia* spp., is polygenic. Gadzhiev (1986) concluded that resistance to *R. solani* was under polygenic control because of the small number of resistant potato (*Solanum* spp.) plants in the F_2 generations of *S. tuberosum* x *S. phureja* and *S. tuberosum* x *S. vernei* crosses. As Wastie (1994) noted, this also could be explained by a few major recessive genes controlling Rhizoctonia resistance in potato. However, in either case, polygenic control of Rhizoctonia resistance in potato is suggested. Resistance to *R. solani* is polygenic in flax (*Linum usitatissimum* L.), and relatively high broad-sense heritabilities and moderate narrow-sense heritabilities have been calculated (Omarn *et al.*, 1968; Vest and Comstock, 1968). There was a negative correlation between host plant resistance and yellow seed-coat color, which is desired in oil seed flax cultivars (Vest and Comstock, 1968; Groth *et al.*, 1970). Studies in flax (Bolkan and Butler, 1974; Anderson and Stretton, 1978) also were used to elucidate genetic control of pathogenicity in *R. solani*. Polygenic control of resistance to *R. solani* also was found for cotton (*Gossypium* spp.) (Poswal *et al.*, 1986), tomato (Barksdale, 1974; Werner *et al.*, 1980), and rice (*Oryza sativa* L.) (Marchetti, 1991).

III. AVAILABLE RESISTANT GERMPLASM

A. Grasses

Rhizoctonia solani, *R. zeae* Voorhees, *R. cerealis* van der Hoeven, and *R. oryzae* Ryker are reported to cause diseases (brown patch, pre- and post-emergence seedling damping-off, crown rot, yellow patch, foliar blight) on over 100 grass species (Couch, 1973; Smith *et al.*, 1989a). In perennial and self-seeding annual grass species, selection for resistance to diseases caused by *Rhizoctonia* spp. provides sustainability of the stand over time. These species include tall fescue (*Festuca arundinacea* Schreb.), red fescue (*Festuca rubra* L.), zoysiagrass (*Zoysia japonica* Steud.), perennial ryegrass (*Lolium perenne* L.), Kentucky bluegrass (*Poa pratensis* L.), creeping bentgrass (*Agrostis palustris* Huds.), common bermudagrass [*Cynodon dactylon* (L.) Pers.], centipedegrass [*Eremochloa ophiuroides* (Munro.) Hack.], and St. Augustinegrass [*Stenotaphrum secundatum* (Walt.) Kuntze] (Burpee, 1980; Hurd and Grisham, 1983; Martin, and Lucas, 1983, 1984; Martin, 1987).

Immunity to disease caused by these *Rhizoctonia* spp. does not occur (Vargas, 1994). However, progress has been made in selection for increased Rhizoctonia resistance in many species (Table 1 and Smith *et al.*, 1989b). Selection for resistance to *Rhizoctonia* has commonly been phenotypic recurrent selection or single-plant progeny testing under natural infection in environments conducive to disease development (Clarke *et al.*, 1985; Meyer and Funk, 1989).

B. Rice

R. solani (AG 1-IA) causes sheath blight in rice, which can cause serious yield losses (Lee and Rush, 1983; Marchetti, 1991; Basu and Gupta, 1992 see also chapter V.9). There are few reports of progress in breeding for resistance to rice sheath blight, both because there are few cultivars with more than moderate resistance and the difficulty in screening germplasm for resistance. However, cultivars with moderate resistance can be used in conjunction with fungicides to suppress the disease and reduce the number of fungicide applications necessary for management (Mew, 1991). Genetic control of resistance appears to be polygenic, with a strong environmental influence on expression in field plots (Marchetti, 1991). There were several unsuccessful attempts to screen for resistance in the greenhouse, using a low number of infection cushions as the selection criterion (Marshall and Rush, 1980a, b; Groth 1992). Most resistant material released was selected in the field (Table 1).

C. Wheat

Rhizoctonia root rot and sharp eyespot are diseases of wheat (*Triticum aestivum* L.) caused by *R. solani* and *R. cerealis,* respectively (Bockus, 1987). These can be serious problems in areas of the United States where soft red winter wheat is grown, especially on acid, sandy, dry soils, under continuous cropping or reduced tillage systems (Bockus, 1987). The cultivar 'Bare', developed by Camas Wheat Breeding, has been used as a source of Rhizoctonia resistance (Pope, 1992; Pope and McProud, 1993). Resistant soft red winter wheat cultivars have been developed and released (Table 1 and Miskin *et al.*, 1991; 1995).

D. Sugar beet

R. solani is endemic in all sugar beet growing areas and causes root- and crown-rot (AG 2-2) and seedling damping-off (primarily AG 4). John Gaskill began a resistance breeding program in the late 1950's. He released the first resistant germplasm in 1966 (Gaskill, 1968), and research has continued since then (Hecker and Ruppel, 1977) with over 15 germplasm lines released in the last 10 years (Table 1). Resistance remains one of the most important means of managing damage by this disease (see also chapter V.10).

E. Solanaceous Plants

1. Potato *R. solani* causes stem canker and black scurf (or "Rhizoc") of potato; these can cause serious yield losses when environmental conditions favor disease development (Burton, 1989; Leach and Webb, 1993). A major source of infection may be the presence of sclerotia on the seed tubers (Burton, 1989). Because different parts of the plant differ in susceptibility, it has been difficult to resolve the genetics of resistance (Wastie, 1994). Leach and Martin (1993) point out that resistance to *R. solani* has not been considered to be important in potato-clone development until recently. It has been difficult for breeders to identify or enhance sources of resistance in the tetraploid *Solanum* species (Leach and Webb, 1993); therefore, wild diploid species have been screened (Osipova and Mikhevva, 1976; Nazarova and Turuleva, 1987; Nazarova, 1991). Screening tetraploid *Solanum* clones for resistance to *R. solani* has resulted in varied degrees of resistance to *Rhizoctonia*, which suggests that, even though not specifically selecting for resistance, breeders have incorporated some resistance by selecting away from damage caused by this pathogen (Leach and Webb, 1993).

2. Tomato Rhizoctonia fruit rot and soil rot in tomato is caused by *R. solani*. Tomato germplasm has been screened and sources of resistance have been identified (Barksdale, 1968; Moustafa and Khafagi, 1992). Both polygenic (Barksdale, 1974; Johnson *et al.*, 1981) and monogenic genetic control (Werner *et al.*, 1980; Johnson *et al.*, 1981) of host plant resistance have been hypothesized.

3. Pepper Although there are no Rhizoctonia-resistant cultivars of lc Chile (*Capsicum annuum* L.), Muhyi and Bosland (1995) described screening techniques and reported sources of resistance.

F. Legumes

1. Pea *R. solani* causes seed rot, tip blight, and seedling stem rot of peas (*Pisum sativum* L.). Some screening techniques and sources of Rhizoctonia resistance in peas and lentils (*Lens culinaris* Medic.) were reviewed by Kraft *et al.* (1994). Attempts to correlate greenhouse testing with field performance have been made and resistant sources have been identified (Shehata and Davis, 1981; McCoy and Kraft, 1984; Kraft *et al.*, 1994). The importance of this germplasm was review by Ali *et al.* (1994).

2. Dry and Snap Bean *R. solani* may cause many diseases in dry beans, including seed decay, seedling damping-off, root rots, and stem canker. Both snap and dry bean germplasm (*P. vulgaris*) have been screened for resistance to *R. solani* (McLean *et al.*, 1968; Prasad and Weigle, 1969; 1970; Tu and Park, 1993). Highly heritable forms of resistance are available and easily transferred to commercial cultivars (Deakin and Dukes, 1975; Muyolo *et al.*, 1993; Tu and Park, 1993). A highly tolerant wax-bean cultivar (Goldcast) and two breeding lines were released in 1970 from the US Vegetable Breeding Laboratory, Charleston, SC (Zaumeyer and Meiners, 1975). However, this resistance has not been incorporated into any dry bean cultivar available from any public/private firm (H. Schwartz, pers. comm.). Resistance most likely is polygenic, controlled by three genes with additive gene action (Silva and Hartmann, 1982).

3. Other Legumes Lima bean (*Phaseolus lunatus* L.) germplasm has been evaluated for resistance to *R. solani* (Warren *et al.*, 1972). Muyolo *et al.* (1993) tested dry bean, lima bean, and soybean (*Glycine max* Merr.) for resistance to Rhizoctonia root and hypocotyl rot and web blight and found a range of resistance in all three species. A range of host plant resistance and pathogen virulence also was found in faba bean (*Vicia faba* L.) germplasm when challenged by nine isolates of *R. solani* (Rashid and Bernier, 1993). At least one cultivar of mung bean [*Vigna radiata* (L.) Wilczek] with Rhizoctonia blight resistance was released (Singh and Singh, 1994). Resistant germplasm also was identified in black gram [*Vigna mungo* (L.) Hepper] (Singh *et al.*, 1991), chickpea (*Cicer arietinum* L.) (Khaleeque *et al.*, 1993), and peanut (*Arachis hypogaea* L.) (Woodard and Jones, 1980; Branch and Brenneman, 1993). Peanut germplasm, resistant to limb rot caused by *R. solani*, has been released as has Rhizoctonia-resistant alfalfa (*Medicago sativa* L.) and birdsfoot trefoil (*Lotus corniculatus* L.) (Table 1).

F. Vegetable Crops

1. Cucumber *R. solani* causes seedling damping-off and belly or fruit rot of cucumber. Yield losses up to 40% have occurred when the environment favored disease development (Sumner and Smittle, 1976). Resistance to these two diseases is not correlated (Booy *et al.*, 1987). Heritability of 0.24 was estimated for resistance to fruit rot, indicating incorporation of resistance into commercial cultivars is possible.

Selection via progeny testing was recommended by Sloane and co-workers (1985) to improve fruit rot resistance.

2. Other Vegetables Radish (*Raphanus sativus* L.) germplasm resistant to *R. solani*, which causes scurfy root, has been identified and greenhouse screening methods established (Humaydan *et al.*, 1976). Methods for screening carrot (*Daucus carota* L.) also have been described (Howard and Williams, 1976). Cabbage (*Brassica oleracea* L.) cultivars were screened and shown to possess varied resistance to *R. solani* (Bocak and Rod, 1992).

E. Oilseeds

1. Cotton Cotton (*Gossypium* spp.) germplasm has been screened for resistance to *R. solani*, but highly resistant sources have not been identified (Mathre and Otta, 1967; Stanton *et al.*, 1994). The Multi-Adversity Resistance program reported improvement in resistance to seedling disease pathogens (Thaxton *et al.*, 1991), but highly resistant cultivars are not available (Stanton *et al.*, 1994). Host plant resistance was reported to be polygenic with low heritability (Poswal *et al.*, 1986).

2. Rapeseed (Canola) Pre-emergence damping-off, postemergence seedling rot, and brown girdling root rot, all caused by *R. solani,* are important diseases of oilseed rape and canola (*Brassica napus* L. and *B. rapa* L.) (Yang and Verma, 1992) that may limit production in some areas (Huber *et al.*, 1992). There have been efforts to screen for resistance, and resistant cultivars were identified (Acharya *et al.*, 1984; Yang and Verma, 1992), but all currently grown cultivars are susceptible to *R. solani* (Yang and Verma, 1992).

3. Flax Root rot and seedling blight of flax (*L. usitatissimum*) is caused by *R. solani* (Islam, 1992). Resistant cultivars have been identified (Anderson, 1977) and resistance was correlated with seed-coat color (Vest and Comstock, 1968; Groth *et al.*, 1970). Resistance was polygenic in nature (Omarn *et al.*, 1968; Vest and Comstock, 1968).

F. Other Crops

Black root rot complex of strawberry (*Fragaria chiloensis* Buchesne var. *ananassa* Bailey) has been associated with *R. fragariae* Husain & McKeen, *Pythium* spp., and *Pratylenchus penetrans* (Cobb) Filip. and Stek. Wing and co-workers showed varied levels of resistance among strawberry cultivars in field screening trials. Methods of screening strawberry germplasm, which resulted in successful selection of germplasm resistant to *R. solani* and *R. fragariae*, have been reported (Nipoti *et al.*, 1989). Geranium (*Pelargonium graveolens* L. Herit) grown for oil production is susceptible to root rot and wilt caused by *R. solani*. Field and laboratory screening methods identified Rhizoctonia resistance among geranium clones (Kalra *et al.*, 1992).

IV. Summary

In an increasing number of plant species, selection has provided germplasm with resistance to diseases caused by *Rhizoctonia* spp. In most cases, the resistance is not immunity, but an ability of the host plant to continue growth in the presence of the pathogen. In a few crops, genetic control of this resistance has been studied. With few exceptions, resistance is governed by more than one gene. In the two crop species where single-gene resistance is hypothesized (Werner *et al.*, 1980; Johnson *et al.*, 1981; Clark and Block, 1984), resistance was due to toughness of the fruit epidermis, thereby providing a mechanical resistance to infection.

Host plant resistance to *Rhizoctonia* has proved durable in many crop species. As regulation of chemical control becomes more stringent, increasing resistance, coupled with sound management practices, will become an even more important method for managing plant diseases caused by *Rhizoctonia* spp. Strong host plant resistance will allow continued food and fiber production in areas where *Rhizoctonia* is present.

V. ACKNOWLEDGMENTS

We thank all of our many colleagues who were so helpful in providing information for this chapter. We would also welcome reprints or references from researchers working with any plant species to understand the genetic control of resistance to *Rhizoctonia*.

VI. REFERENCES

Acharya SN, Verma PR, Dueck J & Downey RK (1984) Screening rapeseed/canola for resistance to damping-off and seedling root rot caused by *Rhizoctonia solani*. Can. J. Plant Pathol. 6:325-328.

Ali SM, Sharma B & Ambrose MJ (1994) Current status and future strategy in breeding pea to improve resistance to biotic and abiotic stresses. Euphytica 73:115-126.

Anderson NA (1977) Evaluation of the *Rhizoctonia* complex in relation to seedling blight of flax. Plant Dis. Rep. 61:140-142.

Anderson NA & Stretton HM (1978) Genetic control of pathogenicity in *Rhizoctonia solani*. Phytopathology. 68:1314-1317.

Bailey RH, Brilman LA, Jacklin AW, Brede AD, Green BKII & Funk CR (1989) Registration of 'Mesa' tall fescue. Crop Sci. 29:1570-1571.

Bailey RH, Kopec DM, White RH & Funk CR (1990) Registration of 'Dandy' perennial ryegrass. Crop Sci. 30:955-956.

Barksdale TH (1968) Rhizoctonia soil rot and buckeye rot of tomatoes: observational differences in susceptibility. Plant Dis. Rep. 52:284-286.

Barksdale TH (1974) Evaluation of tomato fruit for resistance to Rhizoctonia soil rot. Plant Dis. Rep. 58:406-408.

Basu A & Gupta PKS (1992) Loss in yield and seed infection in promising genotypes of rice (*Oryza sativa*) due to sheath-blight disease caused by *Rhizoctonia solani*. Indian J. Agr. Sci. 62:570-571.

Beuselinck PR (1994) Registration of 'Dawn' birdsfoot trefoil. Crop Sci. 34:540.

Beuselinck PR & Steiner JJ (1994) Registration of CAD birdsfoot trefoil germplasm selected for drought resistance. Crop Sci. 34:543.

Bocak L & Rod J (1992) Evaluation of susceptibility of cabbage assortment to the damage caused by flea beetles (*Phyllotreta* spp.) and fungus *Rhizoctonia solani* Kühn. Zahradnictvi 19:19-23.

Bockus WW (1987) Diseases of roots, crowns, and lower stems. Chap. 8A In: Heyne EG (ed.) Wheat and Wheat Improvement. 2nd Edition, (pp 510-527) ASA-CSSA-SSSA, Madison.

Bolkan HA & Butler EE (1974) Studies on heterokaryosis and virulence of *Rhizoctonia solani*. Phytopathology 64:513-522.

Bollich CN, Webb BD, Marchetti MA & Scott JE (1991) Registration of 'Maybelle' rice. Crop Sci. 31:1090.

Booy G, Wehner TC & Jenkins SFJ (1987) Resistance of cucumber lines to *Rhizoctonia solani* damping-off: not related to fruit rot resistance. HortScience 22:105-108.

Branch WD (1994) Registration of 'Georgia Browne' peanut. Crop Sci. 34:1125-1126.

Branch WD & Brenneman TB (1993) White mold and Rhizoctonia limb rot resistance among advanced Georgia peanut breeding lines. Peanut Sci. 20:124-126.

Brauen SE, Goss RL & Brede AD (1993) Registration of 'Putter' creeping bentgrass. Crop Sci. 33:1100.

Brilman LA, Brede AD, Kemp ML, Duell RW & Funk CR (1991) Registration of 'Wrangler' tall fescue. Crop Sci. 31:480-481.

Burpee L (1980) *Rhizoctonia cerealis* causes yellow patch of turfgrasses. Plant Dis. 64:1114-1116.

Burton WG (1989) The Potato. 3rd Edition, Longman Scientific & Technical, New York. 742 p.

Clark RL & Block CC (1984) Belly rot resistance in *Cucumis sativus*. Phytopathology 74:819.

Clarke BB, Funk CR & Halisky PM (1985) Development of *Festuca arundinacea* Schreb. cultivars with improved resistance to *Rhizoctonia solani* Kühn. Chap. 57 In: Proceedings of the Fifth International Turfgrass Research Conference. (pp 641-645) Institut National de la Recherche Agronomique, Paris.

Couch HB (1973) Diseases of Turfgrass. 2nd Edition, Robert E. Krieger Publishing Co., Huntington, NY. 348 p.

Deakin JR & Dukes PD (1975) Breeding snap beans for resistance to diseases caused by *Rhizoctonia solani* Kühn. HortScience 10:269-271.

Dickson MH & Boettger MA (1977) Breeding multiple root rot resistance in snap beans. J. Amer. Soc. Hort. Sci. 102:373-377.

Dudley JW & Moll RH (1969) Interpretation and use of estimates of heritability and genetic variances in plant breeding. Crop Sci. 9:257-261.

Engelke MC, Lehman VG, Kneebone WR, Colbaugh PF, Reinert JA & Knoop WE (1995) Registration of 'Crenshaw' creeping bentgrass. Crop Sci. 35:589.

Farr DF, Bills GF, Chamuris GP & Rossman AY (1989) Fungi On Plants and Plant Products in the United States. APS Press, St. Paul. Minnesota 1252 pp.

Ford TM, Bara RF, Sun S & Funk CR (1995) Registration of 'Affinity' perennial ryegrass. Crop Sci. 35:1220-1221.

Fraser ML, Jones SK, Jones GW & Funk CR (1994) Registration of 'Hubbard 87' tall fescue. Crop Sci. 34:304.

Funk CR, White RH, Bara RF, Pepin GW, Jacob JO & Herb DJ (1989) Registration of 'Pinnacle' perennial ryegrass. Crop Sci. 29:1569-1570.

Funk CR, White RH, Pepin GW, Green II BK & Kopec DM (1990) Registration of 'Sherwood' perennial ryegrass. Crop Sci. 30:742-743.

Gadzhiev NM (1986) The nature of the inheritance of black scurf resistance in interspecific potato hybrids. Selektsiya i Semenovodstvo, USSR 1:18-19.

Gaskill JO (1968) Breeding for Rhizoctonia resistance in sugarbeet. J. Am. Soc. Sugar Beet Technol. 15:105-119.

Gravois KA, Moldenhauer AK, Lee FN, Norman RJ, Helms RS, Wells BR, Dilday RH, Rohman PC & Blocker MM (1994) Registration of 'Adair' rice. Crop Sci. 34:1123.

Groth DE (1992) Selection for resistance of rice sheath blight through number of infection cushions and lesion type. Plant Dis. 76:721-723.

Groth JV, Comstock VE & Anderson NA (1970) Effect of seed color on tolerance of flax to seedling blight caused by *Rhizoctonia solani*. Phytopathology 60:379-380.

Hecker RJ & Ruppel EG (1975) Inheritance of resistance to Rhizoctonia root-rot in sugarbeet. Crop Sci. 15:487-490.

Hecker RJ & Ruppel EG (1976) Polyploid and maternal effects on Rhizoctonia root rot resistance in sugarbeet. Euphytica 25:419-423.

Hecker RJ & Ruppel EG (1977) Rhizoctonia root-rot resistance in sugarbeet: breeding and related research. J. Am. Soc. Sugar Beet Technol. 19:246-256.

Hecker RJ & Ruppel EG (1985) Registration of Rhizoctonia root rot resistant sugarbeet germplasms FC701/6, FC702/7, and FC705/1. Crop Sci. 25:374.

Hecker RJ & Ruppel EG (1986) Registration of Rhizoctonia root rot resistant sugarbeet germplasm FC712. Crop Sci. 26:213-214.

Hecker RJ & Ruppel EG (1987) Registration of Rhizoctonia root rot resistant sugarbeet germplasm FC707(4*x*). Crop Sci. 27:822.

Hecker RJ & Ruppel EG (1988) Registration of Rhizoctonia root rot resistant sugarbeet germplasm FC709. Crop Sci. 28:1039-1040.

Hecker RJ & Ruppel EG (1991) Registration of Rhizoctonia root rot resistant sugarbeet germplasm FC710. Crop Sci. 31:494.

Howard RJ & Williams PH (1976) Methods for detecting resistance to Pythium and Rhizoctonia root diseases in seedling carrots. Plant Dis. Rep. 60:151-156.

Huber DM, Christmas E, Herr LJ, McCay-Buis TS & Baird R (1992) Rhizoctonia crown rot of canola in Indiana. Plant Dis. 76:1251-1253.

Humaydan HS, Williams PH, Jacobsen BJ & Bissonnette HL (1976) Resistance in radish to *Aphanomyces raphani* and *Rhizoctonia solani*. Plant Dis. 60:156-160.

Hurd B & Grisham MP (1983) *Rhizoctonia* spp. associated with brown patch of Saint Augustinegrass. Phytopathology 73:1661-1665.

Hurley RH, Clarke BB, Meyer WA, Kopec DM & Funk CR (1987a) Registration of 'Rebel II' tall fescue. Crop Sci. 27:1086.

Hurley RH, Meyer WA, Rose-Fricker CA & Funk CR (1987b) Registration of 'Cowboy' perennial ryegrass. Crop Sci. 27:1086-1087.

Hurley RH, Jones SK, Jones GW, Clarke BB & Funk CR (1988) Registration of 'Tara' perennial ryegrass. Crop Sci. 28:192.

Hurley RH, Pompei ME, Mazur G, White RH, Kemp ML & Funk CR (1990) Registration of 'Tribute' tall fescue. Crop Sci. 30:957-958.

Hurley RH, Meyer WA, Rose-Fricker CA, Kemp ML & Funk CR (1992) Registration of 'Rebel Jr.' tall fescue. Crop Sci. 32:831.

Hurley RH, Lehman VG, Murphy JA & Funk CR (1994a) Registration of 'Southshore' creeping bentgrass. Crop Sci. 34:1124-1125.

Hurley RH, Meyer WA, Rose-Fricker CA, Lehman VG & Funk CR (1994b) Registration of 'Prelude II' perennial ryegrass. Crop Sci. 34:1129.

Hurley RH, Lehman VG, Meyer WA, Rose-Fricker CA, Fraser ML, Bara RF, Sun S & Funk CR (1995) Registration of 'Rebel 3D' tall fescue. Crop Sci. 35:1215.

Islam MR (1992) Control of flax diseases through genetic resistance. Z. PflKrankh. PflanzenSch. (J. Pl. Dis. Protect.) 99:550-557.

Johnson CE, Moore EL & Batson WE (1981) Inheritance of Rhizoctonia soil rot resistance in tomato. MAFES Res. Rpt., vol. 5 no. 15.

Kalita UC, Baruah DK & Upadhaya LP (1993) Integrated germplasm improvement-rainfed lowland. Internl. Rice Res. Notes 18:22.

Kalra A, Parameswaran TN & Ravindra NS (1992) Influence of planting date on plant losses and yield responses of geranium (*Pelargonium gaveolens*) to root rot and wilt. J. Agric. Sci., Cambridge 118:309-314.

Khaleeque MI, Inam-ul-Haq M, Iqbal M & Alam A (1993) Response of chickpea cultivars to *Rhizoctonia solani* infection. Pakistan J. Phytopathol. 5:18-20.

Kraft JM, Haware MP, Jimenez-Diaz RM, Bayaa B & Harrabi M (1994) Screening techniques and sources of resistance to root rots and wilts in cool season food legumes. Euphytica 73:27-39.

Leach LD & Garber RH (1970) Control of *Rhizoctonia*. In: Parmeter J John R. J (ed.) *Rhizoctonia Solani*: Biology and Pathology. (pp 189-199) University of California Press, Berkeley.

Leach SS & Webb RE (1993) Evaluation of potato cultivars, clones and a true seed population for resistance to *Rhizoctonia solani*. Am. Potato J. 70:317-329.

Lee FN & Rush MC (1983) Rice sheath blight: a major rice disease. Plant Dis. 67:829-832.

Marchetti MA (1991) Quantifications of the relationship between sheath blight severity and yield loss in rice. Plant Dis. 75:773-775.

Marshall DS & Rush MC (1980a) Infection cushion formation of rice sheaths by *Rhizoctonia solani*. Phytopathology. 70:947-950.

Marshall DS & Rush MC (1980b) Relation between infection by *Rhizoctonia solani* and *R. oryzae* and disease severity in rice. Phytopathology 70:941-946.

Martin B (1987) Rapid tentative identification of *Rhizoctonia* spp. associated with diseased turfgrasses. Plant Dis. 71:47-49.

Martin SB & Lucas LT (1983) Pathogenicity of *Rhizoctonia zeae* on tall fescue and other turfgrasses. Plant Dis. 67:676-678.

Martin SB & Lucas LT (1984) Characterization and pathogenicity of *Rhizoctonia* spp. and binucleate Rhizoctonia-like fungi from turfgrasses in North Carolina. Phytopath. 74:170-175.

Mathre DE & Otta JD (1967) Sources of resistance in the genus *Gossypium* to several soilborne pathogens. Plant Dis. Rep. 51:864-867.

McCoy RJ & Kraft JM (1984) Comparison of techniques and inoculum sources in evaluating peas (*Pisum sativum*). Plant Dis. 68:53-55.

McKenzie KS, Rush MC & Groth DE (1986) Registration of two disease-resistant germplasm lines of rice. Crop Sci. 26:839-840.

McLean DM, Hoffman JC & Brown GB (1968) Greenhouse studies on resistance of snap beans to *Rhizoctonia solani*. Plant Dis. Rep. 52:486-488.

Mew TM (1991) Rice diseases. Chap. 5 In: Luh BS (ed.) Rice Production. 2nd Edition, Vol. 1 (pp 187-236) Van Nostrand Reinhold, New York.

Meyer WA & Funk CR (1989) Progress and benefits to humanity from breeding cool-season grasses for turf. In: Sleper DA, Asay KH and Pederson JF (eds.) Contributions from Breeding Forage and Turf Grasses. (pp 31-48) Crop Science Society of America, Inc., Madison, (CSSA Special Publication Number 15).

Meyer WA & Rose-Fricker C (1990a) Registration of 'Birdie II' perennial ryegrass. Crop Sci. 30:1158-1159.

Meyer WA & Rose-Fricker C (1990b) Registration of 'Bonanza' tall fescue. Crop Sci. 30:1157.

Meyer WA, Rose-Fricker CA, Funk CR & Rose BL (1987) Registration of 'Citation II' perennial ryegrass. Crop Sci. 27:815-816.

Meyer WA, Robinson MF, Bara RF, Dickson WK & Funk CR (1988a) Registration of 'SR 4100' perennial ryegrass. Crop Sci. 28:1021.

Meyer WA, Rose-Fricker C, Rose BL & Zajac J (1988b) Registration of 'Jaguar' tall fescue. Crop Sci. 28:191.

Meyer WA, Rose-Fricker CA, Robinson MR & Funk CR (1989) Registration of 'SR-4000' perennial ryegrass. Crop Sci. 29:826-827.

Meyer WA, Rose-Fricker CA, Pepin GW & Funk CR (1990) Registration of 'Jazz' perennial ryegrass. Crop Sci. 30:743.

Meyer WA, Rose-Fricker CA & Funk CR (1991) Registration of 'Apache' tall fescue. Crop Sci. 31:1088.

Miskin K, Beazer C, Fogleman B, Ridge E, Bruns R, Moffatt J, Smith J & Reeder J (1991) Agripro Biosciences, Inc. Ann. Wheat Newsl. 27:16-17. (The AWN is available as in the previous Ref.).

Miskin KE, Beazer C, Glover E & Scruggs D (1995) Agripro Seeds, Inc. - Midwest soft red winter wheat research. Ann. Wheat Newsl. 41 (The AWN is available electronically at the graingenes gopher site at cornell or WWW address http://wheat.pw.usda.gov/graingenes.html).

Moldenhauer KAK, Lee FN, Morman RJ, Helms RS, Wells BR, Dilday RH, Rohman PC & Marchetti MA (1990) Registration of 'Katy' rice. Crop Sci. 30:747.

Mosijidis JA & Donnelly ED (1989) Registration of 'Au Donnelly' sericea lespedeza. Crop Sci. 29:237-238.

Moustafa SES & Khafagi YS (1992) Reaction of certain tomato cultivars to fusarium wilt and root-rot disease caused by *Fusarium oxysporum* f.sp. *lycopersici* and *Rhizoctonia solani* respectively. Assiut J. Agr. Sci. 23:199-206.

Muhyi R & Bosland PW (1995) Evaluation of *Capsicum* germplasm for sources of resistance to *Rhizoctonia solani*. HortScience 30:341-342.

Muyolo NG, Lipps PE & Schmitthenner AF (1993) Reactions of dry bean, lima bean, and soybean cultivars to Rhizoctonia root and hypocotyl rot and web blight. Plant Dis. 77:234-238.

Nazarova LP (1991) An immunological characteristic of interspecific potato hybrids by resistance to Rhizoctonia disease. Nauchno-tekhnicheskii byulleten' Vsesoyuznogo ordena Lenina i ordena druzhby narodov nauchno-issledovatel'skogo instituta rastenievodstva imeni N.I. Vavilova 214:34-37.

Nazarova LP & Turuleva LM (1987) Resistance of North American wild species of potato to Rhizoctonia. Sbornik Nauchnykh Trudov po Prikladnoi Botanike, Genetike i Seletektsii 115:103-106.

Nelson RR (1973) The use of resistance genes to curb population shifts in plant pathogens. Chap. 5 In: Nelson RR (ed.) Breeding Plants for Disease Resistance. (pp. 49-66) The Pennsylvania State University Press, University Park.

Nelson RR & MacKenzie DR (1973) The detection and stability of disease resistance. Chap. 3 In: Nelson RR (ed.) Breeding Plants for Disease Resistance. (pp. 12-39) The Pennsylvania State University Press, University Park.

Nipoti P, Finessi L, Manzali D & Gavina F (1989) Methods for estimating the resistance of strawberry plants to *Verticillium dahliae* Kleb., *Rhizoctonia fragariae* Husain and McKeen and *Rhizoctonia solani* Kühn. Acta Hort. 265:609-613.

Nowick EM & Groth DE (1989) Registration of PR 6555 rice germplasm line with moderate resistance to sheath blight. Crop Sci. 29:1096-1097.

Omarn AO, Frederiksen RA & Atkins IM (1968) Heritability of seedling disease characteristics in flax. Crop Sci. 8:750-753.

Osipova EA & Mikhevva LP (1976) Resistance to *Rhizoctonia* in seedlings of diploid cultivated species of potato after artificial infection. Referativnyi Zhurnal 208:179-183.

Panella LW, Ruppel EG & Hecker RJ (1995) Registration of four multigerm sugarbeet germplasms resistant to rhizoctonia root rot: FC716, FC717, FC718, and FC719. Crop Sci. 35:291-292.

Pepin GW, Wiley WK, King DE, Petersen CJ & Funk CR (1989) Registration of 'Fiesta II' perennial ryegrass. Crop Sci. 29:1328.

Peterson RJ, Johnson-Cicalese JM, Sun S, Funk DC & Funk CR (1991) Registration of 'Seville' perennial ryegrass. Crop Sci. 31:483-484.

Peterson RJ, Clarke BB, Fraser ML, Murphy JA, Sun S & Funk CR (1993) Registration of 'Shenandoah' tall fescue. Crop Sci. 33:1404-1404.

Pierson VG & Gaskill JO (1961) Artificial exposure of sugar beets to *Rhizoctonia solani*. J. Am. Soc. Sugar Beet Technol. 11:574-590.

Pope WK (1992) Camas wheat breeding Moscow, ID. Ann. Wheat Newsl. 28:204-205. (The AWN is availableas in a previously detailed Ref.).

Pope WK & McProud W (1993) Camas wheat breeding, Moscow ID plant breeds 1, Moscow ID. Ann. Wheat Newsl. 39:267-268. (The AWN is availableas in a previously detailed Ref.).

Poswal MAT, El-Zik KM & Bird LS (1986) Gene action and inheritance of resistance to *Rhizoctonia solani* and *Pythium ultimum* in cotton seedlings. Phytopathology. 76:1107.

Prasad K & Weigle JL (1969) Resistance to *Rhizoctonia solani* in *Phaseolus vulgaris* (snap beans). Plant Dis. Rep. 53:350-352.

Prasad K & Weigle JL (1970) Screening for resistance to *Rhizoctonia solani* in *Phaseolus vulgaris*. Plant Dis. Rep. 54:40-44.

Rashid KY & Bernier CC (1993) Genetic diversity among isolates of *Rhizoctonia solani* and sources of resistance in *Vicia faba*. Can. J. Pl. Pathol. 15:23-28.

Reeves AF, Porter GA, Manzer FE, Work TM, Davis AA & Plissey ES (1994) Mainechip: a new chipping potato variety for cool storage processing. Am. Potato J. 71:237-247.

Robinson MF, Clarke BB, Duell RW & Funk CR (1990) Registration of 'Titan' tall fescue. Crop Sci. 30:956-957.

Rose-Fricker CA & Meyer WA (1991a) Registration of '246' perennial ryegrass. Crop Sci. 31:1089.

Rose-Fricker CA & Meyer WA (1991b) Registration of 'Omega II' perennial ryegrass. Crop Sci. 31:1383.

Rose-Fricker CA & Meyer WA (1993) Registration of 'Olympic II' tall fescue. Crop Sci. 33:202.

Rose-Fricker CA & Meyer WA (1994) Registration of 'Eldorado' tall fescue. Crop Sci. 34:1129-1130.

Rose-Fricker CA, Meyer WA & Zajac J (1991) Registration of 'Saturn' perennial ryegrass. Crop Sci. 31:1089.

Rose-Fricker CA, Meyer WA & Van Wuk JP (1993a) Registration of 'Charger' perennial ryegrass. Crop Sci. 33:206.

Rose-Fricker CA, Meyer WA & Van Wuk JP (1993b) Registration of 'Maximize' tall fescue. Crop Sci. 33:203.

Rose-Fricker CA, Meyer WA, Bara RF & Funk CR (1995a) Registration of 'Brightstar' perennial ryegrass. Crop Sci. 35:1508-1509.

Rose-Fricker CA, Meyer WA & Funk CR (1995b) Registration of 'Safari' tall fescue. Crop Sci. 35:1204.

Rose-Fricker CA, Meyer WA, Funk CR & Wick A (1995c) Registration of 'Legacy' perennial ryegrass. Crop Sci. 35:1210.

Rose-Fricker CA, Meyer WA, Funk CR, Zajac JJ & Bara RF (1995d) Registration of 'Prizm' perennial ryegrass. Crop Sci. 35:1208-1209.

Ruppel EG & Hecker RJ (1994) Rhizoctonia root rot on sugarbeet cultivars having varied degrees of resistance. J. Sugar Beet Res. 31:135-142.

Ruppel EG, Schneider CL, Hecker RJ & Hogaboam GJ (1979) Creating epiphytotics of *Rhizoctonia* root rot and evaluating for resistance to *Rhizoctonia solani* in sugarbeet field plots. Plant Dis. Rep. 63:518-522.

Ruppel EG, Hecker RJ & Panella LW (1995) Registration of two sugarbeet germplasms resistant to Rhizoctonia root rot: FC715 and FC715CMS. Crop Sci. 35:290.

Russel RA, McVeigh KJ, Pepin GW, Brilman LA & Funk CR (1987) Registration of 'All*Star' perennial ryegrass. Crop Sci. 27:1308-1309.

Schneider CL, Ruppel EG, Hecker RJ & Hogaboam GJ (1982) Effect of soil deposition in crowns on development of *Rhizoctonia* root rot in sugar beet. Plant Dis. 66:408-410.

Shaner GE, Ohm HW, Foster JE, Patterson FL, Gallun RL, Huber DM, Buechley GC, Safranski GG & Hertel JM (1986) Registration of 'Adder' wheat. Crop Sci. 26:201.

Shehata MA & Davis DW (1981) Screening peas for resistance to stem rot caused by *Rhizoctonia solani*. Plant Dis. 65:417-419.

Sherf AF & MacNab AA (1986) Vegetable Diseases and Their Control. Wiley, New York. 728 p.

Silva GH & Hartmann RW (1982) Inheritance of resistance to *Rhizoctonia solani* Kühn in snap bean (*Phaseolus vulgaris* L.). J. Amer. Soc. Hort. Sci. 107:653-657.

Singh DP & Singh BB (1994) A new variety of mung bean - Narendra Mung 1. Ind. Farming 43:3-4.

Singh DP, Singh BB & Gupta RP (1991) Evaluation and classification of blackgram germplasm. Nar. Deva. J. Agr. Res. 6:138-141.

Sloane JT, Wehner TC & Jenkins Jr. SF (1985) Inheritance of resistance to Rhizoctonia fruit rot in cucumber. HortSci. 20:1119-1120.

Smith JD, Jackson N & Woolhouse AR (1989a) Rhizoctonia foliar disease of established turf. Chap. 25 In: Fungal Diseases of Amenity Turf Grasses. 3rd Edition, (pp 266-275) E. & F.N. Spon Ltd., London.

Smith JD, Jackson N & Woolhouse AR (1989b) Species and cultivars of turf grasses in relation to disease management. Chap. 3 In: Fungal Diseases of Amenity Turf Grasses. 3rd Edition, (pp 15-17) E. & F.N. Spon Ltd., London.

Stanton MA, Rothrock CS & Stewart JMcD (1994) Response of A-genome cotton germplasm to the seedling disease pathogens, *Rhizoctonia solani* and *Pythium ultimum*. Genet. Res. Crop Evol. 41:9-12.

Sumner DR & Smittle DA (1976) Etiology and control of fruit rot of cucumber in single harvesting for pickles. Plant Dis. 60:304-307.

Thaxton PM, El-Zik KM & Kirkpatrick RK (1991) Genetic gains in resistance to pests in the MAR germplasm. In: Herber DJ and Richter DA (eds.) Proceedings of the Cotton Belt-Wide Producers Research Conference. (pp 151-156) Cotton Council of America, Memphis, Tenessee.

Tseng ST & Oster JJ (1994) Registration of 87-Y-550, a rice germplasm line resistant to stem rot disease. Crop Sci. 34:314.

Tu JC & Park SJ (1993) Root-rot resistance in common bean. Can. J. Pl. Sci. 73:365-367.

Vargas J J M (1994) Management of Turfgrass Diseases. 2nd Edition, Lewis Publishers (CRC Press, Inc.), Boca Raton. 294 pp

Vest G & Comstock VE (1968) Resistance in flax to seedling blight caused by *Rhizoctonia solani*. Phytopathology 58:1161-1163.

Warren HL, Helfrich RM & Blount VL (1972 Evaluation of lima bean varieties and selections for resistance to *Rhizoctonia solani*. Plant Dis. Rep. 56:268-270.

Wastie RL (1994) Inheritance of resistance to fungal diseases of tubers. Chap. 18 In: Bradshaw JE and Mackay GR (eds.) Potato Genetics. (pp 411-428) University Press, Cambridge.

Watson CEJR, Linscombe SD, Mclean SD, Meurer CA, Smith MA, Wofford DS & Baswell JW (1995) Registration of 'DeSoto' tall fescue. Crop Sci. 35:939.

Werner RA, Sanders DC & Henderson WR (1980) Inheritance of tolerance to Rhizoctonia fruit rot of tomato. J. Amer. Soc. Hort. Sci. 105:819-822.

Wing KB, Pritts MP & Wilcox WF (1995) Field resistance of 20 strawberry cultivars to black root rot. Fruit Var. J. 49:94-98.

Woodard KE & Jones BL (1980) Screening peanut plant introductions in controlled environment chambers for resistance to *Rhizoctonia solani*. Plant Dis. 64:949-950.

Xie QJ, Linscombe SD, Rush MC & Jodari-Karimi F (1992) Registration of LSBR-33 and LSBR-5 sheath blight-resistant germplasm lines of rice. Crop Sci. 32:507.

Yang J & Verma PR (1992) Screening genotypes for resistance to pre-emergence damping-off and postemergence seedling root rot of oilseed rape and canola caused by *Rhizoctonia solani* AG 2-1. Crop Protection 11:443-448.

Zaumeyer WJ & Meiners JP (1975) Disease resistance in beans. Annu. Rev. Phytopathol. 13:313-334.

VI.C2 STRATEGIES FOR MOLECULAR RESISTANCE BREEDING (& TRANSGENIC PLANTS)

BEN J.C. CORNELISSEN[1], MIRJAM P. DOES[1] AND LEO S. MELCHERS[2]

[1]*Section of Plant Pathology Institute for Molecular Cell Biology BioCentrum Amsterdam Kruislaan 318 1098 SM Amsterdam,*
[2]*MOGEN International nv Einsteinweg 97 2333 CB Leiden,*
THE NETHERLANDS

I. Introduction

Despite the enormous input over the years in the development of control measures, diseases caused by fungi are still a major problem in agriculture. The epidemic spread of those diseases is controlled by a) agricultural management techniques, such as crop rotation and avoiding the spread of infested soil and diseased plant materials, b) the breeding of specific resistance traits into crops and c) the application of chemical compounds. Although breeders have successfully produced cultivars of crops resistant against specific fungal diseases, they are confronted continually with the evolvement of new races of fungal pathogens imposing new problems. Therefore, farmers still rely on the use of agrochemicals. However, the use of chemicals is costly and potentially damaging to the environment, and over time they become less efficient due to adaptation of the pathogen. High production costs together with a growing public concern about the environment have stimulated research into new alternatives in crop protection. One such an alternative is biological control, another the application of gene transfer techniques. During recent years a number of strategies to control fungal diseases with transgenic plants has been proposed. These strategies include the expression of genes encoding proteins able to inhibit growth of fungi, the fungus-inducible production of compounds toxic to either the invading fungus or plant cells, the production of inhibitors of fungal enzymes and others (reviewed by Lamb *et al.*, 1992; Cornelissen and Melchers, 1993). New approaches are still emerging.

Diseases caused by *Rhizoctonia* species can be controlled by cultural methods (chapter VI.A), but as only means of control these methods are usually insufficient. Germplasm for resistance is hardly available (chapter VI.C1), hence a breeding approach to control the epidemic spread of *Rhizoctonia* is not feasible. Therefore, control of Rhizoctonia diseases relies on chemicals, either by treatment of seed and/or soil, or by foliar applications. As to the development of new alternatives, the state of the art in biological control is discussed in chapter VI.B. In this chapter we will review the progress in the development of transgenic plants resistant to *Rhizoctonia*. So far the approaches to develop transgenic plants resistant to *Rhizoctonia* have been limited to the expression of genes encoding antifungal proteins. Therefore, in this chapter the focus will be on this approach.

B. Sneh et al. (eds.),
Rhizoctonia Species: Taxonomy, Molecular Biology, Ecology, Pathology and Disease Constrol, 529–536.

II. Antifungal proteins

Proteins with the ability to inhibit the growth of fungi *in vitro* are abundantly present in nature. Whether they are involved in the defence of the producer of these proteins against fungal infections *in vivo* is not known. Nevertheless, genes encoding such antifungal proteins can be used to render plants resistant to fungi. Proteins able to inhibit the growth of *Rhizoctonia in vitro* include enzymes able to degrade chitin and ß-1,3-glucan, both major structural components of the cell walls of all fungi (Wessels JGH and Sietsma JH, 1981) except the oomycetes. Cell walls of this group of fungi lack detectable amounts of chitin, but like the wall of plant cells they contain cellulose instead. Recent data even suggest that oomycetes are lower plants rather than fungi (Baldauf and Palmer, 1993). A third group of proteins able to inhibit the growth of *Rhizoctonia in vitro* is formed by the ribosome-inactivating proteins.

A. Chitinases

Chitinase catalyzes the hydrolysis of chitin, an insoluble linear ß-1,4-linked homopolymer of *N*-acetylglucosamine. Both endo- and exo-chitinases are present in nature. Endo-chitinases are found in plants and fungi, whereas exo-chitinases are of bacterial origin exclusively.

1. Classification. Plant chitinase genes are classified into four families, Chi-a to Chi-d (Table 1), some of which are divided into two or more subfamilies (Meins *et al.*, 1994, adjusted by Neuhaus, 1995). Within a family genes show high degrees of homology with each other; virtually no homology is found between genes from different families. Genes Chi-a1, Chi-a2 and Chi-a4 encode chitinases that are known in literature as class I, II and IV chitinases, respectively; Chi-a5 encodes the stinging nettle (*Urtica dioica*) lectin precursor. From this precursor an N-terminal signal peptide and a C-terminal chitinase domain is cleaved off to yield the mature lectin (Lerner and Raikhel, 1992). Mature stinging nettle lectin (UDA) consists of two small, cysteine-rich domains both able to bind chitin. These domains are very homologous to the primary structure of hevein, a protein of 43 amino acids from the rubber tree (*Hevea brasiliensis*), and to the so-called chitin-binding domain of many mature chitinases (Beintema, 1994). Chi-a6 encodes a chitinase from beet with an unusually long proline-rich domain of 131 amino acids (Berglund *et al.*, 1995). Chi-b genes code for enzymes formerly known as class III chitinases, and gene Chi-c encodes a tobacco endo-chitinase showing significant sequence similarity to a number of bacterial exo-chitinases (Melchers *et al.*, 1994). Finally, Chi-d genes code for PR 4 and PR 4-like proteins (including *win* proteins), with (Chi-d1) or without (Chi-d2) a chitin-binding domain (Linthorst *et al.*, 1991; Hejgaard *et al.*, 1992; Ponstein *et al.*, 1994). Hevein is encoded by a Chi-d gene as well. This protein is the processing product of a preproprotein from which a N-terminal signal peptide and a C-terminal portion is cleaved off (Lee *et al.*, 1991).

Endo-chitinase activities have been found in both culture filtrates and extracts of many fungi. However, there is only a small number of reports on the purification and characterization of fungal chitinases, and the primary structures of only two or three of them have been published. Because of this lack of data a classification of fungal chitinases has not been proposed as yet.

Based on their primary structures prokaryotic chitinases have been classified into three groups (Watanabe *et al.*, 1993). Group A contains proteins homologous (at least 32% identity) to the catalytic domain of chitinase A1 from *Bacillus circulans*, group B enzymes are homologous (32% identity or more) to the catalytic domain of *B. circulans* chitinase D and group C chitinases do not show homology to proteins of either group. Group A chitinases show homology to the tobacco endo-chitinase encoded by Chi-c (Melchers *et al.*, 1994).

TABLE 1. Classification of plant chitinase genes as proposed by Meins *et al.* (1994) and as adjusted by Neuhaus (1995)

Family	Subfamily	Gene product known as	Other characteristics of gene product
Chi-a	Chi-a1	class I chitinase	
	Chi-a2	class II chitinase	
	Chi-a4	class IV chitinase	
	Chi-a5	UDA	
	Chi-a6		long proline-rich domain present
Chi-b		class III chitinase	
Chi-c			homologous to bacterial exo-chitinases
Chi-d	Chi-d1	class I PR-4 proteins, *win* proteins, hevein	chitin-binding or hevein domain present
	Chi-d2	class II PR-4 proteins	chitin-binding or hevein domain not present

2. Antifungal properties Thus far only a limited number of chitinases have been purified and tested for *in vitro* antifungal activity. It appeared that not all chitinases display antifungal activity *in vitro* and that the sensitivity of fungi for purified chitinases varies greatly. All Chi-a1 plant chitinases tested thus far, including one from rice that was found to be inhibitory for *R. solani*, the causal agent of the sheath blight disease of rice (Anuratha *et al.*, 1992), seem to be able to inhibit growth *in vitro* of at least some fungi (Schlumbaum *et al.*, 1986; Mauch *et al.*, 1988; Broekaert *et al.*, 1988; Verburg and Huynh, 1991; Sela-Buurlage *et al.*, 1993). A Chi-a2 chitinase from tobacco does not show antifungal activity (Sela-Buurlage *et al.*, 1993). In contrast, a purified Chi-a2 chitinase from barley inhibits growth *in vitro* of a number of fungi including *R. solani* (Leah *et al.*, 1991). The Chi-a4 chitinases tested thus far display antifungal activity toward a few fungi only (Huynh *et al.*, 1992). UDA inhibits several fungi *in vitro* (Broekaert *et al.*, 1989). *R. solani* as well shows sensitivity to UDA (Cammue *et al.*, 1992). The antifungal action of UDA differs from the action of enzymes with chitinase activity (Broekaert *et al.*, 1989). The only tested Chi-b chitinase does not inhibit fungal growth *in vitro* (Vogelsang and Barz, 1993), whereas the tobacco endo-chitinase that shows homology to bacterial chitinases displays only limited antifungal activity (Melchers *et al.*, 1994). Finally, all tested proteins encoded by Chi-d chitinase genes, hevein included, have the ability to inhibit growth of at least some fungi (Van Parijs *et al.*, 1991; Hejgaard *et al.*, 1992; Ponstein *et al.*, 1994).

Interestingly, some (plant) chitinases have been observed to inhibit fungal growth *in vitro* in a synergistic way when mixed with the products of other chitinase genes. A Chi-a1 chitinase has been found to act synergistically together with UDA, the mature product of a Chi-a5 chitinase gene (Broekaert *et al.*, 1989). Likewise, Chi-a1 chitinases from both barley and tobacco inhibit fungal growth *in vitro* in a synergistic way when mixed with Chi-d genes encoded products from either barley or tobacco (Hejgaard *et al.*, 1992; Ponstein *et al.*, 1994). Neither UDA nor the barley Chi-d gene products display chitinase activity themselves, despite the fact that they are encoded by chitinase genes. The observation that products of chitinase genes may act synergistically confirms the idea (Broekaert *et al.*, 1989) that the modes of action of these antifungal proteins are different.

So far fungal endo-chitinase have been purified mainly from mycoparasitic fungi like *Trichoderma* and *Gliocladium.* In one report three chitinases from *Trichoderma harzianum* were found to be unable to inhibit growth *in vitro* of *Botrytis cinerea*, despite the fact that the enzymes displayed a lytic activity on purified cell wall of the fungus (De la Cruz *et al.*, 1992). However, in another report a chitinase from *T. harzianum* has been described showing growth inhibiting activity on a number of phytopathogenic fungi including *B. cinerea* (Lorito *et al.*, 1993). In addition, combining this particular chitinase with a chitobiosidase from the same origin resulted in a synergistic increase in antifungal activity. Fungal growth inhibiting activity of an endo-chitinase from *Gliocladum virens* was found to be very low (Di Pietro *et al.*, 1993).

Bacterial chitinases have been reported to lack antifungal activity *in vitro* (Roberts and Selitrennikoff, 1988). However, later it was found that growth of at least *R. solani* could be inhibited by a purified chitinase from *Serratia marcescens* (Jach *et al.*, 1992).

B. ß-1,3-glucanases

ß-1,3-endoglucanases catalyse the hydrolysis of ß-1,3-glucan. These enzymes are found both in plants and fungi. In plants three classes (I-III) have been described (Ward *et al.*, 1991). Class I ß-1,3-glucanases from various plant sources display growth inhibiting activity *in vitro* to some fungi (Mauch *et al.*, 1988; Sela-Buurlage *et al.*, 1993). A class II ß-1,3-glucanase from barley has been shown to inhibit growth *in vitro* as well (Leah *et al.*, 1991). However, a tobacco class II ß-1,3-glucanases lacks antifungal activity. The combination of a class I ß-1,3-glucanase with either a Chi-a1 chitinase (Mauch *et al.*, 1988; Sela-Buurlage *et al.*, 1993), a Chi-c chitinase (Melchers *et al.*, 1994) or a Chi-d chitinase (Ponstein *et al.*, 1994), both type of hydrolases in quantities insufficient to show an effect by themselves, results in a very strong antifungal activity, indicating that these enzymes act synergistically. Also the barley class II ß-1,3-glucanase acts synergistically with a barley Chi-a2 chitinase (Leah *et al.*, 1991).

C. Ribosome-inactivating proteins

Plant ribosome-inactivating proteins (RIPs) inhibit protein synthesis by modifying 28S rRNA (reviewed by Stirpe *et al.*, 1992). RIPs do not affect ribosomes of plants in which they are produced and show various degrees of specificity towards ribosomes of other plants. Fungal ribosomes are targets of RIPs as well. A RIP isolated from barley was shown to exhibit antifungal activity *in vitro* to a number of plant pathogenic fungi, including *R. solani* (Leah *et al.*, 1991). When mixed with either a Chi-a2 chitinase or a class II ß-1,3-glucanase from barley, a strong synergy in antifungal activity was observed.

III. ENHANCED RESISTANCE IN TRANSGENIC PLANTS

The occurrence of proteins with the ability to inhibit fungal growth *in vitro* has led to the idea that the constitutive or pathogen-inducible expression of the corresponding genes in transgenic plants may render these plants fungus resistant. In the last five years several studies have been published that support this idea. In most of these studies *R. solani* has been used to challenge-inoculate transgenic plants to determine levels of resistance. Broglie and co-workers (Broglie *et al.*, 1991) were the first to show enhanced fungal resistance brought about by the expression of an antifungal transgene. They introduced a Chi-a1 chitinase gene from bean under the control of the cauliflower mosaic virus 35S promoter into both tobacco and canola. Compared to control plants, transgenic lines exhibiting elevated levels of chitinase activity were found to be more resistant to *R. solani*: more vigorous growth and lower seedling mortality during growth in inoculated soil were seen. Results of ultra structural and cytochemical studies on the *R. solani* inoculated, transgenic canola plants confirmed the idea that the produced chitinase is at least in part responsible for the observed enhanced resistance (Benhamou *et al.*, 1993). Enhanced protection against *R. solani* was also observed in transgenic rice plants constitutively expressing a rice Chi-a1 chitinase gene (Lin *et al.*, 1995), in *Nicotiana sylvestris* plants transgenic for a Chi-a1 chitinase gene from tobacco (Vierheilig *et al.*, 1993) and in tobacco plants constitutively expressing a ChiA exo-chitinase gene from *S. marcescens* (Jach *et al.*, 1992; Howie *et al.*, 1994).

The observation that chitinases and ß-1,3-glucanases may act synergistically in inhibiting fungal growth *in vitro*, has prompted the production and testing of transgenic plants co-expressing genes for both hydrolases. The constitutive co-expression in tomato of a Chi-a1 chitinase gene and a class I ß-1,3-glucanase, both from tobacco, resulted in increased resistance against *Fusarium oxysporum*, whereas plants expressing either one of the genes did not show enhanced resistance (Van den Elzen *et al.*, 1993; Jongedijk *et al.*, 1995). This result indicates that synergy between antifungal proteins occurs *in planta* as well as *in vitro*. The constitutive expression in *N. sylvestris* of a tobacco Chi-a1 chitinase gene did not alter the susceptibility of the plants to the fungal leaf pathogen *Cercospora nicotianae*, despite the fact that *Cercospora* species are sensitive to chitinases *in vitro* (Neuhaus *et al.*, 1991). Likewise, only some protection against *C. nicotianae* was found in tobacco plants strongly expressing either a Chi-a1 chitinase genes from rice or a class I ß-1,3-glucanase gene from alfalfa. However, upon co-expression of both genes in transgenic tobacco significant levels of resistance were observed (Zhu *et al.*, 1994). As *in vitro,* synergy between a barley Chi-a2 chitinase and a class II ß-1,3-glucanase was observed *in planta* as well: after challenge-inoculation with *R. solani*,up to 60% reduction in disease development was found in transgenic tobacco plants co-expressing the two corresponding genes (Jach *et al.*, 1995).

Progenies of transgenic tobacco with the gene encoding the antifungal RIP from barley seeds under the control of a wound-inducible promoter showed an increased resistance to *R. solani* (Logemann *et al.*, 1992; Jach *et al.*, 1995). Levels of resistance found in these plants were higher than reported earlier for transgenic tobacco plants constitutively expressing the bacterial ChiA exo-chitinase gene. However, the best results were obtained by the constitutive co-expressing of the barley RIP gene and Chi-a2 chitinase gene in tobacco: up to 55% reduction in disease development was found after inoculation with *R. solani* (Jach *et al.*, 1995).

Thus far there is only one report on enhanced protection against fungal diseases brought about by the constitutive expression of a gene encoding an antifungal protein in field grown plants. Over two consecutive years Howie and co-workers (1994) have performed field trials to assess the effect of the constitutive expression of a *S. marcescens ChiA* chitinase gene in transgenic tobacco on infection by *R. solani*. Consistently, disease tolerance was observed in transgenic plants regardless the accumulation of the chitinase intra- or extracellularly (Howie *et al.* 1994).

IV. CONCLUDING REMARKS

For the development of transgenic plants resistant to *Rhizoctonia* only one approach has been employed thus far, notably the expression of genes encoding antifungal proteins. Products of several chitinase genes, glucanases and a barley ribosome-inactivating protein have been found to inhibit growth of *Rhizoctonia in vitro* to a more or lesser extent. Most if not all antifungal proteins seem to inhibit fungal growth in a synergistic way when mixed with a second antifungal protein. This synergy is observed both *in vitro* and *in planta* when two genes encoding antifungal proteins are co-expressed. Enhanced levels of *Rhizoctonia* resistance in transgenic tobacco plants brought about by the accumulation of antifungal proteins are not only observed in greenhouse experiments but also in the only fieldtrial published today. In greenhouse experiments up to 60% reduction in disease development has been observed upon inoculation of transgenic tobacco co-expressing two antifungal genes with *R. solani*. However, for the antifungal-protein-strategy followed thus far to be successful at the long term, levels of resistance in transgenic plants should be enhanced. This might be achieved by the application of either new genes encoding proteins with higher antifungal activity or new combinations of genes. Alternatively the strategy of the constitutive expression of genes encoding antifungal proteins could be used in combination with other approaches.

V. ACKNOWLEDGEMENTS

We thank Dr Christoph Maas (Max-Planck Institute für Züchtungsforschung, Cologne, Germany) for providing us with a manuscript prior to publication.

VI. REFERENCES

Anuratha CS, Huang JK, Pingali A & Muthukrishnan S (1992) Isolation and characterization of a chitinase and its cDNA clone from rice. J. Plant Biochem. & Biotech. 1:5-10.

Baldauf SL & Palmer JD (1993) Animals and fungi are each other's closest relatives:Congruent evidence from multiple proteins . Proc. Natl. Acad. Sci. USA 90:11558-11562.

Beintema JJ (1994) Structural features of plant chitinases and chitin-binding proteins. FEBS Lett. 350:159-163.

Benhamou N, Broglie K, Chet I & Broglie R (1993) Cytology of infection of 35S chitinase transgenic canola plants by *Rhizoctonia solani*:cytochemical aspects of chitin breakdown *in vivo*. Plant J. 4:295-305.

Berglund L, Brunstedt J, Nielsen KK, Chen Z, Mikkelsen JD & Marcker KA (1995) A proline-rich chitinase from *Beta vulgaris*. Plant Mol. Biol. 27:211-216.

Broekaert WF, Van Parijs J, Allen AK & Peumans WJ (1988) Comparison of some molecular, enzymatic and antifungal properties of chitinases from thorn-apple, tobacco and wheat. Physiol. Mol. Plant Pathol. 33:319-331.

Broekaert WF, Van Parijs J, Leyns F, Joos H & Peumans WJ (1989) A chitin-binding lectin from stinging nettle rhizomes with antifungal activity. Science 245:1100-1102.

Broglie K, Chet I, Holliday M, Cressman R, Biddle P, Knowlton S, Mauvais CJ & Broglie R (1991) Transgenic plants with enhanced resistance to the fungal pathogen *Rhizoctonia solani*. Science 254:1194-1197.

Cammue BPA, DeBolle MFC, Terras FRG, Proost P, Van Damme J, Rees SB, Vanderleyden J & Broekaert WF (1992) Isolation and characterization of a novel class of plant antimicrobial peptides from *Mirabilis jalapa* seeds. J. Biol. Chem. 267:2228-2233.

Cornelissen BJC & Melchers LS (1993) Strategies for control of fungal diseases with transgenic plants. Plant Physiol. 101:709-712.

De la Cruz J, Hidalgo-Gallego A, Lora JM, Benitez T, Pintor-Toro JA & Llobell A (1992) Isolation and characterization of three chitinases from *Trichoderma harzianum*. Eur. J. Biochem. 206:859-867.

Di Pietro A, Lorito M, Hayes CK, Broadway RM & Harman GE (1993) Endochitinase from *Gliocladium virens*:isolation, characterization, and synergistic antifungal activity in combination with gliotoxin. Phytopathology 83:308-313.

Hejgaard J, Jacobsen S, Bjorn SE & Kragh KM (1992) Antifungal activity of chitin-binding PR-4 type proteins from barley grain and stressed leaf. FEBS Lett. 307:389-392.

Jach G, Logemann S, Wolf G, Oppenheim A, Chet I, Schell J & Logemann J (1992) Expression of a bacterial chitinase leads to improved resistance of transgenic tobacco plants against fungal infection. Biopract. 1:33-40.

Jach G, Grönhardt B, Mundy J, Logemann J, Pinsdorf E, Leah R, Echell J & Maas C (1995) Enhanced quantitative resistance against fungal disease by combinatorial expression of different barley antifungal proteins in transgenic tobacco. Plant J. 8:97-109.

Howie W, Joe L, Newbigin E, Suslow T & Dunsmuir P (1994) Transgenic tobacco plants which express the *chi*A gene from *Serratia marcescens* have enhanced tolerance to *Rhizoctonia solani*. Trans. Res. 3:90-98.

Huynh QK, Hironaka CM, Levine EB, Smith CE, Borgmeyer JR & Shah DM (1992) Antifungal proteins from plants:purification, molecular cloning, and antifungal properties of chitinases from maize seed. J. Biol. Chem. 267:6635-6640.

Jongedijk E, Tigelaar H, van Roekel JSC, Bres-Vloemans SA, Dekker I, van den Elzen PJM, Cornelissen BJC & Melchers LS (1995) Synergistic activity of chitinases and ß-1,3-glucanases enhances fungal resistance in transgenic tomato plants. Euphytica 85:173-180.

Lamb CJ, Ryals JA, Ward ER & Dixon RA (1992) Emerging strategies for enhancing resistance to microbial pathogens. Bio/Techn. 10:1436-1445.

Leah R, Tommerup H, Svendsen I & Mundy J (1991) Biochemical and molecular characterization of three barley seed proteins with antifungal properties. J. Biol. Chem. 266:1464-1573.

Lee H, Broekaert WF & Raikhel NV (1991) Co- and post-translational processing of the hevein preproprotein of latex of the rubber tree (*Hevea brasiliensis*). J. Biol. Chem. 266:15944-15948.

Lerner DR & Raikhel NV (1992) The gene for stinging nettle lectin (*Urtica dioica* agglutinin) encodes both a lectin and a chitinase. J. Biol. Chem. 267:11085-11091.

Lin W, Anuratha CS, Datta K, Potrykus I, Muthukrishnan S & Datta SK (1995) Genetic engineering of rice for resistance to sheath blight. Bio/Technol. 13:686-691.

Linthorst HJM, Danhash N, Brederode FT, Van Kan JAL, De Wit PJGM & Bol JF (1991) Tobacco and tomato PR proteins homologous to *win* and pro-hevein lack the 'hevein' domain. Mol. Plant-Microbe Interact. 4:586-592.

Logemann J, Jach G, Tommerup H, Mundy J & Schell J (1992) Expression of a barley ribosome inactivating protein leads to increased fungal protection in transgenic tobacco plants. Bio/Technol. 10:305-308.

Lorito M, Harman GE, Hayes CK, Broadway RM, Tronsmo A, Woo SL & Di Pietro A (1993) Chitinolytic enzymes produced by *Trichoderma harzianum*:antifungal activity of purified endochitinase and chitobiosidase. Phytopathology 83:302-307.

Mauch F, Mauch-Mani B & Boller T (1988) Antifungal hydrolases in pea tissue. II Inhibition of fungal growth by combination of chitinase and ß-1,3-glucanase. Plant Physiol. 88:936-942.

Melchers LS, Apotheker-De Groot M, van der Knaap JA, Ponstein AS, Sela-Buurlage MB, Bol JF, Cornelissen BJC, van den Elzen PJM & Linthorst HJM (1994) A new class of tobacco chitinases homologous to bacterial exo-chitinases displays antifungal activity. Plant J. 5:469-480

Meins F, Fritig B, Linthorst HJM, Mikkelsen JD, Neuhaus J-M & Ryals J (1994) Plant chitinase genes. Suppl. to Plant Mol. Biol. Rep. 12:S22-S28.

Neuhaus J-M, Ahl-Goy P, Hinz U, Flores S & Meins F (1991) High-level expression of a tobacco chitinase gene in *Nicotiana sylvestris*. Susceptibility of transgenic plants to *Cercospora nicotianae* infection. Plant Mol. Biol. 16:141-151.

Neuhaus J-M (1995) Proposal presented at the '4th International workshop on pathogenesis-related proteins in plants:biology and biotechnological potential', held September 3-7, 1995, Irsee, Germany.

Ponstein AS, Bres-Vloemans SA, Sela-Buurlage MB, Van den Elzen PJM, Melchers LS & Cornelissen BJC (1994) A novel pathogen- and wound-inducible tobacco (*Nicotiana tabacum*) protein with antifungal activity. Plant Physiol. 104:109-118.

Roberts WK & Selitrennikoff CP (1988) Plant and bacterial chitinases differ in antifungal activity. J. Gen. Microbiol. 134:169-176.

Schlumbaum A, Mauch F, Vogeli U & Boller T (1986) Plant chitinases are potent inhibitors of fungal growth. Nature 324:365-367.

Sela-Buurlage MB, Ponstein AS, Bres-Vloemans SA, Melchers LS, van den Elzen PJM & Cornelissen BJC (1993) Only specific tobacco (*Nicotiana tabacum*) chitinases and ß-1,3-glucanases exhibit antifungal activity. Plant Physiol. 101:857-863.

Stirpe F, Barbieri L, Battelli LG, Soria M & Lappi DA (1992) Ribosome-inactivating proteins from plants:present status and future prospects. Bio/Technol. 10:405-412.

Van den Elzen PJM, Jongedijk E, Melchers LS & Cornelissen BJC (1993) Virus and fungal resistance:from laboratory to field, Phil. Trans. R. Soc. London B 342:835-838.

Van Parijs J, Broekaert WF, Goldstein I.J & Peumans WJ (1991) Hevein:an antifungal protein from rubber-tree (*Hevea brasiliensis*) latex. Planta 183:258-264.

Verburg JG & Huynh QK (1991) Purification and characterization of an antifungal chitinase from *Arabidopsis thaliana*. Plant Physiol. 95:450-455.

Vierheilig H, Alt M, Neuhaus JM, Boller T & Wiemke A. (1993) Colonization of transgenic *Nicotiana sylvestris* plants expressing different forms of *Nicotiana tabacum* chitinase, by the root pathogen *Rhizoctonia solani* and by the mycorrhizal symbiont *Glomus mosseae*. Mol. Plant-Microbe Interact. 6:261-264.

Vogelsang R & Barz W (1993) Purification, characterization and differential hormonal regulation of a ß-1,3-glucanase and two chitinases from chickpea (*Cicer arietinum* L.). Planta 189:60-69.

Ward ER, Payne GB, Moyer MB, Williams SC, Dincher SS, Sharkey KC, Beck JH, Taylor HT, Ahl-Goy P, Meins F, Ryals JA (1991) Differential regulation of ß-1,3-glucanase messenger RNAs in response to pathogen infection. Plant Physiol 96:390-397.

Watanabe T, Kobori K, Miyashita K, Fujii T, Sakai H, Uchida M & Tanaka H (1993) Identification of glutamic acid 204 and aspartic acid 200 in chitinase A1 of *Bacillus circulans* WL-12 as essential residues for chitinase activity. J. Biol. Chem. 268:18567-18572.

Wessels JGH & Sietsma JH (1981) Fungal cell walls:a survey. In:W. Tanner and F.A. Loewus (Eds.) Encyclopedia of Plant Physiology, New series, vol. 13B:Plant Carbohydrates. (pp 352-394) Springer Verlag, Berlin.

Zhu Q, Maher EA, Masoud S, Dixon RA & Lamb CJ (1994) Enhanced protection against fungal attack by constitutive co-expression of chitinase and glucanase genes in transgenic tobacco. Bio/Technol. 12:807-812.

VI.D CHEMICAL CONTROL OF *RHIZOCTONIA* SPECIES

HANS R. KATARIA[1] AND ULRICH GISI[2]

[1]*International Rice Research Institute, IRRI-India Liaison Office, C-18 Friends Colony (East), New Delhi 110 065, INDIA,*
[2]*Agrobiological Research Station, Sandoz Agro Limited, CH-4108 Witterswil, SWITZERLAND*

I. INTRODUCTION

Ever since the destructive potential of *Rhizoctonia* pathogens on crop plants was realized, application of synthetic fungicides has been the most popular disease control measure employed by the farmers on various crops and the use of fungicides is practiced widely throughout the world. Even today many farmers will select disease-prone and high-yielding varieties and rely totally on fungicide treatments to maintain crop health. Disease control with fungicides has been largely successful against *Rhizoctonia* pathogens on many different crops in both temperate and tropical agro-ecological regions, attributable to targeted development of a variety of potent fungicides of diverse chemical nature and appropriate application methods against various Rhizoctonia diseases. But disease control with recommended anti-*Rhizoctonia* fungicides has not always been successful in some situations, and failures to control Rhizoctonia diseases on the same or different crop in an agro-ecological region are also documented in the literature, contrary to general belief that a potent fungicide shall curb the pathogen under all situations. The reasons for such diversity in results for satisfactory disease control are many and intricate, but have hardly been reasoned out scientifically. The purity of fungicides, their formulations and their improper usage have often been the alibi for extension workers to console the cultivators for poor disease control with the tested anti-*Rhizoctonia* fungicides. This contention is strongly discounted by the information generated from studies on quadrangular interactions of host-pathogen-fungicide-environment in the control of Rhizoctonia diseases. Recent progress in taxonomy and ecological diversity of population structures of *Rhizoctonia* pathogens in major cropping systems has greatly helped to understand the possible practical reasons for the effectiveness or ineffectiveness of fungicides to control *Rhizoctonia* in the fields.

Chemical control of *Rhizoctonia* has been the subject of hundreds of research papers and reports published since 1913 when steam-formalin treatment was first recommended to control potato black scurf (Winston, 1913). Rather than a general literature review, this chapter is intended to highlight strategic considerations in using fungicides to achieve commercially desired control of Rhizoctonia diseases, with a viewpoint as to how best we can control *Rhizoctonia* with the available highly potent fungicides, using as lower amounts as possible and as much as necessary, with target-oriented fungicide treatments in conjunction with the other plant protection chemicals and technologies.

II. FUNGICIDE TREATMENTS AGAINST RHIZOCTONIA DISEASES

Rhizoctonia diseases are controlled by seed or soil treatments or both, or foliar applications of a variety of fungicides of diverse chemical groups, *viz.*, aromatic

B. Sneh et al. (eds.),
Rhizoctonia Species: Taxonomy, Molecular Biology, Ecology, Pathology and Disease Constrol, 537–547.

hydrocarbons (quintozene, chloroneb, etridiazole, tolclofos-methyl), carboxamides (carboxin, benodanil, flutolanil, furmecyclox, pyracarbolid), benzimidazoles (thiabendazole, carbendazim, benomyl, thiophanate-methyl), dicarboximides (iprodione, dichlozolin, vinclozolin), triazoles (propiconazole, cyproconazole, hexaconazole, flusilazole, tebuconazole, flutriafol, triadimefon, triadimenol), morpholine (fenpropimorph), phenylurea (pencycuron), validamycin (validamycin A), phenylpyrroles (fenpiclonil, fludioxonil) and β-methoxyacrylates (ICI A5504, BAS 490F). Among these fungicides, quintozene has been most popular and had its monopoly for *Rhizoctonia* control for a long time (1935 - 1965) until the first systemic fungicide carboxin was commercially launched in 1966 and proved highly effective against *R. solani* damping-off and root rot when used as seed or soil treatment. Currently, tolclofos-methyl and pencycuron are the most popular fungicides for *Rhizoctonia* control; pencycuron has been developed specifically to control rice sheath blight and potato scurf, while tolclofos-methyl outperforms most other fungicides for its excellence in controlling almost all types of Rhizoctonia diseases on a number of crop species in diverse environmental conditions. Triazole fungicides possess the potential to control various Rhizoctonia diseases on stem, foliage and subterranean plant parts at very low application rates, but success in disease control with these fungicides is highly variable in different crops and geographic regions. Validamycin A is specifically effective against certain plant diseases caused by *Rhizoctonia* spp., and has been commercially used for the control of sheath blight of rice in Japan since 1973 but not elsewhere on rice or other crops (Wakae and Matsuura, 1975). Phenylpyrroles and β-methoxyacrylates (strobilurins) are derivatives of natural antifungal compounds produced by antagonistic bacteria and fungi respectively, and offer interesting opportunity to control Rhizoctonia diseases with low amounts (Leadbitter *et al.,* 1995). β-methoxyacrylates exhibit slow systemic mobility in plants, a property very desirable in successful long-term control of Rhizoctonia infections.

Seed treatments provide excellent protection against damping-off and seedling root rot for 2-3 weeks. Soil applications with fungicides (drench, dust or granules) are made for extended protection from soil-borne inoculum for longer duration, sometimes throughout the crop season as in case of potato, sugarbeet, peanut, cereals, flowerbulbs and horticultural crops. Fungicides are sprayed on to the aerial plant parts to control stem and foliage blights caused by *Rhizoctonia* fungi in beans, sugarbeet, rice, corn and turfgrasses. Though most anti-*Rhizoctonia* fungicides can be applied to seed, soil and foliage, they exhibit considerable variations in the level of disease control on subterranean and above-ground plant parts, because of marked differences in the physical properties of fungicides and their formulations in terms of lipophilicity and wettability, distribution and retention in root region, penetrability in foliage, redistribution and translaminar mobility, and systemic activity in soil and plants (Lyr, 1995).

In some crops, specifically sugarbeet and beans, *Rhizoctonia* fungi cause many types of disease symptoms during a season, such as seedling damping-off, leaf spots, petiole blight, pod blight, collar rot and root rot. In such situations, the researchers and cultivators are encountered with the difficult task of selecting appropriate fungicides for seed, soil and foliar applications. The big question is: is there one single fungicide that can be applied to seed and soil as well as to stem and foliage for the control of *Rhizoctonia* infections on seedlings, leaves, pods, stems and roots during the entire crop season, or are different fungicides needed to control different phases of the disease in a crop? Also, there are several plant pathogenic species of *Rhizoctonia* and their anastomosis groups (AGs) capable of infecting various crops. Many plant species, particularly rice, wheat, corn, sugarbeet, legumes, strawberry and turfgrasses are susceptible to several species and AGs of *Rhizoctonia* and are infected by them simultaneously or in succession during different growth stages. Successful disease control against multiple infections by *Rhizoctonia* pathogens requires fungicides with activity against all species and AGs involved in a given disease situation. The question often asked is: which of the currently available fungicides are capable of controlling all major *Rhizoctonia* species and AGs? Therefore, it is of utmost importance to seek information on fungitoxic potential of different anti-*Rhizoctonia* fungicides towards various pathogenic species and AGs.

III. Sensitivity of *Rhizoctonia* species and different AGs to fungicides

A. Aromatic hydrocarbon fungicides

Over the past several decades quintozene treatments against *R. solani* have demonstrated effective but highly variable disease control on various crops in different regions. Isolates of *R. solani* from diseased cotton plants differed significantly in their sensitivity to quintozene (Shatla and Sinclair, 1963). This and other early studies indicated that ecologically different isolates of *R. solani*, as well as isolates from the same field, differ widely in their level of sensitivity to quintozene, and tolerance to quintozene existed in some indigenous pathogenic isolates of *R. solani* in fields. Quintozene shows strong activity towards *R. zeae, R. oryzae*, other *Waitea* isolates, binucleate *Rhizoctonia* isolates (*Ceratobasidium* spp.) and AGs 2-2, 6, 9, and BI of *R. solani*, moderate activity against AGs 1, 3, 5, while AGs 2-1, 4, 7 and 8 are least sensitive (Carling *et al.*, 1990). Smiley *et al.* (1990) noted tolerance to quintozene in *R. oryzae* and AG 8 of *R. solani* infecting wheat. By contrast, tolclofos-methyl (an aromatic hydrocarbon fungicide with phosphorothionate moiety) is very effective against all major pathogenic *Rhizoctonia* species and AGs, and exhibits little or no selectivity within these fungi (Ohtsuki and Fujinami, 1982; Smiley *et al.*, 1990; Kataria *et al.*, 1991a,b; Phillips, 1992; Olaya *et al.*, 1994; Anon, 1995). It provides very strong inhibition at quite low concentrations and sclerotia of *Rhizoctonia* are killed within 30 min after contact with this compound. Insensitivity or tolerance to tolclofos-methyl among indigenous populations of *Rhizoctonia* fungi is very rare (van Bruggen and Arneson, 1984).

B. Carboxamide fungicides

Though all carboxamide fungicides possess anti-*Rhizoctonia* activity, they provide different levels of growth inhibition and disease control owing to differences in structural configuration, systemic functions and metabolic stability in plant tissues and fungal cells (Snel *et al.*, 1970; Lyr *et al.*, 1974; Huppatz *et al.*, 1983; Sumner *et al.*, 1992). Benodanil has strong activity against *C. cornigerum, R. cerealis* and *R. solani* but weak activity against *R. zeae* and *R. oryzae* (Kataria and Gisi, 1989; Kataria *et al.*, 1991a). Carboxin is very active against *R. cerealis* and *R. solani* and much more effective against *R. zeae* than against *R. oryzae, R. sasakii* and *C. cornigerum*. Carboxin and furmecyclox exhibit wide variability in their activity against various AGs of *R. solani*, and though the majority of AGs are sensitive to these two fungicides, furmecyclox exerts higher activity than carboxin against each AG (Kataria *et al.*, 1991b); AG 8 is the most sensitive and AG 3 the least sensitive to both carboxin and furmecyclox. Carboxin shows moderate to weak activity towards AG 2-2 and AG 5. Smiley *et al.* (1990) found an AG 8 isolate to be tolerant to carboxin and moderately tolerant to furmecyclox. Photosensitivity of carboxin restricts its use to seed or soil treatment and is not suitable as foliar spray to control stem and foliage infections. Furmecyclox, flutolanil and benodanil have good photostability and provide excellent protection against *Rhizoctonia* pathogens in soil and leaf surfaces (Huppatz *et al.*, 1983; Martin and Torres, 1986; Hirooka *et al.*, 1989; Kataria *et al.*, 1993).

C. Benzimidazole fungicides

Thiabendazole shows strong activity towards various AGs of *R. solani*, though AG 2-1 and AG 2-2 are less sensitive than the other AGs (Kataria *et al.*, 1991b). Benomyl, carbendazim and thiophanate-methyl show strong activity against all species and AGs of *Rhizoctonia*; isolates of *W. circinata* (*R. zeae* and *R. oryzae*) infecting rice, wheat and barley, and *Thanatephorus cucumeris* (*R. solani* and *R. sasakii*) are more sensitive to benomyl than the isolates of *C. gramineum* (*R. cerealis*) and *C. cornigerum*. However, the isolates of *R. zeae* and *R. oryzae* from turfgrasses are reported to be resistant to benomyl (Christensen, 1979; Martin *et al.*, 1984a,b; Carling *et al.*, 1990). The variable levels of activity against different *Rhizoctonia* populations may cause difficulties when benomyl is used to control Rhizoctonia blights on turfgrasses, rice and cereals.

D. Dicarboximide fungicides

Iprodione is highly effective against *R. cerealis* and *R. solani* but shows poor activity against *R. zeae*, *R. oryzae*, and other isolates of *Waitea* (Carling *et al.*, 1990; Smiley *et al.*, 1990; Kataria *et al.*, 1991a; Olaya *et al.*, 1994). Grisham (1983), Price *et al.* (1983) and Martin *et al.* (1984a,b) reported that iprodione inhibited *R. solani* and *R. cerealis* on various graminaceous hosts but was ineffective against *R. sasakii* infections, and high concentrations were required to control foliar blight of turf caused by *R. zeae*. Iprodione exerts consistently strong activity against all AGs of *R. solani* (Kataria *et al.*, 1991b), but a variable activity against different isolates of binucleate *Rhizoctonia* species (Carling *et al.*, 1990). Though most binucleate *Rhizoctonia* pathogens are sensitive to iprodione, occurrence of resistant isolates is not uncommon in turf fields (Sanders *et al.*, 1978a).

E. Triazole fungicides

Of the several triazole fungicides investigated against *Rhizoctonia* fungi, only cyproconazole and hexaconazole have shown remarkably strong activity against all *Rhizoctonia* species and AGs; they strongly inhibit all AGs of *R. solani*, all binucleate *Rhizoctonia* isolates including *R. cerealis* and *C. cornigerum*, and *R. sasakii*, *R. zeae*, *R. oryzae* and other isolates of *Waitea* (Carling *et al.*, 1990; Kataria *et al.*, 1991a,b). Triadimenol shows strong activity against *R. sasakii*, *R. oryzae*, *R. solani* AGs 1 and 4, moderate activity against *R. cerealis*, *C. cornigerum*, *R. zeae* and *R. solani* AGs 2-1 and 3; AG 8 is tolerant to triadimenol (Smiley *et al.*, 1990). Triadimefon, generally less fungitoxic than triadimenol, shows poor activity against *R. cerealis*, and strongly inhibits *R. solani* AGs at 22°C but not at 27°C (Kataria and Gisi, 1989). Propiconazole and flusilazole show strong activity against *R. zeae* and *R. oryzae*, moderately strong activity against *R. cerealis*, but are generally least effective against *R. solani* AGs with wide inter-isolate variations. In view of contrasting differences in the antifungal spectrum of different triazole fungicides within *Rhizoctonia* fungi and other physico-chemical properties, indiscriminate applications of these fungicides may not always yield satisfactory disease control in cropping systems supporting mixed populations of *Rhizoctonia* pathogens (Cotterill *et al.*, 1989).

F. Morpholine, imidazole and pyrimidine fungicides

Fenpropimorph (morpholine) exerts strong activity towards *R. zeae, R. oryzae, R. sasakii, R. solani* AGs 1, 2-1, 2-2, 3, 6, 7 and 9, moderate activity against AGs 4, 5 and 8, and is largely ineffective against *R. cerealis* and *C. cornigerum*. Imazalil (imidazole) shows strong activity against *R. zeae, R. oryzae, R. cerealis* and *C. cornigerum*, but little or no activity against various AGs of *R. solani*. By contrast, prochloraz (another imidazole) shows an almost complete lack of fungitoxic activity against most *Rhizoctonia* species and AGs. Fenarimol and nuarimol (both pyrimidines) exhibit strong activity against *R. oryzae* and *R. sasakii*, moderate to poor activity against *R. zeae, R. cerealis* and *C. cornigerum*, and highly variable activity towards different AGs of *R. solani*: strong activity against AG 1, moderate activity against AG 7, and poor or no activity against the other AGs (Kataria and Gisi, 1989; Carling *et al.*, 1990; Smiley *et al.*, 1990; Kataria *et al.*, 1991a,b). Imidazole and pyrimidine fungicides are not used primarily for *Rhizoctonia* control but in mixtures with several anti-*Rhizoctonia* fungicides for broad spectrum disease management.

G. Phenylurea fungicide

Among all the fungicides investigated so far against *Rhizoctonia* fungi, the phenylurea fungicide pencycuron has shown the highest level of specificity of action within these fungi, and a complete lack of activity towards other fungi (Yamada *et al.*, 1988). It shows very strong activity against *R. solani* AGs 1, 2-1, 2-2, 3, 6 and 9, but is totally inactive against AGs 5, 7, 8, *R. oryzae, R. cerealis, C. cornigerum, C. oryzae sativae* and other binucleate *Rhizoctonia* species (Roberts and Stephens, 1984; Sumner, 1987; Kataria *et al.*, 1991a,b; Olaya *et al.*, 1994). AG 4 isolates from legumes, cereals, millets, eggplant, cotton and chrysanthemum are reported to be highly sensitive to pencycuron, while a large number of AG 4 isolates infecting canola, radish and sugarbeet are tolerant to pencycuron. Since the majority of *Rhizoctonia* species are tolerant to pencycuron and only some AGs are highly sensitive, treatments with this fungicide for *Rhizoctonia* control have to be targeted very specifically only after precise identification of the pathogenic populations.

H. Phenylpyrrole fungicides

Fludioxonil and fenpiclonil show strong activity against *R. solani* AGs 1, 2-1, 2-2, 3, 4, 5 as well as binucleate *Rhizoctonia* isolates, excellent disease control against damping-off, stem canker, root rot and foliar blights in many different crops, and exhibit promising results in *Rhizoctonia* control with low amounts for seed treatment and foliar application (Olaya *et al.*, 1994; Leadbitter *et al.*, 1995). Concerted efforts should continue to screen these and other novel fungicides against various *Rhizoctonia* pathogens from different cropping systems and climates to develop target-oriented treatments for maximum disease control benefits.

IV. FACTORS AFFECTING DISEASE CONTROL

A. Host factors

Varying rates and ways of fungicide applications are employed in various crops in different climatic zones, and the level of disease control by a recommended fungicide against a *Rhizoctonia* pathogen on different crops is likely to vary and it has. Such variations in disease control are usually attributed to the climate and agronomic factors associated with each crop and rarely to the involvement of the host itself in deciding the level of disease control by fungicides. Using four host species inoculated with the same *R. solani* isolate (AG 4) under identical conditions in the greenhouse, Kataria and Grover (1976) demonstrated that disease control by a fungicide varied greatly on different hosts: benomyl gave over 90% control on mungbean but only 45-50% control on long melon and eggplant; quintozene provided 90% control on sugarbeet but only 40% control on eggplant and mung bean; chloroneb gave 70-80% control on sugarbeet and mungbean but only 10% control on eggplant. Disease control by a potent fungicide may also vary in closely related host cultivars; for example, soil treatment with tolclofos-methyl provided 60-70% control of bulb rot in tulip cvs Giuseppe Verdi, Apeldoorns Elite and Roodkapje but no control at all in cv. Christmas Marvel (Koster and Meer, 1990). Seed-treatment fungicides give much better control against damping-off and seedling root rot in *Brassica napus* cv. Westar than in *B. rapa* cv. Tobin inoculated with isolates of either AGs 2-1 or 4. Disease control with most fungicides is generally poor in *B. oleracea* (most susceptible to AG 2-1), while even low amounts of fungicides on the seed of *B. nigra* and *B. juncea* (both moderately resistant to AG 2-1) provide excellent disease control (Kataria and Verma, 1990; Kataria *et al.*, 1993). Varying levels of susceptibility to *Rhizoctonia* pathogens have been noted in cultivars of sugarbeet, potato, pea, alfalfa, beans, mustard, rice and turfgrasses. On moderately resistant hosts *Rhizoctonia* pathogens form only a few and less well organized infection structures (Hofman and Jongebloed, 1988), and low amounts of fungicides should be able to immobilize weakened infective propagules and prevent infection and spread for a longer duration (Ruppel and Hecker, 1987; Burpee, 1993). These examples highlight the practical usefulness of combining low rates of fungicides with host defense and should be exploited wherever possible.

B. Soil factors

Characteristics of soil play an important role in governing the efficacy of seed and soil treatments with fungicides in *Rhizoctonia* control. Disease control is invariably better in light-textured sandy or sandy loam soils, while higher amounts of fungicides are needed for satisfactory disease control in heavy-textured clay or clay loam soils due to adsorption of fungicides on soil colloids, clay particles and organic matter (Rushdi and Jeffers, 1956; Alexander, 1957; Wright, 1968; Kataria and Grover, 1976; Hans *et al.*, 1981). However, not all fungicides are influenced by soil texture and adsorptive capacity: thiabendazole and chloroneb provide good disease control against *R. solani* in various soils regardless of texture (Kataria and Sunder, 1987). Fungicides have a specific range of pH and temperature for maximum efficiency against *Rhizoctonia*, i.e., carbendazim, benomyl and thiophanate-methyl provide maximum control against *R. solani* in neutral to alkaline pH and cool temperature, while quintozene and chloroneb perform better at an acidic pH and in a warm climate. Chemical control in potato fields against stem canker and black scurf with low rates of pencycuron and tolclofos-methyl is more effective in neutral marine loam soils than in slightly acid sandy soils (Jager *et al.*, 1991). Triadimenol is equally effective in controlling wheat root rot (AG 4) at low and high temperatures (Mathieson and Rush, 1991). Soils have to be adequately moist for optimum disease control with fungicides; in water-deficient soils even increased dosages of fungicides do not necessarily overcome the loss in activity due to water stress (Munnecke, 1972; Kataria and Grover, 1976; Hans *et al.*, 1981). In addition, several other soil characteristics and components including compactness, chemical and microbial activities, nutrient status, salinity/sodicity and crop residues may influence disease control directly or indirectly.

Therefore, in making recommendations of fungicides for seed or soil treatment, it is axiomatic to optimize the application rates in view of the soil type in a target field.

C. Fertilizers and manures

Micronutrients and N, P, K fertilizers, applied to intensively cropped agricultural soils, may also interact with fungicides and modify disease control in various ways. Nitrogen fertilization often reduces the efficacy of fungicides by increasing host susceptibility to *Rhizoctonia* pathogens, while disease control is generally better in soils with adequate amounts of potassic and phosphatic fertilizers. Micronutrients like Fe, Zn and Cu severely reduce efficacy of some fungicides (particularly benzimidazoles); boron and molybdenum rather improve disease control by quintozene, chloroneb, carboxin and carbendazim (Kataria *et al.*, 1981; Kataria and Sunder, 1985). Obviously the interactions of fertilizers and fungicides are very complex, and at present it is difficult to reliably predict the circumstances under which fungicidal control of Rhizoctonia diseases will increase or will be reduced after fertilizer application. However, it is well documented that enrichment of soil with peat and organic manures decreases fungicidal control of Rhizoctonia diseases (Munnecke, 1972; Sanders *et al.*, 1978b; Bandyopadhyay *et al.*, 1982; Kataria and Sunder, 1987). Humic and fulvic acids, as predominant constituents of organic manures, sequester substantial amounts of fungicides and render them inaccessible for disease control activities in soil. Manures freed of humic and fulvic acids have little or no effect on disease control by fungicides. Moreover, carboxin and chloroneb defy inactivation by organic manures and humic acid, and provide good disease control also in soils rich in organic matter.

D. *Rhizobium* and other biologicals

In legume crops fungicides are applied to the seed along with nitrogen-fixing Rhizobia. Fungicides and *Rhizobium* are so close to each other on the seed surface that *Rhizobium* can influence the efficacy of fungicides in disease control and the fungicides can influence the ability of *Rhizobium* to form functional nodules (Kataria *et al.*, 1985). Thiabendazole provides excellent disease control both with and without *Rhizobium* on the seedand also supports good nodule density. VA-mycorrizal fungi, implied to reduce root infection by *Rhizoctonia* pathogens, may also be used along with a selective fungicide like pencycuron to augment disease control.

Disease control by fungicides may be augmented in the presence of antagonistic soil bacteria and fungi encouraged by soil amendments or applied directly as biocontrol preparations, and low rates of fungicides might be sufficient for enhancement and greater efficiencies in suppressing Rhizoctonia diseases even in crops with season-long susceptibility. Several studies have highlighted practical usefulness of combining lower rates of fungicides with strong biocontrol agents (*Bacillus, Pseudomonas, Trichoderma, Verticillium*) for efficient and extended control of *Rhizoctonia* in fields (details in chapter 'Integrated Control'). Only fungicides with strong and selective action against the target *Rhizoctonia* pathogens will be reliable partners in combined formulations with biocontrol agents mostly for low to moderate disease pressure situations only. In severely infected crops, fungicides will form major component of effective disease-control strategies (De Waard *et al.*, 1993).

E. Insecticides and herbicides

A variety of insecticides and herbicides are applied also in *Rhizoctonia*-infested fields, often in mixtures with fungicides on seed (insecticides), tank-mixed with fungicides (herbicides), or sequentially before or during seeding. Besides their individual effects on disease severity, insecticides and herbicides greatly influence disease control with fungicides with all sorts of possible interactions (additive, synergistic, or antagonistic). Organophosphorus insecticides, particularly aromatic phosphorothioates, themselves can reduce Rhizoctonia infections and improve disease control with fungicides by additive or synergistic interactions (Kataria *et al.*, 1989; Kataria and Verma, 1993). These insecticides are potent inhibitors of cutinase released by fungal infection structures and certain fungicides strongly inhibit the development of infection cushions by *R. solani* (Kataria & Grover, 1975; Hirooka *et al.*, 1989). Therefore, the chances of successful infections are lowered by fungicide-insecticide mixtures because of their joint action at different stages during the infection process. Carbamate insecticides (carbofuran, aldicarb, cloethocarb) aggravate Rhizoctonia infections and reduce efficiency of fungicides by antagonistic interactions (Kataria and Verma, 1993), and hence should be avoided in *Rhizoctonia*-infested fields.

Pinckard and Standifer (1966) observed that preplant soil application of trifluralin increased *R. solani* damping-off of cotton, but when quintozene was applied along with the herbicide there was no disease in plots which received combined treatments. Similarly disease control against *R. solani* is greatly improved when quintozene and pencycuron are applied in mixture with alachlor, fluchloralin, paraquat, or simazine which themselves have varying effects on the disease (Kataria and Dodan, 1982; Awadalla and El-Refaie, 1994). Cereal herbicides such as ioxynil, bromoxynil, dicamba and DNOC interact synergistically with cyproconazole in controlling *R. cerealis* on wheat (sharp eyespot) and allow substantial reduction in the amount of fungicide required for 90% disease control (Kataria and Gisi, 1990). In situations where herbicide treatments aggravate the disease, it is very essential to use an appropriate fungicide not only to nullify the adverse effects of herbicides but to possibly achieve better disease control by synergistic interaction in combined applications. However, the nature of interactions between herbicides and fungicides differ greatly depending on the growing conditions and the crop on which the mixtures are used (Moustafa-Mahmoud *et al.*, 1993), and it is not possible to generalize on the nature of likely interactions for all crops and seasons.

V. CONCLUSIONS AND FUTURE APPROACHES

We have achieved a great deal in chemical control of Rhizoctonia diseases, and learned that disease control with fungicides depends not only on the intrinsic fungitoxicity and other properties of the fungicides, but it is an interplay of several other factors which govern the biological availability and activity of fungicides and influence the final outcome of disease control by fungicide treatments. Therefore, in targeting chemical control of Rhizoctonia diseases, the key points are 'precision' and 'integration'. Precision involves 'intelligent' selection of appropriate fungicides for a given disease situation, and integration involves 'compatible' combinations of fungicides with other disease control measures as well as with other plant protection chemicals. Because the registration and recommendation of fungicides are determined by economic and commercial factors, the choices that growers have with some crops may be limited. In some parts of the world, fungicides with efficacy against *Rhizoctonia* may not be available on minor crops.

Major precision criteria for selecting fungicides against *Rhizoctonia* center around the pathogen population, host species and cultivars, and the environmental factors. To control a target disease in an area, selection of fungicides is to be judged in view of strong activity against various *Rhizoctonia* species and AGs causing the disease and the suitability of the fungicides for seed, soil and foliar application, as the case may be, on crop species and cultivars on which they are to be used and well-suited for local soil and other environmental conditions in the target area. The selected fungicides have to be compatible with insecticides and herbicides likely to be used in the same field, possibly with synergistic interactions to allow efficient disease control with lower amounts of fungicides. More emphasis will be put on research towards biologicals, as far as they prove active, reliable and economically viable. However, since these developments will be transformed into practical solutions only stepwise and slowly, their intelligent combination with other inputs, including chemicals, offers the most promising strategy for the foreseeable future. Synergistic combinations of reduced rates of fungicides with appropriate biocontrol agents represent realistic means of effective disease control for longer duration. Far more lasting will be combined use of chemicals and host genetic defenses in partnership in sustainable disease-control systems. More field experimentation must be encouraged, so that soundly based judgment on the best strategies for particular circumstances can be achieved. There is much more to learn, of great practical and scientific interest, in *Rhizoctonia* control. Chemical control will be needed for many years, with the objectives of targeted, ecologically safe and economically sound concepts of plant protection. If only plant breeders and biotechnologists could find the genetic equivalent of tolclofos-methyl or cyproconazole, at least in stability and breadth of effect!

VI. References

Alexander PM (1957) Etiology and control of *Poinsettia* root and stem rot. Phytopathology 47: 1.

Anon. (1995) Control of a broad spectrum of anastomosis groups of Rhizoctonia solani in ornamentals, potatoes and vegetable crops with Rizolex. Abstract P-9-17. International Symposium on *Rhizoctonia*. Noordwijkerhout, The Netherlands, June 27-30, 1995.

Awadalla OA & El-Refaie IM (1994) Effect of herbicides on the toxicity of fungicides against *Rhizoctonia solani* causing damping-off of cotton. J. Phytopath. 140:187-192.

Bandyopadhyay R, Yadav JPS, Kataria HR & Grover RK (1982) Fungicidal control of *Rhizoctonia solani* in soil amended with organic manures. Ann. Appl. Biol. 101:251-259.

Burpee LL (1993) A guide to integrated management of turfgrass diseases. Vol 1 Cool season turfgrasses. GCSAA Press, Lawrence, Kansas.

Carling DE, Helm DJ & Leiner RH (1990) *In vitro* sensitivity of *Rhizoctonia solani* and other multinucleate and binucleate *Rhizoctonia* to selected fungicides. Plant Dis. 74:860-863.

Christensen JJ (1979) *Rhizoctonia* species associated with diseased turfgrasses in New Zealand. NZ J. Agr. Res. 22:627-629.

Cotterill PJ, Ballinger DJ & Kollmorgen JF (1989) Use of three screening techniques for the evaluation of fungicides to control Rhizoctonia root rot of wheat. Ann. Appl. Biol. 115:229-235.

De Waard MA, Georgopoulos SG, Hollomon DW, Ishii H, Leroux P, Ragsdale NN & Schwinn FJ (1993) Chemical control of plant diseases:problems and prospects. Annu. Rev. Phytopath. 31:403-421.

Grisham MP (1983) Differential responses of *Rhizoctonia solani* isolates *in vitro* to selected fungicides. Phytopathology 73:501.

Hans JK, Tyagi PD, Kataria HR & Grover RK (1981) The influence of soil and other physical factors on antifungal activity of carbendazim against *Rhizoctonia solani*. Pestic. Sci. 12:425-432.

Hirooka T, Miyagi Y, Araki F & Kunoh H (1989) Biological mode of action of flutolanil in its systemic control of rice sheath blight. Phytopathology 79:1091-1094.

Hofman TW & Jongebloed PHJ (1988) Infection process of *Rhizoctonia solani* on *Solanum tuberosum* and effects of granular nematicides. Neth. J. Plant Path. 94:243-252.

Huppatz JL, Phillips JN & Witrzens B (1983) Laboratory and glasshouse studies of the activity of carboxamide derivatives against *Rhizoctonia solani* in cotton. Plant Dis. 67:45-47.

Jager G, Velvis H, Lamers JG, Mulder A & Roosjen J (1991) Control of *Rhizoctonia solani* in potato by biological, chemical and integrated measures. Potato Res. 34:269-284.

Kataria HR & Dodan DS (1982) The influence of two herbicides on the antifungal activity of some fungicides against *Pythium butleri* and *Rhizoctonia solani* causing damping-off of cowpea. Pestic. Sci. 13:583-588.

Kataria HR & Gisi U (1989) Recovery from soil and sensitivity to fungicides of *Rhizoctonia cerealis* and *R. solani*. Mycol. Res. 92:458-462.

Kataria HR & Gisi U (1990) Interactions of fungicide-herbicide combinations against plant pathogens and weeds. Crop Prot. 9:403-409.

Kataria HR & Grover RK (1975) Inhibition of infection structures of *Rhizoctonia solani* by fungitoxicants. Z. Pflkrank. PflSchutz 82:226-232.

Kataria HR & Grover RK (1976) Some factors affecting the control of *Rhizoctonia solani* by systemic and non-systemic fungicides. Ann. Appl. Biol. 82:267-278.

Kataria HR & Sunder S (1985) Effect of micronutrients on the efficacy of fungicides against *Rhizoctonia solani* on cowpea seedlings. Pestic. Sci. 16:453-456.

Kataria HR & Sunder S (1987) Fungicidal control of *Rhizoctonia solani* in relation to soil texture, organic matter and clay minerals. Ann. Appl. Biol. 110:43-51.

Kataria HR & Verma PR (1990) Efficacy of fungicidal seed treatments against pre-emergence damping-off and post-emergence seedling root rot of growth chamber grown canola caused by *Rhizoctonia solani* AG 2-1 and AG 4. Can. J. Plant Path. 12:409-416.

Kataria HR & Verma PR (1993) Interactions of fungicide and insecticide combinations against *Rhizoctonia* damping-off and root rot in canola. Ann. Appl. Biol. 123:233-246.

Kataria HR, Yadav JPS & Grover RK (1981) Interactions of chemical fertilizers with seed dressing fungicides in controlling *Rhizoctonia solani*. Z. Pflkrank. PflSchutz 88:641-650.

Kataria HR, Yadav JPS, Garg FC & Grover RK (1985) Inactivation of seed-treatment fungicides by *Rhizobium*. Pestic. Sci. 16:337-340.

Kataria HR, Singh H & Gisi U (1989) Interactions of fungicide-insecticide combinations against *Rhizoctonia solani in vitro* and in soil. Crop Prot. 8:399-404.

Kataria HR, Hugelshofer U & Gisi U (1991a) Sensitivity of *Rhizoctonia* species to different fungicides. Plant Path. 40:203-211.

Kataria HR, Verma PR & Gisi U (1991b) Variability in the sensitivity of *Rhizoctonia solani* anastomosis groups to fungicides. J. Phytopath. 133:121-133.

Kataria HR, Verma PR & Rakow G (1993) Fungicidal control of damping-off and seedling root rot in *Brassica* species caused by *Rhizoctonia solani* in the growth chamber. Ann. Appl. Biol. 123:247-256.

Koster AThJ & Meer LJvd (1990) Control of *Rhizoctonia solani* in the field with a low dosage of tolclofos-methyl or flutolanil. Acta Hort. 266:419-424.

Leadbitter NJ, Zang L & Mantel B (1995) Control of *Rhizoctonia* using the phenylpyrroles. Abstract P-9-11. International Symposium on *Rhizoctonia*. Noordwijkerhout, The Netherlands, June 27-30, 1995.

Lyr H (1995) Modern Selective Fungicides. Gustav Fischer Verlag, Jena, Stuttgart & New York.

Lyr H, Ritter G & Banasiak L (1974) Detoxification of carboxin. Z. Allg. Mikrobiol. 14:313-320.

Martin BC & Torres H (1986) Fungicides for the control of *Rhizoctonia solani* damping-off in seedlings derived from true potato seed. Fitopatologia 21:74-80.

Martin SB, Lucas LT & Campbell CL (1984a) Comparative sensitivity of *Rhizoctonia solani* and *Rhizoctonia*-like fungi to selected fungicides in vitro. Phytopathology 74:778-781.

Martin SB, Campbell, CL & Lucas LT (1984b) Response of *Rhizoctonia* blights of tall fescue to selected fungicides in the greenhouse. Phytopathology 74:782-785.

Mathieson JT & Rush CM (1991) Influence of temperature and five fungicides on Rhizoctonia root rot of hard red winter wheat. Plant Dis. 75:983-986.

Moustafa-Mahmoud SM, Sumner DR, Ragab MM & Ragab MM (1993) Interaction of fungicides, herbicides, and planting date with seedling disease of cotton caused by *Rhizoctonia solani* AG 4. Plant Dis. 77:79-86.

Munnecke DE (1972) Factors affecting the efficacy of fungicides in soil. Annu. Rev. Phytopath. 10:375-398.

Ohtsuki S & Fujinami A (1982) Rizolex (tolclofos-methyl). Japan Pestic. Inf. 41:21-25.

Olaya G, Abawi GS & Barnard J (1994) Response of *Rhizoctonia solani* and binucleate *Rhizoctonia* to five fungicides and control of pocket rot of table beets with foliar sprays. Plant Dis. 78:1033-1037.

Phillips AJL (1992) A comparison of dust and acetone infusion applications of tolclofos-methyl to bean seeds for the control of *Rhizoctonia solani*. Plant Path. 41:35-40.

Pinckard JA & Standifer LC (1966) An apparent interaction between cotton herbicidal injury and seedling blight. Plant Dis. Reptr 50:172-174.

Price KR, Wyatt AM & Davies WP (1983) Variation in *Rhizoctonia cerealis* to fungicides. Proc. 10th Internat. Conf. Plant Prot. Brighton, UK. 3:642.

Roberts DL & Stephens CT (1984) Sensitivity of *Rhizoctonia solani* to experimental fungicide NTN 19701. Phytopathology 74:854.

Ruppel EG & Hecker RJ (1987) Prelayby applications of experimental fungicides for suppressing Rhizoctonia root rot in sugar beet. Plant Dis. 71:694-698.

Rushdi M & Jeffers WF (1956) Effect of some soil factors on efficiency of fungicides in controlling *Rhizoctonia solani*. Phytopathology 46:88-90.

Sanders PL, Burpee LL & Cole H Jr (1978a) Preliminary studies on binucleate turfgrass pathogens that resemble *Rhizoctonia solani*. Phytopathology 68:145-148.

Sanders PL, Burpee LL & Cole H Jr (1978b) The effect of soil composition on initial and residual efficacy of four systemic fungicides on pot-grown creeping bentgrass. Plant Dis. Reptr 62:545-549.

Shatla MN & Sinclair JB (1963) Tolerance of pentachloronitrobenzene among cotton isolates of *Rhizoctonia solani*. Phytopathology 53:1407-1411.

Smiley RW, Wilkins DE & Klepper EL (1990) Impact of fungicide seed treatments on *Rhizoctonia* root rot, take-all, eyespot, and growth of winter wheat. Plant Dis. 74:782-787.

Snel M, Schmeling B von & Edgington LV (1970) Fungitoxicity and structure-activity relationships of some oxathiin and thiazole derivatives. Phytopathology 60:1164-1169.

Sumner DR (1987) Efficacy of pencycuron against isolates representing different anastomosis groups of *Rhizoctonia solani* and *Rhizoctonia*-like binucleate fungi. Plant Dis. 71:515-518.

Sumner DR, Lewis JA & Gitaitis RD (1992) Chemical and biological control of *Rhizoctonia solani* AG-4 in snap bean double-cropped with corn. Crop Prot. 11:121-126.

Van Bruggen AHC & Arneson PA (1984) Resistance in *Rhizoctonia solani* to tolclofos-methyl. Neth. J. Plant Path. 90:95-106.

Wakae O & Matsuura K (1975) Characteristics of validamycin as a fungicide for *Rhizoctonia* disease control. Rev. Plant Prot. Res. 8:81-92.

Winston JR (1913) Effect of the steam-formalin treatment on certain soil organisms. Phytopathology 3:74.

Wright NS (1968) Evaluation of terraclor and terraclor super-x for control of *Rhizoctonia* on potato in British Columbia. Can. Plant Dis. Surv. 48:77-81.

Yamada Y, Saito J & Takase I (1988) Development of a new fungicide, pencycuron. J. Pestic. Sci. 13:375-387.

VI.E INTEGRATED CONTROL OF *RHIZOCTONIA* SPECIES.

MARK W. SWEETINGHAM

Agriculture Western Australia, Baron-Hay Court, South Perth, 6151, AUSTRALIA.

I. Introduction

The term integrated control began to be used in the 1950's and from it evolved the term integrated pest management (IPM). Initially it was a concept applied mostly to insect control, born by a desire to rationalize pesticide inputs (Kendrick, 1988). The IPM philosophy has grown to imply more than the application of multiple control methods. Disease monitoring integrated with economic injury thresholds and a detailed knowledge of the pathosystem should be used to plan the rational application of management strategies. Papavizas and Lewis (1979) commented that the literature on integrated management of *Rhizoctonia* in the field was practically non-existent. Since then many integrated systems have been researched and some are being implemented at the farm level.

Integrated control is considered desirable for a number of reasons. Many strategies used in isolation provide an inadequate or variable level of control depending on environmental conditions. This is particularly true for biological control agents. Combined strategies are often additive and sometimes synergistic, leading to more effective and reliable disease control. Decreased pesticide inputs are increasingly sought to reduce environmental and consumer concerns (real and perceived) and are providing a greater incentive to look at integrated approaches to disease management. In the case of Rhizoctonia diseases, chemical control has not been successful in many situations and host resistance is rare, so integrated control has often been pursued as the only means of achieving reliable and economic disease reduction.

Control strategies should be optimized with other aspects of crop management and complimentary to the profitability and sustainability of the farming system. Without this, they are unlikely to be adopted no matter how well they control disease.

II. Combining different control methods

Control of Rhizoctonia diseases may be achieved by biological, chemical, or cultural means as well as in some instances the use of moderately resistant cultivars. Integrated control often employs combinations of these approaches.

A. Augmenting biological methods

For most commercial and experimental biological control agents, integration with additional strategies is increasingly recommended to enhance control.

Bacillus subtilis GB03 is registered as a seed inoculant for the suppression of pathogens including *R. solani* and is applied to large areas of field crops in the USA. Early results showed variable control of *R. solani* AG 4 seedling diseases of cotton and peanut. Fortunately GB03 is compatible with most common seed-treatment fungicides such as quintozene and carboxin and research now focuses on combined formulations (Backman *et al.*, 1994). Fungicides appear to give early seedling

B. Sneh et al. (eds.),
Rhizoctonia Species: Taxonomy, Molecular Biology, Ecology, Pathology and Disease Constrol, 549–558.

protection whilst the bacterium provides longer term protection. Mahaffee and Backman (1993) showed evidence that fungicides increase root colonization by *B. subtilis*, possibly by clearing niches of fungicide sensitive microorganisms. Root colonization is also influenced by cultivar, temperature and edaphic factors such as organic matter and clay content (Backman *et al.*, 1994). Such knowledge may lead to improved application strategies and seed treatment formulations.

Bacterial antagonists readily lend themselves to formulation with fungicides and other examples include the combination of *B. cereus* and pencycuron for the control of Rhizoctonia stem rot of chrysanthemum in Taiwan (Wu *et al.*, 1990).

The integration of fungal antagonists with fungicides offers a greater challenge. To increase their compatibility, *Trichoderma* and *Gliocladium* can be genetically manipulated to increase tolerance to the MBC fungicides (Papavizas and Lewis, 1983). Newer fungicides such as tolclofos-methyl, pencycuron, and flutolanil have high activity against *Rhizoctonia* but are more selective in their antifungal spectrum. These should be more easily integrated with biological control fungi. Jager *et al.* (1991) found that inoculating potato tubers with the mycoparasite *Verticillium biguttatum* in conjunction with 25% of the recommended rate of pencycuron gave control of black scurf and stolon rot equal to pencycuron at the full rate.

The effectiveness of introduced antagonists can also be enhanced by cultural practices. Lewis and Papavizas (1980) showed that a *Trichoderma* soil treatment used in conjunction with moldboard plowing gave greater reduction in Rhizoctonia fruit rot of cucumber than either treatment used alone. In Israel, control of *R. solani* in a range of greenhouse and field crops has been achieved by the integration of soil solarization with the antagonist *Trichoderma harzianum* (Elad *et al.*, 1980, 1981). Similar results were obtained on potato in Turkey by Bicici and Erkilic (1986) who combined a *T. harzianum* tuber treatment with solarization and iprodione.

B. Cultural practices and fungicides

R. solani AG 3, which causes black scurf and stem canker on potatoes, is a specific pathogen of potato and thus crop rotation is an important control strategy . However, infection can originate from tuber-borne sclerotia and hyphae as well as from soil, therefore scurf-free tubers are recommended for seed. Hide and Read (1991) found that with reduced rotation period, seed treatment with tolclofos-methyl became less effective. The decline in fungicide efficacy evidently resulted as the soil became the predominant source of inoculum, highlighting the need to integrate fungicide and crop rotation. Given knowledge of the host-pathogen interaction in many environments, time of sowing, sowing depth, delayed hilling and irrigation scheduling recommendations can also be made (Anon, 1986). In the Netherlands, Dijst (1985) found black scurf was increased by chemical haulm destruction. Consequently, the process of green crop harvesting is recommended as another component of an integrated control program (Luttikholt *et al.*, 1995).

After 4 years of field experimentation Lewis *et al.* (1983) published work on the integrated control of root and hypocotyl rot of snap beans in Maryland caused by *R. solani* AG 4. The major component of the strategy was to moldboard plow infected crop residues to a depth below the infection court of the subsequent crop. High CO_2 levels at depth also reduce survival of the pathogen. Chloroneb plus metalaxyl seed treatments used in conjunction with plowing gave greater yield than either component used individually. Similarly, with Rhizoctonia fruit rot of cucumbers, the controlling effect of moldboard plowing and fungicide effects were additive (Lewis and Papavizas, 1980).

C. Integration with resistance

There are no examples of major gene resistance conferring a high level of resistance to any Rhizoctonia diseases. However, there are reports of moderate resistance in rice, pea, sugar beet, and turf grass genotypes (see Chapter VI.C1.). These levels of resistance are probably inadequate alone but should be useful if integrated with other control strategies. For example, Burpee (1993) recommended tolerant tall fescue cultivars in conjunction with fungicide, reduced nitrogen fertilizer, short mowing height and irrigation practices that limit the duration of foliar wetness for the control of Rhizoctonia blight in lawns. Interestingly, cultivars vary in the extent to which nitrogen fertilizer and mowing height influence disease development and severity (Burpee, 1995).

III. Integration with strategies for control of associated pathogens

A. Fungal pathogens

Seedling damping-off problems involving *R. solani* are often also associated with other fungi, particularly *Pythium* and ideally control strategies should be effective against all pathogens present. Cubeta and Echandi (1991) found that seed treatment with metalaxyl in combination with a non pathogenic binucleate *Rhizoctonia* gave protection against *Rhizoctonia* and Pythium damping-off of cucumbers. Treating seeds with metalaxyl or captan alone controlled *Pythium* but induced a higher incidence of Rhizoctonia damping-off. Similarly, Lewis *et al.* (1983) controlled *Pythium* and *R. solani* on snap bean by combining a seed treatment of metalaxyl and chloroneb with moldboard plowing. Rothrock (1990), dealing with a cotton seedling disease complex of *R. solani*, *Pythium* spp., *Fusarium oxysporum* and *Thielaviopsis basicola,* reported that best stands were obtained with a combined seed treatment of metalaxyl + quintozene and a commercial formulation of *B. subtilis*.

Severe outbreaks of lupin hypocotyl rot in Western Australia, caused by *R. solani* AG 11 (Carling *et al.*, 1994), are usually associated with sowing depths greater than 6cm. Shallower sowing reduces disease by exposing less hypocotyl to infection (Sweetingham and Jarvis, 1993). By contrast the severity of lupin root rot caused by *Pleiochaeta setosa* is increased by shallow sowing (Sweetingham, 1991). *P. setosa* and *R. solani* often co-exist, creating a problem for optimal sowing depth. Increasing the rotational break between lupin crops will reduce inoculum of the host specific *P. setosa,* so that shallow sowing can be employed. Lupin hypocotyl rot can be further suppressed by iprodione seed treatment (W.J. MacLeod and M.W. Sweetingham, unpublished).

B. Nematodes

Crop rotation can be used for both potato cyst nematode (PCN) *Globodera* spp. and *Rhizoctonia* control in potato. However, if crops are grown in short rotations a nematicide or a PCN resistant cultivar will be required (Hide and Read, 1991). Unfortunately, some nematicides including oxamyl, ethoprophos and aldicarb can cause marked increases in Rhizoctonia stem canker (Scholte, 1987). The mechanisms are unclear but it appears that the nematicides are not causing a direct effect on *R. solani* or the mycoparasite *V. biguttatum* (Hofman and Bollen, 1987).

Cereal cyst nematode (CCN) (*Heterodera avenae*) commonly occurs in the same fields as Rhizoctonia bare patch in southern Australia. Direct drilling wheat results in less CCN damage than sowing after tillage (Roget and Rovira, 1985). However, direct drilling increases bare patch. Fortunately, the nematode can be managed by rotation with a legume or oilseed crop and by resistant cultivars. Roget *et al.* (1996) optimized control of both *Rhizoctonia* and CCN in experimental plots through the use of a mild tillage treatment involving a modified seeding point (section IV D).

IV. INTEGRATION WITH CROP MANAGEMENT AND THE FARMING SYSTEM

Beneficial or detrimental effects can arise where control strategies interact with other agronomic practices.

A. Interactions with *Rhizobium* in legumes

Rhizobium can act as an antagonist towards *Rhizoctonia* with disease suppression reported from soil or seed treatments (Farzana *et al.*, 1991; Ehteshamul-Haque and Ghaffar, 1993). Unfortunately, there is potential incompatibility where both *Rhizobium* inoculants and fungicides are required on seed. Kataria *et al.* (1985), working with Rhizoctonia root rot of cowpea, found that some fungicides, including captafol, chloroneb, carbendazim, and thiophanate-methyl were considerably reduced in efficacy when applied to seed with *Rhizobium.* Additionally, fungicides such as thiram are toxic to *Rhizobium* and when applied together on seed can result in reduced nodulation (Evans *et al.*, 1986). On the positive side, *Bacillus* seed treatments applied to peanuts to suppress *Rhizoctonia* increased nodulation (Turner and Backman, 1991).

Genetic engineering of *Rhizobium meliloti* has been used to improve biological control activity by the introduction of a chitinase gene. Sitrit *et al.* (1993) demonstrated the lysis of *R. solani* cells treated with extracts from alfalfa root nodules produced by the modified *R. meliloti*.

B. Insecticide and herbicide interactions

Kataria *et al.* (1989) found that fungicides including carboxin and thiabendazole gave increased control of Rhizoctonia root rot of cowpea when applied to seeds with insecticides such as endosulfan or dimethoate. Synergistic interactions were detected in some fungicide and insecticide combinations. Similar benefits were found for Rhizoctonia damping-off and root rot in canola where combinations of fungicide seed dressings and the insecticides gamma-HCH or terbufos increased control (Kataria and Verma, 1993).

Herbicides can also influence the efficacy of fungicides. Cyproconazole sprayed onto leaves in combination with the herbicides dicamba or bromoxynil had a synergistic effect on the control of *R. cerealis* on wheat seedlings in controlled environment tests (Kataria and Gisi, 1990). Also in pots, Awadalla and El-Refaie (1994) found that fungicide seed treatments on cotton gave better control of *R. solani* (AG 4) when soil was treated with either simazine or paraquat. However in field trials, several herbicides (but not simazine or paraquat) reduced the efficacy of a range of fungicide seed treatments on cotton (Moustafa *et al.*, 1993).

Increases in soil-borne diseases due to herbicides has been reviewed by Altman and Rovira (1989) who cite several examples relating to *R. solani*. In Australia, Rovira and McDonald (1986) observed increased root damage in barley caused by *R. solani* AG 8 in field plots one year after application of chlorsulfuron. Wong *et al.* (1993) found this did not occur on acid soils where chlorsulfuron is less residual.

Clearly, all insecticide and herbicide interactions are strongly influenced by environmental conditions and although incompatibility warnings may be appropriate, it remains to be seen if positive effects can be reliably exploited.

C. Nutrition

Increased levels of nitrogen have been shown to reduce the severity of patches caused by *R. solani* AG 8 in cereals. MacNish (1985a) measured a 50% reduction in patch area by the application of inorganic nitrogen in zero-tilled barley. The effect of nitrogen and control by tillage appeared additive but this could not be demonstrated statistically owing to the low incidence of disease in the tillage treatments. In controlled environment experiments, Wall *et al.* (1994) found that nitrogen levels had no effect on root infection levels and growth of diseased plants responded less to nitrogen than healthy plants. Brennan (1991) concluded from field trials that the rate of applied nitrogen required for optimum wheat yield was higher in the presence of bare patch. Thus it seems under field conditions that nitrogen boosts diseased plants sufficiently for them to find more moisture which is the factor most limiting yield.

Wheat deficient in zinc has also been shown to be more susceptible to *R. solani* AG 8 (Thombai *et al.*, 1993). However, increasing zinc above levels adequate for plant growth did not produce further reductions in disease.

D. Integration with conservation farming practices

Rhizoctonia bare patch has been shown to be consistently more damaging where zero or reduced tillage crop establishment methods are used (Neate, 1988). MacNish (1985a) showed that "scarifying" with a tined cultivator to a depth of 12cm prior to sowing considerably reduced disease severity. Scarifying does not invert the soil and bury *Rhizoctonia* inoculum but decompacts the soils, mineralizes nitrogen and increases the growth rate of seedling roots (Hamblin and Tennant, 1979). Neate (1991) also showed how disruption to the mycelial network of *R. solani* AG 8 in soil reduced its inoculum potential. There is no evidence that tillage reduces disease by enhancing crop residue decomposition as suggested by Chung *et al.* (1988).

Pre-disposing soils to wind erosion has been a major obstacle to adoption of the scarifying treatment by Australian farmers. Another impediment has been the increased risk of CCN damage (section III.B).

To integrate tillage control of *Rhizoctonia* into a conservation farming system, modification of seeding systems has been necessary. Jarvis and Brennan (1986) found that a combine drill modified to cultivate 10cm deep on front tines arranged directly in line with rear tines which sowed at 3cm reduced Rhizoctonia patch to the same extent as scarifying pre-seeding. This single-pass system still left the soil surface erodible, as the relatively wide working points fully disturbed the inter-row surface, but only for the brief period between sowing and seedling establishment. Roget *et al.* (1996) researched narrow blade points cutting a slot 10mm wide and 25mm below the seed depth. This minor amount of soil disturbance resulted in less disease than sowing with a zero disturbance disc drill. Sweetingham and Jarvis (1993) suggested a refinement specifically for sowing lupins into cereal trash. A combine drill is modified to place phosphate fertilizer 10cm deep behind a narrow point whilst seed is directed from a back-angled tube to 5cm depth. Half the tines are removed to double row spacing, providing the means to sow into greater levels of trash for better protection against wind erosion. The deep working point controls *Rhizoctonia* while the deep placed fertilizer improves lupin vigor.

The adoption of reduced tillage farming means an increased reliance on chemical weed control. Rhizoctonia patch severity can be reduced if volunteer weed growth is chemically removed for 3 to 6 weeks prior to sowing (Roget *et al.*, 1987; Wong *et al.*, 1993). The mechanisms behind this effect are not fully resolved and the effect is not consistent in all environments (MacNish and Fang, 1987). Nevertheless, the additive benefits of the chemical fallow and modified sowing points have been shown at one site (Roget *et al.*, 1996). However, for reasons pointed out earlier, chlorsulfuron and possibly other sulfonyl-urea herbicides should be avoided.

At present, there are no biological strategies employed for bare patch control, although *V. biguttatum* has reduced *R. solani* AG 8 damage on barley in controlled environment studies (Morris *et al.*, 1993). Stephens *et al.* (1993) found that the earthworm *Aporrectodea trapezoides* artificially inoculated into field soils could reduce root damage in wheat caused by *R. solani* AG 8. The density of *A. trapezoides* tested was similar to that which can build up in Australian soils under pasture. Collembola, also a potential biological control for *Rhizoctonia,* build up greater populations in reduced tillage systems (Rickerl *et al.*, 1989) as do earthworms. However, the benefits of scarifying would clearly outweigh the benefit of such organisms in reducing bare patch. Nevertheless, it may be possible to build up useful populations in reduced soil disturbance and trash retention systems.

In Maryland, Lewis *et al.* (1983) noted that although moldboard plowing could successfully used to reduce root and hypocotyl rot of snap beans, there was continuing pressure to adopt conservation tillage practices in order to minimize the risk of soil erosion. Moldboard plowing can be used to control Rhizoctonia root and hypocotyl rot of cotton in Georgia (Sumner *et al.*, 1981). This is particularly important on farms where cotton is continuously cropped with winter fallow. However, by switching to a double-cropped winter cereal system, with a cotton-triticale-soybean-triticale rotation, conservation tillage practices can be employed without serious disease losses (Sumner *et al.*, 1995).

V. Rational control response

A. Crop monitoring and epidemic modeling

In integrated control programs it is accepted that some damage to the plant is economically tolerable. Crops should be monitored and disease levels compared with thresholds at which control is warranted. This concept is most useful where prescriptive seasonal tactics, such as fungicide application can be employed (Fick and Power, 1992).

Rice sheath blight (*R. solani* AG 1-IA), has a multicyclic epidemic and the tactical application of chemicals or biologicals within the season is feasible. Several computer simulation models to forecast yield losses have been developed (Hashiba and Ijiri, 1989; Saikia and Baruah, 1990; Ji *et al.*, 1991). Key variables include planting date, nitrogen application, irrigation scheduling, early season infection levels and meteorological conditions. Groth *et al.* (1992) found that early season infection levels were a good indicator of late season disease severity in Louisiana. They defined threshold levels, for spraying fungicide, of 5-10% infected tillers for susceptible and 10-15% for more resistant cultivars.

Most Rhizoctonia diseases have a monocyclic epidemic where intervention after infection is not appropriate and preventative strategies must be employed. Even so, disease thresholds may be important to plan strategies for future seasons. Herr (1987) found that monitoring soil populations of *R. solani* the season prior to sugar beet cropping did not provide a basis for forecasting disease. But in this study all isolates of *R. solani,* without testing, were assumed to be pathogenic to sugar beet. Soon it may be possible to implement preventative strategies after pre-sowing risk assessment based on molecular techniques which quantify inoculum of specific *R. solani* pathotypes in soil. Schumann *et al.* (1994) working with Rhizoctonia blight on turf grass, demonstrated that the use of a field immunoassay kit together with an environmental based forecasting model could reduce the number of fungicide applications compared to calendar schedules whilst providing an acceptable level of disease control.

B. Importance of precise pathogen identification

Anastomosis groups of *R. solani* vary in their host range and the diseases they cause. It is not uncommon to isolate more than one AG from the same field or even the same diseased plant. An example of the need for precise identification is the varied sensitivity of different AGs to fungicides (Kataria *et al.* 1991).

Identification of zymogram (ZG) groups within an anastomosis group is important in Western Australia. Within AG 2-1, ZG 6 causes root and hypocotyl rot of legumes whereas ZG 5, which is often isolated from lupins is not pathogenic to legumes but causes hypocotyl rot of canola. Within AG 2-2, ZG 4 is highly virulent on legumes whilst ZG 10 is only weakly virulent (MacNish *et al.*, 1994) and does not require control.

A patch disease of lupins and barley in Western Australia with very similar symptoms to Rhizoctonia bare patch has recently been shown to be caused by a binucleate *Rhizoctonia* sp. Unlike bare patch, this disease cannot be controlled by tillage but it can be reduced by crop rotation (MacLeod and Sweetingham, 1995).

C. Economic analysis

Implementation of control strategies, particularly those involving crop rotation, need to consider implications for the whole-farm profitability and sustainability. The optimal enterprise mix and rotation are dictated by many economic factors as well as limitations imposed by the physical and biological environment.

The Australian grain belt is a low input - low output mixed cropping and sheep grazing system. The need for production efficiency is paramount and disease control is important but farmers have a limited budget to spend on control measures. In Western Australia a whole-farm bio-economic model (MIDAS) is commonly used to optimize the cropping and grazing activities (Kingwell and Pannell, 1987). Given the varied reaction of cereal species to *R. solani* AG 8 (MacNish, 1985b), decisions on the profitability of growing barley (most susceptible), wheat and oats (most resistant) in the presence of bare patch disease can be resolved.

Farmers involved in the production of high value crops such as potatoes or tulips have a high capital investment in land and specialized machinery. Closer rotations may be most profitable despite high disease pressure from *R. solani,* as more costly control strategies can be considered.

VI. References

Altman J & Rovira AD (1989) Herbicide-pathogen interactions in soil-borne root diseases. Can. J. Plant Path. 11:166-172.

Anon. (1986) Rhizoctonia stem and stolon canker. In: Flint ML (ed.) Integrated Pest Management for Potatoes in the Western United States. (pp 74-75) University of California Division of Agriculture and Natural Resources Oakland, California.

Awadalla OA & El-Refaie IM (1994) Effect of herbicides on the toxicity of fungicides against *Rhizoctonia solani* damping-off of cotton. J. Phytopath. 140:187-192.

Backman PA, Brannen PM & Mahaffee WF (1994) Plant response and disease control following seed inoculation with *Bacillus subtilis*. In: Ryder MH, Stephens PH & Bowen GD (eds.) Improving Plant Productivity with Rhizosphere Bacteria. (pp 3-8) CSIRO, Australia.

Bicici M & Erkilic A (1986) Integrated control of black scurf and stem canker caused by *Rhizoctonia solani* Kühn. on potatoes. Doga Tarim-ve-Ormancilik 10:149-173.

Brennan RF (1991) Effect of nitrogen and residual copper on the occurrence of Rhizoctonia bare patch in wheat grown near Esperance, Western Australia. Aust. J. Exp. Agric. 31:259-262.

Burpee LL (1993) A guide to integrated management of turfgrass diseases. Vol. 1 Cool Season Turfgrasses. GCSAA Press, Lawrence, Kansas.

Burpee LL (1995) Interactions among mowing height, nitrogen fertility and variety affect severity of Rhizoctonia blight of tall fescue. Plant Dis. 79:721-726.

Carling DE, Rothrock CS, MacNish GC, Sweetingham MW, Brainard KA & Winters SW (1994) Characterisation of anastomosis group 11 (AG 11) of *Rhizoctonia solani*. Phytopathology 84:1387-1393.

Chung YR, Hoitink HAH & Lipps PE (1988) Interactions between organic-matter decomposition level and soilborne disease severity. Agric. Ecosys. Environ. 24:183-193.

Cubeta MA & Echandi E (1991) Biological control of *Rhizoctonia* and *Pythium* damping-off of cucumber:an integrated approach. Biol. Control 1:227-236.

Dijst G (1985) Investigations on the effect of haulm destruction and additional root cutting on black scurf on potato tubers. Neth. J. Plant Pathol. 91:153-162.

Elad Y, Katan J & Chet I (1980) Physical, biological and chemical control integrated for soilborne diseases in potatoes. Phytopathology 70:418-422.

Elad Y, Chet I, Henis Y (1981) Biological control of *Rhizoctonia solani* in strawberry fields by *Trichoderma harzianum*. Plant Soil 60:245-254.

Ehteshamul -Haque S & Ghaffar A (1993) Use of rhizobia in the control of root rot diseases of sunflower, soybean and mungbean. J. Phytopath. 138:157-163.

Evans J, O'Connor GE, Seymour AR & Carmichael A (1986) Sensitivity of lupin root nodulation to iprodione fungicide. Austral. Plant Path. 15:66-67.

Farzana A, Ghaffar A & Ali F (1991) Effect of seed treatment with biological antagonists on rhizosphere mycoflora and root infecting fungi of soybean. Pak. J. Bot. 23:183-188.

Fick GW & Power AG (1992) Pests and integrated control. In: Pearson CD (ed.) Ecosystems of the World 18. Field Crop Ecosystems. (pp 59-83) Elsevier, Amsterdam.

Groth DE, Rush MC & Hollier CA (1992) Prediction of rice sheath blight severity and yield loss based on early season infection. Louisiana Agriculture 35:20-23.

Hamblin AP & Tennant D (1979) Interactions between soil type and tillage level in a dryland situation. Aust. J. Soil Res. 17:177-189.

Hashiba T & Ijiri T (1989) Estimation of yield loss and computerized forecasting system (BLIGHTCAST) for rice sheath blight disease. Trop. Agric. Res. Ser. 22:163-172.

Herr LJ (1987) Populations of *Rhizoctonia solani* in soil under crops in rotation with sugar beet. Ann. Appl. Biol. 110:17-24.

Hide GA & Read PJ (1991) Effects of rotation length, fungicide treatment of seed tubers and nematicide on diseases and the quality of potato tubers. Ann. Appl. Biol. 119:77-87.

Hofman TW & Bollen GJ (1987) Effects of granular nematicides on growth and microbial antagonism to *Rhizoctonia solani*. Neth. J. Plant Pathol. 93:201-214.

Jager G, Velvis H, Lamers JG, Mulder A & Roosjen J (1991) Control of *Rhizoctonia solani* in potato by biological, chemical and integrated measures. Potato Res. 34:269-284.

Jarvis RJ and Brennan RF (1986) Timing and intensity of surface cultivation and depth of cultivation affect rhizoctonia patch and wheat yield. Aust. J. Exp. Agric. 26:703-708.

Ji BH, Ding KJ, Tan GJ, Wu J, Zhang CQ, Li XM, Tang XD, Lui JC, Cheng SM, Gao BZ, Zhou YH & Ji MF (1991) Simulation model RSPM-1 for forecast and management of rice sheath blight. Scientia-Agricultura-Sinica 24:65-73.

Kataria HR, Yadav JS, Garg FC & Grover RK (1985) Inactivation of seed treatment fungicides by *Rhizobium*. Pestic. Sci. 16:337-340.

Kataria HR Singh H & Gisi U (1989) Interactions of fungicide-insecticide combinations against *Rhizoctonia solani in vitro* and in soil. Crop Prot. 8:399-404.

Kataria HR & Gisi U (1990) Interactions of fungicide-herbicide combinations against plant pathogens and weeds. Crop Prot. 9:403-409

Kataria HR & Verma PR & Gisi U (1991) Variability in the sensitivity of *Rhizoctonia solani* anastomosis groups to fungicides. J. Phytopathology 133:21-133.

Kataria HR & Verma PR (1993) Interactions of fungicide and insecticide combinations against Rhizoctonia damping-off and root rot in canola. Ann. Appl. Biol. 123:233-246.

Kendrick JB (1988) A viewpoint on integrated pest management. Plant Dis. 72:647.

Kingwell RS & Pannell DJ (1987) MIDAS, a bioeconomic model of a dryland farming system. (197pp) Pudoc, Wageningen.

Lewis JA & Papavizas GC (1980) Integrated control of Rhizoctonia fruit rot of cucumber. Phytopathology 70:85-89.

Lewis JA, Lumsden RD, Papavizas GC & Kantzes JG (1983) Integrated control of snap bean diseases caused by *Pythium* spp. and *Rhizoctonia solani*. Plant Dis. 67:1241-1244.

Luttikholt AJG, Ridder JK & van den Boogert, PHJF (1995) Green-crop harvesting, a mechanical haulm destruction method with potential for black scurf control in potato. Abstract P-8-18. International Symposium on *Rhizoctonia*. Noordwijkerhout, The Netherlands.

MacNish GC (1985a) Methods of reducing Rhizoctonia patch of cereals in Western Australia. Plant Pathol. 34:175-181.

MacNish GC (1985b) Mapping Rhizoctonia patch in consecutive cereal crops in Western Australia. Plant Pathol. 34:165-174.

MacNish GC & Fang, CS (1987) Effect of short chemical fallow on rhizoctonia bare patch and root rot of wheat at Esperance, Western Australia. Aust. J. Exp. Agric. 27:671-677.

MacNish GC, Carling DE, Sweetingham MW & Brainard KA (1994) Anastomosis group (AG) affinity off pectic isozyme (zymogram) groups (ZG) of *Rhizoctonia solani* from the Western Australian cereal-belt. Mycol. Res. 98:1369-1375.

MacLeod WJ & Sweetingham MW (1995) Coming to grips with eradu-patch. J. Agric. West. Aust. 36:66-70.

Mahaffee WF & Backman PA (1993) Effects of seed factors on spermosphere and Rhizosphere colonization of cotton by *Bacillus subtilis* GB03. Phytopathology 83:1120-1125.

Morris RAC, Coley-Smith JR & Whipps JM (1993) Effects of the mycoparasite *Verticillium biguttatum* on barley stunt disease caused by *Rhizoctonia solani* anastomosis group 8 in a model system. Plant Pathol. 42:915-922.

Moustafa-Mahmoud SM, Sumner DR & Ragab MM (1993) Interaction of fungicides, herbicides, and planting date with seedling disease of cotton caused by *Rhizoctonia solani* AG 4. Plant Dis. 77:79-86.

Neate SM (1988) The effect of tillage on disease of cereals caused by *Gaeumannomyces graminis*, *Rhizoctonia solani* and *Heterodera avenae*:A review. Plant Prot. Quart. 3:5-7.

Neate SM (1991) Some mechanisms of action of a narrow sowing point which influences Rhizoctonia root disease and yield of cereals. Abstr. In :8th Australasian Plant Pathology Society Conference October 1991, University of Sydney.

Papavizas GC & Lewis JA (1979) Integrated control of *Rhizoctonia solani*. In: Schippers B & Gams W (eds.). Soil-borne Plant Pathogens. (pp 415-424) Academic Press London.

Papavizas GC & Lewis JA (1983) Physiological and biocontrol characteristics of stable mutants of *Trichoderma viride* resistant to MBC fungicides. Phytopathology 73:407 411.

Rickerl DH, Curl EA & Touchton JT (1989) Tillage and rotation effects on collembola populations and *Rhizoctonia* infestation. Soil Tillage Res. 15:41-49.

Roget DK & Rovira AD (1985) Effect of tillage on *Heterodera avenae* in wheat. In: Paker CA, Rovira AD, Moore KJ, Wong PT & Kollmorgen JF (eds.) Ecology and Management of Soilborne Plant Diseases. (pp 252-254) Inkata Press, Melbourne.

Roget DK, Venn NR & Rovira AD (1987) Reduction of Rhizoctonia root rot of direct-drilled wheat by short-term chemical fallow. Aust. J. Exp. Agric. 27:425-430.

Roget DK, Neate SM & Rovira AD (1996). The effect of sowing point design and tillage practise on the incidence of Rhizoctonia root rot, take-all and cereal cyst nematode. Aust. J. Exp. Agric. (in press).

Rothrock CS (1990) The cotton seedling disease complex in Arkansas. Special Report. Agricultural Experiment Station, Division of Agriculture, University of Arkansas. No. 144, 4pp.

Rovira AD & McDonald HJ (1986) Effects of the herbicide chlorsulfuron on Rhizoctonia bare patch and take-all of barley and wheat. Plant Dis. 70:879-882.

Saikia L & Baruah BP (1990) Linear model for predicting yield loss in rice due to sheath blight (*Rhizoctonia solani*) incidence. Indian J. Agric. Sci. 60:360-361.

Scholte K (1987) The effect of crop rotation and granular nematicides on the incidence of *Rhizoctonia solani* in potato. Potato Res. 30:187-199.

Schumann GL, Clarke BB, Rowley LV & Burpee LL (1994) Use of environmental parameters and immunoassays to predict Rhizoctonia blight and schedule fungicide applications on creeping bentgrass. Crop Prot. 13:211-218.

Sitrit Y, Barak Z, Kapulnik Y, Oppenheim AB & Chet I (1993) Expression of *Serratia marcescens* chitinase gene in *Rhizobium meliloti* during symbiosis on alfalfa roots. MPMI 6:293-298.

Stephens PM, Davoren CW, Doube BM, Ryder MH, Benger AM & Neate SM (1993) Reduced severity of *Rhizoctonia solani* disease on wheat seedlings associated with the presence of the earthworm *Aporrectodea trapeziodes* (Lumbricidae). Soil Biol. Biochem. 25:1477-1484.

Sumner DR, Doupnik B & Boosalis MG (1981) Effects of reduced tillage and multiple cropping on plant diseases. Annu. Rev. Phytopathol. 19:167-187.

Sumner DR, Dowler CC, Johnson AW & Baker SH (1995) Conservation tillage and seedling diseases in cotton and soybean double cropped with triticale. Plant Dis. 79:372-375.

Sweetingham MW (1991) The effect of inoculum distribution and sowing depth on Pleiochaeta root rot of lupin. Aust. J. Agric. Res. 42:121-128.

Sweetingham MW & Jarvis RJ (1993) Controlling lupin fungal diseases. Australian Grain 3:36

Thombai P, Graham RD, Neate SM & Webb MJ (1993) Interaction between zinc nutritional status of cereals and Rhizoctonia root rot severity. Plant Soil 153:215-222.

Turner JT & Backman PA (1991) Factors relating to peanut yield increases after seed treatment with *Bacillus subtilis*. Plant Dis. 75:347-353.

Wall PC, Neate SM, Graham RD, Reuter DJ & Rovira AD (1994). The effect of Rhizoctonia root disease and applied nitrogen on growth, nitrogen uptake and nutrient concentrations in spring wheat. Plant Soil 163:111-120.

Wong PTW, Dowling PM, Tesoriero L & Nicol HI (1993) Influence of preseason weed management and in-crop treatments on two successive wheat crops. Aust. J. Exp. Agric. 33:173-177.

Wu WS, Kuo MH, Tschen J & Liu SD (1990) Integrated control of chrysanthemum stem rot. Plant Prot. Bull. Taipei. 32:77-90.

GPSR Compliance

The European Union's (EU) General Product Safety Regulation (GPSR) is a set of rules that requires consumer products to be safe and our obligations to ensure this.

If you have any concerns about our products, you can contact us on ProductSafety@springernature.com

In case Publisher is established outside the EU, the EU authorized representative is:

Springer Nature Customer Service Center GmbH
Europaplatz 3
69115 Heidelberg, Germany

Batch number: 10800209

Printed by Printforce, the Netherlands